T0145228

Emergence, Complexity and Computation

Volume 23

About this Series

The Emergence, Complexity and Computation (ECC) series publishes new developments, advancements and selected topics in the fields of complexity, computation and emergence. The series focuses on all aspects of reality-based computation approaches from an interdisciplinary point of view especially from applied sciences, biology, physics, or chemistry. It presents new ideas and inter-disciplinary insight on the mutual intersection of subareas of computation, complexity and emergence and its impact and limits to any computing based on physical limits (thermodynamic and quantum limits, Bremermann's limit, Seth Lloyd limits...) as well as algorithmic limits (Gödel's proof and its impact on calculation, algorithmic complexity, the Chaitin's Omega number and Kolmogorov complexity, non-traditional calculations like Turing machine process and its con-sequences,...) and limitations arising in artificial intelligence field. The topics are (but not limited to) membrane computing, DNA computing, immune computing, quantum computing, swarm computing, analogic computing, chaos computing and computing on the edge of chaos, computational aspects of dynamics of complex systems (systems with self-organization, multiagent systems, cellular automata, artificial life,...), emergence of complex systems and its computational aspects, and agent based computation. The main aim of this series it to discuss the above mentioned topics from an interdisciplinary point of view and present new ideas coming from mutual intersection of classical as well as modern methods of com-putation. Within the scope of the series are monographs, lecture notes, selected contributions from specialized conferences and workshops, special contribution from international experts.

More information about this series at http://www.springer.com/series/10624

Andrew Adamatzky
Editor

Advances in Unconventional Computing

Volume 2: Prototypes, Models and Algorithms

 Springer

Editor
Andrew Adamatzky
Unconventional Computing Centre
University of the West of England
Bristol
UK

ISSN 2194-7287   ISSN 2194-7295   (electronic)
Emergence, Complexity and Computation
ISBN 978-3-319-81632-6   ISBN 978-3-319-33921-4   (eBook)
DOI 10.1007/978-3-319-33921-4

Preface

Unconventional computing is a science in flux. What is unconventional today will be conventional tomorrow. Designs being standard in the past are seen now as a novelty. Unconventional computing is a niche for interdisciplinary science, cross-bred of computer science, physics, mathematics, chemistry, electronic engineering, biology, material science and nanotechnology. The aims were to uncover and exploit principles and mechanisms of information processing in and functional properties of physical, chemical and living systems to develop efficient algorithms, design optimal architectures and manufacture working prototypes of future and emergent computing devices.

I invited world's leading scientists and academicians to describe their vision of unconventional computing and to highlight most promising directions of future research in the field. Their response was overwhelmingly enthusiastic: over fifty chapters were submitted spanning almost all fields of natural and engineering sciences. Unable to fit over one and half thousands pages into one volume, I grouped the chapters as "theoretical" and "practical". By "theoretical", I mean constructs and algorithms which have no immediate application domain and do not solve any concrete problems, yet they make a solid mathematical or philosophical foundation to unconventional computing. "Practical" includes experimental laboratory implementations and algorithms solving actual problems. Such a division is biased by my personal vision of the field and should not be taken as an absolute truth.

The first volume brings us mind-bending revelations from gurus in computing and mathematics. The topics covered are computability, (non-)universality and complexity of computation; physics of computation, analogue and quantum computing; reversible and asynchronous devices; cellular automata and other mathematical machines; P-systems and cellular computing; infinity and spatial computation; and chemical and reservoir computing. As a dessert, we have two vibrant memoirs by founding fathers of the field.

The second volume is a tasty blend of experimental laboratory results, modelling and applied computing. Emergent molecular computing is presented by enzymatic logical gates and circuits, and DNA nanodevices. Reaction–diffusion chemical computing is exemplified by logical circuits in Belousov–Zhabotinsky medium and geometrical computation in precipitating chemical reactions. Logical circuits realised with solitons and impulses in polymer chains show advances in collision-based computing. Photochemical and memristive devices give us a glimpse into hot topics of novel hardware. Practical computing is represented by algorithms of collective and immune-computing and nature-inspired optimisation. Living computing devices are implemented in real and simulated cells, regenerating organisms, plant roots and slime mould. Musical biocomputing and living architectures make the ending of our unconventional journey non-standard.

The chapters are self-contained. No background knowledge is required to enjoy the book. Each chapter is a treatise of marvellous ideas. Open the book at a random page and start reading. Abandon all stereotypes, conventions and rules. Enter the stream of unusual. Even a dead fish can go with the flow. You can too.

Bristol, UK   Andrew Adamatzky
March 2016

Contents

Chapter 1
Implementing Molecular Logic Gates, Circuits, and Cascades Using DNAzymes

Matthew R. Lakin, Milan N. Stojanovic and Darko Stefanovic

Abstract The programmable nature of DNA chemistry makes it an attractive framework for the implementation of unconventional computing systems. Our early work in this area was among the first to use oligonucleotide-based logic gates to perform computations in a bulk solution. In this chapter we chart the development of this technology over the course of almost 15 years. We review our work on the implementation of DNA-based logic gates and circuits, which we have used to demonstrate digital logic circuits, autonomous game-playing automata, trainable systems and, more recently, decision-making circuits with potential diagnostic applications.

1.1 Introduction

The development of electronic digital logic was one of the greatest technological achievements of the 20th century, and exponential increases in the computational power of commercially-available microprocessors meant that electronic computers are now ubiquitous and indispensable in the modern world. Contemporaneous advances in molecular biology made it clear that information processing is a fundamental capability of all biological systems. Subsequent rapid progress in that field progressed in parallel with the development of consumer electronics, and elucidated many of the mechanisms behind biological information processing [7]. Given that the information processing capabilities of biological systems were evolved over millions of years, it is fascinating to consider whether we can construct synthetic molecular

M.R. Lakin (✉) · D. Stefanovic
University of New Mexico, Albuquerque, NM 87131, USA
e-mail: mlakin@cs.unm.edu

D. Stefanovic
e-mail: darko@cs.unm.edu

M.N. Stojanovic
Columbia University, New York, NY 10032, USA
e-mail: mns18@columbia.edu

© Springer International Publishing Switzerland 2017   1
A. Adamatzky (ed.), *Advances in Unconventional Computing*,
Emergence, Complexity and Computation 23,
DOI 10.1007/978-3-319-33921-4_1

computers that may be more compact and more robust than their natural or electronic counterparts.

When we say that molecules compute, what we usually mean is that an assembly of molecules detects certain inputs, typically the presence or absence of other molecules, and responds by producing one or more output signals, which may take the form of the release of an output molecule or the generation of a detectable fluorescent signal. The use of fluorescence is merely a convenient means of detecting the circuit response in an experimental setting, and plays no role in the actual computation. The goal of a molecular computer scientist is to engineer the intervening molecular system so that the pattern of output signals is related to the pattern of input signals by the desired logic function.

From an unconventional computing perspective, the development of molecular computers offers intriguing possibilities to implement extremely low-power computation [8, 88] and to implement autonomous computational systems that can survive and thrive in environments hostile or inaccessible to silicon microprocessors, such as within the bloodstream or within living cells. The compact nature of DNA has been previously exploited to demonstrate high-density information storage [37], but in our context the fact that billions of molecules exist in each experimental system may make it feasible to execute massively parallel computations in a very small volume, or to implement novel computational architectures that compute using the dynamics of interactions between molecular circuit components.

Our experimental work focuses on catalytic nucleic acid chemistry, in particular, DNAzymes (also known as deoxyribozymes), which are DNA-based enzymes that can cleave or combine other nucleic acid strands. DNAzymes are not known to occur in nature, and the known DNAzyme catalytic motifs have been isolated in *in vitro* evolution experiments [12, 77, 81]. We turned DNAzymes into logic gates by augmenting them with up to three input-binding modules that regulated the catalytic activity of the DNAzyme based on the pattern of input strands observed in the solution. The cleavage reaction catalyzed by the DNAzyme served as the reporting channel, and we exploited the combinatorial chemistry of DNA to enable us to build systems that processed multiple signals simultaneously in a single solution, with the different information streams identified by different DNA sequences. Thus, each DNAzyme unit implemented a logic gate with up to three inputs, and we constructed a set of such gates complete for Boolean logic [103].

In this chapter we review our designs for DNAzyme-based molecular computers, their integration in large-scale parallel gate arrays exhibiting sophisticated logical and temporal behaviors, and our recent attempts to diversify into sequential logic cascades. We begin by describing our early approach to molecular computing [20, 47], including the first reported complete set of nucleic acid-based logic gates [103]. We then describe how these gates were used to produce autonomous molecular computing systems that implement well-known logic circuits such as adders [58, 106] and large-scale game-playing automata [66, 82, 107]. This approach has been previously reviewed [108], including in the popular literature [67]. We then discuss how, in recent years, we have further developed this approach to achieve signal propagation in DNAzyme signaling cascades, and how we have begun to apply these new techniques to biodetection applications.

1.2 Developing DNAzyme-Based Logic Gates and Circuits

The historical context for our work was set by the publication of Adleman's seminal paper [1], which demonstrated that combinatorial nucleic acid chemistry could be used to solve a small instance of the Hamiltonian path constraint satisfaction problem. By synthesizing a library of DNA strands to represent the vertices and directed edges of the graph, Adleman's approach relied on combinatorial hybridization of these molecules to produce linear structures that encoded paths of various lengths through the graph. There followed an extensive sequence of purification and analysis of the resulting molecules that encoded the possible paths through the directed graph, to locate any paths of the correct length that visited every vertex. The fundamental insight behind this approach was to parallelize the "generate" phase of the "generate and test" paradigm for solving computationally intractable problems such as Hamiltonian path. However, the laborious nature of the "test" phase limited the practical applicability of this incarnation of molecular computing.

In our early work, we adopted an alternative approach to molecular computation. We were inspired by timely reviews [12, 77] on nucleic acid catalysts and aptamers, which got us thinking about using external inputs to control nucleic acid catalysis: an idea that was ripe for implementation [92, 110]. Thus, rather than using combinatorial chemical reactions to search for solutions to computationally hard problems, we instead used large populations of DNA logic gates to compute Boolean logic functions, using bulk fluorescence readouts to assay the result of the computations. This approach greatly simplified the experimental protocols and enabled us to execute relatively sophisticated computations, with human intervention required only to provide external data inputs.

We constructed molecular logic gates using RNA-cleaving DNAzymes, which are single strands of DNA that can catalyze the cleavage of specific substrate molecules. The various parts of a DNAzyme strand are illustrated in Fig. 1.1a, using the "E6" catalytic motif [13] as the example. The central catalytic core sequence is largely fixed (with the exception of a small central loop in the E6 motif): this sequence is believed to coordinate the binding of metal ion cofactors (here Mg^{2+}) that are required for the cleavage reaction to occur. The catalytic core is flanked by two variable substrate binding arms, which recognize and bind to a complementary substrate molecule and position it correctly so that cleavage may take place. Figure 1.1a shows the means by which a DNAzyme binds to a substrate molecule, cleaves the substrate at the cleavage site (marked by a single RNA base in the DNA strand), and unbinds from the two shorter product molecules. We can monitor the progress of the cleavage reaction by labeling the substrate with a fluorescent tag on one end and a corresponding quencher molecule on the other end: when the substrate is cleaved the fluorophore and quencher are separated, which reduces the efficiency of the quenching reaction

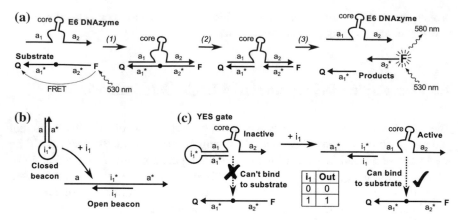

Fig. 1.1 DNAzyme reactions and sensors. **a** DNAzyme structure and DNAzyme-catalyzed cleavage of a substrate molecule. DNAzymes (here the "E6" catalytic motif [13]) consist of conserved catalytic core sequence (labeled core) flanked by two substrate binding arms (labeled a_1 and a_2). The corresponding substrate consists of sequences complementary to the substrate binding arms a_1^* and a_2^*, with a cleavage site in the middle (*denoted by a small disc*). To produce a fluorescent readout of the cleavage reaction, the substrate is labeled with a fluorophore on one end (F) and a quencher on the other (Q). In its uncleaved state, when the fluorophore is excited (here by light of 530 nm wavelength), the energy is transferred to the quencher by Förster resonance energy transfer (FRET). Hence, no output fluorescence is observed. The first step of the cleavage reaction is for the DNAzyme to bind to the substrate (reaction 1). Then, the DNAzyme cleaves the substrate into two shorter product strands (reaction 2). Subsequent unbinding of the products (reaction 3) recycles the active DNAzyme into solution, whereupon it may proceed to interact with further substrates. In the cleaved state, the fluorophore is separated from the quencher, so when the fluorophore is excited, it re-emits the light at its output wavelength (here, 580 nm), which can be observed using standard optical techniques. **b** Molecular beacon reactions. The molecular beacon consists of a stem enclosing a loop whose sequence is complementary to that of the input strand. In the absence of the input, the beacon adopts the energetically favorable hairpin conformation. When the input is added, it binds to the loop and causes the stem to open. **c** A sensor (YES gate) constructed by grafting a molecular beacon input detection module onto a DNAzyme such that the closed stem of the molecular beacon blocks one of the substrate binding arms (a_1). Thus, in the absence of input i_1, the blocked binding arm prevents the DNAzyme from binding to, and cleaving, its substrate, so no fluorescence is observed. However, in the presence of input i_1, the stem of the molecular beacon is opened, exposing the substrate binding arm, so that the DNAzyme can bind and cleave its input, resulting in the generation of a fluorescent output signal. Hence, the YES gate computes the identity function of its input, as shown by the truth table

and causes an increase in observed fluorescence when the fluorophore is excited by a laser. It is important to note that the DNAzyme strand is unchanged by the cleavage reaction and may bind and cleave additional substrates in a "multiple turnover" reaction which provides an innate signal amplification capability. The efficiency of this process is determined by the lengths of the substrate binding arms: too long, and the post-cleavage unbinding reaction is slowed due to increased stability of the DNAzyme-product complex; too short, and the pre-cleavage binding reaction is slowed due to decreased stability of the DNAzyme-substrate complex. In addition to our work on using DNAzymes to construct molecular logic systems, the catalytic properties of DNAzymes have also been exploited to build systems of self-avoiding molecular walkers [65, 74, 83, 97].

There are three parts of an E6 DNAzyme strand that may be independently modified to control its behavior, the two substrate binding arms and the small loop in the catalytic core. We modified these parts of the strand by functionalizing them with *molecular beacons*, which are DNA structures that function as input recognition elements. The basic operation of a molecular beacon is illustrated in Fig. 1.1b. In its native state, it is energetically favorable for the mutually complementary ends of the beacon strand to bind to each other, forming a structure known as a *hairpin*. The single-stranded loop of the hairpin can then serve as an input recognition element: a strand that is complementary to the loop region can bind to the loop and thereby induce a conformational change (by converting the loop from a flexible single-stranded region to a rigid double-stranded region) that opens the hairpin stem. The classical use of molecular beacons in molecular biology is to generate a fluorescent response to the input binding event [115].

However, our interest in molecular beacons was as a means of regulating the catalytic activity of DNAzymes. Thus, our first published result [104] was a logic gate that sensed a single input oligonucleotide and activated a DNAzyme in response (Fig. 1.1c). This logic gate design incorporated a single molecular beacon module that, in the native state, blocks one of the substrate binding arms and thereby prevents the DNAzyme binding to the substrate. However, when the complementary input strand is present, it binds to the molecular beacon module and opens the stem, which exposes the substrate binding arm, enabling the DNAzyme to bind to the reporter substrate and cleave it, which we can detect via fluorescence. We call a gate that senses the input i_1 in this way a YES$_1$ gate [104] (occasionally, a signal detector, sensor, or basic catalytic molecular beacon). Here and henceforth, we represent an input value of 1 by the presence of the corresponding input oligonucleotide and an input value of 0 by its absence. Similarly, we represent an output value of 1 by DNAzyme-catalyzed cleavage of the corresponding substrate molecule (observed via fluorescence) and an output value of 0 by no cleavage taking place. Thus, from a logical perspective, the YES gate simply computes the identity function of its input. Furthermore, by adding a different molecular beacon that blocks the other arm of the YES gate, we obtain a DNAzyme that is only activated when the complementary inputs for *both* molecular beacons are present, as both substrate binding arms must be available for the DNAzyme to bind to the substrate. Thus, such a DNAzyme will function as an AND logic gate, as shown in Fig. 1.2b which implements $i_1 \wedge i_2$. Thus, our DNAzyme-based molecular logic gates are switched by oligonucleotide input signals, just as electronic logic gates are switched by their respective electrical input signals.

Any set of complete Boolean logic gates must include some form of negation, and for this we turned to the third of the potential modification sites on the E6 DNAzyme strand, the small loop within the catalytic core. It turns out that (at least in the E6 catalytic motif) this loop can be enlarged to the same size as the loops from our other molecular beacon control modules without adversely affecting the

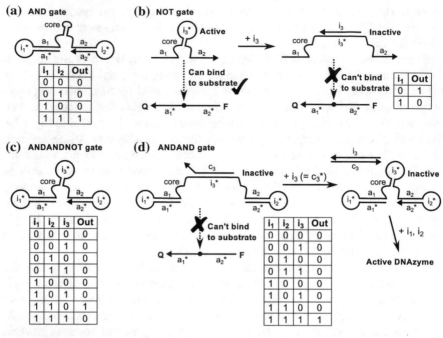

Fig. 1.2 DNAzyme logic gates. **a** AND gate. By extending the YES gate motif with a molecular beacon module on *both* substrate binding arms, we obtain an AND gate. This gate implements AND logic because both substrate binding arms must be exposed for successful binding to the substrate, which is only possible when *both* inputs are present. Thus this gate computes $i_1 \wedge i_2$, as shown in the truth table. **b** NOT gate. Extending the small loop in the catalytic core of the "E6" DNAzyme to a full-size input-binding loop allows us to implement a NOT gate. In the absence of input i_3, the catalytic core structure folds as normal and the DNAzyme is functional. However, when i_3 is added, the loop is opened, which distorts the catalytic core and prevents the DNAzyme from binding to the substrate. Thus, the catalytic activity of the DNAzyme is negatively regulated by the input, which computes $\neg i_3$, as shown in the truth table. **c** ANDANDNOT gate. By placing loops in all three possible positions, we combine the AND gate with the NOT gate to produce an ANDANDNOT gate, which is active only when inputs i_1 and i_2 are present but i_3 is *not* present. Thus, this gate computes $i_1 \wedge i_2 \wedge \neg i_3$, as shown in the truth table. **d** ANDAND gate. By pre-binding the logic gate with strands complementary to the true input strands, we can invert the sense of control of any of the input-binding loops. In this example, the ANDANDNOT gate from **c** is converted into an ANDAND gate by reversing the action of the i_3 input. This is achieved by pre-binding the logic gate with the c_3 strand, which binds to the input-binding loop in the catalytic core and deforms the catalytic core. Then, when input i_3, which is complementary to c_3, is added, it binds to c_3 via the short, single-stranded, exposed "toehold" and removes c_3 from the logic gate via *toehold-mediated strand displacement* [133, 139], which allows the catalytic core to refold. This reaction effectively converts the i_1ANDi$_2$ANDi$_3$ gate into an i_1ANDi$_2$ANDNOTi$_3$, while simultaneously removing input i_3 from solution. The inputs i_1 and i_2 then bind to the other input-binding loops as normal. Thus, the DNAzyme is active only when all three inputs are present, which yields an ANDAND gate that computes $i_1 \wedge i_2 \wedge i_3$, as shown in the truth table

catalytic activity of the DNAzyme. Thus, we were able to use this loop to negatively regulate DNAzyme catalysis, producing a NOT gate that computes $\neg i_3$, as shown in Fig. 1.2b. In this gate design, when an input strand binds to the molecular beacon loop in the catalytic core, the resulting conformational change distorts the catalytic core by pushing the two halves of the catalytic core (and the two substrate binding arms) apart, so that the DNAzyme cannot successfully bind and cleave its substrate. Conversely, the closed state of the molecular beacon forms the correct catalytic core structure so that substrate cleavage may occur, thus, the DNAzyme is active when the input is absent and inactive when the input is present, as required for a NOT gate [103].

As is well known, AND and NOT gates are complete for Boolean logic, assuming that we can form circuits with arbitrary connections between the gates. We will return to the question of gate connectivity when we discuss our more recent work in Sect. 1.4, but here we observe that it is possible to implement more sophisticated information processing using unconnected DNAzyme units. In particular, we constructed three-input logic gates from a single DNAzyme by simultaneously modifying all three potential control sites (the two substrate binding arms and the catalytic core) with molecular beacons. Figure 1.2c shows the basic three-input logic gate, which is an ANDANDNOT gate [58, 103] that computes the logic function $i_1 \wedge i_2 \wedge \neg i_3$. In this case, the catalytic activity of the DNAzyme is positively regulated by the two input binding loops on the substrate binding arms and negatively regulated by the loop in the catalytic core.

The above examples of logic gate design show how direct application of molecular beacon input-binding modules to DNAzymes can be used to implement certain one-, two-, and three-input Boolean logic gates. To broaden our repertoire of logic gates, we employed a strategy of pre-binding gates with blocking strands complementary to certain input binding loops, so that they are initially held open as opposed to folding closed. Then, by adding complementary inputs to strip off the blocking strands via toehold-mediated strand displacement [133, 139], so that the effect of those inputs was negated compared to our previous designs. Strand displacement is an alternative technique for realizing molecular computation, which has been used previously to implement digital logic circuits [89, 96], catalytic cycles [140], artificial neural networks [91], chemical reaction networks [19, 101] and nanomachines [134]. As an example of this approach to DNAzyme logic gate construction, Fig. 1.2d illustrates an ANDAND gate that computes the function $i_1 \wedge i_2 \wedge i_3$, by using a pre-bound blocking strand to reverse the sense of action of the molecular beacon in the catalytic core. Thus, using this technique, a single DNAzyme can implement any Boolean formula that is a conjunction of one, two, or three literals.

It is important to note that we can vary the sequences of substrate binding arms (and of the corresponding substrate), which allows us to produce DNAzymes that cleave different substrates, so we can simultaneously monitor the outputs of different DNAzymes using different fluorophores. Similarly, we can vary the sequences of the inputs (and of the corresponding input-binding loops) without affecting their behavior: this allows us to replicate our logic gate motifs to compute the same logic function for different sets of inputs. The only proviso here is that the chosen sequences must form the correct structures when they are prepared in the test tube, and that there is no unintended cross-reactivity between the sequences. This is a general challenge in the field of DNA nanotechnology, and there are mature algorithms and software packages available to design nucleotide sequences that will fold to produce the desired structures [25–27, 125, 135, 136].

Furthermore, we can implement OR gates implicitly by creating multiple gates with different input patterns that cleave the same substrate. Since the substrate will be cleaved when *any* of the corresponding gates are activated, this gives an implicit OR connection between the gates. We can exploit these properties to construct systems of large numbers of DNAzyme-based logic gates that operate in parallel arrays connected by implicit OR connections, and we can predict the behavior of the ensemble compositionally in terms of the behavior of the original logic gates.

Thus, we can implement any Boolean formula that can be converted into disjunctive normal form (DNF) such that each clause contains at most three literals. This abstraction of biochemical interactions as logic functions is a powerful organizing motif that has since been adopted by many other groups working on solution-phase molecular computation using other molecular computing frameworks [49, 71, 139].

Having developed a collection of elementary logic gates using DNAzymes, we used them to assemble some straightforward demonstration systems. An early circuit that we constructed was a half-adder [106], which took two bits as inputs and produced an output bit for the sum and an output bit for the carry. We used two oligonucleotide inputs i_1 and i_2 to represent the two input bits: the presence of i_1 denoted a value of 1 for the first input bit and its absence denoted a value of 0 the first input bit. Similarly, presence or absence of i_2 encoded values of 1 or 0 for the second input bit, respectively. The output bits were reported via different substrates, cleaved by DNAzymes with different substrate binding arm sequences, which were labeled with fluorescent molecules that emitted different colored light, which we could monitor simultaneously. The collection of logic gates that implement the half adder is presented in Fig. 1.3a: the value of the SUM bit is the XOR of the two input bits, which we implemented using two parallel ANDNOT gates with opposite inputs (SUM $= (i_1 \wedge \neg i_2) \vee (i_2 \wedge \neg i_1)$), and the CARRY bit was generated by a straightforward AND gate (CARRY $= i_1 \wedge i_2$).

We subsequently developed a larger system that implements a full adder [58], which extends the half adder with an extra carry input bit i_3, such that

$$\text{SUM} = (i_1 \wedge i_2 \wedge i_3) \vee (i_1 \wedge \neg i_2 \wedge \neg i_3) \vee (i_2 \wedge \neg i_3 \wedge \neg i_1) \vee (i_3 \wedge \neg i_2 \wedge \neg i_1)$$
$$\text{CARRY} = (i_1 \wedge i_2) \vee (i_1 \wedge i_3) \vee (i_2 \wedge i_3).$$

The logic gates that make up the full adder system are presented in Fig. 1.3b. Since the SUM output bit depends on the two input bits as well as the carry-in bit, this system makes full use of the techniques from Fig. 1.2 for implementing arbitrary three-input logic gates. With further progress on circuit designs that enable information transmission in DNAzyme cascades (see Sect. 1.4), multiple full-adder units could

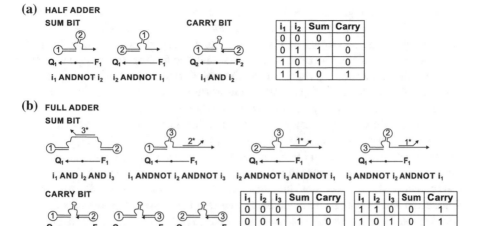

Fig. 1.3 DNAzyme binary adder circuits. **a** A half adder circuit can be constructed using three parallel DNAzyme logic gates. The sum bit is computed by an i_1 ANDNOTi_2 gate and a i_2 ANDNOTi_1 gate which, together, compute the XOR of the two inputs. These gates cleave one output substrate, labeled with fluorophore $\mathbf{F_1}$ and the corresponding quencher $\mathbf{Q_1}$. The carry bit is computed by a single i_1 ANDi_2 gate, which cleaves another output substrate, labeled with a different fluorophore $\mathbf{F_2}$ that emits in a different part of the electromagnetic spectrum from $\mathbf{F_1}$, along with the corresponding quencher $\mathbf{Q_2}$. **b** The full adder circuit extends the half adder design to additionally process the carry input i_3

be chained together with the carry output of one serving as the carry input of the next, producing a molecular system that can add larger binary numbers.

As we shall see below, there are alternatives to using molecular beacons for controlling DNAzyme catalysis. Particularly noteworthy is work on multi-component DNAzymes [10, 30–32, 46, 51, 52, 70, 76, 121] in which the DNAzyme strand is split within the catalytic core to produce two parts that, individually, possess no catalytic activity. Assembly of a catalytically active DNAzyme from these parts is controlled by inputs that can implement logic functions. This idea has also been generalized to self-assembling structures with more than two components [53, 54, 118].

1.3   Towards Wide Circuits Via Parallel Gate Arrays

To move towards implementing larger-scale systems, we decided initially to expand our circuits by adding additional gates with implicit OR connections via shared substrate molecules. This can be viewed as increasing the "width" of the circuit, as opposed to the "depth", i.e., there is no sequential information transfer (or cascading) between DNAzyme gates. Specifically, we worked on developing large-scale arrays of molecular logic gates that function as *molecular automata*, playing games of strategy against human opponents. In this context, a key aspect of circuit design is to render game strategies as Boolean formulae suitable for implementation using the available molecular logic gates [102]. As the resulting formulae tend to be large and complex, the challenge designing these automata was a good test of the extent to which molecular logic arrays may be engineered to implement large formulae.

We have built three generations of game-playing automata, **MAYA-I–III** (originally standing for "Molecular Array of YES and AND gates"). In all three, the human interacts with the automaton by adding input oligonucleotides corresponding to the next human move, and these stimuli cause the logic gates comprising the automaton to change state. These state changes record the history of moves and enable the automaton to signal its move by activating certain DNAzymes to produce a fluorescent response. The original **MAYA-I** automaton [107] played a symmetry-pruned game of tic-tac-toe (Fig. 1.4), and the subsequent **MAYA-II** automaton [66] played an unrestricted version of tic-tac-toe using a richer encoding of inputs (not shown). **MAYA-III** [82] could be trained to play specific strategies in a specially designed simple game, though we will not discuss **MAYA-III** in detail here.

In **MAYA-I**, the tic-tac-toe board is represented by 9 wells in a 3×3 section of a well plate, which we number 1–9 (Fig. 1.4). We assume that the automaton moves first and always claims the center well (well 5), and that the human's first move is either well 1 (if claiming a corner) or well 4 (if claiming a side). This symmetry pruning restricts play to just 19 legal games, which made it feasible to exhaustively test the automaton [107]. The automaton is programmed with an optimal strategy, so that the automaton will win 18 of the 19 possible games, with the human earning a draw only by playing perfectly.

The board is prepared by adding the requisite DNAzyme logic gates (Fig. 1.4a) to each of the 9 wells: there are 23 logic gates in total. The game is initiated by adding the required Mg^{2+} ions to all wells; since well 5 is the only well containing a non-logic-gated DNAzyme, that DNAzyme immediately activates and produces a fluorescent signal indicating that **MAYA** has claimed well 5. Human moves are represented by eight input oligonucleotides, corresponding to the 8 remaining wells, which are added to all wells to signal the next human move. Thus, each well records all moves made by the human, and after each human move each well independently computes whether it will be the next well claimed by the automaton. This is possible because the automaton's strategy is fixed, and the automaton's program is such that a single well will activate in response to each human move. An example of a gameplay sequence for the **MAYA-I** system is presented in Fig. 1.4b, and the Boolean logic functions computed by the logic gates in each of the nine wells are shown in Fig. 1.5.

Our follow-up work on the **MAYA-II** automaton removed the restrictions on the moves available to the human player, so that the initial human move could claim any of the 8 peripheral wells. This increased the number of legal games from 19 for **MAYA-I** to 76 for **MAYA-II**. We also increased the number of inputs from 8 to 32, encoding not just the well number but also the order of selection of that well in the game sequence. Thus, the number of logic gates implementing the game strategy

(a)

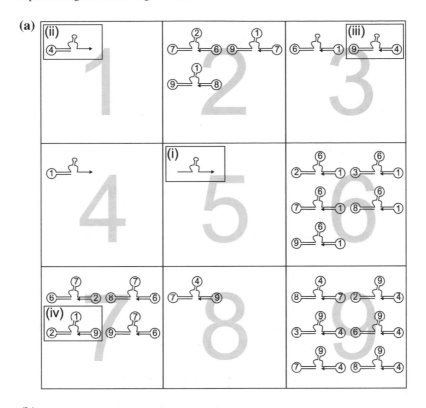

(b)

(i) The automaton moves first in middle well, activated by adding Mg^{2+} to all wells.

(ii) The human player adds input i_4 to all wells, which only activates the YES i_4 gate in well 1 to signal the automaton's response.

(iii) The human player adds input i_9 to all wells to block a diagonal. All wells now contain inputs i_4 and i_9, and thus the i_4 AND i_9 gate in well 3 is activated to signal the automaton's response.

(iv) The human player has no winning move, and adds input i_2 to all wells to block a row. Since all wells now contain inputs i_2, i_4, and i_9, the i_2 AND i_9 ANDNOT i_1 gate in well 7 is activated, which completes a diagonal and the automaton wins the game.

Fig. 1.4 MAYA-I, an automaton that plays a symmetry-pruned game of tic-tac-toe. **a** Distribution of logic gates in wells. The center well (5) contains a DNAzyme without any logic gate attachments, while the other wells contain logic gates. Gates used in the example game from **b** are boxed. **b** Example gameplay for a game in which the human does not play perfectly and therefore loses

Well 1: i_4

Well 2: $(i_7 \wedge i_6 \wedge \neg i_2) \vee (i_9 \wedge i_7 \wedge \neg i_1) \vee (i_9 \wedge i_8 \wedge \neg i_1)$

Well 3: $(i_6 \wedge i_1) \vee (i_9 \wedge i_4)$

Well 4: i_1

Well 5: $\top$

Well 6: $(i_2 \wedge i_1 \wedge \neg i_6) \vee (i_3 \wedge i_1 \wedge \neg i_6) \vee (i_7 \wedge i_1 \wedge \neg i_6) \vee (i_8 \wedge i_1 \wedge \neg i_6)$
$\vee (i_9 \wedge i_1 \wedge \neg i_6)$

Well 7: $(i_6 \wedge i_2 \wedge \neg i_7) \vee (i_8 \wedge i_6 \wedge \neg i_7) \vee (i_2 \wedge i_9 \wedge \neg i_1) \vee (i_9 \wedge i_6 \wedge \neg i_7)$

Well 8: $i_7 \wedge i_9 \wedge \neg i_4$

Well 9: $(i_8 \wedge i_7 \wedge \neg i_4) \vee (i_2 \wedge i_4 \wedge \neg i_9) \vee (i_3 \wedge i_4 \wedge \neg i_9) \vee (i_6 \wedge i_4 \wedge \neg i_9)$
$\vee (i_7 \wedge i_4 \wedge \neg i_9) \vee (i_8 \wedge i_4 \wedge \neg i_9)$

Fig. 1.5 The Boolean logic functions computed by the logic gates in each of the nine wells of the **MAYA-I** automaton

increased from 23 for **MAYA-I** to 96 for **MAYA-II**. We introduced a second set of logic gates that respond to human moves by fluorescing in a second color, using 32 YES gates (4 in each of the 8 peripheral wells). Aside from these changes, the basic mechanism of game playing is the same as for **MAYA-I**, so we do not discuss it further here.

From an information processing perspective, **MAYA-II** is not significantly more capable than **MAYA-I**. However, the main advance in the development of **MAYA-II** was the engineering feat of scaling the system up from 23 gates in **MAYA-I** to a total of 128 gates in **MAYA-II**. In scaling up the system we learned that individual DNAzymes that work perfectly in isolation may fail due to unwanted interference when placed in a parallel OR-gate array with other DNAzymes. Predicting such cross-reactivity and designing to avoid it is a major challenge in the implementation of molecular computing systems, as discussed in Sect. 1.2. In the context of a molecular automaton such as **MAYA-I** or **MAYA-II**, however, each gate must be designed not just against the other gates present in the same well, but also against the constraints imposed by shared sequences that may appear in other wells, such as input binding loops and substrate binding arms. Recent work on computational optimization of nucleic acid structures [25, 125, 135, 136] may aid future work in this direction, although it seems likely that designing the large number of gates present in the **MAYA-II** system would remain challenging even with the assistance of such computational tools.

The **MAYA-I** and **MAYA-II** automata were hardwired to play a fixed strategy in the game of tic-tac-toe. In contrast, for the **MAYA-III** automaton [82] we started with a blank slate and invented a simple retributive game, which we called tit-for-tat, which is played on a 2×2 board, and for which there are 81 different winning strategies. Our goal was to demonstrate a molecular computing system that could be "trained by example" to play a particular strategy in the tit-for-tat game. We will not go into the details of the tit-for-tat game and of the **MAYA-III** implementation that plays it, except to say that the system consisted of 4 YES gates and 12 AND gates implementing the game logic, that were each augmented with an additional input binding loop that detects the training inputs. This allows a non-expert user to select the automaton's strategy in a training phase that mimicks the real gameplay but using the training inputs instead of the gameplay inputs, which "teaches" the system by only activating the required gates in individual wells of the automaton to enable the correct responses for the intended strategy. From a computational standpoint, the training inputs are simply an instance of "staging" the inputs to the system, although in the **MAYA-III** design the training inputs were designed with an additional overhanging toehold, making it possible to retrain a trained system (before it has been used for gameplay) by removing the training inputs via strand displacement. This capability relies on the fact that the input-binding loops will refold into their "closed" hairpin conformation once the training input has been stripped away.

1.4 Towards Deep Circuits Via Signaling Cascades

Our experience developing the **MAYA** series of molecular automata was valuable in that we learned that engineering large-scale assemblies of parallel DNAzyme logic gates is possible, although sometimes challenging. A particular limitation of the **MAYA** approach is that parallel arrays of the available molecular logic gates cannot implement all possible Boolean formulae: any formula whose DNF representation contains a clause with four or more literals cannot be implemented using our available logic gates using a parallel OR-gate array. This is a limitation of our DNAzyme-based framework, as we only had three available locations for functionalization of the DNAzyme with input detection modules. For formulae that are not in DNF, functional completeness [124] can only be achieved by multi-layer circuits, which require arbitrarily many layers to implement arbitrary formulae, in general. The natural remedy for this limitation is to develop methods to connect DNAzyme logic gates via signal propagation cascades, which would enable us to connect DNAzymes into "deep circuits" so that we could, at least in principle, construct a circuit to implement any Boolean logic formula.

Contemporaneously with the work described in Sect. 1.3, we carried out some initial investigations into connecting DNAzymes into signaling cascades and multi-layer logic circuits. We initially developed a two-layer logic cascade in which an upstream ligase DNAzyme (which joins two short substrates into a longer product) activated a downstream phosphodiesterase (substrate cleaving) DNAzyme [105]. This demonstrated signal transmission but precluded building circuits deeper than two levels, as a different kind of DNAzyme was used in the two levels. We also explored the construction of networked DNAzymes attached to microspheres [132], whereby substrate molecules were cleaved from a microsphere once the nearby DNAzyme logic gates were activated, allowing the substrates to diffuse to another microsphere and serve as inputs for the DNAzyme logic gates attached to the second microsphere, thereby achieving signal transmission. This provided a more promising framework for the implementation of deep circuits, as well as offering the opportunity for sequence reuse on different microspheres; however, signal attenuation was significant because the DNAzymes and their substrates were both attached to fixed points on the microsphere, reducing the amount of turnover by activated DNAzymes and hence limiting the potential for scaling up the system.

Thus, we looked for an alternative approach to implement robust, scalable mechanisms for signal transmission in deep circuits of DNAzyme logic gates. We began from the observation that, if we are to implement signal transmission from one substrate-cleaving DNAzyme to another, then the substrate cleavage reaction must enable some downstream reaction involving the cleavage products that was not possible with the uncleaved substrate molecule. Since the cleavage products will both be sub-sequences of the longer uncleaved substrate, this is a non-trivial engineering problem. We addressed this problem by developing substrate molecules that were structured, as opposed to the linear, unstructured substrates used in our prior work on DNAzyme logic gates (Sect. 1.2). Thus, the full sequence of the downstream activator strand is present in the uncleaved substrate structure but is folded up in the structure and is therefore prevented from reacting with any other DNAzymes in the system.

The design of structured DNAzyme substrates is technically challenging because the structure must balance pre-cleavage stability (to prevent undesired activation without cleavage of the substrate) with post-cleavage instability (to promote rapid activation when the substrate *has* been cleaved). We worked on a number of potential designs for these structured substrates, which we summarized previously [55], before settling on a design for a substrate that provided a workable compromise between these concerns [15]. Our *structured chimeric substrate* (SCS) design is summarized in Fig. 1.6a, and consists of a dual stem-loop design. The inner stem and loop sequester the sequence that will activate the downstream DNAzyme, and the outer stem and loop contain the binding and cleavage sites for the upstream DNAzyme. We found that the enhanced stability conferred by the combination of the two stems was vital to maintain stability of the SCS structure prior to cleavage.

The mechanism by which the upstream DNAzyme binds and cleaves the structured substrate is illustrated in Fig. 1.6b. The upstream DNAzyme binds to the SCS via the external toehold and initiates a strand displacement reaction with one of its substrate binding arms, thereby opening the outer stem of the SCS (reaction 1). The second arm of the upstream DNAzyme can then bind to the outer loop and position the SCS cleavage site correctly with respect to the catalytic core of the upstream DNAzyme (reaction 2). Following cleavage of the SCS strand (reaction 3), the upstream DNAzyme unbinds (reaction 4), leaving a short waste strand along with the remainder of the SCS strand, minus the outer stem, which we call the *activator* strand. The structure of the activator is significantly weaker than that of the uncleaved SCS, therefore, the downstream effector sequence contained within the activator structure is made available to interact with downstream DNAzyme gates. Thus the catalytic activity of the upstream DNAzyme causes a covalent modification to the SCS molecule which alters its structure, causing signal propagation to a downstream logic gate by release of the effector sequence.

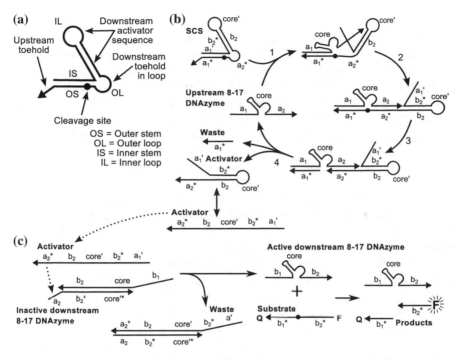

Fig. 1.6 Design of structured chimeric substrates for deep DNAzyme circuits. **a** The design of our structured chimeric substrate (SCS) molecule consists of an inner stem and loop and an outer stem and loop, and an external toehold. The inner stem and loop sequester the downstream activator sequence, and the outer stem and loop comprise the upstream DNAzyme binding and cleavage sites. The outer loop also contains the toehold that will enable the downstream activator to bind to the downstream DNAzyme. We write core′ for a sequence consisting of *part* of the catalytic core sequence of the 8-17 DNAzyme, rather than the full core sequence. **b** The mechanism of the SCS cleavage reaction. Binding of the upstream 8-17 DNAzyme to the SCS initially displaces the outer stem (reaction 1), opening the outer loop so that the upstream DNAzyme can bind to that part of the SCS (reaction 2). This positions the catalytic core of the DNAzyme correctly with respect to the SCS cleavage site, so that the DNAzyme can cleave the SCS (reaction 3). Unbinding of the DNAzyme from the cleaved SCS recycles the DNAzyme and produces a short waste strand and a downstream activator (reaction 4). This activator is structurally weaker than the SCS because it only contains a single stem, and can therefore interconvert into a linearized form. **c** The linearized activator strand interacts with a downstream DNAzyme that has been inhibited by hybridization with a partially complementary inhibitor strand. The activator binds to a complementary toehold on the downstream inhibitor strand and initiates a strand displacement reaction that displaces an active downstream DNAzyme and produces an inert waste complex. The displaced DNAzyme can then cleave its own substrate, which may be a fluorescently-labeled linear readout substrate (as shown here) or another SCS molecule that enables further signal propagation

We used the SCS design from Fig. 1.6 to implement several "deep" circuits that were beyond the capabilities of our previous framework of parallel logic gates (Sect. 1.3). In this context we used a different design for DNAzyme logic gates that were inhibited via direct hybridization with an inhibitor strand and activated via toehold-mediated strand displacement [16], as shown in Fig. 1.6c. The practical reason for this was that using toehold-mediated strand displacement allows us to control the binding pathway in the activation reaction: this information was helpful in determining which parts of the activator sequence to protect, and how, when designing the SCS molecules [55]. We also based these logic gates on the Zn^{2+}-dependent "8–17" DNAzyme motif [93], which is more compact than E6 and has a higher catalytic rate.

As a proof-of-concept for depthwise scaling of DNAzyme circuits using our SCS design, we implemented multi-stage linear signaling cascades up to five layers deep, as outlined in Fig. 1.7a. The cascade consists of a series of inactive DNAzymes, which are activated in turn by a cleaved SCS in a strand displacement reaction, and proceed to cleave their own SCS molecule to activate the DNAzyme in the next layer. This system was inspired by protein signaling cascades such as the MAP kinase phosphorylation cascades [85, 95]. Experimental results from this system for two-, three-, four-, and five-layer variants of the cascade are shown in Fig. 1.7b, c. This cascade comprised only strand-displacement "YES" gates: we have also used the SCS approach to connect DNAzymes controlled via molecular beacons, such as those from Sect. 1.2—see the Supporting Information from [15] for details. This work demonstrates that multi-layer DNAzyme networks may be implemented provided that the information transmission interfaces between DNAzymes are designed carefully, and opens the possibility of scaling up the parallel OR-gate arrays discussed above by connecting the logic gates in multi-layer circuits.

Furthermore, to demonstrate the inclusion of logical processing into SCS circuits, we designed a two-layer, three-input AND circuit using two strand displacement-based AND gates that require two inputs to activate each DNAzyme [16] via a cooperative strand displacement reaction in which the two input strands simultaneously displace part of the sequestered DNAzyme strand [138], as shown in Fig. 1.8a. As a demonstration of our systems' potential applicability for virus detection, we used the circuit template from Fig. 1.8a to implement four logic circuits to detect and distinguish all four serotypes of dengue virus [9, 100]. We chose four target sequences unique to the four serotypes and two generic sequences conserved across the four serotypes, and designed four three-input AND systems that require both generic inputs and a particular serotype-specific input to be present before the fluorescent output is triggered. Experimental results for these circuits are summarized in Fig. 1.8b, showing that the AND logic functions correctly. Furthermore, each version of the circuit is only sensitive to one serotype—see the Supporting Information from [15] for details.

Our approach to depthwise scaling of DNAzyme circuits using the SCS approach was successful because of the use of the SCS as an intermediary between the communicating DNAzymes. This removed the need for direct interaction between the DNAzymes, which allowed us to standardize the SCS design and enabled simpler scaling of the circuit. We believe that this approach could be deployed to implement

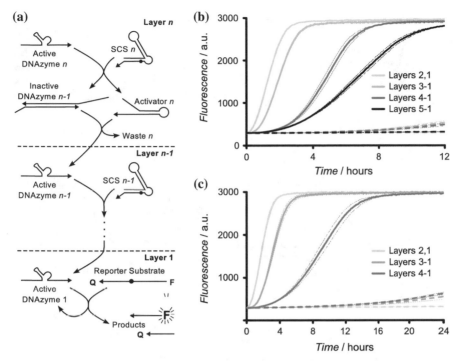

Fig. 1.7 Multi-layer DNAzyme signaling cascades. **a** Signaling cascade reactions. In each layer of the cascade, an active DNAzyme cleaves the corresponding SCS, producing an activator that activates the downstream DNAzyme via a strand displacement reaction, thereby propagating the activating signal to the next layer of the cascade. In the final layer (layer 1), the activated DNAzyme cleaves a linear fluorescent reporter substrate to generate an output signal. **b** Experimental data from DNAzyme signaling cascades. The mean fluorescence signal (solid lines) from multi-layer DNAzyme signaling cascades with equal concentrations (100 nM) of each DNAzyme from each layer. The dashed lines represent the same reaction without the top-layer active DNAzyme, which measures the non-specific activation (leakage) of the cascade. **c** Further experimental data from multi-layer DNAzyme signaling cascades with increasing DNAzyme concentrations in each layer (25 nM in layer 4, 50 nM in layer 3, 75 nM in layer 2, and 100 nM in layer 1), to demonstrate signal amplification. In both **b** and **c**, the *dotted lines* represent the 95 % confidence interval from three replicate experiments

a range of interesting dynamical behaviors such as DNAzyme feedback systems. An interesting direction for future work would be to engineer inhibitory connections between DNAzymes, which would permit the implementation of more kinds of circuit derived from gene networks explored in systems biology [7], such as DNAzyme oscillators [33, 38].

We are gratified that other groups have also begun to explore DNAzyme-based computation cascades. For instance, the Kolpashchikov group has published [35] a design similar to our structured substrate molecule. In that work, cleavage of the structured substrate directly released a downstream DNAzyme that generates an output signal via a color change in the solution, which limits the possibilities for scaling

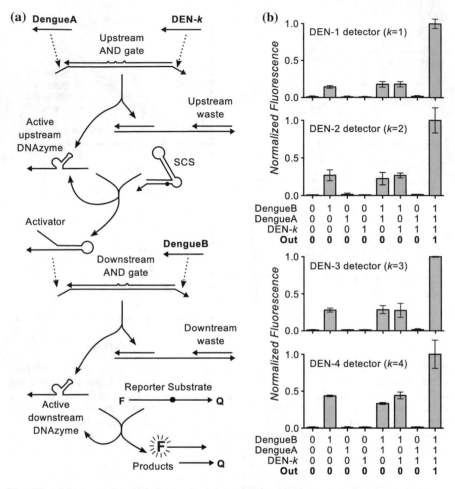

Fig. 1.8 Multi-layer diagnostic logic circuits for the detection of sequences from the genomes of all four dengue serotypes. **a** Circuit design template. The diagnostic circuit for serotype DEN-k ($k \in \{1, 2, 3, 4\}$) requires the presence of two conserved sequences from the dengue genomes (which we call "DengueA" and "DengueB") and one sequence specific to the serotype of interest (which we call "DEN-k"). The circuit consists of two DNAzyme-based AND gates, whose inhibitors contain mismatched bases to promote rapid activation [16]. When both upstream inputs are present, the upstream DNAzyme is displaced by a cooperative strand displacement reaction involving both input strands simultaneously [138]. The active upstream DNAzyme cleaves an SCS molecule, producing an activator that serves as one input to the downstream logic gate. If the second conserved dengue sequence is also present, it serves as the second input to the downstream logic gate. When the downstream DNAzyme is activated, it generates a fluorescent output signal as before. Thus, all three inputs must be present to produce an output, which would increase confidence in the diagnosis in a practical application. We derived detection circuits for all four dengue serotypes (DEN-1 − 4) by modifying part of the upstream logic gate and the SCS molecule. **b** Experimental data for all four dengue serotyping circuits, showing correct operation of all four instantiations of circuit template using all eight combinations of the two conserved sequences and the correct serotype-specific sequence. Error bars represent the 95 % confidence interval from three replicate experiments

up to larger systems. Additionally, the Willner group has developed logic cascades based on DNAzymes which used a two-strand substrate structure [30]: see [121] and citations therein. Finally, other workers have explored DNAzyme cascades in which each DNAzyme directly activates the next by cleaving away an inhibitor strand that was holding the DNAzyme in an inactive, quasi-circular conformation [10], and cross-catalytic amplification cycles in which circularized DNAzymes are linearized by the cleavage reaction [61].

1.5   Towards Applications in Biodetection

In the previous section, we described a two-layer, three-input AND circuit that detected synthetic oligonucleotide targets with sequences corresponding to segments of the four dengue virus genomes. In this section, we briefly review recent work, by ourselves and others, on moving DNAzyme-based logic systems toward practical biodetection applications.

We used as a basis for our biodetection work the strand displacement-based DNAzyme YES gates described in Sect. 1.4. We extended this gate design by incorporating a separate "input binding" module that, when activated, releases a secondary toehold from a loop so that a secondary "fuel" strand can complete the displacement of an active DNAzyme from the complex, as shown in Fig. 1.9a. This two-step process allowed us to retain the strand displacement mechanism of activation while separating the input sequence from that of the DNAzyme itself, so that each may be changed independently of the other. We used this gate design to demonstrate detection of oligonucleotides and of small molecules (such as ATP) via aptamer binding, whereby the small molecule target binds to a particular DNA subsequence that is known to have high binding affinity for the target molecule. We also used this sensor platform to detect genes on DNA extracted from bacteria [14]. Some results from the detection of genetic elements on bacterial DNA are shown in Fig. 1.9b, c, and demonstrate sequence specificity as well as the possibility of detecting a technically challenging double-stranded target. This is a realistic model biodetection assay: most viable protocols for pathogen detection will include similar sample preparation steps.

The use of DNAzymes for biodetection and other analytical chemistry applications has also been explored extensively by other groups, as the innate catalytic activities of DNAzymes make them useful for implementing both detection and output functionalities. The Willner group has published a range of papers on analytical applications of DNAzymes and DNAzyme logic [31, 64, 112, 118–120], as have the Lu group [44, 57, 60, 63, 123, 126–128, 130, 141], the Li group [2–6, 40, 68, 113, 114], the Kolpashchikov group [35, 36, 51, 52], and the Liu group [41–43]. A good example of the output capabilities of DNAzymes is the peroxidase-mimicking DNAzyme which, when activated, turns the solution from clear to colorless. This DNAzyme has been used to produce a visual output from several molecular logic systems [28, 35]. In addition, one of us (D.S.) has worked on applying the molecular beacon-based DNAzyme motifs from Sect. 1.2 to the problem of virus detection [86],

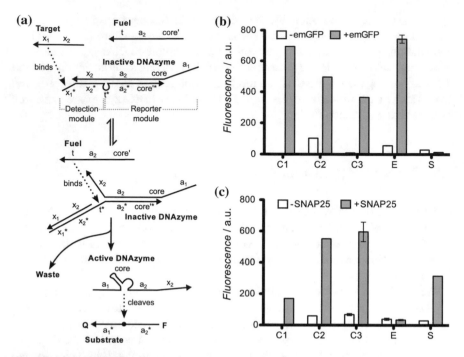

Fig. 1.9 Modular, DNAzyme-based biosensor design and experimental data. **a** Our biosensor design consists of orthogonal "detection" and "reporter" modules. DNA detection targets, such as single-stranded oligonucleotides or denatured double-stranded DNA, bind to the detection module by toehold-mediated strand displacement and, in doing so, expose the secondary toehold in the reporter module. In both cases, this allows the fuel strand to bind and complete displacement of the DNAzyme strand from the inhibitor, which can then fold into a catalytically active conformation and generate an amplified fluorescent output by cleaving fluorescently-labeled substrates. **b** and **c** Detection of genes on denatured plasmid DNA extracted from bacteria. As a demonstration, we designed five biosensors using a common reporter module but varying the detection module, to detect five genetic sequences from plasmids encoding GFP-fusion protein variants: a commercially available Emerald GFP plasmid ("emGFP") and a Pinpoint Xa plasmid containing a SNAP25-GFP fusion protein [94] ("SNAP25"). Three biosensors targeted genetic sequences common to both plasmids (which we called C1–3), one targeted a sequence specific to emGFP (which we called E), and the final biosensor targeted a sequence specific to SNAP25 (which we called S). Results from detecting the emGFP and SNAP25 plasmids using the five biosensors individually are shown in **b** and **c**, respectively. In **b** we observe a strong response from C1–3 and E, and a weak response from S, as expected. Similarly, in **c** we observe a strong response from C1–3 and S, and a weak response from E. Thus, our biosensors are specific to their detection target sequences and can be used to detect bacterial DNA in realistic assay scenarios

demonstrating the use of well-plates as fluorescent seven-segment displays for multiplexed detection of oligonucleotide inputs corresponding to representative viral sequences.

1.6 Conclusions

In conclusion, we have reviewed almost fifteen years of work in our laboratories on the development of DNAzyme-based logic systems. Our early research on solution-phase molecular logic systems was among the first work in this direction, and this approach to molecular logic has now become broadly accepted in the community. That said, recent work [73, 90, 111] hints at a move towards implementing entire molecular computing systems on a single surface. This line of research aims to combine the advantages of autonomous molecular logic, as described here, with the additional benefits conferred by confinement to a surface, such as the possibility for direct sequence reuse in different parts of the circuit. Our work on large-scale molecular automata, which we characterize here as a move towards "wide" circuits of gates operating in parallel, demonstrated that molecular systems can implement non-trivial interactive systems, and we are hopeful that there will be future incarnations of the **MAYA** series of automata, subject to the difficulty of rendering game strategies as Boolean formulae [102].

Our subsequent work on DNAzyme cascades and multi-layer logic circuits, which we characterize here as a move towards "deep" circuits, illustrates the challenges involved in connecting substrate-cleaving DNAzymes into sequential logic circuits. Inspired by protein cascades, our chosen mechanism relied on the secondary structure of the uncleaved substrate molecule to sequester the downstream activator sequence prior to cleavage. However, there may be other potential approaches: in particular, our previous work on DNAzyme ligase logic gates [105] suggests a possible way forward, in which DNAzyme processing units assemble activators (or inhibitors) from short, inactive strands to regulate other DNAzymes. The difficulty with this approach is that turnover in ligase systems is severely restricted by the stability of the bond between the DNAzyme and the ligated product strand. These issues notwithstanding, the sheer variety of chemical reactions that can be catalyzed by DNAzymes [11, 17, 18, 34, 39, 59, 72, 78, 87, 98, 99, 122, 131] offers intriguing possibilities for DNAzyme information processing units as components of a "synthetic metabolism" in artificial living systems [109, 129]; here we may draw inspiration from fascinating work on self-replicating ribozymes [45, 48, 56, 62, 75, 79, 80, 116, 117, 137]. Ribozymes are RNA enzymes, which are similar to DNAzymes but occur naturally, and which may have formed a key component of prebiotic "life".

Our work, and the work of others, on DNAzyme biosensors highlights a key practical advantage of DNAzyme technology: despite the wide array of reactions that DNAzymes catalyze, they are still just short, single strands of DNA and are therefore relatively simple to design, very cheap to synthesize, and robust to degradation in the laboratory. In particular, the efficient RNA-cleaving activity of many DNAzymes makes them attractive for use in gene silencing applications, as they can cleave the RNA molecules that serve as an intermediary in the gene expression process, thereby rendering them inoperable. Preliminary work has demonstrated that DNAzymes can function in the intracellular chemical environment [46]: potential therapeutic applications of DNAzymes include cancer therapy [21–24, 29, 50, 69] and antiviral applications [84]. This work offers a direct route to the development of logic-based molecular therapeutics that perform non-trivial information processing based on the observed intracellular chemical environment before making an informed decision about whether to activate its therapeutic gene-silencing DNAzyme payload. This offers the potential for highly targeted therapy, so that well-defined cell types (such as tumor cells) may be eliminated with minimal side effects. Thus, there is great potential for future research in DNAzyme-based computing systems in a wide range of fields, from logic and artificial life to biomedical diagnostics and therapeutics.

Acknowledgments We acknowledge our other experimental collaborators, in particular, Joanne Macdonald, Sergei Rudchenko, Steven Graves, and Carl Brown, III. This material is based upon work supported by the National Science Foundation under grants 1027877, 1028238, and 1318833. M.R.L. gratefully acknowledges support from the New Mexico Cancer Nanoscience and Microsystems Training Center.

References

1. Adleman, L.M.: Molecular computation of solutions to combinatorial problems. Science **266**, 1021–1024 (1994)
2. Aguirre, S.D., Ali, M.M., Kanda, P., Li, Y.: Detection of bacteria using fluorogenic DNAzymes. J. Vis. Exp. **63**, e3961 (2012). doi:10.3791/3961
3. Aguirre, S.D., Ali, M.M., Salena, B.J., Li, Y.: A sensitive DNA enzyme-based fluorescent assay for bacterial detection. Biomolecules **3**, 563–577 (2013). doi:10.3390/biom3030563
4. Ali, M.M., Aguirre, S.D., Lazim, H., Li, Y.: Fluorogenic DNAzyme probes as bacterial indicators. Angew. Chem. Int. Ed. **50**, 3751–3754 (2011)
5. Ali, M.M., Aguirre, S.D., Mok, W.W.K., Li, Y.: Developing fluorogenic RNA-cleaving DNAzymes for biosensing applications. In: Hartig, J.S., (ed.) Ribozymes, Methods in Molecular Biology, vol. 848, pp. 395–418. Springer, New York (2012)
6. Ali, M.M., Li, Y.: Colorimetric sensing using allosteric DNAzyme-coupled rolling circle amplification and a peptide nucleic acid-organic dye probe. Angew. Chem. Int. Ed. **48**, 3512–3515 (2009)
7. Alon, U.: An Introduction to Systems Biology: Design Principles of Biological Circuits. Chapman & Hall/CRC, Boca Raton (2007)
8. Bennett, C.H.: The thermodynamics of computation–a review. Int. J. Theor. Phys. **21**(12), 905–940 (1982)
9. Bhatt, S., Gething, P.W., Brady, O.J., Messina, J.P., Farlow, A.W., Moyes, C.L., Drake, J.M., Brownstein, J.S., Hoen, A.G., Sankoh, O., Myers, M.F., George, D.B., Jaenisch, T., Wint, G.R.W., Simmons, C.P., Scott, T.W., Farrar, J.J., Hay, S.I.: The global distribution and burden of dengue. Nature **496**, 504–507 (2013). doi:10.1038/nature12060
10. Bone, S.M., Hasick, N.J., Lima, N.E., Erskine, S.M., Mokany, E., Todd, A.V.: DNA-only cascade: A universal tool for signal amplification, enhancing the detection of target analytes. Anal. Chem. **86**(18), 9106–9113 (2014). doi:10.1021/ac501811r

11. Brandsen, B.M., Velez, T.E., Sachdeva, A., Ibrahim, N.A., Silverman, S.K.: DNA-catalyzed lysine side chain modification. Angew. Chem. Int. Ed. **53**(34), 9045–9050 (2014). doi:10. 1002/anie.201404622

12. Breaker, R.R.: In vitro selection of catalytic polynucleotides. Chem. Rev. **97**(2), 371–390 (1997)

13. Breaker, R.R., Joyce, G.F.: A DNA enzyme with Mg^{2+}-dependent RNA phosphoesterase activity. Chem. Biol. **2**(10), 655–656 (1995)

14. Brown III, C.W., Lakin, M.R., Fabry-Wood, A., Horwitz, E.K., Baker, N.A., Stefanovic, D., Graves, S.W.: A unified sensor architecture for isothermal detection of double-stranded DNA, oligonucleotides, and small molecules. ChemBioChem **16**, 725–730 (2015). doi:10. 1002/cbic.201402615

15. Brown III, C.W., Lakin, M.R., Horwitz, E.K., Fanning, M.L., West, H.E., Stefanovic, D., Graves, S.W.: Signal propagation in multi-layer DNAzyme cascades using structured chimeric substrates. Angew. Chem. Int. Ed. **53**(28), 7183–7187 (2014). doi:10.1002/anie.201402691

16. Brown III, C.W., Lakin, M.R., Stefanovic, D., Graves, S.W.: Catalytic molecular logic devices by DNAzyme displacement. ChemBioChem **15**, 950–954 (2014). doi:10.1002/cbic. 201400047

17. Chandra, M., Sachdeva, A., Silverman, S.K.: DNA-catalyzed sequence-specific hydrolysis of DNA. Nat. Chem. Biol. **5**(10), 718–720 (2009)

18. Chandrasekar, J., Silverman, S.K.: Catalytic DNA with phosphatase activity. Proc. Natl. Acad. Sci. U.S. Am. **110**(14), 5315–5320 (2013). doi:10.1073/pnas.1221946110

19. Chen, Y.J., Dalchau, N., Srinivas, N., Phillips, A., Cardelli, L., Soloveichik, D., Seelig, G.: Programmable chemical controllers made from DNA. Nat. Nanotechnol. **8**, 755–762 (2013). doi:10.1038/nnano.2013.189

20. Credi, A.: Molecules that make decisions. Angew. Chem. Int. Ed. **46**(29), 5472–5475 (2007)

21. Dass, C.R.: Deoxyribozymes: cleaving a path to clinical trials. Trend. Pharmacol. Sci. **25**(8), 395–397 (2004)

22. Dass, C.R., Choong, P.F., Khachigian, L.M.: DNAzyme technology and cancer therapy: cleave and let die. Mol. Cancer Ther. **7**(2), 243–251 (2008). doi:10.1158/1535-7163.MCT-07-0510

23. Dass, C.R., Galloway, S.J., Choong, P.F.: Dz13, a c-jun DNAzyme, is a potent inducer of caspase-2 activation. Oligonucleotides **20**(3), 137–146 (2010). doi:10.1089/oli.2009.0226

24. Dass, C.R., Saravolac, E.G., Li, Y., Sun, L.Q.: Cellular uptake, distribution, and stability of 10–23 deoxyribozymes. Antisense Nucl. Acid Drug Dev. **12**, 289–299 (2002)

25. Dirks, R.M., Bois, J.S., Schaeffer, J.M., Winfree, E., Pierce, N.A.: Thermodynamic analysis of interacting nucleic acid strands. SIAM Rev. **49**, 65–88 (2007)

26. Dirks, R.M., Pierce, N.A.: A partition function algorithm for nucleic acid secondary structure including pseudoknots. J. Comput. Chem. **24**, 1664–1677 (2003)

27. Dirks, R.M., Pierce, N.A.: An algorithm for computing nucleic acid base-pairing probabilities including pseudoknots. J. Comput. Chem. **25**, 1295–1304 (2004)

28. Eckhoff, G., Codrea, V., Ellington, A.D., Chen, X.: Beyond allostery: catalytic regulation of a deoxyribozyme through an entropy-driven DNA amplifier. J. Syst. Chem. **1**, 13 (2010). doi:10.1186/1759-2208-1-13

29. Elahy, M., Dass, C.R.: Dz13: c-jun downregulation and tumour cell death. Chem. Biol. Drug Des. **78**, 909–912 (2011). doi:10.1111/j.1747-0285.2011.01166.x

30. Elbaz, J., Lioubashevski, O., Wang, F., Remacle, F., Levine, R.D., Willner, I.: DNA computing circuits using libraries of DNAzyme subunits. Nat. Nanotechnol. **5**(6), 417–422 (2010)

31. Elbaz, J., Moshe, M., Shlyahovsky, B., Willner, I.: Cooperative multicomponent self-assembly of nucleic acid structures for the activation of DNAzyme cascades: A paradigm for DNA sensors and aptasensors. Chemistry–A. Eur. J. **15**, 3411–3418 (2009)

32. Elbaz, J., Wang, F., Remacle, F., Willner, I.: PH-programmable DNA logic arrays powered by modular DNAzyme libraries. Nano Lett. **12**, 6049–6054 (2012). doi:10.1021/nl300051g

33. Farfel, J., Stefanovic, D.: Towards practical biomolecular computers using microfluidic deoxyribozyme logic gate networks. In: Carbone, A., Pierce, N.A. (eds.) Proceedings of the 11th International Meeting on DNA Computing, Lecture Notes in Computer Science, vol. 3892, pp. 38–54. Springer, Heidelberg (2006)

34. Flynn-Charlebois, A., Wang, Y., Prior, T.K., Rashid, I., Hoadley, K.A., Coppins, R.L., Wolf, A.C., Silverman, S.K.: Deoxyribozymes with 2'-5' RNA ligase activity. J. Am. Chem. Soc. **125**, 2444–2454 (2003). doi:10.1021/ja028774y

35. Gerasimova, Y.V., Cornett, E.M., Edwards, E., Su, X., Rohde, K.H., Kolpashchikov, D.M.: Deoxyribozyme cascade for visual detection of bacterial RNA. ChemBioChem **14**, 2087–2090 (2013). doi:10.1002/cbic.201300471

36. Gerasimova, Y.V., Kolpashchikov, D.M.: Folding of 16S rRNA in a signal-producing structure for the detection of bacteria. Angew. Chem. Int. Ed. **52**, 10586–10588 (2013). doi:10.1002/anie.201303919

37. Goldman, N., Bertone, P., Chen, S., Dessimoz, C., LeProust, E.M., Sipos, B., Birney, E.: Towards practical, high-capacity, low-maintenance information storage in synthesized DNA. Nature **494**, 77–80 (2013). doi:10.1038/nature11875

38. Goudarzi, A., Lakin, M.R., Stefanovic, D.: DNA reservoir computing: a novel molecular computing approach. In: Soloveichik, D., Yurke, B. (eds.) Proceedings of the 19th International Conference on DNA Computing and Molecular Programming, Lecture Notes in Computer Science, vol. 8141, pp. 76–89. Springer, Heidelberg (2013). doi:10.1007/978-3-319-01928-4_6

39. Gu, H., Furukawa, K., Weinberg, Z., Berenson, D.F., Breaker, R.R.: Small, highly active DNAs that hydrolyze DNA. J. Am. Chem. Soc. **135**, 9121–9129 (2013). doi:10.1021/ja403585e

40. He, S., Qu, L., Shen, Z., Tan, Y., Zeng, M., Liu, F., Jiang, Y., Li, Y.: Highly specific recognition of breast tumors by an RNA-cleaving fluorogenic DNAzyme probe. Anal. Chem. **87**(1), 569–577 (2014). doi:10.1021/ac5031557

41. Huang, P.J.J., Liu, M., Liu, J.: Functional nucleic acids for detecting bacteria. Rev. Anal. Chem. **32**(1), 77–89 (2013). doi:10.1515/revac-2012-0027

42. Huang, P.J.J., Vazin, M., Liu, J.: In vitro selection of a new lanthanide-dependent DNAzyme for ratiometric sensing lanthanides. Anal. Chem. **86**(19), 9993–9999 (2014). doi:10.1021/ac5029962

43. Huang, P.J.J., Vazin, M., Matuszek, Z., Liu, J.: A new heavy lanthanide-dependent DNAzyme displaying strong metal cooperativity and unrescuable phosphorothioate effect. Nucleic Acid. Res. **43**(1), 461–469 (2015). doi:10.1093/nar/gku1296

44. Hwang, K., Wu, P., Kim, T., Lei, L., Tian, S., Wang, Y., Lu, Y.: Photocaged DNAzymes as a general method for sensing metal ions in living cells. Angew. Chem. Int. Ed. **53**, 13798–13802 (2014)

45. Hayden, J.E., Riley, C.A., Burton, A.S., Lehman, N.: RNA-directed construction of structurally complex and active ligase ribozymes through recombination. RNA **11**(11), 1678–1687 (2005). doi:10.1261/rna.2125305

46. Kahan-Hanum, M., Douek, Y., Adar, R., Shapiro, E.: A library of programmable DNAzymes that operate in a cellular environment. Sci. Rep. **3**, 1535 (2013)

47. Katz, E., Privman, V.: Enzyme-based logic systems for information processing. Chem. Soc. Rev. **39**(5), 1835–1857 (2010)

48. Kim, D.E., Joyce, G.F.: Cross-catalytic replication of an RNA ligase ribozyme. Chem. Biol. **11**, 1505–1512 (2004). doi:10.1016/j.chembiol.2004.08.021

49. Kim, J., Winfree, E.: Synthetic in vitro transcriptional oscillators. Mol. Syst. Biol. **7**, 465 (2011). doi:10.1038/msb.2010.119

50. Kim, S.H., Dass, C.R.: Induction of caspase-2 activation by a DNA enzyme evokes tumor cell apoptosis. DNA Cell Biol. **31**(1) (2012). doi:10.1089/dna.2011.1323

51. Kolpashchikov, D.M.: A binary deoxyribozyme for nucleic acid analysis. ChemBioChem **8**, 2039–2042 (2007)

52. Kolpashchikov, D.M.: Binary probes for nucleic acid analysis. Chem. Rev. **110**, 4709–4723 (2010)

53. Kolpashchikov, D.M., Gerasimova, Y.V., Khan, M.S.: DNA nanotechnology for nucleic acid analysis: DX motif-based sensor. ChemBioChem **12**, 2564–2567 (2011)

54. Lake, A., Shang, S., Kolpashchikov, D.M.: Molecular logic gates connected through DNA four-way junctions. Angew. Chem. Int. Ed. **49**, 4459–4462 (2010). doi:10.1002/anie.200907135

55. Lakin, M.R., Brown III, C.W., Horwitz, E.K., Fanning, M.L., West, H.E., Stefanovic, D., Graves, S.W.: Biophysically inspired rational design of structured chimeric substrates for DNAzyme cascade engineering. PLoS ONE **9**(10), e110986 (2014). doi:10.1371/journal. pone.0110986
56. Lam, B.J., Joyce, G.F.: Autocatalytic aptazymes enable ligand-dependent exponential amplification of RNA. Nat. Biotechnol. **27**(3), 288–292 (2009). doi:10.1038/nbt.1528
57. Lan, T., Furuya, K., Lu, Y.: A highly selective lead sensor based on a classic lead DNAzyme. Chem. Commun. **46**, 3896–3898 (2010)
58. Lederman, H., Macdonald, J., Stefanovic, D., Stojanovic, M.N.: Deoxyribozyme-based three-input logic gates and construction of a molecular full adder. Biochemistry **45**(4), 1194–1199 (2006)
59. Lee, C.S., Mui, T.P., Silverman, S.K.: Improved deoxyribozymes for synthesis of covalently branched DNA and RNA. Nucleic Acid. Res. **39**(1), 269–279 (2011). doi:10.1093/nar/gkq753
60. Lee, J.H., Wang, Z., Liu, J., Lu, Y.: Highly sensitive and selective colorimetric sensors for uranyl (UO_2^{2+}): development and comparison of labeled and label-free DNAzyme-gold nanoparticle systems. J. Am. Chem. Soc. **130**, 14217–14226 (2008)
61. Levy, M., Ellington, A.D.: Exponential growth by cross-catalytic cleavage of deoxyribozymogens. Proc. Natl. Acad. Sci. U.S. Am. **100**(11), 6416–6421 (2003). doi:10.1073/pnas. 1130145100
62. Lincoln, T.A., Joyce, G.F.: Self-sustained replication of an RNA enzyme. Science **323**, 1229–1232 (2009)
63. Liu, J., Cao, Z., Lu, Y.: Functional nucleic acid sensors. Chem. Rev. **109**, 1948–1998 (2009). doi:10.1021/cr030183i
64. Lu, C.H., Wang, F., Willner, I.: Zn^{2+}-ligation DNAzyme-driven enzymatic and nonenzymatic cascades for the amplified detection of DNA. J. Am. Chem. Soc. **134**(25), 10651–10658 (2012)
65. Lund, K., Manzo, A.J., Dabby, N., Michelotti, N., Johnson-Buck, A., Nangreave, J., Taylor, S., Pei, R., Stojanovic, M.N., Walter, N.G., Winfree, E., Yan, H.: Molecular robots guided by prescriptive landscapes. Nature **465**, 206–209 (2010)
66. Macdonald, J., Li, Y., Sutovic, M., Lederman, H., Pendri, K., Lu, W., Andrews, B.L., Stefanovic, D., Stojanovic, M.N.: Medium scale integration of molecular logic gates in an automaton. Nano Lett. **6**(11), 2598–2603 (2006)
67. Macdonald, J., Stefanovic, D., Stojanovic, M.N.: DNA computers for work and play. Sci. Am. **299**(5), 84–91 (2008)
68. McManus, S.A., Li, Y.: Turning a kinase deoxyribozyme into a sensor. J. Am. Chem. Soc. **135**(19), 7181–7186 (2013)
69. Mitchell, A., Dass, C.R., Sun, L.Q., Khachigian, L.M.: Inhibition of human breast carcinoma proliferation, migration, chemoinvasion and solid tumour growth by DNAzymes targeting the zinc finger transcription factor EGR-1. Nucleic Acid. Res. **32**(10), 3065–3069 (2004). doi:10. 1093/nar/gkh626
70. Mokany, E., Bone, S.M., Young, P.E., Doan, T.B., Todd, A.V.: MNAzymes, a versatile new class of nucleic acid enzymes that can function as biosensors and molecular switches. J. Am. Chem. Soc. **132**, 1051–1059 (2010). doi:10.1021/ja9076777
71. Montagne, K., Plasson, R., Sakai, Y., Fujii, T., Rondelez, Y.: Programming an in vitro DNA oscillator using a molecular networking strategy. Mol. Syst. Biol. **7**, 466 (2011). doi:10.1038/ msb.2011.12
72. Mui, T.P., Silverman, S.K.: Convergent and general one-step DNA-catalyzed synthesis of multiply branched DNA. Organ. Lett. **10**(20), 4417–4420 (2008)
73. Muscat, R.A., Strauss, K., Ceze, L., Seelig, G.: DNA-based molecular architecture with spatially localized components. In: ISCA '13: Proceedings of the 40th Annual International Symposium on Computer Architecture, pp. 177–188. ACM, New York (2013)
74. Olah, M.J., Stefanovic, D.: Superdiffusive transport by multivalent molecular walkers moving under load. Phys. Rev. E **87**, 062713 (2013). doi:10.1103/PhysRevE.87.062713
75. Olea, Jr. C., Horning, D.P., Joyce, G.F.: Ligand-dependent exponential amplification of a self-replicating LRNA enzyme. J. Am. Chem. Soc. **134**, 8050–8053 (2012). doi:10.1021/ ja302197x

76. Orbach, R., Remacle, F., Levine, R.D., Willner, I.: DNAzyme-based 2:1 and 4:1 multiplexers and 1:2 demultiplexer. Chem. Sci. **5**, 1074–1081 (2014)
77. Osborne, S.E., Ellington, A.D.: Nucleic acid selection and the challenge of combinatorial chemistry. Chem. Rev. **97**(2), 349–370 (1997)
78. Parker, D.J., Xiao, Y., Aguilar, J.M., Silverman, S.K.: DNA catalysis of a normally disfavored RNA hydrolysis reaction. J. Am. Chem. Soc. **135**(23), 8472–8475 (2013). doi:10. 1021/ja4032488
79. Paul, N., Joyce, G.F.: A self-replicating ligase ribozyme. Proc. Natl. Acad. Sci. U.S. Am. **99**(20), 12733–12740 (2002). doi:10.1073/pnas.202471099
80. Paul, N., Joyce, G.F.: Minimal self-replicating systems. Curr. Opin. Chem. Biol. **8**, 634–639 (2004). doi:10.1016/j.cbpa.2004.09.005
81. Paul, N., Springsteen, G., Joyce, G.F.: Conversion of a ribozyme to a deoxyribozyme through in vitro evolution. Chem. Biol. **13**, 329–338 (2006). doi:10.1016/j.chembiol.2006.01.007
82. Pei, R., Matamoros, E., Liu, M., Stefanovic, D., Stojanovic, M.N.: Training a molecular automaton to play a game. Nat. Nanotechnol. **5**, 773–777 (2010)
83. Pei, R., Taylor, S.K., Stefanovic, D., Rudchenko, S., Mitchell, T.E., Stojanovic, M.N.: Behavior of polycatalytic assemblies in a substrate-displaying matrix. J. Am. Chem. Soc. **128**(39), 12693–12699 (2006)
84. Peracchi, A.: Prospects for antiviral ribozymes and deoxyribozymes. Rev. Med. Virol. **14**, 47–64 (2004). doi:10.1002/rmv.415
85. Plotnikov, A., Zehorai, E., Procaccia, S., Seger, R.: The MAPK cascades: Signaling components, nuclear roles and mechanisms of nuclear translocation. Biochimica et Biophysica Acta **1813**, 1619–1633 (2011)
86. Poje, J.E., Kastratovic, T., Macdonald, A.R., Guillermo, A.C., Troetti, S.E., Jabado, O.J., Fanning, M.L., Stefanovic, D., Macdonald, J.: Visual displays that directly interface and provide read-outs of molecular states via molecular graphics processing units. Angew. Chem. Int. Ed. **53**(35), 9222–9225 (2014)
87. Purtha, W.E., Coppins, R.L., Smalley, M.K., Silverman, S.K.: General deoxyribozyme-catalyzed synthesis of native 3'-5' RNA linkages. J. Am. Chem. Soc. **127**, 13124–13125 (2005). doi:10.1021/ja0533702
88. Qian, L., Soloveichik, D., Winfree, E.: Efficient Turing-universal computation with DNA polymers. In: Sakakibara, Y., Mi, Y. (eds.) Proceedings of the 16th International Conference on DNA Computing and Molecular Programming, Lecture Notes in Computer Science, vol. 6518, pp. 123–140. Springer, New York (2011)
89. Qian, L., Winfree, E.: Scaling up digital circuit computation with DNA strand displacement cascades. Science **332**, 1196–1201 (2011). doi:10.1126/science.1200520
90. Qian, L., Winfree, E.: Parallel and scalable computation and spatial dynamics with DNA-based chemical reaction networks on a surface. In: Murata, S., Kobayashi, S. (eds.) Proceedings of the 20th International Conference on DNA Computing and Molecular Programming, Lecture Notes in Computer Science, vol. 8727, pp. 114–131. Springer, New York (2014)
91. Qian, L., Winfree, E., Bruck, J.: Neural network computation with DNA strand displacement cascades. Nature **475**, 368–372 (2011). doi:10.1038/nature10262
92. Robertson, M.P., Ellington, A.: In vitro selection of an allosteric ribozyme that transduces analytes to amplicons. Nat. Biotechol. **17**(1), 62–66 (1999)
93. Santoro, S.W., Joyce, G.F.: A general purpose RNA-cleaving DNA enzyme. Proc. Natl. Acad. Sci. U.S. Am. **94**, 4262–4266 (1997)
94. Saunders, M.J., Edwards, B.S., Zhu, J., Sklar, L.A., Graves, S.W.: Microsphere-based flow cytometry protease assays for use in protease activity detection and high-throughput screening. In: Robinson, J.P. (ed.) Current Protocols in Cytometry Unit 13.12. Wiley, Hoboken (2010). doi:10.1002/0471142956.cy1312s54
95. Schaeffer, H.J., Weber, M.J.: Mitogen-activated protein kinases: specific messages from ubiquitous messengers. Mol. Cell. Biol. **19**(4), 2435–2444 (1999)
96. Seelig, G., Soloveichik, D., Zhang, D.Y., Winfree, E.: Enzyme-free nucleic acid logic circuits. Science **314**, 1585–1588 (2006). doi:10.1126/science.1132493

97. Semenov, O., Olah, M.J., Stefanovic, D.: Cooperative linear cargo transport with molecular spiders. Nat. Comput. **12**(2), 259–276 (2013). doi:10.1007/s11047-012-9357-2

98. Silverman, S.K.: Deoxyribozymes: DNA catalysts for bioorganic chemistry. Organ. Biomol. Chem. **2**, 2701–2706 (2004)

99. Silverman, S.K.: Catalytic DNA (deoxyribozymes) for synthetic applications—current abilities and future prospects. Chem. Commun. pp. 3467–3485 (2008). doi:10.1039/b807292m

100. Simmons, C.P., Farrar, J.J., van Vin Chau, N., Wills, B.: Dengue. New Engl. J. Med. **366**, 1423–1432 (2012)

101. Soloveichik, D., Seelig, G., Winfree, E.: DNA as a universal substrate for chemical kinetics. Proc. Natl. Acad. Sci. U.S. Am. **107**(12), 5393–5398 (2010). doi:10.1073/pnas.0909380107

102. Stefanovic, D., Stojanovic, M.N.: Computing game strategies. In: Computability in Europe: The Nature of Computation, pp. 383–392. Milano (2013)

103. Stojanovic, M.N., Mitchell, T.E., Stefanovic, D.: Deoxyribozyme-based logic gates. J. Am. Chem. Soc. **124**, 3555–3561 (2002)

104. Stojanovic, M.N., de Prada, P., Landry, D.W.: Catalytic molecular beacons. ChemBioChem **2**, 411–415 (2001)

105. Stojanovic, M.N., Semova, S., Kolpashchikov, D., Macdonald, J., Morgan, C., Stefanovic, D.: Deoxyribozyme-based ligase logic gates and their initial circuits. J. Am. Chem. Soc. **127**, 6914–6915 (2005). doi:10.1021/ja043003a

106. Stojanovic, M.N., Stefanovic, D.: Deoxyribozyme-based half adder. J. Am. Chem. Soc. **125**(22), 6673–6676 (2003)

107. Stojanovic, M.N., Stefanovic, D.: A deoxyribozyme-based molecular automaton. Nat. Biotechnol. **21**(9), 1069–1074 (2003)

108. Stojanovic, M.N., Stefanovic, D., Rudchenko, S.: Exercises in molecular computing. Acc. Chem. Res. **47**, 1845–1852 (2014)

109. Tabor, J.J., Levy, M., Ellington, A.D.: Deoxyribozymes that recode sequence information. Nucleic Acid. Res. **34**, 2166–2172 (2006)

110. Tang, J., Breaker, R.R.: Rational design of allosteric ribozymes. Chem. Biol. **4**(6), 453–459 (1997)

111. Teichmann, M., Kopperger, E., Simmel, F.C.: Robustness of localized DNA strand displacement cascades. ACS Nano **8**(8), 8487–8496 (2014)

112. Teller, C., Shimron, S., Willner, I.: Aptamer-DNAzyme hairpins for amplified biosensing. Anal. Chem. **81**, 9114–9119 (2009)

113. Tram, K., Kanda, P., Li, Y.: Lighting up RNA-cleaving DNAzymes for biosensing. J. Nucleic Acid. p. 958683 (2012). doi:10.1155/2012/958683

114. Tram, K., Kanda, P., Salena, B.J., Huan, S., Li, Y.: Translating bacterial detection by DNAzymes into a litmus test. Angew. Chem. Int. Ed. **53**(47), 12799–12802 (2014). doi:10. 1002/anie.201407021

115. Tyagi, S., Kramer, F.R.: Molecular beacons: probes that fluoresce upon hybridization. Nat. Biotechnol. **14**(3), 303–309 (1996)

116. Vaidya, N., Manapat, M.L., Chen, I.A., Xulvi-Brunet, R., Hayden, E.J., Lehman, N.: Spontaneous network formation among cooperative RNA replicators. Nature **491**, 72–77 (2012). doi:10.1038/nature11549

117. Vaidya, N., Walker, S.I., Lehman, N.: Recycling of informational units leads to selection of replicators in a prebiotic soup. Chem. Biol. **20**(2), 241–252 (2013). doi:10.1016/j.chembiol. 2013.01.007

118. Wang, F., Elbaz, J., Orbach, R., Magen, N., Willner, I.: Amplified analysis of DNA by the autonomous assembly of polymers consisting of DNAzyme wires. J. Am. Chem. Soc. **133**, 17149–17151 (2011)

119. Wang, F., Elbaz, J., Teller, C., Willner, I.: Amplified detection of DNA through an autocatalytic and catabolic DNAzyme-mediated process. Angew. Chem. Int. Ed. **50**, 295–299 (2011)

120. Wang, F., Elbaz, J., Willner, I.: Enzyme-free amplified detection of DNA by an autonomous ligation DNAzyme machinery. J. Am. Chem. Soc. **134**, 5504–5507 (2012)

121. Wang, F., Lu, C.H., Willner, I.: From cascaded catalytic nucleic acids to enzyme-DNA nanostructures: controlling reactivity, sensing, logic operations, and assembly of complex structures. Chem. Rev. **114**, 2881–2941 (2014). doi:10.1021/cr400354z

122. Wang, Y., Silverman, S.K.: Deoxyribozymes that synthesize branched and lariat RNA. J. Am. Chem. Soc. **125**, 6880–6881 (2003). doi:10.1021/ja035150z

123. Wang, Z., Lee, J.H., Lu, Y.: Label-free colorimetric detection of lead ions with a nanomolar detection limit and tunable dynamic range by using gold nanoparticles and DNAzyme. Adv. Mater. **20**(17), 3263–3267 (2008)

124. Wernick, W.: Complete sets of logical functions. Trans. Am. Math. Soc. **51**, 117–132 (1942)

125. Wolfe, B.R., Pierce, N.A.: Sequence design for a test tube of interacting nucleic acid strands. ACS Synth. Biol. **4**(10), 1086–1100 (2015). doi:10.1021/sb5002196

126. Wu, P., Hwang, K., Lan, T., Lu, Y.: A DNAzyme-gold nanoparticle probe for uranyl ion in living cells. J. Am. Chem. Soc. **135**, 5254–5257 (2013). doi:10.1021/ja400150v

127. Xiang, Y., Lu, Y.: Using personal glucose meters and functional DNA sensors to quantify a variety of analytical targets. Nat. Chem. **3**, 697–703 (2011). doi:10.1038/nchem.1092

128. Xiang, Y., Lu, Y.: Expanding targets of DNAzyme-based sensors through deactivation and activation of DNAzymes by single uracil removal: Sensitive fluorescent assay of uracil-DNA glycosylase. Anal. Chem. **84**, 9981–9987 (2012). doi:10.1021/ac302424f

129. Xiang, Y., Wang, Z., Xing, H., Lu, Y.: Expanding DNAzyme functionality through enzyme cascades with applications in single nucleotide repair and tunable DNA-directed assembly of nanomaterials. Chem. Sci. **4**, 398–404 (2013). doi:10.1039/c2sc20763j

130. Xiang, Y., Wu, P., Tan, L.H., Lu, Y.: DNAzyme-functionalized gold nanoparticles for biosensing. In: Gu, M.B., Kim, H.S. (eds.) Biosensors Based on Aptamers and Enzymes, Advances in Biochemical Engineering/Biotechnology, vol. 140, pp. 93–120. Springer, New York (2014)

131. Xiao, Y., Wehrmann, R.J., Ibrahim, N.A., Silverman, S.K.: Establishing broad generality of DNA catalysts for site-specific hydrolysis of single-stranded DNA. Nucleic Acid. Res. **40**(4), 1778–1786 (2012). doi:10.1093/nar/gkr860

132. Yashin, R., Rudchenko, S., Stojanovic, M.N.: Networking particles over distance using oligonucleotide-based devices. J. Am. Chem. Soc. **129**(50), 15581–15584 (2007)

133. Yurke, B., Mills Jr., A.P.: Using DNA to power nanostructures. Genet. Program. Evol. Mach. **4**, 111–122 (2003). doi:10.1023/A:1023928811651

134. Yurke, B., Turberfield, A.J., Mills Jr., A.P., Simmel, F.C., Neumann, J.L.: A DNA-fuelled molecular machine made of DNA. Nature **406**, 605–608 (2000). doi:10.1038/35020524

135. Zadeh, J.N., Steenberg, C.D., Bois, J.S., Wolfe, B.R., Pierce, M.B., Khan, A.R., Dirks, R.M., Pierce, N.A.: NUPACK: analysis and design of nucleic acid systems. J. Comput. Chem. **32**, 170–173 (2011)

136. Zadeh, J.N., Wolfe, B.R., Pierce, N.A.: Nucleic acid sequence design via efficient ensemble defect optimization. J. Comput. Chem. **32**, 439–452 (2011)

137. Zenisek, S.F.M., Hayden, E.J., Lehman, N.: Genetic exchange leading to self-assembling RNA species upon encapsulation in artificial protocells. Artif. Life **13**(3), 279–289 (2007). doi:10.1162/artl.2007.13.3.279

138. Zhang, D.Y.: Cooperative hybridization of oligonucleotides. J. Am. Chem. Soc. **133**, 1077–1086 (2011). doi:10.1021/ja109089q

139. Zhang, D.Y., Seelig, G.: Dynamic DNA nanotechnology using strand-displacement reactions. Nat. Chem. **3**, 103–113 (2011). doi:10.1038/nchem.957

140. Zhang, D.Y., Turberfield, A.J., Yurke, B., Winfree, E.: Engineering entropy-driven reactions and networks catalyzed by DNA. Science **318**, 1121–1125 (2007)

141. Zhang, X.B., Kong, R.M., Lu, Y.: Metal ion sensors based on DNAzymes and related DNA molecules. Annu. Rev. Anal. Chem. **4**, 105–128 (2011)

Chapter 2
Enzyme-Based Reversible Logic Gates Operated in Flow Cells

Evgeny Katz and Brian E. Fratto

Abstract Reversible logic gates, such as Feynman gate (Controlled NOT), Double Feynman gate, Toffoli gate and Peres gate, with 2-input/2-output and 3-input/3-output channels, were realized using reactions biocatalyzed by enzymes and performed in flow systems. The flow devices were constructed using a modular approach, where each flow cell was modified with one enzyme that biocatalyzed one chemical reaction. Assembling the biocatalytic flow cells in different networks, with different pathways for transporting the reacting species, allowed the multi-step processes mimicking various reversible logic gates. The chapter emphasizes "logic" reversibility but not the "physical" reversibility of the constructed systems. Their advantages and disadvantages are discussed and potential use in biosensing systems, rather than in computing devices, is suggested.

2.1 Introduction

Computer technology presently based on silicon-materials and binary algorithms is coming to the end of its exponential development, being limited not only by further component-miniaturization but also the speed of their operation. Conceptually novel ideas are needed to break through these limitations. In order to reach another level of information processing and to maintain fast progress in computer technology, new and intuitive forms of technology are needed. The quest for novel ideas in information processing has resulted in several exciting directions in the general area of unconventional computing [1], including research in quantum computing and biologically inspired molecular computing. While molecular computing [2–4] is generally motivated by mimicking natural biological information processing, the tools are not necessary based on biological systems and often represented by synthetic molecules with signal-controlled properties. Synthetic molecular systems [5] and nano-species [6] have been designed to mimic the operation of Boolean logic gates and demon-

E. Katz (✉) · B.E. Fratto
Clarkson University, Potsdam, NY, USA
e-mail: ekatz@clarkson.edu

© Springer International Publishing Switzerland 2017
A. Adamatzky (ed.), *Advances in Unconventional Computing*,
Emergence, Complexity and Computation 23,
DOI 10.1007/978-3-319-33921-4_2

strate basic arithmetic functions and memory units. Despite the great progress that has been achieved in the development of molecular computing systems [7, 8], the major challenge in this research area is further increase of their complexity [9]. A new advancement in the development of molecular information systems has been achieved with use of biomolecular species [10, 11], borrowing some ideas from systems biology [12]. The first demonstration of computational processes performed by DNA molecules to solve some combinatorial problems [13] was recently extended to include the use of various biomolecular systems based on DNA/RNA [14–16], oligopeptides [17], proteins [18], enzymes [19] and even whole biological cells [20, 21] for mimicking various information processing steps, Fig. 2.1. One of the obvious advantages of biomolecular systems is their ability to integrate in artificially designed complex reacting processes mimicking multi-step information processing networks. Their operation in biological environment complementing natural biological processes was demonstrated [22]. Multi-step biochemical cascades mimicking electronic circuitries have demonstrated the ability to perform simple arithmetic operations [23–25], play games [26] and make decisions in multi-choice situations [27]. Novel functionalities, supplementary to electronics, achievable in biomolecular systems are the most challenging goals of this research [28, 29]. These systems are still far away from the natural information processing in cells, but are already much more complex than pure synthetic molecular systems [30]. Recent research in unconventional computing [1], particularly using molecular [2–5, 7–9, 31] and biomolecular [10, 11, 14, 15, 19, 32] systems has resulted in artificial (bio)chemical systems mimicking Boolean logic operations, including AND, OR, XOR, NAND, NOR and other logic gates. Reversible [33–36], reconfigurable [37, 38] and resettable [39–41] logic gates for processing of chemical signals have been designed using sophisticated synthetic molecules or complex biomolecular assemblies.

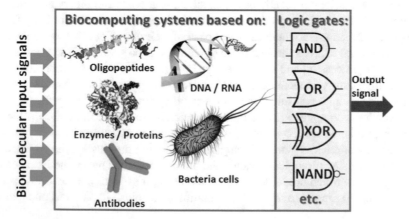

Fig. 2.1 Biocomputing systems based on various biomolecular/biological species can process multiple chemical input signals and generate an output signal according to different logic implemented in the systems (Adapted from Ref. [30] with permission)

Molecular [33–36, 42, 43] and biomolecular [44–47] systems mimicking operation of reversible logic gates (e.g., Toffoli [44], Fredkin [44–46] and Feynman [34, 35, 47] gates) are of particular interest, since they provide unique output patterns for each combination of input signals. It should be noted that the term "reversibility" in the definition of logic operations has the meaning different from commonly used in chemistry. In information processing the "logic reversibility" means the possibility to recover initial information from the processed information. This is not possible for most of trivial Boolean gates such as AND, OR, etc. Indeed, the logic output **0** can be generated by AND gate for different input combinations: **0, 0; 0, 1** and **1, 0**, thus the original pattern of the input signals cannot be recovered when the output signal **0** is generated. Special property of logic reversibility can be achieved when the number of channels for the output and input signals is the same, for example, 2-input/2-output, 3-input/3-output. In such case each combination of the input signals can produce a unique pattern of the output signals that allows recovering the initial inputs. Depending on the signal processing function, which is unique to the specific reversible logic gate, the output signals will change differently when various combinations of the inputs are applied. Using this change, the unknown input signals can always be recovered from the measured output signals if the logic function is known. This inherent property of reversible logic gates is important, particularly for biosensing applications of logic gates [48, 49] when all the information in regards to the inputs is needed. For example, medical analysis of biomarkers signaling on injuries [49] can be performed using AND logic gates combining together two biomarkers of low specificity. The decision about the medical conditions, e.g., specific kind of injury, would be based on the simultaneous presence of two biomarkers appearing at concentrations above specific thresholds. While this is enough for the medical conclusion about the presence of the injury conditions, the negative result (logic output **0**) has no clear meaning because it can originate from various combinations of biomarkers, when one or both of them appear below the thresholds. Knowing exactly the original combination of the biomarkers would be the great advantage for biomedical applications. This could be achievable with the use of reversible logic gates.

While justifying the importance of reversible logic gates another inherent physical property that is frequently emphasized is the ability to save energy. During irreversible logic computations, each bit of information lost generates $kT \ln 2$ joules of heat energy based on thermodynamic consideration [50]. This release of heat results in serious problems for computer engineering, particularly with densely packed nano-sized elements. Theoretically [51, 52] and experimentally [53], in some electronic realizations, reversible information processing should not be accompanied by an entropy increase [54], thus classifying it as an energy saving process. This is considered to be a very important feature of reversible information processing, and will be needed in future electronic computers to allow fast computing in nano-scale elements without generation of heat. However, this is not applicable to chemical systems where energy dissipation is inevitable. Therefore, the molecular/biomolecular realizations of reversible logic gates keep only *logic* reversibility, while the energy savings are illusory goals. This rather trivial and obvious conclusion should not be overlooked when conceiving and/or implementing reversible logic gates in (bio)molecular systems [44].

In order to be useful in a practical sense, reversible logic gates should be integrated in complex information processing networks, which include multi-step Boolean and non-Boolean operations. Unfortunately, some of the experimentally realized systems were all-photonic [43], where the input/output signals were optical signals. This does not allow for easy integration with molecular networks, thus resulting in standalone operations of the gates. Other reversible logic gates used chemical input signals (e.g., metal ions) [33–36], which cannot be easily produced by preceding chemical reactions. This makes integration of the designed reversible gates in complex logic networks very difficult. Although these systems illustrate novel functions, it is certainly not enough for any practical application. Only a few recent experimental examples have demonstrated reversible logic operations with input/output signals that are represented by biomolecules using DNA [44, 46] and enzyme [47] reactions, thus allowing for the extension of information processing and integration of reversible logic gates into complex biomolecular networks that can be designed to mimic biological systems.

Recently designed logic networks based on enzyme-biocatalyzed multi-step reactions [19, 55, 56] are the easiest for the practical realization of systems where logic operations are particularly useful in biomedical sensing [48, 49] as well as diagnostic applications [57, 58]. Despite the fact that complex multi-input/multi-step information processing systems have been successfully realized with enzymatic cascades [19], the requirement of several independently read output signals for realization of reversible logic gates is not a simple task. Cross-talking between enzymatic pathways and chemically produced output signals limit the complexity of the enzyme-based logic systems when they are realized in a homogeneous system. In order to assemble more complex systems for realization of reversible logic gates, "clocking" (temporal control) and spatial separation (compartmentalization) [36] of various steps (e.g., in flow devices) are needed [59, 60]. The present chapter reports on the first experimental realization of reversible logic gates [61] (Feynman gate (Controlled NOT) with 2-input/2-output channels, Double Feynman gate, Toffoli gate and Peres gate with 3-input/3-output channels) using enzyme biocatalytic reactions which are performed in modular flow systems, where the chemical inputs and outputs may be potentially extended to include additional information processing steps, thus allowing for further increase of system complexity. The studied systems were composed of simple single-enzyme-functionalized cells connected in specific networks with a modular design and are presented in the chapter in the order of increasing of their complexity.

2.2 Results and Discussion

Figure 2.2a shows a single flow cell which was used for assembling flow devices mimicking the operation of reversible logic gates. The cells were each modified with one immobilized enzyme, before they were assembled in circuitries to organize

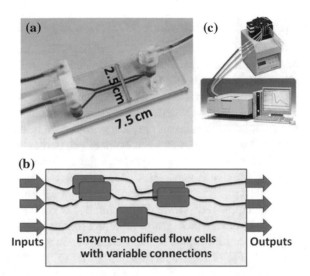

Fig. 2.2 a A single flow cell used as a component of the system. **b** Schematic of the system with many interconnected enzyme-modified cells assembled in a circuitry with variable connections to mimic different logic gates (the scheme does not correspond to any specific logic network; for specific circuitries see Figs. 2.3, 2.6 and 2.9). **c** Spectrophotometer connected to the flow system with help of a multi-channel peristaltic pump (Adapted from Ref. [66] with permission)

biocatalytic cascades composed of several interconnected biocatalytic cells, Fig. 2.2b. For realizing different logic gates, Feynman gate (Controlled NOT; CNOT) Double Feynman gate (DFG), Toffoli gate and Peres gate, the physical layout of the flow circuitries varied depending on the needs of the logic operations. The flow systems that were composed of several interconnected cells had three input and three output channels. The input channels were connected to test-tubes filled with stock solutions of the input signal chemicals, while the output channels were connected to the flow-cuvette in the spectrophotometer for optical reading of the output signals, Fig. 2.2c. The flow of reacting species moving from one enzyme-modified cell to another and finally to the spectrophotometer was maintained with a peristaltic pump.

The following biocatalytic enzymes and substrate-inputs were used to perform reversible logic operations described in the following sections:
Enzymes immobilized in flow cells: alkaline phosphatase (AP; E.C. 3.1.3.1) from bovine intestinal mucosa, glucose-6-phosphate dehydrogenase (G6PDH; E.C. 1.1. 1.49) from *Leuconostoc mesenteroides*, glucose dehydrogenase (GDH; E.C. 1.1.1.47) from *Pseudomonas sp.*, lactate dehydrogenase (LDH; E.C. 1.1.1.27) from porcine heart, glucose oxidase (GOx; E.C. 1.1.3.4) from *Aspergillus niger*, peroxidase from horseradish (HRP; E.C. 1.11.1.7) and diaphorase from *Clostridium kluyveri* (Diaph; E.C. 1.8.1.4). *Substrates/cofactors used as input signals*: p-nitrophenyl phosphate (PNPP), glucose (Glc), glucose-6-phosphate (G6P), pyruvate (Pyr), β-nicotinamide adenine dinucleotide hydrate (NAD^+), β-nicotinamide adenine dinucleotide reduced dipotassium salt (NADH) and hydrogen peroxide (H_2O_2).

2.2.1 Feynman Gate—Controlled NOT (CNOT) Gate

Here we illustrate the realized Feynman gate (Controlled NOT (CNOT) gate) [47], which is the only non-trivial reversible gate with two input and two output signals [62]. The first experimental realization of a CNOT gate was accomplished in 1995 [63]. Since its realization the CNOT gate has been implemented in many physical systems. This implementation demonstrats great importance for future quantum computing [64], however, it was never researched in chemical, particularly in biochemical, systems aiming at its integration with biomolecular logic networks. Figure 2.3a shows the truth table of the CNOT gate. The CNOT gate flips the second Input B (the target input) and directs it to the second Output Q (the target output) if and only if the first Input A (the control input) is **1**, while the first Input A is always copied to Output P (the control output). In other words, Output P represents an Identity gate for Input A, while Output Q is processing Inputs A and B according to a XOR logic operation. The logic scheme illustrating CNOT operation is shown in Fig. 2.3b–d, where Input A is directed to an Identity gate and a XOR gate is activated by both Inputs A and B. In our experimental realization we defined Input A as a Tris-buffer solution (0.1 M, pH 7.1) containing p-nitrophenyl phosphate (PNPP) and pyruvate (Pyr) with concentrations 10 and 1 mM, respectively, to represent logic **1** value. Logic **0** value for Input A was defined as the absence of PNPP and Pyr in the solution. Input B was defined as a Tris-buffer solution (0.1 M, pH 7.1) containing glucose-6-phosphate (G6P), 6 mM, for logic **1** value. The absence of G6P was used to encode logic **0** for Input B. Note that the concentrations of the reacting species were optimized experimentally as will be described later. PNPP representing one of the reacting species in Input A

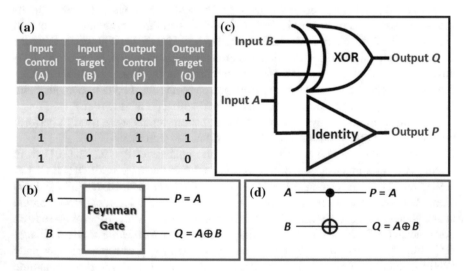

Fig. 2.3 The truth table (**a**), block diagram (**b**), logic circuitry (**c**) and equivalent electronic circuitry (**d**) for Feynman (CNOT) gate

was converted to p-nitrophenol (PNP) in the reaction biocatalyzed by alkaline phosphatase (AP) resulting in the optical absorbance increase ($\lambda_{max} = 420$ nm), Fig. 2.4. This biocatalytic reaction represented the Identity gate, Fig. 2.3c, and the produced absolute values of absorbance changes were used as Output P, which was defined as logic **1** or **0** when $|\Delta Abs|>0.1$ or $|\Delta Abs| < 0.1$, respectively. Input A was also directed to the biochemical system performing a XOR logic operation, Fig. 2.3c.

We designed a XOR gate with two oppositely directed biocatalytic redox reactions [65] activated by pyruvate (Pyr; second reacting species in Input A) and glucose-6-phosphate (G6P; Input B), Fig. 2.4. Input B was defined as logic **0** and **1** for absence of G6P and its presence with the concentration of 6 mM, respectively. The reaction media (all input solutions) also included NADH (0.4 mM) and NAD$^+$ (10 mM) cofactors as a part of the gate "machinery". NADH was oxidized by Pyr in the reaction biocatalyzed by lactate dehydrogenase (LDH) producing NAD $^+$ and decreasing the NADH absorbance at $\lambda_{max} = 340$ nm. NAD $^+$ was reduced by G6P in the reaction biocatalyzed by glucose-6-phosphate dehydrogenase (G6PDH) producing NADH and increasing its absorbance. Output Q produced by the XOR gate, Fig. 2.3c, was measured as absolute values of the absorbance changes at $\lambda_{max} = 340$ nm and defined as logic **1** and **0** for $|\Delta Abs|>0.2$ or $|\Delta Abs| < 0.2$.

The Identity and XOR gates operating in parallel were realized in a flow system schematically outlined in Fig. 2.4. The biocatalytic reactions proceeding in the flow cells are shown schematically in each box representing a single cell, Fig. 2.4. Figure 2.5 shows a photo of the experimental setup for realization of the Feynman gate (CNOT) gate. The solutions containing biochemicals representing Inputs A and B with the variable binary logic values **0** and **1**, as well as the constant composition of the "machinery" represented by mixed NADH/NAD$^+$ were pumped through flow devices containing immobilized enzymes biocatalyzing chemical transformations mimicking logic operations. The flow design of the biochemical device allowed "clocking" (temporal control) and spatial separation of the reaction steps. While the Identity gate realized in a single flow unit performed a simple biocatalytic transformation of PNPP to PNP yielding an optically readable signal, the XOR gate was composed of two flow-through units: one modified with G6PDH that was reducing NAD$^+$ when G6P was available; another modified with LDH was oxidizing NADH

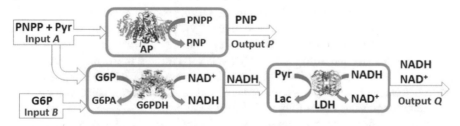

Fig. 2.4 The biocatalytic cascade mimicking the CNOT gate operation (G6PA—6-phosphogluconic acid; Lac—lactate; all other abbreviations and reactions are explained in the text)

Fig. 2.5 The experimental
setup for realization of the
Feynman (CNOT) gate.
Please note different colored
dyes used to illustrate
experimental realization and
mixing where applicable
(Adapted from Ref. [47]
with permission)

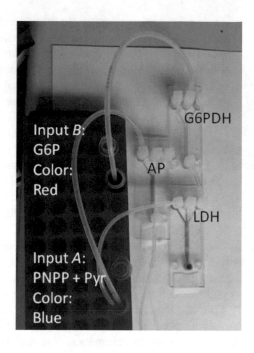

in the presence of Pyr, Fig. 2.4. Therefore, the optical absorbance corresponding
to the NADH concentration was decreasing in the presence of Pyr and absence of
G6P (input combination **1, 0**) and increasing in the absence of Pyr and presence of
G6P (input combination **0, 1**). Since the NADH concentration and the corresponding absorbance were changed in different directions (decreasing and increasing), the
Output Q value **1** was defined as the absolute value of the absorbance change. The
flow cells modified with G6PDH and LDH were optimized (balanced) in such a
way that simultaneous presence of Pyr and G6P (input combination **1, 1**) resulted
in no changes in the concentration of NADH, meaning logic value **0** for Output Q.
Balancing of the biocatalytic reaction rates was achieved by optimization of input
concentrations for Pyr and G6P for the specific activity of the enzymes immobilized in the flow units. Obviously, the absence of Pyr and G6P (input combination
0, 0) did not result in any reaction and preserved the NADH absorbance unchanged
resulting in logic **0** for Output Q. These biocatalytic processes performed in the flow
system allowed operation of the XOR part of the CNOT gate. It should be noted
that the absorbance measurements for the solutions reacted in the flow device were
performed versus the "machinery" solution containing NADH/NAD$^+$ applied to the
reference channel of the spectrophotometer, thus reflecting absorbance difference in
the reacting solutions rather than their full absorbance.

Figure 2.6 shows the experimental data obtained upon application of the input
signals in different logic combinations. Figure 2.6A shows the absorbance changes
observed in the Output P (from the Identity gate). In the absence of PNPP only
minor absorbance changes were observed, meaning an Output P logic **0** value. Each

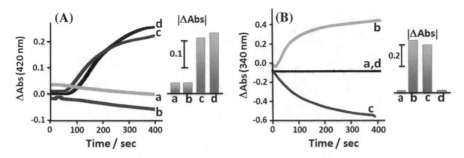

Fig. 2.6 A, B Optical responses of the Identity and XOR parts of the CNOT gate, respectively, upon application of input combinations: *a* **0, 0**; *b* **0, 1**; *c* **1, 0** and *d* **1, 1**. *Insets* show the optical outputs obtained after 400 s of the flow system operation (Adapted from Ref. [47] with permission)

time application of PNPP (regardless presence or absence of any other species) resulted in the absorbance increase at $\lambda_{max} = 420$ nm reflecting the formation of PNP and resulting in the output logic **1**; thus, Input *A* was directly copied to Output *P*. Figure 2.6B shows the XOR gate performance where only unbalanced input signals **0, 1** and **1, 0** resulted in the absorbance changes (output **1**), while the absence of the reacting Pyr and G6P (inputs **0, 0**) or their balanced application (inputs **1, 1**) resulted in no absorbance changes (output **0**).

2.2.2 Double Feynman Gate (DFG) Operation

Figure 2.7a shows the truth table of DFG. DFG operates with three input and three output signals, where Input *A* is copied to Output *P*, while two other output signals *Q* and *R* are the results of two XOR logic functions performed on Inputs *A*, *B* and *A*, *C*, respectively, Fig. 2.7b. In other words, DFG can be represented as the Identity (ID) gate and two XOR gates operating in parallel, Fig. 2.7c, d. Therefore, when realized in a biochemical system [66], the DFG is a very convenient example of a parallel computing system. This convenience is illustrated by the simplistic design of the gate as a flow-through device with parallel channels.

The ID and two XOR gates operating in parallel were realized in a flow system outlined in Fig. 2.8. The solutions containing biochemicals, pyruvate (Pyr), glucose-6-phosphate (G6P) and glucose (Glc) representing Inputs *A*, *B* and *C*, respectively, with the variable binary logic values **0** and **1**, as well as the constant composition of the "machinery" represented by mixed NADH/NAD$^+$ were pumped through the flow devices containing immobilized enzymes biocatalyzing chemical transformations mimicking logic operations. The flow design of the biochemical device allowed "clocking" (temporal control) and spatial separation of the reaction steps. While the ID gate realized in a single flow unit performed a simple biocatalytic transformation (NADH oxidation) yielding an optically readable signal, the XOR gates were

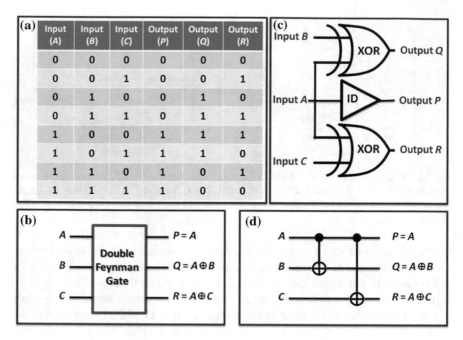

Fig. 2.7 The truth table (**a**), block diagram (**b**), logic circuitry (**c**) and equivalent electronic circuitry (**d**) for Double Feynman gate (DFG) (Adapted from Ref. [66] with permission)

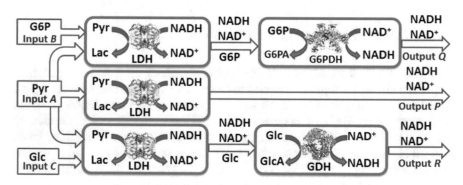

Fig. 2.8 Experimental realization of the biocatalytic Double Feynman gate (DFG) in the flow device (GlcA—gluconic acid, Lac—lactate, G6PA—6-phosphogluconic acid; all other abbreviations and processes are explained in the text). (Adapted from Ref. [66] with permission)

composed of two flow-through units connected in serial: one modified with the enzyme oxidizing NADH to yield NAD^+ and another with the enzyme reducing NAD^+ resulting in the formation of NADH, thus decreasing and increasing optical absorbance corresponding to NADH concentration. The input signals were applied to the 3-channel flow system, where the biocatalytic reactions were performed in parallel, Fig. 2.8. The biocatalytic reactions proceeding in the flow cells are shown schematically in each box representing a single cell, Fig. 2.8.

Input A represented by Pyr was applied to the flow cell functionalized with LDH, Fig. 2.8. Pyr, if it is present in Input A (logic **1** value), resulted in the oxidation of NADH in the reaction biocatalyzed by LDH, thus resulting in the optical absorbance decrease ($\lambda_{max} = 340$ nm). In the absence of Pyr (logic **0** value) the NADH oxidation was not possible and the absorbance was not changed. This biocatalytic reaction represented the ID gate, Fig. 2.7c, and the produced absolute values of absorbance changes were used as Output P, which was defined as logic **1** or **0** when $|\Delta Abs| > 0.2$ or $|\Delta Abs| < 0.2$, respectively. Input A (Pyr) was also directed to biochemical systems performing XOR logic operations, Fig. 2.7c. The first XOR channel was fed with Inputs A and B (Pyr and G6P, respectively). This channel was composed of two flow cells functionalized with LDH and G6PDH operating in sequence, Fig. 2.8. The XOR gate was designed to operate with two oppositely directed biocatalytic redox reactions [65] activated by Pyr and G6P. In the presence of Pyr, NADH was oxidized in the reaction biocatalyzed by LDH in the first flow cell, thus decreasing its absorbance at 340 nm. On the other hand, in the presence of G6P, NAD$^+$ was reduced in the reaction biocatalyzed by G6PDH in the second flow cell, thus increasing absorbance at 340 nm corresponding to the formation of NADH. Output Q in this channel was measured as absolute values of the absorbance changes at $\lambda_{max} = 340$ nm and defined as logic **1** and **0** for $|\Delta Abs| > 0.2$ or $|\Delta Abs| < 0.2$, respectively. The second XOR channel was activated with Inputs A and C (Pyr and Glc, respectively). This channel was composed of two flow cells functionalized with LDH and GDH operating in sequence, Fig. 2.8. This XOR gate was also designed to operate with two opposite biocatalytic reactions. In the presence of Pyr, NADH was oxidized in the reaction biocatalyzed by LDH in the first flow cell, thus decreasing its absorbance at 340 nm. In the presence of Glc, NAD$^+$ was reduced in the reaction biocatalyzed by GDH in the second flow cell, thus increasing absorbance at 340 nm corresponding to the formation of NADH. Output R in this channel was measured as absolute values of the absorbance changes at $\lambda_{max} = 340$ nm and defined as logic **1** and **0** for $|\Delta Abs| > 0.2$ or $|\Delta Abs| < 0.2$, respectively. Figure 2.9 shows the experimental setup for realization of the DFG.

The flow cells modified with enzymes oxidizing NADH and reducing NAD$^+$ were optimized (balanced) in such a way that simultaneous presence of the oxidizing input (Pyr) and reducing input (G6P or Glc) (input combination **1, 1**) resulted in negligible overall changes in the concentration of NADH, meaning logic value **0** for Outputs Q and R. Balancing of the biocatalytic reaction rates was achieved by optimization of input concentrations for Pyr and G6P in one channel and Pyr and Glc in another channel for the specific activity of the enzymes immobilized in the flow units. Obviously, the absence of Pyr and G6P or Glc (input combinations **0, 0**) did not result in any reaction and preserved the NADH absorbance unchanged resulting in logic **0** for Outputs Q and R. The unbalanced input signals (**0, 1** or **1, 0**) applied to the XOR channels resulted in the NADH absorbance changes considered as the output signal **1**. These biocatalytic processes performed in the flow system allowed operation of the XOR parts of DFG. It should be noted that the absorbance

Fig. 2.9 Experimental
realization of DFG (photo of
the flow cell circuitry).
Different colored dyes are
used in this image to
illustrate the experimental
realization including the
mixing of channels where it
is applicable (Adapted from
Ref. [66] with permission)

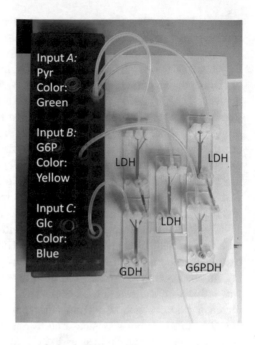

measurements for the solutions reacted in the flow device were performed versus the
"machinery" solution containing a constant amount of NADH/NAD$^+$ applied to the
reference channel of the spectrophotometer, thus reflecting the absorbance difference
in the reacting solutions rather than their full absorbance.

Figure 2.10 shows the experimental data obtained upon application of the input
signals in 8 different logic combinations. Figure 2.10A, B shows the absorbance
changes observed in Output P (from the ID gate). In the absence of Pyr no absorbance
changes were observed, meaning a logic **0** value for Output P. Each application of
Pyr (regardless presence or absence of any other species) resulted in the decrease of
absorbance at $\lambda_{max} = 340$ nm, thus reflecting the oxidation of NADH and resulting
in the output logic **1**. Figure 2.10C, D and E, F show the XOR gate performance
where only unbalanced A, B and A, C input signals (**0**, **1** and **1**, **0**) resulted in the
change of absorbance (output **1**), while the absence of the reacting species (inputs
0, **0**) or their balanced application (inputs **1**, **1**) resulted in no change of absorbance
(output **0**). Overall, the obtained experimental data show the output signal pattern,
Fig. 2.10, corresponding to the truth table, Fig. 2.7a, characteristic of DFG.

2.2.3   Toffoli Gate Operation

Figure 2.11a shows the truth table of Toffoli gate. Toffoli gate operates with three
input and three output signals, where Inputs A and B are directly copied to Outputs

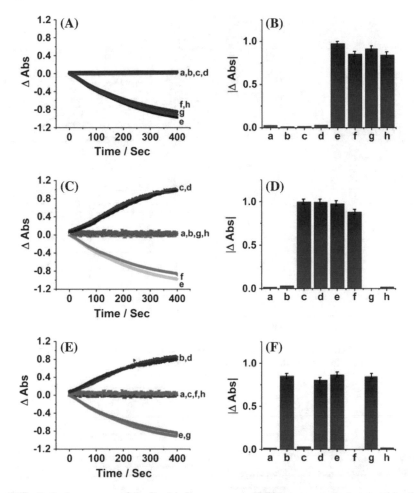

Fig. 2.10 Optical responses of the Double Feynman gate (DFG) system to various combinations of the input signals: *a* **0, 0, 0**; *b* **0, 0, 1**; *c* **0, 1, 0**; *d* **0, 1, 1**; *e* **1, 0, 0**; *f* **1, 0, 1**; *g* **1, 1, 0**; *h* **1, 1, 1**. (Note that the logic values are shown for the input signals in the following order: *A*, *B*, *C*.) **A**, **B** panels show *P* output corresponding to the Identity operation copying *A* input signal. **C**, **D** panels show *Q* output corresponding to the XOR operation on *A* and *B* inputs. **E**, **F** panels show *R* output corresponding to the XOR operation on *A* and *C* inputs. **A**, **C** and **E** plots show kinetics of the signal generation. **B**, **D** and **F** bar charts show the signals obtained after 400 s of the flow system operation. All output signals were read at λ_{max} 340 nm. The data shown in the bar-charts are average of three independent experiments (Adapted from Ref. [66] with permission)

P and *Q*, thus representing two ID gates operating in parallel. In addition to their role in the ID gates, Inputs *A* and *B* are directed to an AND gate to generate an intermediate output which is then directed to a XOR gate together with Input *C* to yield Output *R*, Fig. 2.11b–d. In other words, Toffoli gate can be represented as two ID gates operating in parallel with the AND-XOR concatenated gates, Fig. 2.11c.

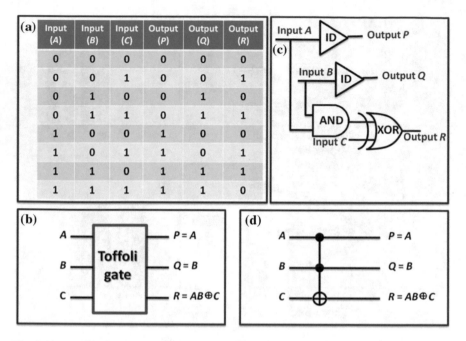

Fig. 2.11 The truth table (**a**), block diagram (**b**), logic circuitry (**c**) and equivalent electronic circuitry (**d**) for Toffoli gate (Adapted from Ref. [66] with permission)

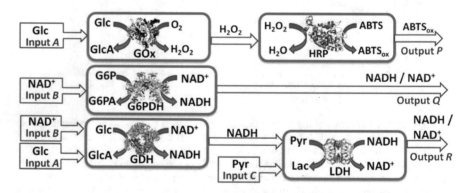

Fig. 2.12 Experimental realization of the biocatalytic Toffoli gate in the flow device (GlcA—gluconic acid, Lac—lactate, G6PA—6-phosphogluconic acid; all other abbreviations and processes are explained in the text) (Adapted from Ref. [66] with permission)

Two ID gates and the AND-XOR concatenated gates operating in parallel were realized [66] in a flow system outlined in Fig. 2.12. Figure 2.13 shows a photo of the experimental setup for realization of the Toffoli gate. The solutions containing biochemicals, Glc, NAD^+ and Pyr, representing Inputs A, B and C, respectively, with the variable binary logic values **0** and **1**, as well as the constant composition of the "machinery" that is represented by a mixture of NADH/G6P/ABTS (note that

Fig. 2.13 Experimental
realization of Toffoli gate
(photo of the flow cell
circuitry). Different colored
dyes are used in this image
to illustrate the experimental
realization including the
mixing of channels where it
is applicable (Adapted from
Ref. [66] with permission)

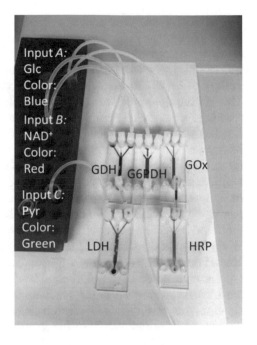

O_2 was always present in the solution in equilibrium with air) were pumped through
flow cells containing immobilized enzymes biocatalyzing chemical transformations
mimicking logic operations. The first ID gate activated by Input A was composed
of two flow cells operating in sequence: the first cell modified with GOx produced
H_2O_2 in the presence of glucose (Glc) (if Input A was applied at logic value **1**).
Then for convenient optical detection, the produced H_2O_2 was reacted with ABTS in
the second flow cell modified with HRP to yield colored oxidized ABTS (ABTS$_{ox}$)
which represented Output P. Obviously, in the absence of glucose (logic value **0** for
Input A) H_2O_2 was not produced and ABTS was not oxidized, thus preserving the
optical absorbance in this channel without changes. Output P in this channel was
measured as the absorbance changes at $\lambda_{max} = 415$ nm characteristic of ABTS$_{ox}$ and
defined as logic **1** and **0** for Δ Abs > 0.2 or ΔAbs < 0.2, respectively. The second
ID gate activated with Input B was represented by a single flow cell modified with
G6PDH. In the presence of NAD$^+$ (logic value **1** for Input B) this cell produced
NADH (note that the reducing species G6P were always present in the "machinery"
solution) which was considered as Output Q. In the absence of NAD$^+$ (logic value
0 for Input B) NADH was not produced, thus keeping the initial absorbance without
changes. It should be noted that NADH was also present in the "machinery" solution,
thus the NADH produced in the flow cells was added to the background amount of
NADH. Output Q in this channel was measured as absolute values of the absorbance
changes at $\lambda_{max} = 340$ nm characteristic of NADH and defined as logic **1** and **0** for
$|\Delta$Abs$| > 0.2$ or $|\Delta$Abs$| < 0.2$, respectively. Overall, Inputs A and B were directly
copied to Outputs P and Q.

The AND gate was activated with a combination of Inputs A and B (Glc and NAD$^+$, respectively). In the presence of both reacting species (inputs **1, 1**), GDH catalytically reduced NAD$^+$ to NADH, which was moving with the flow to the next cell modified with LDH. If either or both reacting species were absent (inputs **0, 1**; **1, 0**; **0, 0**) NADH was not produced, thus demonstrating the AND logic features. The cell functionalized with LDH was additionally fed with Input C (Pyr). If both reacting species (NADH and Pyr) were present (**1, 1, 1** combination for Inputs A, B, C), the reaction biocatalyzed by LDH resulted in oxidation of NADH, thus bringing its concentration down to the original level and keeping the optical absorbance with no changes (Output R **0**). It should be noted that the Pyr concentration was carefully optimized to keep the balance and to compensate the NADH production in the first reacting cell. The same result was achieved if no reactions were activated in both connected flow cells (**0, 0, 0** combination for Inputs A, B, C). When NADH was produced in the first cell but Pyr was not present (**1, 1, 0** combination for Inputs A, B, C) Output R demonstrated increasing absorbance corresponding to the produced NADH (Output R **1**). When NADH was not produced in the first cell, but Pyr was present (**0, 0, 1**; **0, 1, 1**; **1, 0, 1** combinations for Inputs A, B, C) the reaction in the second cell resulted in the consumption of NADH present in the background solution, thus resulting in the absorbance decrease. Output R in this channel was measured as absolute values of the absorbance changes at $\lambda_{max} = 340$ nm characteristic of NADH and defined as logic **1** and **0** for $|\Delta Abs| > 0.2$ or $|\Delta Abs| < 0.2$, respectively.

Figure 2.14 shows the experimental data obtained upon application of the input signals in 8 different logic combinations. Figure 2.14A, B shows the absorbance changes observed in Output P (from the first ID gate). In the absence of Glc no absorbance changes were observed, meaning logic **0** value for Output P. Each application of Glc (regardless presence or absence of any other species) resulted in the increase of absorbance at $\lambda_{max} = 415$ nm, reflecting the oxidation of ABTS and resulting in the output logic **1**. Figure 2.14C, D shows the absorbance changes observed in Output Q (from the second ID gate). In the absence of NAD$^+$ no absorbance changes were observed, meaning logic **0** value for Output Q. When NAD$^+$ was applied (regardless presence or absence of any other species) the absorbance at $\lambda_{max} = 340$ nm was increased corresponding to the production of NADH and resulting in the output logic **1**. Figure 2.14E, F shows the AND-XOR gates performance where only unbalanced A, B and C input signals (**0, 0, 1; 0, 1, 1; 1, 0, 1; 1, 1, 0**) resulted in the change of absorbance (output **1**), while the absence of the reacting species (inputs **0, 0, 0**) or their balanced application (inputs **0, 1, 0; 1, 0, 0; 1, 1, 1**) resulted in no change of absorbance (output **0**). Overall, the obtained experimental data show the output signal pattern, Fig. 2.14, corresponding to the truth table, Fig. 2.11a, characteristic of Toffoli gate.

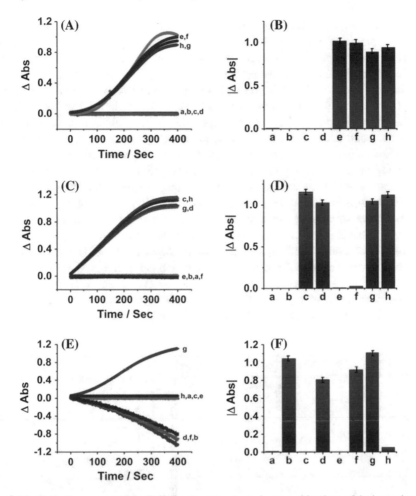

Fig. 2.14 Optical responses of the Toffoli gate system to various combinations of the input signals: *a* **0, 0, 0**; *b* **0, 0, 1**; *c* **0, 1, 0**; *d* **0, 1, 1**; *e* **1, 0, 0**; *f* **1, 0, 1**; *g* **1, 1, 0**; *h* **1, 1, 1**. (Note that the logic values are shown for the input signals in the following order: *A*, *B*, *C*.) **A, B** panels show *P* output corresponding to the Identity operation copying *A* input signal. **C, D** panels show *Q* output corresponding to the Identity operation copying *B* input signal. **E, F** panels show *R* output corresponding to the AND operation on *A* and *B* inputs followed by the XOR operation on the output from the AND gate and *C* input. **A, C** and **E** plots show kinetics of the signal generation. **B, D** and **F** bar charts show the signals obtained after 400 s of the flow system operation. Output *P* was read at λ_{max} 415 nm and Outputs *Q* and *R* were read at λ_{max} 340 nm. The data shown in the bar-charts are average of three independent experiments (Adapted from Ref. [66] with permission)

2.2.4 Peres Gate Operation

Figure 2.15a shows the truth table of Peres gate. Peres gate operates with three input and three output signals, where Input *A* is directly copied to Output *P*, while two other output signals Q and R are the results of complex logic operations. Inputs A and B are processed through a XOR gate to yield Output Q. Inputs A and B are also processed through an AND gate and the resulting intermediate goes to another XOR gate together with Input C to generate Output R at the end of the concatenated logic operations, Fig. 2.15b. In other words, Peres gate can be represented as the ID gate operating in parallel with a XOR gate and an AND-XOR circuit, Fig. 2.15c, d.

The ID, XOR and AND-XOR gates operating in parallel were realized [66] in a flow system outlined in Fig. 2.16. Figure 2.17 shows a photo of the experimental setup for realization of the Peres gate. The solutions containing biochemicals, NADH, Glc and H_2O_2, representing Inputs A, B and C, respectively, with the variable binary logic values **0** and **1**, as well as the constant composition of the "machinery" represented by mixed $K_3[Fe(CN)_6]$, $K_4[Fe(CN)_6]$ and Pyr (note that O_2 was always present in the solution in equilibrium with air) were pumped through flow cells containing immobilized enzymes biocatalyzing chemical transformations mimicking logic operations. The ID gate activated by Input A (NADH) was represented with a flow cell functionalized with Diaph which catalyzed NADH oxidation by $[Fe(CN)_6]^{3-}$. This process resulted in the production of reduced $[Fe(CN)_6]^{4-}$ and absorbance decrease at $\lambda_{max} = 420$ nm characteristic of $[Fe(CN)_6]^{3-}$ when Input A was applied at logic **1** value. In the absence of NADH (logic **0**) the reaction did

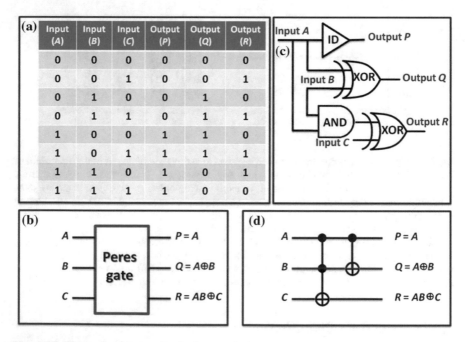

Fig. 2.15 The truth table (**a**), block diagram (**b**), logic circuitry (**c**) and equivalent electronic circuitry (**d**) for Peres gate (Adapted from Ref. [66] with permission)

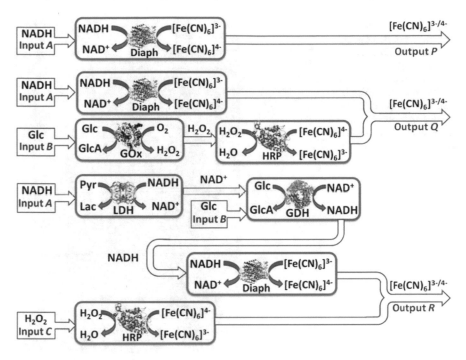

Fig. 2.16 Experimental realization of the biocatalytic Peres gate in the flow device (GlcA—gluconic acid, Lac—lactate; all other abbreviations and processes are explained in the text) (Adapted from Ref. [66] with permission)

not proceed and the absorbance was not changed regardless of logic values of other inputs.

The XOR gate activated with Inputs A and B was realized in the following way. Input A (NADH) was applied to a flow cell functionalized with Diaph and operating in the way similar to described above. This biocatalytic pathway resulted in the consumption of $[Fe(CN)_6]^{3-}$. Input B (Glc) was applied to another flow cell functionalized with GOx. The reaction in this cell resulted in biocatalytic oxidation of glucose and concomitant production of H_2O_2 (note that O_2 was present in the solution). The in situ produced H_2O_2 was directed to the next flow cell modified with HRP where $[Fe(CN)_6]^{4-}$ was oxidized by H_2O_2 to yield $[Fe(CN)_6]^{3-}$. This reaction chain was activated only in the presence of Glc (Input B in logic **1** value). The solutions flowing out of the Diaph-modified cell fed with Input A and out of the GOx-HRP-modified cells fed with Input B were mixed after the biochemical reactions to yield finally Output Q. The overall result showed increasing or decreasing absorbance at $\lambda_{max} = 420$ nm corresponding to the increasing or decreasing concentration of $[Fe(CN)_6]^{3-}$ when Inputs A and B were applied with the combinations **0**, **1** or **1**, **0**, respectively. When Inputs A and B were applied at logic values **0**, **0** or **1**, **1**, the overall absorbance was not changed because the final concentration of $[Fe(CN)_6]^{3-}$ coming from two parallel channels was not changed due to the absent (**0**,

Fig. 2.17 Experimental realization of Peres gate (photo of the flow cell circuitry). Different colored dyes are used in this image to illustrate the experimental realization including the mixing of channels where it is applicable (Adapted from Ref. [66] with permission)

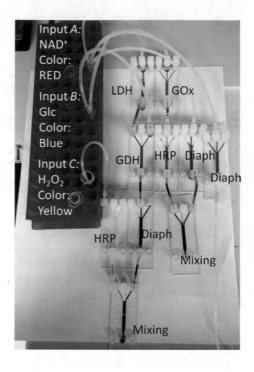

0 inputs) or balanced (**1, 1** inputs) reactions. Output Q in this channel was measured as the absolute values of the absorbance changes at $\lambda_{max} = 420$ nm, characteristic of $[Fe(CN)_6]^{3-}$ and defined as logic **1** and **0** for $|\Delta Abs| > 0.1$ or $|\Delta Abs| < 0.1$, respectively.

The most sophisticated function was realized for Output R, Fig. 2.16. First Input A (NADH) was processed in the cell modified with LDH where NADH was bio-catalytically oxidized with Pyr (note that Pyr was always present in the background solution) to yield NAD^+ which was then directed to the next cell modified with GDH. In that cell, NAD^+ was reduced back to NADH in the presence of Input B (Glc). The two-step biocatalytic cascade was terminated with the production of NADH if both reaction steps were activated (combination **1, 1** for Inputs A and B). When NADH input was absent in the first cell (Input A **0**) the biocatalytic cascade was not activated at all, regardless of the absence or presence of Glc in the second cell (input combinations **0, 0** and **0, 1** for Inputs A and B). In the presence of NADH in the first cell modified with LDH and in the absence of Glc in the second cell modified with GDH (combination **1, 0** for Inputs A and B), the overall result was production of NAD^+ in the first cell which was not returned back to NADH in the second cell. Thus, the final output in the form of the NADH production after two consecutive cells was only in the presence of both Inputs A and B in **1, 1** combination, thus demonstrating the features of the AND function. The produced NADH was applied to the next flow cell modified with Diaph where NADH was used to reduce $[Fe(CN)_6]^{3-}$ yielding $[Fe(CN)_6]^{4-}$. Another reaction operating in parallel was performed in the cell modified with HRP where Input C (H_2O_2) oxidized $[Fe(CN)_6]^{4-}$ to yield $[Fe(CN)_6]^{3-}$. The solution produced in this cell was mixed with the solution coming out of the cascade of reactions realized in the cells modified with LDH-GDH-Diaph. The solution coming out of HRP-modified cell will increase the $[Fe(CN)_6]^{3-}$ concentration if Input C is applied at logic value **1** (meaning presence of H_2O_2). The solution coming out of LDH-GDH-Diaph cells will decrease the $[Fe(CN)_6]^{3-}$ concentration if Inputs A and B are applied at **1, 1** combination. The overall result is the unchanged $[Fe(CN)_6]^{3-}$ concentration in case of **0, 0, 0; 0, 1, 0; 1, 0, 0** and **1, 1, 1** combinations for Inputs A, B and C because of the absence of the corresponding reactions or due to their balancing. All other input combinations (**0, 0, 1; 0, 1, 1; 1, 0, 1; 1, 1, 0**) resulted in the $[Fe(CN)_6]^{3-}$ concentration change, thus resulting in the corresponding absorbance change. These results correspond to the logic operation of the concatenated AND-XOR gates as needed for the part of Peres gate. Output R in this channel was measured as absolute values of the absorbance changes at $\lambda_{max} = 420$ nm characteristic of $[Fe(CN)_6]^{3-}$ and defined as logic **1** and **0** for $|\Delta Abs| > 0.1$ or $|\Delta Abs| < 0.1$, respectively. It should be noted that the absorbance measurements for the solutions reacted in the flow device were performed versus the "machinery" solution that contained a constant concentration of $[Fe(CN)_6]^{3-}/[Fe(CN)_6]^{4-}$ that was applied to the reference channel of the spectrophotometer, thus reflecting absorbance difference in the reacting solutions rather than their full absorbance.

Figure 2.18 shows the experimental data obtained upon application of the input signals in 8 different logic combinations. Figure 2.18A, B shows the absorbance changes observed in Output P (from the ID gate). In the absence of NADH no absorbance changes were observed, meaning logic **0** value for Output P. Each application of NADH (regardless presence or absence of any other species) resulted in the decrease of absorbance at $\lambda_{max} = 420$ nm, reflecting the reduction of $[Fe(CN)_6]^{3-}$ and resulting in the output logic **1**. Figure 2.18C, D shows the absorbance changes observed in Output Q (from the XOR gate). In the absence of NADH and Glc no reactions proceeded and no absorbance changes were observed. The same overall result was observed for the balanced reactions in the presence of both NADH and Glc. Only unbalanced reactions (meaning presence of one of the reacting species, NADH or Glc) resulted in the concentration changes for $[Fe(CN)_6]^{3-}$ and corresponding absorbance changes at $\lambda_{max} = 420$ nm. This channel represented the XOR gate operation activated by Inputs A and B. Figure 2.18E, F shows the absorbance changes in Output R (from AND-XOR concatenated gates). This channel shows the unchanged absorbance at $\lambda_{max} = 420$ nm (meaning the unchanged concentration of $[Fe(CN)_6]^{3-}$) for input combinations **0, 0, 0** (when no reactions proceed) and **0, 1, 0; 1, 0, 0; 1, 1, 1** (when the reactions are balanced), thus resulting in logic **0** for Output R. All other input combinations resulted in the unbalanced reactions and the corresponding optical changes, thus resulting in the logic value **1** for Output R. Overall, the obtained experimental data show the output signal pattern, Fig. 2.18, corresponding to the truth table, Fig. 2.15a, characteristic of Peres gate.

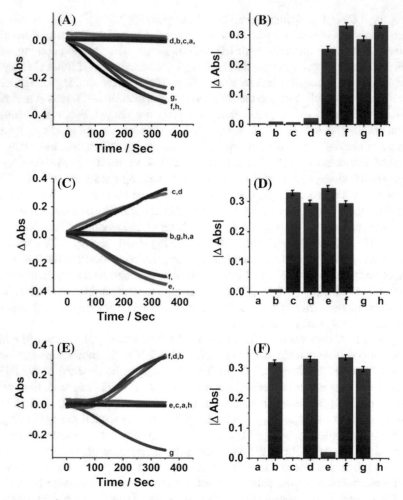

Fig. 2.18 Optical responses of the Peres gate system to various combinations of the input signals: *a* **0, 0, 0**; *b* **0, 0, 1**; *c* **0, 1, 0**; *d* **0, 1, 1**; *e* **1, 0, 0**; *f* **1, 0, 1**; *g* **1, 1, 0**; *h* **1, 1, 1**. (Note that the logic values are shown for the input signals in the following order: *A*, *B*, *C*.) **A**, **B** panels show *P* output corresponding to the Identity operation copying *A* input signal. **C**, **D** panels show *Q* output corresponding to the XOR operation on *A* and *B* inputs. **E**, **F** panels show *R* output corresponding to the AND operation on *A* and *B* inputs followed by the XOR operation on the output from the AND gate and *C* input. **A**, **C** and **E** plots show kinetics of the signal generation. **B**, **D** and **F** bar charts show the signals obtained after 350 s of the flow system operation. All output signals were read at λ_{max} 420 nm. The data shown in the bar-charts are average of three independent experiments (Adapted from Ref. [66] with permission)

2.3   Advantages and Disadvantages of the Developed Approach

2.3.1   Advantages

Biochemical systems of high complexity composed of many operating enzymes biocatalyzing several reactions cannot be realized in a single solution without compartmentalization of the reacting species similarly to the approach used by Nature in biological cell. Therefore, the approach applied in the present study, where each biocatalytic reaction is running in a separate volume in individual reacting cells and communicating via flow moving from one cell to another, is a good solution for increasing the system complexity. While biocatalytic systems organized in a single solution were limited by maximum 3–4 reaction steps [56, 67] and required extremely complicated optimization [68], the reactions separated in individual cell are much simpler for performance and optimization, thus allowing the increased complexity of the systems. Most important, the modular design with individual flow cells, each modified with one enzyme, allows easy combination of them in various networks for performing different logic operations (AND, XOR, etc.) being parts of the complex information processes, as it is demonstrated in the realized reversible logic gates.

If the signal-processing system needs to have several output signals and when it is organized in one solution, the output signals must have different optical properties being represented by different chemical species. For example, one of the recently reported systems with several inputs and three outputs used individually readable absorbance changes corresponding to NADH and ABTS$_{ox}$ (λ_{max} 340 and 420 nm, respectively) and a fluorescence output produced by luciferase-luciferin system in the presence of ATP (λ_{max} 552 nm) [69]. If output signals are read by any other means, for example using redox species analyzed electrochemically, the problem still persists since the produced species should be chemically different to demonstrate different redox potentials that are individually readable by cyclic voltammetry or by any other electrochemical technique. This requirement puts serious limitations on the use of biochemical reactions. The present approach does not have this limitation because the output signals are read from separated channels, thus allowing the outputs to be represented by the same chemical species; for example, NADH can be the output from all logic processes proceeding in parallel in different channels.

Another advantage of the designed biomolecular systems can be found in comparison with the reversible logic systems based on complex synthetic macro-molecules. All-photonic gates activated by light signals and producing light emission as output cannot be easily extended to other chemical information processing steps [43]. On the other hand, the output signals represented by chemical species (e.g., NADH) could be easily connected to the extending information processing steps or even used for chemical actuation processes (e.g., stimulation of drug release) [70, 71].

2.3.2 Disadvantages

In electronic realizations of reversible logic gates the input and output signals are represented by electrical potential/current changes. Therefore, the end of one logic gate can be easily connected to the beginning of another. This common property of electronic systems allows assembling reversible logic gates in complex networks for sophisticated computation. In the present chemical realization, the output signals are different by their nature from the input signals. Thus, connecting the logic elements in high hierarchy systems is difficult or even impossible, at least for some of the presently developed systems. This certainly limits the complexity of possible systems. In other words, the chemical extensions of the present logic systems are possible, but connecting the same components to each other is certainly not as trivial as an electronic system.

It should be noted that the developed systems allow *logic reversibility* meaning the possibility to recover the whole set of input signals by analyzing the pattern of the output signals. The biocatalytic reactions cannot run in the opposite directions, thus the physical reversibility is not possible in these systems. Also, the biochemical systems do not have energy saving properties expected theoretically for electronic realization of the reversible computation.

Obviously, the time-scale of the chemical systems operation is in minutes, thus being incomparably longer than that of electronic systems [72]. This makes the use of chemical systems very problematic for realistic computational applications. Even with these drawbacks, the new systems are very interesting for their possible integration with biosensors for logic processing biosensor inputs. Although, this drawback is not as problematic as it may seem, as it could be at least partially resolved if the systems are scaled down to microfluidic devices operating on a shorter time-scale [73].

Overall, it should be noted that the disadvantages briefly discussed above are characteristic of molecular/biomolecular computing systems in general, rather than being specific for the present approach based on the use of flow devices. On the other hand, the advantages of the developed approach clearly demonstrated the possibility of increasing complexity of the information processing systems which is not achievable in a homogeneous system without compartmentalization of reaction steps.

2.4   Conclusions and Perspectives

The obtained results demonstrated for the first time information processing in reversible logic gates, such as Feynman gate (CNOT), Double Feynman gate (DFG), Toffoli gate and Peres gate, performed in enzyme-based biocatalytic systems. The first results for CNOT, DFG, Toffoli and Peres gates realized in biocatalytic flow systems [47, 66] are promising for integrating the reversible gates into complex biomolecular logic networks. The logic processing of biomolecular signals in the reversible mode will be particularly beneficial for biosensing applications that need each combination of the output signals to correspond with a unique pattern of the input signals, thus allowing restoration of the original input values. The application of logically reversible gates for the analysis of biomedically important biomarker signaling for various physiological dysfunctions, similarly to previously reported injury diagnostics [57, 58], are feasible. Technological realization of the information processing systems in flow devices allows for "clocking" (temporal control) as well as spatial separation of the various steps of multistage biochemical processes, thus providing novel options for their sophistication and functional flexibility. The developed approach allows designing other reversible logic gates of higher complexity, including Fredkin gate and other gates with 3-input/3-output and 4-input/4-output compositions [61, 72]. Still technological realization of reversible logic systems with flow networks requires a lot of additional work. Increasing complexity of the interconnected flow networks may result in difficulties, which require fundamental research on biochemical reactions and transport in complex structured systems [74]. Practical use of the systems discussed in this chapter will certainly require transition from the macro-scale flow devices used in the present preliminary study to microfluidic lab-on-a-chip devices, allowing their miniaturization and potentially faster response operation.

Acknowledgments This work was supported by National Science Foundation (award CBET-1403208).

Appendix

This section is addressed to the readers interested in technical details of the experimental realization of the reversible logic gates described above.

Chemicals and Materials

Enzymes and their substrates used in the biocatalytic reactions were specified in Sect. 2.2. Results and Discussion (second paragraph).

Other chemicals used for the immobilization procedure and being components of reacting solutions: pepsin (E.C. 232.629.3) from porcine gastric mucosa, glutaric dialdehyde, poly(ethyleneimine) solution (PEI) (average M_w ca. 750,000), 2-amino-2-hydroxymethyl-propane-1,3-diol (Tris-buffer), 2, 2'-azino-*bis* (3-ethylbenzothiazoline-6-sulfonic acid) diammonium salt (ABTS), $K_3[Fe(CN)_6]$ and $K_4[Fe(CN)_6]$. The chemicals listed above and other standard inorganic/organic reactants were purchased from Sigma-Aldrich and used as supplied. Ultrapure water

(18.2 MΩ· cm) from NANOpure Diamond (Barnstead) source was used in all of the experiments.

Instruments and Devices

Flow cells (μ-Slide III 3in1 Flow Kit; ibidi GmbH) were used for the biocatalytic reactions. A Shimadzu UV-2450 UV-Vis spectrophotometer with flow-through quartz cuvettes (1 cm optical pathway) connected to the tubing of the flow device was used for all optical measurements. The reacting solutions were pumped through the flow cells and spectrophotometer cuvettes with the help of a peristaltic pump (Gilson Minipuls 3) connected with polyethylene tubing, 1 mm internal diameter.

Immobilization of Enzymes in the Flow Cells

Before any experimental data were realized, the flow cells were flushed with concentrated sulfuric acid to remove residual physical adsorption of PEI left on the cell surface from previous experiments. After this initial preparatory step, all subsequent cleanings were conducted with the following method. The flow cells were washed with a minimum of 10 mL of deionized water and then reacted with pepsin solution, 0.5 mg/mL, in 0.1 M phosphate buffer, pH 2.0, for 1 h. Then, the cells were washed with a minimum of 10 mL of deionized water. These cleaning steps aimed at removing remnant enzymes from previous experiments and prepared the cell surface for adsorption of PEI. Then, the flow cells were treated with a PEI solution (2 % v/v) for 1 h and then, thoroughly washed with 5 mL of deionized water, resulting in physical adsorption of PEI on polystyrene and providing the amino groups needed for the enzyme immobilization. Then, the amino-functionalized surface was reacted with glutaric dialdehyde (5 % v/v) for 1 h; after that, the surface was washed with 5 mL of deionized water to remove non-reacted glutaric dialdehyde. The enzyme solutions were reacted with the flow cells and the wells that were activated with glutaric dialdehyde for 1 h and then, the cells were thoroughly washed with Tris-buffer (0.1 M, pH 7.1) to remove non-reacted enzymes from the cells. The following enzyme concentrations were used for preparing the logic gates: (*i*) Feynman gate: AP ca. 500 units/mL; G6PDH ca. 280 units/mL; LDH ca. 600 units/mL, (*ii*) Double Feynman gate: GDH ca. 46 units/mL; G6PDH ca. 14 units/mL; LDH ca. 30 units/mL, (*iii*) Toffoli gate: GDH ca. 140 units/mL; G6PDH ca. 280 units/mL; LDH ca. 340 units/mL; GOx ca. 140 units/mL; HRP ca. 120 units/mL, (*iv*) Peres gate: GDH ca. 190 units/mL; GOx ca. 665 units/mL; LDH ca. 343 units/mL; HRP ca. 400 units/mL; Diaph ca. 66 units/mL. This procedure resulted in the enzymes covalently bound to the adsorbed PEI through Schiff-base bonds. The flow cell devices with the immobilized enzymes demonstrated reproducible performance for at least two days allowing pumping of the input solutions over long period of time, thus proving stable immobilization of the enzymes and preserving their biocatalytic activity.

Optimization of the Input Concentrations

The input concentrations were experimentally optimized for the specific enzyme activity in the flow cells. The optimization was aimed at the output signals with the comparable intensity upon application of different combinations of the input signals. Balancing output signals for XOR gates, when optimizing the reversible gates, was particularly important.

The following optimized input concentrations were considered as logic **1** values:

Feynman gate: Input A (PNPP + Pyr) 10 mM + 1 mM, respectively, Input B (G6P) 6 mM.
Double Feynman gate: Input A (Pyr) 1.46 mM, Input B (G6P) 2.22 mM, Input C (Glc) 0.6 mM.
Toffoli gate: Input A (Glc) 0.6 mM, Input B (NAD$^+$) 2.75 mM, Input C (Pyr) 1.1 mM.
Peres gate: Input A (NADH) 0.02 mM, Input B (Glc) 10 mM, Input C (H$_2$O$_2$) 0.7 mM.

Logic **0** value for all input signals was defined as the absence of the corresponding chemicals (meaning their zero physical concentration in the background solutions).

Flow Cell Performance and the Output Signal Measurements

The enzyme-modified flow cells were activated with solutions containing the input signals applied in all possible logic combinations (4 variants for Feynman gate and 8 variants for all other gates). The solutions also included non-variable reacting components which had the same initial concentrations for all combinations of the inputs signals. The solution compositions used for different gate are listed below:

Feynman gate

The input signals (represented with PNPP + Pyr and G6P solutions also containing non-variable NADH (0.4 mM) and NAD$^+$ (10 mM) cofactors were pumped through the flow system with the volumetric rate of 50 μL/min. Optical absorbance measurements were performed for the Identity gate channel (Output P) at $\lambda = 420$ nm characteristic of p-nitrophenol (PNP) and for the XOR gate channel (Output Q) at $\lambda = 340$ nm characteristic of NADH. The reference channel (cuvette) of the spectrophotometer was filled with the background ("machinery") solution containing NADH (0.4 mM) and NAD$^+$ (10 mM), thus allowing the absorbance change measurements versus the composition of the background solution.

Double Feynman gate

The input signals (represented with Pyr, G6P and Glc solutions also containing non-variable NADH, 0.4 mM, and NAD^+, 5.0 mM) were pumped through the flow system with the volumetric rate of 50 μL/min. Optical absorbance measurements were performed for the Identity and two XOR gate channels at $\lambda = 340$ nm characteristic of NADH. The reference channel (cuvette) of the spectrophotometer was filled with the background ("machinery") solution containing NADH (0.4 mM) and NAD^+ (5.0 mM), thus allowing the absorbance change measurements versus the composition of the background solution.

Toffoli gate

The input signals (represented with Pyr, Glc and NAD^+ solutions also containing non-variable NADH, 0.375 mM, G6P, 0.4 mM, ABTS, 4.0 mM) were pumped through the flow system with the volumetric rate of 50 μL/min. Optical absorbance measurements were performed for the GOx/HRP Identity gate at $\lambda = 415$ nm characteristic of the oxidized form of ABTS ($ABTS_{ox}$), while the G6PDH Identity gate and the AND/XOR gate channels were measured at $\lambda = 340$ nm characteristic of NADH. The reference channel (cuvette) of the spectrophotometer was filled with the background ("machinery") solution containing NADH (0.375 mM), G6P (0.4 mM) and ABTS (4.0 mM), thus allowing the absorbance change measurements versus the composition of the background solution.

Peres gate

The input signals (represented with NADH, Glc and H_2O_2 solutions also containing non-variable $K_4[Fe(CN)_6]$ 2.0 mM, $K_3[Fe(CN)_6]$ 1.0 mM and Pyr 0.065 mM) were pumped through the flow system with the volumetric rate of 25 μL/min. Optical absorbance measurements were performed at $\lambda = 420$ nm characteristic of $K_3[Fe(CN)_6]$. The reference channel (cuvette) of the spectrophotometer was filled with the background ("machinery") solution containing $K_4[Fe(CN)_6]$ 2.0 mM, $K_3[Fe(CN)_6]$ 1.0 mM and Pyr 0.065 mM, thus allowing the absorbance change measurements versus the composition of the background solution.

References

1. Calude, C.S., Costa, J.F., Dershowitz, N., Freire, E., Rozenberg, G. (eds.): Unconventional Computation. Lecture Notes in Computer Science, vol. 5715. Springer, Berlin (2009)
2. Szacilowski, K.: Infochemistry. Wiley, Chichester (2012)
3. de Silva, A.P.: Molecular Logic-Based Computation. Royal Society of Chemistry, Cambridge (2013)
4. Katz, E. (ed.): Molecular and Supramolecular Information Processing – From Molecular Switches to Unconventional Computing. Willey-VCH, Weinheim (2012)
5. de Silva, A.P.: Molecular logic and computing. Nat. Nanotechnol. **2**, 399–410 (2007)

6. Claussen, J.C., Hildebrandt, N., Susumu, K., Ancona, M.G., Medintz, I.L.: Complex logic functions implemented with quantum dot bionanophotonic circuits. ACS Appl. Mater. Interfaces **6**, 3771–3778 (2014)
7. Pischel, U.: Advanced molecular logic with memory function. Angew. Chem. Int. Ed. **49**, 1356–1358 (2010)
8. Szacilowski, K.: Digital information processing in molecular systems. Chem. Rev. **108**, 3481–3548 (2008)
9. Pischel, U., Andreasson, J., Gust, D., Pais, V.F.: Information processing with molecules - Quo Vadis? ChemPhysChem **14**, 28–46 (2013)
10. Katz, E. (ed.): Biomolecular Computing – From Logic Systems to Smart Sensors and Actuators. Willey-VCH, Weinheim (2012)
11. Benenson, Y.: Biomolecular computing systems: principles, progress and potential. Nat. Rev. Genet. **13**, 455–468 (2012)
12. Alon, U.: An Introduction to Systems Biology. Design Principles of Biological Circuits. Chapman & Hall/CRC Press, Boca Raton (2007)
13. Adleman, L.M.: Molecular computation of solutions to combinatorial problems. Science **266**, 1021–1024 (1994)
14. Stojanovic, M.N., Stefanovic, D., Rudchenko, S.: Exercises in molecular computing. Acc. Chem. Res. **47**, 1845–1852 (2014)
15. Stojanovic, M.N., Stefanovic, D.: Chemistry at a higher level of abstraction. J. Comput. Theor. Nanosci. **8**, 434–440 (2011)
16. Ezziane, Z.: DNA computing: Applications and challenges. Nanotechnology **17**, R27–R39 (2006)
17. Ashkenasy, G., Dadon, Z., Alesebi, S., Wagner, N., Ashkenasy, N.: Building logic into peptide networks: Bottom-up and top-down. Isr. J. Chem. **51**, 106–117 (2011)
18. Unger, R., Moult, J.: Towards computing with proteins. Proteins **63**, 53–64 (2006)
19. Katz, E., Privman, V.: Enzyme-based logic systems for information processing. Chem. Soc. Rev. **39**, 1835–1857 (2010)
20. Rinaudo, K., Bleris, L., Maddamsetti, R., Subramanian, S., Weiss, R., Benenson, Y.: A universal RNAi-based logic evaluator that operates in mammalian cells. Nat. Biotechnol. **25**, 795–801 (2007)
21. Arugula, M.A., Shroff, N., Katz, E., He, Z.: Molecular AND logic gate based on bacterial anaerobic respiration. Chem. Commun. **48**, 10174–10176 (2012)
22. Kahan, M., Gil, B., Adar, R., Shapiro, E.: Towards molecular computers that operate in a biological environment. Phys. D **237**, 1165–1172 (2008)
23. Baron, R., Lioubashevski, O., Katz, E., Niazov, T., Willner, I.: Elementary arithmetic operations by enzymes: a model for metabolic pathway based computing. Angew. Chem. Int. Ed. **45**, 1572–1576 (2006)
24. Stojanovic, M.N., Stefanovic, D.: Deoxyribozyme-based half-adder. J. Am. Chem. Soc. **125**, 6673–6676 (2003)
25. Benenson, Y.: Biocomputing: DNA computes a square root. Nat. Nanotechnol. **6**, 465–467 (2011)
26. Pei, R.J., Matamoros, E., Liu, M.H., Stefanovic, D., Stojanovic, M.N.: Training a molecular automaton to play a game. Nat. Nanotechnol. **5**, 773–777 (2010)
27. Qian, L., Winfree, E., Bruck, J.: Neural network computation with DNA strand displacement cascades. Nature **475**, 368–372 (2011)
28. MacVittie, K., Halámek, J., Privman, V., Katz, E.: A bioinspired associative memory system based on enzymatic cascades. Chem. Commun. **49**, 6962–6964 (2013)
29. Privman, V., Katz, E.: Can bio-inspired information processing steps be realized as synthetic biochemical processes? Phys. Status Solidi A **212**, 219–228 (2015)
30. Katz, E.: Biocomputing - Tools, aims, perspectives. Curr. Opin. Biotechnol. **34**, 202–208 (2015)
31. de Silva, A.P.: Molecular computing - A layer of logic. Nature **454**, 417–418 (2008)
32. Benenson, Y.: Biocomputers: from test tubes to live cells. Mol. BioSyst. **5**, 675–685 (2009)

33. Pérez-Inestrosa, E., Montenegro, J.-M., Collado, D., Suau, R., Casado, J.: Molecules with multiple light-emissive electronic excited states as a strategy toward molecular reversible logic gates. J. Phys. Chem. C **111**, 6904–6909 (2007)
34. Cervera, J., Mafé, S.: Multivalued and reversible logic gates implemented with metallic nanoparticles and organic ligands. ChemPhysChem **11**, 1654–1658 (2010)
35. Remón, P., Ferreira, R., Montenegro, J.-M., Suau, R., Perez-Inestrosa, E., Pischel, U.: Reversible molecular logic: a photophysical example of a Feynman gate. ChemPhysChem **10**, 2004–2007 (2009)
36. Remón, P., Hammarson, M., Li, S., Kahnt, A., Pischel, U., Andréasson, J.: Molecular implementation of sequential and reversible logic through photochromic energy transfer switching. Chem. Eur. J. **17**, 6492–6500 (2011)
37. Sun, W., Xu, C.H., Zhu, Z., Fang, C.J., Yan, C.H.: Chemical-driven reconfigurable arithmetic functionalities within a fluorescent tetrathiafulvalene derivative. J. Phys. Chem. C **112**, 16973–16983 (2008)
38. Fratto, B.E., Roby, L.J., Guz, N., Katz, E.: Enzyme-based logic gates switchable between OR, NXOR and NAND Boolean operations realized in a flow system. Chem. Commun. **50**, 12043–12046 (2014)
39. Sun, W., Zheng, Y.-R., Xu, C.-H., Fang, C.-J., Yan, C.-H.: Fluorescence-based reconfigurable and resettable molecular arithmetic mode. J. Phys. Chem. C **111**, 11706–11711 (2007)
40. Liu, D.B., Chen, W.W., Sun, K., Deng, K., Zhang, W., Wang, Z., Jiang, X.Y.: Resettable, multi-readout logic gates based on controllably reversible aggregation of gold nanoparticles. Angew. Chem. Int. Ed. **50**, 4103–4107 (2011)
41. O'Steen, M.R., Cornett, E.M., Kolpashchikov, D.M.: Nuclease-containing media for resettable operation of DNA logic gates. Chem. Commun. **51**, 1429–1431 (2015)
42. Semeraro, M., Credi, A.: Multistable self-assembling system with three distinct luminescence outputs: prototype of a bidirectional half-subtractor and reversible logic device. J. Phys. Chem. C **114**, 3209–3214 (2010)
43. Andréasson, J., Pischel, U., Straight, S.D., Moore, T.A., Moore, A.L., Gust, D.: All-photonic multifunctional molecular logic device. J. Am. Chem. Soc. **133**, 11641–11648 (2011)
44. Orbach, R., Remacle, F., Levine, R.D., Willner, I.: Logic reversibility and thermodynamic irreversibility demonstrated by DNAzyme-based Toffoli and Fredkin logic gates. Proc. Natl. Acad. USA **109**, 21228–21233 (2012)
45. Roy, S., Prasad, M.: Novel proposal for all-optical Fredkin logic gate with bacteriorhodopsin-coated microcavity and its applications. Opt. Eng. **49**, Article ID 065201 (2010)
46. Klein, J.P., Leete, T.H., Rubin, H.: A biomolecular implementation of logically reversible computation with minimal energy dissipation. Biosystems **52**, 15–23 (1999)
47. Moseley, F., Halámek, J., Kramer, F., Poghossian, A., Schöning, M.J., Katz, E.: An enzyme-based reversible CNOT logic gate realized in a flow system. Analyst **139**, 1839–1842 (2014)
48. Katz, E., Wang, J., Privman, M., Halámek, J.: Multi-analyte digital enzyme biosensors with built-in Boolean logic. Anal. Chem. **84**, 5463–5469 (2012)
49. Wang, J., Katz, E.: Digital biosensors with built-in logic for biomedical applications. Isr. J. Chem. **51**, 141–150 (2011)
50. Landauer, R.: Irreversibility and heat generation in the computing process. IBM J. Res. Develop. **5**, 261–269 (1961)
51. Toffoli, T.: Physics and computation. Int. J. Theor. Phys. **21**, 165–175 (1982)
52. Fredkin, E., Toffoli, T.: Conservative logic. Int. J. Theor. Phys. **21**, 219–253 (1982)
53. Takeuchi, N., Yamanashi, Y., Yoshikawa, N.: Reversible logic gate using adiabatic superconducting devices. Sci. Rep. **4**, Article ID 6354 (2014)
54. Bennett, C.H.: Logical reversibility of computation. IBM J. Res. Develop. **17**, 525–532 (1973)
55. Zhou, J., Arugula, M.A., Halámek, J., Pita, M., Katz, E.: Enzyme-based NAND and NOR logic gates with modular design. J. Phys. Chem. B **113**, 16065–16070 (2009)
56. Privman, V., Arugula, M.A., Halámek, J., Pita, M., Katz, E.: Network analysis of biochemical logic for noise reduction and stability: a system of three coupled enzymatic AND gates. J. Phys. Chem. B **113**, 5301–5310 (2009)

57. Halámková, L., Halámek, J., Bocharova, V., Wolf, S., Mulier, K.E., Beilman, G., Wang, J., Katz, E.: Analysis of biomarkers characteristic of porcine liver injury - From biomolecular logic gates to animal model. Analyst **137**, 1768–1770 (2012)
58. Halámek, J., Windmiller, J.R., Zhou, J., Chuang, M.-C., Santhosh, P., Strack, G., Arugula, M.A., Chinnapareddy, S., Bocharova, V., Wang, J., Katz, E.: Multiplexing of injury codes for the parallel operation of enzyme logic gates. Analyst **135**, 2249–2259 (2010)
59. Toepke, M.W., Abhyankar, V.V., Beebe, D.J.: Microfluidic logic gates and timers. Lab Chip **7**, 1449–1453 (2007)
60. Scida, K., Li, B.L., Ellington, A.D., Crooks, R.M.: DNA detection using origami paper analytical devices. Anal. Chem. **85**, 9713–9720 (2013)
61. Garipelly, R., Madhu Kiran, P., Santhosh Kumar, A.: A Review on reversible logic gates and their implementation. Int. J. Emerging Technol. Adv. Eng. **3**, 417–423 (2013)
62. O'Brien, J.L., Pryde, G.J., White, A.G., Ralph, T.C., Branning, D.: Demonstration of an all-optical quantum controlled-NOT gate. Nature **426**, 264–267 (2003)
63. Monroe, C., Meekhof, D.M., King, B.E., Itano, W.M., Wineland, D.J.: Demonstration of a fundamental quantum logic gate. Phys. Rev. Lett. **75**, 4714–4717 (1995)
64. Siomau, M., Fritzsche, S.: Universal quantum Controlled-NOT gate. Eur. Phys. J. D **60**, 417–421 (2010)
65. Privman, V., Zhou, J., Halámek, J., Katz, E.: Realization and properties of biochemical-computing biocatalytic XOR gate based on signal change. J. Phys. Chem. B **114**, 13601–13608 (2000)
66. Fratto, B.E., Katz, E.: Reversible logic gates based on enzyme-biocatalyzed reactions and realized in flow cells – Modular approach. ChemPhysChem **1**, 1 (2015, in press)
67. Privman, V., Zavalov, O., Halámková, L., Moseley, F., Halámek, J., Katz, E.: Networked enzymatic logic gates with filtering: new theoretical modeling expressions and their experimental application. J. Phys. Chem. B **117**, 14928–14939 (2013)
68. Halámek, J., Bocharova, V., Chinnapareddy, S., Windmiller, J.R., Strack, G., Chuang, M.-C., Zhou, J., Santhosh, P., Ramirez, G.V., Arugula, M.A., Wang, J., Katz, E.: Multi-enzyme logic network architectures for assessing injuries: digital processing of biomarkers. Molec. Biosyst. **6**, 2554–2560 (2010)
69. Guz, N., Halámek, J., Rusling, J.F., Katz, E.: A biocatalytic cascade with several output signals - Towards biosensors with different levels of confidence. Anal. Bioanal. Chem. **406**, 3365–3370 (2014)
70. Mailloux, S., Guz, N., Zakharchenko, A., Minko, S., Katz, E.: Majority and minority gates realized in enzyme-biocatalyzed systems integrated with logic networks and interfaced with bioelectronic systems. J. Phys. Chem. B **118**, 6775–6784 (2014)
71. Mailloux, S., Halámek, J., Katz, E.: A model system for targeted drug release triggered by biomolecular signals logically processed through enzyme logic networks. Analyst **139**, 982–986 (2014)
72. De Vos, A.: Reversible Computing: Fundamentals, Quantum Computing, and Applications. Willey-VCH, Weinheim (2010)
73. Zhao, Y., Chakrabarty, K.: Digital microfluidic logic gates and their application to built-in self-test of lab-on-chip. IEEE Trans. Biomed. Circuits Syst. **4**, 250–262 (2010)
74. Konkoli, Z.: Phys. Rev. E **72**, Article ID 011917 (2005)

Chapter 3
Modeling and Modifying Response of Biochemical Processes for Biocomputing and Biosensing Signal Processing

Sergii Domanskyi and Vladimir Privman

Abstract Processes involving multi-input multi-step reaction cascades are used in developing novel biosensing, biocomputing, and decision making systems. In various applications different changes in responses of the constituent processing steps (reactions) in a cascade are desirable in order to allow control of the system's response. Here we consider conversion of convex response to sigmoid by "intensity filtering," as well as "threshold filtering," and we offer a general overview of this field of research. Specifically, we survey rate equation modelling that has been used for enzymatic reactions. This allows us to design modified biochemical processes as "network components" with responses desirable in applications.

3.1   Introduction

In theoretical rate-equation modeling of chemical and biochemical reactions in several-step cascades that are being investigated for novel biosensing or biomolecular computing applications [1, 8, 9, 11, 16, 17, 26, 27, 40, 42, 47–49, 51, 52, 54, 56, 59, 62, 63, 66, 79, 80, 88, 105, 106, 108, 117, 126, 127], one frequently focuses on the select few primary kinetic pathways [57, 64, 85, 86] for each step (reaction, process). This is done in order to limit the number of adjustable parameters in such systems, for which experimental data are typically noisy [73, 84] and not sufficiently detailed for a more accurate multi-parameter description of all the possible reaction pathways. Here we illustrate this approach by considering two specific recently studied systems [30, 94] relevant to biosensing and biocomputing [1, 17, 52, 68, 107]. However, the illustrated general framework for setting up rate-equation modeling applies to many other chemical, biochemical and biomolecular systems in the biosensing and biomolecular computing (biocomputing) contexts, extensively researched over the past decade [54, 64].

S. Domanskyi · V. Privman (✉)
Department of Physics, Clarkson University, Potsdam, NY 13676, USA
email: privman@clarkson.edu

© Springer International Publishing Switzerland 2017
A. Adamatzky (ed.), *Advances in Unconventional Computing*,
Emergence, Complexity and Computation 23,
DOI 10.1007/978-3-319-33921-4_3

61

We concentrate on processes with multi-input reaction cascades that are used in biosensing, biocomputing, and decision making devices and setups utilizing (bio)chemical processes with well-defined responses [8, 9, 49, 51, 53, 56, 59, 105, 106, 117, 126, 127]. Enzymatic processes are of particular interest because they promise short-term development of new biosensing [21, 40–42, 59, 117, 120] and bioactuating applications [58, 65, 95] with several signal processing steps. Indeed, most biosensing and bioanalytical devices involve enzymatic reactions, which are biocompatible, selective (specific), and also relatively easy to integrate with electronics [125]. For instance, enzyme-based logic systems [9, 55, 56, 68, 105, 106] operating as binary YES/NO biosensors can be interfaced with electrochemical/electronic devices by coupling to electrodes [41] or field-effect transistors [51, 60, 83].

The considered rate equation modelling has been used for cascades of enzymatic reactions [30, 88, 93, 94]. The set of kinetic rate equations describing the key (bio)chemical reaction steps of interconversion of different chemicals as well as the output buildup, is typically solved numerically with finite difference methods. These rate equations model the main reaction steps and enable fitting key process parameters to the extent allowed by limited and/or noisy experimental data. Indeed, the full kinetic description of each enzymatic process would in most cases require numerous parameters (rate constants) for each enzyme. We have developed models [30, 88, 93, 94] that give a reasonable system's response control—and description for potential modifications for applications—with a small number of adjustable parameters.

The use of the rate-equation modeling reviewed here, allows us to "design" modified biochemical processes as "device components" (signal processing steps) with responses desirable in applications. This is illustrated in Fig. 3.1, where panel (a) shows a typical "convex" response of a (bio)chemical process. The output is limited by the input chemical for small inputs, which results in a linear dependence. However, as the input is increased, other chemicals' availability limits the output, and its response to large input values reaches saturation. This can be modified for various applications, as sketched in Fig. 3.1.

In Sect. 3.2, we offer an illustration of a system where the convex response is modified to yield a sigmoid "filter" shape, by intensity-filtering [6, 30, 43, 44, 50, 82, 90, 92, 93, 96, 114, 122–124] whereby the input [96] or output [2, 43, 82] is chemically depleted up to a limited extent. For such filtering, the dashed line, Fig. 3.1, panel (c), illustrates the possibility of signal loss as the price paid for modifying the system's response. Linear response is desirable in many biosensing applications [5, 20, 22, 25, 29, 34, 76, 91, 97, 101–103, 111, 124, 128], see Fig. 3.1, panel (b). However, in certain cases threshold response is preferred [94], as shown in Fig. 3.1, panel (d). In Sect. 3.3, we offer an example of a system where an added enzymatic process accomplishes such a response-modification by an interesting new enzyme-functioning mechanism. Section 3.4 offers a Summary.

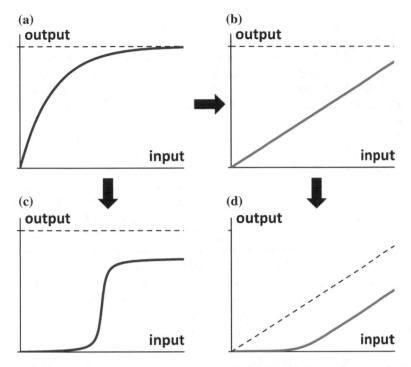

Fig. 3.1 a A typical "convex" response shape for a chemical or biochemical process. **b** Linear response desirable in many biosensing applications. **c** "Binary" sigmoid-shape response of interest in biomolecular computing, desired to be symmetrical and steep at the middle inflection. **d** In certain applications the conversion of a linear response to the threshold one, followed by a linear behavior, (b) → (d), is required. Adapted with permission from Ref. [94]. Copyright © 2014, American Chemical Society

3.2 Sigmoid Response and Its Numerical Rate-Equation Modeling

3.2.1 Sigmoid Response for Noise Reduction in Binary Gate Functioning

Sigmoid response is useful when "binary" input and output values are of interest in processing based on biomolecular reactions, which has recently been investigated for "digital" sensor and actuator design, logic systems, and other novel ideas in interfacing and computing involving biomolecules [52, 53]. Enzyme-catalyzed reactions have been used in such systems, with emphasis on novel diagnostic applications [4, 19, 27, 52, 53, 87]. For example, binary signal differentiation can be useful for future biomedical and diagnostic applications involving analysis of biomarkers indicative of specific illnesses or injury [42, 43, 74, 90]. Processing steps then mimic binary logic gates and their networks. These developments promise new functionalities for analytical purposes, offering a new class of biosensors which can generate a binary output of the alert type: YES/NO, in response to several input signals. These are parts of biosensor-bioactuator "Sense/Act/Treat" systems [57, 71, 115, 117]. The approach has already been used to analyze biomarkers indicative of certain traumas [69, 81]. Binary (digital) in such applications refers to the ability to identify specific values or ranges of values corresponding to **1** or **0** (YES/NO, Act/Don't Act) signals [85]. Standard binary logic gates, including AND, OR, XOR, INHIB, etc. [2, 7, 32, 87, 104, 108], and also few-gate model biomolecular networks [3, 21, 41, 88, 121] were demonstrated, some mimicking simple digital electronics designs.

Control of noise in functioning of biomolecular gates used as network elements is an important topic to consider [3, 85, 87]. An effective approach to noise control has been to modify some of the biomolecular reaction responses in a network of processing steps, according to (a) → (c), per Fig. 3.1, i.e., have the output a sigmoid function of the input(s). This mechanism is also used in natural systems [15, 87, 99]. Sigmoid response then "filters" the output towards the two reference binary values. Such biomolecular filtering based on several mechanisms has been considered, including, the use of allosteric enzymes that have substrates with self-promoter properties [89], "intensity filtering" (defined shortly) by redox transformations [90], pH control by buffering [82], and intensity filtering utilizing competing enzymatic processes [82]. These developments have built on earlier approaches to understand or realize sigmoid/digital (ON/OFF, YES/NO) responses in natural or synthetic biological and biochemical systems [13, 14, 33, 77].

The convex response, Fig. 3.1a, when scaled to the logic **0** to **1** input and output ranges, and assuming that the logic **0**s and at physical **0**s (of the reactants' concentrations), always has slope larger than 1 at the origin, and therefore amplifies the spread of the input(s) due to noise, as it is transmitted to the output. In intensity filtering a fraction of the output [87, 90, 108] signal or that of the input signal(s) is neutralized [96] (or converted into one of the intermediate reagents) by an added chemical process, but only for small values of the signals. The two approaches are interrelated especially when the considered processes are networked: outputs then become inputs to other gates. The partial output removal approach has been successfully applied to systems of interest in applications [43], as well as yielded realizations [44, 123] of double-sigmoid (means, with "filtering" properties with respect to both inputs) AND and OR logic gates. As sketched in Fig. 3.1c, the price paid when using such "intensity filtering" is the potential loss of some of the signal intensity (the spread between the physical values corresponding to the binary **0** and **1**).

Intensity filtering based on partial input neutralization has been theoretically analyzed [30] for optimizing the binary output signal. In the present section we survey this approach as an example. In the next subsection we describe the system for which experimental data were obtained in Ref. [96]. We illustrate how a simplified kinetic description of the enzymatic processes involved can be set up, to limit the number of fitted parameters to key rate constants. Furthermore, ideally the model setup should be done in a way that allows us to identify those chemical or physical parameters of biocatalytic processes which could be adjusted to control the quality of the realized sigmoid response. Quality measures of the sigmoid response should then be optimized, including the steepness and symmetry of the sigmoid curve, as well as the issue of avoiding too much signal intensity loss due to the added filtering.

3.2.2 Sigmoid Response Achieved by Neutralizing a Fraction of the Input

As an example, we analyze a specific system [96] that corresponds to signal transduction: The simplest "identity" logic gate that converts a single input: **0** or **1**, to the same binary value, **0** or **1**, of the output. In principle, the physical "logic values" (or ranges) of inputs that are designated as **0**s or **1**s are determined by the application. In fact, logic **0** needs not necessarily be at the physical zero. In the present case [96] the input is glucose in solution, the initial $t = 0$, where t is the time, concentration of which can be varied. We take the experimental [96] input values 0 mM and 10 mM, for the binary **0** and **1**, respectively.

The signal processing was biocatalyzed by an electrode-immobilized enzyme glucose oxidase, resulting in oxidation of glucose. The output was measured [96] at the "gate time," $t_g = 180$ s, as the current resulting from the transfer of two elementary units of charge per each oxidation cycle. In Fig. 3.2, the current, $I(t_g)$, normalized per its maximum value $I_{max}(t_g)$ for the largest glucose input, $G_{max} = 10$ mM, is plotted versus the glucose input. The data are taken from Ref. [96], whereas the model fit, detailed later, is from Ref. [30].

For evaluating the effects of noise [73, 87] in the signals, we have to consider the shape of the whole response curve, e.g., Fig. 3.2, i.e., the output current versus the input glucose concentration not only near the logic points but also generally over the whole **0** to **1** interval of values and somewhat beyond. As expected, the response curve here is convex. As described earlier, it is useful to convert the response to sigmoid, which offers advantages in noise handling, because small or zero slope at both logic points results in suppression of noise in the input as it is converted to the output.

Here we consider the approach [30] realized in Ref. [96], of neutralizing (consuming) a fraction of the input (glucose) in an added competing chemical process that only can use up a *limited amount* of glucose. Enzyme hexokinase was added to the solution, and adenosine triphosphate (ATP) was introduced in limited amounts as compared to the maximum 10 mM of glucose, to "switch on" the filtering effect. Indeed, the process biocatalyzed by hexokinase consumes glucose but only to the extent that ATP is not used up, without contributing to the output current. This makes the output signal, the current at the electrode, sigmoid. The corresponding experimental points from Ref. [96] and model fit (detailed shortly) are shown in Fig. 3.3.

As mentioned earlier, entirely phenomenological data fitting with properly shaped (convex or sigmoid) curves in not satisfactory, because we want to explicitly identify and model the dependence on those parameters which could be controlled to

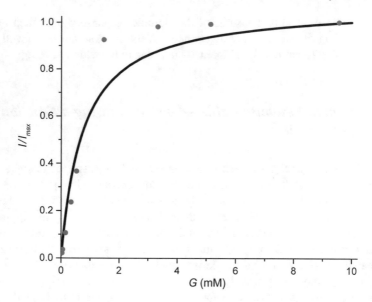

Fig. 3.2 Experimental data [96] (*circles*) and our numerical model (*line*) for the normalized current at time t_g versus the *initial* glucose concentration, G, without the "filter" process, with the fitted parameters as described in the text. Adapted with permission from Ref. [30]. Copyright © 2012, American Chemical Society

optimize the system's response. Phenomenological approaches [46] include the Hill function fitting [96]. Here we survey a different approach based on rate-equation modeling of the key steps of the enzymatic processes. We identify the concentrations of hexokinase and ATP as parameters to change, to significantly improve the sigmoid response.

Due to complexity of most enzymatic reactions, in our modeling we focus on few key processes for each of them. Indeed, as emphasized earlier we want few adjustable parameters, suitable for the noisy data available in this field, e.g., Figs. 3.2 and 3.3. With numerous parameters the specific data set might look better fitted, but the extrapolative power of the model will be lost. Thus, only enough adjustable parameters are kept to have a schematic overall-trend description of the response curves such as those shown in Figs. 3.2 and 3.3.

Let us first consider glucose oxidase (GOx) only, without the added "filtering." We identify the following key process steps and their rates:

$$E + G \xrightarrow{k_1} C \xrightarrow{k_2} E + \cdots . \tag{3.1}$$

Here E denotes the concentration of GOx, and G that of glucose. The intermediate products are produced in the first step, involving concentration C of the modified enzyme. For glucose, unlike some other possible substrates for GOx, the first step (lumping several processes) can usually be assumed practically irreversible [41, 42,

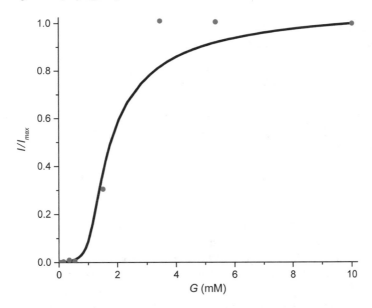

Fig. 3.3 Experimental data [96] (*circles*) and our numerical model (*line*), the same as in Fig. 3.2, but with the filtering process active. The process here is the same as in Fig. 3.2, but with added hexokinase (2 μM). The initial concentration of ATP, 1.25 mM, was a fraction of the maximum initial glucose concentration, 10 mM. Adapted with permission from Ref. [30]. Copyright © 2012, American Chemical Society

45, 46]. The last step is also irreversible. It is important to emphasize that we do not aim at a detailed kinetic study of the enzymatic reactions involved. As pointed out earlier, we seek a simple, few-parameter description of the response curve based on data from Ref. [96]. We ignore the kinetics of all the other reactants, input or product, except for the rate equation for $C(t)$,

$$\frac{dC(t)}{dt} = k_1 G(t)E(t) - k_2 C(t), \tag{3.2}$$

which should be solved with $E(t) = E(0) - C(t)$. Indeed, we need this quantity only, because the current is proportional to the rate of the second step in Eq. (3.1),

$$I(t) \propto k_2 C(t), \tag{3.3}$$

i.e., our output is $I(t_g) \propto C(t_g)$.

Without the hexokinase "filtering" part of the process, we can assume that the GOx reaction at the electrode practically does not consume glucose: $G(t) = G(0) = G$. This assumption is appropriate for electrochemical designs for glucose sensing [39, 112, 113]. We also assume that the oxygen concentration is constant (and therefore is absorbed in a rate constant), ignoring the fact that for the largest glucose concentrations some corrections might possibly be needed due to oxygen depletion at the electrode [96]. With these assumptions, Eq. (3.2) can be solved in closed form,

$$C(t) = \frac{k_1 E(0) G}{k_1 G + k_2} [1 - e^{-(k_1 G + k_2)t}]. \tag{3.4}$$

Here the initial (and later remaining constant) value of G is the input, varying from 0 to G_{max}. For logic-gate-functioning analysis of such processes, we define scaled logic-range variables,

$$x = \frac{G(0)}{G_{max}}, \qquad y = \frac{I(t_g; G(0))}{I_{max}} = \frac{C(t_g; G(0))}{C(t_g; G_{max})}, \tag{3.5}$$

where $I_{max} = I(t_g; G_{max})$. The slope of $y(x)$ near the logic point values $x = 0$ and 1, determines the noise transmission factors [85, 87].

The data in Ref. [96] were given as the values of y for several inputs, $G(0)$. Without the filter process, least-squares fit of these data in our case yielded the estimates $k_1 \cong 80 \, \text{mM}^{-1}\text{s}^{-1}$, $k_2 \cong 60 \, \text{s}^{-1}$. However, these estimates are rather imprecise, as indicated by the numerical fitting procedures. Indeed, these rate constants are large in the sense that the dimensionless combinations $k_2 t_g$ and $k_1 G_{max} t_g$ are both much larger than 1. This is consistent with other estimates of these rate constants for GOx with glucose as a substrate [12, 31, 38, 61]. The dependence of the scaled variable y on $G = G(0)$,

$$y = \frac{\frac{k_1 E(0) G}{k_1 G + k_2} [1 - e^{-(k_1 G + k_2)t_g}]}{\frac{k_1 E(0) G_{max}}{k_1 G_{max} + k_2} [1 - e^{-(k_1 G_{max} + k_2)t_g}]} \approx \frac{G(G_{max} + \frac{k_2}{k_1})}{G_{max}(G + \frac{k_2}{k_1})}, \tag{3.6}$$

is then to a good approximation only controlled by the ratio k_2/k_1, for which a relatively precise estimate is possible, $k_2/k_1 = 0.75 \pm 0.02$ mM. The quality of the fits such as that shown in Fig. 3.2, is not impressive, but this is similar to the situation with the more phenomenological Hill-function fitting [96].

With the filter process added, in the presence of hexokinase (HK), of concentration denoted $H(t)$, and ATP, of concentration $A(t)$, glucose will be depleted. Data are then available [96] for several initial values $A(0)$, all smaller than G_{max}. In order to limit the number of adjustable parameters we will only consider that pathway of the HK biocatalytic process [36, 118] in which glucose is taken in as the first substrate, to form an intermediate product of concentration $D(t)$. We again take a simplified scheme for the HK activity, ignoring a possible reversibility of the complex formation and other details [36, 37, 45, 118],

$$H + G \xrightarrow{k_3} D + \cdots, \qquad D + A \xrightarrow{k_4} H + \cdots. \tag{3.7}$$

This approach yields only two adjustable parameters which enter the rate equations that determine the time-dependence of glucose to use in Eq. (3.2) for calculating $C(t)$,

$$\frac{dG}{dt} = -k_3HG, \tag{3.8}$$

$$\frac{dH}{dt} = -k_3HG + k_4DA,$$

$$\frac{dD}{dt} = k_3HG - k_4DA,$$

$$\frac{dA}{dt} = -k_4DA.$$

Note that the two middle equations can be made into one by using $D(t) + H(t) = H(0)$. The available data were for $H(0) = 2\,\mu M$. The resulting system was solved numerically, and the data available for four initial nonzero ATP concentrations were fitted to yield the estimates $k_3 = 14.3 \pm 0.7\,\mathrm{mM^{-1}s^{-1}}$, $k_4 = 8.1 \pm 0.4\,\mathrm{mM^{-1}s^{-1}}$. The earlier estimate for k_2/k_1 was used to obtain these values.

3.2.3   Sigmoid Response Optimization

For fault-tolerant [3, 32, 88] information processing when gates are connected in a network [35, 116], parameters must be chosen to reduce the analog noise amplification or avoid it, the latter accomplished by filtering. There are various sources of noise in the biochemical reaction processes that affect their performance as binary "gates." Imprecise and/or noisy realization of the expected response curve, $y(x)$, is one such source. There is also noise in the input(s) that is passed to the output. In biochemical environments the noise in the inputs is quite large [23, 28, 52, 53, 85, 87, 98, 110]. Avoiding this "analog noise" being amplified during signal processing is paramount to small-scale network stability. For larger networks, additional consideration of "digital" errors [85] is required, but here we focus on the single gate design.

Unless the input noise levels are very large or the response curve has non-smooth features near the logic point $x = 0$ or 1, then the noise transmission factor is simply given by the slope of the curve $y(x)$ near each of the two logic points. Filtering can make both slopes (at **0** and **1**) much smaller than 1, compare Figs. 3.2 and 3.3. For best results, the filtering response-curve shape should be centered away from **0** or **1** and also steep. However, improvement of the quality of filtering should not be done at the expense of the intensity of the output signal in terms of its actual range of values, here equal I_{max}, as opposed to the scaled variable y. Loss of intensity amplifies the *relative level* of noise from all the sources discussed above.

The inputs are set by the gate usage and typically cannot be adjusted. We can select other parameters values to optimize the filtering quality. Here we formulate quantitative criteria for such optimization. Note that within the assumed regime of functioning, in our model the shape of $y(x)$ does not depend on $E(0)$. However, other "gate machinery" (means, not input or output) initial chemical concentrations can be varied. Here we consider the adjustment of $H(0)$ and $A(0)$. Other modifications can include changing the physical or chemical conditions (which affects the rates of various processes) or limiting the supply of oxygen [72].

To have the response curve as symmetric as possible we consider the position of the peak of the slope, $y'(x)$. In enzymatic processes, sigmoid response-curves are typically not symmetrical with respect to the inflection region; see Fig. 3.3 and also some approximate analytic expressions and their plots in Refs. [92, 93]. We can define the width of the peak of the derivative by the difference $x_2 - x_1$, where $y'(x_{1,2}) = 1$. The middle-point of the peak is defined at $(x_2 + x_1)/2$. Figure 3.4 shows three different illustrative sigmoid response curves, as well as their derivatives calculated in our model, with the parameter values discussed in the preceding subsection. Figure 3.5 presents a contour plot of the *deviation* of the middle-point peak position from 1/2. Our aim is to get it rather close to 1/2 without compromising the other gate-quality criteria. One of these is analyzed in Fig. 3.6, which plots the width of the peak, which we would like to be as small as possible.

A "non-binary" criterion for gate quality is that of avoiding to the extent possible the loss of the signal intensity. Since enzymatic processes usually approach saturation at large inputs, here this can be defined as the fractional loss:

$$1 - \frac{I_{max}(H(0) > 0, A(0) > 0)}{I_{max}(H(0) = 0, A(0) = 0)} = 1 - \frac{I(t_g, G_{max})_{H(0)>0,A(0)>0}}{I(t_g, G_{max})_{H(0)=0,A(0)=0}}. \tag{3.9}$$

This quantity is shown in Fig. 3.7 as the percentage value. Figures 3.5, 3.6 and 3.7 span values safely within the experimentally realizable ranges of the considered control quantities, $H(0)$ and $A(0)$. Consideration of Figs. 3.5 and 3.6 suggests that the peak can be made optimally centrally positioned and narrow, by selecting approximately $H(0) = 4\,\mu M$ and $A(0) = 4$ mM. The optimal choices correspond to the regions marked by the white ovals in the figures. At least some loss of intensity is usually present for this type of filtering. However, it can be tolerated if percentage-wise it is comparable to (or smaller than) the degree of noise otherwise present in the output. The approximately 5 % loss level in the oval-delineated region (see Fig. 3.7) is therefore acceptable.

Our optimal sigmoid response shape and its derivative are shown as curves (a) in Fig. 3.4. While not symmetrical, the response curve is centrally positioned and rather narrow. The derivative of the output signal in regions $0 \leq x \lesssim 0.37$ and $0.63 \lesssim x \leq 1$ is less than 1, see Fig. 3.4: bottom panel, curve (a). In these two regions, each extending $\sim 37\%$ from the logic points **0** and **1**, on the input axis, the noise in the input will not be amplified. The criteria just surveyed are quite general and can be applied to other systems contemplated for information and signal processing or for biosensing with biomolecular processes.

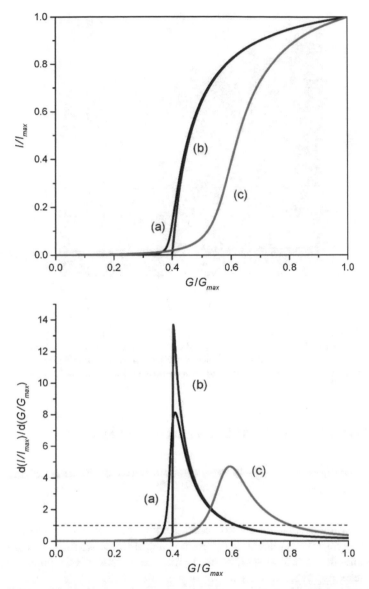

Fig. 3.4 Examples of sigmoid curves (*top panel*) and their derivatives (*bottom panel*) for three different selections of the parameters used to control the response: **a** $H(0) = 4\,\mu M$ and $A(0) = 4$ mM; **b** $H(0) = 8\,\mu M$ and $A(0) = 4$ mM; **c** $H(0) = 3\,\mu M$ and $A(0) = 6$ mM. The values (**a**) correspond to the center of the optimal range as described in the text. The *dashed line* indicates the level at which the width of the peak of the derivative is measured. Adapted with permission from Ref. [30]. Copyright © 2012, American Chemical Society

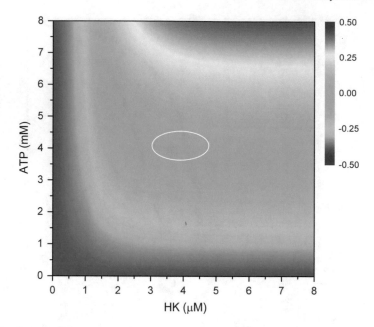

Fig. 3.5 Contour plot for various initial values of HK and ATP, of the deviation of the middle-point of the peak location from 1/2, i.e., $(x_2 + x_1 - 1)/2$. The optimal values are as small as possible (*green color*). The *oval* defines the best choice of the parameters considering the other criteria for optimizing the response: see text. Reprinted with permission from Ref. [30]. Copyright © 2012, American Chemical Society

3.3 Threshold Response in an Enzymatic System

3.3.1 Modifications of Response Functions of Biomolecular Processes

In the preceding section we considered conversion of convex response to sigmoid. However, in various applications different changes in the response function might be desirable. In biosensing applications in many situations it is useful to modify the generic response to make it as linear as possible [5, 20, 22, 25, 29, 34, 76, 91, 97, 101–103, 111, 124, 128], i.e., the conversion (a) → (b) in Fig. 3.1, here also hoping to avoid too much loss in the overall signal intensity. Recently, a model was developed [91, 124] (not reviewed here) and applied to data analysis, of how two enzymatic processes with different nonlinear responses can be combined to yield an extended approximately linear response regime.

Recently, experiments [66] on three-input majority and minority enzymatic gates for biocomputing applications have underscored the importance of another type of "biochemical filtering" as a part of the biochemical post-processing of the output to achieve the desired response. In this case the conversion of a linear response to the

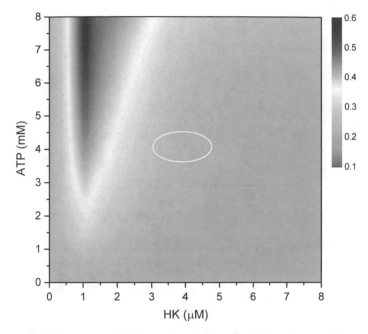

Fig. 3.6 Contour plot of the width of the peak. The optimal values are as small as possible (the *green shades*). The *oval* defines the best choice of the parameters considering the other criteria for optimizing the response: see text. Reprinted with permission from Ref. 30. Copyright © 2012, American Chemical Society

threshold one, followed by a linear behavior, (b) → (d) in Fig. 3.1, is required. Here we review results [94] establishing that such "filtering" mechanism in the reported experiments [66] (and in the earlier work on filtering [67]) utilizing the enzyme malate dehydrogenase (MDH), also called malic dehydrogenase, is a result on an unusual mechanism of enzymatic biocatalytic activity of this enzyme, noted in an early work on the mechanism of action of MDH [100]. This work [100] considered what is called [70] a reversible random-sequential bi bi mechanism of action for MDH, and reported that MDH can undergo a variant of inhibition [100] that results in the slowing-down of the oxidation/reduction of one of the two substrate/product redox couples.

As suggested by this observation, modeling of the filtering effect here is quite different from that for the afore-surveyed [6, 30, 44, 93, 122–124] "intensity filtering." We survey an appropriate description, which was applied [94] to data for a system where the initial linear response is due to the biocatalytic action of another enzyme, glucose dehydrogenase (GDH). We also report (in the next subsection) additional interesting conclusions for "intensity filtering" that was considered in the preceding section.

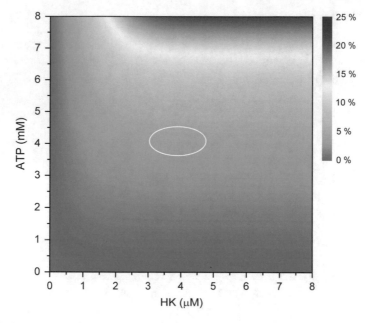

Fig. 3.7 Contour plot of the measure of the loss of the output signal intensity, Eq. (3.9). This measure should be minimized (*green color*) without compromising the other gate-quality criteria. The *oval* defines the best choice of the parameters considering the other two criteria: see text. Reprinted with permission from Ref. 30. Copyright © 2012, American Chemical Society

3.3.2 Signal Transduction Combined with Fast Reversible Deactivation of the Output

The system that is considered here is shown schematically in Fig. 3.8. We already emphasized that the full kinetic description of enzymatic processes requires several rate constants for each enzyme. We will revisit this later (in Sect. 3.3.3). Let us first attempt to use a simple model with a minimal number of parameters in an attempt to describe the effect on a linear response of the type shown in Fig. 3.1b, of an added process that affects the output product, of concentration, $P(t)$, by rapidly interconverting it to and from (equilibrating it with) another compound that is inert as far as contributing to the output signal. Our conclusion will be that this simple description is not adequate for the system of interest [94]. However, the model itself is useful to study because adding fast, reversible processes that affect the product can be done relatively easily in most cases by chemical or biochemical means.

The first enzyme in the cascade, GDH, was utilized in the kinetic regime quite typical for many uses of enzymes, i.e., with both of its input chemicals (substrates), glucose and NAD^+, provided with the initial concentrations large enough to have the products of the reaction generated with a practically constant rate for the times of the experiment. For the product of interest, NADH, we thus assume that its concentration, $P(t)$, varies according to

$$\frac{dP}{dt} = RG, \tag{3.10}$$

$$P(t_g) = RGt_g, \tag{3.11}$$

where R is a rate constant that can be fitted from the data, whereas G is the initial concentration of glucose, which is the input at time $t = 0$, was varied from 0 up to, here, $G_{max} = 8$ mM. Other reagents in the present system have fixed initial concentrations. The linear behavior in time applies for all but the smallest inputs, G, and it breaks down for very short times as well as for very long times on the time-scales of the experiments that went up to 600 s.

The second enzyme, MDH, is also used in the regime of plentiful supply of the initially available substrates (one of the two in each direction of functioning, see Fig. 3.8). Since its functioning is reversible, we could attempt to describe the kinetics of the present system by the effective processes

$$G \xrightarrow{R} P, \qquad P \underset{r_+}{\overset{r_+}{\rightleftharpoons}} M \tag{3.12}$$

We note that MDH oxidizes NADH to NAD^+, which is then our "inert" compound (not contributing to the measured signal obtained by optically detecting the concentration of NADH), but since NAD^+ is already present in the system in a large quantity, the variation of its concentration has little effect on the reverse process. However, malate, denoted, $M(t)$, see Fig. 3.8, not initially present, directly (and for

Fig. 3.8 The schematics of the enzymatic processes in the biocatalytic cascade [94] surveyed in Sect. 3.3. The reactants and biocatalysts that are initially in the system (with filtering) are color-coded *blue*, including β-nicotinamide adenine dinucleotide (NAD^+) and its reduced form (NADH). The *double-arrows* emphasize that the functioning of MDH is reversible. Reprinted with permission from Ref. [94]. Copyright © 2014, American Chemical Society

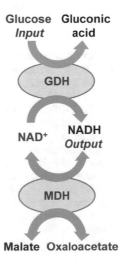

simplicity we assume linearly) affects the reverse process rate. The present model is not accurate, but interesting because the resulting rate equations can be solved in closed form,

$$\frac{dP}{dt} = RG - r_+P + r_-M, \qquad \frac{dM}{dt} = r_+P - r_-M, \qquad (3.13)$$

$$P(t) = RG\left\{ \frac{r_+[1 - e^{-(r_+ + r_-)t}]}{(r_+ + r_-)^2} + \frac{r_-t}{r_+ + r_-} \right\}. \qquad (3.14)$$

One could speculate that an added fast reversible process that deactivates a part of the product, up to a fraction that corresponds to the concentrations of the rate-limiting chemicals for which that reversible process equilibrates, might have some "filtering" effect. But Eq. (3.14) suggests that there is no "filtering" at all. Instead, the dependence of the product $P(t_g)$ on the input, G, remains linear for any fixed "gate time" t_g, with a reduced slope (means, with loss of intensity). The original time-dependence, Eq. (3.11), is linear in both G and t_g. However, with the added process the time dependence is modified. Figure 3.9 illustrates that for small times the rate of the product output is unchanged (the added process is not really active). For large times a reduced rate, $RGr_-/(r_+ + r_-)$, is approached.

Interestingly, the experimentally observed [66] change from the linear to threshold response, (b) → (d) in Fig. 3.1, must therefore be due to more complicated kinetic mechanisms than that summarized in Eq. (3.12). The origin of the observed effect turns out to be connected to an interesting kinetic property of the functioning of MDH, reviewed in the rest of this section. The model just considered, however, suggests that, generally adding a fast, reversible process of deactivation of the input by *equilibrating* it with another species cannot in itself result in threshold type (at low inputs) intensity filtering. Examples [6, 30, 43, 44, 50, 82, 90, 92, 93, 96, 114, 122–124] when such an approach worked have always involved the *absence of equilibration* by kinetic restrictions, for example due to a limitation on how much of

Fig. 3.9 Time dependence of the NADH concentration, $P(t)$, for typical parameter values [94] with (the *red curve*) and without (the *black straight line*) the added fast reversible "output deactivation" process biocatalyzed by MDH. The *dashed line* is the asymptotic rate, $RGr_-/(r_+ + r_-)$, for large times. Adapted with permission from Ref. [94]. Copyright © 2014, American Chemical Society

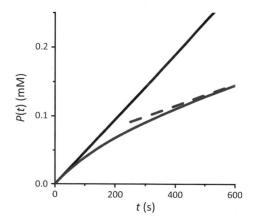

the other species could be produced (imposed by the process requiring some other, limited-supply chemical).

3.3.3 MDH Kinetics with Inhibition

Enzymes have rather complicated kinetic mechanisms. These typically involve the formation of complexes with substrate(s), then follow-up processes involving these complexes, etc., in most cases resulting in the restoration of the enzyme at the end of the cascade, when products are released. Our first enzyme, GDH, has such a standard mechanism of action [10, 78, 109], that would require several rate constants to fully model. The second enzyme, MDH, has a complicated and less common mechanism of action [24, 75, 100, 119], with a number of intermediate complexes. It is in fact not fully studied. MDH can form complexes [100] with all four of the relevant substrates for the direct (NADH and oxaloacetate) or reverse (NAD$^+$ and malate) functioning, and then form triple-complexes in which the actual redox-pair conversions occur. This is sketched in Fig. 3.10a. Modeling [18] of all the processes would require at least 18 rate constants. This illustrates why it is so important to use few-parameter kinetic models for a semi-qualitative description of the response in sensor and biomolecular computing applications. Such approaches [85, 93] usually involve setting up an effective rate equation description that captures the main process pathways.

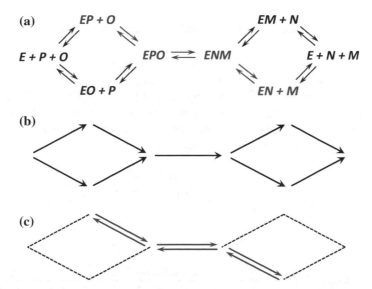

Fig. 3.10 a Mechanism of action of MDH. Here E stands for the enzyme, P for NADH (the product), N for NAD$^+$, whereas malate and oxaloacetate are denoted by M and O, respectively. **b** The "direct" reaction pathways activate at early times. **c** A hypothetical mechanism for a reaction pathway subset that dominates at later times

The output product, NADH, denoted P, generated by the GDH process, activates all the "direct" complex-formation and redox conversion processes of MDH, Fig. 3.10b. The latter not only partially convert NADH back to NAD$^+$, to be denoted N, but these processes also build up the concentration of malate, M. The "reverse" processes of MDH are then also activated, driving the system towards equilibration. However, it has been reported in the literature [100] that, as the concentration of malate is increased relative to oxaloacetate, denoted O, the redox inter-conversion rate NADH $\leftrightarrow$ NAD$^+$ actually slows down, whereas the inter-conversion rate oxaloacetate $\leftrightarrow$ malate increases. This might look paradoxical, but a likely explanation is as shown in Fig. 3.10c. Most of the enzyme, E, becomes trapped in the complexes EP and EN (as well as in complexes, ENM and EPO). The fast inter-conversion oxaloacetate $\leftrightarrow$ malate ($O \leftrightarrow M$) is compensated for by the inter-conversion $EP \leftrightarrow EN$. This interesting mechanism can be either kinetic or caused by malate inhibiting [100] some of the reaction pathways. It is important to emphasize that despite the earlier experimental evidence [100], this mechanism is largely a conjecture. In fact, the observation that this assumption leads to modeling [94] that fits the data provides an additional support to it.

To model this effect with a minimal possible number of parameters, considering that oxaloacetate is supplied in large quantity, we ignore its depletion. We assume that the concentration of malate that would correspond to steady state is M_0. We then write the rate equation of the linear supply of the product, cf. Eq. (3.10), but with the added depletion term,

$$\frac{dP}{dt} = RG - K(M_0 - M)P = -KP^2 - K(M_0 - Rt)P + RG. \tag{3.15}$$

Here K is the rate constant for the decrease in the amount of the product, P, due to the initially active mechanism, Fig. 3.10b, which is gradually replaced by the mechanism involving $EP \leftrightarrow EN$ as M increases from 0 to M_0, Fig. 3.10c. This assumes that the relative rates of the two mechanisms are directly proportional to $M_0 - M$ and M, respectively. The second expression in Eq. (3.15) was obtained by using $M(t) = RGt - P(t)$. This can be solved to yield

$$P(t) = RGt - M_0 + \frac{M_0 e^{-K(\frac{1}{2}RGt - M_0)t}}{1 + KM_0 \int_0^t e^{-K(\frac{1}{2}RG\tau - M_0)\tau} d\tau}, \tag{3.16}$$

or

$$P(t) = RGt - M_0 + \frac{2\sqrt{KRG}M_0 e^{\frac{Kt}{2}(2M_0 - RGt)}}{\sqrt{2\pi}KM_0 e^{\frac{KM_0^2}{2RG}}\left[\operatorname{erf}\left(\sqrt{\frac{K}{2RG}}M_0\right) - \operatorname{erf}\left(\sqrt{\frac{K}{2RG}}(M_0 - RGt)\right)\right] + 2\sqrt{KRG}}. \tag{3.17}$$

This expression provides the dependence of $P(t_g)$ on G, of the type shown in Fig. 3.1d, and was successful in experimental data fitting [94] for a system the

Fig. 3.11 *Top panel* Measured time dependence [94] for input $G = 7$ mM, points practically merged into *solid lines*, and model results, shown as *dashed lines*, for a typical experiment **a** without filtering, and **b** with filtering. *Bottom panel* Measured dependence on the initial glucose concentration [94] for fixed time $t_g = 360$ s, shown as *dots*, for **a** without filtering, and **b** with filtering; model results are shown as *dashed lines*. Adapted with permission from Ref. [94]. Copyright © 2014, American Chemical Society

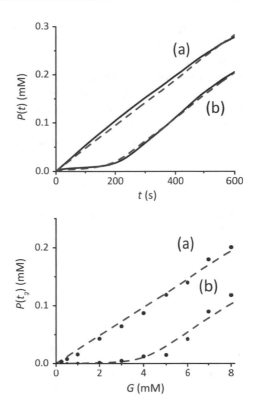

schematic of which is sketched in Fig. 3.8. Figure 3.11 provides an illustration of fitting the experimentally measured [94] time dependence, and also shows an example of data fitting [94] for the response function, which should be compared with Fig. 3.1b and 3.1d.

3.4 Summary

We reviewed the biochemical "intensity filtering," by considering approaches to modeling binary AND gate performance and optimization of its "digital" response. Specifically, we considered the recently introduced approach of a partial input conversion into inactive compounds, which yields sigmoid response of the output, of interest in information/signal processing and in biosensing applications. For selected examples, we established criteria for optimizing such a "binary" response. Different physical or chemical conditions can be changed to impact enzymatic processes, and we demonstrated this by an example of how our system's response changed when the initial concentrations of two "filter process" chemicals were varied. The developed criteria are quite general and can be applied to other systems contemplated for information/signal processing, and for biosensing, with biomolecular processes.

Applying a similar rate-equation modelling approach we then demonstrated that reversible conversion of the product to another compound cannot on its own result in (bio)chemical "filtering." Experimentally observed biochemical "threshold filtering" by a reaction biocatalyzed by an enzyme with an unusual mechanism of action was instead attributed to inhibition of certain process pathways for this enzyme once one of its substrates builds up in concentration. Experimental data analysis supports the model's validity.

Acknowledgments The authors thank their many colleagues [3, 6, 30, 32, 43, 44, 56–59, 63, 73, 82, 84–95, 122–124], notably, Prof. E. Katz, for collaborative teamwork. They also gratefully acknowledge funding of the research work reviewed here by the US National Science Foundation, most recently under award CBET-1403208.

References

1. Adamatzky, A., De Lacy Costello, B., Bull, L., Stepney, S., Teuscher, C.: Unconventional Computing. Luniver Press, UK (2007)
2. Amir, L., Tam, T.K., Pita, M., Meijler, M.M., Alfonta, L., Katz, E.: J. Am. Chem. Soc. **131**, 826 (2009)
3. Arugula, M.A., Halámek, J., Katz, E., Melnikov, D., Pita, M., Privman, V., Strack, G.: IEEE Comp. Soc. Conf. Publ. Serv. **1** (2009)
4. Ashkenazi, G., Ripoll, D.R., Lotan, N., Scheraga, H.A.: Biosensors. Bioelectronics **12**, 85 (1997)
5. Bakker, E.: Anal. Chem. **76**, 3285 (2004)
6. Bakshi, S., Zavalov, O., Halámek, J., Privman, V., Katz, E.: J. Phys. Chem. B **117**, 9857 (2013)
7. Baron, R., Lioubashevski, O., Katz, E., Niazov, T., Willner, I.: Angewandte Chemie International Edition **45**, 1572 (2006)
8. Benenson, Y.: Mol. BioSyst. **5**, 675 (2009)
9. Benenson, Y.: Nat. Rev. Genet. **13**, 455 (2012)
10. Bhaumik, S.R., Sonawat, H.M.: Indian J. Biochem. Biophys. **36**, 143 (1999)
11. Bowsher, C.G.: J. R. Soc. Interface **8**, 186 (2010)
12. Brights, H.J., Appleby, M.: J. Biol. Chem **244**, 3625 (1969)
13. Buchler, N.E., Cross, F.R.: Mol. Syst. Biol. **5**, 272 (2009)
14. Buchler, N.E., Louis, M.: J. Mol. Biol. **384**, 1106 (2008)
15. Buchler, N.E., Gerland, U., Hwa, T.: Proc. Natl. Acad. U. S. A. **102**, 9559 (2005)
16. Buisman, H.J., ten Eikelder, H.M.M., Hilbers, P.A.J., Liekens, A.M.L.: Artif. Life **15**, 5 (2009)
17. Calude, C.S., Costa, J.F., Dershowitz, N., Freire, E., Rozenberg, G.: Unconventional Computation. Lecture Notes in Computer Science, vol. 5715. Springer, Berlin (2009)
18. Cha, S.: J. Biol. Chem. **243**, 820 (1968)
19. Chen, J., Fang, Z., Lie, P., Zeng, L.: Anal. Chem. **84**, 6321 (2012)
20. Chen, X., Zhu, J., Tian, R., Yao, C.: Sens. Actuators B: Chem. **163**, 272 (2012)
21. Chuang, M., Windmiller, J.R., Santhosh, P., Ramirez, G.V., Katz, E., Wang, J.: Chem. Commun. **47**, 3087 (2011)
22. Coche-Guerente, L., Cosnier, S., Labbe, P.: Chem. Mater. **9**, 1348 (1997)
23. Credi, A.: Angewandte Chemie International Edition **46**, 5472 (2007)
24. Cunningham, M.A., Ho, L.L., Nguyen, D.T., Gillilan, R.E., Bash, P.A.: Biochemistry **36**, 4800 (1997)

25. De Benedetto, G.E., Palmisano, F., Zambonin, P.G.: Biosensors. Bioelectronics **11**, 1001 (1996)
26. de Silva, A.P.: Nature **454**, 417 (2008)
27. de Silva, A.P., Uchiyama, S.: Nat. Nanotechnol. **2**, 399 (2007)
28. de Silva, A.P., Uchiyama, S., Vance, T.P., Wannalerse, B.: Coord. Chem. Rev. **251**, 1623 (2007)
29. Delvaux, M., Walcarius, A., Demoustier-Champagne, S.: Biosensors. Bioelectronics **20**, 1587 (2005)
30. Domanskyi, S., Privman, V.: J. Phys. Chem. B **116**, 13690 (2012)
31. Duke, F.R., Weibel, M., Page, D.S., Bulgrin, V.G., Luthy, J.: J. Am. Chem. Soc. **91**, 3904 (1969)
32. Fedichkin, L., Katz, E., Privman, V.: J. Comp. Theor. Nanosci. **5**, 36 (2008)
33. Ferrell Jr., J.E.: Curr. Biol. **18**, R244 (2008)
34. Ferri, T., Maida, S., Poscia, A., Santucci, R.: Electroanalysis **13**, 1198 (2001)
35. Flood, A.H., Ramirez, R.J.A., Deng, W., Muller, R.P., Goddard III, W.A., Stoddart, J.F.: Aust. J. Chem. **57**, 301 (2004)
36. Fromm, H.J., Zewe, V.: J. Biol. Chem. **237**, 1661 (1962)
37. Fromm, H.J., Zewe, V.: J. Biol. Chem. **237**, 3027 (1962)
38. Gibson, Q.H., Swoboda, E.W., Massey, V.: J. Biol. Chem **239**, 3927 (1964)
39. Gough, D.A., Leypoldt, J.K., Armour, J.C.: Diabet. Care **5**, 190 (1982)
40. Guz, N., Halámek, J., Rusling, J., Katz, E.: Anal. Bioanal. Chem. **406**, 3365 (2014)
41. Halámek, J., Windmiller, J.R., Zhou, J., et al.: Analyst **135**, 2249 (2010)
42. Halámek, J., Bocharova, V., Chinnapareddy, S., et al.: Mol. BioSyst. **6**, 2554 (2010)
43. Halámek, J., Zhou, J., Halámková, L., Bocharova, V., Privman, V., Wang, J., Katz, E.: Anal. Chem. **83**, 8383 (2011)
44. Halámek, J., Zavalov, O., Halámková, L., Korkmaz, S., Privman, V., Katz, E.: J. Phys. Chem. B **116**, 4457 (2012)
45. Hammes, G.G., Kochavi, D.: J. Am. Chem. Soc. **84**, 2069 (1962)
46. Heidel, J., Maloney, J.: ANZIAM J. **41**, 83 (1999)
47. Hillenbrand, P., Fritz, G., Gerland, U.: PLoS ONE **8**, e68345 (2013)
48. Jiang, H., Riedel, M.D., Parhi, K.K.: IEEE Des.Test Comput. **29**, 21 (2012)
49. Jia, Y., Duan, R., Hong, F., Wang, B., Liu, N., Xia, F.: Soft Matter **9**, 6571 (2013)
50. Kang, D., Vallée-Bélisle, A., Porchetta, A., Plaxco, K.W., Ricci, F.: Angewandte Chemie International Edition **51**, 6717 (2012)
51. Katz, E.: Isr. J. Chem. **51**, 132 (2011)
52. Katz, E.: Molecular, Supramolecular Information Processing: From Molecular Switches to Logic Systems. Wiley-VCH Verlag GmbH & Co. KGaA, Weinheim (2012)
53. Katz, E.: Biomolecular Information Processing. From Logic Systems to Smart Sensors, Actuators. Wiley-VCH, Weinheim (2012)
54. Katz, E.: Curr. Opin. Biotechnol. **34**, 202 (2015)
55. Katz, E., Minko, S.: Chem. Commun. **51**, 3493 (2015)
56. Katz, E., Privman, V.: Chem. Soc. Rev. **39**, 1835 (2010)
57. Katz, E., Privman, V., Wang, J.: In: Proceedings of the Conference ICQNM 2010, vol. 1 (2010)
58. Katz, E., Bocharova, V., Privman, M.: J. Mater. Chem. **22**, 8171 (2012)
59. Katz, E., Wang, J., Privman, M., Halámek, J.: Anal. Chem. **84**, 5463 (2012)
60. Kramer, M., Pita, M., Zhou, J., Ornatska, M., Poghossian, A., Schöning, M.J., Katz, E.: J. Phys. Chem. C **113**, 2573 (2009)
61. Leskovac, V., Trivić, S., Wohlfahrt, G., Kandrač, J., Peričin, D.: Int. J. Biochem. Cell Biol. **37**, 731 (2005)
62. Liu, Y., Kim, E., White, I.M., Bentley, W.E., Payne, G.F.: Bioelectrochemistry **98**, 94 (2014)
63. MacVittie, K., Halámek, J., Privman, V., Katz, E.: Chem. Commun. **49**, 6962 (2013)
64. Mailloux, S., Katz, E.: Biocatalysis **1**, 13 (2014)
65. Mailloux, S., Halámek, J., Katz, E.: Analyst **139**, 982 (2014)

66. Mailloux, S., Guz, N., Zakharchenko, A., Minko, S., Katz, E.: J. Phys. Chem. B **118**, 6775 (2014)
67. Mailloux, S., Zavalov, O., Guz, N., Katz, E., Bocharova, V.: Biomater. Sci. **2**, 184 (2014)
68. Mailloux, S., Gerasimova, Y.V., Guz, N., Kolpashchikov, D.M., Katz, E.: Angewandte Chemie International Edition **54**, 6562 (2015)
69. Manesh, K.M., Halámek, J., Pita, M., et al.: Biosensors. Bioelectronics **24**, 3569 (2009)
70. Marangoni, A.G.: Enzyme Kinetics. A Modern Approach. Wiley, New York (2003)
71. Margulies, D., Hamilton, A.D.: J. Am. Chem. Soc. **131**, 9142 (2009)
72. McMahon, C.P., Rocchitta, G., Serra, P.A., Kirwan, S.M., Lowry, J.P., O'Neill, R.D.: Anal. Chem. **78**, 2352 (2006)
73. Melnikov, D., Strack, G., Pita, M., Privman, V., Katz, E.: J. Phys. Chem. B **113**, 10472 (2009)
74. Melnikov, D., Strack, G., Zhou, J., et al.: J. Phys. Chem. B **114**, 12166 (2010)
75. Minárik, P., Tomášková, N., Kollárová, M., Antalík, M.: Gen. Physiol. Biophys. **21**, 265 (2002)
76. Nikitina, O., Shleev, S., Gayda, G., Demkiv, O., Gonchar, M., Gorton, L., Csöregi, E., Nistor, M.: Sens. Actuators B: Chem. **125**, 1 (2007)
77. Novick, A., Weiner, M.: Proc. Natl. Acad. U. S. A. **43**, 553 (1957)
78. Ohshima, T., Ito, Y., Sakuraba, H., Goda, S., Kawarabayasi, Y.: J. Mol. Catal. B **23**, 281 (2003)
79. Pischel, U.: Angewandte Chemie International Edition **49**, 1356 (2010)
80. Pischel, U., And réasson, J., Gust, D., Pais, V.F.: ChemPhysChem **14**, 28 (2013)
81. Pita, M., Zhou, J., Manesh, K.M., Halámek, J., Katz, E., Wang, J.: Sens. Actuators B: Chem. **139**, 631 (2009)
82. Pita, M., Privman, V., Arugula, M.A., Melnikov, D., Bocharova, V., Katz, E.: Phys. Chem. Chem. Phys. **13**, 4507 (2011)
83. Poghossian, A., Malzahn, K., Abouzar, M.H., Mehndiratta, P., Katz, E., Schöning, M.J.: Electrochim. Acta **56**, 9661 (2011)
84. Privman, V.: J. Comput. Theor. Nanosci. **8**, 490 (2011)
85. Privman, V.: Isr. J. Chem. **51**, 118 (2011)
86. Privman, V.: Approaches to Control of Noise in Chemical, Biochemical Information, Signal Processing, p. 281. Wiley-VCH Verlag GmbH & Co. KGaA, Weinheim (2012)
87. Privman, V., Strack, G., Solenov, D., Pita, M., Katz, E.: J. Phys. Chem. B **112**, 11777 (2008)
88. Privman, V., Arugula, M.A., Halámek, J., Pita, M., Katz, E.: J. Phys. Chem. B **113**, 5301 (2009)
89. Privman, V., Pedrosa, V., Melnikov, D., Pita, M., Simonian, A., Katz, E.: Biosensors. Bioelectronics **25**, 695 (2009)
90. Privman, V., Halámek, J., Arugula, M.A., Melnikov, D., Bocharova, V., Katz, E.: J. Phys. Chem. B **114**, 14103 (2010)
91. Privman, V., Zavalov, O., Simonian, A.: Anal. Chem. **85**, 2027 (2013)
92. Privman, V., Fratto, B.E., Zavalov, O., Halámek, J., Katz, E.: J. Phys. Chem. B **117**, 7559 (2013)
93. Privman, V., Zavalov, O., Halámková, L., Moseley, F., Halámek, J., Katz, E.: J. Phys. Chem. B **117**, 14928 (2013)
94. Privman, V., Domanskyi, S., Mailloux, S., Holade, Y., Katz, E.: J. Phys. Chem. B **118**, 12435 (2014)
95. Privman, M., Tam, T.K., Bocharova, V., Halámek, J., Wang, J., Katz, E., Appl, A.C.S.: Mater. Interfaces **3**, 1620 (2011)
96. Rafael, S.P., Vallée-Bélisle, A., Fabregas, E., Plaxco, K., Palleschi, G., Ricci, F.: Anal. Chem. **84**, 1076 (2012)
97. Rinken, T., Rinken, P., Kivirand, K.: Biosensors - Emerging Materials, Applications, vol. 3 (2011)
98. Saghatelian, A., Völcker, N.H., Guckian, K.M., Lin, V.S.-Y., Ghadiri, M.R.: J. Am. Chem. Soc. **125**, 346 (2003)
99. Setty, Y., Mayo, A.E., Surette, M.G., Alon, U.: Proc. Natl. Acad. **100**, 7702 (2003)
100. Silverstein, E., Sulebele, G.: Biochemistry **8**, 2543 (1969)

101. Simonian, A.L., Badalian, I.E., Smirnova, I.P., Berezov, T.T.: Biochemical Engineering-Stuttgart, vol. 344. G.Fisher Pub., New York (1991)
102. Simonian, A.L., Khachatrian, G.E., Tatikian, S.S., Avakian, T.M., Badalian, I.E.: Biosensors. Bioelectronics **6**, 93 (1991)
103. Simonian, A.L., Badalian, I.E., Berezov, T.T., Smirnova, I.P., Khaduev, S.H.: Anal. Lett. **27**, 2849 (1994)
104. Sivan, S., Lotan, N.: Biotechnol. Prog. **15**, 964 (1999)
105. Stojanovic, M.N., Stefanovic, D.: J. Comput. Theor. Nanosci. **8**, 434 (2011)
106. Stojanovic, M.N., Stefanovic, D., Rudchenko, S.: Acc. Chem. Res. **47**, 1845 (2014)
107. Stojanovic, M.N., Stefanovic, D., Rudchenko, S.: Acc. Chem. Res. **47**, 1845 (2014)
108. Strack, G., Ornatska, M., Pita, M., Katz, E.: J. Am. Chem. Soc. **130**, 4234 (2008)
109. Strecker, H.J., Korkes, S.: J. Biol. Chem. **196**, 769 (1952)
110. Szacilowski, K.: Chem. Rev. **108**, 3481 (2008)
111. Tian, F., Zhu, G.: Anal. Chim. Acta **451**, 251 (2002)
112. Updike, S.J., Hicks, G.P.: Nature **214**, 986 (1967)
113. Updike, S.J., Shults, M., Ekman, B.: Diabet. Care **5**, 207 (1982)
114. Vallée-Bélisle, A., Ricci, F., Plaxco, K.W.: J. Am. Chem. Soc. **134**, 2876 (2012)
115. von Maltzahn, G., Harris, T.J., Park, J., Min, D., Schmidt, A.J., Sailor, M.J., Bhatia, S.N.: J. Am. Chem. Soc. **129**, 6064 (2007)
116. Wagner, N., Ashkenasy, G.: Chem. Eur. J. **15**, 1765 (2009)
117. Wang, J., Katz, E.: Anal. Bioanal. Chem. **398**, 1591 (2010)
118. Wilson, J.E.: J. Exp. Biol. **206**, 2049 (2003)
119. Wolfe, R.G., Neilands, J.B.: J. Biol. Chem. **221**, 61 (1956)
120. Xia, F., Zuo, X., Yang, R., et al.: J. Am. Chem. Soc. **132**, 8557 (2010)
121. Xu, J., Tan, G.J.: J. Comput. Theor. Nanosci. **4**, 1219 (2007)
122. Zavalov, O., Bocharova, V., Halámek, J., Halámková, L., Korkmaz, S., Arugula, M.A., Chinnapareddy, S., Katz, E., Privman, V.: Int. J. Unconv. Comput. **8**, 347 (2012)
123. Zavalov, O., Bocharova, V., Privman, V., Katz, E.: J. Phys. Chem. B **116**, 9683 (2012)
124. Zavalov, O., Domanskyi, S., Privman, V., Simonian, A.: Proc. Conf. ICQNM **2013**, 54 (2013)
125. Zhang, X., Ju, H., Wang, J.: Electrochemical Sensors, Biosensors, their Biomedical Applications. Academic Press, New York (2008)
126. Zhou, M., Dong, S.: Acc. Chem. Res. **44**, 1232 (2011)
127. Zhou, M., Zhou, N., Kuralay, F., Windmiller, J.R., Parkhomovsky, S., Valdés-Ramírez, G., Katz, E., Wang, J.: Angewandte Chemie International Edition **51**, 2686 (2012)
128. Zhu, L., Yang, R., Zhai, J., Tian, C.: Biosensors. Bioelectronics **23**, 528 (2007)

Chapter 4
Sensing Parameters of a Time Dependent Inflow with an Enzymatic Reaction

Jerzy Gorecki, Joanna N. Gorecka, Bogdan Nowakowski, Hiroshi Ueno,
Tatsuaki Tsuruyama and Kenichi Yoshikawa

Abstract Functionality of living organisms is based on decision making. Chemical reactions stand behind information processing in biological systems. Therefore, it is interesting to consider reaction models that show ability to make decisions by evolving towards significantly different states, depending on conditions at which those reactions proceed. It has been recently demonstrated that a system exhibiting cooperative or sigmoidal response with respect to the input exhibits the potential to function as a discriminator of the amplitude or the frequency of its external periodic perturbation. Here we consider a few models of allosteric enzymatic reactions and discuss their applicability for sensing the frequency or the amplitude of the time dependent input in a form of reagent inflow. The output is coded in a product oscillation type. On the basis of numerical simulations we compare results for a full reaction model with its reduced, easier to analyze version.

J. Gorecki (✉) · B. Nowakowski
Institute of Physical Chemistry, Polish Academy of Sciences,
Kasprzaka 44/52, 01-224 Warsaw, Poland
e-mail: jgorecki@ichf.edu.pl

B. Nowakowski
e-mail: bnowakowski@ichf.edu.pl

J.N. Gorecka
Institute of Physics, Polish Academy of Sciences, al. Lotnikow 32/46, 02-668
Warsaw, Poland
e-mail: gorec@ifpan.edu.pl

H. Ueno · K. Yoshikawa
Faculty of Life and Medical Sciences, Doshisha University, Kyoto 610-0394, Japan
e-mail: uenoh1111@gmail.com

K. Yoshikawa
e-mail: keyoshik@mail.doshisha.ac.jp

T. Tsuruyama
Department of Pathology, Graduate School of Medicine, Kyoto University, Kyoto, Japan
e-mail: tsuruyam@kuhp.kyoto-u.ac.jp

© Springer International Publishing Switzerland 2017
A. Adamatzky (ed.), *Advances in Unconventional Computing*,
Emergence, Complexity and Computation 23,
DOI 10.1007/978-3-319-33921-4_4

4.1 Introduction

The field of unconventional computation is concerned with investigation of computing media as alternatives to the standard semiconductor based electronics. The progress of silicon microprocessor technology, measured by the Moore law, is unprecedented in the history of civilization. However, it is anticipated that this trend should finally terminate. New computational strategies, new types of computing substrates and methods of information coding are needed to ensure the present rate of progress in construction of information processing devices [1–6].

Studies on information processing based on chemical reactions bring a significant contribution to unconventional computations. Some of chemical processes considered for information processing operations involve a few molecules only [7, 8]. This observation is promising because it shows that the future computing devices, based on chemical reactions, can be reduced to the molecular scale.

Many chemical processes with potential applications to computationally oriented tasks can be found in Nature. Among them there are reactions leading to spatial structures formed by replicable molecules like RNA or DNA [9–12] and enzymatic processes characterized by complex, nonlinear chemical kinetics and exhibiting a rich variety of stationary spatio-temporal structures. Studies aimed on integration of these both approaches, for example: an interface that enables communication of otherwise incompatible nucleic-acid and enzyme computational systems, has been recently reported [13].

The enzymatic systems, we are concerned with accept and transform information in the form of chemical substrates with particular properties that meet the binding specificity criteria of a selected enzyme in a lock and key fashion. Enzymatic reactions can process information, coded in concentrations of specific molecules, catalyzing reactions and producing reaction products with different properties than the input reagents. The output information is coded in concentrations of products. The products can then participate in further chemical reactions. The transformations of products into yet other molecules can be regarded as following information processing steps. Studies on networks of connected enzyme-catalyzed reactions, with added chemical and enzymatic processes that incorporate the filtering steps into the functioning of this biocatalytic cascades have been recently published (see [14]). It has been demonstrated that scaled logic variables for the inputs, output, and some intermediate products can be useful in describing enzyme cascade behavior by identifying quantities that offer the most direct control of the network properties. Therefore, using enzymatic reactions we can design information processing cascades, feedback loops, and other complex sequences of operations.

Studies on applications of chemical computing have a long history. It has been noticed that within the Michaelis–Menten kinetics model the shape of the convex line representing the rate of product production as the function of substrate concentration is similar to the relationship between the collector-emitter voltage and collector current in a bipolar transistor. Using this analogy the concepts of transistor based information processing devices can be re-formulated in the language of chemical reactions and reagent flows. The theoretical background to chemistry based logic was presented in a number of publications of Hjelmfelt and Ross and their co-workers [15–18]. They considered a perfectly stirred system and assumed that binary information is coded in stationary concentrations of reagents involved. In the proposed models of chemical neurons the stationary output concentration rapidly switches from low to high values if the concentration of the input reagent exceeds a specific value. In such case the relationship between reagent concentrations and binary logic values is straightforward, high concentrations represent the logic TRUE state, the low ones are interpreted as the logic FALSE. Simulation studies demonstrated how to implement the basic logic operations within the model of a chemical neuron [18, 19].

A convex line representing the relationship between the substrate concentration and the rate of product generation is typical for enzymatic reactions. For such relationship there is an identifiable linear regime, typically near the physical zero concentrations, as well as the saturation regime for larger concentrations. Most experimental studies on (bio)chemical information processing have been focused on the binary logic [20]. Low and high concentrations of selected reagents are interpreted as the logic variables. In this respect, a sigmoid, filter-like relationship between the substrate concentration and the product generation rate seems to be more suited for information processing applications, because it gives a better balance between low and high values of concentrations. Moreover, for information coded in stationary values of concentrations the sigmoidal response has advantage over the convex one because it allows to reduce noise of the output element important for noise-tolerant networking for chemical information processing [21].

It has been observed that sigmoidal kinetics often appears for properly chained enzymatic reactions [22]. Therefore, such kinetics can be designed on demand by coupling an enzymatic process characterized by a convex kinetics with other reactions [23]. Different physical or chemical stimuli can be applied to impact enzymatic processes. For example it was demonstrated that in enzymatic reaction involving glucose oxidase system response depends on initial concentration of hexokinase and ATP [23]. Even more interesting it has been observed that the transition between convex and sigmoidal kinetics can be achieved by system illumination [24].

The sigmoidal kinetics can be also expected for allosteric enzymatic reactions [25–27]. Allosteric regulation is one of many ways in which enzyme activity can be controlled. Enzyme activity is regulated by its conformational dynamics [28]. A typical enzyme contains binding sites where substrate molecules can be attached and the catalytic site where the activation energy is reduced and the reaction, specific for a given enzyme, proceeds. The allosteric enzymes contain another region, separated from the substrate binding site, to which small, regulatory molecules can attach to and and thereby control the catalytic activity. The allosteric regulatory molecules change the conformation or dynamics of the enzyme that is transduced to the active site and thus affect the reaction rate of the enzyme. In this way, allosteric interactions can either inhibit or activate enzymes. As we demonstrate below, the regulating reactions can couple with the main enzymatic process. If the rates for these processes are properly selected then the resulting kinetics of an enzymatic reaction has a sigmoidal form.

The recent studies on applications of enzymatic reactions for computing oriented tasks match experiments with numerical simulations. It has been demonstrated that the simple information processing devices can be constructed with properly selected enzymatic reactions. Biocatalytic system with a double-sigmoid filter (sigmoid with respect to two types of molecules) are especially interesting because they can directly operate on two input signals [28–30]. Enzymatic reactions capable of recognizing two specific molecules or ions in the solution (for example Mg^{2+} and Ca^{2+} [28, 31]) were identified and their applications for construction of the logic gates (the AND binary gate [32], the OR binary gate [33]) have been reported. Enzymatic systems capable of more complex computational tasks like three-input logic gates or molecular full adders were also reported [34, 35].

The studies discussed above were based on information coded in time-independent, stationary concentrations of selected reagents. Here we present a new approach to chemical sensing based on enzymatic reaction with information coded in time dependent evolution of the medium. We discuss an application of chemical computing to time dependent phenomena considering the problem of discrimination of the parameters describing periodic perturbations of a computing medium.

Living organisms have to process information in order to find the optimum environment for their existence under time-dependent environmental change. A significant part of their computational activity is focused on decision making. In a highly organized society life of its members is determined by the choices they make. Some of decision making problems require a binary (YES or NO) answer. Having in mind that information processing activity of living organisms is based on (bio-)chemical reactions we search for and identify chemical reaction models that can be applied for decision making problems in which the input information is coded in periodically changing inflow of reagents. Such inflow can be related to a periodic stimulus forced by cycling change in environment, like the circadian rhythm.

Rational decision making is based on time-dependent information about the problem. Information can be collected by chemical reactions showing different behaviour depending on the conditions they proceeds in. A number of information processing strategies, operating on time dependent inputs have been reported in the literature [1–4]. An answer coming from a chemical sensor can be continuous or discrete. For example, one can get information on the distance to a source of excitations by comparing the frequency of pulses excited in a few identical excitable channels [36, 37]. An excitable chemical medium can be also applied as a sensor of critical changes in time dependent medium parameters. It has been demonstrated [38] that an excitation pulse can propagate in the medium with a slowly decreasing excitability level. On the other hand if exactly the same decrease in system excitability occurs rapidly then the excitation pulse vanishes. Therefore, by observing a distance propagated by a pulse in an excitable medium we are able to determine the rate of temporal changes in the medium excitability. The simplest sensors are binary discriminators (or binary classifiers). They are capable of distinguishing if the conditions at which reactions proceed belong to a given class or not. For example, an excitable medium with a propagating pulse can be regarded as a discriminator if the temporal changes in its excitability are larger than a given value.

The examples presented above illustrate sensing potential of a spatially distributed medium. However, there are reactions that proceeds in a homogeneous medium that can be used as discriminators of the conditions in which it proceeds. In our recent paper [39] we considered a dynamical system with hysteresis as a prototype of a decision making automaton. Systems with hysteresis, commonly observed in physics, chemistry and biology [40, 41], are natural candidates for binary classifiers. In a systems with hysteresis one can distinguish two different classes of stable states S_1 and S_2. System evolution towards a state from a particular class is determined by the value of control parameter λ and the initial condition. Let us assume that the increase in control parameter λ value above the threshold λ_1 triggers the transition from S1 to S2. The reverse transition from S_2 to S_1 occurs if the value of control parameter drops below λ_2. Such system can be obviously used as a discriminator of the control parameter. For example, if the initial state belongs to S_1 and if after some time we observe the system in a state belonging to S_2 then it means that at certain moment of time the value of control parameter necessarily exceeded λ_1. However, if only time monotonic changes in the value of control parameter are considered then the system discrimination ability is reduced to just two values λ_1 and λ_2.

We have recently demonstrated [39] that the applicability of a dynamical system characterized by a sigmoidal kinetics for discrimination oriented tasks increases if a periodic perturbation of inflow is imposed. Periodically perturbed bistable system oscillates. The location of system oscillations in the phase space and the oscillation amplitude depend on the amplitude and frequency of perturbation. In particular we can distinguish oscillations between states in the class S_1 and oscillations between states in the class S_2. The numerical simulations for reaction kinetics defined by a third order rational function revealed a non-trivial property of such system: a marginal change of inflow parameters (amplitude or frequency) can force a sharp transition between oscillations belonging to different classes. The transition occurs in a narrow range of perturbation parameters. The change between oscillation types can be easily distinguished and therefore a chemical reaction exhibiting hysteresis can be used as a discriminator of perturbation properties.

In this report we concentrate on realistic models of enzymatic reactions that can be applied as discriminators of periodically changing stimuli. All models are analyzed assuming that the reaction kinetics follows the mass action law. We consider model enzymatic reactions in which an enzyme is activated by one, two or three identical control molecules. The applicability of such models as sensors of inflow parameters is investigated. We start with the simplest reaction model in which an enzyme is activated by binding with a single molecule and demonstrate that such reaction model does not work as a discriminator of inflow. However, enzymatic reactions involving allosteric activation by two or three control molecules lead to a medium exhibiting bistability if the reaction rates are properly selected. These models lead to kinetic term in a Hill-like form with the Hill coefficient equal to 2 and 3, which value is typical for allosteric reactions characterized by a positive cooperativity [26, 42]. For both reaction models there is a sharp transition in the character of oscillations in product concentration in the phase space of parameters describing amplitude and frequency of periodic inflow of control molecules. Therefore, these enzymatic reactions can be used as binary discriminators of the periodic inflow parameters.

For sigmoidal kinetic we are able to stabilize oscillations in different regions of the phase space and we observe a sharp transition between oscillations of different classes. As discussed in this article such effect is absent for the case of convex kinetics.

4.2 Determination of Inflow Properties with Enzymatic Reactions

In this Section we consider three models of an enzymatic reaction and estimate their applicability for sensing oriented applications. In all models the enzyme E has to bind with a few control molecules X to perform its function. The complex of enzyme and X transforms the reactant B into the product Y. We assume that the complex disintegrates after the reaction and all molecules of X are detached at the same time.

4.2.1 Enzyme Activation with a Single Control Molecule

At the beginning we consider a model in which a single molecule of X activates the enzyme:

$$E + X \underset{k_{-1}}{\overset{k_1}{\rightleftharpoons}} EX$$

$$B + EX \overset{k_p}{\longrightarrow} X + Y + E$$

It is easy to notice that the sum of concentration of E (e) and of EX (f) does not change in time.

In all models discussed in this Section we assume that reagents X and Y are involved in the following processes: transformation of product Y into reactant X:

$$C + Y \overset{k_c}{\longrightarrow} X + products_C$$

and the decomposition of reactant X:

$$D + X \overset{k_d}{\longrightarrow} products_D$$

We assume that $products_C$ and $products_D$ are inert for all reactions listed above, so there is no need to specify them. The last of listed reactions removes X from the system, so the concentration of X (x) tends to zero at long times. In order to have other, more interesting behaviour of the system we assume that there is a time dependent inflow of the reagent X to the system described by a time dependent function $I(t)$:

$$inflow \xrightarrow{I(t)} X$$

For the following analysis we assume that concentrations of reagents B,C and D (denoted as b, c and d, respectively) are much larger than of the other reagents and we can assume that they remain constant in time. Therefore, the complete description of the system is given by three kinetic equations describing the time evolution of the concentrations of EX ($f(t)$), X ($x(t)$) and Y ($y(t)$) read:

$$\frac{dx(t)}{dt} = -k_1 ex(t) + k_{-1}f(t) + k_p bf(t) - k_d dx(t) + k_c cy(t) + I(t) \qquad (4.1)$$

$$\frac{df(t)}{dt} = k_1 ex(t) - k_{-1}f(t) - k_p bf(t) \qquad (4.2)$$

$$\frac{dy(t)}{dt} = k_p bf(t) - k_c cy(t) \qquad (4.3)$$

If we assume that the reactions involving enzyme and its complex are fast, and those leading to $products_C$ and $products_D$ are slow then the concentration of EX complex can be regarded as quasistationary one. In such case the system time evolution description reduces to two kinetic equations for $x(t)$ and $y(t)$:

$$\frac{dx(t)}{dt} = -k_d dx(t) + k_c cy(t) + I(t) \qquad (4.4)$$

$$\frac{dy(t)}{dt} = \frac{k_1 k_p be_0 x(t)}{k_{-1} + k_p b + k_1 x(t)} - k_c cy(t) \qquad (4.5)$$

Here the symbol e_0 denotes the total concentration of enzyme in its free E and complexed forms EX ($e_0 = e(t) + f(t) = const.$).

If the inflow is constant $I(t) = A_0$ then the analysis of nullclines for the dynamical system Eqs. (4.4) and (4.5) shows that for each set of model parameters (rate constants and concentrations of reagents that do not change in time) the system has a single, stable, stationary state. Therefore, the enzymatic reaction with activation by a single control molecule does not function as a discriminator of time dependent inflow parameters according to the strategy described in [39].

4.2.2 Enzyme Activation with Two Control Molecules

Hysteresis is observed in a slightly more complex reaction model in which the enzyme molecule E has to combine with two molecules of reagent X in order to transform the reactant B into the final product Y:

$$E + X \underset{k_{-1}}{\overset{k_1}{\rightleftharpoons}} EX$$

$$EX + X \underset{k_{-2}}{\overset{k_2}{\rightleftharpoons}} EX_2$$

$$B + EX_2 \xrightarrow{k_p} 2X + Y + E$$

As in the previous model we consider the transformation of product Y into reactant X:

$$C + Y \xrightarrow{k_c} X + products_C$$

and the decomposition of reactant X:

$$D + X \xrightarrow{k_d} products_D$$

The complete description of such reaction model is given by four kinetic equations describing the time evolution of $f(t)$, $x(t)$, $y(t)$ and the concentrations of EX_2 ($g(t)$):

$$\frac{dx(t)}{dt} = -k_1 e(t)x(t) + k_{-1}f(t) - k_2 f(t)x(t) + k_{-2}g(t)$$
$$+ 2k_p bg(t) - k_d dx(t) + k_c cy(t) + I(t) \tag{4.6}$$

$$\frac{df(t)}{dt} = k_1 e(t)x(t) - k_{-1}f(t) - k_2 f(t)x(t) + k_{-2}g(t) \tag{4.7}$$

$$\frac{dg(t)}{dt} = k_2 f(t)x(t) - k_{-2}g(t) - k_p bg(t) \tag{4.8}$$

$$\frac{dy(t)}{dt} = k_p bg(t) - k_c cy(t) \tag{4.9}$$

If we assume that the reactions involving enzyme and its complexes are faster than those leading to $products_C$ and $products_D$ then the concentrations of complexes EX and EX_2 can be regarded as quasistationary ones. In such case the time evolution of the system is described by two kinetic equations for $x(t)$ and $y(t)$:

$$\frac{dx(t)}{dt} = Q(x(t), y(t), t)$$
$$= -k_d dx(t) + k_c cy(t) + I(t) \tag{4.10}$$

$$\frac{dy(t)}{dt} = R(x(t), y(t))$$
$$= \frac{k_1 k_2 k_p be_0 x^2(t)}{k_{-1}k_{-2} + k_1 k_p b + k_1 k_{-2}x(t) + k_1 k_p bx(t) + k_2 k_p bx(t) + k_1 k_2 x^2(t)}$$
$$- k_c cy(t)$$

$$\tag{4.11}$$

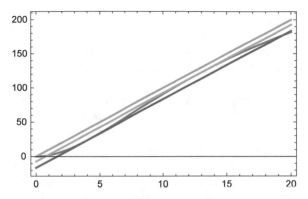

Fig. 4.1 Positions of nullclines for the dynamical system defined by Eqs. (4.10, 4.11) in the phase space (x, y). The nullcline $R(x, y) = 0$ is plotted with the *blue line*. The nullcline $Q(x, y, t) = 0$ is shown for a few cases: $I(t) \equiv 0$ (*the orange line*), $I(t) \equiv 0.731284$ (*the green line*) and $I(t) \equiv 1.64789$ (*the red line*). For the selected parameters of the model, the concentrations at tangential points are $(x_1, y_1) = (4.12112, 24.7323)$ and $(x_2, y_2) = (12.9281, 121.968)$

where e_0 denotes the total concentration of enzyme in its free and all complexed forms ($e_0 = e(t) + f(t) + g(t) = const.$).

For this reaction model one can select the values of parameters such that the nullcline of Eq. (4.11) has a sigmoidal form, so the system exhibits hysteresis. For example, such behaviour is observed when $b = c = d = e_0 = 1$, the rate constants for fast reactions are $k_1 = 10, k_{-1} = 200, k_2 = 20, k_{-2} = 160, k_p = 40$ and the rate constants for the slow reactions leading to products are $k_c = 0.1, k_d = 1$. Figure 4.1 illustrates the nullclines for these values of parameters. The nullcline $Q(x, y, t) = 0$ is shown or a few values of inflow: $I(t) \equiv 0$ (no inflow of X), $I(t) \equiv 0.731284$ and $I(t) \equiv 1.64789 = A_1$. The last two values of inflow correspond to cases when the nullcline $Q(x, y, t)$ is tangential to the nullcline $R(x, y)$. At the beginning we consider an inflow that remains constant after it is switched on at $t = 0$, i.e.: $I(t) = I_0 \cdot \Theta(t)$, where $\Theta(t)$ is the Heaviside step function. If the inflow $I_0 < 0.731284$ then the system has a single, stable stationary state (x_s, y_s) for which $x_s < 4.12112$ and $y_s < 24.7323$. For $I_0 > 1.64789$ the system has a single, stable stationary state (x_s, y_s) for which $x_s > 12.928$ and $y_s > 121.968$. If $0.731284 < I_0 < 1.64789$ then the system is bistable and shows hysteresis. Having in mind results of our recent paper [39] it is expected that a reaction in which an enzyme is activated by two control molecules can function as a discriminator of frequency and amplitude of a time dependent inflow.

In the following we consider periodic inflow described by the formula $I(t) = A \cdot (\sin(2\pi \nu t + \phi_0) + 1) \cdot \Theta(t)$. It is characterized by frequency ν and the initial (for $t = 0$) phase ϕ_0. Such inflow term is always non-negative. The Heaviside step function describes the case where there is no inflow for $t < 0$ and it is switched on at $t = 0$. If $\phi_0 = 3 \cdot \pi/2$ then $I(t)$ is a continuous function. In this case $I(t = 0) = 0$, next it increases and finally oscillates. For any other phase the inflow term is not continuous at $t = 0$; for example if $\phi_0 = \pi/2$ then $I(t = 0) = 2 \cdot A$, next

$I(t)$ decreases and oscillates later. For $t > 0$ the time average of $I(t)$ equals to A and it is independent of the frequency and the initial phase. If the inflow amplitude $A = 0$ than $(x = 0, y = 0)$ is a only steady state of Eqs. (4.10, 4.11) and it is stable. We assume that this stable state is the initial state for the system evolution studied in numerical simulations.

Observations of changes in the character of oscillations in concentrations of reagents can be used to determine the frequency and amplitude of the inflow. Figure 4.2 illustrates the time evolution of concentration of Y for the fixed amplitude $A = 1.2$ and a few different frequencies $v = 0.0022, 0.00306, 0.00307, 0.00309, 0.0031$. The solid line shows the time evolution calculated from the full reaction model described by Eqs. (4.6)–(4.9). The blue dashed line plots the time evolution for the reduced, two-variable model based on Eqs. (4.10) and (4.11). The full model of reaction and the reduced model are in a full agreement for low and high frequencies. Like in our previous paper, for the lowest input frequency the oscillations of $y(t)$ extend over a large range of concentrations. For high inflow frequencies the system oscillates below the tangential value $y_2 = 121.968$. In a narrow interval of frequencies the character of evolution significantly changes. Oscillations observed in $y(t)$ switch from large values exceeding y_2 to oscillations below this value. For the considered amplitude the full model predicts such transition for frequencies in the interval $[0.00307, 0.00309]$. The reduced model predicts that this transition occurs at slightly higher frequencies, in the interval $[0.00309, 0.0031]$. Therefore, as concluded in our previous paper, from the observation of time evolution of concentration in the considered enzymatic system one can discriminate the frequency of the inflow, it the amplitude of inflow is known.

Results shown in Fig. 4.2 illustrate that around the frequency that separates different oscillation types the medium needs some time before it reaches the steady form of oscillations. For the enzyme activation with two molecules we apply classification of the oscillation type is based on the minimum and maximum of $y(t)$ observed in a long time interval for which the evolution has presumably reached its stationary character. Here we use $t_{min} = 8000$ and $t_{max} = t_{min} + 2000$ We introduce:

$$y_{min} = min_{t \in [t_{min}, t_{max}]} \, y(t) \qquad (4.12)$$

and

$$y_{max} = max_{t \in [t_{min}, t_{max}]} \, y(t) \qquad (4.13)$$

We distinguish oscillations on the upper stable branch (USB) of $R(x, y) = 0$ nullcline if $y_{min} > y_1 = 24.7323$ and $y_{max} > y_2$ from oscillations on the lower stable branch (LSB). For those $y_{min} < y_1$ and $y_{max} < y_2$.

The line separating USB from LSB oscillations in the phase space v, A is shown in Fig. 4.3. The solid line corresponds to the full model, whereas symbols give results of the reduced model. For the selected values of reaction parameters the agreement between both models is very good. Our discrimination method is based on location of a given oscillation type with respect to the line separating USB for LSB oscillations. Let us denote points on this line as (v_c, A_c). The frequency v_c treated as a function of

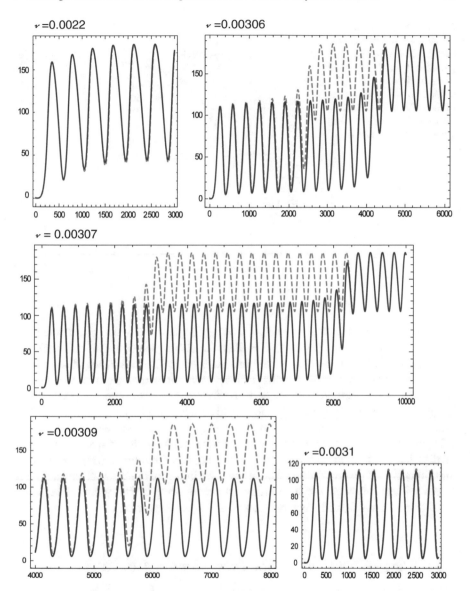

Fig. 4.2 The concentration of Y as a function of time (given on the horizontal axis) for the fixed amplitude $A = 1.2$ and a few different frequencies ($\nu = 0.0022, 0.00306, 0.00307, 0.00309, 0.0031$). The *solid line* shows the time evolution calculated from the full reaction model Eqs. (4.6) and (4.7). The *blue dashed line* plots the time evolution for the reduced, two-variable model based on Eqs. (4.10) and (4.11)

Fig. 4.3 The phase diagram showing the oscillation type as a function of inflow parameters (ν, A). The *solid line* corresponds to the full model, symbols give results of the reduced model. The *horizontal dashed lines* mark positions of $A_1/2$ and A_1

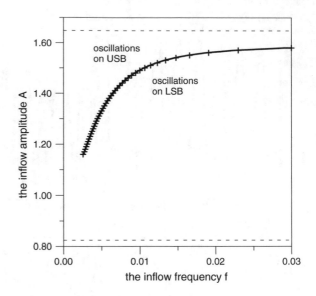

A_c is an increasing function $\nu_c = Z(A_c)$. For example if we like to determine if the frequency of inflow ν is higher than ν_0 then we should select the inflow amplitude equal to $A = Z^{-1}(\nu_0)$ and observe the character of oscillations. If we observe USB oscillations then $\nu < \nu_0$. Observation of LSB oscillations indicates that $\nu > \nu_0$. The sensitivity of the method changes with the frequency ν_0. In order to detect a high frequency inflow, the inflow amplitude should be selected with much higher precision than for $0.005 \leq \nu_c \leq 0.01$. One can also use the transition between class LSB and USB oscillations to determine the inflow amplitude. Unlike for frequency, the range of discriminated amplitudes does not extend outside the interval $[A_1/2, A_1]$. If the amplitude $A > A_1$ than, at a high frequency ν the inflow of control molecules is close to the average inflow A. For such inflow the steady state of the system is located at USB and no transition between oscillation character is expected.

4.2.3 Enzyme Activation with Three Control Molecules

Finally, let us consider yet more complex reaction model in which the enzyme molecule E has to combine with three molecules of reagent X in order to transform the reactant B into the final product Y:

$$E + X \underset{k_{-1}}{\overset{k_1}{\rightleftharpoons}} EX$$

$$EX + X \underset{k_{-2}}{\overset{k_2}{\rightleftharpoons}} EX_2$$

$$EX_2 + X \underset{k_{-3}}{\overset{k_3}{\rightleftharpoons}} EX_3$$

$$B + EX_3 \overset{k_p}{\longrightarrow} 3X + Y + E$$

As in the previous case we assume that these reactions involving the enzyme E and its complexes are fast and lead to quasistationary concentrations of E, EX, EX_2 and of EX_3 (denoted as $h(t)$). Moreover, like in the cases discussed before these reactions combine with slow processes:

$$C + Y \overset{k_c}{\longrightarrow} X + products_C$$

and

$$D + X \overset{k_d}{\longrightarrow} products_D$$

Now the complete description of the system is given by five kinetic equations describing the time evolution of $f(t)$, $g(t)$, $h(t)$, $x(t)$ and $y(t)$:

$$\frac{dx(t)}{dt} = -k_1 e(t)x(t) + k_{-1}f(t) - k_2 f(t)x(t) + k_{-2}g(t) - k_3 g(t)x(t)$$
$$+ k_{-3}h(t) + 3k_p bh(t) - k_d dx(t) + k_c cy(t) + I(t) \tag{4.14}$$

$$\frac{df(t)}{dt} = k_1 ex(t) - k_{-1}f(t) - k_2 f(t)x(t) + k_{-2}g(t) \tag{4.15}$$

$$\frac{dg(t)}{dt} = k_2 ex(t) - k_{-2}g(t) - k_3 g(t)x(t) + k_{-3}h(t) \tag{4.16}$$

$$\frac{dh(t)}{dt} = k_3 g(t)x(t) - k_{-3}h(t) - k_p bh(t) \tag{4.17}$$

$$\frac{dy(t)}{dt} = k_p bh(t) - k_c cy(t) \tag{4.18}$$

If we assume that concentrations of EX, EX_2 and EX_3 complexes can be regarded as quasistationary ones then the model can be reduced to two kinetic equations for the time evolution of $x(t)$ and $y(t)$:

$$\frac{dx(t)}{dt} = Q(x(t), y(t), t)$$
$$= -k_d dx(t) + k_c cy(t) + I(t) \tag{4.19}$$

$$\frac{dy(t)}{dt} = S(x(t), y(t))$$

$$= k_p bh(t) - k_c cy(t)$$

$$= -k_c cy(t) + k_1 k_2 k_3 k_p b^2 e_0 x^3(t)/$$

$$(k_{-1}k_{-2}k_{-3} + k_{-1}k_{-2}k_p b + k_1 k_{-2}k_{-3} bx(t) + k_{-1}k_3 k_p bx(t) + k_1 k_{-2}k_p b^2 x(t)$$

$$+ k_1 k_2 k_{-3} x^2(t) + k_1 k_2 k_p b^2 x^2(t) + k_1 k_3 k_p b^2 x^2(t) + k_2 k_3 k_p bx^2(t)$$

$$+ k_1 k_2 k_3 b x^3(t)) \tag{4.20}$$

where e_0 denotes the total concentration of enzyme in its free and all complexed forms ($e_0 = e(t) + f(t) + g(t) + h(t) = const.$).

For the full reaction model it seems difficult to guess the values of parameters for which the model exhibits hysteresis and can be used to discriminate the amplitude or frequency of the inflow. It is easier to consider the reduced model and find reaction parameters leading to the sigmoidal shape of $S(x, y) = 0$ nullcline. For example, such behaviour is observed when $b = c = d = e_0 = 1$, the rate constants for fast reactions are $k_1 = 10, k_{-1} = 20, k_2 = 10, k_{-2} = 40, k_3 = 10, k_{-2} = 20, k_p = 15$ and the rate constants for the slow reactions leading to products $k_c = 1, k_d = 1$. Figure 4.4 illustrates the nullclines for these values of parameters. The nullcline $Q(x, y, t) = 0$ is shown for a few values of inflow: $I(t) \equiv 0, I(t) \equiv 0.583234$ and $I(t) \equiv 0.914414 = A_2$. The last two values of inflow correspond to cases when the nullcline $Q(x, y, t)$ is tangential to the nullcline $S(x, y)$. If the inflow $I(t) < 0.583234$ then the system has a single, stable stationary state (x_s, y_s) for which $x_s < 1.71818$ and $y_s < 0.803767$. If the inflow $I(t) > 0.914414$ then the system has a single, stable stationary state (x_s, y_s) for which $x_s > 4.39961$ and $y_s > 3.81637$. If the inflow does not depend on time after it is switched on at $t = 0$ ($I(t)) = I_0 \cdot \Theta(t)$ and $0.583234 < I_0 < 0.914414$ then the

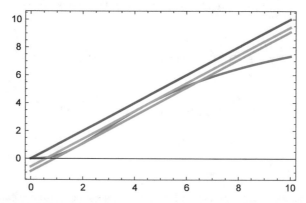

Fig. 4.4 Positions of nullclines for the dynamical system defined by Eqs. (4.19, 4.20) in the phase space (x, y). The nullcline $S(x, y) = 0$ is plotted with the *blue line*. The nullcline $Q(x, y, t) = 0$ is shown for a few cases: $I(t) \equiv 0$ (*the red line*), $I(t) \equiv 0.583234$ (*the green line*) and $I(t) \equiv 0.914414$ (*the orange line*). For the selected parameters of the model, the concentrations at tangential points are $(x_1, y_1) = (1.71818, 0.803767)$ and $(x_2, y_2) = (4.39961, 3.81637)$

system is bistable and shows hysteresis effect. Therefore, the reaction in which the enzyme is activated by three molecules can also function as a discriminator of the inflow parameters.

Like in the previous case we consider the periodic inflow of X with the frequency v and the initial phase ϕ_0 described by the expression: $I(t) = A \cdot (\sin(2\pi v t + \phi_0) + 1) \cdot \Theta(t)$. The initial state for numerical simulations is always $(x = 0, y = 0)$, because it is the only stable state when $I(t) \equiv 0$. In order to classify oscillations we calculate y_{min} and y_{max} according to Eqs. (4.12, 4.13) for $t_{min} = 1000$ and $t_{max} = 2000$. Yet again we distinguish states on the upper stable branch (USB) of $S(x, y) = 0$ nullcline if $y_{min} > y_2 = 3.81637$ and oscillations on the lower stable branch (LSB) for which $y_{max} < y_2$.

Figure 4.5 illustrates the time evolution of concentration of Y for the fixed amplitude $A = 0.8$ and six different frequencies $v = 0.01, 0.015, 0.023, 0.024, 0.0282$ and 0.0284. Like in Fig. 4.2 for low frequencies the oscillations of $y(t)$ extend over a large interval of concentrations. For higher frequencies the system exhibit USB oscillations. If the frequency of inflow v exceeds a critical value LSB oscillations appear. The LSB oscillations remain stable at high frequencies. Qualitatively the same scenario is seen for both full and the reduced model. The fact that the transition between two types of oscillations occurs in a narrow range of frequencies indicates that the considered enzymatic reaction can be also applied for discrimination of the inflow parameters.

Unlike for the enzyme activated by two molecules of X we observe significant quantitative differences between the full and the reduced models. For $A = 0.8$ the time evolutions predicted by both models are in perfect agreement for $v = 0.01$, however at higher frequencies results differ. The full model predicts the transition between USB and LSB oscillations in the interval $[0.023, 0.24]$, whereas the reduced models places this transition at higher frequencies in the interval $[0.0282, 0.284]$. We believe that the difference comes from a high inflow frequency for which quasi-stationarity of the enzyme complexes does not hold. We expect that the difference becomes more important in systems with higher cooperativity.

Figure 4.6 illustrates the regions of parameters (v, A) for which a given oscillation pattern is observed. The solid line corresponds to the full model and symbols give results of the reduced model. The agreement between models at low frequencies is good but the results diverge when v increases. Therefore, the reduced model can be used for qualitative studies of transition between different oscillation types and the full reaction model has to be used to calculate the line separating different oscillation types. Like in the previous case the accuracy of frequency determination decreases as a function of frequency.

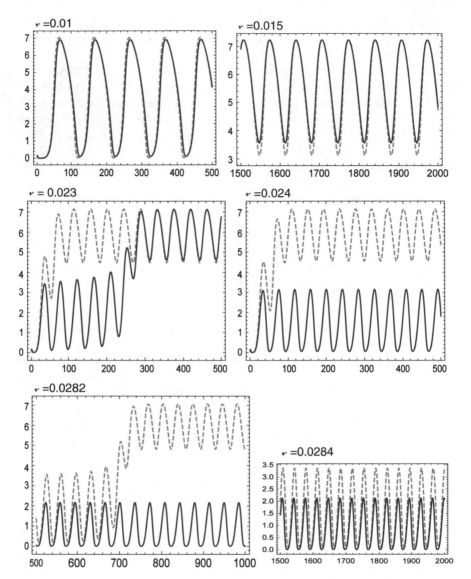

Fig. 4.5 The concentration of Y as a function of time (given on the horizontal axis) for the model in which the enzyme molecule is activated by three molecules of reagent. Results are presented for a fixed amplitude $A = 0.8$ and a few selected inflow frequencies ν, given above corresponding figures. The initial phase is $\phi_0 = 3\pi/2$ for all cases. The *red solid line* shows the time evolution calculated from the full reaction model (Eqs. (4.14–4.18)). The *blue dashed line* plots the time evolution for the reduced, two-variable model based on Eqs. (4.19, 4.20)

Fig. 4.6 The phase diagram showing the oscillation type a the function of inflow parameters (ν, A). The *solid line* corresponds to the full model; the symbols give results of the reduced model. The *horizontal dashed lines* mark positions of $A_2/2$ and A_2

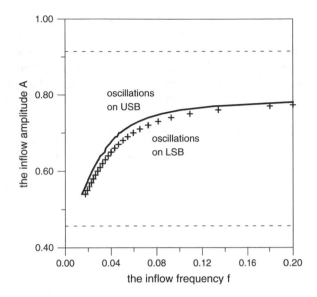

4.3 Conclusions

In this report we presented two models of an enzymatic reaction that can be applied for unconventional information processing as discriminators of input parameters. We have demonstrated that reaction models in which the enzyme is activated by two or three control molecules show hysteresis if the model parameters are properly selected. Therefore, the reactions can be used for a novel discrimination strategy described in [39]. This discrimination strategy has been originally introduced for a formal model of dynamical system. According to it the information about the amplitude and frequency of the inflow can be obtained by observation of oscillations in reagent concentration. Here we confirmed that the strategy applies to realistic models of chemical reactions.

There are many parameters (rate constants, concentrations of reagents in excess) in the considered reaction models. Depending on the model a state of system is described by concentrations of 4 or 5 reagents. Therefore, a straightforward identification of parameter values, at which the reaction can be used as a discriminator seems difficult. Here we restricted our attention to the cases in which reactions involving the enzyme and its complexes are faster than the other processes. We reduced the full reaction model to two kinetic equations. In the reduced model one of the nullclines has a sigmoidal form. Using the arguments given in [39] one can easily estimate the range of input frequency or amplitude in which a sharp transition between different forms of oscillations, necessary for discrimination, is observed. The comparison of results shows a qualitative agreement between the full reaction model and the reduced one. For the reaction with enzyme activation by two control molecules we have also found a good quantitative agreement between the models. In the other case (enzyme activation by three control molecules) we observed significant differences in the inflow amplitude and frequency at which the transition between different forms of oscillations occurs. Therefore, although the strategy of discrimination can be easily explained for two-variable model with a sigmoidal nullcline, precise calculations of discriminator characteristics should be based on the detailed reaction model, which provides more realistic interpretation to the dynamic response.

Acknowledgments This work was supported by KAKENHI Grants-in-Aid for Scientific Research (15H02121, 25103012). Two of the authors (JG and BN) thanks for the financial support from the PAN-JSPS joint research project: Spontaneous creation of chemical computing structures based on interfacial interactions.

References

1. Henson, A., Parrilla Gutierrez, J.M., Hinkley, T., Tsuda, S., Cronin, L.: Towards heterotic computing with droplets in a fully automated droplet-maker platform, Phil. Trans. R. Soc. A **373**, 20140221 (2015)
2. Gorecki, J., Gizynski, K., Guzowski, J., Gorecka, J.N., Garstecki, P., Gruenert, G., Dittrich, P.: Chemical computing with reaction-diffusion processes. Phil. Trans. R. Soc. A **373**, 20140219 (2015)
3. Nehaniv, C.L., Rhodes, J., Egri-Nagy, A., Dini, P., Morris, E.R., Horváth, G., Karimi, F., Schreckling, D., Schilstra, M.J.: Symmetry structure in discrete models of biochemical systems: natural subsystems and the weak control hierarchy in a new model of computation driven by interactions. Phil. Trans. R. Soc. A **373**, 20140223 (2015)
4. Adamatzky, A.: Slime mould processors, logic gates and sensors. Phil. Trans. R. Soc. A **373**, 20140216 (2015)
5. Katz, E. (ed.): Biomolecular Information Processing: From Logic Systems to Smart Sensors and Actuators. Wiley, Hoboken (2012)
6. Katz, E. (ed.): Molecular and Supramolecular: From Molecular Switches to Logic Systems Information Processing. Wiley, Hoboken (2012)
7. Szacilowski, K.: Digital Information Processing in Molecular Systems. Chem. Rev. **108** 3481–3548 (2008)
8. Pischel, U., Andreasson, J., Gust, D., Pais, V.F.: Information Processing with Molecules – Quo Vadis? ChemPhysChem **14**, 28–46 (2013)
9. Head, T., Rozenberg, G., Bladergroen, R.S., Breek, C.K.D., Lommerse, P.H.M., Spaink, H.P.: Computing with DNA by operating on plasmids. BioSystems **57**, 87–93 (2000)
10. Head, T., Chen, X., Yamamura, M., Gal, S.: Aqueous computing: a Survey with an Invitation to Participate. J. Comput. Sci. Technol. **17**, 672–681 (2002)
11. Benenson, Y.: Biocomputers: from test tubes to live cells. Mol. BioSyst. **5**, 675–685 (2009)
12. M. N. Stojanovic, D. Stefanovic, S. Rudchenko: Exercises in Molecular Computing, Acc. Chem. Res. **47**, 1845–1852 (2014). doi:10.1021/ar5000538
13. Mailloux, S., Gerasimova, Y.V., Guz, N., Kolpashchikov, D.M., Katz, E.: Bridging the two worlds: a universal interface between enzymatic and dna computing systems. Angew. Chem. Int. Ed. **54**, 6562–6566 (2015)
14. Privman, V., Zavalov, O., Halamkova, L., Moseley, F., Halamek, J., Katz, E.: Networked enzymatic logic gates with filtering: new theoretical modeling expressions and their experimental application. J. Phys. Chem. B **117**, 14928–14939 (2013). doi:10.1021/jp408973g
15. Hjelmfelt, A., Weinberger, E.D., Ross, J.: Chemical implementation of neural networks and Turing machines. Proc. Natl. Acad. Sci. U.S.A. **88**, 10983–10987 (1991)

16. Hjelmfelt, A., Weinberger, E.D., Ross, J.: Chemical implementation of finite—state machines. Proc. Natl. Acad. Sci. U.S.A. **89**, 383–387 (1992)
17. Hjelmfelt, A., Ross, J.: Mass-coupled chemical systems with computational properties. J. Phys. Chem. **97**, 7988–7992 (1993)
18. Hjelmfelt, A., Ross, J.: Implementation of logic functions and computations by chemical kinetics. Phys. D **84**, 180–193 (1995)
19. Arkin, A., Ross, J.: Computational functions in biochemical reaction networks. Biophys. J. **67**, 560–578 (1994)
20. Katz, E., Privman, V.: Enzyme-based logic systems for information processing. Chem. Soc. Rev. **39**, 1835–1857 (2010)
21. Privman, V., Halamek, J., Arugula, M.A., Melnikov, D., Bocharova, V., Katz, E.: Biochemical filter with sigmoidal response: increasing the complexity of biomolecular logic. J. Phys. Chem. B. **114**, 14103–14109 (2010)
22. Pita, M., Privman, V., Arugula, M.A., Melnikov, D., Bocharova, V., Katz, E.: Towards biochemical filters with a sigmoidal response to pH changes: buffered biocatalytic signal transduction. Phys. Chem. Chem. Phys. **13**, 4507–4513 (2011)
23. Domanskyi, S., Privman, V.: Design of digital response in enzyme-based bioanalytical systems for information processing applications. J. Phys. Chem. B. **116**, 13690–13695 (2012)
24. Privman, V., Fratto, B.E., Zavalov, O., Halamek, J., Katz, E.: Enzymatic AND Logic gate with sigmoid response induced by photochemically controlled oxidation of the output. J. Phys. Chem. B. **117**, 7559–7568 (2013). doi:10.1021/jp404054f
25. Laskowski, R.A., Gerick, F., Thornton, J.M.: The structural basis of allosteric regulation in proteins. FEBS Lett. **583**, 1692–1698 (2009)
26. Tsuruyama, T., Nakamura, T., Jin, G., Ozeki, M., Yamada, Y., Hiai, H.: Constitutive activation of Stat5a by retrovirus integration in early pre-B lymphomas of SL/Kh strain mice. Proc. Natl. Acad. Sci. **99**, 8253–8258 (2002)
27. Tsuruyama, T.: A model of cell biological signaling predicts a phase transition of signaling and provides mathematical formulae. PLoS One **9**, e102911 (2014)
28. Zauner, K.P., Conrad, M.: Enzymatic Computing. Biotechnol. Prog. **17**, 553–559 (2001)
29. Rabinowitz, J.D., Hsiao, J.J., Gryncel, K.R., Kantrowitz, E.R., Feng, X.-J., Li, G., Rabitz, H.: Dissecting enzyme regulation by multiple allosteric effectors: nucleotide regulation of aspartate transcarbamoylase. Biochemistry. **47**, 5881–5888 (2008)
30. Bakshi, S., Zavalov, O., Halamek, J., Privman, V., Katz, E.: Modularity of biochemical filtering for inducing sigmoid response in both inputs in an enzymatic AND gate J. Phys. Chem. B. **117**, 9857–9865 (2013). doi:10.1021/jp4058675
31. Barrett, H.C.: Enzymatic computation and cognitive modularity. Mind Lang. **20**, 259–287 (2005)
32. Halamek, J., Zavalov, O., Halamkova, L., Korkmaz, S., Privman, V., Katz, E.: Enzyme-based logic analysis of biomarkers at physiological concentrations: and gate with double-sigmoid "Filter" response. J. Phys. Chem. B. **116**, 4457–4464 (2012). doi:10.1021/jp300447w
33. Zavalov, O., Bocharova, V., Privman, V., Katz, E.: Enzyme-based logic: OR Gate with Double-sigmoid filter response. J. Phys. Chem. B. **116**, 9683–9689 (2012). doi:10.1021/jp305183d
34. Lederman, H., Macdonald, J., Stefanovic, D., Stojanovic, M.N.: Deoxyribozyme-based Three-input logic gates and construction of a molecular full adder. Biochemistry. **45**, 1194–1199 (2006)
35. Baron, R., Lioubashevski, O., Katz, E., Niazov, T., Willner, I.: Elementary arithmetic operations by enzymes: a model for metabolic pathway based computing. Angew. Chem., Int. Ed. **45**, 1572–1576 (2006)
36. Gorecki, J., Gorecka, J.N., Yoshikawa, K., Igarashi, Y., Nagahara, H.: Sensing the distance to a source of periodic oscillations in a nonlinear chemical medium with the output information coded in frequency of excitation pulses. Phys. Rev. E. **72**, 046201 (2005)
37. Yoshikawa, K., Nagahara, H., Ichino, T., Gorecki, J., Gorecka, J.N., Igarashi, Y.: On chemical methods of direction and distance sensing. Int. J. Unconventional Comput. **5**, 53–65 (2008)

38. Tanaka, M., Nagahara, H., Kitahata, H., Krinsky, V., Agladze, K., Yoshikawa, K.: Survival versus collapse: abrupt drop of excitability kills the traveling pulse, while gradual change results in adaptation. Phys. Rev. E. **76**, 016205 (2007)
39. Ueno, H., Tsuruyama, T., Nowakowski, B., Gorecki, J., Yoshikawa, K.: Discrimination of time-dependent inflow properties with a cooperative dynamical system. Chaos **25**, 103115 (2015)
40. Murray, J.D.: Mathematical Biology. I. An Introduction, 3rd edn. Springer, Berlin (2002)
41. Krasnoselsky, M.A., Pokrowsky, A.W.: Sistemy s gisteresisom (Systems with Hysteresis, in Russian). Nauka, Moscow (1983)
42. Goutelle, S., Maurin, M., Rougier, F., Barbaut, X., Bourguignon, L., Ducher, M., Maire, P.: The Hill equation: a review of its capabilities in pharmacological modelling. Fundam. Clin. Pharmacol. **22**, 633 (2008)

Chapter 5
Combinational Logic Circuit Based on BZ Reaction

Mingzhu Sun and Xin Zhao

Abstract As a basic unit of large scale integration, combinational logic circuit is very important in the development of digital computer. Chemical computation should have the ability to replicate the basic function of combinational logic circuit in BZ medium, in order to realize chemical computer. In this chapter, we design and implement different types of combinational logic circuits from two perspectives. On one hand, based on the basic chemical processors and logic gates, the cascade method is applied to achieve the functions of multi-bit combinational logic by using low-bit logic circuits. On the other hand, a universal method is put forward to construct combinational logic circuits according to their sum-of-products expressions. Simulation results demonstrate the effectiveness of the two construction methods, as well as all the combinational logical circuits designed in this chapter. We believe that the realization of combinational logical circuits will be helpful to fulfil other logic and arithmetic functions, and ultimately can bring great potential applications for the implementation of chemical computer and other intelligent systems.

5.1 Introduction

Belousov–Zhabotinsky (BZ) reaction is an important chemical oscillating reaction discovered by Boris Belousov and further developed by Zhabotinsky [17]. It has been proposed that some types of photosensitive BZ reactions can implement various computational operations [6, 16, 27]. As an unconventional strategy of information processing, this kind of computation, which is so called chemical computation,

M. Sun · X. Zhao (✉)
Nankai University, No. 94 Weijin Road, Tianjin, China
e-mail: zhaoxin@nankai.edu.cn

M. Sun
e-mail: sunmz@nankai.edu.cn

© Springer International Publishing Switzerland 2017
A. Adamatzky (ed.), *Advances in Unconventional Computing*,
Emergence, Complexity and Computation 23,
DOI 10.1007/978-3-319-33921-4_5

relies on geometrically constrained excitable chemical medium, and uses the change of concentrations of the BZ reagents to transmit information and realize computation [4, 8].

The first successful application of chemical computation was image processing in 1989 [13]. In 1995, the properties of BZ excitable medium were exploited to find minimum-length paths in complex labyrinths [24]. Since then, many different prototypes of chemical computation have been designed and tested on a variety of tasks, for example, optimal path search in a labyrinth [20, 21], image processing [19], robot navigation [1], direction detection [18], Voronoi diagram solving [2], and so on, which indicates that chemical computation has the ability and high efficiency for some kinds of complex computations.

Beyond that, one benchmark of chemical computation is capable of replicating components used in conventional computation, such as logic circuit [10]. According to the construction of digital computer, logic circuit based on BZ reaction can be connected to create chemical computer, which is a type of non-classical computers and can perform universal computation. In the past 20 years, the chemical implementations of logic circuit have attracted the attention of many researchers. Since the Showalter Laboratory realized the first logic gates in BZ medium in 1994 [28, 29], researchers have built several logic circuits, such as chemical diode [3, 7, 12], Boolean logic gates [14, 15, 23, 25], adders [5, 11, 30], decoders [26], counters [9], memory cells [14], and so on. Many of them have already been verified in simulations or in chemical experiments.

As a basic unit of digital circuit system, combinational logic circuit is the precondition for the development of digital computer. Though adders and decoders belong to combinational logic circuit, no one has discussed completely how to construct different types of combinational logic circuits based on BZ reaction until now. In this chapter, we focus on the design and simulation of combinational logic circuits in BZ medium. Based on the basic chemical signal processors implemented before, we systematically analysis the design of combinational logic circuits, and use the conventional method and universal method to fulfill construction respectively. On one hand, logic gates are designed based on existing structures and applied to construct low-bit logic circuits according to their logic expressions; then multi-bit combinational logic circuits, including binary adder, binary comparator, binary encoder and binary decoder, are designed and implemented in simulation by the cascade method. On the other hand, a universal method is put forward to design logic circuits, since the logical function of any combinational logic circuit can be expressed by sum-of-products expressions. Universal structure of sum-of-products expressions and multi-input multi-output structure are formed and tested respectively for logical arithmetic and signal transmission, then universal structure of combinational logic circuits is designed and verified.

The rest of the chapter is organized as follows: we present the details of the two-variable Rovinsky–Zhabotinsky (RZ) model of BZ reaction for simulation in Sect. 5.2; Sect. 5.3 outlines some basic chemical signal processors, which have been verified before and are used as components of combinational logic circuits. Sections 5.4 and 5.5 show the construction and simulation results of two types of combinational logic circuits. At last, the chapter is concluded in Sect. 5.6.

5.2 The RZ Model of BZ Reaction

We employ the RZ model of the BZ reaction to calculate the propagation of the pulses [9, 26, 27, 30]. The RZ model, which is derived from the Field–Koros–Noyes (FKN) reaction mechanism [17], has two variables, x and z, corresponding to dimensionless concentrations of activator $HBrO_2$ and of catalyst $Fe(phen)_3^{3+}$. In the active regions, which contain the catalyst, the time evolution of the concentrations of x and z is described by Eq. (5.1).

$$\frac{\partial x}{\partial \tau} = \frac{1}{\varepsilon}\left[x(1-x) - (2q\,\alpha\,\frac{z}{1-z} + \beta)\frac{x-\mu}{x+\mu}\right] + \nabla^2 x,$$
$$\frac{\partial z}{\partial \tau} = x - \alpha\,\frac{z}{1-z}. \tag{5.1}$$

In the passive regions, where catalyst is absent, the concentrations of x and z evolve according to Eq. (5.2).

$$\frac{\partial x}{\partial \tau} = -\frac{1}{\varepsilon}\left[x^2 + \beta\frac{x-\mu}{x+\mu}\right] + \nabla^2 x,$$
$$z = 0. \tag{5.2}$$

In numerical calculations, Eqs. (5.1) and (5.2) are solved numerically using Euler method with a five-node Laplace operator for the diffusion term. The time step $\Delta\tau$ is 0.0001, and the distance between each grid point $\Delta\rho$ is 0.3301 [26, 30]. The other parameters are the same values as considered in Refs. [9, 26, 30]: $\varepsilon = 0.1176$, $q = 0.5$, $\alpha = 0.068$, $\beta = 0.0034$, and $\mu = 0.00051$. For these values of parameters, the stationary concentrations of x and z in the active region, which are the stationary solution of Eq. (5.1), are: $x = 7.27 \times 10^{-4}$, $z = 1.06 \times 10^{-2}$; and the stationary concentrations in the passive region, which are the stationary solution of Eq. (5.2), are: $x = 5.12 \times 10^{-4}$, $z = 0$.

Based on the geometrical configuration of the channels, we can implement many kinds of computational devices by the model. In simulations, the pulses in the active regions are initiated by increasing the value of x to 0.1 at the end of signal channels, and then the excitation waves will propagate inside the channels. In simulation results, the black color and the white color show the distribution of the active and passive regions respectively. A high concentration of activator x is marked as a gray wave in the active regions.

5.3 Basic Chemical Signal Processors Based on BZ Reaction

The combinational logic circuits designed here are constructed by using some kinds of basic chemical signal processors, such as, T-shaped coincidence detector, chemical diode, time delay unit and crossover structure, which are constituted with simple straight line boundaries. Since the basic structures of these processors have already been verified in numerical simulations and chemical experiments [3, 8, 9, 14], the combinational logic circuits will work well in BZ medium, if they can be implemented and verified in simulations. In order to keep the structures of combinational logic circuits uniform and compact, we set the width of the channels to 20 grid points in all simulations.

5.3.1 T-Shaped Coincidence Detector

A T-shaped coincidence detector is used to detect whether two pulses meet in the proper area of the channel. As shown in Fig. 5.1, the device has two parts: a horizontal bar above as the signal channel, and a T-shaped structure below to detect coincidence [9]. The distance between the above bar and the horizontal bar on the T (detector bar) is very important. If two pulses propagate simultaneously from two sides of the above bar, they will meet in the middle and annihilate. With proper distance between the two bars, a pulse appears in the detector bar, and an output pulse is sent through the vertical part of the T. However, when a single pulse propagates inside the bar, it dies at the other end and does not excite the detector bar.

In Fig. 5.1, the distance between the above bar and the horizontal bar on "T" is set to 10 grid points for a 20-grid point wide channel. This structure can detect the meeting areas of the pulses and will be employed to construct logic gates and combinational logic circuits in Sect. 5.4.

5.3.2 Chemical Diode

Chemical diode is used for unidirectional signal transmission. The structure shown in Fig. 5.2 can achieve this purpose. The structure is formed by a straight channel on

Fig. 5.1 A classical T-shaped coincidence detector [9]. Grid size is 100 × 90, the channels are 20 grid points wide and the gap is 10 grid points wide

Fig. 5.2 A classical chemical signal diode [3]. Grid size is 120 × 60, the channels are 20 grid points wide and the gap is 8 grid points wide

the left and a triangular-tipped channel on the right [3]. The asymmetric geometry in BZ medium results in an asymmetric pulse propagation, such as unidirectional propagation. By selecting the proper width of the passive gap between two channels, pulse propagates through the gap from left to right, but not vice versa.

In Fig. 5.2, the distance between the straight channel and the triangular-tipped channel is set to 8 grid points for a 20-grid point wide channel. This structure is used to implement binary encoder in Sect. 5.4.

5.3.3 Time Delay Unit

In numerical simulations, the propagation speed of the pulse is a constant, if the parameters of simulation model are fixed. The time delay unit can reduce the propagation speed due to the penetration property of the BZ reaction, so that the synchronized input pulses can be generated. In BZ reaction, penetration means a pulse propagating in the active region can penetrate into a passive part and disappears after some distance [8]. The pulse can excite the active region behind the passive stripe, if the stripe is narrow enough [22]. The pulse propagates more quickly in active region than passive region, so penetration can be used for pulse delay.

The time delay unit is shown in Fig. 5.3. With a 20-grid point wide channel and an 8-grid point wide passive stripe, the time delay unit leads to 13-grid point delay under the simulation condition. This property will be applied to implement binary adder and binary decoder in Sect. 5.4.

(a)   (b)

Fig. 5.3 Pulse delay by penetration property of the BZ reaction. Grid size is 120 × 80, the channels are 20 grid points wide and the gap is 8 grid points wide. It leads to 13-grid point delay after penetration. **a** Pulses propagation before penetration. **b** Pulses propagation after penetration

5.3.4 Crossover Structure

Crossover structure is used for cross propagation of the pulses. This structure utilizes another property of the BZ reaction: angle dependent penetration of passive region [8]. When a pulse passes through the passive region, it is said that the maximum width of the passive stripe depends on the angle between the wave vector of the pulse and the normal to the stripe. A pulse with the wave vector perpendicular to the stripe can pass a wider stripe than a pulse propagates along the strip. Thus, in a junction of two channels, the interactions can be excluded by adding passive stripes with proper width.

When channel is 20 grid points wide, we cannot find a proper width of the passive stripe under the simulation condition. While for 40-grid point wide channel, the width of the stripe can be set to 10 grid points. Figure 5.4a illustrates the crossover structure, in which, a junction of two 40-grid point wide channels are separated by four 10-grid point wide passive stripes, and connected to 20-grid point wide channels. This device permits the pulse to propagate through channel AA' or BB' without outputs in channel BB' or AA'. For unidirectional propagation, we reduce the length of 40-grid point wide channels to 10 grid points, as shown in Fig. 5.4b. The pulse from O_1 or O_2 cannot pass the passive stripe as it cannot penetrate so long in an almost 20-grid point wide channel.

This crossover structure, which realizes single input single output cross propagation, will be employed to construct binary adder and binary encoder in Sect. 5.4. When both input channels of the crossover structure are excited simultaneously, there's no output pulse in any output channels, since the pulses meet in the center and annihilate. This property is used to achieve NOT operation in Sect. 5.5.

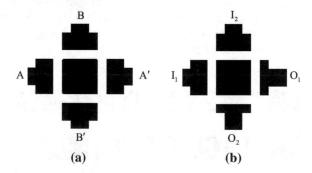

(a)   **(b)**

Fig. 5.4 Crossover structures for single input and single output. Grid size is 120×120, the cross channels and the propagation channels are 40 and 20 grid points wide respectively; the width of the passive stripes is 10 grid points. **a** A bidirectional crossover structure. **b** A unidirectional crossover structure. The length of the 40-grid point wide cross channels is 10 grid points. The pulse can only propagate from I_1 to O_1 or from I_2 to O_2

5.4 Classic Combinational Logic Circuits

In digital electronics, combinational logic circuits are constructed by combining logic gates directly, or by extending low-bit logic circuits by cascade method. Similarly, in this section, we design combinational logic circuits in BZ medium in these two ways. For one type of combinational logic circuit, we first build 1-bit or 2-bit logic circuit based on logic gates; then multi-bit combinational logic circuit is constructed by cascade method. Generally, An n-bit ($n \geqslant 2$) combinational logic circuit can be built by using one or two ($n - 1$)-bit logic circuits.

In this section, we build logic gates based on basic chemical signal processors firstly, then discuss the construction of binary adder, binary comparator, binary encoder, binary decoder, which are four types of classic combinational logic circuits.

5.4.1 Logic Gates

The logic gates used here are described as follows:

1. AND gate: In AND gate, a HIGH output (1) results only if both inputs are 1. If neither or only one input is 1, a LOW output (0) results. A T-shaped coincidence detector is regarded as chemical realization of AND gate, as shown in Fig. 5.5a, c, d. With the presence of two input pulses, the detector becomes excited and the output is 1; in other cases, the detector is not excited and the output is 0.
2. OR gate: In OR gate, if one or both inputs are 1, the output is 1. If neither input is 1, the output is 0. In BZ reaction, OR gate is realized by linking channels directly

Fig. 5.5 The structures of logic gates based on the BZ reaction. **a** AND gate. **b** OR gate. **c** NOT gate. **d** XOR gate

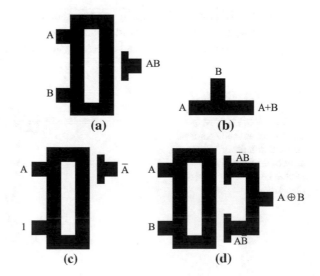

(Fig. 5.5b). The output is 0 only if no pulse appears at channel A or B; otherwise, an output pulse is sent through the channel, which gives 1 as the output.

3. NOT gate: NOT gate outputs the opposite logic level to its input. As shown in Fig. 5.5c, NOT gate is replaced by AND-NOT gate with a constant auxiliary 1 input. With the absence of pulse at channel A, the auxiliary 1 input propagates through the channel, and the output is 1; while if there is a pulse at A, the two pulses will meet and annihilate, as a result, the output is 0.

4. XOR gate: In XOR gate, if one and only one input is 1, the output is 1. If both inputs are both 0 or 1, the output is 0. Figure 5.5d shows the structure of XOR gate, which combines AND, OR and NOT gates.

Note that the pulse from one input will propagate into the other in the structures in Fig. 5.5, but it has no influence on the outputs. In the rest of this chapter, we do not add additional chemical diodes in the input channels for simplicity, if the outputs do not interfere with the inputs.

5.4.2 Binary Adder

5.4.2.1 One-Bit Binary Adder

Binary adder performs addition of binary numbers. A one-bit half-adder adds two single binary digits A_i and B_i, with two outputs sum S_i and carry C_i. The outputs are expressed as:

$$S_i = A_i \oplus B_i = \bar{A}_i B_i + A_i \bar{B}_i,$$
$$C_i = A_i B_i. \tag{5.3}$$

From Eq. 5.3, one-bit half-adder is constructed by combining an AND gate and an XOR gate directly. As shown in Fig. 5.6, we put carry channel C_i on the top of the structure, so the output channel of AND gate $A_i B_i$ should cross the channel $\bar{A}_i B_i$.

Fig. 5.6 The structure of one-bit half-adder. Grid size is 240 × 260. The devise consists of an AND gate, an XOR gate and a crossover structure, which is used to exchange the order of $A_i B_i$ and $\bar{A}_i B_i$

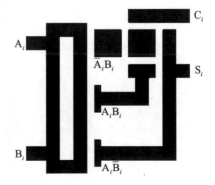

Since A_iB_i and $\bar{A}_iB_i$ cannot be 1 at the same time, a modified crossover structure is applied to achieve channel crossing.

For one-bit half-adder, we should consider four cases of the two inputs:

1. $A_i = 0$, $B_i = 0$: With the absence of the two pulses, the BZ medium stays at the stable stationary state. There is no pulse in the output channels, so the outputs C_i and S_i are 0 and 0.
2. $A_i = 1$, $B_i = 0$: Pulse from channel A_i propagates through the channel, generating a new pulse in channel $A_i\bar{B}_i$. Finally, the pulse is sent through channel S_i, so the outputs are 0 and 1.
3. $A_i = 0$, $B_i = 1$: Similarly, pulse from channel B_i generates a new pulses in channel $\bar{A}_iB_i$. The pulse passes through the crossover structure, giving 0 and 1 as the outputs.
4. $A_i = 1$, $B_i = 1$: Two pulses from channel A_i and B_i meet in the middle, and generates a pulse in channel A_iB_i. Through the crossover structure, this pulse is sent through channel C_i. As a result, the final outputs are 1 and 0.

Figure 5.7 shows the simulation results of one-bit half-adder in BZ reaction. Table 5.1 lists all possible inputs and the corresponding outputs of half-adder. The results indicates that this device realizes the addition of two one-bit binary numbers.

A full-adder adds binary numbers and accounts for values carried in as well as carried out. A one-bit full-adder adds three one-bit numbers, A_i, B_i and C_{i-1}, A_i and B_i are the operands, and C_{i-1} is the carry from the previous significant bit position. We can build a one-bit full-adder by using two half-adders: the first half-adder adds

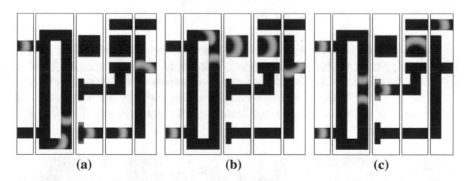

Fig. 5.7 Simulation results of one-bit half-adder when inputs (A_i and B_i) are **a** 1, 0. **b** 0, 1. **c** 1, 1

Table 5.1 Inputs and corresponding outputs of the structure of half-adder

Input		Output	
A_i	B_i	S_i	C_i
0	0	0	0
0	1	1	0
1	0	1	0
1	1	0	1

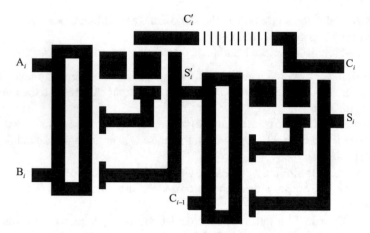

Fig. 5.8 The structure of one-bit full-adder. Grid size is 460 × 300. One-bit full-adder uses two one-bit half-adders to add three one-bit numbers

A_i and B_i, with outputs S_i' and C_i'; the second half-adder adds S_i' and C_{i-1} to generate final sum output S_i, and the final carry output C_i is produced by ORing carries of two half-adders. The outputs can be expressed as Eq. (5.4). The structure of one-bit full-adder is shown in Fig. 5.8, some time delay units are involved to keep the outputs synchronized with different inputs.

$$S_i = S_i' \oplus C_{i-1} = A_i \oplus B_i \oplus C_{i-1},$$
$$C_i = C_i' + S_i'C_{i-1} = A_iB_i + (A_i \oplus B_i)C_{i-1}. \tag{5.4}$$

Figure 5.9 shows the simulation results when previous carry input C_{i-1} is 1. In Fig. 5.9a, A_i and B_i are both 1, the outputs C_i' and S_i' of first half-adder are 1 and 0. Consider the previous carry input C_{i-1}, the final outputs C_i and S_i are 1 and 1. In Fig. 5.9b, A_i is 1 and B_i is 0, the first half-adder outputs 0 and 1 for C_i' and S_i'. In the second half-adder, two 1s add together, the final outputs C_i and S_i are 1 and 0.

(a)   (b)

Fig. 5.9 Simulation results of one-bit full-adder when inputs (A_i, B_i, C_{i-1}) are **a** 1, 1, 1. **b** 1, 0, 1

The simulation results demonstrate that this structure fulfils the function of one-bit full-adder.

5.4.2.2   Two-Bit Binary Adder

Two-bit binary adder adds two binary numbers A_1A_0 and B_1B_0, giving S_1, S_0 as sum outputs and C_1 as carry output. We can extend the structure of one-bit binary adder to achieve two-bit binary adder, which combines a one-bit half-adder and a one-bit full-adder. The half-adder achieves the addition of two lower bits A_0 and B_0, the sum outputs as the lower sum S_0. The full-adder achieves the addition of two higher bits A_1 and B_1, as well as the carry output from the half-adder, the outputs of the full-adder are set as higher sum S_1 and carry C_1 of the two-bit adder. Figure 5.10 shows the structure of two-bit adder, in which, time delay units are used in the structure for pulse delay.

There are sixteen kinds of inputs for two-bit binary adder. Figure 5.11 shows two typical simulation results. In Fig. 5.11a, the inputs A_1A_0 and B_1B_0 are both 11, so the carry and sum of the half-adder are 1 and 0; three 1s add together in the full-adder, suggesting that the outputs $C_1S_1S_0$ of two-bit adder are 110. In Fig. 5.11b, the inputs A_1A_0 and B_1B_0 are 11 and 10, so the carry and sum of the half-adder are 0 and 1; the full-adder adds two 1s, the final outputs are 101. We have performed simulations for all 16 input combinations. The simulation results indicate that this structure realizes the addition function of two-bit binary numbers.

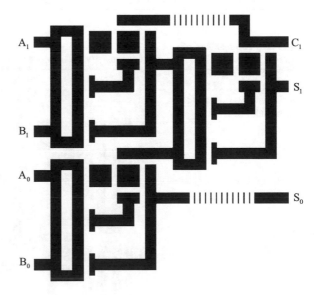

Fig. 5.10 The structure of two-bit binary adder. Grid size is 460 × 500. The structure consists of a one-bit full-adder and a one-bit half-adder

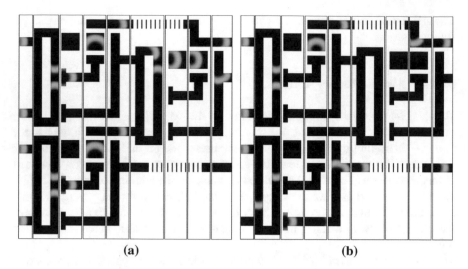

Fig. 5.11 Simulation results of two-bit binary adder when inputs (A_1A_0 and B_1B_0) are **a** 11, 11. **b** 11, 10

5.4.2.3 Three-Bit Binary Adder

It is easy to design three-bit binary adder by the similar cascade method. Three-bit binary adder consists of a two-bit binary adder at the bottom and a one-bit full-adder on the top. As shown in Fig. 5.12, A_2, A_1, A_0 and B_2, B_1, B_0 are the input channels for the two three-bit binary numbers; S_2, S_1, S_0 are the sum output channels, and C_2

Fig. 5.12 The structure of three-bit binary adder. Grid size is 680 × 760. The structure consists of a two-bit binary adder and a one-bit full-adder

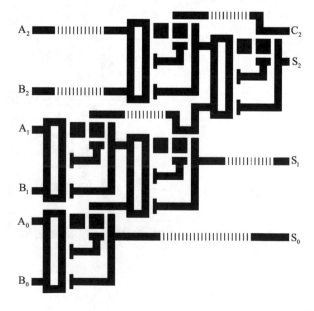

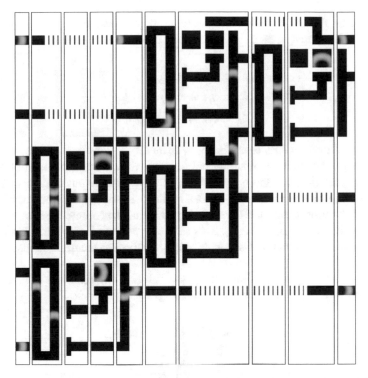

Fig. 5.13 Simulation result of three-bit binary adder when inputs ($A_2A_1A_0$ and $B_2B_1B_0$) are 110 and 011

is the carry output channel. We get two lower sums S_1 and S_0 from the two-bit adder, and add two highest bits of the inputs and the carry of the two-bit adder to generate the highest sum S_2 and the carry C_2 of the three-bit adder. Figure 5.13 shows one of the simulation results. When inputs ($A_2A_1A_0$ and $B_2B_1B_0$) are 110 and 011, the corresponding outputs ($C_2S_2S_1S_0$) are 1001. The simulations of all 64 cases of input combinations have been done to indicate that this structure can achieve the function of three-bit binary adder.

Using this cascade method and the same basic structure, higher bit binary adders can be implemented easily. An n-bit ($n \geq 2$) binary adder can be constructed by combining a ($n - 1$)-bit adder and a one-bit full-adder.

5.4.3 Binary Comparator

5.4.3.1 One-Bit Binary Comparator

Binary comparator is used to compare binary numbers. It takes two binary numbers as input and determines whether one number is greater than, less than or equal to the other number. A one-bit binary comparator compares two single bits A_i and B_i, and outputs the comparison result, which can be expressed as Eq. (5.5).

$$(A_i > B_i) = A_i\bar{B}_i,$$
$$(A_i = B_i) = \bar{A}_i\bar{B}_i + A_iB_i, \qquad\qquad (5.5)$$
$$(A_i < B_i) = \bar{A}_iB_i.$$

In BZ medium, it is easy to realize $A_i > B_i$ and $A_i < B_i$ by using basic logic gates. In our design, the binary comparator only outputs the comparison result of "greater than" and "less than". If there is no pulse in these two output channels, it means that the two numbers are equal ($A_i = B_i$). Figure 5.14 shows the structure of one-bit binary comparator, which uses two T-shaped coincidence detectors to realize the logical functions of $A_i\bar{B}_i$ and $\bar{A}_iB_i$. This structure is vertically symmetrical, so that the pulses from different input channels propagate through the output channels at the same time.

For one-bit binary comparator, we consider all four combinations of inputs listed below. The simulation results are shown in Fig. 5.15.

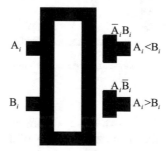

Fig. 5.14 The structure of one-bit binary comparator. Grid size is 150×220. Two T-shaped coincidence detectors are used in the structure to realize the logical functions of $A_i\bar{B}_i$ and $\bar{A}_iB_i$

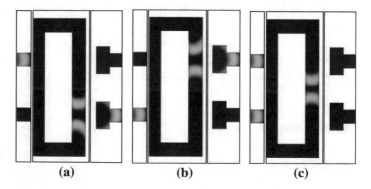

Fig. 5.15 Simulation results of one-bit binary comparator when inputs (A_i and B_i) are **a** 1, 0. **b** 0, 1. **c** 1, 1

Table 5.2 Inputs and corresponding outputs of the structure of one-bit binary comparator

Input		Output		
A_i	B_i	$A_i > B_i$	$A_i < B_i$	$A_i = B_i$
0	0	0	0	1
0	1	0	1	0
1	0	1	0	0
1	1	0	0	1

1. $A_i = 0, B_i = 0$: Similar to one-bit half-adder, the BZ medium stays at the stable stationary state without stimulation. There is no pulse in the output channels, so the outputs ($A_i > B_i$ and $A_i < B_i$) are 0 and 0, which means $A_i = B_i$.
2. $A_i = 1, B_i = 0$: As shown in Fig. 5.15a, pulse from channel A_i propagates through the channel, generating a new pulse in channel $A_i \bar{B_i}$. So the outputs are 1 and 0, which means the comparison result is $A_i > B_i$.
3. $A_i = 0, B_i = 1$: Opposite to 2, pulse from channel B_i generates a new pulses in channel $\bar{A_i} B_i$ in Fig. 5.15b. So the outputs are 0 and 1, which means the comparison result is $A_i < B_i$.
4. $A_i = 1, B_i = 1$: In Fig. 5.15c, two pulses from channel A_i and B_i propagate through the channels, meet in the middle and annihilate. There is no new pulse generated in output channels, so the outputs are 0, 0, which means the comparison result is $A_i = B_i$.

Table 5.2 lists all the outputs of one-bit binary comparator with different input combinations. Though the comparison result $A_i = B_i$ can not be obtained in the structure directly, it can be expressed by $A_i > B_i$ and $A_i < B_i$ indirectly. When $A_i > B_i$ and $A_i < B_i$ are both 0, it means $A_i = B_i$. Therefore, we conclude that the device shown in Fig. 5.14 realizes the comparison operation of two one-bit numbers.

5.4.3.2   Two-Bit Binary Comparator

In two-bit binary comparator, the two binary numbers to be compared are $A = A_1 A_0$, $B = B_1 B_0$, and the comparison result is represented by $A > B$ or $A < B$. Same as one-bit comparator, when two outputs are both 0, it means $A = B$. We implement two-bit binary comparator based on the structure of one-bit binary comparator. The structure of two-bit comparator includes two parts: the left part, which consists of two one-bit comparators, achieves comparison bit by bit; the right part synthesizes the by-bit comparison results to get the final result. Since the final result of binary comparator is mainly decided by the comparison result of the higher bits, the outer one-bit comparator compares the higher bit of the input data on the left part of the two-bit comparator. The structure of two-bit binary comparator is shown in Fig. 5.16, in which, A_1, A_0, B_1, B_0 are four input channels for two two-bit numbers, and $A > B, A < B$ are two output channels.

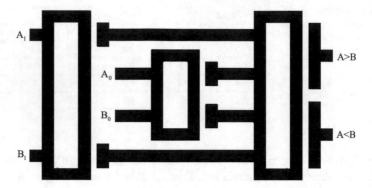

Fig. 5.16 The structure of two-bit binary comparator. Grid size is 500×300. The structure includes two parts: the left part achieves comparison bit by bit, the right part synthesizes the by-bit comparison results to get the final result

(a)   (b)

Fig. 5.17 Simulation results of two-bit binary adder when inputs (A_1A_0 and B_1B_0) are **a** 10, 01. **b** 10, 11

For this structure, we performed simulations for all sixteen input combinations to indicate that the structure can realize the comparison of two-bit binary numbers. Figure 5.17 shows two typical simulation results. In Fig. 5.17a, when inputs A_1A_0 and B_1B_0 are 10, 01, the final result is decided by the comparison result of the higher bits. So the outputs are 10, which means $A > B$. In Fig. 5.17b, when inputs A_1A_0 and B_1B_0 are 11 and 10, two higher bits are equal, the final result is decided by the comparison result of the lower bits, which means that the outputs are 01 and $A < B$.

5.4.3.3 Three-Bit Binary Comparator

Three-bit binary comparator compares two three-bit binary numbers, which are $A = A_2A_1A_0$ and $B = B_2B_1B_0$, and gives information of $A > B$ and $A < B$ as the result. Similar to two-bit comparator, three-bit comparator consists of the by-bit comparison unit and the comprehensive comparison unit. In by-bit comparison unit, three one-bit comparators on the left compare each bit of the two numbers; while

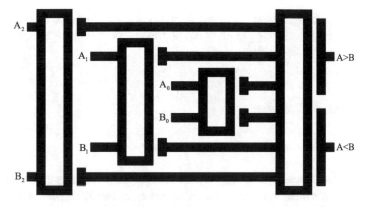

Fig. 5.18 The structure of three-bit binary comparator. Grid size is 680×430. There are three one-bit comparators on the left and a comprehensive comparison unit on the right

the comprehensive comparison unit on the right with six inputs synthesizes the comparison results. As shown in Fig. 5.18, the structure of three-bit comparator has six input channels $A_2, A_1, A_0, B_2, B_1, B_0$ and two output channels $A > B, A < B$. We have performed all the simulations to ensure that this structure realizes the comparison function of three-bit numbers. One of the simulation results is shown in Fig. 5.19. When inputs $A_2A_1A_0$ and $B_2B_1B_0$ are 101, 010, the outputs are 10, which mean $A > B$.

We can build higher bit binary comparator by the same cascade method. An n-bit $(n \geqslant 2)$ binary comparator consists of n one-bit binary comparators and a comprehensive comparison unit with $2n$ inputs. n one-bit binary comparators with decreasing size place one inside the other. The outer one is bigger and compares higher order of

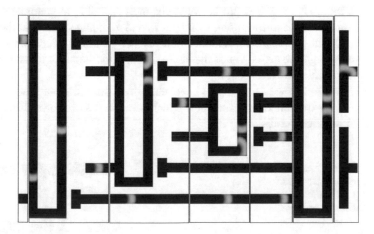

Fig. 5.19 Simulation result of three-bit binary comparator when inputs $(A_2A_1A_0$ and $B_2B_1B_0)$ are 101 and 010

the input data, whereas inner one is smaller and compares lower order. The comprehensive comparison unit collects all the comparison results of every bit and compares them to achieve the final result.

5.4.4 Binary Encoder

5.4.4.1 One-Bit and Two-Bit Binary Encoder

Binary encoder compresses 2^n (or fewer) binary inputs into n outputs. In one-bit (2 to 1) binary encoder, the inputs and output are I_1, I_0 and Y_0 respectively, and Y_0 is equal to I_1. In BZ medium, we can build a one-bit encoder by connecting the input channel I_1 to the output channel Y_0 directly.

For two-bit (4 to 2) binary encoder, the Boolean expressions can be represented as:

$$Y_1 = I_2 + I_3,$$
$$Y_0 = I_1 + I_3. \qquad (5.6)$$

where I_3, I_2, I_1, I_0 are four inputs, Y_1 and Y_0 are two outputs. Moreover, a group signal (GS) is provided as an additional output for cascade. GS is 1 when one or more of the inputs are active, which can be expressed as:

$$GS = I_0 + I_1 + I_2 + I_3. \qquad (5.7)$$

The related inputs are ORed to produce all three outputs. In BZ medium, this device can be constructed by using OR gates and chemical diodes directly. In the simple case, we ignore the order of inputs and outputs in the structure. As shown in Fig. 5.20, input channel I_3 is put between channels I_1 and I_2 to generate outputs Y_0 and Y_1 conveniently; furthermore, channels Y_0 and Y_1 intersect, so that the pulse from channel I_1 can propagate to channel GS. Since two pulses from channels Y_0 and Y_1 never go through the crossover at the same time, a modified crossover structure is employed to achieve crossing in this structure.

Since one and only one of four inputs is 1 at each time, there are four kinds of input combinations for two-bit binary encoder, which are listed as follows. Figure 5.21 shows the simulation results.

Fig. 5.20 The structure of two-bit binary encoder. Grid size is 220 × 190. A modified crossover structure is used to exchange the order of Y_0 and Y_1

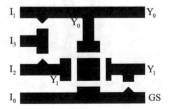

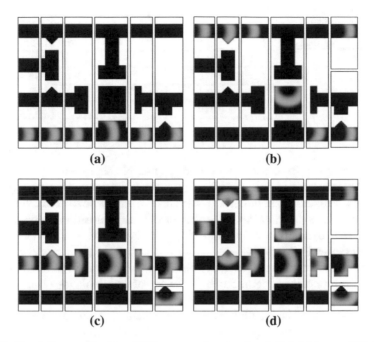

Fig. 5.21 Simulation results of two-bit binary encoder when inputs ($I_3I_2I_1I_0$) are **a** 0001, **b** 0010, **c** 0100, **d** 1000

1. $I_3I_2I_1I_0 = 0001$: As shown in Fig. 5.21a, pulse from channel I_0 propagates and outputs through channel GS. So the outputs ($GS\ Y_1Y_0$) are 100.
2. $I_3I_2I_1I_0 = 0010$: When channel I_1 has pulse (Fig. 5.21b), a single pulse passes the crossover and sends through channels GS and Y_0. So the outputs are 101.
3. $I_3I_2I_1I_0 = 0100$: Similar to 2, pulse from channel I_2 pulse passes the crossover and sends through channels GS and Y_1, as shown in Fig. 5.21c. So the outputs are 110.
4. $I_3I_2I_1I_0 = 1000$: The pulse from channel I_3 generates two pulses in channels Y_0 and Y_1, so two pulses propagate into the crossover. The pulse from channel Y_1 arrives at the crossover earlier than the pulse from channel Y_0, as shown in Fig. 5.21d, then the pulse from channel Y_0 cannot pass the crossover. As a result, there is still one output pulse in channel GS, the outputs are 111.

Table 5.3 summarizes all possible inputs and the corresponding outputs. It is clear that this device implements information encoding of two bits.

5.4.4.2 Three-Bit Binary Encoder

For three-bit (8 to 3) binary encoder, $I_7, I_6, I_5, I_4, I_3, I_2, I_1, I_0$ are eight inputs; Y_2, Y_1, Y_0 are three outputs and GS is the additional output for cascade. Three-bit

Table 5.3 Inputs and corresponding outputs of the structure of two-bit binary encoder

Input				Output		
I_3	I_2	I_1	I_0	GS	Y_1	Y_0
0	0	0	1	1	0	0
0	0	1	0	1	0	1
0	1	0	0	1	1	0
1	0	0	0	1	1	1

binary encoder can be constructed based on two-bit binary encoder by cascade method. The structure of three-bit encoder consists of two two-bit binary encoders, in which the outputs of individual two-bit encoders are ORed to produce Y_0, Y_1 and GS, respectively; and GS of the higher bit encoder is regarded as output Y_2. Three additional crossover structures are introduced in the structure to produce the final outputs, as shown in Fig. 5.22a.

For the structure of three-bit binary encoder, we performed simulations for all eight input combinations. Figure 5.22b shows one of the simulation results, in which the inputs $(I_7I_6I_5I_4I_3I_2I_1I_0)$ are 1000000. When I_7 is 1, the outputs of the higher bit encoder are 111. These output pulses pass three additional crossover structures, giving 1111 as the final outputs $(GS\ Y_2Y_1Y_0)$. The simulation results indicate that this structure realizes the encoding of three-bit binary numbers.

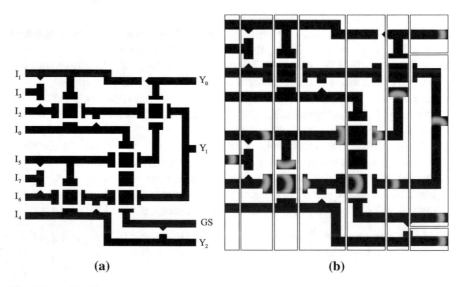

(a)   (b)

Fig. 5.22 **a** The structure of three-bit binary encoder. Grid size is 460×490. This structure consists of two two-bit binary encoders and three additional crossover structures. **b** Simulation result of three-bit binary encoder when inputs $(I_7I_6I_5I_4I_3I_2I_1I_0)$ are 1000000

It is easy to form higher bit binary encoders based on BZ reaction from the lower bit encoders and crossover structures. An n-bit ($n \geqslant 2$) binary encoder can be constructed by cascading two ($n - 1$)-bit encoders.

5.4.5 Binary Decoder

5.4.5.1 One-Bit and Two-Bit Binary Decoder

A decoder is a digital circuit that performs the inverse operation of an encoder. It has n input lines and a maximum of 2^n unique output lines. The logic expressions of one-bit (1 to 2) binary decoder are given by Eq. (5.8), where I_0 is the input, and Y_1, Y_0 are two outputs. It is easy to build one-bit binary decoder in BZ medium by combining an AND gate and a NOT (AND NOT) gate with an auxiliary 1 input. Figure 5.23 shows the structure, in which the auxiliary 1 input is put on the top.

$$Y_1 = I_0 = I_0 1,$$
$$Y_0 = \bar{I}_0 = \bar{I}_0 1. \tag{5.8}$$

A two-bit (2 to 4) binary decoder has two inputs I_1, I_0, and four outputs Y_3, Y_2, Y_1, Y_0. The Boolean expressions of two-bit decoder are expressed as:

$$Y_3 = I_1 I_0, \quad Y_2 = I_1 \bar{I}_0,$$
$$Y_1 = \bar{I}_1 I_0, \quad Y_0 = \bar{I}_1 \bar{I}_0. \tag{5.9}$$

Substituting Eq. (5.8) into Eq. (5.9), we have:

$$Y_3 = I_1 Y_1', \quad Y_2 = I_1 Y_0',$$
$$Y_1 = \bar{I}_1 Y_1', \quad Y_0 = \bar{I}_1 Y_0'. \tag{5.10}$$

where Y_1', Y_0' are the outputs of one-bit binary decoder. Equation (5.10) links two-bit decoder with one-bit decoder. A cascade method is adopt here to build two-bit decoder. Based on the structure of one-bit decoder, two-bit decoder can be divided into two parts: a one-bit decoder with lower bit input I_0 and the auxiliary 1 input

Fig. 5.23 The structure of one-bit binary decoder. Grid size is 130×160. It is constructed by combining an AND gate and a NOT (AND NOT) gate with an auxiliary 1 input

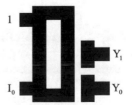

Fig. 5.24 The structure of
two-bit binary decoder. Grid
size is 260×270. This
structure consists of a one-bit
binary decoder and a
rectangular ring structure

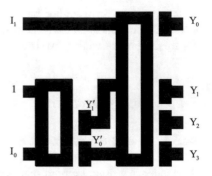

as inputs, and a rectangular ring structure with outputs of one-bit decoder Y_1', Y_0'
and higher bit input I_1 as inputs. Figure 5.24 shows the structure of two-bit binary
decoder, in which, the length of the output channels of one-bit decoder is adjusted
to keep the pulses in pace.

Figure 5.25 shows all simulation results of two-bit binary decoder. There are four
kinds of input combinations for two-bit decoder, which are listed as follows:

1. $I_1 I_0 = 00$: As shown in Fig. 5.25a, with the absence of inputs I_1 and I_0, a pulse
   from the auxiliary 1 input channel propagates in the structure of one-bit decoder,
   generating a new pulse in channel Y_0'. This pulse continues propagating in the
   rectangular ring structure, and send through the output channel Y_0, which means
   that the final outputs $(Y_3 Y_2 Y_1 Y_0)$ are 0001.
2. $I_1 I_0 = 01$: When channel I_0 has pulse (Fig. 5.25b), two pulses propagate through
   the input channels, generating and sending a new pulse through channel Y_1'. Then
   the pulse from channel Y_1' generates another new pulse in output channel Y_1,
   suggesting the outputs are 0010.
3. $I_1 I_0 = 10$: Similar to 1, channel Y_0' has an output pulse in one-bit decoder. The
   pulses from input channel I_1 and channel Y_0' collide and generate a new pulse in
   output channel Y_2, as shown in Fig. 5.25c. So the outputs are 0100.
4. $I_1 I_0 = 11$: In Fig. 5.25d, with the presence of inputs I_1, I_0 and auxiliary 1, the
   pulses are sent through Y_1' and Y_3 in one-bit and two-bit decoders respectively, so
   the outputs are 1000.

All possible inputs and the corresponding outputs are listed in Table 5.4. The
simulation results and the truth table demonstrate that this structure realizes the
decoding function of two bits.

5.4.5.2 Three-Bit Binary Decoder

Three-bit binary decoder (3 to 8 decoder) has three inputs I_2, I_1, I_0 and eight outputs $Y_7, Y_6, Y_5, Y_4, Y_3, Y_2, Y_1, Y_0$. Similarly, The logic expressions of three-bit binary
decoder can be represented by highest bit input I_2 and the outputs of two-bit decoder:

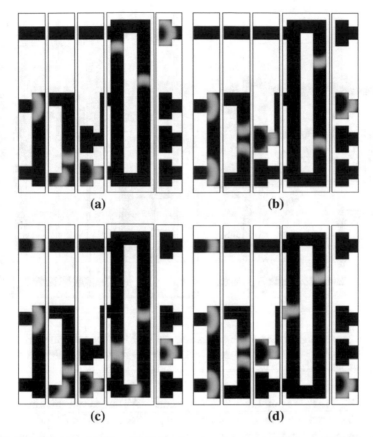

Fig. 5.25 Simulation results of two-bit binary decoder when inputs ($I_1 I_0$) are **a** 00, **b** 01, **c** 10, **d** 11

Table 5.4 Inputs and corresponding outputs of the structure of two-bit binary decoder

Input		Output			
I_1	I_0	Y_3	Y_2	Y_1	Y_0
0	0	0	0	0	1
0	1	0	0	1	0
1	0	0	1	0	0
1	1	1	0	0	0

$$Y_7 = I_2 Y_3', \quad Y_6 = I_2 Y_2',$$
$$Y_5 = I_2 Y_1', \quad Y_4 = I_2 Y_0',$$
$$Y_3 = \bar{I}_2 Y_3', \quad Y_2 = \bar{I}_2 Y_2',$$
$$Y_1 = \bar{I}_2 Y_1', \quad Y_0 = \bar{I}_2 Y_0' \tag{5.11}$$

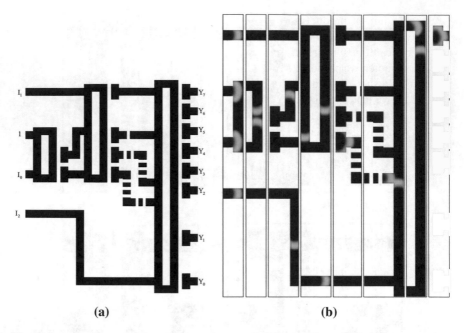

Fig. 5.26 **a** The structure of three-bit binary decoder. Grid size is 440×560. This structure consists of a two-bit binary decoder and a rectangular ring structure. **b** Simulation result of three-bit binary decoder when inputs $(I_2 I_1 I_0)$ are 111

where Y_3', Y_2', Y_1', Y_0' are outputs of two-bit decoder. We build three-bit binary decoder by the similar cascade method. As shown in Fig. 5.26a, three-bit decoder consists of a two-bit decoder and a rectangular ring structure, where time delay units are used for pulse delay. In order to arrange the output channels in numerical order, the highest bit input channel I_2 is set to the bottom of the rectangular ring structure.

There are eight input combinations for three-bit binary decoder. We have performed all the simulations to ensure that this structure fulfils three-bit numbers decoding. Figure 5.26b shows one of the simulation results with inputs $(I_2 I_1 I_0)$ 111. With the presence of all inputs, the pulses is sent through Y_3' and Y_7 in two-bit and three-bit decoders. As a result, the final outputs $(Y_7 Y_6 Y_5 Y_4 Y_3 Y_2 Y_1 Y_0)$ are 10000000.

Using the same basic structure and principle, higher bit decoders can be implemented easily. An n-bit ($n \geqslant 2$) binary decoder can be constructed by combining a $(n-1)$-bit decoder and a rectangular ring structure.

5.5 Universal Combinational Logic Circuits

In digital electronics, the logical function of any combinational logic circuit can be expressed by sum-of-products expressions. Therefore, the universal combinational logic circuits can be constructed based on BZ reaction, if we can achieve the function of universal sum-of-products expressions. Moreover, the connection structure should be involved in combinational logic circuits, so that the input signals can be transmitted to arithmetical unit synchronously.

In this chapter, we propose a universal method to build combinational logic circuits based on their sum-of-products expressions. Firstly, we design the universal structure of sum-of-products expression to realize the arithmetical function of combinational logic circuits; secondly, we design multi-input multi-output connection structure to fulfill the transmission function. At last, the universal combinational logic circuits are constructed based on these two parts.

5.5.1 Universal Structure of Sum-of-Products Expression

Generally, sum-of-products expression is formed as:

$$Y = A_0A_1A_2A_3 + A_4A_5 + A_6A_7A_8 + \cdots + A_{n-1}A_n \tag{5.12}$$

where, $A_0, A_1, \ldots, A_n$ are binary inputs and Y is binary output. Equation 5.12 is comprised of multi-bit AND, multi-bit OR and NOT operations. These three basic structures are discussed below.

5.5.1.1 Multi-Bit AND Operation

In multi-bit AND operation, only if all the inputs are 1, the output is 1. In Sect. 5.4.1, T-shaped coincidence detector is used to achieve AND operation of two binary numbers; multi-bit AND structure can be built by using multiple T-shaped coincidence detectors. We need $n(n-1)/2$ T-shaped coincidence detectors for n-bit AND operation, which will occupy a lot of space.

In this section, we modify T-shaped coincidence detectors to straight channels separated by passive stripes, and design a multi-bit AND structure, which is simple and easily extensible. The multi-bit AND structure has two part: the inputs are on the left, while the operation and the output are on the right. In the left part of the structure, horizontal channels are applied to receive and transmit input pulses; in the right part, channels are arranged vertically for coincidence detection, the rightmost vertical channel is set as the output channel. The numbers of horizontal channels and vertical channels are both the same as the inputs. Taking 5-bit AND expression $Y = A_0A_1A_2A_3A_4$ as an example, Fig. 5.27 shows the structure. In the figure, the width of the channels and the passive stripes is 20 and 10 grid points respectively; the width of the input channels is added to 40 grid points at the end to fulfill penetration.

When input pulse penetrates to the right part from its input channel, the pulse will fork in the first vertical channel. Two forked pulses generated by different input pulses meet in the first vertical channel, generating a new pulse in the second vertical channel. Similarly, the new pulse forks in the second vertical channel, meets another

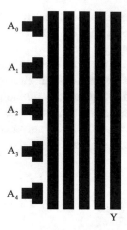

Fig. 5.27 The structure of multi-bit AND operation. Grid size is 200 × 400. This structure consists of two parts: the input part on the left, the operation and output part on the right

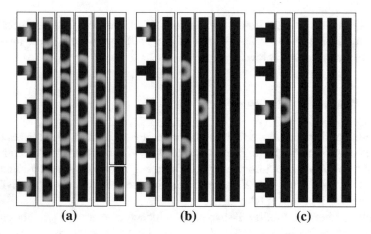

Fig. 5.28 Simulation results of multi-bit AND structure when inputs ($A_4A_3A_2A_1A_0$) are **a** 11111, **b** 10101, **c** 00100

forked pulse, exciting third vertical channel, and so on. Figure 5.28 shows some simulation results of 5-bit AND structure, in which, there are five, three and one input pulses. Simulation results indicate that the number of excited vertical channels is decided by the number of input pulses. If and only if all the inputs are 1, the rightmost vertical channel is excited, which means the output is 1.

5.5.1.2 Multi-Bit OR Operation

In multi-bit OR operation, if one or more than one inputs are 1, the output is 1. Similar to multi-bit AND structure, multi-bit OR structure also contains two parts: the left part is the input part, each input pulse enters the structure from a horizontal input channel; the right part have only one vertical channel, which achieves OR operation and outputting. Figure 5.29a demonstrates the structure of 5-bit OR expression $Y = A_0 + A_1 + A_2 + A_3 + A_4$, in which, the width of the channels is 20 grid points. With the presence of one or more than one input pulses, the vertical output channel is excited, which means the output is 1. The simulation results with five and one input pulses are shown in Fig. 5.29b, c respectively.

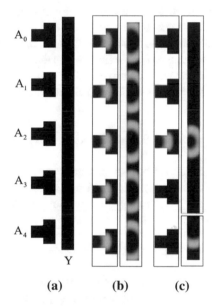

(a)   (b)   (c)

Fig. 5.29 **a** The structure of multi-bit OR operation. Grid size is 80×400. Similar to multi-bit AND structure, this structure also consists of two parts. **b** Simulation result of multi-bit OR structure when inputs $(A_4A_3A_2A_1A_0)$ are 11111. **c** Simulation result of multi-bit OR structure when inputs $(A_4A_3A_2A_1A_0)$ are 00100

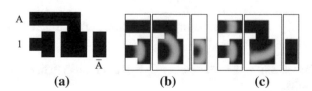

(a)   (b)   (c)

Fig. 5.30 **a** The structure of NOT operation. Grid size is 130×90. This structure is modified from crossover structure. **b** Simulation result of NOT structure when input (A) is 0. **c** Simulation result of NOT structure when input (A) is 1

5.5.1.3 NOT Operation

NOT operation is realized in BZ medium by rectangular ring structure in Sect. 5.4.1. In this section, we design a new NOT structure based on modified crossover structure, in order to connect it to the straight channels in multi-bit AND structure and multi-bit OR structure conveniently.

Similar to Sect. 5.4.1, an auxiliary 1 input is needed, except input A of NOT operation. Taking A and auxiliary 1 as two inputs of the crossover structure, we can obtain $\bar{A}$ in the output channel opposite to the input channel of 1. Furthermore, we connect input channel A to square in the middle of the crossover structure, as shown in Fig. 5.30a, so that two input channels are horizontal, and two input pulses can arrive the center square at the same time.

Figure 5.30b, c show the simulation results. When input A is 0, the pulse from the auxiliary 1 input channel penetrates two passive stripes, generating a new pulse in channel $\bar{A}$. So the output $\bar{A}$ is 1. When input A is 1, the pulse from the auxiliary 1 input channel penetrates one passive stripe, meets the input pulse in the center of the modified crossover structure. The mixed pulse can not pass the second passive stripe, there is no pulse in channel $\bar{A}$, giving 0 as the output.

5.5.1.4 Universal Structure of Sum-of-Products Expression

It is easy to achieve the function of sum-of-products expression based on BZ reaction, by combining multi-bit AND structure, multi-bit OR structure and NOT structure. In this section, we take the following logic expression (Eq. 5.13) as an example, and discuss the design of the universal structure of sum-of-products expression.

$$Y = A_0 \bar{A_1} A_2 + \bar{A_3} A_4 \tag{5.13}$$

Equation 5.13 is comprised of two multi-bit AND, two NOT and one multi-bit OR operations. So the structure of Eq. 5.13 includes two multi-bit AND structures, two NOT structures and one multi-bit OR structure. The structure is constructed in BZ medium by the following steps:

1. Build the multi-bit AND structures of $A_0 A_1 A_2$ and $A_3 A_4$ respectively.
2. Connect NOT structures to the inputs of multi-bit AND structures to achieve the logical functions of $A_0 \bar{A_1} A_2$ and $\bar{A_3} A_4$.
3. Connect the rightmost channels of two multi-bit AND structures to achieve OR operation of $A_0 \bar{A_1} A_2$ and $\bar{A_3} A_4$.

The structure of Eq. 5.13 is shown in Fig. 5.31a. Figure 5.31b, c demonstrate two simulation results.

When inputs $(A_4 A_3 A_2 A_1 A_0)$ are 11111, the input pulses from channels A_1 and A_3 pass NOT structure and annihilate. There are two input pulses in multi-bit AND structure $A_0 A_1 A_2$, which excites two vertical channels. Similarly, there is one input

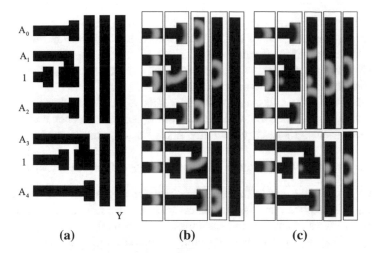

Fig. 5.31 a Universal structure of sum-of-products expression of Eq. 5.13. Grid size is 190×390. **b** Simulation result when inputs $(A_4A_3A_2A_1A_0)$ are 11111. **c** Simulation result when inputs $(A_4A_3A_2A_1A_0)$ are 10101

pulse in multi-bit AND structure A_3A_4, so only the leftmost vertical channel is excited. As a result, no new pulse is generated in the rightmost channel, which means the output is 0. When inputs are 10101, there are three and two input pulses in two multi-bit AND structures $A_0A_1A_2$ and A_3A_4 respectively, so the rightmost vertical channel is excited twice, giving 1 as the final output.

5.5.2 Multi-input Multi-output Connection Structure

The logical function of sum-of-products expression, such as Eq. 5.13, has been realized based on BZ reaction in previous section. However, in the case that one input is involved in more than one multi-bit AND operations, for example, logic expression $Y = A_0A_1 + A_0A_2$, we should solve the crossing problem of different inputs by designing a new connection structure.

Supposing the input pulses pass the connection structure from left to right synchronously; meanwhile, the pulses output below for logical operation. Furthermore, we try to keep the symmetry of the structure, so that the output pulses leave the structure from right side and underside at the same time. In this section, we extend the crossover structure to build the multi-input multi-output connection structure. The construction steps are listed below:

1. Arrange 40-grid point wide square structures vertically as the input channels of the pulses.

2. Extend crossover structures in the plane by connecting center square of each crossover. The numbers of squares in the horizontal and vertical directions are the same as that of the inputs.
3. Connect the input squares to the crossover array horizontally, so that the pulses can pass through the structure from left to right.
4. Lengthen the proper squares to 90 grid points vertically, so that the pulses can propagate below, since the wave vectors of the pulses have been changed.
5. Lengthen the lengthened squares horizontally for diagonal symmetry of the structure.
6. Fill the blanks in the structure with squares to ensure the pulse propagation.

Taking three inputs as an example, the connection structure is shown in Fig. 5.32, in which, I_0, I_1, I_2 are three input channels, O_0, O_1, O_2 and T_0, T_1, T_2 are output channels for logical operation and transmission respectively.

In Fig. 5.32, when pulses pass through the marked regions horizontally, these regions are in the refractory period of the BZ medium, so the pulses in the upside input channels, such as I_0 and I_1, can not propagate vertically in these regions. We can solve this problem by lengthening the rectangular region above the marked region, until the time interval between two pulses pass the mark region is greater than the refractory period. However, this structure will occupy huge space. For simplicity, we change the concentrations of activator and catalyst to stationary concentrations in the marked regions and the regions below in the process of simulation, so that the pulses can propagate in the structure normally.

Figure 5.33 demonstrates the simulation results with one and three inputs. In the horizontal direction, the pulse penetrates the passive stripes and outputs from the right side. In the vertical direction, when the pulse passes the rectangular region, the wave vector of the pulse changes, so the pulse can penetrate the passive stripes and output below. In Fig. 5.33a, the pulse from input channel I_1 passes through the structure and outputs from channels O_1 and E_1 at the same time. In Fig. 5.33b, three pulses pass their crossover structures, and output synchronously from six output channels.

Fig. 5.32 The multi-input multi-output connection structure. Grid size is 360×310. This structure is constructed by extending the crossover structure

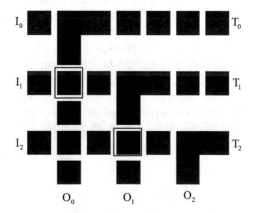

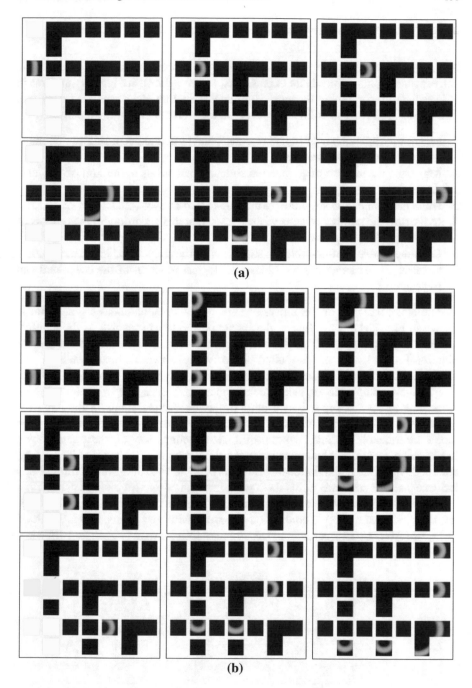

Fig. 5.33 Simulation results of multi-input multi-output connection structure with **a** one input, **b** three inputs

5.5.3 Universal Structure of Combinational Logic Circuits

The logical function of any combinational logic circuit can be converted to the corresponding sum-of-products expressions, so we achieve its logical function in BZ medium by combining the sum-of-products structure and connection structure. The universal method to build combinational logic circuit is described below:

1. Make sure the number of inputs in connection structure according to the inputs of sum-of-products expressions.
2. Make sure the number of connection structures according to the sum-of-products expressions.
3. Build the connection structure by the steps listed in Sect. 5.5.2.
4. Make sure the number of sum-of-products structures according to the outputs of sum-of-products expressions.
5. Construct every sum-of-products structures by the steps listed in Sect. 5.5.1.4.
6. Connect all the structures to achieve the logical function of the combinational logic circuit.

The block diagram of the universal structure of combinational logic circuit is shown in Fig. 5.34. In the figure, the inputs on the left propagate to the sum-of-products structures below through connection structures, then the results of the combinational logic circuit are outputted below. We take two-bit binary encoder and one-bit binary adder as two examples to verify this universal construction method.

In two-bit binary encoder, there are four inputs I_3, I_2, I_1, I_0 and two outputs Y_1, Y_0. According to Eq. 5.6, input I_0 is useless to the logic expressions. In the universal structure of two-bit binary encoder, we need two connection structures with three inputs for transmission and two multi-bit OR structures for calculation, as shown in Fig. 5.35a. Figure 5.35b demonstrates the simulation result when inputs $(I_3 I_2 I_1 I_0)$ are 1000. We also do simulations with inputs 0100 and 0010, the simulation results indicate that this structure fulfils the logical function of two-bit binary encoder.

In one-bit binary adder, there are two inputs A_i, B_i and two outputs C_i, S_i. According to Eq. 5.3, we need two connection structures with two inputs, two NOT structures, two AND structures and one OR structure to achieve $S_i = \bar{A}_i B_i + A_i \bar{B}_i$; while

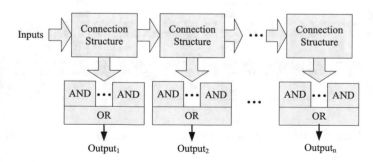

Fig. 5.34 Block diagram of the universal structure of combinational logic circuits

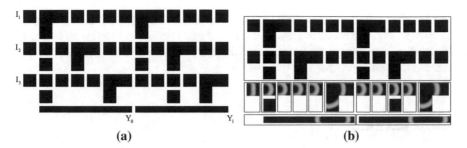

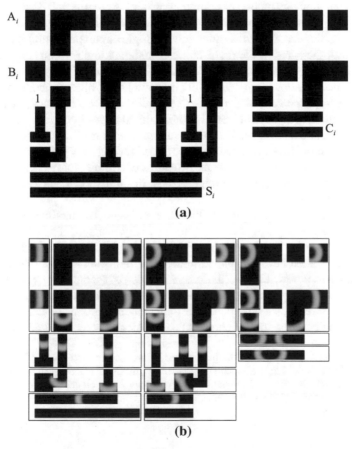

Fig. 5.35 **a** Universal structure of two-bit binary encoder. Grid size is 660×340. **b** Simulation result of two-bit encoder when inputs $(I_3 I_2 I_1 I_0)$ are 1000

Fig. 5.36 **a** Universal structure of one-bit binary adder. Grid size is 660×390. **b** Simulation result of one-bit adder when inputs (A_i, B_i) are 1, 1

another connection structure with two inputs and an AND structure are used for $Ci = A_i B_i$. Figure 5.36 shows the structure and one simulation result with inputs 11. With all the simulations for four input combinations, we conclude that this structure achieve the adding function of two bits.

5.6 Conclusion

It is very important to realize combinational logic circuit based on BZ reaction, since logical functions are the basis for chemical computation. In this chapter, we focus on the design and implementation of combinational logic circuits in BZ medium in two different perspectives. On one hand, we use the conventional method to construct classical combinational logic circuits: low-bit logic circuits are build based on the structure of logic gates, while multi-bit combinational logic circuits are build base on low-bit logic circuits by the cascade method. Four types of classic combinational logic circuits, including binary adder, binary comparator, binary encoder and binary decoder, are implemented and verified in simulation. On the other hand, we propose a universal method to achieve universal combinational logic circuits according to their sum-of-products expressions. Universal structure of sum-of-products expressions and multi-input multi-output structure are formed respectively for logical arithmetic and signal transmission. The universal structures of two-bit binary encoder and one-bit binary adder are designed and implemented in simulation to verify the universal construction method.

The combinational logic circuits based on conventional method have compact structures and can be expended to multi-bit easily, but their fixed structures can only be used to solve fixed problems; While the combinational logic circuits based on universal method have flexible structures, but the structures are huge, and their implementation is restricted by the properties of BZ medium, such as refractory period. By combining these two methods, we can construct different types of combinational logic circuits, which expands the logical functions of chemical computation, facilitating the constructions of chemical computer and other intelligent systems.

Acknowledgments We are grateful to the National Natural Science Foundation of China (NSFC) (61327802, 61273341), Tianjin Research Program of Application Foundation and Advanced Technology (14JCQNJC04700), the National High Technology Research and Development Program (863 Program) of China (2013AA041102).

References

1. Adamatzky, A., Arena, P., Basile, A., Carmona-Galan, R., Costello, B.D.L., Fortuna, L., Frasca, M., Rodriguez-Vazquez, A.: Reaction-diffusion navigation robot control: from chemical to vlsi analogic processors. IEEE Trans. Circuits Syst. I: Regul. Pap. **51**, 926–938 (2004)

2. Adamatzky, A., Costello, B.D.L.: On some limitations of reaction-cdiffusion chemical computers in relation to voronoi diagram and its inversion. Phys. Lett. A **309**, 397–406 (2003)
3. Agladze, K., Aliev, R.R., Yamaguchi, T., Yoshikawa, K.: Chemical diode. J. Phys. Chem. **100**, 13895–13897 (1996)
4. Asai, T., Adamatzky, A., Amemiya, Y.: Towards reaction-diffusion computing devices based on minority-carrier transport in semiconductors. Chaos, Solitons Fract. **20**, 863–876 (2004)
5. Costello, B.D.L., Adamatzky, A., Jahan, I., Zhang, L.: Towards constructing one-bit binary adder in excitable chemical medium. Chem. Phys. **381**, 88–99 (2011)
6. Gorecka, J., Gorecki, J.: Multi-argument logical operations performed with excitable chemical medium. J. Chem. Phys. **124**, 084101 (2006)
7. Gorecka, J.N., Gorecki, J., Igarashi, Y.: One dimensional chemical signal diode constructed with two nonexcitable barriers. J. Phys. Chem. A **111**, 885–889 (2007)
8. Gorecki, J., Gorecka, J.N., Igarashi, Y.: Information processing with structured excitable medium. Nat. Comput. **8**, 473–492 (2009)
9. Gorecki, J., Yoshikawa, K., Igarashi, Y.: On chemical reactors that can count. J. Phys. Chem. A **107**, 1664–1669 (2003)
10. Holley, J., Adamatzky, A., Bull, L., Costello, B.D.L., Jahan, I.: Computational modalities of belousov-zhabotinsky encapsulated vesicles. Nano Commun. Net. **2**, 50–61 (2011)
11. Holley, J., Jahan, I., Costello, B.D.L., Bull, L., Adamatzky, A.: Logical and arithmetic circuits in belousov-zhabotinsky encapsulated disks. Phys. Rev. E **84**, 056110 (2011)
12. Ichino, T., Igarashi, Y., Motoike, I.N., Yoshikawa, K.: Different operations on a single circuit: field computation on an excitable chemical system. J. Chem. Phys. **118**, 8185 (2003)
13. Kuhnert, L., Agladze, K.I., Krinsky, V.I.: Image processing using light-sensitive chemical waves. Nature **337**, 244–247 (1989)
14. Motoike, I., Yoshikawa, K.: Information operations with an excitable field. Phys. Rev. E **59**, 5354–5360 (1999)
15. Motoike, I.N., Adamatzky, A.: Three-valued logic gates in reaction-diffusion excitable media. Chaos, Solitons Fract. **24**, 107–114 (2005)
16. Motoike, I.N., Yoshikawa, K.: Information operations with multiple pulses on an excitable field. Chaos, Solitons Fract. **17**, 455–461 (2003)
17. Murray, J.D.: Mathematical Biology: I. An Introduction. Springer, Berlin (2002)
18. Nagahara, H., Ichino, T., Yoshikawa, K.: Direction detector on an excitable field: Field computation with coincidence detection. Phys. Rev. E **70**, 036221 (2004)
19. Rambidi, N.G., Shamayaev, K.E., Peshkov, G.Y.: Image processing using light-sensitive chemical waves. Phys. Lett. A **298**, 375–382 (2002)
20. Rambidi, N.G., Yakovenchuk, D.: Finding paths in a labyrinth based on reaction-diffusion media. BioSystems **51**, 67–72 (1999)
21. Rambidi, N.G., Yakovenchuk, D.: Chemical reaction-diffusion implementation of finding the shortest paths in a labyrinth. Phys. Rev. E **63**, 026607 (2001)
22. Sielewiesiuk, J., Gorecki, J.: Logical functions of a cross junction of excitable chemical media. J. Phys. Chem. A **105**, 8189–8195 (2001)
23. Steinbock, O., Kettunen, P., Showalter, K.: Chemical wave logic gates. J. Phys. Chem. **100**, 18970–18975 (1996)
24. Steinbock, O., Toth, A., Showalter, K.: Navigating complex labyrinths: optimal paths from chemical waves. Science **267**, 868–871 (1995)
25. Stevens, W.M., Adamatzky, A., Jahan, B.D.L., Costello, I.: Time-dependent wave selection for information processing in excitable media. Phys. Rev. E **85**, 066129 (2012)
26. Sun, M.Z., Zhao, X.: Multi-bit binary decoder based on Belousov-Zhabotinsky reaction. J. Chem. Phys. **138**, 114106 (2013)
27. Szymanski, J., Gorecka, J.N., Igarashi, Y., Gizynski, K., Gorecki, J., Zauner, K., Planque, M.: Droplets with information processing ability. Int. J. Unconv. Comput. **7**, 185–200 (2011)
28. Toth, A., Gaspar, V., Showalter, K.: Signal transmission in chemical systems: propagation of chemical waves through capillary tubes. J. Phys. Chem. **98**, 522–531 (1994)
29. Toth, A., Showalter, K.: Logic gates in excitable media. J. Chem. Phys. **103**, 2058 (1995)
30. Zhang, G.M., Wong, I., Chou, M., Zhao, X.: Towards constructing multi-bit binary adder based on belousov-zhabotinsky reaction. J. Chem. Phys. **136**, 164108 (2012)

Chapter 6
Associative Memory in Reaction-Diffusion Chemistry

James Stovold and Simon O'Keefe

Abstract Unconventional computing paradigms are typically very difficult to program. By implementing efficient parallel control architectures such as artificial neural networks, we show that it is possible to program unconventional paradigms with relative ease. The work presented implements correlation matrix memories (a form of artificial neural network based on associative memory) in reaction-diffusion chemistry, and shows that implementations of such artificial neural networks can be trained and act in a similar way to conventional implementations.

6.1 Introduction

Whilst, in theory at least, unconventional computing paradigms offer significant advantages [33] over the traditional Turing/von Neumann approach, there remain a number of concerns [32] regarding larger-scale applicability, including the appropriate method for programming and controlling such unconventional approaches to computation. The massive parallelism obtainable through many of these paradigms is both the basis for, and a source of problems for, much of this power. Given how quickly these approaches could overtake conventional computing methods, large-scale implementations would be incredibly exciting, and would allow for a larger proportion of possible computations to be computable [12]. The power of such parallelism can be seen by the recent advent of general-purpose graphic processing unit (GPGPU) technology, where huge speed-ups are gained by parallelising repetitive tasks. However, translating between the traditional, algorithmic approach to problem solving that is ubiquitous in computer science, and the complex,

J. Stovold · S. O'Keefe (✉)
Department of Computer Science, University of York,
York, North Yorkshire YO10 5GE, England
e-mail: simon.okeefe@york.ac.uk

J. Stovold
e-mail: jhs503@york.ac.uk

© Springer International Publishing Switzerland 2017
A. Adamatzky (ed.), *Advances in Unconventional Computing*,
Emergence, Complexity and Computation 23,
DOI 10.1007/978-3-319-33921-4_6

dynamical, nonlinear environment that most unconventional paradigms exploit, can be a highly non-trivial task.

Conrad [11] proposed the principle of a tradeoff between programmability, efficiency and evolvability of a computational system. This principle states that the price paid for having such a highly-programmable system is in terms of efficiency, evolvability or both. A corollary to this is that, in order to increase the efficiency of a computational construct, either the programmability or the evolvability must be sacrificed.

This principle underlies the field of unconventional computing. By considering alternative, non-standard approaches to computation, there is the possibility that the efficiency of the system can be increased with a minimal loss of programmability.

There are many examples of unconventional approaches to computation, each with their potential advantages, all with a notable decrease in programmability. For example, Adleman [5] showed experimentally that DNA could be used as a computational substrate by computing a solution to the Hamiltonian Path Problem; Nakagaki [28] showed that the plasmodial slime mould *Physarum polycephalum* could compute the shortest path through a maze; and Laplante [24] showed how a mass-coupled chemical reaction could implement a Hopfield network. While these are all clearly much harder to program than traditional computational substrates, the potential for much more efficient computation is present in all of them through their intrinsic parallelism. By exploring many alternate paths of computation simultaneously, certain unconventional substrates have the effect of being able to implement a non-deterministic UTM, allowing them to potentially solve NP problems in polynomial time.

The problem faced by researchers in the unconventional computing field is that in order to get access to this huge potential power, the system must be programmed, and in order to use the substrate as a general-purpose computer, the system must be reliable. These considerations are necessarily non-trivial. The reliability problem is inherent in the non-determinism, as there is no guarantee that the substrate will perform the computation intended. The programmability problem is that of finding 'algorithms' that can be used to perform meaningful computation on the substrate.

Conrad [12] quantified the idea that general-purpose machines are not particularly general-purpose, and that only certain problems can be solved using conventional computational methods. From this, it becomes clear that different approaches to the same problem—whilst being Turing equivalent—may be different from other perspectives. For example, consider *Turing tarpit* languages such as 'OISC' [15] and 'brainfuck' [*sic*] [27]. Although these languages are Turing complete in the same sense as languages such as C and Java, they have particularly obfuscated programs. These systems serve to elucidate the difference between being Turing computable and being *usably* computable. While an extreme example, 'brainfuck' shows the importance of programmability in general-purpose computers. If this idea is extended to computability in general, just because a system is labelled as Turing complete does not mean that it should be explicitly used as a Turing machine.

Recent approaches have attempted to utilise parallelism to increase the number of computable functions. In conventional computing paradigms, this consists of many UTMs (Universal Turing Machines) working in parallel on the same function. The use of GPGPU processing has been the primary result of this, but the problem with using multiple forms of the same computational construct is that each computational unit has the same limitations. Field-programmable gate arrays (FPGAs) help with this, in that processing units can be implemented alongside bespoke hardware, to speed up regularly-used operations, but problems arise in terms of the time taken to reprogram the devices. Because of this, the reprogrammability—and hence the programmability—of FPGAs as a computational construct is reduced in response to the efficiency gained.

If unconventional computing paradigms are considered as a complement to UTMs, the number of computable functions is likely to be increased, as the limitations of each system may offset those of the other. This is because there will inevitably be certain functions that UTMs are better suited to, and other functions that unconventional approaches are better suited to.

In this paper, we present an implementation of an artificial neural network (a correlation matrix memory) in an unconventional paradigm (reaction-diffusion chemistry). This implementation will serve as a method of simplifying the process of programming the unconventional paradigm (we already know how to program correlation matrix memories, and it's simpler than programming reaction-diffusion reactors).

The paper is organised as follows: in Sect. 6.2 we introduce diffusive computation and reaction-diffusion chemistry, and show how this can be used for computation; in Sect. 6.3 we describe associative memory and correlation matrix memories; Sect. 6.4 discusses the methods used to get our results; Sects. 6.5, 6.6, 6.7, and 6.8 detail the process of designing and testing the correlation matrix memories in reaction-diffusion chemistry. Finally, Sect. 6.9 presents our conclusions.

6.2   Diffusive Computation and Reaction-Diffusion Chemistry

We define *diffusive computation* as an unconventional approach to computing that harnesses the diffusion of particles as a representation of data. The power of diffusive computing systems come from their highly parallel architectures, with the emergent behaviour of interactions between diffusing particles dictating the operation of the system. A simple example of a diffusive computing system is a cellular automaton. While nothing physical diffuses in a cellular automaton, the emergent behaviour of structures such as gliders gives rise to the diffusion of information. A simple rule for the cellular automaton can be used to give complex behaviour that is shown to be similar to that of reaction-diffusion systems [4].

One of the primary benefits of considering this unconventional computing paradigm is the potential it offers compared to existing systems. The properties of diffusive computation compared with general-purpose machines make it a much better candidate for producing systems that have similar complexity to that of the brain [29].

There are many different forms of diffusive computation, such as silicon-based diffusion processors [3], slime mould [2] and cellular automata [4]. In this paper, we consider a fundamental form of diffusive computation: reaction-diffusion chemistry.

Reaction-diffusion chemistry (RD chemistry) is a form of chemical reaction that changes state in such a way that wavefronts of reagent appear to flow (or diffuse) across the solution. The most widely-used models are based on the Belousov–Zhabotinsky reaction [9, 40], which may display single waves of reaction or an oscillation between two (observable) states, depending on the setup used.

6.2.1 Belousov–Zhabotinsky Reaction

Belousov [9] discovered a visibly periodic chemical reaction based on Bromate and citric acid. The reaction cycled through a series of colour changes on a timescale of many seconds—making it very easy for humans to observe the changes in state. Zhabotinsky [40] then enhanced the effect of the colour change by replacing the citric acid used by Belousov with malonic acid. This reaction was soon termed the *Belousov–Zhabotinsky* (BZ) reaction.

The chemical reaction can be distributed across a wide area, such as in a Petri dish, with the depth of the liquid reduced to less than 2 mm. This can also be achieved by binding the catalyst to a chemical gel, with a thin layer of chemicals over the top. This 'thin-layer' reactor allows the system to be considered as a pseudo-two-dimensional system. As the reaction oscillates, each point in the reactor will react only with the chemicals available locally to it, meaning that a small disturbance in the reaction—such as an inhomogeneity (gas bubble, speck of dust etc.)—will cause a wave of colour change to propagate across the reactor [41], and can result in some very interesting effects (such as those in Fig. 6.1).

There are many potential catalysts for BZ reactions, each with different properties. Ruthenium bipyridyl complex $(Ru(bipy)_3^{2+})$ is particularly interesting as a catalyst in BZ reactions as it is photosensitive. This in an incredibly helpful property, as the rate with which ruthenium catalyses the reaction is inversely proportional to the level of illumination. This can be harnessed as a method of controlling the reaction

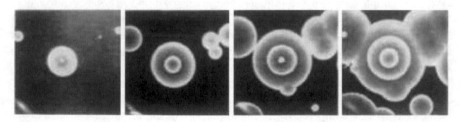

Fig. 6.1 A 'leading centre' pattern, observed in a thin-layer Belousov–Zhabotinsky reaction (from [39])

for computation [23]. By disabling different regions of the reactor, either through illumination or by constraining the distribution of catalyst in a chemical gel, circuits can be constructed that allow logic gates and other computation to be performed [6].

6.2.2   Applications

The different applications of diffusive chemical reactions have been studied in great depth since it was proposed as a basis for morphogenesis by [37]. Between 1986 and 1989, a series of seminal papers proposed a method of using RD chemistry as a computational medium, showing that it can be used for a number of image processing operations: in particular, contrast modification and contour discernment [21–23]. These papers sparked much research interest, particularly given the CPU-intensive nature of sequential image processing algorithms and how intrinsically parallelised the process becomes through the use of RD chemistry.

Since then, a large amount of work has been invested examining the computational properties of RD chemistry. RD chemistry was shown to have the intrinsic ability to find the shortest path through a maze, with a time complexity that is independent of the complexity of the maze, only dependent on the distance between start and finish [31]. This is because the propagating waves explore every reachable path simultaneously, but at each junction, only the first wave to arrive can continue (because of the refractory period of the waves, which will inhibit later waves).

There have been many other applications of RD chemistry, for example, construction of a Voronoi diagram [35]; a chemical diode [6]; and the involute for a static object [25].

The computational universality of RD chemistry (and other nonlinear media) has been discussed in depth [1], and many authors have suggested different approaches to constructing logic gates. The use of capillary tubes to constrain the reaction has been considered, using the diameter of the tube to control the behaviour of the wave, which can be used to implement logic gates (including a universal set) [36]. Various logic gates, a memory cell and a coincidence detector have been constructed using simulations of the diffusion dynamics when constrained to channels [26].

Gorecki et al. [16] take this idea a step further, constructing the coincidence detector, which implements an AND gate, both in simulation and in experiments. The coincidence detector (shown in Fig. 6.2) allows a single wave to pass through in either direction, but the presence of two waves travelling towards each other raises the activator concentration between the colliding waves higher than either of the single waves would alone. This higher concentration then diffuses farther into the gap in catalytic material compared to the concentration from the single waves. The lower part of the gate, which acts as the output, can then be stimulated by this diffusion. The same paper then presents an extension of this idea by constructing a cascadable binary counter.

If a series of waves arrive at a gap in chemical medium within a certain time period, the concentration on the far side of the gap will rise higher than with a single

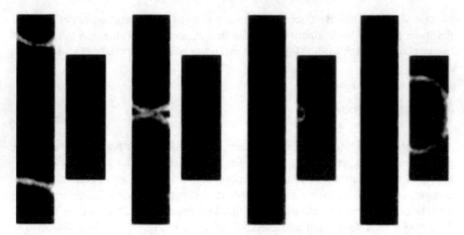

Fig. 6.2 Four photographs of a chemical coincidence detector. Each photo has two sections of active substrate, the *left-most* acting as inputs, the *right-most* as output. The waves are inputted (from the *top* and *bottom*), and as they collide, the activator concentration between them rises higher than if only one wave was present. This results in the activator diffusing farther into the gap between input and output section, producing a wave on the output section (from [16])

wave, because of the diffusion of chemicals across the gap [17]. If this stimulation is sufficient, it can cause an excitatory response on the far side of the gap [13], which is directly analogous to the behaviour of spiking neurons [34].

6.3 Associative Memory

Associative memory is a form of memory that associates stimuli with responses. There are two forms of associative memory, *autoassociative* memory and *heteroassociative* memory. In autoassociative memory, the associated, complete pattern can be retrieved from a small cue: for example, given the cue 'lived happily ever...' most people would immediately think of '...after'. However, with heteroassociative memory, other patterns that are associated with the cue are retrieved, but not necessarily the completed pattern from the cue. The brain works in a similar way to a combination of these forms of memory, as when presented with the same cue as above, most people would still complete the associated pattern, but subsequently think of a fairytale or children's story.

From a computational perspective, this could be implemented as an artificial neural network [18] that behaves in a similar fashion, where a noisy pattern or subset of the pattern could be used as the stimulus to the network in order to retrieve the complete, clean pattern.

The basic idea of an artificial associative memory is to train a fully-connected, two layer artificial neural network (see Fig. 6.3) in such a way that, when presented

Fig. 6.3 Two-layer artificial neural network showing the relationship between input–output pairs and the correlation matrix $\mathcal{M}$, as given in matrix (6.1), below

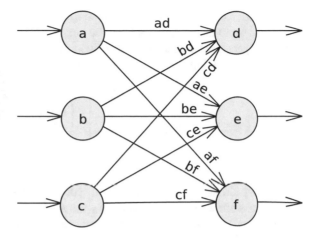

with the stimulus vector x, the connections between the input and output layers will produce the associated output vector y [10]. In training such a network, the weights on these connections are crucial, as they are what define the association. The correlation weight matrix, w, stores associations representing p patterns, and is produced via:

$$w_{ij} = \sum_{1}^{p} x_i^{(p)} \cdot y_j^{(p)}$$

Because of the importance of the correlation weight matrix, one form of artificial associative memory uses it as the entire basis for the computational construct. This is the *Correlation Matrix Memory*.

6.3.1   Correlation Matrix Memories

Willshaw [38] proposed a potential method for the brain to store memories inspired by 'correlograms,' produced by shining light through two pieces of card with pinholes on. The resulting spots of light were captured on a third piece of card, and pertained to the correlation between the patterns on the first two cards. By capturing the correlation present on the third card, an inverted pattern could be used to retrieve the original pattern. Willshaw [38] then proposed a discretised form of correlogram, called an 'associative net' that captured this idea, but without the imperfections of an experimental setup. By presenting a series of input patterns and output patterns to the associative net simultaneously, the associations can be built up within the net and the patterns later retrieved by presenting one of the original patterns.

The idea of an associative net later developed into the binary Correlation Matrix Memory (CMM) [20]. The CMM is a matrix-based representation of a fully-connected two-layer neural network (one input layer, one output layer), where binary values in the matrix represent the binary weights on the connections between the two layers. As such, the neural network in Fig. 6.3 would be represented by the CMM, $\mathcal{M}$, with k input–output pairs $\mathcal{I}$ and $\mathcal{O}$:

$$\begin{pmatrix} \mathcal{I}_k \end{pmatrix} \begin{pmatrix} \mathcal{M} \end{pmatrix} \Bigg| \begin{pmatrix} a \\ b \\ c \end{pmatrix} \begin{pmatrix} ad & ae & af \\ bd & be & bf \\ cd & ce & cf \end{pmatrix}$$

$$\begin{pmatrix} \mathcal{O}_k \end{pmatrix} \Bigg| \begin{pmatrix} d & e & f \end{pmatrix} \tag{6.1}$$

Before training, the initial matrix $\mathcal{M}$ would be filled with zeros (as there are no associations stored in the network). As the k binary-valued input–output pairs are presented to the network, the associations are built up in the matrix.

Upon recall, we get [18]:

$$\mathcal{O} = \mathcal{M}\mathcal{I}_r$$

where $\mathcal{I}_r$ is the input pattern, and $\mathcal{O}$ is the output from the network trained with associations stored in $\mathcal{M}$. The desired output pattern, $\mathcal{O}_r$ is currently combined with noise from the other patterns stored in the network, e_r, hence:

$$\mathcal{O} = \mathcal{O}_r + e_r$$
$$e_r = \sum_{\substack{k=1 \\ k \neq r}}^{N} (\mathcal{I}_k^T \mathcal{I}_r) \, \mathcal{O}_k$$

The output vector $\mathcal{O}$ can be thresholded to a appropriate level, which (depending on how saturated the network is) should leave the desired output vector $\mathcal{O}_r$. This process can be used as a method of generalisation, allowing the network to retrieve complete patterns from a noisy or distorted cue.

Because the association matrix, $\mathcal{M}$, is a binary matrix (only allowing 0 or 1), the network can also be represented by a grid of wires, with connections between horizontal and vertical wires representing the 1 s in $\mathcal{M}$, and hence the associations stored in the network (see Fig. 6.4).

The binary nature of the CMM lends itself to be efficiently implemented using RD chemistry. The propagating waves in RD chemistry can be used to implement the binary signals (0/1 encoded as absence/presence of a wave) that are required to perform training and recall in the CMM. Interactions between the input waves and correlation matrix can allow recall to occur without external influence. Section 6.5 discusses how the correlation matrix, $\mathcal{M}$, may be stored in an RD memory structure.

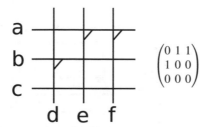

Fig. 6.4 A CMM can be viewed as an electrical circuit, with associations between input–output pairs represented by the connections between input and output wires (*left*). The matrix representation of this diagram is given as the correlation matrix on the *right*

6.4   Methods

Given the speed of the chemical reaction, we have simulated the dynamics of the BZ reaction instead of using a wet lab setup. A large number of authors have taken this approach previously, including some of the early papers by Rovinsky [30] who give a mathematical model of the dynamics alongside the chemical definition.

6.4.1   Simulation

We simulate a diffusive chemical reaction using the Oregonator model [14]:

$$\frac{\partial u}{\partial t} = \frac{1}{\varepsilon}\left[u - u^2 - (fv + \phi) \cdot \frac{u - q}{u + q}\right] + D_u\nabla^2 u$$

$$\frac{\partial v}{\partial t} = u - v$$

(6.2)

with parameter values as given in Table 6.1. The parameters (ε, f, ϕ, q, D_u) are described in detail in [19], but (briefly): ε is a scaling factor, f is the stoichiometric coefficient (a conservation of mass parameter), q is a propagation scaling factor, ϕ is representative of the illumination, and D_u is the diffusion coefficient for the solution. These are fixed over the course of the reaction and reactor, other than ϕ, which varies spatially across the reactor in order to construct 'circuits.'

The diffusion term ($D_x\nabla^2 x$ for chemical x) is provided for the activator u, but not for inhibitor v. This is because the simulation assumes that v is bound to an immobile substrate, such as a chemical gel. Equation 6.2 are integrated using an explicit forward Euler integration, with timestep $\delta t = 0.001$ and five-node discrete Laplace operator with grid spacing $\delta x = 0.25$. Waves are started through the simultaneous stimulation of 15 adjacent pixels by setting their activator concentration (u) to 1.0.

Table 6.1 Parameter values for excitable Belousov–Zhabotinsky simulation

Parameter	Value
ε	0.0243
f	1.4
ϕ_{active}	0.054
$\phi_{passive}$	0.0975
q	0.002
D_u	0.45
δt	0.001
δx	0.25

6.4.2 Extracting Results

Two different methods are used to extract results from the simulation. The first is to use time-lapse images produced by the simulation to visually check the presence or absence of a wave in a given channel; the second is through the use of 'pixel traces'.

Time-lapse images were produced by thresholding the activator concentration at $u = 0.04$ every 1000 simulated time-steps. All points above $u = 0.04$ were marked as white, while the rest left as they were (see Fig. 6.5).

The idea behind pixel traces, on the other hand, is to record the concentration values of a small set of pixels throughout the entire run of the simulation. This record of chemical concentrations can then be used to produce a graph, showing how the concentrations vary over time. Alternatively, if many runs are being performed, and the arrival of a single wave is all that matters, the activator concentration can be thresholded to an appropriate value, and used to determine the approximate arrival time of the wave. The graph presented in Fig. 6.6 shows an example pixel trace extracted from the simulator, and the threshold point. All time values with concentration above this threshold are then extracted, and the centre of the 'bump' is calculated, as an approximation of the waves arrival time at the specified pixel. For the work presented, this threshold value is held constant at $u = 0.54$.

Fig. 6.5 Example time-lapse output from the simulator. The activator concentration is thresholded at $u = 0.04$ every 1000 simulated time-steps, and marked as *white*. The illuminated (passive) regions of reactor (high ϕ) are signified by *grey* background, and non-illuminated (active) regions of reactor (low ϕ) are *black*

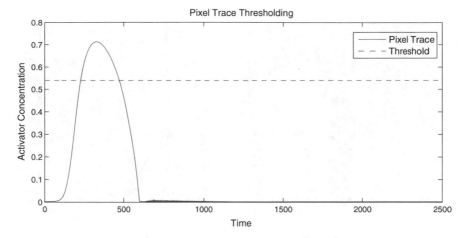

Fig. 6.6 Graph showing the pixel trace values over time (*solid line*) and threshold point (*dashed line*) for approximating wave arrival time

6.5   Isolated CMM Neuron

While the basic structure of the CMM is a matrix of binary neurons, we first consider the construction of an individual binary neuron and its connections with other neurons in the matrix. There already exist a number of physical implementations, including the use of a simple RAM device [8]. This use of a memory device to store the neural network was the starting point for the construction of the CMM presented here.

Motoike [26] showed how memory cells can be constructed in RD chemistry using a unidirectional ring of catalytic channels. By arranging the catalyst such that an external signal can excite the channels, or annihilate an existing excitatory wavefront, the memory cell can perform set/reset operations. Gorecki [17] proposed a method of separating the reset and read operations that were previously linked. An early design for this is given in Fig. 6.7, where the ring round the edge contains the state of the memory cell and the 'S' shape in the centre can be used as a reset signal.

A side-effect of this particular design of memory cell is that by taking an output from one point in the ring, it can be used to produce a periodic signal (as the wave travels round the ring, it will pass the output point at a rate proportional to the internal perimeter of the ring).

By storing the value of the CMM neuron in a memory cell in this way (i.e. if there is a connection in the matrix at that point, the memory cell is set to 1), then training a CMM is just a case of setting the corresponding memory cells.

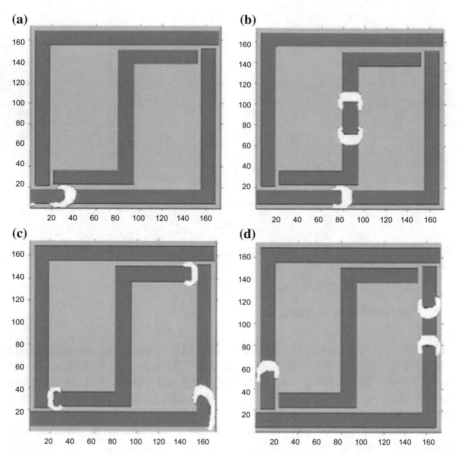

Fig. 6.7 Basic (simulated) memory cell. The wave propagates round the outer ring, in one direction only, but can be cleared by stimulating the S-shaped structure in the centre. From [17]

6.5.1 Requirements

The following requirements were identified as the basis for the CMM design:

Req. 1: The system must implement functionality of a CMM neuron.

Req. 1a: The neuron must maintain an internal state, $s_i \in \{0, 1\}$.

Req. 1b: The neuron must output $s_i \wedge x$ for some binary input $x \in \{0, 1\}$.

Req. 1c: The neuron must allow the state, s_i to be set to 1 (for training).

Req. 2: The neuron must allow the input (output) of one neuron to be passed to the next as input horizontally (output vertically).

6.5.2   Design

Figure 6.8 shows the initial logical design of the circuit. The circuit consists of a single memory cell, used to store the internal state (s_i) of the neuron. This fulfills Req. 1a. It is also used to provide a periodic pulse to coincidence detector (a), when trained to value 1. The second input to coincidence detector (a) is the input signal x, which corresponds to an individual binary element of the input vector ℑ. The output of detector (a) is then $s_i \wedge x$, and so fulfills Req. 1b. Finally, on coincidence of input signal x and training signal z (which corresponds to an individual binary element of the output vector ℴ) the second coincidence detector (b) sets the memory cell to 1, satisfying Req. 1c. From the fulfilling of all three sub-requirements, the system can then be said to fulfill Req. 1.

The cascading of inputs horizontally is achieved by splitting the x input value so that it feeds into both the coincidence detectors in the neuron and to the next neuron in the row. The z input in the design is representative of the vertical cascade functionality. This is used to pass the training signals through to each neuron and also to pass through the outputs from those neurons above this in the column to the bottom of the matrix. These points fulfill Req. 2.

In order to implement the vertical (train/output) cascading without interfering with the logic of the neuron, it has been moved from the centre to the side, leaving space for the memory cell to be compactly placed in the centre of the neuron (see Fig. 6.9).

During training, only an input–output pair of (1, 1) will load the memory cell, because coincidence detector (b), which is used for training, implements a logical-AND operation.

During recall, given an input of $x = 1$, if the memory cell is loaded (i.e. if there exists a link at this point in the matrix), then this cell will produce an output wave. This is then fed down to the output channel, marked z. This output channel feeds into the top of the cell below it in the matrix, as a means of propagating all the values from that column down to the thresholding logic that will be present at the bottom of the matrix. Because all the waves in any particular column will be produced

Fig. 6.8 Diagram showing the logical design of a single CMM neuron. The loop on the right implements a memory cell (with state $s_i \in \{0, 1\}$), the input (output) path is represented by x (z), and is used for both training and recall, and the original value of x (z) is passed through to the next neuron in the row (column)

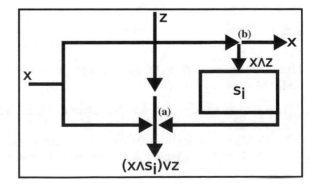

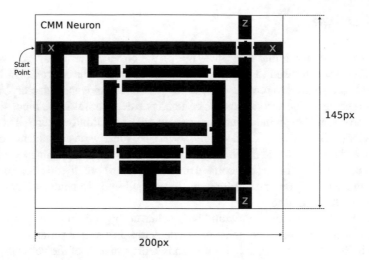

Fig. 6.9 Chemical layout for single CMM neuron. There are two coincidence detectors (AND-gates) in this design, the first is above the memory cell and is used during training ($x \wedge z$, detector (b) in Fig. 6.8). The second is below the memory cell and is used during recall ($x \wedge s_i$, detector (a) in Fig. 6.8)

simultaneously, the waves in the z-channel will not interfere with each other in any way. They also cannot interfere with the x-channels, as by the time an output wave is generated, the x-value will already have propagated to the next column.

Because of the way the neuron has been designed, we need to include a further requirement to ensure the design is scalable:

Req. 2a: The periodicity of the memory cell output must match the time to cascade horizontally.

This requirement means that the cascaded inputs will match with the periodic output of the memory cell across the entire row of neurons (i.e. the inputs to each neuron will arrive at the same point in the memory cell's periodic cycle).

6.5.3 Testing

Three tests were identified to ensure the proposed design was sufficient for use as a CMM neuron:

Test 1: Test periodicity of memory cell output is equal to time for cascading input horizontally.
Test 2: Test recall logic works for two neurons in each direction.
Test 3: Test training logic works for two neurons in each direction.

Test 1 requires precise measurements of the periodicity of the memory cell output and of the cascading input. To implement this test, two neurons were wired up horizontally, and a periodic boundary implemented on the far-side of the second neuron. This means that when the input propagates through the second neuron, it will reappear at the input of the first neuron. In doing so, the input becomes a periodic signal that can be compared to the signals from the memory cell output.

Test 2 can be implemented alongside Test 1, as a side effect of having a memory cell loaded and providing an input signal. Provided the first neuron produces output at every input signal, and the second produces no output, then the logic is correct.

Tests 1 and 2 were measured through pixel traces at locations $(98, 44)$ and $(98, 121)$ and time-lapse images.

Test 3 required training signals to be provided to the neurons from both the top and left. As this cannot be implemented alongside the other tests, it was set up separately. The test was looking for whether the coincidence of training signals will load the memory cell, and also tested that the output of said memory cell will coincide with the input from a recall signal. On top of this, the test needed to check for whether another training signal will 'overload' the memory cell, i.e. loading the cell with two waves instead of one.

Test 3 was measured using time-lapse images.

6.5.4   Results

Figure 6.10 shows the pixel traces for both memory cell and input cascade, for Test 1. The times that the waves pass the measurement points line up exactly for every other memory cell output—the semi-periodic nature of the x signal is due to there being two neurons wired up, so the signal takes twice as long to reach the measurement point.

The recall logic of the CMM neuron worked as expected, as can be seen from the time-lapse image produced towards the start of the run, in Fig. 6.11.

Finally, the training logic of the CMM also worked as expected, as can be seen from the time-lapse images in Fig. 6.12.

At this point, all three tests have been shown to be a success and the CMM neuron can now be considered as fulfilling its requirements.

6.6   CMM Thresholding

The next stage in the development of a complete CMM network is the ability to threshold the output. One of the main strengths of associative memory is the ability to generalise from a noisy or incomplete input pattern and returning the complete pattern. As described in Sect. 6.3.1, the output pattern, $\mathcal{O}_r$ should contribute the most

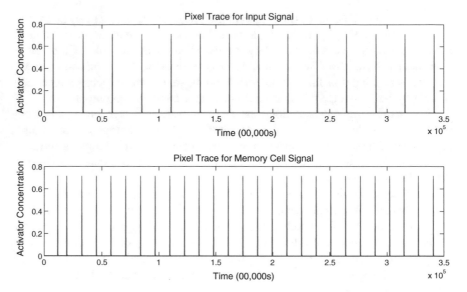

Fig. 6.10 *Top* graph shows semi-periodic input of x signal, *bottom* graph shows periodic output of memory cell. The spikes at the start of the *bottom* graph result from the initial stimulations

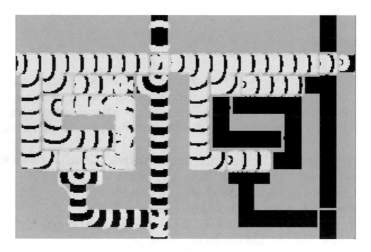

Fig. 6.11 Time-lapse image showing correctly implemented recall logic in CMM neuron. *Left* Neuron is trained to value 1, and upon recall outputs value 1. *Right* Neuron is trained to value 0, and upon recall outputs value 0

to the output of the network, with just a comparatively small noise term that needs to be cancelled out.

One method for achieving this is to input a series of waves (corresponding to the integer value of the threshold) up the output channels of the matrix. This will have the effect of annihilating that number of waves in each channel. Any waves remaining will propagate to the output, and should represent $\mathcal{O}_r$.

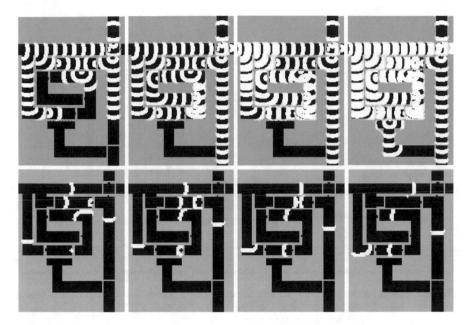

Fig. 6.12 Time-lapse images (*top*) showing CMM neuron training, and failure to 'overload' the neuron (where more than one value is stored in the memory cell at the same time (images are multiples of 10,000 simulated time steps). Images on the *bottom* show snapshots (at 500 simulated time step intervals) of the same behaviour, showing how the refractory period of the wave in the cell prevents a further wave from starting

6.6.1 Requirements

The following were identified as requirements for an appropriate thresholding circuit:

Req. 1: The system must generate the specified number (θ) of waves as output.
Req. 1a: The system must maintain an internal state representing the number of waves outputted.
Req. 2: The system must allow extraneous (i.e. $>\theta$) output waves from the CMM to pass through uninhibited.

6.6.2 Design

Figure 6.13 shows the basic logical design proposed to fill these requirements. The memory cell is used to produce the periodic output required from the circuit, while the binary counter keeps track of how many waves have been produced, fulfilling Req. 1a. Once the binary counter reaches threshold, it inhibits the production of further waves by clearing the memory cell, fulfilling Req. 1.

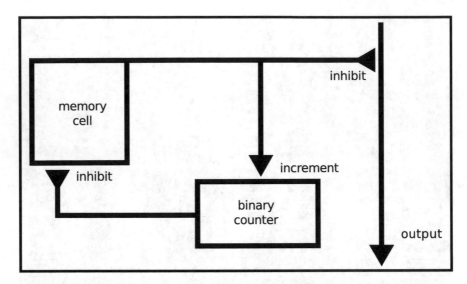

Fig. 6.13 Design for thresholding logic with threshold value θ. The circuit should produce θ waves, which are used to inhibit the output line. Once the counter reaches θ, it produces a single wave that inhibits the memory cell, stopping the output of waves

Figure 6.14 shows the chemical design (for threshold $\theta = 4$). By inserting a diode junction [6] on the output channel, the circuit fulfills Req. 2, as the thresholding waves will propagate up the output channel, until the appropriate threshold has been reached, at which point the output waves will propagate past the thresholding logic to the output.

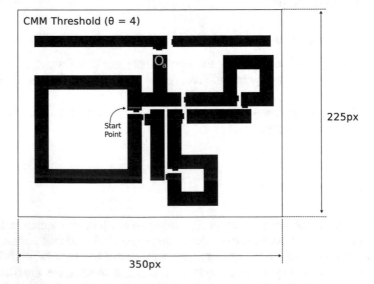

Fig. 6.14 Reaction layout for thresholding logic with threshold value $\theta = 4$. The two memory cells on the *right* and *bottom* are the binary counter, and the large memory cell on the *left* produces the output waves to inhibit the output line (across the *top*)

6.6.3   Testing

Test 1: Test number of waves outputted from thresholding logic matches threshold value, θ.

Test 2: Test number of waves outputted from CMM after thresholding matches number produced less threshold value.

Both tests can be performed simultaneously, by hooking up a series of five CMM neurons to a threshold circuit with $\theta = 4$. If both tests are successful, the circuit should output a single wave after thresholding. These tests can be measured using pixel traces, counting the number of waves produced on the output of the threshold logic, and then visually counting the number of waves produced on the post-threshold output channel.

Figure 6.15 shows the pixel trace for the threshold circuit output, clearly showing four waves at regular intervals, as expected. Figure 6.16 shows a time-lapse image with a single output wave on the output channel. At this stage, each test has been passed successfully, and the thresholding logic can now be considered as fulfilling its requirements.

6.7   CMM Training

The final checks to perform on a column of CMM neurons is to ensure that the training mechanism functions correctly. While Req. 1c in Sect. 6.5 showed that an individual neuron can be set to 1, this did not show how multiple neurons could be trained to a specific binary pattern.

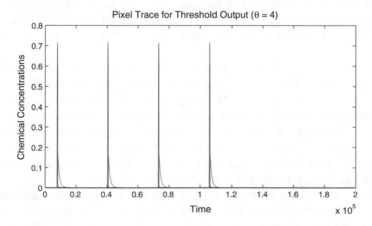

Fig. 6.15 Graph showing correct output from threshold circuit, with four periodic waves produced

Fig. 6.16 Time-lapse image showing single output wave on output channel, after thresholding. *Note* The image has been rotated from the original definition of a CMM, so this represents a single column of neurons

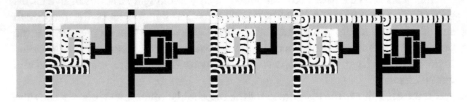

Fig. 6.17 A single column (rotated) of a correlation matrix, successfully trained to the binary value of 21 (10110)

The first test will be to check that a column of neurons is able to be trained. In order to achieve this, a column of five CMM neurons have been wired up, as in Fig. 6.16, but without the threshold circuit (for simplicity). These will be trained to the binary pattern that encodes 21 (10110).

Five waves are sent down the z-channel of the column, with a periodicity such that the waves arrive at the neurons at the same time as each other (one wave every 15,750 simulated timesteps). When all five had been sent, the x-channels of the column were stimulated with the binary pattern to be stored (at simulated time $t = 77,000$). For the training to be successful, the memory cells pertaining to the 1s in the stored pattern will be set, and no others.

As can be seen from the time-lapse image in Fig. 6.17, the first, third and fourth memory cells (from the top of the column) have been set to 1, and no others. This shows that the training stage for a single pattern in a single column has been successful.

The next step in testing the single column is to store a second pattern in the pre-trained column of CMM neurons. At present, three of the neurons (1, 3, 4) have been set. By subsequently storing the binary pattern encoding 5 (00101), the column should only set the final (fifth) neuron in the column, so that the final neurons set are (1, 3, 4, 5). This pertains to the logical OR of the first and second patterns (10111). If any of the pre-set neurons are affected, then this test will have failed.

As can be seen from the time-lapse image in Fig. 6.18, the training has been successful and the appropriate neurons are set, as anticipated.

The next stage in the training testing is to ensure that multiple columns can be trained with different patterns. This will be achieved by connecting up two columns of CMM neurons, and training the first (left-most) column with the binary pattern encoding 25 (11001) and the second column with the binary pattern encoding 21 (10110, as before).

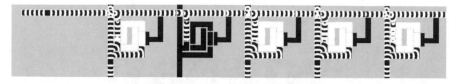

Fig. 6.18 A single column (rotated) with binary value of 21 (10110) already stored, is successfully trained with the binary value of 5 (00101), resulting in the superposed value 10111 stored in the column

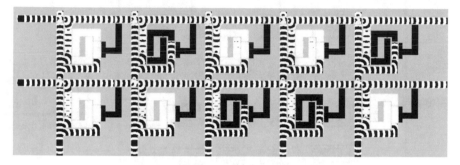

Fig. 6.19 Two columns (rotated) are successfully trained with distinct patterns, without interference or crosstalk between the patterns. First pattern (*bottom*) is the binary value of 25 (11001); second pattern (*top*) is the binary value of 21 (10110)

As can be seen from Fig. 6.19, the multi-column training was successful, with each column successfully trained without interference from the other column.

At this point, it is evident that the mechanisms provided for training a matrix of CMMs are sufficient. The next stage is to check the recall from a trained network of CMMs.

6.8   Full CMM Networks

Before we can be sure we have constructed associative memory, we need to ensure a network of CMM neurons will perform the computation we expect it to. The design decisions made when considering the individual neuron, and the testing that has already been performed on the thresholding and training aspects of the network, ensure that this is straightforward.

The testing in this section will consist of all the major stages required to construct a fully-functional CMM network. The network will be trained on a number of patterns. It will then be presented with a noisy version of one of the patterns, and the response thresholded to retrieve the answer. These tests will be performed on a 6×4 matrix of CMM neurons, with threshold $\theta = 2$. The patterns stored in the matrix will be

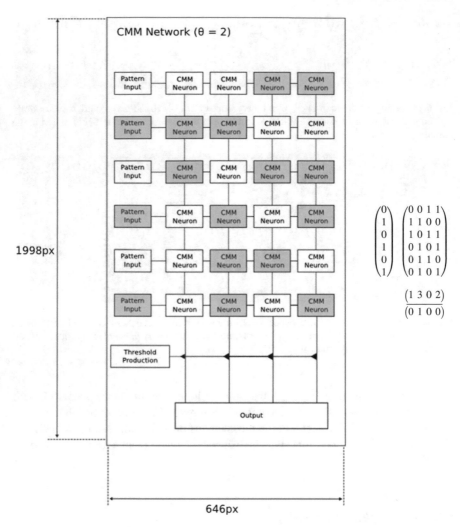

Fig. 6.20 *Left* Logical construction of a trained 6 × 4 CMM, with four patterns (011000, 010111, 101010 and 101101) stored, and input pattern 010101). *Right* Matrix representation of the same CMM, including expected results after thresholding at $\theta = 2$

the binary representations of 24, 23, 42 and 45. Upon presentation of the binary representation of 21, the network should respond with 23.

Figure 6.20 shows the logical (trained) construction of this network, where the grey and white boxes represent 1 and 0 respectively.

As is evident from Fig. 6.21, the training phase successfully stored the four patterns in the network.

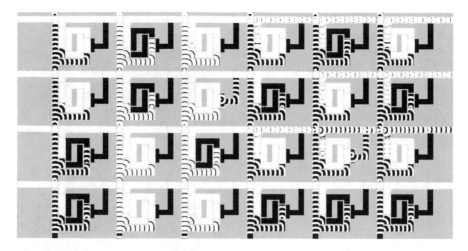

Fig. 6.21 Time-lapse image of the 6 × 4 CMM network (rotated) after training with the four patterns (011000, 010111, 101010 and 101101)

The final phase is to check the recall and generalisation of the network. The threshold circuit is started, and the input pattern (010101) presented to the network. As is evident from Fig. 6.22, the network responded with the second column, which corresponds to the number 23. From this, it can be seen that the circuit successfully implements a CMM-based associative memory.

6.9   Conclusions and Future Work

Reaction-diffusion chemistry and diffusive computation could offer a viable alternative to traditional approaches to computer science, but are generally very difficult to program. By providing a method of implementing different forms of neural network (spiking neurons previously [34]) and associative memory herein, we offer different methods of encoding problems in RD chemistry using paradigms that are more well-known.

The work presented implements a binary correlation matrix memory, and shows that the memory exhibits that same behaviour in RD chemistry as it would in traditional substrates. One of the main benefits of using binary CMMs is that training the network doesn't require altering the circuit in any way, just requires the setting of a series of memory cells. This is much simpler than training the form of spiking neuron we proposed previously [34], although a simpler spiking neuron implementation has been proposed [29] but requires a more implicit representation to that proposed by the current authors.

We suggest that diffusive computation, such as reaction-diffusion chemistry, has many further applications, and through the use of silicon-based diffusive proces-

Fig. 6.22 Final time-lapse
of the 6 × 4 CMM network.
The time-lapse was cleared
at simulated time 575000, to
make the results clearer. The
input pattern (010101) gives
a single output after
thresholding (second
column) which pertains to
the value 010111

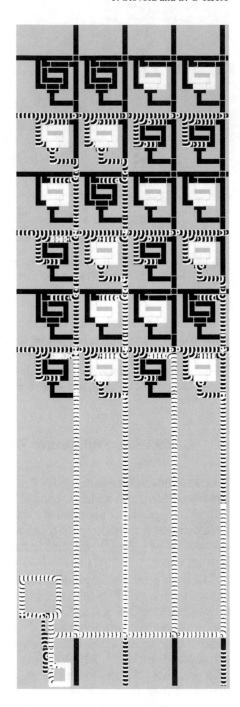

sors [7] these systems could potentially be used alongside traditional computing systems, such that those problems that are amenable to solution by diffusive computation can be offloaded to the diffusive processor for efficient processing.

References

1. Adamatzky, A.: Computing in Nonlinear Media and Automata Collectives. IoP Publishing, Bristol, UK (2001)
2. Adamatzky, A.: If BZ medium did spanning trees these would be the same trees as Physarum built. Phys. Lett. A **373**(10), 952–956 (2009)
3. Adamatzky, A., De Lacy Costello, B., Asai, T.: Reaction-Diffusion Computers. Elsevier Science, Amsterdam (2005)
4. Adamatzky, A., Wuensche, A., De Lacy Costello, B.: Glider-based computing in reaction-diffusion hexagonal cellular automata. Chaos, Solitons Fractals **27**(2), 287–295 (2006)
5. Adleman, L.: Molecular computation of solutions to combinatorial problems. Science **266**(5187), 1021–1024 (1994)
6. Agladze, K., Aliev, R.R., Yamaguchi, T., Yoshikawa, K.: Chemical diode. J. Phys. Chem. **100**, 13,895–13,897 (1996)
7. Asai, T., De Lacy Costello, B., Adamatzky, A.: Silicon implementation of a chemical reaction-diffusion processor for computation of Voronoi diagram. Int. J. Bifurc. Chaos **15**(10), 3307–3320 (2005)
8. Austin, J., Stonham, T.J.: Distributed associative memory for use in scene analysis. Image Vis. Comput. **5**(4), 251–260 (1987)
9. Belousov, B.P.: A Periodic Reaction and Its Mechanism. Med Publ, Moscow (1959)
10. Carpenter, G.A.: Neural network models for pattern recognition and associative memory. Neural Netw. **2**(4), 243–257 (1989)
11. Conrad, M.: The brain-machine disanalogy. Biosystems **22**(3), 197–213 (1989)
12. Conrad, M., Zauner, K.: Conformation-based computing: A rationale and a recipe. In: Sienko, T., Adamatzky, A., Rambidi, N., Conrad, M. (eds.) Molecular Computing, chap 1. MIT Press, Massachusetts, pp. 1–31 (2003)
13. Dolnik, M., Marek, M., Epstein, I.R.: Resonances in periodically forced excitable systems. J. Phys. Chem. **96**(8), 3218–3224 (1992)
14. Field, R.J., Janz, R.D., Vanecek, D.J.: Composite double oscillation in a modified version of the Oregonator model of the Belousov-Zhabotinsky reaction. J. Chem. Phys. **73**(7), 3132–3138 (1980)
15. Gilreath, W.F., Laplante, P.A.: Historical review of OISC. In: Computer Architecture: A Minimalist Perspective. The Kluwer International Series in Engineering and Computer Science, vol. 730, pp. 51–54. Springer, US (2003)
16. Gorecki, J., Yoshikawa, K., Igarashi, Y.: On chemical reactors that can count. J. Phys. Chem. A **107**(10), 1664–1669 (2003)
17. Gorecki, J., Gorecka, J., Igarashi, Y.: Information processing with structured excitable medium. Nat. Comput. **8**, 473–492 (2009)
18. Haykin, S.: Neural Networks: A Comprehensive Foundation, 1st edn. Prentice Hall PTR, Upper Saddle River (1994)
19. Holley, J., Adamatzky, A., Bull, L., De Lacy Costello, B., Jahan, I.: Computational modalities of Belousov–Zhabotinsky encapsulated vesicles. ArXiv e-prints (2010)
20. Kohonen, T.: Correlation matrix memories. IEEE Trans. Comput. **C-21**(4), 353–359 (1972)
21. Kuhnert, L.: A new optical photochemical memory device in a light-sensitive chemical active medium. Nature **319**(6052), 393–394 (1986)
22. Kuhnert, L.: Photochemische manipulation von chemischen wellen (in German). Naturwissenschaften **73**, 96–97 (1986)

23. Kuhnert, L., Agladze, K.I., Krinsky, V.I.: Image processing using light-sensitive chemical waves. Nature **337**(6204), 244–247 (1989)
24. Laplante, J., Pemberton, M., Hjelmfelt, A., Ross, J.: Experiments on pattern recognition by chemical kinetics. J. Phys. Chem. **99**(25), 10,063–10,065 (1995)
25. Lázár, A., Noszticzius, Z., Farkas, H., Försterling, H.D.: Involutes: the geometry of chemical waves rotating in annular membranes. Chaos **5**(2), 443–447 (1995)
26. Motoike, I., Yoshikawa, K.: Information operations with an excitable field. Phys. Rev. E **59**, 5354–5360 (1999)
27. Müller, U.: Brainfuck—an eight-instruction Turing-complete programming language. http://www.muppetlabs.com/~breadbox/bf/ (1993)
28. Nakagaki, T., Yamada, H., Tóth, Á.: Path finding by tube morphogenesis in an amoeboid organism. Biophys. Chem. **92**(1–2), 47–52 (2001)
29. Rambidi, N.: Chemical-based computing and problems of high computational complexity: the reaction-diffusion paradigm. In: Sienko, T., Adamatzky, A., Rambidi, N., Conrad, M. (eds.) Molecular Computing, chap 4, pp. 91–152. MIT Press, Massachusetts (2003)
30. Rovinsky, A.B., Zhabotinsky, A.M.: Mechanism and mathematical model of the oscillating bromate-ferroin-bromomalonic acid reaction. J. Phys. Chem. **88**(25), 6081–6084 (1984)
31. Steinbock, O., Tóth, Á., Showalter, K.: Navigating complex labyrinths: optimal paths from chemical waves. Science **267**(5199), 868–871 (1995)
32. Stepney, S.: Unconventional computer programming. In: Symposium on Natural/Unconventional Computing and Its Philosophical Significance (2012)
33. Stepney, S., Abramsky, S., Adamatzky, A., Johnson, C., Timmis, J.: Grand challenge 7: journeys in non-classical computation. In: Visions of Computer Science, London, UK, September 2008, pp. 407–421 (2008)
34. Stovold, J., O'Keefe, S.: Simulating neurons in reaction-diffusion chemistry. In: Lones, M., Smith, S., Teichmann, S., Naef, F., Walker, J., Trefzer, M. (eds.) Information Processing in Cells and Tissues. Lecture Notes in Computer Science, vol. 7223, pp. 143–149. Springer, Berlin (2012)
35. Tolmachiev, D., Adamatzky, A.: Chemical processor for computation of Voronoi diagram. Adv. Mater. Opt. Electron. **6**(4), 191–196 (1996)
36. Tóth, Á., Showalter, K.: Logic gates in excitable media. J. Chem. Phys. **103**(6), 2058–2066 (1995)
37. Turing, A.M.: The chemical basis of morphogenesis. Philos. Trans. R. Soc. Lond. Ser. B, Biol. Sci. **237**(641), 37–72 (1952)
38. Willshaw, D.J., Buneman, O.P., Longuet-Higgins, H.C.: Non-holographic associative memory. Nature **222**(5197), 960–962 (1969)
39. Zhabotinsky, A., Zaikin, A.: Autowave processes in a distributed chemical system. J. Theor. Biol. **40**(1), 45–61 (1973)
40. Zhabotinsky, A.M.: Periodic course of the oxidation of malonic acid in a solution (studies on the kinetics of Belousov's reaction). Biofizika **9** (1964)
41. Zhabotinsky, A.M.: A history of chemical oscillations and waves. Chaos: Interdiscip. J. Nonlinear Sci. **1**(4), 379–386 (1991)

Chapter 7
Calculating Voronoi Diagrams Using Chemical Reactions

Ben De Lacy Costello and Andrew Adamatzky

Abstract This chapter overviews work on the use of simple chemical reactions to calculate Voronoi diagrams and undertake other related geometric calculations. This work highlights that this type of specialised chemical processor is a model example of a parallel processor. For example increasing the complexity of the input data within a given area does not increase the computation time. These processors are also able to calculate two or more Voronoi diagrams in parallel. Due to the specific chemical reactions involved and the relative strength of reaction with the substrate (and cross-reactivity with the products) these processors are also capable of calculating Voronoi diagrams sequentially from distinct chemical inputs. The chemical processors are capable of calculating a range of generalised Voronoi diagrams (either from circular drops of chemical or other geometric shapes made from adsorbent substrates soaked in reagent), skeletonisation of planar shapes and weighted Voronoi diagrams (e.g. additively weighted Voronoi diagrams, multiplicatively weighted crystal growth Voronoi diagrams). The chapter also discusses some limitations of these processors. These chemical processors constitute a class of pattern forming reactions which have parallels with those observed in natural systems. It is possible that specialised chemical processors of this general type could be useful for synthesising functional structured materials.

7.1 Introduction

Voronoi diagrams are used to partition space into spheres of influence in many branches of science and can be observed in numerous natural systems. They find uses in many fields such as astronomy (galaxy cluster analysis [1]), biology (modelling tumour cell growth [2]), chemistry (modelling crystal growth [3]), ecology

B. De Lacy Costello (✉) · A. Adamatzky
Unconventional Computing Centre, University of the West of England, Bristol, UK
e-mail: ben.delacycostello@uwe.ac.uk

A. Adamatzky
e-mail: andrew.adamatzky@uwe.ac.uk

© Springer International Publishing Switzerland 2017 167
A. Adamatzky (ed.), *Advances in Unconventional Computing*,
Emergence, Complexity and Computation 23,
DOI 10.1007/978-3-319-33921-4_7

(modelling competition [4]) and computational geometry (solving nearest neighbour problems [5]). In addition to standard and generalised Voronoi diagrams weighted diagrams are commonly used for example to model crystal growth [6]. In simple terms an additively weighted Voronoi diagram (AWVD) is where all sources grow at the same rate but some start at different times. A multiplicatively weighted Voronoi diagram (MWVD) is where all sources start growing at the same time but some grow with different rates. The multiplicatively weighted crystal growth Voronoi diagram (MWCGVD) [6] is an extension of the MWVD and overcomes the problem of disconnected regions which are not possible in growing crystals (or expanding chemical fronts). Therefore, in the MWCGVD when regions with large growth rates meet regions with lower growth rates they can "wrap around" these regions to form a connected Voronoi diagram of the plane. The bisectors of weighted Voronoi diagrams are hyperbolic segments rather than straight lines. For a more in depth description and mathematical treatment of Voronoi diagrams and weighted derivatives see Refs. [6, 7].

In addition to their many uses in modeling, Voronoi diagrams can be observed in nature, for example, in animal coat markings [8] between interacting bacterial, fungal or slime mould colonies [9, 10] between growing crystals and even in universal structures such as gravitational caustics [11]. They were also observed in gas discharge systems [12]. The formation of Voronoi diagrams in chemical systems with one reagent and one substrate was first reported in [13]. This and similar reactions were utilised as chemical processors for the computation of the shortest obstacle free path [14], skeletonisation of a planar shape [15, 16] and construction of a prototype XOR gate [17]. The formation of weighted Voronoi diagrams from two reagents reacted on one substrate (potassium ferrocyanide gel) were first reported in [18]. In the same year the calculation of three Voronoi diagrams in parallel was reported in a chemical system based on the reaction of two reagents on a mixed substrate gel (potassium ferrocyanide and potassium ferricyanide) [19]. More recently complex tessellations of the plane have been reported where one of two binary reagents exhibited limited or no reactivity with the gel substrate [20]. It was shown that re-useable processors for calculation of Voronoi diagrams could be constructed using a crystallisation process [21]. However, these processors must be reset completely prior to carrying out sequential calculations. Subsequent to this it was found that by carefully selecting reagents it was possible to calculate two or more Voronoi diagrams sequentially on the same substrate. Sequential calculations are useful in comparing overlapping Voronoi regions.

A new class of pattern forming inorganic reaction based on the reaction of sodium hydroxide with copper chloride loaded gels was reported [22]. These reactions exhibited a range of self-organised patterns such as cardioid and spiral waves only observed previously in more complex chemical reactions such as the Belousov–Zhabotinsky (BZ) reaction. This is highly significant given the very simple nature of the chemical reactants.

More recently another inorganic system has been identified based on aluminium chloride gels reacted with sodium hydroxide [23]. This system is particularly remarkable as unlike the system mentioned above [22] the wave evolution can be easily observed in real-time.

In previous work we were able to exert control over another class of inorganic reaction with the same primary wave splitting mechanism [24]. The reaction involved the addition of copper chloride to potassium ferricyanide immobilised in an agarose gel. At high concentrations of potassium ferricyanide circular (cone shaped) waves are spontaneously generated by heterogeneities and expand. Where these waves collide they form a natural tessellation of the plane similar to those obtained in crystallisation processes. By reducing the concentration of potassium ferricyanide then the point and time at which circular waves are initiated could be controlled by marking the gel with a very fine glass needle. By using this methodology user defined precipitation patterns could be created that equate to Voronoi tessellations of the plane. In previous work [25] involving the aluminium chloride sodium hydroxide reaction we adopted the same strategy and were able to show that initiation of travelling waves was possible. However, marking the gel prior to initiation of the reaction produced complex travelling waves made up of multiple spiral wave fragments. In further work [26] it was established that marking the gel after initiation (pouring on the outer electrolyte) allowed selection of cardioid or circular waves, depending on the time after initiation. Therefore, the construction of Voronoi diagrams and additively weighted Voronoi diagrams in this system proved possible.

Section 7.2 of the chapter relates to the construction of Voronoi diagrams and related geometric calculations in stable chemical systems. Section 7.3 relates to the construction of Voronoi diagrams in unstable chemical systems. Section 7.4 compares the systems from Sects. 7.2 and 7.3 with other unconventional parallel processors, both chemical and biological. Section 7.5 gives some general conclusions.

7.2 Construction of Voronoi Diagrams in Stable Systems

7.2.1 Potassium Ferricyanide or Ferrocyanide Loaded Gels

For the purposes of this chapter we will deal mainly with reactions based on the reaction of potassium ferrocyanide or ferricyanide loaded gels reacted with various metal salts. However, a huge variety of inorganic reactions studied display this phenomena in the right concentration ranges (usually low substrate concentration and high outer electrolyte concentration). To undertake the reactions agar Gel (Sigma, St Louis, USA, 0.3 g) was added to deionised water (30 ml) and heated with stirring to 70 °C and then removed from the heat. Potassium ferricyanide (Fisher Scientific, Leicestershire, UK, 75 mg (7.59 mM)) or potassium ferrocyanide (BDH chemicals Ltd, Poole, UK, 75 mg (5.91 mM)) was added with stirring. The solution was then poured into five 9 cm diameter Petri dishes and left to set for one hour. All possible binary combinations of the metal ions listed below were reacted on both potassium ferrocyanide and ferricyanide gels. The metal ions used in construction of the Voronoi tessellations were as follows: iron (III) nitrate [300 mg/ml, 1.24 M] (can be substituted with chloride salt), iron (II) sulphate [300 mg/ml, 1.97 M], silver (III) nitrate [300 mg/ml, 1.76 M], cobalt (II) chloride hexahydrate [300 mg/ml, 1.26 M], lead (II) chloride [300 mg/ml, 1.07 M], manganese (II) chloride tetrahydrate [300 mg/ml, 1.51 M], chromium (III) nitrate nonahydrate [300 mg/ml, 1.26 M], nickel (II) chloride [300 mg/ml, 2.31 M] and copper (II) chloride [300 mg/ml, 2.23 M]. To form simple Voronoi diagrams drops of these metal ion solutions are carefully placed on the surface of the gel.

Figure 7.1a shows a Voronoi diagram constructed when ferric chloride was reacted with potassium ferrocyanide. The drops of ferric chloride were placed in three different regular patterns to highlight the versatility of this approach. Figure 7.1b shows a simple Voronoi diagram calculated in the reaction between cupric chloride and potassium ferrocyanide. It should be noted that in these chemical systems bisector width is proportional to the distance between initiation sources.

7.2.2 Palladium Chloride Loaded Gels

For the purposes of these experiments a reaction-diffusion processor based on palladium chloride was used. A gel of agar (2 % by weight) containing palladium chloride (in the range 0.2–0.9 % by weight, for the images displayed the lower range of 0.2 % was favoured) was prepared by mixing the solids in warm deionised water. The mixture was heated with a naked flame until it boiled with constant stirring to ensure full dissolution of palladium chloride and production of a uniform gel (on cooling). The liquid was then rapidly transferred to Petri dishes. The outer electrolyte used was a saturated solution (at 20 °C) of potassium iodide. Figure 7.2 shows a generalised Voronoi diagram constructed in the palladium chloride potassium iodide chemical processor.

7.2.3 Generalised Voronoi Diagram Construction

Actually all Voronoi diagram construction using chemical inputs in stable systems are generalised as the input is not a point source. However, circular drops of reagent construct an almost identical Voronoi diagram to that expected for a point source. In this section generalised refers to an input which is not circular.

It was found that filter paper soaked in the outer electrolyte could be used to initiate reaction fronts rather than drops of the outer electrolyte. Therefore, generalized Voronoi diagrams could be constructed, i.e. where the sources of wave generation are geometric shapes rather than point sources. In this case the bisectors are curved rather than straight lines. This kind of computation is not trivial for conventional computer processors. However, for the parallel chemical processors the complexity

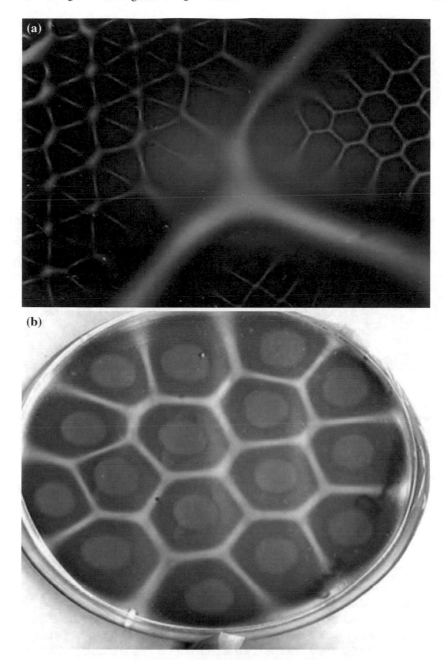

Fig. 7.1 Voronoi diagrams formed in stable chemical systems via the application of drops of reagents. **a** Voronoi diagram formed when ferric chloride was reacted with a potassium ferrocyanide loaded gel. The reactant drops (ferric chloride) were placed in various array patterns in different sections of the Petri dish. The bisectors with low precipitate concentration are clearly visible. **b** Voronoi diagram formed when copper chloride was reacted with a potassium ferrocyanide loaded gel. In this case the position of the original drops and the bisector formation can be observed

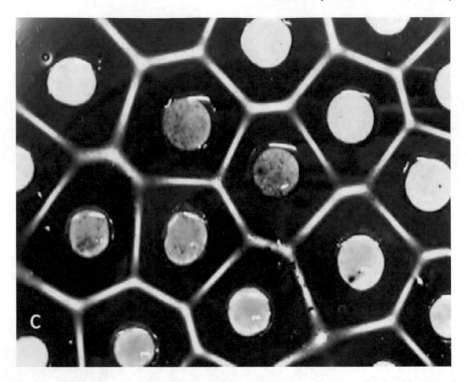

Fig. 7.2 Voronoi diagrams formed in stable chemical systems via the application of drops of reagents. Voronoi diagram formation when potassium iodide was reacted with a palladium chloride loaded gel. In this case the original positions of the drops and the bisectors appear as low precipitate regions, probably due to re-dissolution of the primary product in the outer electrolyte (in contrast to the ferricyanide/ferrocyanide based systems)

of the computational task is almost irrelevant. Figure 7.3 shows an example of a generalized Voronoi diagram formed on a potassium ferrocyanide gel.

7.2.4   Skeletonisation of a Planar Shape

This reaction can be used to calculate the internal Voronoi diagram (skeleton) of a contour. This is a useful data reduction method for image recognition etc. Figure 7.4 shows the skeletonisation of a geometric contour. The skeleton of a pentagon as calculated by the chemical reaction is a five pointed star. This is because as mentioned the distance between the opposed initiation fronts controls bisector width.

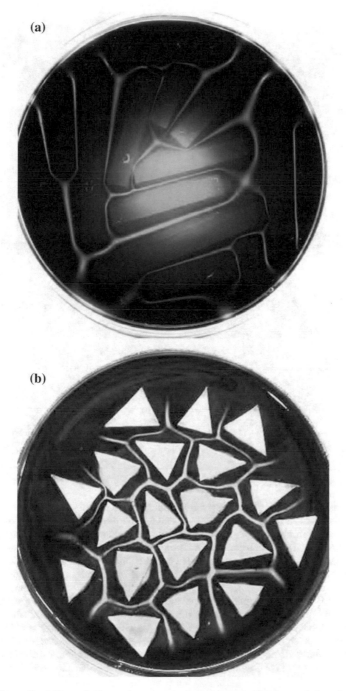

Fig. 7.3 Generalised Voronoi diagrams constructed from geometric shapes (other than circles). **a** Generalised Voronoi diagram formed when ferric chloride soaked filter paper was reacted with a potassium ferrocyanide loaded gel. **b** Generalised Voronoi diagram when potassium iodine soaked filter paper was reacted with a palladium chloride loaded gel

(a)   (b)

(c)   (d)

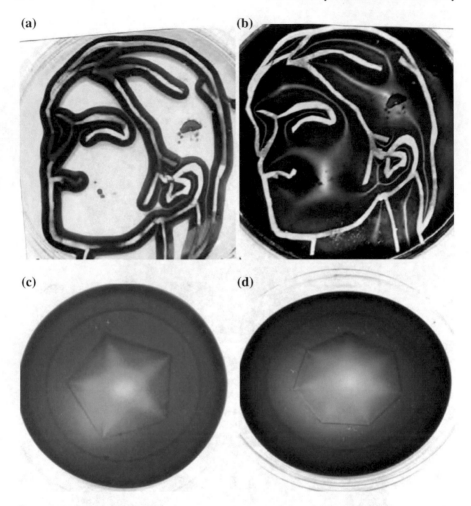

Fig. 7.4 Skeletonisation of planar shapes or contours using a similar technique to that used to form generalised voronoi diagrams in Fig. 7.3. **a** and **b** Show a contour soaked in potassium iodide reacted on a palladium chloride gel. The bisectors represent a data reduction of the original contour. As the chemical reactions give bisectors which vary in width according to the distance separating the reaction fronts of the contour then additional information is encoded in the computed skeleton. This is apparent in (**c**) and (**d**) which show skeletons formed when ferrous sulphate soaked filter paper was reacted with a potassium ferricyanide loaded gel. In this case the skeleton of a pentagon is a five pointed star and of a heptagon a 7 pointed star. It is apparent that the chemical processor more successfully constructs the skeleton of the pentagon. This is because eventually a geometric shape with more sides would approximate a circle in which case the skeleton would be a circular region of reduced precipitate. Thus increasing the angle between two sides and reducing the length of individual features affects the accuracy of the calculation. This is because chemical fronts do not maintain the shape of the original boundary over long distances but adopt a distance dependent curvature

7.2.5   Mixed Cell Voronoi Diagrams

7.2.5.1   Balanced Binary Voronoi Tessellations

Two or more different metal ion solutions can be reacted with the substrate loaded gel
in parallel to create a series of colourful tessellations of the plane. If the reagents have
the same diffusion rate and react fully with the substrate gel then a simple Voronoi
diagram is constructed (albeit that the cells are different colours). Figure 7.5 shows an
example of a mixed cell Voronoi diagram of this type. If the reagents had equal reactivity with the gel but differing diffusion rates then a multiplicatively weighted crystal growth Voronoi diagram (MWCGVD) would be constructed. Figure 7.6 shows

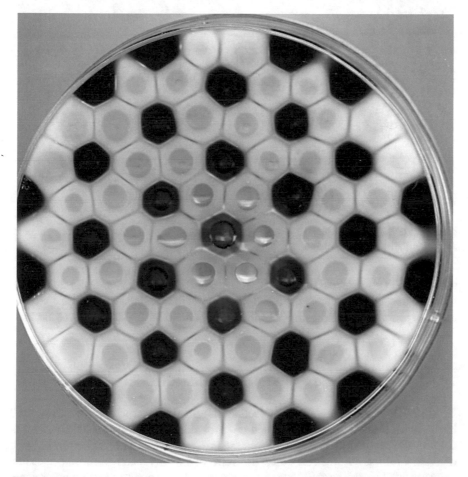

Fig. 7.5 Mixed cell Voronoi diagram formed when ferrous sulphate and manganese chloride were
reacted with a potassium ferrocyanide loaded gel

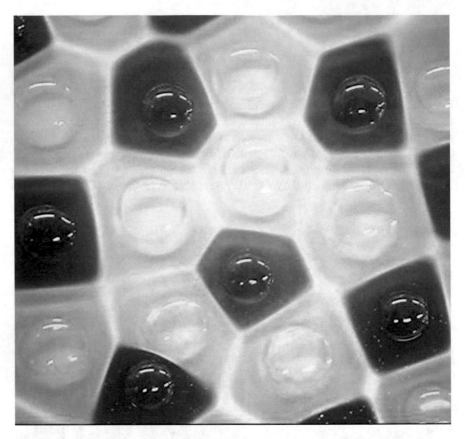

Fig. 7.6 Multiplicatively weighted crystal growth Voronoi diagram formed when cupric chloride and ferric chloride were reacted on potassium ferrocyanide gel

a MWCGVD constructed in this type of reaction. The bisectors are slightly curved between the cells formed via the reaction of cupric ions and the cells formed via the reaction of ferric ions indicating that the diffusion rate of the cupric ions is slightly faster.

7.2.5.2   Unbalanced Binary Voronoi Tessellations

Figure 7.7 shows an example of the tessellation formed when a partially reactive metal ion solution (ferric chloride) is reacted on a gel alongside a fully reactive metal ion solution (cobaltous chloride). In this reaction fronts are initiated from drops of both reactant solutions but when they meet the fronts of the fully reactive reactant cross the fronts formed by the partially reactive reagent and only annihilate (in a precipitate free region) when they meet fronts emanating from another fully reactive source. Therefore, the resulting tessellation is a Voronoi diagram of all initiation sites and a

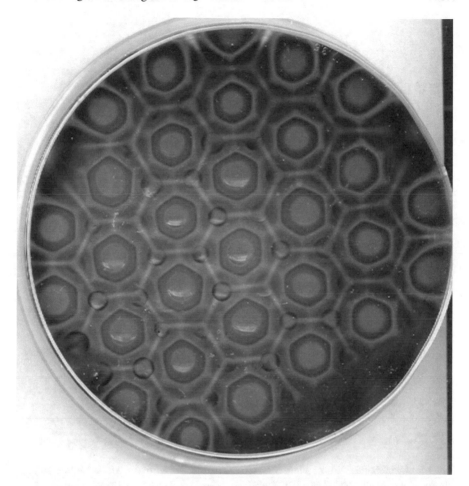

Fig. 7.7 Complex tessellation created via the reaction of ferric chloride and cobaltous chloride on potassium ferricyanide gel

Voronoi diagram of only sites containing the fully reactive reagent. So two Voronoi diagrams can be calculated in parallel using these simple chemical reactions.

7.2.5.3 Cross Reactive Binary Voronoi Tessellations

Figure 7.8 shows a fully cross reactive tessellation in progress. In this case both the primary reactants produce coloured cells and a simple Voronoi diagram. However, where the fronts meet both primary products are cross reactive and a secondary precipitate is formed where the fronts overlap. Figure 7.9 Shows the completed reaction where only a white precipitate remains across most of the Petri dish. However, what is remarkable is that the precipitate has a highly complex and ordered tessellation

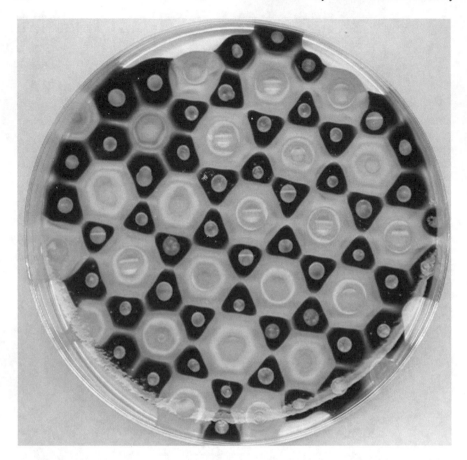

Fig. 7.8 A tessellation formed when two reagents with cross reactive products (ferrous sulphate and silver nitrate) are reacted on a potassium ferricyanide gel. The reaction is still in progress

made up of high and low precipitate regions. It is possible that this type of reaction may be useful in directed materials synthesis if the mechanism can be better understood.

7.2.5.4 Balanced and Unbalanced Tertiary Voronoi Diagrams

The number of distinct reagents reacted on the same substrate gel can be increased to three or more. Unlike the binary case where the limit of possibilities has been fully mapped, tertiary Voronoi tessellations have only been partially classified [27]. However, they fit into the same general types, balanced (all reagents are reactive equally with the gel), unbalanced (one or more reagent is partially reactive) and cross reactive. These tertiary tessellations could be computing MWCGVD depending on the relative diffusion and reaction rates of the primary reagents. What is certain is that

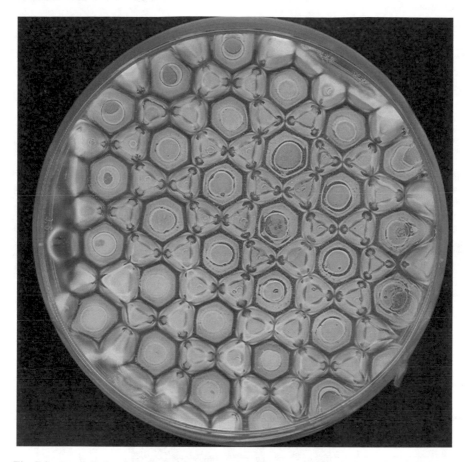

Fig. 7.9 A tessellation formed when two reagents with cross reactive products (ferrous sulphate and silver nitrate) are reacted on a potassium ferricyanide gel. Completed reaction

they produce an amazing array of colourful tessellations of the plane. Any design can be recreated provided that it can be subdivided into symmetrical regions which are spaced evenly (and allowing for partial and cross-reactivity, or incorporating them into the design). Figure 7.10 shows examples of this type of tessellation.

7.2.6   Tessellations Formed by Combining Two Exclusive Chemical Couples

Ferric ions do not react with potassium ferricyanide and ferrous ions do not react with potassium ferrocyanide. By using a mixed gel of potassium ferricyanide and ferrocyanide and reacting this with drops of ferric ions and ferrous ions a complex tessellation can be obtained, see Fig. 7.11. This tessellation equates to three separate

(a) (b)

(c) (d)

(e)

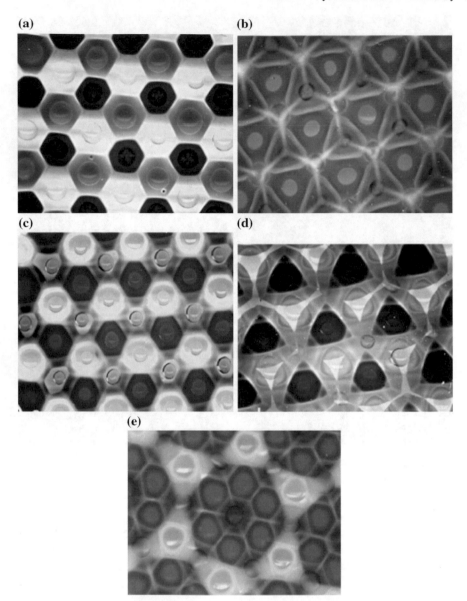

Fig. 7.10 Showing a range of tertiary tessellations (3 metal ion reagents reacted simultaneously on the substrate gel) of the plane. **a** Showing a balanced tessellation when ferrous sulphate (*blue*), nickel (II) chloride (*yellow*) and cobalt (II) chloride (*red*) was reacted on potassium ferricyanide gel. **b** Showing a tessellation formed where cobalt (II) chloride (*red*), iron (III) nitrate (partially reactive) and chromium (III) nitrate (unreactive were reacted on potassium ferricyanide gel. **c** Showing tessellation when ferric chloride (partially reactive), nickel (II) chloride and cobalt chloride were reacted on potassium ferricyanide gel. **d** Showing a cross reactive formed when iron (III) nitrate, copper (II) chloride and iron (II) sulphate were reacted on potassium ferricyanide gel. **e** Tessellation where cobalt chloride, nickel chloride and ferric chloride were reacted on potassium ferricyanide gel

Fig. 7.11 Tessellation formed when ferrous ions and ferric ions are reacted on a mixed potassium ferricyanide and potassium ferricyanide gel

Voronoi diagrams constructed in parallel. One pertaining to all reactant drops and two which are exclusive to each set of different reactant drops. These are calculated where both sets of fronts cross and only annihilate in a precipitate free region where they meet another front from their exclusive chemical couple.

7.2.7　Sequential Voronoi Diagram Calculation

Previously it had been assumed that these chemical processors were single use in terms of sequential parallel geometric computations. However, the careful selection of primary and secondary reactants based on their reactivity with the substrate gel, allows for sequential calculations to be undertaken (see Fig. 7.12). Thus a reagent with

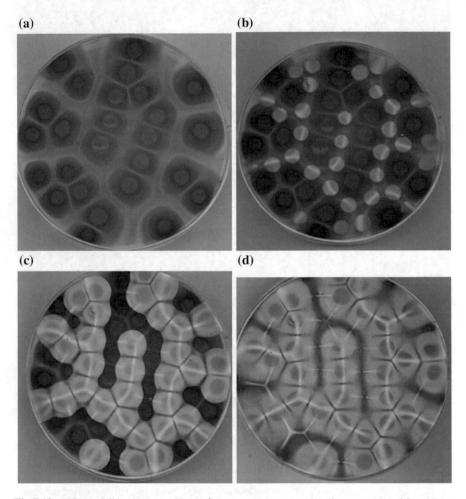

Fig. 7.12　a Voronoi diagram formed by Ag^+ ions reacted on potassium ferrocyanide gel. **b** Addition of secondary Co^{2+} reactant to the gel substrate containing the original Voronoi diagram. **c** It can be seen that a physical precipitation reaction is initiated c. 45 min into the reaction, Co^{2+} diffusing fronts interact to form secondary Voronoi diagram bisectors (precipitate free regions). **d** Final reaction showing formation of distinctive interacting Voronoi diagrams. The bisectors formed in these chemical reactions inherently code for distance. Therefore, points/sites separated by larger distances produce wide bisectors. This is useful when used in optimal path planning and minimal data reduction via the skeletonisation of planar shapes

partial reactivity with the substrate is first reacted to give a Voronoi diagram at which point subsequent points are added using a reactant with full reactivity (alternatively use of reagents whose products have mutual cross reactivity can be undertaken to achieve a similar result).

7.2.8 Speed of Computation

In these stable reactions the speed of Voronoi diagram calculation is relatively slow at the scales we have presented. The reason for conducting these experiments in this manner was to ensure a visual output to the computation, and ease of adding the reagents in parallel to the substrate. However, as mentioned the complexity of adding three or more reagents or calculating up to three Voronoi diagrams in parallel has no real affect on the computation time within a defined area. Furthermore, if the reagent density is increased and the input size decreased then the computation is much faster (of the order of a few seconds). This could presumably be reduced further by optimisation of the gel system and identifying a method such as inkjet printing to add nanolitre sized droplets in parallel to the substrate surface. Figure 7.13 shows a Voronoi diagram computed using a manually applied high drop density. The drop volume is calculated to be in the nanolitre range.

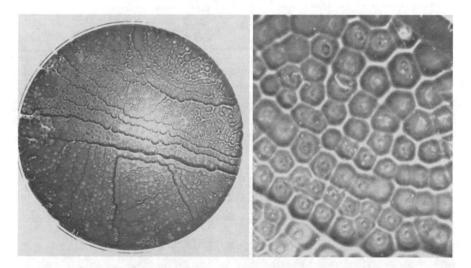

Fig. 7.13 Voronoi diagram calculated from hundreds of nanolitre sized droplets of copper chloride reacted on a potassium ferrocyanide gel. The computation was complete in under 30 s. This compares to a time of an hour or more for larger drops spaced according to the tessellations previously presented in the chapter

7.3   Formation of Voronoi Diagrams in Unstable Systems

7.3.1   Failure to Compute a Voronoi Diagram

Figure 7.14 shows the results of raising the potassium ferrocyanide concentration from 2.5 mg/ml (5.91 mM) to 7.5 mg/ml (17.73 mM). At this concentration the circular fronts can be observed to spontaneously split meaning that a complete Voronoi diagram is not formed. This splitting of the reaction front is qualitatively similar to the primary splitting mechanism described by Hantz [22] for the copper chloride sodium hydroxide reaction. In this phase of the reaction extended pattern formation

Fig. 7.14 Failure to compute Voronoi diagram due to chemical instability which disrupts the circular fronts

can be observed. Interestingly, Fig. 7.14 shows Voronoi bisectors calculated at points where no fronts meet (due to the instability in the front, causing the progress of one or more reaction fronts to cease) but should have met if the fronts had maintained stability. This does highlight the long range forces that must be acting in advance of the fronts.

7.3.2 Spontaneous Voronoi Diagram Formation in Cupric Chloride Potassium Ferro/Ferricyanide Systems

If the concentration of inner electrolyte is raised further to a concentration circa 30 mg/ml (70.92 mM) then circular reaction fronts do not form but where solution has been applied to the gel a number of small expanding circular waves can be observed. Because these waves represent a splitting of the front in three dimensions they are actually expanding cone shaped regions enclosing unreacted substrate. Where these cone shaped regions collide they form tessellations which when observed in 2D equate to Voronoi diagrams of the original points of instability, see Fig. 7.15.

7.3.3 Controlled Voronoi Diagram Formation in Cupric Chloride Potassium Ferro/Ferricyanide Systems

By reducing the concentration of potassium ferrocyanide or potassium ferricyanide in the gel a reaction where spontaneous wave formation was minimised was obtained. However, when the gel was pre-marked with a thin glass needle prior to initiation of the reaction this provided a large enough heterogeneity in the system after initiation (pouring on outer electrolyte) to selectively initiate a cone shaped wave from this point. Therefore, programmable Voronoi diagrams could be constructed in these unstable systems see, Fig. 7.16. For experimental details refer to [24]. Figure 7.17 shows a schematic of how Voronoi diagrams are formed from interacting precipitation fronts.

7.3.4 Controlling Wave Generation in the Aluminium Chloride Sodium Hydroxide Reaction

Lagzi and co-workers described the generation of moving self organised structures in a simple precipitation system [23]. Structures such as circular waves, cardioid and spiral waves can be observed at different concentration ranges of the reaction. For details of the experimental procedures used see [23–26]. We constructed a phase diagram for the reaction in order to identify whether there was a controllable phase in

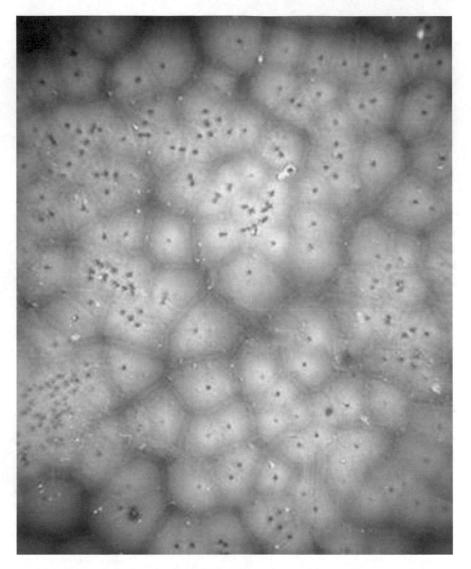

Fig. 7.15 Spontaneous "Voronoi diagram" formation when copper chloride solution 300 mg/ml (2.24 M) is poured over a potassium ferrocyanide gel (30 mg/ml, 70.92 mM)

this reaction. By choosing concentration ranges in the controllable region but close to the spontaneous pattern generation region of the phase diagram it was possible to initiate disorganised spiral waves from specific locations by marking the gel with a fused silica needle (diameter 0.25 mm) prior to initiation of the reaction, see Fig. 7.18. It did not prove possible to initiate circular waves selectively at any concentration ranges within this region using this technique. Therefore, we marked the gel at specific time intervals after the initiation of the reaction. Figure 7.19 shows a Petri dish

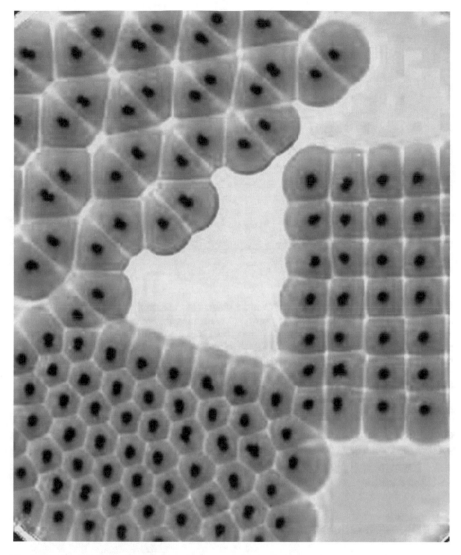

Fig. 7.16 Controlled Voronoi diagram construction in an unstable system based on the reaction of cupric chloride (2.93 M) with a potassium ferricyanide gel (0.09 M). The gel was pre-marked with the pattern using a glass needle (diameter 0.5 mm). Then the outer electrolyte solution of cupric chloride was poured over the gel to initiate the 3D precipitate waves

where a set of three points have been marked at increasing time intervals after the reaction has been initiated. What this shows is that there is a tendency towards single circular wave generation with increasing time interval after initiation. Therefore, if the gel is marked two minutes after initiation single circular waves can be generated consistently. Interestingly, it was possible to generate cardioid waves fairly repro-

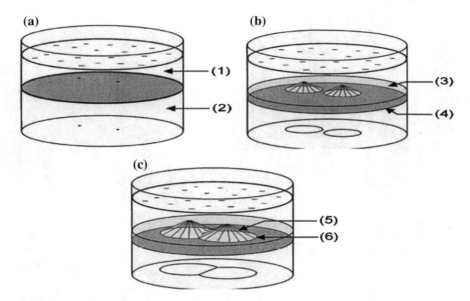

Fig. 7.17 Schematic representation of Voronoi diagram formation through the "regressing edge mechanism." *1* Outer electrolyte, *2* Agarose gel containing the inner electrolyte, *3* Precipitate, *4* Reaction front active border, *5* Passive border, *6* Regressing edge. The contours of the patterns are drawn on the bottom of the vessel. **a** The precipitation does not start at some points of the gel surface. **b** These points expand into precipitate-free cones as the front progresses. **c** The regressing edges meet on the line that is an equal distance from the tip of the empty cones to form a Voronoi diagram (in a 2D view)

ducibly if the gel was marked 60 s after initiation. In addition we observed that in this system the formation of single circular waves was due to the failure of an expanding wave fragment to form a cardioid wave. Thus the spiral tips were annihilated in the collision resulting in a seemingly circular expanding wave.

7.3.4.1 Selective Initiation of Circular Waves

Therefore, by marking the gel at time intervals greater than 2 min circular waves rather than disorganised waves (obtained if gel was marked prior to initiation) can be reproducibly obtained, see Fig. 7.20a. Re-marking the gel at the centre of the circular front results in the initiation of a second circular front, see Fig. 7.20b.

7.3.4.2 Construction of a Voronoi Diagram

Therefore, if circular waves can be initiated reproducibly and the circular fronts annihilate in straight line bisectors then a Voronoi diagram can be constructed, see Fig. 7.21.

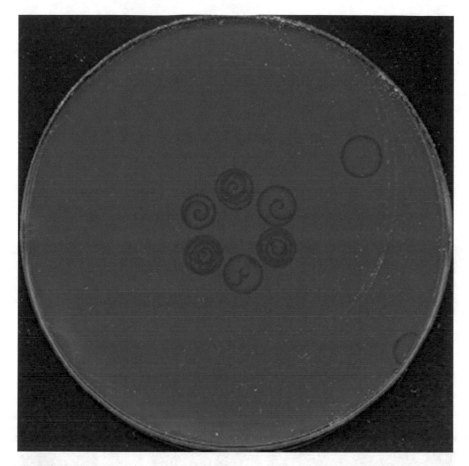

Fig. 7.18 Controlled evolution of six complex waves in the aluminium chloride (0.26 M) and sodium hydroxide reaction (5 M). The gel was marked once at each of six positions in a hexagonal pattern. After marking the gel the sodium hydroxide was poured onto the gel to initiate the precipitation front

7.3.4.3 Construction of an Additively Weighted Voronoi Diagram

A further step is to construct weighted diagrams. In this system it is relatively easy to construct an Additively weighted Voronoi diagram by marking the gel at different locations at various time intervals after the initiation of the reaction, see Fig. 7.22. The curved bisectors between the smaller cells initiated at longer time intervals after the start of the reaction and the other larger cells signifies correct calculation of the weighted diagram.

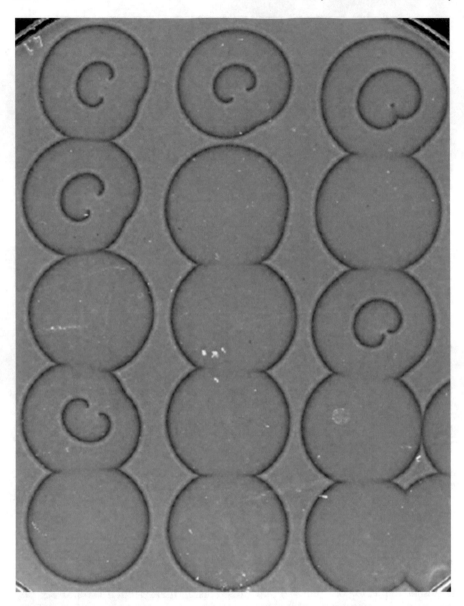

Fig. 7.19 The selective initiation of waves in the aluminium chloride (0.26 M) sodium hydroxide (5 M) reactions. The gel was marked (from *top* to *bottom*) 60 s after initiation and subsequently at 15 s intervals with the final set of three points marked 2 min after initiation

(a)   (b)

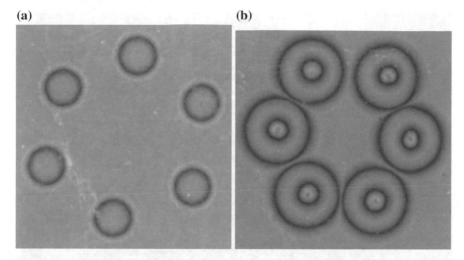

Fig. 7.20   a The selective initiation of circular waves in the aluminium chloride (0.26 M) sodium hydroxide (5 M) reaction. The gel was marked 5 min after initiation. **b** Initiation of nested circular waves in the aluminium chloride (0.26 M) sodium hydroxide (5 M) reaction

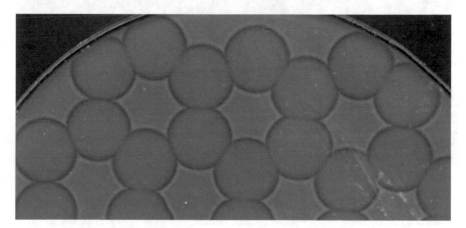

Fig. 7.21   Construction of a Voronoi diagram in the aluminium chloride (0.26 M) sodium hydroxide (5 M) reaction. Reaction in progress

7.4   Discussion

7.4.1   Stable Versus Unstable Systems

It has been demonstrated that Voronoi diagrams can be formed in both stable systems where input is in the form of drops of reagents (such as metal ions) which diffuse and react with a substrate loaded gel [13–16, 18–20, 27, 28] and unstable systems [24–26]

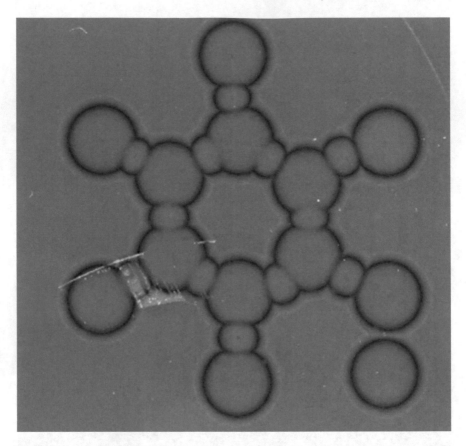

Fig. 7.22 Additively weighted Voronoi diagram under construction in the aluminium chloride (0.26 M) sodium hydroxide (5 M) reaction

either spontaneously (presumably at heterogeneities in the system which trigger wave evolution) or in controlled systems (where concentration is held just below the level likely to result in natural wave evolution and large physical heterogeneities are applied to the system in order to trigger waves). It is interesting to note that stable and unstable Voronoi diagram formation can occur in two distinct phases of the same reactions (i.e. the reagents have not changed only the concentration of the reagents). It is also interesting to note that Voronoi diagram formation (generalized) is computed by both stable and unstable systems but additively weighted diagrams are not computed correctly by stable systems (studied so far) and multiplicatively weighted diagrams (crystal growth type) are not computed correctly by unstable systems (studied so far). This is in contrast to crystal growth whereby crystals within the same domain can grow with different speeds and start growing at different times relative to each other. Although unstable systems are similar in some way to crystallization processes, they rely predominantly on the diffusion rate of the outer electrolyte through the gel to control growth rate of the waves. Thus at each point in the evolution of the reaction the growth rate of the initiated waves will be almost identical.

The advantages of unstable systems versus stable systems are:

- Voronoi diagram formation is from point source (not circular region)—accuracy of calculation increased
- Additively weighted diagram can easily be calculated—by marking the gel before and after initiation (adding the reagent). In stable systems additively weighted diagrams cannot be calculated correctly as fronts seem to exert attractive forces.
- Expanding precipitate waves exert repulsive forces leading to correct bisector calculation in AWVD
- Ease of data input. This is because inputs for calculation are added prior to initiation of the reaction.
- Potential to design functional materials in 3D
- Faster computation (potentially) and on smaller length scales

The disadvantages of unstable systems versus stable systems are:

- Methods of multiple parallel computations yet to be established.
- Programmable multiplicatively weighted Voronoi diagram cannot be computed simply—as wave expansion velocity hard coded to the reaction (whereas in stable systems diffusion and reaction rates of different reactants which can be reacted on the same substrate naturally differ and this info can be utilised to undertake specific computations.)
- Cannot be used for sequential calculations.
- The computational output is not as obvious in some reactions and may require post image processing to extract data.

7.4.2 Computational Limitations

As mentioned in the previous sections, these chemical processors have some potential limitations from a computational perspective. These are dealt with in some detail in [28]. The accuracy of computational output is limited by the diffusion and reaction velocity of the chemical species and the distance between sources. This is because fronts are subject to natural curvature which increases with time. This limits the accuracy of the calculated bisectors especially if the source of waves is a planar shape with acute angles (as these will not be reproduced in the growing front). In the calculation of a skeleton another problem exists, and in fact acute angles allow more accurate reconstruction than obtuse angles where the fronts lose definition and approximate to a circular wavefront. Also in the calculation of a skeleton the data reduction is limited. This is because the fronts do not just interact to form output at maximal points of interaction but continuously across the substrate resulting in a 3D data reduction. Although this is useful for reconstruction of the exact contour, data reduction is limited per se. In the previous paper [28] the problem of inverting a Voronoi diagram was studied and it was found that due to the inherent mechanism

(a)   (b)

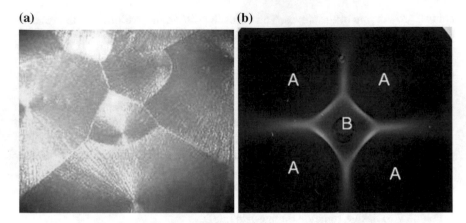

Fig. 7.23 This shows the correct calculation of the bisectors in an additively weighted Voronoi diagram computed **a** naturally (i.e. without initiation) in growing crystals and **b** incorrect calculation in a stable chemical processor, the bisectors are have inverted curvature, suggesting an attractive force rather than a repulsive force is responsible for bisector formation

of the chemical construction it proved impossible to recreate the original data point set from a calculated Voronoi diagram.

There are specific problems in calculating weighted Voronoi diagrams in stable and unstable systems. As mentioned the stable systems easily calculate a MWCGVD but fail to compute an AWVD. This is because even though it is easy to add point sources which diffuse at the same speed at distinct time points in the reactions evolution, bisector construction is not correct. Figure 7.23 shows an example of one such calculation. This shows that the bisectors are not computed correctly according to the theoretical output. In fact the bisectors are inverted completely from those expected theoretically. This suggests that an attractive force is responsible for the mechanism of bisector formation rather than a repulsive force which would be required for correct calculation. In unstable systems it is the opposite problem calculation of an additively weighted diagram is facile, but calculation of a multiplicatively weighted diagram is problematic because at a given point in the reaction evolution all growing precipitation fronts in the unstable system are controlled only by diffusion and reaction rate and these are uniform at a given point in the progress of the reactant front.

7.4.3 Precipitating Systems Versus Other Unconventional Parallel Processors for Computation of Voronoi Diagrams

The obvious initial comparison is with the work of Adamatzky on the "hot ice computer" for calculation of a Voronoi diagram [21]. The advantage of the hot ice approach is fast evolution compared to both unstable and stable systems (although all systems could be optimized further). The hot ice computer is reversible and thus it is possible to perform additional (but not sequential) calculations. They all share a permanent output which is advantageous as the results do not have to be reconstructed, and could be used directly to interface with another unconventional processor for shortest path calculations, robot control etc. The advantage of the stable systems over the hot ice computer are the ability to perform sequential calculations, the ability to compute multiple Voronoi diagrams in parallel (up to three to date). The ability of the unstable systems over the hot ice approach are the ease of inputs, it is possible to input 100 s of points in parallel and because this is done prior to initiating the reaction (simply pouring on the reagent), this is not as difficult as positioning large arrays of input devices. As the hot ice computer is so fast due to the spontaneous nature of the crystallization process then it is difficult to control the evolution of weighted Voronoi diagrams, due to the difficult of inputting two sets of data with a time delay (e.g. for AWVD calculation).

The next obvious comparison is with work undertaken by Adamatzky and de Lacy Costello utilizing the Belousov–Zhabotinsky (BZ) reaction to compute a Voronoi diagram [29]. This approach involved controlling the excitability of the BZ reaction to limit natural wave evolution and then initiation of single circular waves using arrays of silver wires. In this respect it shares similarities with the approach used for initiation of the hot ice computer, and also the approach for making controllable waves in unstable precipitating reactions i.e. limiting the natural reactivity to obtain a "controllable "substrate. How does this approach compare? It is still slower than the hot ice computer but has the ability to be optimized. It is marginally faster than the precipitating reactions if reacted within the same area with same input density. It does not give a permanent output and so image analysis is required to reconstruct the Voronoi diagram calculated by the interacting wave fronts (this is because wave fronts in BZ reaction must annihilate). It is inherently difficult to control the inputs beyond a certain number, as with the hot ice computer (but this may simply be an engineering challenge rather than a computational limitation).

As mentioned Voronoi diagrams can be calculated in gas discharge systems [12, 30]. The computation is fast and reversible, but data input can be problematic and the approach is relatively expensive compared to precipitation methods. Also in gas discharge systems it is not apparent how certain weighted diagrams could be computed or indeed multiple diagrams in parallel.

Voronoi diagrams and skeletonisation has also been undertaken using Physarum polycephalum [31], bacterial and fungal cultures [9, 10]. These systems are probably slower than the computation in precipitation systems (although again this could be optimized). They can only compute one Voronoi diagram in parallel and results are not reversible although the computation can be altered during calculation which is a useful function. The bisectors are not formed in the same way as the chemical systems in that they do not leave permanent outputs of the calculation (at least not as clearly). The accuracy of calculation is probably lower than the chemical case because growing fronts tend to be discontinuous. They should be able to compute weighted diagrams, for example if smaller amounts of inoculum are added in combination with larger sources then programmable MWCGVD should be created as the distinct sources should grow with different rates.

In summary there are many unconventional approaches to calculating Voronoi diagrams, all of which have some benefits but also some drawbacks. One thing they all possess is the advantage of parallel computation which is especially beneficial for geometric calculations. However, another thing they all share is slow speed. Additionally they all possess varying degrees of accuracy and some calculations have to be extracted using conventional processing power. However, they are capable of easily solving complex computational problems which some conventional processors would find challenging. However, conventional processors although inefficient for certain geometric calculations possess incredibly high speed and huge numbers of elementary processing units.

7.5   Conclusions

We have shown that Voronoi diagrams can be constructed in "stable" precipitating reactions via the addition of drops of an outer electrolyte to a gel containing an inner electrolyte. The bisectors of the Voronoi diagram are low precipitate areas separating coloured cells. A mechanism of substrate competition between the advancing fronts is suggested as the reason for bisector formation.

A range of complex and colourful tessellations may be obtained if two or more reagents are reacted with the substrate loaded gel. This is particularly true if one or more of the reagents has limited reactivity with the gel. In this case two Voronoi diagrams may be constructed in parallel where the additional diagram corresponds to the original positions of the reactive drops. Even more complex tessellations may be constructed if the products are cross-reactive. This may have some use in materials synthesis as the complex tessellation involves precipitate and precipitate free areas. If two exclusive chemical couples (ferric ions/ferrocyanide gel and ferrous ions ferricyanide gel) are combined then three Voronoi diagrams may be constructed in parallel. One corresponds to the position of all drops and one corresponds to the position of ferric or ferrous ions.

If the concentration of substrate in the gel is raised then the chemical reactions become unstable and the fronts spontaneously split. Thus Voronoi diagrams are not formed and the controllable pattern formation is lost. If the substrate level is raised even further then the reactions become unstable in three dimensions leading to the formation of growing conical waves. As the front advances through the thickness of the gel these conical waves grow and eventually collide where they annihilate in the formation of a Voronoi diagram. If the concentration is reduced slightly then these conical waves are not formed unless the gel is pre-marked with a glass needle. By using this method Voronoi diagrams could be calculated accurately and fairly rapidly (dependent on heterogeneity density). The remarkable feature of these controllable chemical systems is that the higher the density of information the faster the calculation is completed (upto the theoretical maximum).

We tried to implement the same approach with the newly discovered aluminum chloride sodium hydroxide reaction. However, if the gel was marked prior to initiation disorganized waves were generated albeit at specific locations. Therefore, the gel was marked at various time intervals after initiation of the reaction. It was found that wave type could be fairly reproducibly selected. For example if the gel was marked one minute after initiation cardioids waves were the predominant wave initiated. However, if the gel was marked two minutes after initiation then circular waves predominate. Therefore, we were able to construct a range of simple Voronoi diagrams. We also found that marking the gel repeatedly art different time intervals after initiation resulted in the selective initiation of circular waves. Therefore, target like waves, weighted Voronoi diagrams and other user define patterns could be constructed. This work could be very useful in the synthesis of functional materials. This would particularly be true if the effect was reproduced in a number of other inorganic and polymerization type reactions. Also a better method of control rather than physical marking of the gel would be desirable.

This chapter has highlighted the wide range of precipitating chemical reactions which are capable of geometric calculations and thus can be classed as specialized chemical parallel processors. It would be desirable to gain a better understanding of the underlying chemical and physical mechanism responsible for the pattern formation in terms of exerting better control over the reactions especially at smaller scales. If this were possible this type of wave interaction computing could be used to design functional materials. Better methods of data input and truly reversible adaptable systems could also be realized.

References

1. Pastzor, L., Csillag, F.: In: Crabtree, D.R., Hanisch, R.J., Barnes, J. (eds.) Astronomical data analysis software and systems. ASP Conference Series, vol. 61, p. 331 (1994)
2. Blackburn, C.G., Dunckley, L.: The application of Voronoi tessellations in the development of 3D stochastic models to represent tumour growth. Zeitschrift fur Angewandte Mathematik und Mechanik **76**, 335 (1996)
3. Hargittai, I. (ed.): Symmetry: Unifying Human Understanding. Pergamon Press, Oxford (1986)
4. Byers, J.A.: Dirichlet tessellation of bark beetle spatial attack points. J. Anim. Ecol. **61**, 759 (1992)
5. Graham, R., Yao, F.: A whirlwind tour of computational geometry. Am. Math. Mon. **97**, 687 (1990)
6. Schaudt, B.F., Drysdale, R.L.: Multiplicatively weighted crystal growth Voronoi diagrams. In: Proceedings of the 7th Annual Symposium on Computational Geometry, p. 214 (1999)
7. Preparata, F.P., Shamos, M.I.: Computational Geometry: An Introduction. Springer, New York (1985)
8. Walter, M., Fournier, A., Reimers, M.: Clonal mosaic models for the synthesis of mammalian coat patterns. In: Proceedings of Graphics Interface, vol. 118, p. 82 (1998)
9. Aurenhammer, F.: Voronoi diagrams - a survey of a fundamental geometric data structure. ACM Comput. Surv. **23**, 345 (1991)
10. Muller, S.C., Mair, T., Steinbock, O.: Travelling waves in yeast extract and in cultures of dictyostelium discoieum. Biophys. Chem. **72**, 37 (1998)

11. Stevens, P.S.: Patterns in Nature. Penguin Books Ltd., London (1976)
12. Zanin, A.L., Liehr, A.W., Moskalenko, A.S., Purwins, H.-G.: Voronoi diagrams in barrier gas discharge. Appl. Phys. Lett. **81**, 3338–3340 (2002)
13. Tolmachiev, D., Adamatzky, A.: Chemical processors for computation of Voronoi diagram. Adv. Mater. Opt. Electron. **6**, 191 (1996)
14. Adamatzky, A., de Lacy Costello, B.: Reaction diffusion path planning in a hybrid reaction diffusion and cellular automaton processor. Chaos, Solitons Fractals **16**, 727 (2003)
15. Adamatzky, A., Tolmachiev, D.: Chemical processor for computation of skeleton and planar shape. Adv. Mater. Opt. Electron. **7**, 135 (1997)
16. Adamatzky, A., de Lacy Costello, B., Ratcliffe, N.: Experimental reaction diffusion pre-processor for shape recognition. Phys. Lett. A **297**, 344 (2002)
17. Adamatzky, A., de Lacy Costello, B.: Experimental logic gates in a reaction-diffusion medium: the XOR gate and beyond. Phys. Rev. E **66**, 046112 (2002)
18. De Lacy Costello, B.P.J.: Constructive chemical processors - experimental evidence that shows this class of programmable pattern forming reactions exist at the edge of highly nonlinear region. Int. J. Bifurc. Chaos **13**, 1561 (2003)
19. De Lacy Costello, B.P.J., Adamatzky, A.I.: On multitasking in parallel processors: experimental findings. Int. J. Bifurc. Chaos **13**, 521 (2003)
20. De Lacy Costello, B.P.J., Jahan, I., Adamatzky, A., Ratcliffe, N.M.: Chemical tessellations. Int. J. Bifurc. Chaos **19**(2), 619 (2009)
21. Adamatzky, A.: Hot ice computer. Phys. Lett. A **374**, 264–271 (2009)
22. Hantz, P.: Pattern formation in the NaOH + CuCl$_2$ Reaction. J. Phys. Chem. B **104**, 4266 (2000)
23. Volford, A., Izsak, F., Ripszam, M., Lagzi, I.: Pattern formation and self-organisation in a simple precipitation system. Langmuir **23**(3), 961–964 (2007)
24. De Lacy Costello, B.P.J., Ratcliffe, N.M., Hantz, P.: Voronoi diagrams formed by regressing edges of precipitation fronts. J. Chem. Phys. **120**(5), 2413–2416 (2004)
25. De Lacy Costello, B.P.J.: Control of complex travelling waves in simple inorganic systems - the potential for computing. Int J. Unconv. Comput. **4**, 297 (2008)
26. De Lacy Costello, B.P.J., Armstrong, J., Jahan, I., Ratcliffe, N.M.: Fine control and selection of travelling waves in inorganic pattern forming reactions. Int. J. Nanotechnol. Mol. Comput. **1**(3), 26 (2009)
27. de Lacy Costello, B.P.J., Jahan, I., Hambidge, P., Locking, K., Patel, D., Adamatzky, A.: Chemical tessellations-results of binary and tertiary reactions between metal ions and ferricynide or ferrocyanide loaded gels. Int. J. Bifurc. Chaos **20**(7), 2241–2252 (2010)
28. Adamatzky, A., de Lacy Costello, B.: On some limitations of reaction-diffusion chemical computers in relation to Voronoi diagram and its inversion. Phys. Lett. A **309**(5–6), 397–406 (2003)
29. Adamatzky, A., de Lacy Costello, B.: Collision free path planning in the Belousov-Zhabotinsky medium assisted by a cellular automaton. Die Natuwissenschaften **89**, 474 (2002)
30. de Lacy Costello, B.P.J., et al.: The formation of Voronoi diagrams in chemical and physical systems: experimental findings and theoretical models. Int. J. Bifurc. Chaos **14**, 2187–2210 (2004)
31. Shirakawa, T., Adamatzky, A., Gunji, Y.-P., Miyake, Y.: On simultaneous construction of Voronoi diagram and Delaunay triangulation by Physarum Polycephalum. Int. J. Bifurc. Chaos **19**, 3109–3117 (2009)

Chapter 8
Light-Sensitive Belousov–Zhabotinsky Computing Through Simulated Evolution

Larry Bull, Rita Toth, Chris Stone, Ben De Lacy Costello
and Andrew Adamatzky

Abstract Many forms of unconventional computing, i.e., massively parallel non-linear computers, can be realised through simulated evolution. That is, the behaviour of non-linear media can be controlled automatically and the structural design of the media optimized through the nature-inspired machine learning approach. This chapter describes work using the Belousov–Zhabotinsky reaction as a non-linear chemical medium in which to realise computation. A checkerboard image comprising of varying light intensity cells is projected onto the surface of a catalyst-loaded gel resulting in rich spatio-temporal chemical wave behaviour. Cellular automata are evolved to control the chemical activity through dynamic adjustment of the light intensity, implementing a number of Boolean functions in both simulation and experimentation.

8.1 Introduction

Excitable and oscillating chemical system—the Belousov–Zhabotinsky (BZ) reaction [35]—have been used to solve a number of computational tasks such as implementing logical circuits [3, 31], image processing [22], shortest path problems [30] and memory [27]. In addition chemical diodes [4], coincidence detectors [15] and transformers where a periodic input signal of waves may be modulated by the barrier

L. Bull (✉) · C. Stone · A. Adamatzky
Department of Computer Science, University of the West of England, Bristol, UK
e-mail: larry.bull@uwe.ac.uk

A. Adamatzky
e-mail: andrew.adamatzky@uwe.ac.uk

B. De Lacy Costello
Institute of BioSensing Technology, University of the West of England, Bristol, UK
e-mail: ben.delacycostello@uwe.ac.uk

R. Toth
High Performance Ceramics, EMPA, Dubendorf, Switzerland
e-mail: Rita.Toth@empa.ch

© Springer International Publishing Switzerland 2017
A. Adamatzky (ed.), *Advances in Unconventional Computing*,
Emergence, Complexity and Computation 23,
DOI 10.1007/978-3-319-33921-4_8

199

into a complex output signal depending on the gap width and frequency of the input [28] have all been demonstrated experimentally.

A number of experimental and theoretical constructs utilising networks of chemical reactions to implement computation have been described. These chemical systems act as simple models for networks of coupled oscillators such as neurons, circadian pacemakers and other biological systems [21]. Ross and co-workers [5] produced a theoretical construct suggesting the use of "chemical" reactor systems coupled by mass flow for implementing logic gates neural networks and finite-state machines. In further work Hjelmfelt et al. [16, 17] simulated a pattern recognition device constructed from large networks of mass-coupled chemical reactors containing a bistable iodate-arsenous acid reaction. They encoded arbitrary patterns of low and high iodide concentrations in the network of 36 coupled reactors. When the network is initialized with a pattern similar to the encoded one then errors in the initial pattern are corrected bringing about the regeneration of the stored pattern. However, if the pattern is not similar then the network evolves to a homogenous state signalling non-recognition.

In related experimental work Laplante et al. [23] used a network of eight bistable mass coupled chemical reactors (via 16 tubes) to implement pattern recognition operations. They demonstrated experimentally that stored patterns of high and low iodine concentrations could be recalled (stable output state) if similar patterns were used as input data to the programmed network. This highlights how a programmable parallel processor could be constructed from coupled chemical reactors. This described chemical system has many properties similar to parallel neural networks. In other work Lebender and Schneider [24] described methods of constructing logical gates using a series of flow rate coupled continuous flow stirred tank reactors (CSTR) containing a bistable nonlinear chemical reaction. The minimal bromate reaction involves the oxidation of cerium(III) (Ce^{3+}) ions by bromate in the presence of bromide and sulphuric acid. In the reaction the Ce^{4+} concentration state is considered as "0" ("False") and "1" ("True") if a given steady state is within 10 % of the minimal (maximal) value. The reactors were flow rate coupled according to rules given by a feedforward neural network run using a PC. The experiment is started by feeding in two "true" states to the input reactors and then switching the flow rates to generate "true"-"false", "false"-"true" and "false"-"false". In this three coupled reactor system the AND (output "true" if inputs are both high Ce^{4+}, "true"), OR (output "true" if one of the inputs is "true"), NAND (output "true" if one of the inputs is "false") and NOR gates (output "true" if both of the inputs are "false") could be realised. However to construct XOR and XNOR gates two additional reactors (a hidden layer) were required. These composite gates are solved by interlinking AND and OR gates and their negations. In their work coupling was implemented by computer but they suggested that true chemical computing of some Boolean functions may be achieved by using the outflows of reactors as the inflows to other reactors, i.e., serial mass coupling.

As yet no large scale experimental network implementations have been undertaken mainly due to the complexity of analysing and controlling many reactors. That said there have been many experimental studies carried out involving coupled oscillating and bistable systems (e.g., see [6, 10, 18, 32]). The reactions are coupled together either physically by diffusion or an electrical connection or chemically, by having two oscillators that share a common chemical species. The effects observed include multistability, synchronisation, in-phase and out of phase entrainment, amplitude or "oscillator death", the cessation of oscillation in two coupled oscillating systems, or the converse, "rhythmogenesis", in which coupling two systems at steady state causes them to start oscillating [11].

Alongside the development of unconventional computers has been the growing use of machine learning techniques to aid their design and programming (see [25] for an overview). Since techniques such as evolutionary computing (e.g., [12]) have been shown able to handle various complex tasks effectively, the aim is to apply them to harness the as yet only partially understood intricate dynamics of non-linear media to perform computations more effectively than with traditional architectures. Previous theoretical and experimental studies have shown that reaction-diffusion chemical systems are capable of information processing. As such, we have been exploring the use of simulated evolution to design such chemical systems which exploit collision-based computing (e.g., [1]). We use a spatially-distributed light-sensitive form of the BZ reaction in gel which supports travelling reaction-diffusion waves and patterns. Exploiting the photoinhibitory property of the reaction, the chemical activity (amount of excitation on the gel) can be controlled by an applied light intensity, namely it can be decreased by illuminating the gel with high light intensity and vice versa. In this way a BZ network is created via light and controlled using (heterogeneous) cellular automata (CA) designed using simulated evolution [8, 9, 14]. This architecture is adapted from the system described in [34] and experimental chemical computers have been realised, as will be described.

8.2 Simulated Media

We use two-variable Oregonator equation [13] adapted to a light-sensitive Belousov–Zhabotinsky (BZ) reaction with applied illumination [7]:

$$\frac{\partial u}{\partial t} = \frac{1}{\varepsilon}(u - u^2 - (fv + \phi)\frac{u - q}{u + q}) + D_u\nabla^2 u$$

$$\frac{\partial v}{\partial t} = u - v \tag{8.1}$$

The variables u and v represent the instantaneous local concentrations of the bromous acid autocatalyst and the oxidized form of the catalyst, HBrO$_2$ and tris (bipyridyl) Ru (III), respectively, scaled to dimensionless quantities. The rate of the photo-induced bromide production is designated by ϕ, which also denotes the excitability of the system in which low simulated light intensities facilitate excitation while high intensities result in the production of bromide that inhibits the process, experimentally verified by [19]. The system was integrated using the Euler method with a five-node Laplacian operator, time step $\Delta t = 0.001$ and grid point spacing $\Delta x = 0.62$. The diffusion coefficient, D_u, of species u was unity, while that of species v was set to zero as the catalyst is immobilized in the gel. The kinetic parameters were set to $\varepsilon = 0.11, f = 1.1$ and $q = 0.0002$. The medium is oscillatory in the dark which made it possible to initiate waves in a cell by setting its simulated light intensity to zero. At different ϕ values the medium is excitable, sub-excitable or non-excitable.

8.2.1 Cellular Automata

We have used cellular automata (CA) to control such chemical systems (Fig. 8.1), i.e., finite automata are arranged in a two-dimensional lattice with aperiodic boundary conditions (an edge cell has five neighbours, a corner cell has three neighbours, all other cells have eight neighbours each). Use of a CA with such a two-dimensional topology is a natural choice given the spatio-temporal dynamics of the BZ reaction. Each automaton updates its state depending on its own state and the states of its neighbours. States are updated in parallel and in discrete time. In standard CA all cells have the same state transition function (rule), whereas in this work the CA is heterogeneous, i.e., each cell/automaton has its own state transition function. The transition function of every cell is evolved by a simple evolutionary process. This approach is very similar to that presented in [29]. However, his reliance upon each

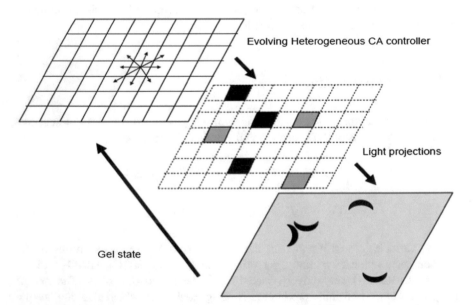

Fig. 8.1 Relationship between the CA controller, applied grid pattern and chemical system comprising one process control cycle. Modified with permission from J. Chem. Phys. 129, 184708 (2008). Copyright 2008, AIP Publishing LLC

cell having access to its own fitness means it is not applicable in the majority of chemical computing scenarios we envisage. Instead, fitness is based on emergent global phenomena in our approach (as in [26], for example). Thus, following [20], we use a simple approach wherein each automaton of the two-dimensional CA controller is developed via a simple genetics-based hillclimber. After fitness has been assigned, some proportion of the CA genes are randomly chosen and mutated. Mutation is the only variation operator used here to modify a given CA cell's transition rule to allow the exploration of alternative light levels for the grid state. For a CA cell with eight neighbours there are 29 possible grid state to light level transitions, each of which is a potential mutation site. After the defined number of such mutations have occurred, an evolutionary generation is complete and the simulation is reset and repeated. The system keeps track of which CA states are visited since mutation. On the next fitness evaluation (at the end of a further 25 control cycles) mutations in states that were not visited are discarded on the grounds that they have not contributed to the global fitness value and are thus untested. We also performed control experiments with a modified version to determine the performance of an equivalent random CA controller. This algorithm ignored the fitness of mutants and retained all mutations except those from unvisited states.

8.2.2 Controller

For a given experiment, a random set of CA rules is created for a two-dimensional array of size 10-by-10 cells. The rule for each cell is represented as a gene in a genome, which at any one time takes one of the discrete light intensity values used in the experiment. As previously mentioned, the grid edges are not connected (i.e., the grid is planar and does not form a toroid) and the neighborhood size of each cell is of radius 1; cells consider neighborhoods of varying size depending upon their spatial position, varying from three in the corners, to five for the other edge cells, and eight everywhere else. In the model each of the 100 cells consists of 400 (20-by-20) simulation points for the reaction. The reaction is thus simulated numerically by a lattice of size 200-by-200 points, which is divided into the 10-by-10 grid.

To begin examining the potential for the evolution of controllers for such temporally dynamic structures in the continuous, non-linear 2D media described we have designed a simple scheme to create a number of two-input Boolean logic gates. Excitation is fed in at the bottom of the grid into a branching pattern. To encode a logical '1' and '0' either both branches or just one branch of the two "trees" shown in Fig. 8.2 are allowed to fill with excitation, i.e., the grid is divided into two for the inputs (Fig. 8.3). These waves were channelled into the grid and broken up into 12 fragments by choosing an appropriate light pattern as shown in Fig. 8.2. The black area represents the excitable medium whilst the white area is non-excitable. After initiation three light levels were used: one is sufficiently high to inhibit the reaction; one is at the sub-excitable threshold such that excitation just manages to propagate; and the other low enough to fully enable it. The modelled chemical system was

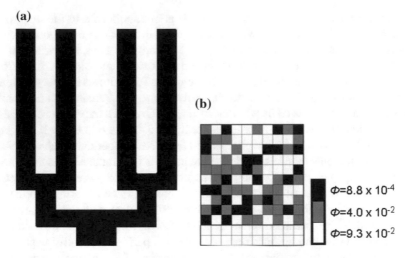

Fig. 8.2 Showing initiation pattern **a** and a typical example of a coevolved light pattern **b**. Modified with permission from J. Chem. Phys. 129, 184708 (2008). Copyright 2008, AIP Publishing LLC

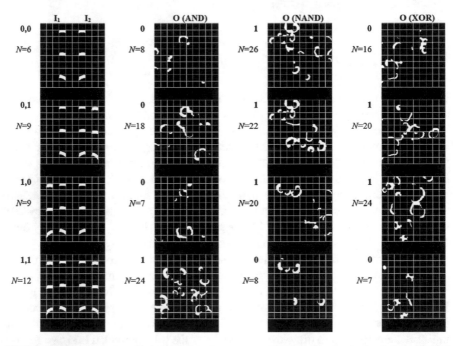

Fig. 8.3 Typical examples of solutions of AND, NAND and XOR logic gates in simulation, required active cells: 20, N: actual number of active cells. Input states I1, I2 for the logic gates are shown on the left and consist of two binary digits, spatially encoded using left and right "initiation trees" (Fig. 8.7). Input values of '0' are encoded using a single branch of the relevant tree resulting in 3 fragments, while binary '1' is encoded using both branches of the tree resulting in 6 fragments. Evolution found a solution in 56 (AND), 364 (NAND) and 16556 (XOR) generations

run for 600 iterations of the simulator. This value was chosen to produce network dynamics similar to those obtained in experiment over 10 s of real time.

8.3 Simulated Experiments

A colour image was produced by mapping the level of oxidized catalyst at each simulation point into an RGB value. Image processing of the colour image was necessary to determine chemical activity. This was done by differencing successive images on a pixel by pixel basis to create a black and white thresholded image. Each pixel in the black and white image was set to white (corresponding to excitation) if the intensity of the red or blue channels in successive colour images differed by more than 5 out of 256 pixels (1.95 %). Pixels at locations not meeting this criterion were set to black. An outline of the grid was superimposed on the black and white images to aid visual analysis of the results.

The black and white images were then processed to produce a 100-bit description of the grid for the CA. In this description each bit corresponds to a cell and it is set to true if the average level of activity within the given cell is greater than a pre-determined threshold of 10 %. Here, activity is computed for each cell as the fraction of white pixels in that cell. This binary description represents a high-level depiction of activity in the BZ network and is used as input to the CA. Once cycle of the CA is performed whereby each cell of the CA considers its own state and that of its neighbours (obtained from the binary state description) to determine the light level to be used for that grid cell in the next time step. Each grid cell may be illuminated with one of three possible light levels. The CA returns a 100-digit trinary action string, each digit of which indicates whether high ($\phi = 0.093023$), sub-excitable threshold ($\phi = 0.04$) or low ($\phi = 0.000876$) intensity light should be projected onto the given cell. The progression of the simulated chemical system, image analysis of its state and operation of the CA to determine the set of new light levels comprises one control cycle of the process. A typical light pattern generated by the CA controller is shown in Fig. 8.2b. Another 600 iterations are then simulated with those light-levels projected, etc. until 25 control cycles have passed. The number of active cells in the grid, that is those with activity at or above the 10 % threshold, is used to distinguish between a logical '0' and '1' as the output of the system. For example, in the case of XOR, the controller must learn to keep the number of active cells below the specified level for the 00 and 11 cases but increase the number for the 01 and 10 cases.

Figure 8.3 shows typical examples of each of the three logic gates learned using the simulated chemical system. Each of the four possible input combinations is presented in turn—00 to 11—and for each input combination the system is allowed to develop for 25 control cycles. Fitness for each input pattern is evaluated after the 25 control cycles with each correct output scoring 1, resulting in a maximum possible fitness of 4. Figure 8.4 shows the fitness averaged over ten runs for AND and NAND tasks with mutation rate 4000, and similar results for XOR are shown for mutation rate

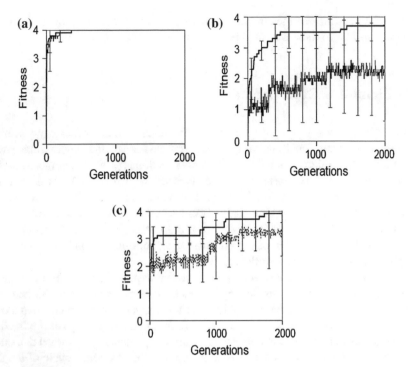

Fig. 8.4 Showing the performance of evolving CA controllers for the three logic gate tasks considered. Dashed lines show the equivalent performance of random search. **a** AND gate. **b** NAND gate. **c** XOR gate

6000. Favourable comparisons to an equivalent random controller are also shown in each case.

8.4 Laboratory Experiments

The success of our simulated experiment encouraged us to build an experimental setup (Fig. 8.5) and perform the same tasks in the real light-sensitive BZ medium. We immobilised the light-sensitive (Ru(byp)$_3$$^{2+}$ catalyst in a thin layer of silica gel which was bathed in the catalyst free BZ reagents. All chemicals were purchased from Aldrich (U.K.) and used as received unless stated otherwise. Ru(bpy)$_3$SO$_4$ was recrystallised from Ru(bpy)$_3$Cl$_2$ using sulphuric acid. For the silica gel 222 mL of the purchased 27 % sodium silicate solution (stabilized in 4.9 M sodium hydroxide) was acidified by adding 57 mL of 2 M sulphuric acid and 187 mL of deionised water. Then 0.6 mL of 0.025 M Ru(bpy)$_3$SO$_4$ and 0.65 mL of 1.0 M sulphuric acid solutions were added to 2.5 mL of the acidified silicate solution. This solution was used to prepare the silica gel in a custom designed 0.3 mm deep Perspex mould. After 3 hours gelation

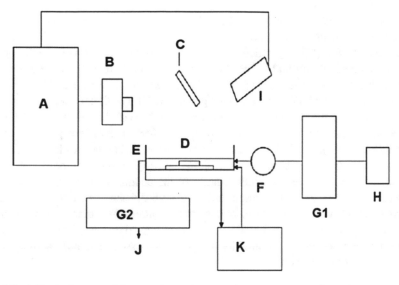

Fig. 8.5 A block diagram of the experimental setup where **a**: computer, **b**: projector, **c**: mirror, **d**: microscope slide with the catalyst-loaded gel, **e**: thermostated Petri dish, **f**: CSTR, **g1** and **g2**: pumps, **h**: stock solutions, **i**: camera, **j**: effluent flow, **k**: thermostated water bath

time the 26 mm × 26 mm × 300 μm gel layers were removed from the mould, carefully washed and stored in water until use. The experiments were performed in a thermostated (22 C°) open reactor containing the catalyst loaded silica gel and the catalyst free BZ solution (0.42 M sodium bromate, 0.19 M malonic acid, 0.64 M sulphuric acid and 0.11 M bromide). This reactor was fed by a continuously-fed stirred tank reactor (CSTR) which freshly mixed the BZ reagents to keep the system far from its equilibrium state. The flow between the two reactors and the removal of the effluents was maintained by two peristaltic pumps. An InFocus Model LP820 Projector was used to shine a computer generated 10-by-10 cell checkerboard grid pattern (with a size of 20 mm × 20 mm) on the surface of the gel through a 455 nm narrow bandpass interference filter, 100/100 mm focal length lens pair and mirror assembly (Fig. 8.5). Three light intensity levels were used in the checkerboard image representing excitable, subexcitable threshold, and non-excitable domains, with 0.35, 1.6 and 3.5 mW cm^{-2}, respectively. Images of the chemical wave fragments on the gel were captured using a Lumenera Infinity2 USB 2.0 scientific digital camera. To improve visibility and enable subsequent image processing images were captured while a uniform grey level of 3.5 mW cm^{-2} was projected on the gel for 10 ms instead of the checkerboard grid. Captured images were processed to identify activity in the same way as for the model.

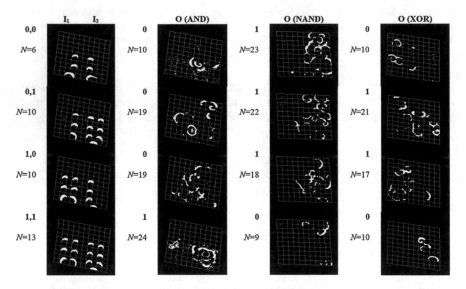

Fig. 8.6 Typical examples of solutions of AND, NAND and XOR logic gates in chemical experiment, required number of active cells: 15 (20 for AND), N: actual number of active cells. Input states I1, I2 for the logic gates are shown on the left and consist of two binary digits, spatially encoded using left and right "initiation trees" (Fig. 8.7a). Input values of '0' are encoded using a single branch of the relevant tree resulting in 3 or 4 fragments, while binary '1' is encoded using both branches of the tree resulting in 6 or 7 fragments. The simulated evolution—seeded with a CA evolved during the simulated runs—found a solution in 16 generations in each case

Figures 8.6 and 8.7 show how similar performance is possible on the real chemical system for each of the three logic functions. In order to produce working XOR and NAND gates from these experiments, it was necessary to use a value of 15 for the

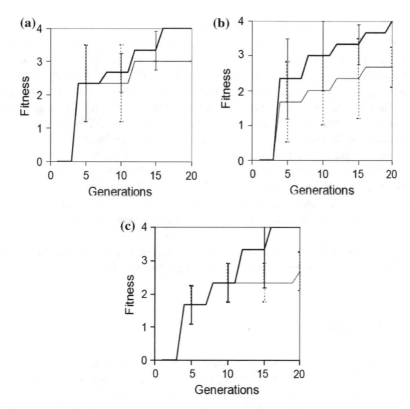

Fig. 8.7 Showing the performance of evolving CA controllers for the three logic gate tasks considered. *Dashed lines* show the equivalent performance of random search. **a** AND gate. **b** NAND gate. **c** XOR gate

required number of active cells due to the relative difficulty of these tasks. All other parameters were the same as those used for numerical simulation.

Because of the limited lifetime of the medium these runs were seeded with CA evolved during the simulated runs presented in Fig. 8.3. Runs using random initial controllers were also explored on the real chemical system (dashed lines on Fig. 8.7), but no successful runs were found over the 40 generations. This is not surprising considering that the average generations needed to find a good solution was higher than 40 in the simulations because of the relative increase in difficulty. Nevertheless the two systems are very similar since the runs with seeded CA evolved during the simulations found solutions in a very short time, namely in 16 or 20 generations. If a solution had been found in four generations it would have meant that the initial states of the simulated and real chemical system are perfectly identical. However, since there is a noticeable difference between the initial states, the solution found by the evolutionary algorithm in simulation was very close to the solution needed for the real chemistry, but a few generations of evolution were needed to adapt to the difference between the two systems. These results show that the approach is capable of adapting to small changes in its environment and finding a solution very quickly when presented with domain-specific knowledge obtained from modeling.

8.5 Conclusion

Excitable and oscillating chemical systems have previously been used to solve a number of simple computational tasks. However, the experimental design of such systems has typically been non-trivial. In this chapter we have presented results from a methodology by which to achieve the complex task of designing such systems—through the use of simulated evolution. We have shown using both simulated and real systems that it is possible in this way to control dynamically the behavior of the BZ reaction, and to design the topology of a network-based approach to chemical computing. As discussed in [25], evolution can also be used to pre-configure programmable/changeable elements of an unconventional medium, such as voltages within liquid crystal, before computation occurs. We have used a similar approach for the gel-based system described above, enabling evolution to predefine where fragment waves of excitation can enter a central area of collision/computation (e.g.,

[33]). Which of these approaches is best able to exploit the properties of non-linear media for computation—or whether their use in combination is possible—remains open to future exploration.

References

1. Adamatzky, A. (ed.): Collision-based Computing. Springer, London (2002)
2. Adamatzky, A., De Lacy Costello, B., Asai, T.: Reaction-Diffusion Computers. Elsevier, Amsterdam (2005)
3. Adamatzky, A., Holley, J., Bull, L., De Lacy Costello, B.: On computing in fine-grained compartmentalised Belousov–Zhabotinsky medium. Chaos, Solitons Fractals **44**(10), 779–790 (2011)
4. Agladze, K., Aliev, R.R., Yamaguhi, T., Yoshikawa, K.: Chemical diode. J. Phys. Chem. **100**, 13895–13897 (1996)
5. Arkin, A., Ross, J.: Computational functions in biochemical reaction networks. Biophys. J. **67**(2), 560–578 (1994)
6. Bar-Eli, K., Reuveni, S.: Stable stationary-states of coupled chemical oscillators: Experimental evidence. J. Phys. Chem. **89**, 1329–1330 (1985)
7. Beato, V., Engel, H.: Pulse propagation in a model for the photosensitive Belousov-Zhabotinsky reaction with external noise. In: Proceedings of the SPIE's First International Symposium on Fluctuations and Noise. International Society for Optics and Photonics (pp. 353-362). (May, 2003)
8. Bull, L.: Evolving Boolean networks on tuneable fitness landscapes. IEEE Trans. Evol. Comput. **16**(6), 817–828 (2012)
9. Bull, L.: Using genetical and cultural search to design unorganised machines. Evol. Intell. **5**(1), 23–33 (2012)
10. Crowley, M.F., Field, R.J.: Electrically coupled Belousov–Zhabotinsky oscillators 1: experiments and simulations. J. Phys. Chem. **90**, 1907–1915 (1986)
11. Dolnik, M., Epstein, I.R.: Coupled chaotic oscillators. Phys. Rev. E **54**, 3361–3368 (1996)
12. Eiben, A., Smith, J.: Introduction to Evolutionary Computing. Springer, Heidelberg (2003)
13. Field, R.J., Noyes, R.M.: Oscillations in chemical systems. IV. Limit cycle behavior in a model of a real chemical reaction. J. Chem. Phys. **60**(5), 1877–1884 (1974)
14. Fogel, L. J., Owens, A.J., Walsh, M.J. Artificial intelligence through a simulation of evolution. In M. Maxfield et al. (Eds.) Biophysics and Cybernetic Systems: Proceedings of the 2nd Cybernetic Sciences Symposium. pp. 131–155. Spartan Books (1965)
15. Gorecki, J., Yoshikawa, K., Igarashi, Y.: On chemical reactors that can count. J. Phys. Chem. A **107**, 1664–1669 (2003)
16. Hjelmfelt, A., Weinberger, E.D., Ross, J.: Chemical implementation of neural networks and turing machines. PNAS **88**, 10983–10987 (1991)
17. Hjelmfelt, A., Ross, J.: Mass-coupled chemical systems with computational properties. J. Phys. Chem. **97**, 7988–7992 (1993)
18. Holz, R., Schneider, F.W.: Control of dynamic states with time-delay between 2 mutually flow-rate coupled reactors. J. Phys. Chem. **97**, 12239 (1993)
19. Kadar, S., Amemiya, T., Showalter, K.: Reaction mechanism for light sensitivity of the Ru(bpy)$_3^{2+}$ -catalyzed Belousov-Zhabotinsky reaction. J. Phys. Chem. A **101**(44), 8200–8206 (1997)
20. Kauffman, S.A.: The Origins of Order. Oxford Press, Oxford (1993)
21. Kawato, M., Suzuki, R.: Two coupled neural oscillators as a model of the circadian pacemaker. J. Theor. Biol. **86**, 547–575 (1980)
22. Kuhnert, L., Agladze, K.I., Krinsky, V.I.: Image processing using light sensitive chemical waves. Nature **337**, 244–247 (1989)

23. Laplante, J.P., Pemberton, M., Hjelmfelt, A., Ross, J.: Experiments on pattern recognition by chemical kinetics. J. Phys. Chem. **99**, 10063–10065 (1995)
24. Lebender, D., Schneider, F.W.: Logical gates using a nonlinear chemical reaction. J. Phys. Chem. **98**, 7533–7537 (1994)
25. Miller, J., Harding, S., Tufte, G.: Evolution in-matrio: evolving computation in materials. Evol. Intell. **7**(1), 49–67 (2014)
26. Mitchell, M., Hraber, P., Crutchfield, J.: Revisiting the edge of chaos: evolving cellular automata to perform computations. Complex Syst. **7**, 83–130 (1993)
27. Motoike, I.N., Yoshikawa, K., Iguchi, Y., Nakata, S.: Real time memory on an excitable field. Phys. Rev. E **63**, 1–4 (2001)
28. Sielewiesiuk, J., Gorecki, J.: Passive barrier as a transformer of chemical frequency. J. Phys. Chem. A **106**, 4068–4076 (2002)
29. Sipper, M.: Evolution of Parallel Cellular Machines. Springer, Heidelberg (1997)
30. Steinbock, O., Toth, A., Showalter, K.: Navigating complex labyrinths: optimal paths from chemical waves. Science **267**, 868–871 (1995)
31. Steinbock, O., Kettunen, P., Showalter, K.: Chemical wave logic gates. J. Phys. Chem. **100**, 18970–18975 (1996)
32. Stuchl, I., Marek, M.: Dissipative structures in coupled cells: experiments. J. Phys. Chem. **77**, 2956–2963 (1982)
33. Toth, R., Stone, C., De Lacy Costello, B., Adamatzky, A., Bull, L.: Simple collision-based chemical logic gates with adaptive computing. J. Nanotechnol. Mol. Comput. **1**(3): 1–16 (2009)
34. Wang, J., Kádár, S., Jung, P., Showalter, K.: Noise driven avalanche behavior in subexcitable media. Phys. Rev. Lett. **82**, 855–858 (1999)
35. Zaikin, A.N., Zhabotinsky, A.M.: Concentration wave propagation in two-dimensional liquid-phase self-oscillating system. Nature **225**, 535–537 (1970)

Chapter 9
On Synthesis and Solutions of Nonlinear Differential Equations—A Bio-Inspired Approach

Ivan Zelinka

Abstract This chapter discusses an alternative approach for mathematical-physical problems solution by means of bio-inspired methods, especially by evolutionary algorithms. Two different approaches are demonstrated here. The first one is the use of evolutionary algorithms on design, parameter estimation and control of the chemical reactor that is represented by 5 nonlinear and mutually joined differential equations, the second one is the use of analytic programming (method of the same class as genetic programming or grammatical evolution) to solve two different differential equations (4th and 2nd order), that represent problems from civil engineering by appropriate function synthesis. Theoretical background as well as applications are discusses here.

9.1 Introduction

Evolutionary computation is a sub-discipline of computer science belonging to the 'bio-inspired' computing area. The main ideas of evolutionary computation have been published for example in [1] and widely introduced to the scientific community [6, 8]. The most well known evolutionary techniques are Genetic Algorithms (GA) introduced by J. Holland [6, 8] based on ideas of A.M Turing [7] and based on first computer experiments by Barricelli in [2], Evolutionary Strategies (ES) by Schwefel [21] and Rechenberg [18] and Evolutionary Programming (EP) by Fogel [5] for example.

The main idea is that every individual of a species can be characterized by its features and abilities that help it to cope with its environment in terms of survival and reproduction. These features and abilities can be termed by its fitness and are inheritable via its genome. In the genome the features/abilities are encoded. The

I. Zelinka (✉)
Faculty of Electrical Engineering and Computer Science, Department of Computer Science,
VSB-TU, 17. Listopadu 15, Ostrava-Poruba, Czech Republic
e-mail: ivan.zelinka@vsb.cz
URL:http://www.ivanzelinka.eu

© Springer International Publishing Switzerland 2017   213
A. Adamatzky (ed.), *Advances in Unconventional Computing*,
Emergence, Complexity and Computation 23,
DOI 10.1007/978-3-319-33921-4_9

code in the genome can be viewed as a kind of description that allows to store, process and transmit the information needed to build the individual. So, the fitness coded in the parent's genome can be handed over to new descendants and support the descendants in performing in the environment. The evolutionary principles are transferred into computational methods in a simplified form that will be outlined now. If the evolutionary principles are used for the purposes of calculations, the following procedure is used, as reported in [26]:

1. Specification of the EA parameters: For each algorithm, parameters that control the run of algorithm or terminate it regularly must be defined, if the termination criterion defined in advance are fulfilled (for example, the number of cycles— generations). Part of this point is the definition of the cost function (objective function) or, as the case may be, what is called fitness—a modified return value of the objective function). The objective function is usually a mathematical model of the problem, whose minimization or maximization leads to the solution of the problem.
2. Generation of the initial population (generally $N \times M$ matrix, where N is the number of parameters of an individual—D. Depending on the number of optimized arguments of the objective function and the user's criterions, the initial population of individuals is generated. An individual is a vector of numbers having such a number of components as the number of optimized parameters of the objective function. These components are set randomly and each individual thus represents one possible specific solution of the problem. The set of individuals is called population.
3. All the individuals are evaluated through a defined objective function and to each one of them fitness value is assigned, which is a modified (usually normalized) value of the objective function, or directly the value of the objective function.
4. Now parents are selected according to their quality (fitness, value of the objective function) or, as the case may be, also according to other criterions.
5. Descendants are created by crossbreeding the parents. The process of crossbreeding is different for each algorithm. Parts of parents are changed in classic genetic algorithms, in a differential evolution, crossbreeding is a certain vector operation, etc.
6. Every descendant is mutated by presence of randomness. In other words, a new individual is changed by means of a suitable random process. This step is equivalent to the biological mutation of the genes of an individual.
7. Every new individual is evaluated in the same manner as in step 3.
8. The best individuals are selected.
9. The selected individuals fill a new population.
10. The old population is eliminated and is replaced by a new population; step 4 represents further continuation.

Steps 4–10 are repeated until the number of evolution cycles (generations etc.) specified before by the user is reached or if the required quality of the solution is not achieved. The principle of the evolutionary algorithm outlined above is general and might more or less differ in specific cases. For more detailed overview about evolutionary algorithms we strongly recommend to read [26].

Together with, say classical evolutionary algorithms, an alternative approach how to use evolution has been developed (since 90's) so that instead of numerical solutions (i.e. unknown parameters estimation) of given problem, symbolic solutions have been derived in the form of mathematical formulas, electronic circuits etc. Generally this approach can be called as "symbolic regression" and the most popular and classic approaches are for example Genetic Programming (GP) [11] or Grammatical Evolution (GE) [15]. Another interesting research was carried out by Artificial Immune Systems (AIS) or/and systems, which do not use tree structures like linear GP and other similar algorithm like Multi Expression Programming (MEP), etc.

In this chapter, a different method called Analytic Programming (AP) [27], is used. AP is a grammar free algorithm—structure, which can be used by any programming language and also by any arbitrary evolutionary algorithm (EA) or another class of numerical optimization method.

The term *symbolic regression* represents a process during which measured data sets are fitted, thereby a corresponding mathematical formula is obtained in an analytical way. An output of the symbolic expression could be, for example, $\sqrt[N]{x^2 + \frac{y^3}{k}}$, and the like. For a long time, symbolic regression was a domain of human calculations but in the last few decades it involves computers for symbolic computation as well.

For closer description of the theoretical principles of AP it is recommended to study [27]. Comparative studies with selected well known case examples from GP as well as applications on synthesis of: controller, systems of deterministic chaos, electronics circuits, etc. are described there. For simulation purposes, AP has been co-joined with EA's like Differential Evolution (DE) [22], Self-Organizing Migrating Algorithm (SOMA) [3], Genetic Algorithms (GA) [6] and Simulated Annealing (SA) [4, 10]. All case studies are described, mentioned and referenced there.

The initial idea of symbolic regression by means of a computer program was proposed in GP [11]. The other approach of GE was developed in [15] and AP in [27]. Another interesting investigation using symbolic regression was carried out in [20] on AIS and Probabilistic Incremental Program Evolution (PIPE), which generates functional programs from an adaptive probability distribution over all possible programs. Yet another new technique is the so called *Transplant Evolution*, see [23] or [24] which is closely associated with the conceptual paradigm of AP, and modified for GE. GE was also extended to include DE by [14]. Symbolic regression is schematically depicted in Fig. 9.1. Generally speaking, it is a process which combines, evaluates and creates more complex structures based on some elementary and non-complex objects, in an evolutionary way. Such elementary objects are usually simple mathematical operators $(+, -, \times, \ldots)$, simple functions (*sin, cos, And, Not, ...*), user-defined functions (simple commands for robots—MoveLeft, TurnRight, ...), etc. An output of symbolic regression is a more complex "object" (formula, function, command, ...), solving a given problem like data fitting of the so-called Sextic and Quintic problem described by Eq. (9.1) [12], randomly synthesized

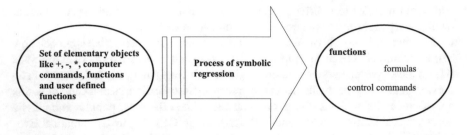

Fig. 9.1 Symbolic regression—schematic view

function by Eq. (9.2), Boolean problems of parity and symmetry solution (basically logical circuits synthesis) by Eq. (9.3) [11, 29], or synthesis of quite complex robot control command by Eq. (9.4) [12, 16]. Equations (9.1)–(9.4) mentioned here are just a few samples from numerous repeated experiments done by AP, which are used to demonstrate how complex structures can be produced by symbolic regression in general for different problems.

$$x\left(K_1 + \frac{\left(x^2 K_3\right)}{K_4\left(K_5 + K_6\right)}\right) * \left(-1 + K_2 + 2x\left(-x - K_7\right)\right) \tag{9.1}$$

$$\sqrt{t}\left(\frac{1}{\log(t)}\right)^{\sec^{-1}(1.28)} \log^{\sec^{-1}(1.28)}\left(\sinh\left(\sec\left(\cos\left(1\right)\right)\right)\right) \tag{9.2}$$

$$
\begin{aligned}
&Nor[(Nand[Nand[B \,||\, B,\ B \,\&\&\, A],\ B]) \,\&\&\, C \,\&\&\, A \,\&\&\, B,\\
&Nor[(\,!C \,\&\&\, B \,\&\&\, A \,||\, !A \,\&\&\, C \,\&\&\, B \,||\, !C \,\&\&\, !B \,\&\&\, !A) \,\&\&\,\\
&(\,!C \,\&\&\, B \,\&\&\, A \,||\, !A \,\&\&\, C \,\&\&\, B \,||\, !C \,\&\&\, !B \,\&\&\, !A) \,||\,\\
&A \,\&\&\, (\,!C \,\&\&\, B \,\&\&\, A \,||\, !A \,\&\&\, C \,\&\&\, B \,||\, !C \,\&\&\, !B \,\&\&\, !A),\\
&(C \,||\, !C \,\&\&\, B \,\&\&\, A \,||\, !A \,\&\&\, C \,\&\&\, B \,||\, !C \,\&\&\, !B \,\&\&\, !A) \,\&\&\, A]]
\end{aligned}
\tag{9.3}
$$

$$
\begin{aligned}
&\text{Prog2[Prog3[Move, Right, IfFoodAhead[Left, Right]],}\\
&\quad\text{IfFoodAhead[IfFoodAhead[Left, Right], Prog2[IfFoodAhead[}\\
&\quad\text{IfFoodAhead[IfFoodAhead[Left, Right], Right], Right],}\\
&\quad\text{IfFoodAhead[Prog2[Move, Move], Right]]]]}
\end{aligned}
\tag{9.4}
$$

Lets briefly discuss the main ideas of GP, GE and AP to give better overview of SR background in order to understand how it has been used on examples here.

9.1.1   Genetic Programming

GP was the first tool for symbolic regression carried out by means of computers instead of humans. The main idea comes from GA, which was used in GP [11, 12]. Its ability to solve very difficult problems is well proven; for example, GP performs so well that it can be applied to synthesize highly sophisticated electronic circuits, etc.

The main principle of GP is based on GA, which is working with populations of individuals represented in the LISP programming language. Individuals in a canonical form of GP are not binary strings, different from GA, but consist of LISP symbolic objects (commands, functions, …), etc. These objects come from LISP, or they are simply user-defined functions. Symbolic objects are usually divided into two classes: functions and terminals. Functions were previously explained and terminals represent a set of independent variables like x, y, and constants like π, 3.56, etc.

The main principle of GP is usually demonstrated by means of the so-called trees (basically graphs with nodes and edges, as shown in Figs. 9.2 and 9.3, representing individuals in LISP symbolic syntax). Individuals in the shape of a tree, or formula like $0.234\,Z + X - 0.789$, are called programs. Because GP is based on GA,

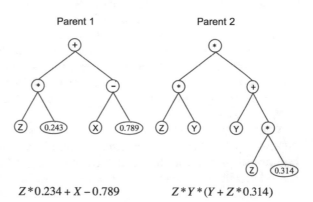

Fig. 9.2   Parental trees

$$Z*0.234 + X - 0.789 \qquad Z*Y*(Y + Z*0.314)$$

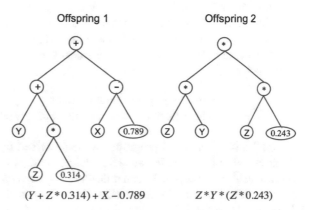

Fig. 9.3   Offsprings

$$(Y + Z*0.314) + X - 0.789 \qquad Z*Y*(Z*0.243)$$

evolutionary steps (mutation, crossover, …) in GP are in principle the same as GA. As an example, GP can serve two artificial parents—trees on Figs. 9.2 and 9.3, representing programs $0.234 Z + X - 0.789$ and $ZY(Y + 0.314Z)$. When crossover is applied, for example, subsets of trees are exchanged. Resulting offsprings of this example are shown on Fig. 9.3.

Subsequently, the offspring fitness is calculated, such that the behavior of the just-synthesized and evaluated individual-tree should be as similar as possible to the desired behavior. The desired behavior can be regarded as a measured data set from some process (a program that should fit them as well as possible) or like an optimal robot trajectory, i.e., when the program is evaluating a sequence of robot commands (TurnLeft, Stop, MoveForward,...) leading as close as possible to the final position. This is basically the same for GE.

For detailed description of GP, see for example classical books [11, 12].

9.1.2 Grammatical Evolution

GE is another program developed in [15] which performs a similar task to that of GP. GE has one advantage over GP, which is the ability to use any arbitrary programming language, not only LISP as in the case of the canonical version of GP. In contrast to other EA's, GE was used only with a few search strategies, and with a binary representation of the populations. The last successful experiment with DE applied on GE was reported in [14]. GE in its canonical form is based on GA, thanks to few important changes it has in comparison with GP. The main difference is in the individual coding.

While GP manipulates in LISP symbolic expressions, GE uses individuals based on binary strings. These are transformed into integer sequences and then mapped into a final program in the Backus-Naur Form (BNF) [15], as explained by the following artificial example. Let $T = \{+, -, \times, /, x, y\}$ be a set of operators and terminals and let F = {epr, op, var} be the so-called nonterminals. In this case, the grammar used for final program synthesis is given in Table 9.1. The rule used for individuals transforming into a program is based on Eq. (9.5) below. GE is based on binary chromosome with a variable length, divided into the so-called codons (range of integer values, 0–255), which is then transformed into an integer domain according to Table 9.2.

$$\text{unfolding} = \text{codon mod rules}$$
$$\text{where rules is number of rules for given nonterminal} \tag{9.5}$$

Synthesis of an actual program can be described by the following. Start with a nonterminal object expr. Because the integer value of Codon 1 (see Table 9.2) is 40, according to Eq. (9.5), one has an unfolding of *expr* = *op expr expr* (40 mod 2, 2 rules for *expr*, i.e., 0 and 1). Consequently, Codon 2 is used for the unfolding of *op* by * (162 mod 4), which is the terminal and thus the unfolding for this part of program is

Table 9.1 Grammatical evolution—rules

Nonterminals	Unfolding	Index
expr	::= op expr expr	0
	var	1
op	::= +	0'
	-	1'
	*	2'
	/	3'
var	:: X	0"
	Y	(1")

Table 9.2 Grammatical evolution—codon

Chromozone	Binary	Integer	BNF index
Codon 1	101000	40	0
Codon 2	11000011	162	2'
Codon 3	1100	67	1
Codon 4	10100010	12	0"
Codon 5	1111101	125	1
Codon 6	11100111	231	1"
Codon 7	10010010	146	Unused
Codon 8	10001011	139	Unused

closed. Then, it continues in unfolding of the remaining nonterminals (*expr expr*) till the final program is fully closed by terminals. If the program is closed before the end of the chromosome is reached, then the remaining codons are ignored; otherwise, it continues again from the beginning of the chromosome. The final program based on the just-described example is in this case $x \cdot y$ (see Fig. 9.4). For a fully detailed description of GE principles, see [15].

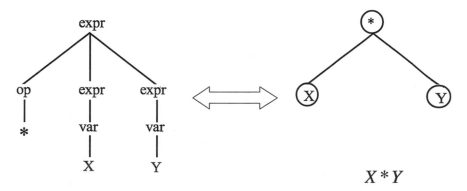

Fig. 9.4 Final program by GE

9.1.3 Analytic Programming

The final method described here and used for experiments in this chapter is called AP, which has been compared to GP with very good results (see, for example, [16, 27, 29]).

The basic principles of AP were developed in 2001 [27] and it is also based on the set of functions, operators and terminals, which are usually constants or independent variables alike, for example:

- **functions**: $sin, tan, tanh, And, Or, \ldots$
- **operators**: $+, -, \times, /, dt, \ldots$
- **terminals**: 2.73, 3.14, $t, \ldots$

All these objects create a set, from which AP tries to synthesize an appropriate solution. Because of the variability of the content of this set, it is called a general functional set (GFS). The structure of GFS is nested, i.e., it is created by subsets of functions according to the number of their arguments (Fig. 9.5). The content of GFS is dependent only on the user. Various functions and terminals can be mixed together. For example, GFS_{all} is a set of all functions, operators and terminals, GFS_{3arg} is a subset containing functions with maximally three arguments, GFS_{0arg} represents only terminals, etc. (Fig. 9.5).

AP, as further described later, is a mapping from a set of individuals into a set of possible programs. Individuals in population and used by AP consist of non-numerical expressions (operators, functions, ...), as described above, which are represented by their integer position indexes in the evolutionary process (Figs. 9.6 and 9.7, see also Chap. 2). This index then serves as a pointer into the set of

Fig. 9.5 Hierarchy in GFS

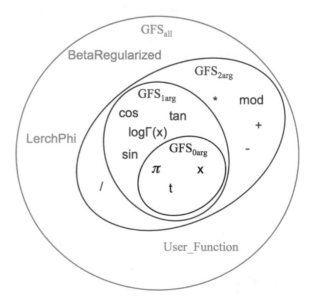

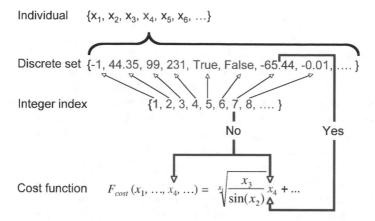

Individual $\{x_1, x_2, x_3, x_4, x_5, x_6, \ldots\}$

Discrete set $\{-1, 44.35, 99, 231, \text{True}, \text{False}, -65.44, -0.01, \ldots\}$

Integer index $\{1, 2, 3, 4, 5, 6, 7, 8, \ldots\}$

No   Yes

Cost function $F_{cost}(x_1, \ldots, x_4, \ldots) = x_1\sqrt{\dfrac{x_3}{\sin(x_2)}} x_4 + \ldots$

Fig. 9.6 DSH-Integer index, see Chap. 2

Fig. 9.7 Principle of mapping from GFS to programs

Individual parameters $\{1, 6, 7, 8, 9, 9\}$ are used by AP like pointers into GFS and through serie of mappings m1 - m5 final formula $\sin(\tan(t)) + \cos(t)$ is created.

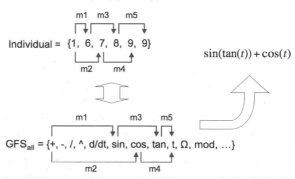

m1   m3   m5

Individual $= \{1, 6, 7, 8, 9, 9\}$ $\sin(\tan(t)) + \cos(t)$

m2   m4

m1   m3   m5

$GFS_{all} = \{+, -, /, \wedge, d/dt, \sin, \cos, \tan, t, \Omega, \text{mod}, \ldots\}$

m2   m4

expressions and AP uses it to synthesize the resulting function-program for cost function evaluation.

Figure 9.7 demonstrates an artificial example how a final function is created from an integer individual via Discrete Set Handling (DSH). Number 1 in the position of the first parameter means that the operator $+$ from GFS_{all} is used (the end of the individual is far enough). Because the operator $+$ must have at least two arguments, the next two index pointers 6 (*sin* from GFS) and 7 (*cos* from GFS) are dedicated to this operator as its arguments. The two functions, *sin* and *cos*, are one-argument functions, so the next unused pointers 8 (*tan* from GFS) and 9 (*t* from GFS) are dedicated to the *sin* and *cos* functions. As an argument of *cos*, the variable *t* is used, so this part of the resulting function is closed (*t* is zero-argument) in its AP development. The one-argument function *tan* remains, and because there is one unused pointer 9, *tan* is mapped on *t* which is on the 9th position in GFS.

To avoid synthesis of pathological functions, a few security *tricks* are used in AP. The first one is that GFS consists of subsets containing functions with the same or a smaller number of arguments. The nested structure (see also Fig. 9.5) is used in the special security subroutine, which measures how far the end of an individual is and, according to this, mathematical elements from different subsets are selected to avoid pathological functions synthesis. More precisely, if more arguments are desired then a possible function (the end of the individual is near) will be replaced by another function with the same index pointer from the subset with a smaller number of arguments. For example, it may happen that the last argument for one function will not be a terminal (zero-argument function). If the pointer is longer than the length of subset, e.g., a pointer is 5 and is used GFS_0, then the element is selected according to the rule: element = pointer_value mod number_of_elements_in_GFS_0. In this example, the selected element would be the variable t (see GFS_0 in Fig. 9.5).

GFS need not be constructed only from clear mathematical functions as demonstrated above, but may also be constructed from other user-defined functions, e.g., logical functions, functions which represent elements of electrical circuits or robot movement commands, linguistic terms, etc.

9.1.3.1 Versions

AP was evaluated in three versions. All three versions utilize the same set of functions for program synthesis, terminals, etc., as in GP [11, 12]). The second version labelled as AP_{meta} (the first version, AP_{basic}) is modified in the sense of constant estimation. For example, the so-called sextic problem was used in [12] to randomly generate constants, whereas AP uses only one, called K, which is inserted into the formula (9.6) below at various places by the evolutionary process. When a program is synthesized, all K's are indexed as K_1, K_2, ..., K_n to obtain (9.7) the formula, and then all K_n are estimated by using a second EA, the result of which can be, for example, (9.8). Because EA (slave) "works under" EA (master), i.e., $EA_{master} \rightarrow$ program $\rightarrow K$ indexing $\rightarrow EA_{slave} \rightarrow$ estimation of K_n, this version is called AP with metaevolution, denoted as AP_{meta}.

$$\frac{x^2 + K}{\pi^K} \tag{9.6}$$

$$\frac{x^2 + K_1}{\pi^{K_2}} \tag{9.7}$$

$$\frac{x^2 + 3.56}{\pi^{-229}} \tag{9.8}$$

Due to this version being quite time-consuming, AP_{meta} was further modified to the third version, which differs from the second one in the estimation of K. This is accomplished by using a suitable method for nonlinear fitting (denoted AP_{nf}). This method has shown the most promising performance when unknown constants are present. Results of some comparative simulations can be found in [27]. AP_{nf} was the method chosen for the simulations described in this chapter.

9.1.3.2   Analytic Programming Subroutines

AP described above is in full detail explained in [27]. Special procedures, that ensure that AP is stable, fast and efficient algorithm are described there. Reader who is interested in this topic can get an explanation what structure is created by basic building block elements, how is individual (i.e. the vector of the real number representation) mapped into space of possible programs/solutions (i.e. math formulas, electronic circuits etc.), how crossover and mutation are done, what is reinforced AP evolution as well as how, so called security procedures, are used to ensure that AP will generate "non-pathological" solutions (i.e. closed solutions—no divide by 0, synthesized function has all arguments as needed etc.). Also similarities and differences between AP, GP and GE are discussed in [27].

9.2   Selected Applications

This section briefly describes some selected applications of classical EA and AP use, which have been conducted during the past few years. The most representative from our own research are

1. Evolutionary control of the chemical reactor control and synthesis of its geometrical structure and parameters estimation.
2. Civil engineering problem solution.

The first case has been solved by classical evolutionary approach and the second one by means of SR with AP and SOMA use.

9.2.1   Chemical Reactor—Predictive Control and Design

Chemical process control requires intelligent monitoring due to the dynamic nature of the chemical reactions and the non-linear functional relationship between the input and output variables involved. Chemical reactors are one of the major processing units in many chemical, pharmaceutical and petroleum industries as well as in environmental and waste management engineering. In spite of continuing advances in optimal solution techniques for optimization and control problems, many of such problems remain too complex to be solved by the known techniques. In chemical engineering, evolutionary optimization has been applied by the author and others to

system identification a model of a process is built and its numerical parameters are found by error minimization against experimental data. Evolutionary optimization has been widely applied to the evolution of neural networks models for use in control applications. The area of reactor network synthesis currently enjoys a proliferation of contributions in which researchers from various perspectives are making efforts to develop systematic optimization tools to improve the performance of chemical reactors. The contributions reflect on the increasing awareness that textbook knowledge and heuristics [13], commonly employed in the development of chemical reactors, are now deemed responsible for the lack of innovation, quality, and efficiency that characterizes many industrial designs, [13]. The main aim of this participation is to show that evolutionary algorithms (EAs) are capable of optimization on chemical engineering processes. The ability of EAs to successfully work with at investigation on optimization and predictive control of chemical reactors. Firstly, a nonlinear mathematical model is required to describe the dynamic behavior of batch process; this justifies the use of evolutionary method to deal with this process, for static optimization of a chemical continuous stirred tank reactor. Consequently, it is used to design geometry technique equipment for chemical reaction. In the next part, we have used EAs to predictive control of chemical process of reactor, too. The optimized reactor and predictive control were used in a simulation with optimization by evolutionary algorithms and the results are presented in graphs. The use of evolutionary algorithms in optimization and control of chemical technologies is very important today and many researchers are working in that field. They are using classical as well as evolutionary algorithms for those purposes, lets mention for example [13, 17] for more classical approach to solve problems of chemical technologies, [9]. Surprisingly, many problems can be defined as optimization problems, e.g. the optimal trajectory of robot arms; the optimal thickness of steel in pressure vessels; the optimal set of parameters for controllers; optimal relations or fuzzy sets in fuzzy models; and so on. Solutions to such problems are usually more or less hard to arrive at, their parameters usually including variables of different types, such as real or integer variables. Evolutionary algorithms are quite popular because they allow the solution of almost any problem in a simplified manner, because they are able to handle optimizing tasks with mixed variables—including the appropriate constraints, as and when required. This paper explains SOMA's use on design and predictive control of given chemical reactor.

9.2.1.1 Reactor Description

Model of the reactor as depicted in Fig. 9.8 inside which above-mentioned reactions can be realized was given by the system of equations (9.9). This is set of 5 nonlinear differential equations that are mutually joined and coupled. Exact solution of such kind of equation system is in an analytic way usually very hard or impossible. In the system (9.9) there are also many free, i.e. adjustable parameters. The set of adjustable parameters is in Table 9.3. Some of them are given by kind of chemical reactions running inside reactor, some of them can be set 'arbitrary' based on expert knowledge

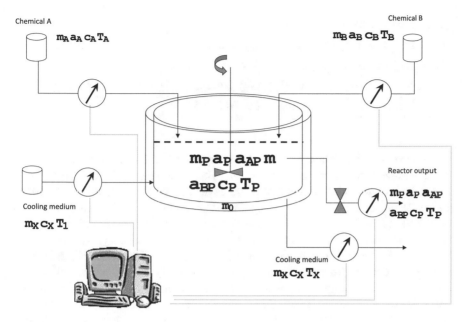

Fig. 9.8 Reactor scheme

Table 9.3 Chemical reactor adjustable parameters

Value	Units in SI	Description
m	Kg	Chemical mass inside reactor
m_A	Kgs^{-1}	Chemical input A
m_B	Kgs^{-1}	Chemical input B
a_A	%	Concentration of chemical o input A
a_B	%	Concentration of chemical on input B
T_A	K	Input temperature of chemical on input A
T_B	K	Input temperature of chemical on input B
T_1	K	Input temperature of cooling medium
m_P	Kgs^{-1}	Output of chemical product
S	m^2	Cooling surface
m_0	Kgs^{-1}	I/O cooling medium
m_X	Kg	Cooling medium inside reactor

or/and by selected numerical methods of optimization. For our purposes the set of adjustable parameters is in Table 9.3. The chemical reactions in this reactor belong to the class of exothermic reactions (releasing heat) so reactor needs to be cooled, as mentioned on the Fig. 9.8 (double reactor wall for cooling medium). For more about chemical reactors and its control it is recommended to see for example [13, 17].

$$\dot{a}_{AP} = \frac{a_A m_A}{m} - \frac{a_{AP}(t)m_P}{m} - Ae^{-\frac{E}{RT_P(t)}} a_{AP}(t)a_{BP}(t)$$

$$\dot{a}_{BP} = \frac{a_B m_B}{m} - \frac{a_{BP}(t)m_P}{m} - Ae^{-\frac{E}{RT_P(t)}} a_{AP}(t)a_{BP}(t)$$

$$\dot{a}_P = Ae^{-\frac{E}{RT_P(t)}} a_{AP}(t)a_{BP}(t) - \frac{a_P(t)m_P}{m}$$

$$\dot{T}_P = \frac{T_A m_A c_A}{mc_P} - \frac{T_B m_B c_B}{mc_P} + \frac{Ae^{-\frac{E}{RT_P(t)}} a_{AP}(t)a_{BP}(t)\Delta \dot{H}_r}{c_P} - \frac{Skp_1 T_P(t)}{mc_P} - \frac{m_P T_P(t)}{m} + \frac{Skp_1 T_x(t)}{mc_P}$$

$$\dot{T}_x = \frac{T_1 m_x}{m_0} + \frac{Skp_1 T_P(t)}{m_0 c_X} - \frac{T_P(t)m_X}{m_0} - \frac{Skp_1 T_X(t)}{m_0 c_X}$$

$$(9.9)$$

Certain class of chemical reactions like enzymatic dechromation technology, etc. can be realized inside this kind of reactor. The advantage of the enzymatic reaction is the production of protein hydrolyzates of relatively good quality and chrome sludge. Using organic bases to form alkaline reaction mixture increases the quality of both its products. A partial regeneration of organic base when diluted protein hydrolyzates undergo concentration cuts the operating costs of enzymatic hydrolysis. In commercial application, the greatest volume of protein hydrolyzate is channelled into agriculture. Hydrolyzate, as an organic nitrogenous fertilizer, not only equals the combined urea-ammonium nitrate fertilizer in crop yield, but also surpasses its manifolds in the foodstuff value of consumer's greens. The content of nitrates is as much as 200 times lower on average. Hydrolyzate is also used in the manufacture of biodegradable foil, especially for producing sowing tape. Application of protein hydrolyzates appears to have good potential in the building industry and in manufacture of modified amino-plastic for adhesive compounds of zero, or at least very low, free formaldehyde content. A serious problem, and at present one lacking a completely satisfactory solution, is the development of recycling technology for chrome from so called filter cake. The main obstacle for the utilization of the chrome sludge is a relatively high content of proteins in the dry substance of cake. In closing, it may be said that enzymatic hydrolysis has a place in the treatment of chromium containing tannery waste and the funds expended on this field of research have brought satisfactory results.

9.2.1.2 Experiment Design and Used Algorithms

For our experiments, focused on predictive control of reactor (9.9), described here, reactor has been tested for its behavior under expert setting and evolutionary estimated and set parameters. Before predictive control has been tested on this reactor, behavior of reactor was investigated in order to select the best configuration for predictive control experiments. This was done in [28] and is only mentioned here. Expert parameters were used for initial—original setting. They come from experiences of experts and literature and were partly obtained during visit in laboratory, Resine and Composite for Forest Products' in Sainte-Foy, Canada. This set of parameters consisted of two kind of parameters i.e. parameters of chemical materials and physical parameters of reactor under consideration, see system of (9.9). An initial conditions $(a_{AP0}, a_{BP0}, a_{P0}, T_{P0}, T_{X0},)$ used in following simulations are also described in Eq. (9.9). As the result shows [28] the reactor under expert parameters produce unsatisfactory behavior [28]. Reactor production stabilizes itself (i.e. without control procedure) after 27 h on 15 % concentration of output chemical. From that point of view above-mentioned parameters were regarded as unsatisfactory. Used evolutionary algorithm was SOMA [3, 25]. It is a stochastic optimization algorithm that is modeled based on the social behavior of competitive—cooperating individuals. It was chosen because it has been proven that this algorithm has the ability to converge towards the global optimum. Before predictive control a few static optimizations by SOMA algorithm were consequently done in [28]. They were done in following steps:

1. Optimization without restrictions.
2. Optimization with restrictions applied exactly in given time.
3. Optimization with penalty applied during time interval.
4. Optimization with penalty applied during time interval and sub optimization of cooling surface.

The last static optimization is the important one. Each of four optimization cases was 10× repeated and consequently depicted [28]. From all 10 simulations the best reactor was finally chosen. In all four cases evolutionary optimization was focused on parameters (Table 9.3) estimation. According to reactor setting based on the Table 9.3 a global extreme was found in 13th dimensional configuration space. Last 13th dimension was cost value of cost function. In case of the last optimization (optimization with penalty applied during time interval and sub-optimization of cooling surface) searching for global extreme had run in 11th dimensional space because of relations among some parameters. Based on this, we decided that for predictive control (again done by evolutionary algorithms) will be used reactor whose structure was optimized by evolutionary algorithms. As mentioned in [28], reactor under expert parameters produce unsatisfactory behavior. Reactor production stabilizes itself (i.e. without control procedure) after 27 h on 15 % concentration of output chemical. Optimization of the parameters and reactor structure gives configuration, in which reactor production stabilizes itself (i.e. without control procedure) after 8 min on 50 % concentration of output chemical. It is 0.009 of the original time from expert set reactor. Thus improvement of the reactor is more than significant.

Differences between the two reactors are best seen in Table 9.5. Parameters expertly designed reactor and parameters obtained by static optimization are reported there. What is clear from both sets of parameters, the size of the reactor, which was the most successful optimization tied to the volume and surface cooling reactor is not clear from the Table 9.5. The terms r_S radius r_m and Δr (for the optimal case, the reactor is completely filled) are reported there. The parameter r_s is the radius of the reactor in the event that this relates to the cooling surface S and the radius r_m related to the total reactor volume derived from the m. In the original reactor there is a visible the gap between the two radii, which means that the reactor would need either additional auxiliary cooling surface or part of the cooling surface should not to be used. This is not the case of the optimized reactor. The parameter Δr is the difference between the outer and inner radius because it is a double-walled reactor. The original reactor is the distance between outer and inner casing $\Delta r = 0.08$ m while the optimized $\Delta r = 0.14$ m. Remaining parameters are clearly seen from Tables.

The cost value which minimization leads to the optimal control setting was done by Eq. (9.10) which consist of two parts. The first one was output product concentration a_P and the second output product temperature T_P. In fact during whole study has been used different cost functions (as mentioned in [28]) like Eq. (9.10) (basic optimization of the reactor structure), (9.11) (basic control on fixed output concentration of chemical and temperature a_p), (9.12) and (9.13) advanced cost functions for predictive control without and with penalization of output quality.

$$f_{cost} = 0.6 - \sum_{t=500}^{1000} a_P(t) \tag{9.10}$$

$$f_{cost} = \sum_{i=0}^{3000} a_p(i) + T_p(i) \quad i = \{0, 10, 20, \ldots, 3000\} \tag{9.11}$$

$$f_{cost} = |0.6 - a_P(t)| + |a_{AP}(\tau)| + |a_{BP}(\tau)| + |T_P(\tau)| + |T_X(\tau)|$$

$$where \quad \begin{matrix} t = 1200 \\ \tau = 100 \end{matrix}$$

under conditions

$$a_{AP}(\tau) = \begin{cases} 0 & if \quad a_{AP}(\tau) \in < 0, \ 1 > \\ 100 \, a_{AP}(\tau) & else \end{cases}$$

$$a_{BP}(\tau) = \begin{cases} 0 & if i \quad a_{BP}(\tau) \in < 0, \ 1 > \\ 100 \, a_{BP}(\tau) & else \end{cases} \tag{9.12}$$

$$T_P(\tau) = \begin{cases} 0 & if \quad T_P(\tau) \in < 273.15, \ 273.15 + 150 > \\ T_P(\tau) & else \end{cases}$$

$$T_X(\tau) = \begin{cases} 0 & if \quad T_X(\tau) \in < 273.15, \ 273.15 + 500 > \\ T_X(\tau) & else \end{cases}$$

$$f_{cost} = 0.6 - \sum_{t=500}^{1000} a_P(t) + |a_{AP}(\tau)| + |a_{BP}(\tau)| + |T_P(\tau)| + |T_X(\tau)|$$

under conditions for $\tau \in < 500, \ 1000 >$

$$a_{AP}(\tau) = \begin{cases} 0 & if \quad Max(a_{AP}(\tau)) \ \& \ Min(a_{AP}(\tau)) \in < 0, \ 1 > \\ 100 \, Max(a_{AP}(\tau)) & else \end{cases}$$

$$a_{BP}(\tau) = \begin{cases} 0 & if \quad Max(a_{BP}(\tau)) \ \& \ Min(a_{BP}(\tau)) \in < 0, \ 1 > \\ 100 \, Max(a_{BP}(\tau)) & else \end{cases}$$

$$T_P(\tau) = \begin{cases} 0 & if \quad Max(T_P(\tau)) \ \& \ Min(T_P(\tau)) \in < 273.15, \ 273.15 + 150 > \\ Max(T_P(\tau)) & else \end{cases}$$

$$T_X(\tau) = \begin{cases} 0 & if \quad Max(T_X(\tau)) \ \& \ Min(T_X(\tau)) \in < 273.15, \ 273.15 + 150 > \\ Max(T_X(\tau)) & else \end{cases}$$

$$\tag{9.13}$$

9.2.1.3   Predictive MIMO Evolutionary Control

The main aim of this experiment was to provide predictive control of chemical reactor (9.9), see also [13, 17]. In this case control was considered as the MIMO (multiple input–multiple output) control. Control process was done by evolutionary algorithms use instead of classical controller. Selected algorithm was SOMA [3] and was used like predictive controller estimating control parameters for reactor in order to reach desired values a_{AP} in the shortest time for different values of $a_{AP} = 0.35, 0.25, 0.1,$ 0.45(output concentration) and constant T_P (product temperature). The cost function (the predictive control functional) that has to be minimized by SOMA, in order to follow a_{AP} and T_P was given by Eq. (9.14).

$$J(N_1, N_2, N_u) = \sum_{i=1}^{L}\sum_{j=1}^{N_2} \eta_i(j)[y(k+j) - w(k+j)]^2 + \sum_{i=1}^{M}\sum_{j=1}^{N_u} \lambda_i(j)[\Delta u(k+j-1)]^2$$

(9.14)

In this functional there is variable penalty of overshooting (controlled trajectory can overshoot desired value w and thus needs to be penalized) and in fact there are 2 of them ($L = 2$, penalization of two outputs). The dimension of inputs was 5: $\lambda = \Delta m_A, \Delta m_B, \Delta T_A, \Delta T_B, \Delta T_1 = [400, 200, 20, 20, 20]$, i.e. MIMO 5:2. Controlled variables ap and Tp were penalized when controlled values overstepped bigger value than $\eta = a_p, T_P = [400, 20]$. These parameters were chosen so that the effects of the two controlled variables on the resulting value of the functional approximately equal. Penalization of a_p was selected in the bigger range than of T_p. Thanks to this the impact of penalization on the final value of the functional was at least approximately the same level. Optimization would run of course even for the badly set penalization or also without it. Used algorithm would then put more emphasis on changing the values on which it was appropriate functional more sensitive, i.e. those that contribute more to change its cost value. The above values have been set approximate based on trial and error approach. Those that are numerically shown above, this seemed to be the most suitable for control of the proposed reactor above (9.9). Compared to classical approach of the predictive control, special feature of these simulations was that it did not carry out the calculation of the new controller output in each k_{th} step, but only if system behavior was changed by some external error or by change in the desired values. Given that, control actions were calculated based on the knowledge of future behavior (according to model reactor), it is still predictive control despite modification of the standard conventions for calculating control actions at each step k. This predictive control has been taken for the variable of mass flow rates and temperatures of both input products including coolant temperature. There were done two kinds of predictive control, without and with penalization of the output-controlled values [28], see also Figs. 9.9 and 9.10.

Fig. 9.9 Controlled
concentration

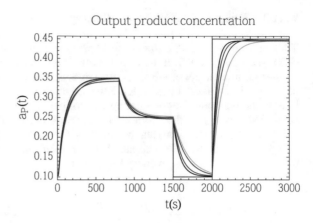

Fig. 9.10 Controlled
temperature

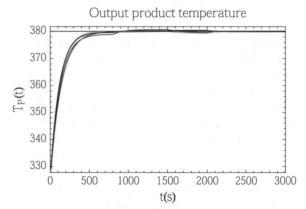

9.2.1.4 Results

As discussed in the previous section, MIMO 5:2 was used in chemical reactor to control two kinds of predictive control, without and with penalization of the output-controlled values. Selected Figs. 9.9 and 9.10 from our experiments show both kinds of simulations that have been 4 times repeated and all 4 simulations were depicted in one figure to show its quality and diversibility. Figures 9.9 and 9.10 show how output variable a_p followed changes of desired value and how T_p stand on desired value for the control with penalization. Results are also reported in the Tables 9.4 and 9.5 and are calculated according to Eqs. (9.10)–(9.13). As seen from figures (Figs. 9.9 and 9.10) and tables, penalization plays important role in such kind of optimization and control.

Table 9.4 Optimal reactor behavior estimated by evolution

Parameter	a_p	T_p
Max	11.46	767.36
Avg	9.05	711.24
Min	6.99	654.15

Table 9.5 Difference between selected reactor parameters

Parameters	Expert setting	SOMA setting
r_s	11.46	767.36
r_m	9.05	711.24
Δr	6.99	654.15

9.2.2 Symbolic Solution of the Nonlinear ODEs

Another use of evolution, in this case the use of the AP on ordinary differential equations (ODE) solution synthesis was used here, especially on

1. ODE solving, exactly $u''(t) = cos(t), u(0) = 1, u(\pi) = -1, u'(0) = 0, u'(\pi) = 0$, see [19], 100× repeated
2. ODE solving, exactly $((4 + x)u''(x))'' + 600u(x) = 5000(x - x2), u(0) = 0, u(1) = 0, u''(0) = 0, u''(1) = 0$, see [19], 5× repeated

In this case we used SR—AP in order to search for such functions u (of course different for each case) that satisfied equations in both cases. Cost value of cost function for these simulations were difference (see Figs. 9.14 and 9.15, size of the filed area between original and just founded function) between actually generated and expected function.

9.2.2.1 On the Hilbert Space and the Classical Approach

Hilbert space widely used in physics and applied mathematics can be viewed like special case of analytic programming (AP). Lets see it in more detailed view. Hilbert space is usually defined like functional space with following properties [19]

1. Completeness.
2. Compactness.
3. Orthogonality.
4. Orthonormality (if no, Schmidt orthonormalization can be used, see [19].
5. …

These demands allow to solve various problems (i.e. PDE etc.) usually in a simple manner. In a geometrical point of view Hilbert space can be depicted like mutually orthogonal axes. Each axis represents one base function, usually periodic functions like sinus or cosine. Position of each point in such so-called *functional space* can be described by its co-ordinates on all axes, or on the contrary, composition of

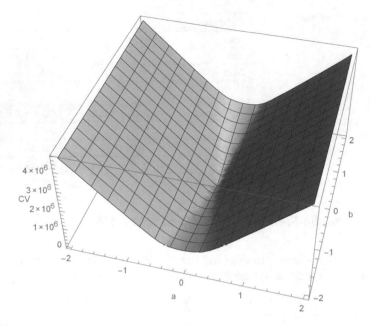

Fig. 9.11 Space of possible $u(x)$ solutions. Each point on surface is a value of the area between curves as on Fig. 9.14

all vector components destined by co-ordinates on all axes, give resulting function – vector from origin to discussed point in the Hilbert space. Solution of given problem, which is usually represented by ODE or PDE, is done in Hilbert space so that there is selected proper functional base of orthogonal functions, inside this base is build resulting *functional* and by means of suitable methods are estimated its unknown parameters. For example in [19] is solved ODE (see Case 2) whose cost function surface is depicted in Fig. 9.11 and some solutions on Fig. 9.14. Surface from Fig. 9.11 is depicted so, that z axis (each point of surface) in Fig. 9.11 is size of area between desired and actual solution depicted in Fig. 9.14. The aim of the numerical methods applied in the Hilbert space is to minimize this area, i.e. to find minimum on the cost function surface from Fig. 9.11.

9.2.2.2 Synthesis of the Case 1

This set of simulations was focused on ODE solving. In [19] it is solved by means of Ritz or Galerkin method on a priori selected functional base, which was orthogonal. Here it was solved by AP without any a priori demands on mixed functional base (sin, cos, ...). The results were fully identical as with classical methods, i.e. $u(t) = -cos(t)$. This example was of course quite simple. More interesting example is the Case 2: the solution synthesis of the $((4 + x)u''(x))'' + 600u(x) = 5000(x - x2)$, $u(0) = 0, u(1) = 0, u''(0) = 0, u''(1) = 0$.

9.2.2.3   Synthesis of the Case 2

This simulation was focused on solution of quite complicated ODE (Case 2) which come from civil engineering. Original solution obtained by means of Ritz or Galerkin method [19], based on Hilbert functional space that consisted of sinus functions and original solution, obtained by means of Ritz or Galerkin method, was according to [19] $u(x) = 1.243 sin(\pi x) + 0.0116 sin(2\pi x) + 0.00154 sin(3\pi x)$.

AP with SOMA was used in two ways. In the first one SOMA was used to estimate only parameters a, b, c of founded $u(x)$, i.e. $u(x) = a \sin(\pi x) + b \sin(2\pi x) + c \sin(3\pi x)$. In the original solution all three coefficients were calculated by means of quite complicated Ritz or Galerkin method. SOMA was able to find all three coefficients as is depicted on Fig. 9.13. This problem was basically classical optimization because functions were a priori known. This simulation was 100 times repeated and in all cases has lead to the same results. The second use of SOMA here was not focused only on parameter estimation. AP with SOMA was used on complex set of functions $(sin, cos, \ldots)$ operators $(+, -, /, \ldots)$ and constants (a, b, c) to find their best combination i.e. to build up function fitting function u as well as possible. The best result are shown on Fig. 9.12. On Figs. 9.14 and 9.15 are snapshots of the temporary solutions recorded during evolution. Through such solutions has AP/SOMA passed through all to the best one.

Fig. 9.12 Acceptable solution by AP (*blue dotted curve*)
$u(x) = 1.243 \sin(\pi x) + 0.0116 \sin(2\pi x)$

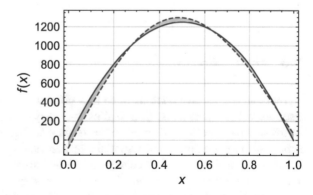

Fig. 9.13 The best solution (*blue dotted curve*)
$u(x) = 1.23413 \sin(\pi x) + 0.00922643 \sin(2\pi x) + 0.000997829 \sin(3\pi x)$
obtained during evolution

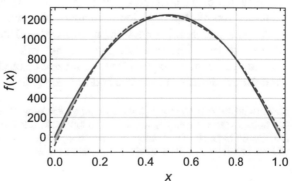

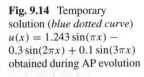

Fig. 9.14 Temporary solution (*blue dotted curve*) $u(x) = 1.243 \sin(\pi x) - 0.3 \sin(2\pi x) + 0.1 \sin(3\pi x)$ obtained during AP evolution

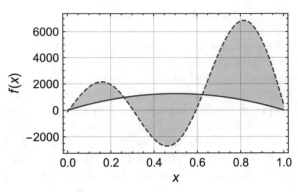

Fig. 9.15 Another temporary solution (*blue dotted curve*) $u(x) = 1.243 \sin(\pi x) + 1 \sin(2\pi x) + 1 \sin(3\pi x)$ obtained during AP evolution

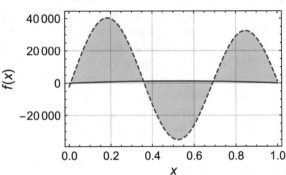

9.3 Conclusion

In this chapter we have demonstrated how different evolutionary approaches can be used in real life problems. Classical evolutionary algorithms as well as symbolic regression have been used in chemical reactor design and control. This system is represented by 5 nonlinear differential equations and evolution has been solving its geometrical design, chemical concentrations and also its control. In the second part we have introduced application of the symbolic regression on the unknown function (solution of the differential equation) synthesis and compared it with classical solution given by Ritz or Galerkin methods in functional spaces. As clearly visible from figures, all simulations show results with satisfactorily quality. This and also other results from the area of bio-inspiring algorithms clearly show that bio-inspired methods are definitely powerful and can be often used to solve very hard problems from various domains of science and/or technology.

Acknowledgments The following grants are acknowledged for the financial support provided for this research: Grant Agency of the Czech Republic–GACR P103/15/06700S, VSB-TU internal grant SGS 2016/175 and by The Ministry of Education, Youth and Sports from the National Programme of Sustainability (NPU II) project "IT4Innovations excellence in science–LQ1602".

References

1. Back, T., Fogel, B., Michalewicz, Z.: Handbook of Evolutionary Computation. Institute of Physics, London (1997)
2. Barricelli, N.: Esempi numerici di processi di evoluzione. *Methodos*, pp. 45–68 (1954)
3. Davendra, D.D., Zelinka, I.: Self-Organizing Migrating Algorithm Methodology and Implementation. Springer, Heidelberg (2016)
4. Černý, V.: Thermodynamical approach to the traveling salesman problem: an efficient simulation algorithm. J. Opt. Theory Appl. **45**(1), 41–51 (1985)
5. Fogel, G., Corne, D.: Evolutionary Computation in Bioinformatics. Bioinformatics artificial intelligence. Morgan Kaufmann, Burlington (2003)
6. Holland, J.H.: Adaptation in Natural and Artificial Systems. The University of Michigan Press, Ann Arbor (1975)
7. Holland, J.H.: Intelligent machinery, unpublished report for national physical laboratory (1975)
8. Holland, J.H.: Adaptation in Natural and Artificial Systems: An Introductory Analysis with Applications to Biology. Control and Artificial Intelligence. MIT Press, Cambridge (1992)
9. Hildebrandt, D., Hopley, F., Glasser, D.: Optimal reactor structures for exothermic reversible-reactions with complex kinetics. Chem. Eng. Sci. **51**(10), 1533–2520 (1996)
10. Kirkpatrick, S., Gelatt, C.D., Vecchi, M.P.: Optimization by simulated annealing. Science **220**(4598), 671–680 (1983)
11. Koza, J.R.: Genetic Programming: On the Programming of Computers by Means of Natural Selection. MIT Press, Cambridge (1992)
12. Koza, J.R., Andre, D., Bennett, F. H., Keane, M.A.: Genetic Programming III: Darwinian Invention and Problem Solving, 1st edn. Morgan Kaufmann Publishers Inc., San Francisco (1999)
13. Luyben, W.L.: Chemical Reactor Design and Control. Wiley-Interscience, 1 edn, (August 2007)
14. O'Neill, M., Brabazon, A.: Grammatical differential evolution. In: Arabnia, H.R (ed.) Proceedings of the 2006 International Conference on Artificial Intelligence, ICAI 2006, vol. 1, pp. 231–236, CSREA Press, Las Vegas, Nevada, USA (2006)
15. O'Neill, Michael, Ryan, Conor: Grammatical Evolution: Evolutionary Automatic Programming in an Arbitrary Language. Kluwer Academic Publishers, Norwell (2003)
16. Oplatková, Z., Zelinka, I.: Investigation on artificial ant using analytic programming. In Proceedings of the 8th Annual Conference on Genetic and Evolutionary Computation, GECCO '06, pp. 949–950, ACM, New York, NY, USA (2006)
17. Perry, R.H., Green, D.W. (eds): Perry's Chemical Engineering Handbook. 6th edn. McGraw-Hill, New York (1984)
18. Rechenberg, I.: Evolutionsstrategie: optimierung technischer systeme nach prinzipien der biologischen evolution. Frommann-Holzboog (1973)
19. Rektorys, K.: Variational methods in Engineering Problems and Problems of Mathematical Physics, vol. 1. Academia, Prague (1999)
20. Rafal, S., Jürgen, S.: Probabilistic incremental program evolution. Evol. Comput. **5**(2), 123–141 (June 1997)
21. Schwefel, H.P.: Numerische Optimierung von Computer-Modellen mittels der Evolutionsstrategie. ISR, vol. 26. Birkhaeuser, Basel/Stuttgart (1977)
22. Storn, R., Price, K.: Differential evolution-a simple and efficient heuristic for global optimization over continuous spaces. J. Glob. Opt. **11**(4), 341–359 (Dec 1997)
23. Weisser, R., Osmera, P.: Two-level tranpslant evolution. In Proceedings of the 17th Zittau Fuzzy Colloquium (2010)
24. Weisser, R., Osmera, P., Matousek, R.: Transplant evolution with modified schema of differential evolution : optimization structure of controllers. In Proceedings of the International Conference on Soft Computing, MENDEL, Brno, Czech Republic (2010)
25. Zelinka, I.: Soma-self organizing migrating algorithm. In: Onwubolu, G.C., Babu, B. (eds.) New Optimization Techniques in Engineering, Springer, New York, pp. 167–218 (2004). ISBN 3-540-20167X

26. Zelinka, I., Celikovský, S., Richter, H., Chen, G. (eds.): Evolutionary Algorithms and Chaotic Systems. Studies in Computational Intelligence, vol. 267. Springer, Heidelberg (2010)
27. Zelinka, I., Davendra, D., Senkerik, R., Jasek, R., Oplatkova, Z.: Analytical Programming-a Novel Approach for Evolutionary Synthesis of Symbolic Structures. InTech (2011)
28. Zelinka, I., Davendra, D.D., Šenkeřík, R., Pluháček, M.: Investigation on evolutionary predictive control of chemical reactor. J. Appl. Log. **13**(2 Part A):156–166, 2015
29. Zelinka, I., Oplatkova, Z., Nolle, L.: Analytic programming-symbolic regression by means of arbitrary evolutionary algorithms. Int. J. Simul. Syst. Sci. Technol. **6**(9):44–56, aug 2005. Special Issue on: Intelligent Systems

Chapter 10
Marangoni Flow Driven Maze Solving

Kohta Suzuno, Daishin Ueyama, Michal Branicki, Rita Tóth,
Artur Braun and István Lagzi

Abstract Algorithmic approaches to maze solving problems and finding shortest paths are generally NP-hard (Non-deterministic Polynomial-time hard) and thus, at best, computationally expensive. Unconventional computational methods, which often utilize non-local information about the geometry at hand, provide an alternative to solving such problems much more efficiently. In the past few decades several chemical, physical and other methods have been proposed to tackle this issue. In this chapter we discuss a novel chemical method for maze solving which relies on the Marangoni flow induced by a surface tension gradient due to a pH gradient imposed between the entrance and exit of the maze. The solutions of the maze problem are revealed by paths of a passive dye which is transported on the surface of the liquid in the direction of the acidic area, which is chosen to be the exit of the maze. The shortest path is visualized first, as the Marangoni flow advecting the dye particles is the most intense along the shortest path. The longer paths, which also solve the maze, emerge subsequently as they are associated with weaker branches of the chemically-induced Marangoni flow which is key to the proposed method.

K. Suzuno · D. Ueyama
Graduate School of Advanced Mathematical Sciences, Meiji University,
Tokyo, Japan

M. Branicki
School of Mathematics, University of Edinburgh, Edinburgh, UK

R. Tóth · A. Braun
Laboratory for High Performance Ceramics, Empa, Swiss Federal Laboratories
for Materials Science and Technology, Dübendorf, Switzerland

I. Lagzi (✉)
Department of Physics, Budapest University of Technology and Economics,
Budafoki út 8, Budapest 1111, Hungary
email: istvanlagzi@gmail.com

© Springer International Publishing Switzerland 2017
A. Adamatzky (ed.), *Advances in Unconventional Computing*,
Emergence, Complexity and Computation 23,
DOI 10.1007/978-3-319-33921-4_10

10.1 Introduction

Mazes and the ability to find a way through them have an intriguing and mysterious appeal to humans. They have been enshrined in the human culture for millennia, from ornaments and mythologies (e.g., the story of Theseus and the Minotaur in the Greek mythology), to contemporary fairytales, books and movies (e.g., Maze runner by James Dashner). The motif of a maze in human culture is unsurprisingly associated with the task of solving a complex problem with potentially many viable answers which cannot be distinguished in their entirety by a local observer. It is thus not surprising that the geometric and topological complexity of a maze and its solutions (i.e., one or more paths leading from the entrance to the exit) serves as a model configuration in many areas of science and technology (e.g., logistics, robot control, neuroscience, etc.). It has been shown that besides humans, animals, and computer algorithms, some amoeboid organisms [1–3], and even nonliving, synthetic constructs are 'able' to solve mazes [1–12]. Such chemical, physical or biological systems are initially in a non-equilibrium thermodynamic state with a spatial gradient of some thermodynamic variable, e.g., temperature, chemical potential, pressure, electric or magnetic field, which induces a flow of matter (momentum) or energy within the system to reach its equilibrium state. Some of the most prominent approaches are briefly mentioned next. Microfluidic networks are often solved by imposing a pressure gradient across the corresponding maze [4] between the entrance and the exit so that the pressure-induced flow has the largest amplitude along the shortest path. An electric field gradient was used to induce a glow discharge in gas-filled microchannels and to identify the shortest path in mazes or urban city maps [5, 6]; in a medium conducting electric current the shortest path is characterized by the largest gradient of the electric field which ionizes a gas and induces a plasma glow. Maze solving by a network of memristors is also based on the presence of an electric potential gradient [7]. Chemical and electric potential wave propagation along a dendritic tube of a single cell organism is the most commonly employed setup for identifying the shortest path between two food sources in a biological system [1–3]. Finally, in chemical systems a chemical potential gradient created at the beginning of the experiment induces a flow of matter which highlights the shortest path in a maze in a number of distinct ways [8–13]. A silver ion gradient initiates the propagation of a chemical wave in the Belousov–Zhabotinsky solution along the paths of a maze with the fastest wave corresponding to the shortest path [8, 9]. The concentration gradient of sodium acetate initiates the propagation of a supersaturation front in a complex structure of a hot ice computer [10]. A pH gradient is responsible for the movement of a surfactant covered organic droplet in a maze filled with an alkaline solution [11]. When a surfactant is in the system, the formation of a pH gradient has more intricate consequences on the resulting macroscopic dynamics in the maze. Surfactants reduce surface tension and the concentration of the fatty acid surfactant depends on the pH of the medium. Therefore, the pH gradient creates a difference between the surface tension of the two sides of the droplet facing the acid and base, making it move in the direction of acid.

In this chapter we show that the so called Marangoni flow induced by a pH gradient can be used as an operator for efficient maze solving. For the practical realization of such a chemical computer it is necessary to mention that we have to deal with a liquid chemistry environment [12, 14].

10.2 Experimental

First, mazes with various topological complexity and spatial extent were designed and fabricated from polydimethylsiloxane (PDMS) using photo- and laserlithography (with thickness and depth of 1.4 and 1 mm, respectively). In a typical experiment the maze was filled with a 0.05 M alkaline solution of potassium hydroxide (KOH, Sigma-Aldrich) containing 0.2 % of 2-Hexyldecanoic acid (Sigma Aldrich) (2-HDA). 2-HDA is a fatty acid and by itself is not soluble in water. However, in alkaline solution the head group is deprotonated and becomes soluble. Consequently, the fatty acid molecules are oriented at the liquid-air interface and the deprotonated form of 2-HDA acts as surfactant (reducing the surface tension at the liquid-air interface). An acidic hydrogel (Agarose, Sigma-Aldrich) block ($\sim 1 \times 1 \times 1$ mm) was placed at the exit of the maze. After addition of an acidic block, a small amount (~ 0.3 mg) of dry Phenol Red dye powder was placed at the liquid-air interface at the starting point (the other entrance of the maze). With this technical set-up of the maze and the necessary chemical reaction and diffusion partners, we can run a time resolved experiment where we observe and track the spatial transient of the colorization of the paths through the maze.

10.3 Results and Discussion

In the presented experimental setup the dye particles traveled passively at the liquid-air interface towards the acidic hydrogel block, i.e., the region of low pH. The dye particles transported by the pH-induced Marangoni flow gradually dissolved in the water phase and the color showed their paths through the maze. Figure 10.1a shows maze solving experiments in various mazes filled with an alkaline solution of a fatty acid. The symmetry in the system is broken by an acidic hydrogel. In a typical experiment, the shortest path can be found and visualized within ~ 10 s (Fig. 10.1a).

The Marangoni flow facilitating the maze solving is induced by the non-uniform distribution of the surface tension at the liquid-air interface, and it drives transport of the top fluid layer towards the higher surface tension regions from the low surface tension regions. The intensity of the fluid flow is propositional to the gradient of the surface tension. Addition of an acidic block to the maze filled with an alkaline solution of a fatty acid changes the surface tension of the solution at the liquid-air interface. This surface tension difference (surface tension gradient) creates and maintains a fluid flow in the liquid phase which is commonly referred to as the Marangoni flow.

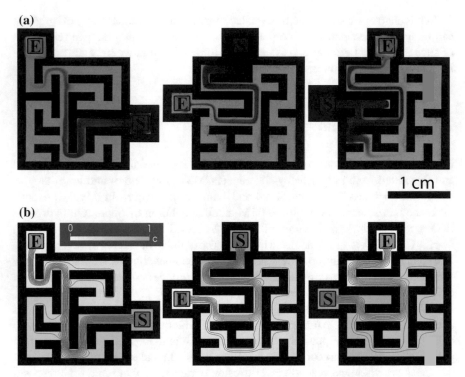

Fig. 10.1 Maze solving and finding the shortest path in various mazes, experiments (**a**) and numerical simulations (**b**). Position of the gel soaked with acid (exit) is indicated by letter *E*. Letter *S* shows the entrances of the maze (starting point), where Phenol Red dye particles (passive tracer) are added. Reprinted from [12] under Creative Commons Attribution 4.0 International Public License

The most intense fluid flow is established along the shortest path in a maze, where the gradient of the surface tension at the liquid-air interface is the highest. Thus, the most of the tracer particles are caught up in the dominant flow branch along the shortest path which is thus characterized by the most intense color contrast of the dissolved dye. However, it should be noted that in a relatively complex maze all possible (and not only the shortest) solution paths can be explored by the Marangoni flow provided that a sufficiently long time is allowed. In our setup it is ∼60 s instead of 10 s which was sufficient for exploring just one path (the shortest path).

The existence of the gradient of the surface tension in a maze can be explained by the effect of pH on the protonation rate of fatty acid molecules. Fatty acid molecules can be protonated/deprotonated by different extent depending on the pH of the medium. Therefore, pH can be used as a technical control parameter to create and maintain a Marangoni flow in a channel network. We performed a range of numerical simulations in mazes to verify the experimental concept and obtain the shortest paths (Fig. 10.1b). Figure 10.2 shows the numerically simulated analogue of the determination of the shortest path between two arbitrary points in a complex maze (downtown

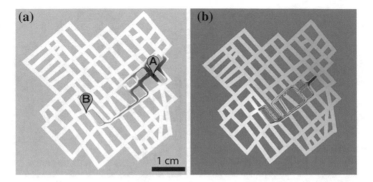

Fig. 10.2 Finding the shortest path between two points in a channel network (made from PDMS) based on the street map of downtown of Budapest, experiments (**a**) and numerical simulations showing major streamlines (**b**). Position of the gel soaked with acid (end point) is indicated by letter B. Letter A shows the starting point, where Phenol Red dye particles are added. Figure 10.2a is reprinted from [12] under Creative Commons Attribution 4.0 International Public License

of Budapest in this case). The streamlines associated with the possible paths were calculated from the gradient of a stationary concentration $c(x, y)$ between two points (marked by letter A and B in Fig. 10.2) which is calculated form the time-invariant solutions of the diffusion equation in two-dimensions given by

$$\frac{\partial c}{\partial t} = D \left(\frac{\partial^2 c}{\partial x^2} + \frac{\partial^2 c}{\partial y^2} \right), \tag{10.1}$$

with Dirichlet boundary conditions, $c = 1$ at A, and $c = 0$ at B. The solutions procedure involved the ADI (Alternating Direction Implicit) method to solve (10.1) on a 1000×1000 uniform rectangular grid. Finally, the streamlines were calculated form the gradient field of the concentration. The simulation results capture the main features of the experiments. Importantly, the shortest and the second-shortest paths found in the chemical computer experiments are in good agreement with the paths detected in the numerical simulations. Several other solution paths visible in the simulations are less pronounced in the experiments. This discrepancy could arise from the finite viscosity of the fluid in the maze (the dynamics of the fluid is not yet considered in the purely diffusive simulations of (10.1)) and the finite time duration of the experiment. In experiments with channels where the surface tension gradient is weak enough, the fluid viscosity suppresses the generation of the convective flow. Moreover, the elapsed time in the real experiment is finite, whereas the numerical solution represents the stationary, time-asymptotic solution. One additional aspect pertaining to this method deserves a mention. Theoretically, the path that connects the bottom-right branch of the site A to B (see arrow in Fig. 10.3) should be the third-shortest path, although it is not found and observed neither experimentally or numerically.

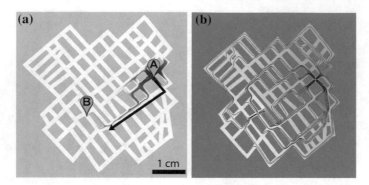

Fig. 10.3 Finding the shortest path between two points in a channel network based on the street map of downtown of Budapest, experiments (**a**) and numerical simulations showing all streamlines (**b**). Figure 10.3a is modified from [12] under Creative Commons Attribution 4.0 International Public License

Hence, the geometrically-third-shortest path is not detected in this system. It would require more conceptual and experimental efforts to raise this kind of orthogonal solution as well. Interestingly, this property is both disadvantageous and advantageous. The path-finding method presented here does not guarantee identification of geometrically shortest paths, since it is based on the real physicochemical state of the system. Consequently, the experimental outcome could be disturbed by the fluid resistance, environmental perturbations and the system size, etc. On the other hand, the system can be treated as a solver that handles the path-finding problem with realistic constraints, which might be very inefficient to solve 'on-the-fly' algorithmically. The presented method allows for detecting those solution paths which are the physicochemically-permitted shortest ones, including the restriction of the channel capacity, rather than the geometrically-optimal paths. This property would be important when we consider an application, for example the chemically-induced transport of chemical entities (e.g., drug delivery).

10.4 Conclusions

In this chapter we presented a novel approach to maze solving which is achieved by generating a pH induced Marangoni flow in the considered channel network. Our method is based on the simple fact that pH change can affect the surface tension of a fatty acid solution, and it generates a surface tension difference at the liquid-air interface. The resulting global Marangoni flow can transport passive soluble dye particles from the starting point (high pH region) to the exit of the maze (low pH region) and in the process highlight the possible solution paths to the maze problem. It is worth pointing out that other non-local phenomena affecting the surface tension of liquids or solutions can be also utilized for maze solving (e.g., temperature, irradiation). These might represent for some communities the technically more welcome control mechanisms in practice depending on the experimental constraints.

Acknowledgments Authors acknowledge the financial support of the Hungarian Scientific Research Fund (OTKA K104666). A.B. is grateful for financial support from the Swiss National Science Foundation project no° #200021-137868. R. T. is grateful for financial support from the Swiss National Science Foundation Marie Heim Vögtlin Fellowship No. PMPDP2-139689/1. K.S., D.U. and I.L. gratefully acknowledge the financial support of the National Research, Development and Innovation Office of Hungary (TÉT_12_JP-1-2014-0005).

References

1. Nakagaki, T., Yamada, H., Tóth, A.: Maze-solving by an amoeboid organism. Nature **407**, 470–470 (2000)
2. Nakagaki, T., Yamada, H., Tóth, A.: Path finding by tube morphogenesis in an amoeboid organism. Biophys. Chem. **92**, 47–52 (2001)
3. Adamatzky, A.: Slime mold solves maze in one pass, assisted by gradient of chemo-attractants. IEEE Trans. Nanobiosci. **11**, 131–134 (2012)
4. Fuerstman, M.J., Deschatelets, P., Kane, R., Schwartz, A., Kenis, P.J.A., Deutch, J.M., Whitesides, G.M.: Solving mazes using microfluidic networks. Langmuir **19**, 4714–4722 (2003)
5. Reyes, D.R., Ghanem, M.M., Whitesides, G.M., Manz, A.: Glow discharge in microfluidic chips for visible analog computing. Lab Chip **2**, 113–116 (2002)
6. Dubinov, A.E., Maksimov, A.N., Mironenko, M.S., Pylayev, N.A., Selemir, V.D.: Glow discharge based device for solving mazes. Phys. Plasmas **21**, 093503 (2014)
7. Pershin, Y.V., Di Ventra, M.: Solving mazes with memristors: a massively parallel approach. Phys. Rev. E **84**, 046703 (2011)
8. Steinbock, O., Tóth, A., Showalter, K.: Navigating complex labyrinths: optimal paths from chemical waves. Science **267**, 868–871 (1995)
9. Steinbock, O., Kettunen, P., Showalter, K.: Chemical wave logic gates. J. Phys. Chem. **100**, 18970–18975 (1996)
10. Adamatzky, A.: Hot ice computer. Phys. Lett. A **374**, 264–271 (2009)
11. Lagzi, I., Soh, S., Wesson, P.J., Browne, K.P., Grzybowski, B.A.: Maze solving by chemotactic droplets. J. Am. Chem. Soc. **132**, 1198–1199 (2010)
12. Suzuno, K., Ueyama, D., Branicki, M., Tóth, R., Braun, A., Lagzi, I.: Maze solving using fatty acid chemistry. Langmuir **30**, 9251–9255 (2014)
13. Cejkova, J., Novak, M., Stepanek, F., Hanczyc, M.M.: Dynamics of chemotactic droplets in salt concentration gradients. Langmuir **30**, 11937–11944 (2014)
14. Braun, A., Tóth, R., Lagzi, I.: Künstliche intelligenz aus dem chemiereaktor. Nachr. Chem. **63**, 445–446 (2015)

Chapter 11
Chemotaxis and Chemokinesis of Living and Non-living Objects

Jitka Čejková, Silvia Holler, To Quyen Nguyenová,
Christian Kerrigan, František Štěpánek and Martin M. Hanczyc

Abstract One of the fundamental properties of living organisms is the ability to sense and respond to changes in their environment by movement. If a motile cell senses soluble molecules and follows along a concentration gradient to the source, or if it moves away from a source of undesirable chemicals, e.g. repellent, toxin, it is displaying a directional movement called positive or negative chemotaxis, respectively. This phenomenon is well-known to biologists and intensively studied in living systems. In contrast chemokinesis is a change in movement due to environmental input but the resulting movement is non-vectorial and can be considered directionally random. Recently, in the last ten years, few laboratories started to focus on the movement properties of artificial constructs, including the directional movement of non-living objects in chemical gradients. This chapter will focus on chemotaxis and chemokinesis of natural and synthetic systems that may provide chemical platforms for unconventional computing.

J. Čejková (✉) · T.Q. Nguyenová · F. Štěpánek
Department of Chemical Engineering, University of Chemistry
and Technology Prague, Prague, Czech Republic
e-mail: jitka.cejkova@vscht.cz

T.Q. Nguyenová
e-mail: nguyentq@vscht.cz

F. Štěpánek
e-mail: frantisek.stepanek@vscht.cz

S. Holler · M.M. Hanczyc
Laboratory for Artificial Biology, Centre for Integrative Biology (CIBIO),
University of Trento, Trento, Italy
e-mail: martin.hanczyc@unitn.it

S. Holler
e-mail: silvia.holler@unitn.it

C. Kerrigan
Architect (ARB, RIBA), 200 Year Continuum, Astudio Architecture, London, UK
e-mail: christiankerrigan@gmail.com

© Springer International Publishing Switzerland 2017
A. Adamatzky (ed.), *Advances in Unconventional Computing*,
Emergence, Complexity and Computation 23,
DOI 10.1007/978-3-319-33921-4_11

11.1   Cellular Movement in Biological Systems

Living cells can physically move through several mechanisms. Due to the length scales of most living cells, viscosity will dominate over inertia. Therefore at such low Reynolds numbers, motion of the cell will require the expenditure of energy. However, some cells rely entirely upon passive flotation and Brownian motion for dispersal. Under the microscope non-motile (and also dead) cells seem to move in a purposeful way, though they may frequently change direction, but this is due to random molecular bombardment of cells by the molecules of the solvent. This type of undirected motion demands no energy nor sophisticated machinery from the cells and therefore it is necessary to distinguish the effects of passive Brownian motion from active cellular motility [12]. There are several primary mechanisms of active cell motion: swimming using rotating flagella in prokaryotes, e.g. *E. coli*, or cilia in eukaryotes, *Tetrahymena*; cytoskeleton polarization, e.g. *Dictyostelium*; and gliding, cyanobacteria. Overall, cell motility can be divided into three types [25]:

1. basal random motility, which takes place in the absence of chemical stimuli,
2. chemokinesis, which corresponds to increased random motility in response to chemical stimuli,
3. chemotaxis, which corresponds to stimulated migration towards an increasing (or decreasing) chemical gradient.

Cells can move independently or collectively, and they can migrate as single unattached cells or multicellular groups (see Table 2 in [60]). For example, single tumor cells can perform amoeboid migration or mesenchymal migration. Some carcinoma cells with an amoeboid morphology can move at high speeds inside the tumors ($\sim 4\,\mu$m/min), whereas mesenchymal migration is characterized by an elongated cell morphology with established cell polarity and relatively low speeds of cell migration (0.1–1 μm/min). Directed multicellular migration can be either collective migration, in which the cells are in tight contact with each other (also known as cohort migration) or cell streaming, in which the coordinated cell migration involves cells that are not always in direct physical contact.

11.2   Sensing in Biology

Most organisms (even bacteria) can sense hearing, light, pressure, gravity and chemicals and as a response to them can display specific movements (phototaxis-response to light [11], magnetotaxis-response to magnetic field [58], thermotaxis-movement along thermal gradient [50], galvanotaxis-movement along the electric field [5], haptotaxis-movement along the gradient of adhesion sites [15], and gravitaxis-movement along the direction of the gravitational force [59]). Intelligent animals can find food by coordinating their senses (smell, sight, and sound) with their central nervous system. Small organisms with no sense-specific organs or central nervous system find food by chemotaxis. The bacterium *E. coli* as a well-studied organism can sense and respond to changes in temperature, osmolarity, pH, noxious chemicals, DNA-damaging agents, mineral abundance, energy sources, electron acceptors, metabolites, chemical signals from other bacteria, and parasites [49]. *E. coli* has five chemotaxis sensors: Trg, for ribose and galactose; Tar, for aspartate; Tsr, for serine; Tap, for peptides; and the newly discovered Aer, which may be a redox detector [31].

Depending on the type and size of a cell, there are two modes of chemical gradient sensing: temporal and spatial. Bacteria are very small organisms and due to their length scale it is thought that the cells are not able to detect gradients spatially, therefore they detect the chemicals temporally. In other words chemotaxis of bacteria involves memory. Bacteria detect the gradient by comparing their currently sensed chemical environment with that sensed in recent past. If the conditions are improving, then the cell will continue with running along its current path. However, if the conditions are not improving because the cell is moving away from the source of nutrients, then the cell will tumble and try to find a new path that yields a more favorable trajectory. By this trial-and-error approach, the bacterial cell would eventually move up the gradient. This mechanism is based on sensory adaptation as the bacterium responds only to temporal changes in signal molecule concentration rather than their absolute values [19].

Eukaryotic cells detect differences in the chemicals concentrations across their cell body because of different receptor occupancy on opposite sides resulting on polarization of the molecular distributions in the cell [38]. Such a spatial gradient sensing mechanism leads to more or less straight line migration, as opposed to the running and tumbling of flagellated bacteria. One of the best-studied examples of eukaryotic chemotactic system is the migration of amoeba *Dictyostelium discoideum* along an increasing concentration of cyclic adenosine-3′,5′-monophosphate (cAMP) [43, 61]. A *Dictyostelium* cell can measure about 1 % of concentration difference over its total length (10–$20\,\mu m$) in a spatial cAMP gradient. In the absence of cAMP, the cells are not at rest, but perform a random type of motion with the average motility $4.19\,\mu m/min$ [62]. Cells show no directional response to negligible stationary, linear cAMP concentration gradients (of less than $10^{-3}\,nM/\mu m$) and exhibited a constant basal motility. In steeper gradients, cells move up the gradient with an average motility of $\sim 9\,\mu m/min$ directly towards the cAMP source. In very steep gradients (above $10\,nM/\mu m$) the cells lose directionality and the motility returned to random motion (see Fig. 6 in [62]). From this it is evident that the chemotaxis can occur only in a certain range of chemoattractant concentration. Extremely low or conversely too high concentrations do not trigger the chemotactic migration.

The true slime mold *Physarum polycephalum* in the plasmodial stage also shows the chemotactic behavior [6]. Several works focused on the effect of various compounds on *Physarum* chemotaxis. For example, carbohydrates [14, 18, 44], amino acids and nucleotides [18], and volatile organic chemicals [20] have been found as chemoattractants, whereas inorganic salts [3] were found to be chemorepellents.

11.3   Importance of Chemotaxis in Multicellular Organisms

In multicellular organisms, chemotaxis is very important in physiological processes, such as the recruitment of inflammatory cells to sites of infection and in organ development during embryogenesis [26]. Neuronal and embryonic cells migrate during development. During angiogenesis, endothelial cells undergo chemotaxis to form blood vessels, while epithelial cells and fibroblasts chemotaxis during wound healing [48]. In addition, chemotaxis governs the motion of sperm towards the egg during mammalian fertilization [24].

However in addition to a role in obviously beneficial processes in multicellular organisms, chemotaxis is also involved in each crucial step of tumor cell dissemination and metastatic progression (invasion, intravasation and extravasation). Many cancer cells such as breast cancer cells are known to preferentially metastasize into certain tissues and organs. This preference is correlated with the production of chemoattractants by the target tissues and organs and up-regulation of chemoattractant receptors in the cancer cells [60, 71].

11.4   Chemotaxis Versus Chemokinesis

There are two different terms in biology that link chemical input with motion output: chemotaxis and chemokinesis [56]. Whereas chemotaxis results in oriented movement, chemokinesis occurs through the increase or decrease of speed, but the directionality is not biased. Both types of movement have been shown in both healthy and malignant cells. For example, multiple growth factors stimulate chemotaxis and/or chemokinesis in malignant mesothelioma cells [47]. For healthy neurons and epithelial cells, micromolar concentrations of gamma-aminobutyric acid (GABA) induce chemokinesis in rat cortical neurons during development, while femtomolar concentrations induce chemotaxis [10]. Furthermore, vascular endothelial growth factor A (VEGFA) was shown, using microfluidic chemotactic chambers, to increase chemotaxis of epithelial cells while decreasing chemokinesis [37]. Although both chemotaxis and chemokinesis may play essential roles in biology, modes of cellular motion employing one or both mechanisms will be different.

11.5   Artificial Chemotaxis and Chemokinesis

The spontaneous motion of liquid droplets, solid particles, and gels under nonequilibrium conditions has been investigated both experimentally and theoretically in several works. Self-propelled systems are studied in broad range of scales and can be called nano-/micro-robots [21, 45], nano-/micro-motors [72], nano-/micro-swimmers [23, 57], and some such systems can be studied at the macroscale [73].

The self-propelled movement of non-biological objects can mimic the chemotactic or chemokinetic behavior of living cells. Usually such examples are based on mechanisms completely different from biological movement, but a link is established between chemical energy and mechanical work.

Camphor grains in water represent the classic non-living chemical system studied for autonomous motion [67] and one of the most coherent examples of chemokinesis. Starting far from equilibrium, the solid camphor spreads as a surfactant at the air water interface and at the same time sublimates. Therefore surface tension gradients are established in the system and fluid can flow through Marangoni-type motion. Due to sublimation of the camphor from the interface, the system will not become blocked with surfactant but the motion can be sustained as long as the camphor grains persist.

Lateral motion of oil or alcohol droplets in aqueous solution containing surfactants are reported in several papers [8, 34, 65, 68]. The droplets are usually propelled by unbalanced surface tension resulting in fluid dynamics through Marangoni-type instabilities. This behavior could be compared to the chemokinesis of living cells, because droplets perform a movement without any orientation towards the source of chemoattractant or from the chemorepellent.

As with motile organelles and cells that exhibit polarity, spontaneous symmetry breaking is needed for self-propelled movement of non-biological objects. A droplet on a surface of a substrate could move when the underlying surface is asymmetrically patterned creating a difference in the interfacial energy between the leading edge and the trailing one of the droplet [22, 63]. Furthermore a droplet could break symmetry through a coupled chemical reaction that occurs at the interface between the droplet and its surrounding medium. The chemical reaction produces a symmetry breakage due to the accumulation and release of the products [35] and the droplets swim through the aqueous media without the need of an air-water or solid-liquid interface.

We have presented several works describing a system of oil droplet movement based on fatty acid chemistry. In [34], the oil containing oleic anhydride precursor was introduced into an aqueous environment that contains oleate micelles. The products of the precursor hydrolysis were coupled to the movement of the oil droplet and the production of waste vesicles. The oil droplet successfully moves away from this waste product into fresh unmodified solution and even displays a primitive form of chemotaxis. This example mimics the cells that move away from their metabolic products and waste into regions with fresh nutrients. In another system, the catalyst solvated in the anhydrous droplet reacts with surfactant substrate in the surrounding aqueous environment that results in symmetry breakage and fuelled droplet motion [68].

Several trajectory patterns of self-propelling droplets that mimic chemokinesis can be observed: irregular [53], circular [52], back-and-forth, knot-forming, and irregular [66]. They may or may not depend on the shape of the boundary. The center of trajectory either corresponds to the dish center (global trajectory) or it does not correspond to the dish center (local trajectories) [66]. We have studied the movement of oleic anhydride-loaded nitrobenzene droplets in oleate micelle solution and we have found that droplet size (volume) correlates with various behavioral modes, namely fluctuating, directional, vibrational, and circular [36].

Research on asymmetric bimetallic catalytic rods or spheres in the concentration gradient of a chemoattractant shows oriented movement of artificial particles that could be compared to chemotaxis. Typically, bimetal nanorod contains one segment made of a metal that functions as a catalyst for the decomposition of H_2O_2.

During the decomposition, oxygen is produced at the surface of the metallic segment according to the reaction $2H_2O_2(l) \rightarrow O_2(g) + 2H_2O\,(l)$ and provides the driving force for nanorod propulsion. The other segment can be made of any material that is not reactive towards hydrogen peroxide, such as an inorganic, organic or polymeric metal, semiconductor or insulator; often this segment is made of gold. Bidoz et al. [27] studied Ni-Au nanorods in the H_2O_2 gradient. The group of Ayusman Sen [55] studied the artificial chemotaxis of platinum-gold rods ($2\,\mu m$ in length, $370\,nm$ in diameter) in the H_2O_2 gradient. Sundararajan with colleagues [64] used also Pt-Au rods in H_2O_2, in this case connected with the polystyrene bead cargo. Howse et al. [37] prepared polystyrene $1.6\,\mu m$ microspheres half-coated in platinum and studied their movement in the H_2O_2 gradient, later Ke et al. [40] performed similar experiments with silica $1\,\mu m$ particles half-coated in Pt by providing a chemical mechanism of motion and imposed asymmetric design, such systems display or may be capable of artificial chemotaxis.

In several works, it was shown that the diffusio-phoretic phenomenon shares similarities with chemotaxis [1]. When a rigid colloidal particle is placed in a solution which is not uniform in the concentration of a solute that interacts with the particle, the particle will be propelled in the direction of lower concentration of the solute. The resulting locomotion is called diffusio-phoresis [7].

Artificial systems can be created to mimic also other kind of induced directional movement, such as magnetotaxis. Dreyfus et al. [23] showed that a linear chain of colloidal magnetic particles linked by DNA and attached to a red blood cell can act as a flexible artificial flagellum. The filament was aligned with an external uniform magnetic field and was readily actuated by oscillating a transverse field. They found that the actuation induced a beating pattern that propels the structure, and that the external fields can be adjusted to control the velocity and the direction of motion, reproducing one magnetotaxis like behavior. Chaturvedi et al. [17] fabricated microscale catalytic motors using particles of various material compositions, including metal and magnetic polymer. Such particles have shown autonomous movement, in random directions, in the presence of H_2O_2 and UV light, but if an external magnetic field was also present, they showed directed movement, forming exclusion regions around them. The magneto-phototactic response shown by the magnetic colloidal chains opens a new path for their use as delivery vehicles.

Several methods are used in biology to study chemotaxis. In the most direct method, cells and their trails can be directly visualized with a camera and then analyzed with image analysis software. Chemotaxis of populations of cells is often measured by counting cells at the attractant source over time [39]. Similar methods are used for the non-biological chemotaxis studies. Chemotaxis studies require a way to deliver chemicals to the observed object (cell, particle, droplet) in a controlled way. A wide range of techniques is available to evaluate chemotactic activity of cells, such as generation of gradient with micropipette [70], using of special chambers (Boyden chamber [13], Dunn chamber [74], Zigmond chamber [75]), and microfluidic devices. Microfluidic devices are ideal tools for creating and controlling chemotactic gradients and are widely used and popular in recent times. Microfabrication methods allow arbitrary microchannel designs to be created through which cells can move. Microchannel dimensions can be created small enough that diffusion occurs in minutes to seconds, thus reducing the waiting time for a gradient to become established [69]. Other methods for studying chemotaxis have been reviewed in [30, 32, 41]. For the study of artificial chemotaxis the most common methods are direct addition of signal molecules by pipette [15, 34], addition of gels soaked with chemoattractant [35, 46], and microfluidics devices [9, 29].

11.6 Motile Systems in Mazes

Living cells and non-living objects can move in mazes towards chemoattractant targets. Many examples of directional movement of single cells in natural maze-like environments can be found. For example, leukocytes, in the case of inflammation, move towards the inflamed site through the intricate maze of the circulatory system following chemoattractants released from the injured tissue [28]. During axonal growth, neurons move through the intricate maze of the brain and create new synapses responding to specific neutrophic factors gradients [51].

More complex organisms, such as slime molds [54], solve mazes doped with chemoattractants. For example, in [4] slime mould is inoculated in a maze's peripheral channel and an oat flake (source of attractants) in a the maze's central chamber. The slime mould grows toward target oat flake and connects the flake with the site of original inoculation with a pronounced protoplasmic tube. The protoplasmic tube represents a path in the maze. The plasmodium solves maze in one pass because it is assisted by a gradient of chemo-attractants propagating from the target oat flake.

The method for solving the maze involves the search of several alternative solutions in parallel. When one or more of the searches finds the source of food then these successful paths are reinforced in time and the non-successful branches in the slime mold are diminished. Currently *Physarum* is used for unconventional computation and for fabrication of novel and emerging non-silicon computers and robots. How to create experimental *Physarum* machines—a green and environmentally friendly unconventional computers—has been shown in Adamatzky's book titled 'Physarum Machines: Computers from Slime Mould' [2].

In our recent work we have shown that even nonliving objects such as decanol droplets in decanoate solutions are able to move chemotactically in salt gradients [16]. It has been demonstrated that this artificial chemotactic system bears many qualitative similarities with natural chemotactic systems, namely:

1. the ability to perform chemotaxis repeatedly when the chemoattractant gradients are recreated,
2. to perform chemotaxis in topologically complex environments (Fig. 11.1),

source of salt

decanol droplet

Fig. 11.1 Maze solving decanol droplet. (Supplementary movie in youtube http://youtu.be/
P5uKRqJIeSs). Reprinted with permission from [16]. ©2014, American Chemical Society

3. to select the chemotaxis direction based on the relative strength of alternative
   chemoattractant sources,
4. to rest in a dormant state and later respond to a stimuli-responsive chemoattractant
   release, and finally,
5. to deliver a reactive payload at the chemoattractant source.

Lagzi and his colleagues [46] showed a droplet of mineral oil and 2-hexyldecanoic
acid in a labyrinth with a pH gradient, where the droplets were able to find the shortest
path away from the maze. They prepared a maze and filled it with a KOH solution. Due
to the presence of a gel soaked with HCl in the 'goal', a pH gradient was established
in the maze channels. Then a 1 μl droplet of mineral oil and 2-hexyldecanoic acid
placed into the 'start' position moved toward the source of HCl. Such droplets have
shown chemotactic movement in the pH gradient (pH-taxis).

11.7   Chemotactic Decanol Droplets as 'Chemo-Taxi'

Chemotactic droplets could serve as transporters for chemicals or small objects, a
sort of 'chemo-taxi'. In our previous work [16], it was shown that the decanol droplet
can carry a reactive payload and deliver it towards the source of chemoattractant (see
Fig. 9 in [16]). Here it will be shown the use of decanol droplets for transport of small
objects.

Generally, a taxi is a type of vehicle that conveys passengers between locations of
their choice. Let us consider that our decanol droplet could serve as a taxi and as our
model passenger a dead fly was chosen. Figure 11.2 represents the transport of a fly
from the right hand side of the glass slide to the left hand side of the glass into the place
where in the beginning of experiment small amount of salt was added. In principle
this kind of chemo-taxi could serve for the transport of any other small object.
Further examples with piece of paper, polystyrene and hair are shown in Fig. 11.2.
In addition, a piece of conductive wire can be transported through chemotaxis. Then
the environmental conditions can be changed such that the droplet releases the cargo
on demand, in this case by adding 25 mM decanoate in water (pH 12) to the system.
See movie [78] and Fig. 11.3.

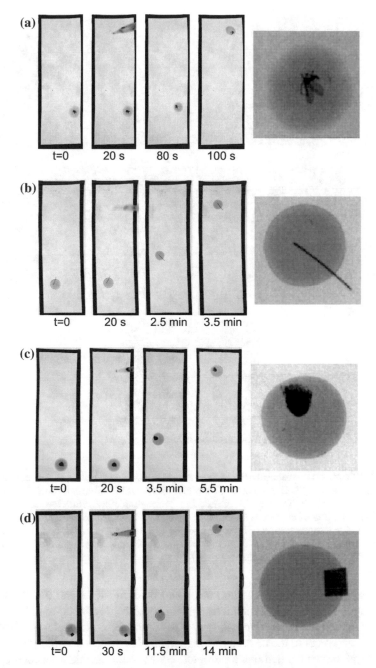

Fig. 11.2 'Chemo-taxi'-decanol droplets transporting small objects in salt gradients. **a** fly, **b** hair, **c** polystyrene and **d** paper. The size of a rectangular slide is 25×75 mm

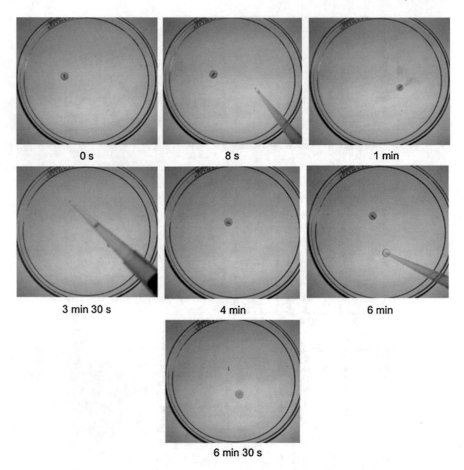

Fig. 11.3 'Chemo-taxi'—decanol droplet transporting and dropping a piece of conductive copper wire of 1.5 mm in length. The diameter of the glass petri dish is 60 mm. 0 s: droplet with wire placed in experiment; 8 s: salt gradient added; 15 s–1 min: droplet migration towards salt source; 1 min–3 min 30 s: 25mM decanoate addition; 3 min 30 s: second salt addition; 3 min 30 s–4 min: droplet migration towards second salt source; 4–6 min 25 mM decanoate addition to trigger release of wire from droplet; 6 min third salt addition; 6 min–6 min 30 s migration towards third salt source and wire left behind. See movie at 8× real time [78]

The idea to have such a 'chemo-taxi' could be used for several potential applications in the future. This chemotactic system could be developed to help machine engineers maintain their equipment, because droplets could track down pH, salt or heat gradients to get to the desired destination and lubricate a specific joint or axle. Another application of chemotactic droplets could be for chemical reaction control. The motile reaction containers in the form of the oil droplets could move to another place or closer to other reaction container via the concentration gradient. Gradients are thus one way of controlling reactions and for manipulation of reactants or products. Chemotactic droplets can fulfil various tasks in response to chemical signals in hazardous areas or limited spaces beyond the control of external power sources.

11.8 Chemotactic Technology with Design Applications

The pattern of movement of the chemotactic decanol system depends on the environmental context. Usually we perform such experiments by placing a decanol droplet with dye in a glass petri dish or glass slide containing 10 mM decanoate pH 12 surfactant. We then see chemotaxis of the droplet towards the high salt source with little or no observable distortion in droplet shape. However if we use a more challenging environmental context, we see a different pattern. For example, when we saturate glass fiber paper with the decanoate solution, then place a droplet of decanol with dye on the surface, the decanol will migrate towards the salt source by producing multiple protruding ends through the fiber matrix. Those ends that successfully manage the fiber network and find the salt will be reinforced over ends that find little or no salt. In this way we see a pattern of migration and chemotaxis similar in phenomenology to the searching behavior of slime mold [76]. Figure 11.4 shows such an experiment. Two droplets of decanol are placed with a space of about 1 cm between them on decanoate saturated glass fiber paper. Concentrated NaCl is added in the gap between them. After about one minute the salt gradient reaches the decanol droplets and the initial protuberances form (see Fig. 11.4, arrows A and B), These protuberances extend, bifurcate and they migrate through the glass fiber matrix. The figure demonstrates the chemotactic movement with the leading edges of the droplets growing towards the chemoattractant and the trailing edges retracting, resulting the net migration towards the source.

The decanol system on fiber glass is a novel system utilizing the design potential of chemotactic technology as its principle criterion [33]. The decanol system moves to towards a salt source over the substrate [77]. Three design conditions which establish the success of the movement are gravity, buoyancy of the decanol system and the design of the structural matrix of the glass fibre substrate. By infusing the decanol system with dyes the scale of the droplets is enhanced in order to observe its interaction across a substrate. It was observed that its manoeuvrability around obstacles solve complex paths to reach the salt solution.

This process shares similar problem solving criteria to the design of architecture, a theme also explored through the 200 Year Continuum [42]. By observing the decanol system, facets of architectural design are developed using site-specific environmental conditions affecting the outcome of successful architectural solutions. By designing the decanol system to move through the substrate we observe behavioral characteristics in relation to its environment-that is, moving between fiber glass matrixes or navigating across or around grains of sand etc. [77]. The initial conditions are set and the experiment unfolds through movement to explore and solve a complex chemical and physical landscape.

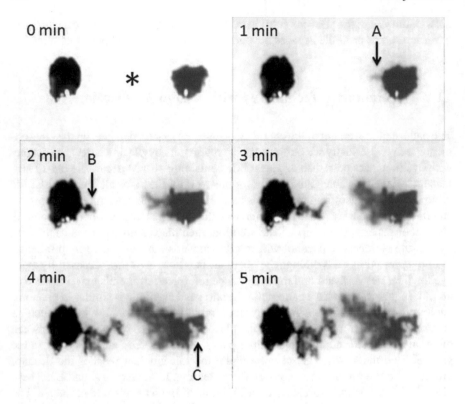

Fig. 11.4 Chemotaxis of decanol through glass fiber paper. Time series including initial state, then every minute time point post NaCl addition. *Star* indicates point of NaCl addition. *Arrows A, B* indicate initial protrusions from red and black droplets, respectively, shown here in grey scale as *black (right)* and *grey (left)*. *Arrow C* indicates retraction of red droplet. Area of glass fiber paper shown is $3 \times 1.8\,cm$ [76]

11.9 Conclusion

We have discussed the role of movement in natural cellular and artificial systems. Cellular movement is important in single cell nutrition and survival but also in the larger macroscopic healthy and disease states in multicellular organisms. Sensing of the immediate environment by cells affects the rate of motion through chemokinesis and/or the direction of motion through chemotaxis. Some cells are capable of both types of motion which often depends on the amount of chemical stimulant present. Interest in the design of self-moving artificial systems on the same size scale as cells is rapidly developing. Such systems either with imposed asymmetry or spontaneously breaking symmetry are already capable of chemokinesis and chemotaxis and are sensitive to different chemical signals such as salt concentration, light, and pH. In some cases, as with droplets, the artificial system can move and leave a chemical pattern on the environment but also pick up and transfer physical cargo. Basic computational aspects of such motile artificial systems are being demonstrated in maze solving tasks. The future development of self-motile artificial systems towards applications in unconventional computing is intriguing and worthy of further study and experimentation.

Acknowledgments This work was supported by the European Commission FP7 Future and Emerging Technologies Proactive: 318671 (MICREAgents) and 611640 (EVOBLISS).

References

1. Abecassis, B., Cottin-Bizonne, C., Ybert, C., Ajdari, A., Bocquet, L.: Boosting migration of large particles by solute contrasts. Nature Mater. **7**, 785–789 (2008)
2. Adamatzky, A.: Physarum Machines: Computers from Slime Mould. Series on Nonlinear Science Series A, vol. 74. World Scientific, Singapore (2010)
3. Adamatzky, A.: Routing Physarum with repellents. Eur. Phys. J. **31**, 403–410 (2010)
4. Adamatzky, A.: Slime mold solves maze in one pass, assisted by gradient of chemo-attractants. IEEE Trans. NanoBioscience **11**(2), 131–134 (2012)
5. Adler, J., Shi, W.: Galvanotaxis in a bacteria. Cold Spring Harbor Symp. Quant. Biol. **53**, 23–25 (1988)
6. Aldrich, H.: Cell Biology of Physarum and Didymium V1: Organisms, Nucleus, and Cell Cycle. Elsevier Science, Amsterdam (2012)
7. Anderson, J.L., Prieve, D.C.: Diffusiophoresis: migration of colloidal particles in gradients of solute concentration. Sep. Purif. Methods **13**, 67–103 (1984)
8. Ban, T., Yamagami, T., Nakata, H., Okano, Y.: pH-Dependent motion of self-propelled droplets due to marangoni effect at neutral pH. Langmuir **29**, 2554–2561 (2013)
9. Baraban, L., Harazim, S.M., Sanchez, S., Schmidt, O.G.: Chemotactic behavior of catalytic motors in microfluidic channels. Angewandte Chemie International Edition **52**, 5552–5556 (2013)
10. Behar, T.N., Li, Y.X., Tran, H.T., Ma, W., Dunlap, V., Scott, C., Barker, J.L.: GABA stimulates chemotaxis and chemokinesis of embryonic cortical neurons via calcium-dependent mechanisms. J. Neurosci. **16**(5), 1808–1818 (1996)
11. Bennett, R.R., Golestanian, R.: A steering mechanism for phototaxis in chlamydomonas. J. R. Soc. Interface **12**, (2015)
12. Berg, H.C.: Random Walks in Biology. Princeton University Press, Princeton (1993)
13. Boyden, S.: The chemotactic effect of mixtures of antibody and antigen on polymorphonuclear leucocytes. J. Exp. Med. **115**, 453–466 (1962)
14. Carlile, M.J.: Nutrition and chemotaxis in the myxomycete Physarum polycephalum: the effect of carbohydrates on the plasmodium. J. Gen. Microbiol. **63**, 221–226 (1970)
15. Carter, S.B.: Haptotaxis and the mechanism of cell motility. Nature **213**, 256–260 (1967)
16. Čejková, J., Novák, M., Štěpánek, F., Hanczyc, M.M.: Dynamics of chemotactic droplets in salt concentration gradients. Langmuir **30**, 11937–11944 (2014)
17. Chaturvedi, N., Hong, Y., Sen, A., Velegol, D.: Magnetic enhancement of phototaxing catalytic motors. Langmuir **26**, 6308–6313 (2010)
18. Chet, I., Naveh, A., Henis, Y.: Chemotaxis of Physarum polycephalum towards carbohydrates, amino acids and nucleotides. J. Gen. Microbiol. **102**, 145–148 (1977)
19. Collin, M., Schuch, R.: Bacterial Sensing and Signaling. Karger, Switzerland (2009)
20. de Lacy Costello, B.P.J., Adamatzky, A.I.: Assessing the chemotaxis behavior of Physarum polycephalum to a range of simple volatile organic chemicals. Commun. Integr. Biol. **6**, (2013)

21. Dey, K.K., Bhandari, S., Bandyopadhyay, D., Basu, S., Chattopadhyay, A.: The pH taxis of an intelligent catalytic microbot. Small (2013)
22. Dos Santos, F.D., Ondarcuhu, T.: Free-running droplets. Phys. Rev. Lett. **75**, 2972–2975 (1995)
23. Dreyfus, R., Baudry, J., Roper, M.L., Fermigier, M., Stone, H.A., Bibette, J.: Microscopic artificial swimmers. Nature **437**, 862–865 (2005)
24. Eisenbach, M.: Sperm chemotaxis. Rev. Reprod. **4**, 56–66 (1999)
25. Eisenbach, M.: Chemotaxis, vol. 499. Imperial College Press, London (2004)
26. Entschladen, F., Zänker, K.S.: Cell Migration: Signalling and Mechanisms. Karger, Switzerland (2009)
27. Fournier-Bidoz, S., Arsenault, A.C., Manners, I., Ozin, G.A.: Synthetic self-propelled nanorotors. Chem. Commun., 441–443 (2005)
28. Fox, R.B., Hoidal, J.R., Brown, D.M., Repine, J.E.: Pulmonary inflammation due to oxygen toxicity: involvement of chemotactic factors and polymorphonuclear leukocytes. Am. Rev. Respir. Dis. **123**, 521 (1981)
29. Francis, W., Fay, C., Florea, L., Diamond, D.: Self-propelled chemotactic ionic liquid droplets. Chem. Commun. **51**, 2342–2344 (2015)
30. Genzer, J., Bhat, R.R.: Surface-bound soft matter gradients. Langmuir **24**, 2294–2317 (2008)
31. Grebe, T.W., Stock, J.: Bacterial chemotaxis: the five sensors of a bacterium. Curr. Biol. **8** (1998)
32. Guan, J.L.: Cell migration: developmental methods and protocols. Methods in Molecular Biology. Humana, New York (2004)
33. Hanczyc, M.M., Ikegami, T.: Protocells as smart agents for architectural design. Technoetic Arts J. **7**(2), 117–120 (2009)
34. Hanczyc, M.M., Toyota, T., Ikegami, T., Packard, N., Sugawara, T.: Fatty acid chemistry at the oil-water interface: self-propelled oil droplets. J. Am. Chem. Soc. **129**, 9386–9391 (2007)
35. Hiroki, M., Hanczyc, M.M., Ikegami, T.: Self-maintained movements of droplets with convection flow. Progress in Artificial Life. Springer, Berlin (2007)
36. Horibe, N., Hanczyc, M.M., Ikegami, T.: Mode switching and collective behavior in chemical oil droplets. Entropy **13**, 709–719 (2011)
37. Howse, J.R. et al. Self-motile colloidal particles: From directed propulsion to random walk. Phys. Rev. Lett. **99** (2007)
38. Jeon, K.W.: International Review of Cytology: A Survey of Cell Biology. Elsevier Science, Amsterdam (2007)
39. Jin, T., Hereld, D.: Chemotaxis: Methods and Protocols. Humana Press, New York (2009)
40. Ke, H., Ye, S., Carroll, R.L., Showalter, K.: Motion analysis of self-propelled Pt-Silica particles in hydrogen peroxide solutions. J. Phys. Chem. A **114**, 5462–5467 (2010)
41. Keenan, T.M., Folch, A.: Biomolecular gradients in cell culture systems. Lab Chip **8**, 34–57 (2008)
42. Kerrigan, C.: The 200 Year Continuum. Leonardo, Massachusetts Institute of Technology Press, **42**(4), 314–323 (2009)
43. Kessin, R.H.: Dictyostelium: Evolution, Cell Biology, and the Development of Multicellularity. Cambridge University Press, Cambridge (2001)
44. Knowles, D.J.C., Carlile, M.J.: The chemotactic response of plasmodia of the myxomycete physarum polycephalum to sugars and related compounds. J. Gen. Microbiol. **108**, 17–25 (1978)
45. Lagzi, I.: Chemical robotics-chemotactic drug carriers. Cent. Eur. J. Med., 1–6 (2013)
46. Lagzi, I., Soh, S., Wesson, P.J., Browne, K.P., Grzybowski, B.A.: Maze solving by chemotactic droplets. J. Am. Chem. Soc. **132**, 1198–1199 (2010)
47. Liu, Z., Klominek, J.: Chemotaxis and chemokinesis of malignant mesothelioma cells to multiple growth factors. Anticancer Res. **24**, 1625–1630 (2004)
48. Martin, P.: Wound healing-aiming for perfect skin regeneration. Science **276**, 75–81 (1997)
49. Meyers, L.A., Bull, J.J.: Fighting change with change: adaptive variation in an uncertain world. Trends Ecol. Evol. **17**, 551–557 (2002)

50. Mori, I., Ohshima, Y.: Neural regulation of thermotaxis in Caenorhabditis elegans. Nature **376**, 344–348 (1995)
51. Mortimer, D., Fothergill, T., Pujic, Z., Richards, LJ., Goodhill, GJ.: Growth cone chemotaxis. Trends Neurosci. **31**, 90–98 (2008)
52. Nagai, K., Sumino, Y., Kitahata, H., Yoshikawa, K. Mode selection in the spontaneous motion of an alcohol droplet. Phys. Rev. **71** (2005)
53. Nagai, K.H., Takabatake, F., Ichikawa, M., Sumino, Y., Kitahata, H., Yoshinaga N.: Rotational motion of a droplet induced by interfacial tension. Phys. Rev. **87** (2013)
54. Nakagaki, T., Yamada, H., Toth, A.: Intelligence: maze-solving by an amoeboid organism. Nature **407**, 470–470 (2000)
55. Paxton, W.F., Kistler, K.C., Olmeda, C.C., Sen, A., St Angelo, S.K., Cao, Y., Mallouk, T.E., Lammert, P.E., Crespi, V.H.: Catalytic nanomotors: autonomous movement of striped nanorods. J. Am. Chem. Soc. **126**, 13424–13431 (2004)
56. Petrie, R.J., Doyle, A.D., Yamada, K.M.: Random versus directionally persistent cell migration. Nat. Rev. Mol. Cell Biol. **10**, 538–549 (2009)
57. Peyer, K.E., Zhang, L., Nelson, B.J.: Bio-inspired magnetic swimming microrobots for bio-medical applications. Nanoscale **5**, 1259–1272 (2013)
58. Polyakova, T., Zablotskii, V.: Magnetization processes in magnetotactic bacteria systems. J. Magn. Magn. Mater. **293**, 365–370 (2005)
59. Roberts, A.M.: Mechanisms of gravitaxis in chlamydomonas. Biol. Bull. **210**, 78–80 (2006)
60. Roussos, E.T., Condeelis, J.S., Patsialou, A.: Chemotaxis in cancer. Nat. Rev. Cancer **11**, 573–587 (2011)
61. Ševčíková, H., Čejková, J., Krausová, L., Přibyl, M., Štěpánek, F., Marek, M.: A new traveling wave phenomenon of Dictyostelium in the presence of cAMP. Phys. D: Nonlinear Phenom. **239**, 879–888 (2010)
62. Song, L., Nadkarnia, S.M., Boedeker, H.U., Beta, C., Bae, A., Francka, C., Rappele, W.J., Loomisf, W.F., Bodenschatz, E.: Dictyostelium discoideum chemotaxis: threshold for directed motion. Eur. J. Cell Biol. **85**, 981–989 (2006)
63. Sumino, Y., Magome, N., Hamada, T., Yoshikawa, K.: Self-running droplet: emergence of regular motion from non equilibrium noise. Phys. Rev. Lett. **94** (2005)
64. Sundararajan, S., Lammert, P.E., Zudans, A.W., Crespi, V.H., Sen, A.: Catalytic motors for transport of colloidal cargo. Nano Lett. **8**, 1271–1276 (2008)
65. Takabatake, F., Magome, N., Ichikawa, M., Yoshikawa, K.: Spontaneous mode-selection in the self-propelled motion of a solid/liquid composite driven by interfacial instability. J. Chem. Phys. **134** (2011)
66. Tanaka, S., Sogabe, Y., Nakata, S.: Spontaneous change in trajectory patterns of a self-propelled oil droplet at the air-surfactant solution interface. Phys. Rev. **91** (2015)
67. Tomlinson, C.: On the motions of camphor on the surface of water. Proc. R. Soc. Lond. **11**, 575–577 (1860)
68. Toyota, T., Maru, N., Hanczyc, M.M., Ikegami, T., Sugawara, T.: Self-propelled oil droplets consuming "fuel" surfactant. J. Am. Chem. Soc. **131**, 5012–5013 (2009)
69. Walker, G.M., Sai, J., Richmond, A., Stremler, M., Chung, C.Y., Wikswo, J.P.: Effects of flow and diffusion on chemotaxis studies in a microfabricated gradient generator. Lab Chip **5**, 611–618 (2005)
70. Wang, F., Herzmark, P., Weiner, O.D., Srinivasan, S., Servant, G., Bourne, H.R.: Lipid products of PI(3)Ks maintain persistent cell polarity and directed motility in neutrophils. Nat. Cell Biol. **4**, 513–518 (2002)
71. Wu, D.: Signaling mechanisms for regulation of chemotaxis. Cell Res. **15**, 52–56 (2005)
72. Yamamoto, D., Shioi, A.: Self-propelled nano/micromotors with a chemical reaction: under-lying physics and strategies of motion control. KONA Powder Part. J. **32**, 2–22 (2015)
73. Zhao, G., Pumera, M.: Macroscopic self-propelled objects. Chem.- Asian J. **7**, 1994–2002 (2012)
74. Zicha, D., Dunn, G.A., Brown, A.F.: A new direct-viewing chemotaxis chamber. J. Cell Sci. **99**, 769–775 (1991)

75. Zigmond, S.H.: Ability of polymorphonuclear leukocytes to orient in gradients of chemotactic factors. J. Cell Biol. **75**, 606–616 (1977)
76. https://vimeo.com/36049652
77. https://vimeo.com/47461384
78. http://youtu.be/l3aAjdjQ0m0
79. Hong, Y., Blackman, N.M.K., Kopp, N.D., Sen, A., Velegol, D.: Chemotaxis of nonbiological colloidal rods. Phys. Rev. Lett. **99** (2007)

Chapter 12
Computing with Classical Soliton Collisions

Mariusz H. Jakubowski, Ken Steiglitz and Richard Squier

Abstract We review work on computing with solitons, from the discovery of solitons in cellular automata, to an abstract model for particle computation, information transfer in collisions of optical solitons, state transformations in collisions of vector solitons, a proof of the universality of blinking spatial solitons, and the demonstration of multistable collision cycles and their application to state-restoring logic. We conclude by discussing open problems and the prospects for practical computing applications using optical soliton collisions in photo-refractive crystals and fibers.

12.1 Introduction

In most present-day conceptions of a "computer," information travels between logical elements fixed in space. This is true for abstract models like Turing machines, as well as real silicon-chip-based electronic computers. This paper will review work over the past several years that views computation in an entirely different way: information is carried through space by particles, computation occurs when these particles collide.

Much of this review will focus on our own work, but of course will necessarily touch on related work by many others. Our intent is to describe the progression of ideas which has brought the authors to a study of the computational power of the Manakov system and its possible practical implementation, and in no way do we intend to minimize important contributions by others to the growing field of

M.H. Jakubowski (✉) · K. Steiglitz
Computer Science Department, Princeton University,
Princeton, NJ 08544, USA
e-mail: Mariusz.Jakubowski@microsoft.com

K. Steiglitz
e-mail: ken@cs.princeton.edu

R. Squier
Computer Science Department, Georgetown University,
Washington, DC 20057, USA
e-mail: squier@cs.georgetown.edu

© Springer International Publishing Switzerland 2017
A. Adamatzky (ed.), *Advances in Unconventional Computing*,
Emergence, Complexity and Computation 23,
DOI 10.1007/978-3-319-33921-4_12

embedded computation, and nonstandard computation in general. We will cite such related work as we are aware of, and will appreciate readers bringing omissions to our attention.

This paper is written more or less historically, and we will trace the development in the following stages:

- solitons in automata,
- the particle machine model and embedded arithmetic,
- information transfer in collisions of continuum solitons,
- the Manakov system and state transformations in collisions,
- universality of time-gated spatial Manakov solitons,
- multistable collision cycles and state-restoration.

We conclude with a discussion of open problems.

12.2   Computation in Cellular Automata

Our own story begins in a sense with influential work of S. Wolfram in the mid-1980s [1], and in particular with the seminal study [2], where he observed that the behavior of a simple but representative type of cellular automata (CA) can be partitioned into four classes, the most interesting of which he conjectured to be Turing universal. At this time S. Wolfram was in Princeton, at the Institute for Advanced Study, and his work led one of us (KS) to begin experimentation with CA. It was subsequently discovered that CA in a new class, the *parity rule filter automata* (PRFA) [3], support particles which behave remarkably like solitons in continuum systems. (We review the essential features of continuum solitons in physical systems in Sect. 12.4.)

The PRFA differ from conventional CA in that the update rule uses new values as soon as they become available, scanning to update from left to right. They are thus analogous to so-called infinite impulse response (IIR) digital filters, while ordinary CA correspond to finite impulse response (FIR) digital filters [4]. We note that although the operations of FA and CA are different, the two classes of automata are equivalent, as shown in [3]. However, to our knowledge it is an open question whether the subclass of PRFA are computationally universal. The term "parity rule" refers to the algorithm for refreshing site values: to update at a particular site, the total number of ones in a window centered at that site is found (using new values to the left), and the new site value is set to one if that sum is even but not zero, and to zero otherwise. Figure 12.1 shows a typical soliton-like collision in a PRFA. Notice that the bit pattern of both particles is preserved in the collision, and that both particles are displaced from their pre-collision paths.

The subsequent paper [5] showed that a carry-ripple adder can be realized in a PRFA. This construction was not simple, and we were not able to "program" more complex computation in these one-dimensional structures. However, several

Fig. 12.1 Typical
soliton-like collision in a
PRFA. The *dots* are ones
against a field of zeros and
the time evolution progresses
downward. Notice the
displacements caused by the
collision, quite similar to
what happens in continuum
systems. This example uses a
window radius of five and is
from [3]

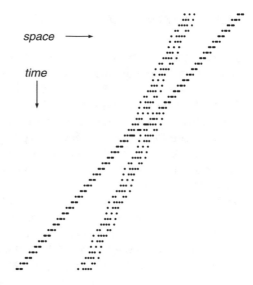

researchers contributed to the study of embedded computation in automata using soliton-like particles in PRFA and related CA in the 1990s, including Goldberg [6], Fokas, Papadopoulou, and Saridakis [7, 8], Ablowitz, Keiser, and Takhtajian [9], Bruschi, Santini and O. Ragnisco [10], and Siwak [11–13]. Since then there has been much more work on particles in CA in general, and the reader is referred to the *Journal of Cellular Automata*, which was founded in 2006, for a key to the rapidly growing literature in the field.

12.3   Particle Machines (PMs)

The *particle machine* (PM) was studied in [5, 14] and is intended to capture the notion of computation by propagating and colliding particles. It was a natural outgrowth of the study of the computational power of the PRFA. The PM can do general computation [14], because it can simulate a Turing machine (TM), and also operates in discrete time and space. However, while the TM's tape, read-write head, and uniprocessing operations hint at mechanical, man-made origins, the PM's particle interactions and fine-grain parallelism are reminiscent of natural physical systems.

We will review the PM model and mention several efficient algorithms that have been encoded in the model. We define a PM as a cellular automaton (CA) with states that represent idealized particles, and with an update rule that encodes the propagation and collisions of such particles. While PMs can have any number of dimensions, we concentrate here on one-dimensional PMs, which are nevertheless powerful enough to support efficient implementations of arithmetic and convolution.

12.3.1   Characteristics of PMs

Quite apart from their use as an abstract model of computation, PMs can be viewed as a way to incorporate the parallelism of systolic arrays [15] in hardware that is not application-specific and is easy to fabricate. A PM can be realized easily in VLSI and the resultant chips are locally connected, very regular (being CA), and can be concatenated with a minimum of glue logic. Thus, many identical VLSI chips can be strung together to provide a very long PM, which can then support many computations in parallel. What computation takes place is determined entirely by the stream of injected particles: There are no multipliers or other fixed arithmetic units in the machine, and the logic supports only particle propagation and collisions. While many algorithms for a PM mimic systolic arrays and achieve their parallelism, these algorithms are not hard-wired, but are "soft," or "floating," in the sense that they do not determine any fixed hardware structures.

An interesting consequence of this flexibility is that the precision of fixed-point arithmetic is completely arbitrary and determined at run time by the user. In [14] the authors show that FIR filtering (convolution) of a continuous input stream, and arbitrarily nested combinations of fixed-point addition, subtraction, and multiplication, can all be performed in one fixed CA-based PM in time linear in the number of input bits, all with arbitrary precision. Later in this section we complete this suite of parallel arithmetic operations with a linear-time implementation of division that exploits the PM's flexibility by changing precision during computation.

12.3.2   The PM Model

We define the PM formally as follows:

Definition 1 A *Particle Machine (PM)* is a CA with an update rule designed to support the propagation and collision of logical *particles* in a one-dimensional homogeneous medium. Each particle has a distinct identity, which includes the particle's velocity. We think of each cell's state in a PM as a *binary occupancy vector*, in which each bit represents the presence or absence of one of n particle types (the same idea is used in lattice gasses; see, for example, [16]). The state of cell i at time $t + 1$ is determined by the states of cells in the *neighborhood* of cell i, where the neighborhood includes the $2r + 1$ cells within a distance, or *radius*, r of cell i, including cell i. In a PM, the radius is equal to the maximum velocity of any particle, plus the maximum displacement that any particle can undergo during collision.

Although this definition is explicitly in one-dimension, it can be generalized easily to higher dimensions.

In summary, a PM is a CA with an update rule modeling propagation and collision of logical particles that are encoded by the state values in one cell, or in a number of adjacent cells. Particles propagate with constant velocities. Two or more particles

Fig. 12.2 The basic conception of a particle machine

may collide; a set of *collision rules*, which are encoded by the CA update rule, specifies which particles are created, which are destroyed, and which are unaffected in collisions. A PM begins with a finite *initial configuration* of particles and evolves in discrete time steps.

12.3.3 Simple Computation with PMs

Figure 12.2 shows the general arrangement of a 1-d PM. Particles are injected at one end of the one-dimensional CA, and these particles move through the medium provided by the cells. When two or more particles collide, new particles may be created, existing particles may be annihilated, or no interaction may occur, depending on the types of particles involved in the collision.

Figure 12.3 illustrates some typical collisions when binary addition is implemented by particle collisions. This particular method of addition is only one of many possibilities. The basic idea here is that each addend is represented by a stream of particles containing one particle for each bit in the addend, one stream moving left and the other moving right. The two addend streams collide with a carry-ripple adder particle where the addition operation takes place. The carry-ripple particle keeps track of the current value of the carry between collisions of subsequent addend-bit particles as the streams collide least-significant-bit first. As each collision occurs, a new right-moving result-bit particle is created and the two addend particles are annihilated. Finally, a trailing "reset" particle moving right resets the carry-ripple to zero and creates an additional result-bit particle moving right.

12.3.4 Algorithms

Arithmetic Addition and subtraction on a PM are relatively straightforward to implement, and both can operate in linear time and with arbitrary precision. A multiplication algorithm with similar properties is not much more difficult to obtain, but a linear-time, arbitrary-precision division algorithm is somewhat more involved. We briefly describe some arithmetic algorithms, and we refer the reader to [14, 17, 18] for details.

Figure 12.4 shows the particle arrangement for fixed-point multiplication. This mirrors a well known systolic array for the same purpose [15], but of course the structure is "soft" in the sense that it represents only the input stream of the PM that

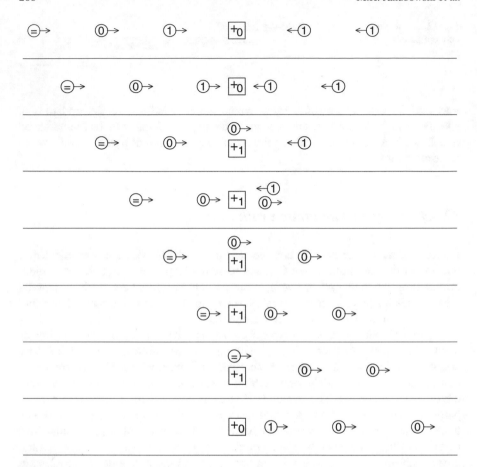

Fig. 12.3 An example illustrating some typical particle collisions, and one way to perform addition in a particle machine. What is shown is actually the calculation $01_2 + 11_2 = 100_2$, implemented by having the two operands, one moving *left* and the other moving *right*, collide at a stationary "carry-ripple" particle. When the leading, least-significant bits collide (in the third row from the top of the figure), the carry-ripple particle changes its identity so that it encodes a carry bit of 1, and a right-moving sum particle representing a bit value of 0 is created. The final answer emerges as the right-moving stream 100_2, and the carry-ripple particle is reset by the "equals" particle to encode a carry of 0. The bits of the two addends are annihilated when the sum and carry bits are formed. Notice that the particles are originally separated by empty cells, and that all operations can be effected by a CA with a neighborhood size of 3 (a radius of 1)

accomplishes the operation. Figure 12.5 shows a simulation of this multiplication scheme for the product $11_2 * 11_2$. In that figure, the particles depicted by R and L represent right- and left-moving operand bits, respectively; p represents stationary "processor" particles in the computation region where the product is formed; c represents "carry" particles propagated during computation; and 0 and 1 represent stationary bits of the product. The top of the figure shows two operands

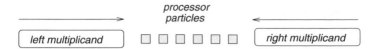

Fig. 12.4 Multiplication scheme, based on a systolic array. The processor particles are stationary and the data particles collide. Product bits are stored in the identity of the processor particles, and carry bits are stored in the identity of the data particles, and thereby transported to neighbor bits

```
.R    .    .R    .    .p   .p   .p   .p    .    .    .L    .    .L    .

  .    .R   .    .R   .p   .p   .p   .p    .    .L   .    .L    .    .
  .    .R   .    .R   .p   .p   .p   .p    .    .L   .    .L    .    .

  .    .    .R   .    .Rp  .p   .p   .p    .L   .    .L   .    .    .
  .    .    .R   .    .Rp  .p   .p   .p    .L   .    .L   .    .    .

  .    .    .    .R   .p   .Rp  .p   .Lp   .    .L   .    .    .    .
  .    .    .    .R   .p   .Rp  .p   .Lp   .    .L   .    .    .    .

  .    .    .    .    .Rp  .p   .RLp .p    .L   .    .    .    .    .
  .    .    .    .    .Rp  .p   .RL1 .p    .L   .    .    .    .    .

  .    .    .    .    .p   .RLp .1   .RLp  .    .    .    .    .    .
  .    .    .    .    .p   .RL1 .1   .RL1  .    .    .    .    .    .

  .    .    .    .    .Lp  .1   .RL1 .1    .R   .    .    .    .    .
  .    .    .    .    .Lp  .1   .RL0c .1   .R   .    .    .    .    .

  .    .    .    .L   .p   .L1c .0   .R1   .    .R   .    .    .    .
  .    .    .    .L   .p   .L0c .0   .R1   .    .R   .    .    .    .

  .    .    .L   .    .Lpc .0   .0   .1    .R   .    .R   .    .    .
  .    .    .L   .    .L1  .0   .0   .1    .R   .    .R   .    .    .

  .    .L   .    .L   .1   .0   .0   .1    .    .R   .    .R   .    .
  .    .L   .    .L   .1   .0   .0   .1    .    .R   .    .R   .    .

.L    .    .L    .    .1   .0   .0   .1    .    .    .R   .    .R   .
.L    .    .L    .    .1   .0   .0   .1    .    .    .R   .    .R   .
```

Fig. 12.5 Simulator output for PM multiplication. The *top line* in the figure gives the initial state of the PM's medium, representing the multiplication problem $11_2 * 11_2$, as described in the text. Each successive pair of lines depicts the state of the medium after the propagation and collision phases of each time step. The *bottom line* in the picture shows the stationary answer, 1001_2, in the central computation region, along with the unchanged operands moving away from the region

(11_2 and 11_2) on their way toward collisions in the central computation region containing stationary "processor" particles; the bottom of the figure shows the same operands emerging unchanged from the computation, with the answer (1001_2) remaining in the central computation region.

For division, a PM can implement a linear-time, arbitrary-precision algorithm based on Newtonian iteration and described by Leighton [19]. Figure 12.6 shows a simulation of such an algorithm running on a PM. Details of this algorithm can be found in [17, 18].

Convolution, Filtering, and Other Systolic-Array Algorithms   As mentioned above, arithmetic operations can be nested to achieve the pipelined parallelism inherent in systolic arrays [14]. This leads to highly parallel, pipelined particle machine implementations of convolution, filtering, and other common digital signal

Fig. 12.6 Output generated by a simulation of the division implementation. Each cell is represented by a small *circle* whose shading depends on which particles are present in that cell. For clarity, only every seventh generation is shown. The example shown is actually the division 1/7

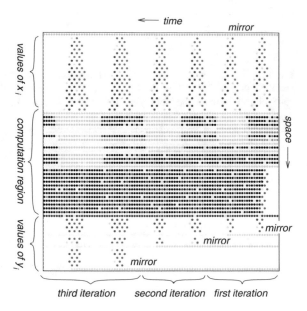

processing algorithms, such as the FFT. Similar non-numerical algorithms with systolic implementations also fit in this category (see, for example, [20]) and are amenable to the same soft realizations.

12.3.5   Comment on VLSI Implementation

Whether it is actually advantageous to implement applications like these in VLSI is an interesting question. The tradeoff is clearly between the efficiencies of using fixed, modular, concatenated chips in a very long pipeline on the one hand, and the inefficiencies in transferring all the problem coding to a very low-level "program" in terms of particles. Where that tradeoff ends up depends a great deal on technology and economies of production scale.

We will not pursue this question of VLSI implementation further in this chapter, but rather follow the road that leads to particles as solitons in nonlinear optical materials.

12.3.6   Particles in Other Automata

Before we leave CA, we mention some related early work on CA with particles. Crutchfield, Das, D'haeseleer, Hanson, Hordijk, Mitchell, Nimwegen, and others report intriguing work in which CA are evolved to perform some simple computational tasks (see [21], for example). Particles appear in these evolved CAs quite spontaneously, suggesting that they may be a very natural way to embed computation in regular structures and materials. Boccara, Nasser, and Roger [22] describe a wide variety of particles observed in a conventional CA. Takahashi [23] presents an appealingly simple box-and-ball model for a CA that supports solitons. Santini [24] extends the concept of integrability to algebraic and functional equations, as well as CA, including the joint work with Bruschi and Ragnisco [10]. Adamatzky [25, 26] described particle-like waves in an excitable medium.

Finally, we also mention some additional, historically significant work: the very simple universal model using ideal elastically colliding billiard balls in the plane [27, 28]; the collection edited by Wolfram [1]; the exhaustive study of universal dynamic computations by Adamatzky [29]; the very early example of pipelined computation in a one-dimensional CA by Atrubin [30]; Conway's universal Game of Life in 2 + 1 dimensions [31]; perhaps the simplest known universal CA in 2 + 1 dimensions [32]; and the well known book by Wolfram [33], in which the universal Rule 110 1-d CA plays a central role. As mentioned before, work in this field has exploded in the last decade, and a good key to current developments can be found in the *Journal of Cellular Automata*.

12.4   Solitons and Computation

12.4.1   Scalar Envelope Solitons

It is a natural step from trying to use solitons in CA to trying to use "real" solitons in physical systems, and the most promising candidates appear to be optical solitons in nonlinear media such as optical fibers and photorefractive crystals. We can envision such computation as taking place via collisions inside a completely uniform medium and aside from its inherent theoretical interest might ultimately offer the advantages of high speed, high parallelism, and low power dissipation. We cannot review here in any detail the fascinating development of soliton theory and its application to optical solitons, but the reader is referred to [34], still a classic beginning reference, and the general book [35] for further reading. We will, however, review the essential features of *envelope* solitons, which are most relevant to our work.

For our purposes a soliton can be defined as in [36]:

Definition 2 A *soliton* is a solitary wave which asymptotically preserves its shape and velocity upon nonlinear interaction with other solitary waves, or, more generally, with another (arbitrary) localized disturbance.

We focus on solitons that arise in systems described by the nonlinear Schrödinger (NLS) equation

$$iu_t + Du_{xx} + N(|u|)u = 0, \qquad (12.1)$$

Fig. 12.7 An envelope
soliton

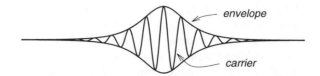

where D is a real number, N an arbitrary operator on $|u|$, and the subscripts denote
partial differentiation. This describes nonlinear wave propagation in a variety of
physical systems, most notably certain optical fibers, where to first order $N(|u|) =
|u|^2$, and in certain (so-called saturable) photorefractive crystals, where $N(|u|) =
m + k|u|^2/(1 + |u|^2)$, where k and m are real constants. In the former instance we
will call the equation the *cubic NLS* (3-NLS), and in the latter, the *saturable NLS* (sat-
NLS). The solitons that result in these systems are called *envelope solitons*, and as
illustrated in Fig. 12.7, they are wave packets, consisting of a *carrier wave* moving at
a characteristic *phase velocity*, modulated by an *envelope*, moving at a characteristic
group velocity.

12.4.2 Integrable and Nonintegrable Systems

There is a crucial difference in behavior between *integrable* and *nonintegrable* systems. Omitting technical details, the integrable systems we consider are analytically
solvable, and collisions between solitons are perfectly *elastic*. That is, solitons emerge
from collisions with all their original energy. Collisions in nonintegrable systems are
characterized by *radiation*—energy that is lost from the solitons in the form of waves
radiating away from the collision site. Such unavoidable energy loss means that collisions cannot be cascaded in many stages, and that any useful computational system
must entail restoration of full-energy solitons.

Clearly, some particle-like behavior is sacrificed in nonintegrable systems, and
in fact purists (generally the mathematical physicists), reserve the term *soliton* for
integrable systems only. Particle physicists on the other hand are more forgiving, and
we will follow their lead in using the term *soliton* more loosely [37, 38].

12.4.3 The Cubic NLS

The most obvious candidate for a useful soliton system is the integrable equation,
3-NLS. This is one of the two or three best-studied soliton equations, and the resultant sech-shaped solitons have been observed experimentally in real optical fibers
for many years. To proceed, we need to identify some soliton parameters as *state
variables* that can be used to carry information. Of the possible parameters, the
amplitude and velocity can be ruled out because they are unaffected by collisions.
The remaining parameters are the carrier phase and positional phase (location). Now

what happens in 3-NLS collisions is very disappointing from the point of view of computation: the values of the state variables that can change do not have any effect on the results of subsequent collisions. This rules out communication of information from soliton to soliton and effectively rules out useful computation in 3-NLS.

12.4.4   Oblivious and Transactive Collisions

We next introduce two definitions that allow us to state the preceding argument somewhat more precisely.

Definition 3   For a given system define the *state* of a soliton to be a set of selected parameters that can change during collisions.

Definition 4   Collisions of solitons in a given system are termed *transactive* if some changes in the state of one colliding soliton depend on the state of the other. If collisions are not transactive, they are termed *oblivious*.

We also call systems themselves transactive or oblivious. We see therefore that 3-NLS is oblivious. The key problem then becomes finding a transactive system.

12.4.5   The Saturable NLS

At the time this obstacle was encountered it seemed to us that all integrable systems are oblivious, and we began looking at some nonintegrable systems, which strictly speaking do not support solitons, but which in fact support near-solitons [39]. At this point M. Segev brought sat-NLS to our attention, an equation that describes the recently discovered $1 + 1$-dimension (one space and one time dimension) photorefractive optical spatial solitons in steady state [40–42], and optical spatial solitons in atomic media in the proximity of an electronic resonance [43]. A numerical study revealed definite transactivity [44]. But the observed effect is not dramatic, and it comes at the cost of unavoidable radiation.

At this point it appeared that transactivity and elastic collisions were somehow antagonistic properties, and that integrable systems were doomed to be oblivious. A pleasant surprise awaited us.

12.5   Computation in the Manakov System

The surprise came in the form of the paper by R. Radhakrishnan, M. Lakshmanan and J. Hietarinta [45], which gave a new bright two-soliton solution for the *Manakov* system [46], and derived explicit asymptotic results for collisions. The solutions were more general than any given previously, and were remarkable in demonstrating what amounts to pronounced transactivity in perfectly integrable equations. The

Manakov system consists of two coupled 3-NLS equations, and models propagation of light in certain materials under certain circumstances. The two coupled components can be thought of as orthogonally polarized. Manakov solitons were observed experimentally in [47].

The Manakov system is less well known than 3-NLS or sat-NLS, so we will describe it in some detail, following [48].

12.5.1 The Manakov System and Its Solutions

As mentioned, the Manakov system consists of two coupled 3-NLS equations,

$$iq_{1t} + q_{1xx} + 2\mu(|q_1|^2 + |q_2|^2)q_1 = 0,$$
$$iq_{2t} + q_{2xx} + 2\mu(|q_1|^2 + |q_2|^2)q_2 = 0, \tag{12.2}$$

where $q_1 = q_1(x, t)$ and $q_2 = q_2(x, t)$ are two interacting optical components, μ is a positive parameter, and x and t are normalized space and time. Note that in order for t to represent the propagation variable, as in Manakov's original paper [46], our variables x and t are interchanged with those of [45]. The system admits single-soliton solutions consisting of two components,

$$q_1 = \frac{\alpha}{2} e^{-\frac{R}{2} + i\eta_I} sech\left(\eta_R + \frac{R}{2}\right),$$

$$q_2 = \frac{\beta}{2} e^{-\frac{R}{2} + i\eta_I} sech\left(\eta_R + \frac{R}{2}\right), \tag{12.3}$$

where

$$\eta = k(x + ikt), \tag{12.4}$$

$$e^R = \frac{\mu(|\alpha|^2 + |\beta^2|)}{k + k^*}, \tag{12.5}$$

and α, β, and k are arbitrary complex parameters. Subscripts R and I on η and k indicate real and imaginary parts. Note that $k_R \neq 0$. Solitons with more than one component, like these, are called *vector* solitons.

12.5.2 State in the Manakov System

The three complex numbers α, β, and k (with six degrees of freedom) in Eq. 12.3 characterize bright solitons in the Manakov system. The complex parameter k is unchanged by collisions, so two degrees of freedom can be removed immediately

Fig. 12.8 A general
two-soliton collision in the
Manakov system. The
complex numbers ρ_1, ρ_L, ρ_2,
and ρ_R indicate the variable
soliton states; k_1 and k_2
indicate the constant soliton
parameters

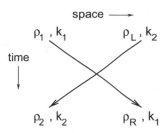

from an informational state characterization. We note that Manakov [46] removed
an additional degree of freedom by normalizing the polarization vector determined
by α and β by the total magnitude $(\alpha^2 + \beta^2)^{1/2}$. However, it is a remarkable fact that
the single complex-valued polarization state $\rho = \alpha/\beta$, with only two degrees of
freedom [49], suffices to characterize two-soliton collisions when the constants k of
both solitons are given [48].

We use the tuple (ρ, k) to refer to a soliton with variable state ρ and constant
parameter k:

- $\rho = q_1(x, t)/q_2(x, t) = \alpha/\beta$: a complex number, constant between collisions;
- $k = k_R + ik_I$: a complex number, with $k_R \neq 0$.

We use the complex plane extended to include the point at infinity.

Consider a two-soliton collision, and let k_1 and k_2 represent the constant soliton
parameters. Let ρ_1 and ρ_L denote the respective soliton states before impact. Suppose
the collision transforms ρ_1 into ρ_R, and ρ_L into ρ_2 (see Fig. 12.8). We will always
associate k_1 and ρ_1 with the right-moving particle, and k_2 and ρ_L with the left-moving
particle. To specify these state transformations, we write

$$T_{\rho_1, k_1}(\rho_L, k_2) = \rho_2, \tag{12.6}$$

$$T_{\rho_L, k_2}(\rho_1, k_1) = \rho_R. \tag{12.7}$$

The soliton velocities are determined by k_{1I} and k_{2I}, and are therefore constant.

It turns out that the state change undergone by each colliding soliton takes on the
very simple form of a linear fractional transformation (LFT) (also called *bilinear* or
Möbius transformation). The coefficients are simple functions of the other soliton in
the collision. Explicitly, the LFTs are

$$\rho_2 = \frac{[(1 - g)/\rho_1^* + \rho_1]\rho_L + g\rho_1/\rho_1^*}{g\rho_L + (1 - g)\rho_1 + 1/\rho_1^*}, \tag{12.8}$$

where

$$g(k_1, k_2) = \frac{k_1 + k_1^*}{k_2 + k_1^*}. \tag{12.9}$$

and

$$\rho_R = \frac{[(1 - h^*)/\rho_L^* + \rho_L]\rho_1 + h^*\rho_L/\rho_L^*}{h^*\rho_1 + (1 - h^*)\rho_L + 1/\rho_L^*}, \tag{12.10}$$

where

$$h(k_1, k_2) = \frac{k_2 + k_2^*}{k_1 + k_2^*}. \tag{12.11}$$

We assume here, without loss of generality, that $k_{1R}, k_{2R} > 0$.

Several properties of these transformations are derived in [48], including characterization of inverse operators, fixed points, and implicit forms. In particular, when viewed as an operator every particle has an *inverse*, and the two traveling together constitute an *inverse pair*. Collision with an inverse pair leaves the state of every particle intact.

12.5.3   Particle Design for Computation

In any particle collision we can view one of the particles as an "operator" and the other as "data." In this way we can hope to find particles and states that effect some useful computation. We give some examples that illustrate simple logical operations.

An i Operator A simple nontrivial operator is pure rotation by $\pi/2$, or multiplication by i. This changes linearly polarized solitons to circularly polarized solitons, and vice versa. A numerical search yielded the useful transformations

$$T_{\rho_L}(\rho) = T_{0, 1-i}(\rho, 1 + i) = (1 - h^*(1 + i, 1 - i))\rho = \frac{1}{\sqrt{2}}e^{-\frac{\pi}{4}i}\rho, \tag{12.12}$$

$$T_{\rho_L}(\rho) = T_{\infty, 5-i}(\rho, 1 + i) = \frac{\rho}{1 - h^*(1 + i, 5 - i)} = \sqrt{2}e^{\frac{3\pi}{4}i}\rho, \tag{12.13}$$

which, when composed, result in the transformation

$$U(\rho, 1 + i) = i\rho. \tag{12.14}$$

Here we think of the data as right-moving and the operator as left-moving. We refer to U as an i operator. Its effect is achieved by first colliding a soliton $(\rho, 1 + i)$ with $(0, 1 - i)$, and then colliding the result with $(\infty, 5 - i)$, which yields $(i\rho, 1 + i)$.

A -1 Operator (NOT Processor) Composing two i operators results in the -1 operator, which with appropriate encoding of information can be used as a logical NOT processor. Figure 12.9 shows a NOT processor with reusable data and opera-

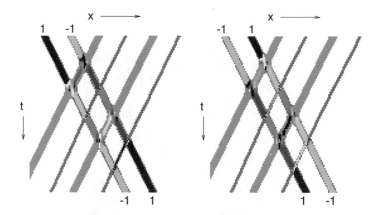

Fig. 12.9 Numerical simulation of a NOT processor implemented in the Manakov system. These graphs display the color-coded phase of ρ for solitons that encode data and processors for two cases. In the initial conditions (top of graphs), the two leftmost (data) solitons are an inverse pair that can represent a 0 in the *left graph*, and a 1 in the *right graph*. In each graph, these solitons collide with the four rightmost (processor) solitons, resulting in a soliton pair representing a 1 and a 0, respectively. The processor solitons are unchanged. These graphs were obtained by numerical simulation of Eq. 12.2 with $\mu = 1$

tor solitons. The two right-moving particles represent data and are an inverse pair, and thus leave the operator unchanged; the left-moving group comprise the four components of the -1 operator. This figure was obtained by direct numerical simulation of the Manakov system, with initial state that contains the appropriate data and processor solitons.

This NOT processor switches the phase of the (right-moving ± 1) data particles, using the energy partition of the (left-moving 0 and ∞) operator particles. A kind of dual NOT gate exists, which switches the energy of data particles using only the phase of the operator particles. In particular, if we use the same k's as in the phase-switching NOT gate, code data as 0 and ∞, and use a sequence of four ± 1 operator particles, the effect is to switch 0 to ∞ and ∞ to 0—that is, to switch all the energy from one component of the data particles to the other (see Fig. 12.10).

A "Move" Operator Figure 12.11 depicts a simple example of information transfer from one particle to another, reminiscent of an assembly-language MOVE instruction. In the initial conditions of each graph, a "carrier" particle C collides with the middle particle; this collision transfers information from the middle particle to C. The carrier particle then transfers its information to another particle via a collision. The appropriate particles A, B, and C for this operation were found through a numerical search, as with the particles for our NOT gate.

Note that "garbage" particles arise as a result of this "move" operation. In general, because the Manakov system is reversible, such "garbage" often appears in computations, and needs to be managed explicitly or used as part of computation, as with conservative logic [27]. Of course reversibility does not necessarily limit

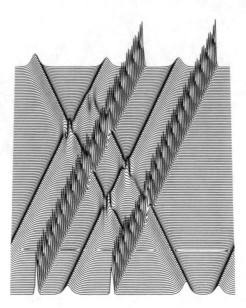

Fig. 12.10 Numerical simulation of an energy-switching NOT processor implemented in the Manakov system. These graphs display the magnitude of one component, for the same two cases as in the previous figure. In this gate the right-moving (data) particles are the inverse pair with states ∞, 0 (*left*), or 0, ∞ (*right*) and the first component is shown. As before, the left-moving (operator) particles emerge unchanged, but here have initial and final states ± 1

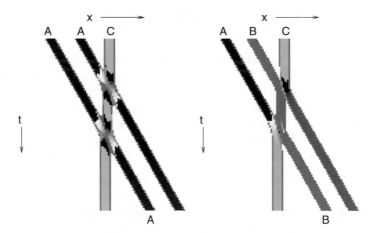

Fig. 12.11 Numerical simulation of a "move" operation implemented in the Manakov system. These graphs display the color-coded phase of ρ. In each graph, the information contained in the middle particle in the initial conditions (*top of graphs*) is moved to the middle particle in the final conditions (*bottom of graphs*). The information transfer is effected by the "carrier" particle C. These graphs were obtained by numerical simulation of Eq. 12.2 with $\mu = 1$

the computational power of the Manakov system, since reversible systems can be universal [50].

12.6   Time-Gated Spatial Manakov Solitons Are Universal

To carry forward our program of embedding general computation in a homogeneous medium, we next sketch the construction of a system of collisions of ideal Manakov solitons that is Turing-equivalent. The approach is straightforward: we will show that we can, in effect, interconnect a universal set of logic gates in an arbitrary manner. Keep in mind that we use the term *gate* to mean a prearranged sequence of soliton collisions that effect a given logical operation, and not, as is in the usual usage, an isolated physical device. We will also use other hardware terms, such as *wiring* and *memory* to refer to the corresponding embedded processing. By *wiring* we will mean moving information from one place to another, and by *memory* we will mean storing it for future use. We will proceed by first describing basic gates that can be used for COPY and FANOUT. The same basic configuration can be adapted for NOT and NAND gates. To complete the computer we will then show how time gating can be used to lay out an arbitrary interconnection of these gates, thus showing universality. The details are reported in [51].

In this section we will use *spatial* solitons, which can be visualized as beams in two spatial dimensions, as opposed to the space-time picture of a pulse, or temporal soliton, traveling down a fiber. The existence and stability of spatial solitons have been well established both theoretically and experimentally in a variety of materials [52]. As pointed out in [52], bright spatial Kerr solitons are stable only in $(1 + 1)$-dimensional systems—that is, systems where the beam can diffract in only one dimension as it propagates. Such solitons are realized in slab waveguides, and are robust with respect to perturbations in both width and intensity.

12.6.1   The General Plan

The main obstacle to implementing what amounts to an arbitrary wiring diagram is the problem of crossing wires without having their signals interfere with one another. We solve this problem by time-gating spatial solitons, so the beams "blink" to avoid unwanted collisions.

It is an interesting question, open to the authors' knowledge, whether a trick like time-gating is necessary for Manakov collision systems to be universal, or whether, as in certain one-dimensional CA like Wolfram's Rule 110 CA [33], arbitrary computation can be embedded in the original, natural space of the underlying medium. But the result with time-gating is physically realizable, and also provides some evidence that the unadorned Manakov collision system may also be rich enough to be Turing-universal.

Fig. 12.12 The general physical arrangement considered in the construction of a universal collision-based computer. Time-gated beams of spatial Manakov solitons enter at the top of the medium, and their collisions result in state changes that reflect computation. Each *solid arrow* represents a beam segment in a particular state

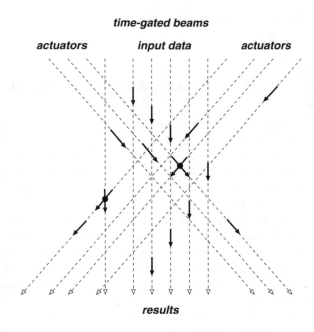

Fig. 12.13 Colliding spatial solitons

Fig. 12.14 Convenient representation of colliding spatial solitons

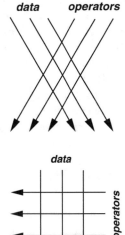

The general arrangement is shown in Fig. 12.12. The usual picture of colliding solitons for computation is shown in Fig. 12.13, but to make it easier to visualize, we will rotate the axes and change the scale so that the data beams travel down and the operator beams travel horizontally as shown in Fig. 12.14.

For the binary states we will use two complex vector soliton states, and it turns out to be possible to use complex state 0 and 1 to represent logical 0 and 1, respectively,

Fig. 12.15 COPY gate

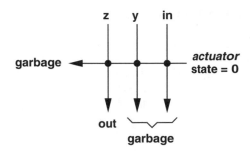

which is convenient but not necessary. The complex soliton states 0 and 1 and logical 0 and 1 will be used interchangeably without risk of confusion.

12.6.2 The COPY and FANOUT Gates

We construct the FANOUT gate by starting with a COPY gate, implemented with collisions between three down-moving, vertical solitons and one left-moving horizontal soliton. This was anticipated by the two-collision "MOVE" gate described in Sect. 12.5.3 and originally in [18]. The use of three collisions and a fixed actuator now makes more flexible gates possible. Figure 12.15 shows the arrangement.

The soliton state labeled *in* will carry a logical value, and so be in one of the two states 0 or 1. The left-moving soliton labeled *actuator* will be in the fixed state 0, as will be the case throughout this construction. The plan is to adjust the (so far) arbitrary states z and y so that $out = in$, justifying the name COPY. It is reasonable to expect that this might be possible, because there are four degrees of freedom in the two complex numbers z and y, and two complex equations to satisfy: that out be 1 and 0 when in is 1 and 0, respectively. Values that satisfy these four equations in four unknowns were obtained numerically. We will call them z_c and y_c. It appears that it is not always possible to solve these equations, and just when they do and do not have solutions remains a subject for future study. However, explicit solutions have been found for all the cases used in this construction.

To be more specific about the design problem, write Eq. 12.8 as the left-moving product $\rho_2 = L(\rho_1, \rho_L)$, and similarly write Eq. 12.10 as $\rho_R = R(\rho_1, \rho_L)$. The successive left-moving products in Fig. 12.15 are $L(in, 0)$ and $L(y, L(in, 0))$. The *out* state is then $R(z, L(y, L(in, 0)))$. The stipulation that 0 maps to 0 and 1 maps to 1 is expressed by the following two simultaneous complex equations in two complex unknowns

$$R(z, L(y, L(0, 0))) = 0 \tag{12.15}$$
$$R(z, L(y, L(1, 0))) = 1$$

Fig. 12.16 FANOUT gate

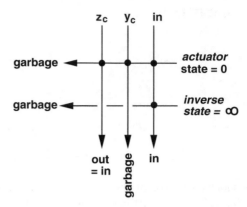

Using the symbolic manipulation program Maple it turns out to be possible to solve for z as a function of y and then eliminate z from the equations, yielding one complex equation in the one complex unknown y, which can be easily solved numerically.

To make a FANOUT gate, we need to recover the input, which we can do using a collision with a soliton in the state which is the inverse of 0, namely ∞ [48]. Figure 12.16 shows the complete FANOUT gate. Notice that we indicate collisions with a dot at the intersection of paths, and require that the continuation of the inverse soliton not intersect the continuation of z that it meets. We indicate that by a break in the line, and postpone the explanation of how this "wire crossing" is accomplished. It is actually immaterial whether the continuation of the inverse operator hits the continuation of y, because neither is used later. We call solitons that are never used again, like the continuation of the inverse operator, *garbage* solitons.

12.6.3   NOT and ONE Gates

In the same way we designed the complex pair of states (z_c, y_c) to produce a COPY and FANOUT gate, we can find a pair (z_n, y_n) to get a NOT gate, mapping 0 to 1 and 1 to 0; and a pair (z_1, y_1) to get ONE gate, mapping both 0 and 1 to 1.

We should point out that the ONE gate in itself, considered as a one-input, one-output gate, is not invertible, and could never be achieved by using the continuation of one particular soliton through one, or even many collisions. This is because such transformations are always nonsingular linear fractional transformations, which are invertible [48]. The algebraic transformation of state from the input to the continuation of z is, however, much more complicated and provides the flexibility we need to get the ONE gate. It turns out that this ONE gate will give us a row in the truth table of a NAND, and is critical for realizing general logic.

Fig. 12.17 A NAND gate, using converter gates to couple copies of one of its inputs to its z and y parameters

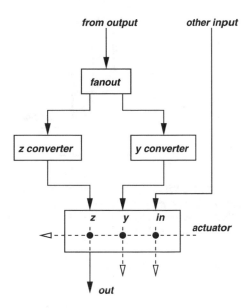

12.6.4   Output/Input Converters and a NAND Gate

To perform logic of any generality we must of course be able to use the output of one operation as the input to another. To do this we need to convert logic (0/1) values to some predetermined z and y values, the choice depending on the type of gate we want. This enables us to construct two-input, one-output gates.

As an important example, here's how a NAND gate can be constructed. We design a z-converter that converts 0/1 values to appropriate values of z, using the basic three-collision arrangement shown in Fig. 12.15. For a NAND gate, we map 0 to z_1, the z value for the ONE gate, and map 1 to z_n, the z value for the NOT gate. Similarly, we construct a y-converter that maps 0 to y_1 and 1 to y_n. These z- and y-converters are used on the fanout of one of the inputs, and the resulting two-input gate is shown in Fig. 12.17. Of course these z- and y-converters require z and y values themselves, which are again determined by numerical search.

The net effect is that when the left input is 0, the other input is mapped by a ONE gate, and when it is 1 the other input is mapped by a NOT gate. The only way the output can be 0 is if both inputs are 1, thus showing that this is a NAND gate. Another way of looking at this construction is that the 2×2 truth table of (left input) $\times$ (right input) has as its 0 row a ONE gate of the columns (1 1), and as its 1 row a NOT gate of the columns (1 0).

The importance of the NAND gate is that it is *universal* [53]. That is, it can be used with interconnects and fanouts to construct any other logical function. Thus we have shown that with the ability to "wire" we can implement any logic using the Manakov model.

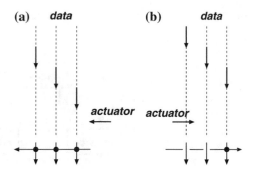

Fig. 12.18 **a** When entered from the right and properly timed, the actuator pulse hits all three data pulses, as indicated in the projection at the bottom; **b** When entered from the left and properly timed, the actuator pulse misses two data pulses and hits only the rightmost data pulse, as indicated in the projection at the bottom

12.6.5 Time Gating

We next take up the question of interconnecting the gates described above, and begin by showing how the continuation of the input in the COPY gate can be restored without affecting the other signals. In other words, we show how a simple "wire crossing" can be accomplished in this case.

The key flexibility in the model is provided by assuming that input beams can be time-gated; that is, turned on and off. When a beam is thus gated, a finite segment of light is created that travels through the medium. We can think of these finite segments as finite light pulses, and we will call them simply *pulses*.

Figure 12.18a shows the basic three-collision gate implemented with pulses. Assuming that the actuator and data pulses are appropriately timed, the actuator pulse hits all three data pulses, as indicated in the projection below the space-space diagram. The problem is that if we want a later actuator pulse to hit the rightmost data pulse (to invert the state, for example, as in the FANOUT gate), it will also hit the remaining two data pulses because of the way they must be spaced for the earlier three collisions.

We can overcome this difficulty by sending the actuator pulse from the left instead of the right. Timing it appropriately early it can be made to miss the first two data pulses, and hit the third, as shown in Fig. 12.18b. It is easy to check that if the velocity of the right-moving actuator solitons is algebraically above that of the data solitons by the same amount that the velocity of the data solitons is algebraically above that of the left-moving actuator solitons, the same state transformations will result.

12.6.6 Wiring

Having shown that we can perform FANOUT and NAND, it remains only to show that we can "wire" gates so that any outputs can be fed to any inputs. The basic

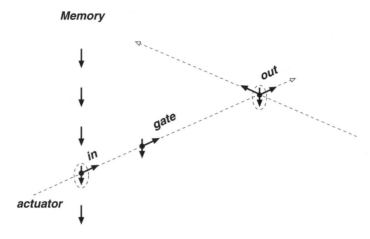

Fig. 12.19 The frame of this figure is moving down with the data pulses on the left. A data pulse in memory is operated on with a three-collision gate actuated from the left, and the result deposited to the upper right

method for doing this is illustrated in Fig. 12.19. We think of data as stored in the down-moving pulses in a column, which we can think of as "memory". The observer moves with this frame, so the data appears stationary.

Pulses that are horizontal in the three-collision gates shown in previous figures will then appear to the observer to move upward at inclined angles. It is important to notice that these upward diagonally moving pulses are evanescent in our picture (and hence their paths are shown dashed in the figure). That is, once they are used, they do not remain in the picture with a moving frame and hence cannot interfere with later computations. However, all vertically moving pulses remain stationary in this picture.

Once a diagonal trajectory is used for a three-collision gate, reusing it will in general corrupt the states of all the stationary pulses along that diagonal. However, the original data pulse (gate input) can be restored with a pulse in the state inverse to the actuator, either along the same diagonal as the actuator, provided we allow enough time for the result (the gate output, a stationary z pulse) to be used, or along the other diagonal.

It turns out that there is one problem remaining with this general idea: we run out of usable diagonals so that, for example, it becomes impossible to fan out the output of a gate output. A simple solution to this problem is to introduce another speed, using velocities ± 0.5, say, in addition to ± 1. This effectively provides four rather than two directions in which a pulse can be operated on, and allows true FANOUT and general interconnections. Figure 12.20 shows such a FANOUT; the data pulse at the lower left is copied to a position above it using one speed, and to another position, above that, using another.

Fig. 12.20 The introduction
of a second speed makes true
FANOUT possible. For
simplicity data and operator
pulses are indicated by *solid
dots*, and the *y* operator
pulses are not shown. The
paths of actuator pulses are
indicated by *dashed lines*

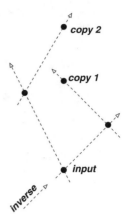

We refer the reader to [51] for more details, including specific gate designs for
the NAND gate.

12.6.7 Universality

It should be clear now that any sequence of three-collision gates can be implemented
in this way, copying data out of the memory column to the upper left or right, and
performing NAND operations on any two at a time in the way shown in the previous
section. The computation can proceed in a breadth-first manner, with the results of
each successive stage being stored above the earlier results. Each additional gate can
add only a constant amount of height and width to the medium, so the total area
required is no more than proportional to the square of the number of gates.

The "program" consists of down-moving *y* and *z* operator pulses, entering at the
top with the down-moving data, and actuator pulses that enter from the left or right at
two different speeds. In the frame moving with the data, the data and operator pulses
are stationary and new results are deposited at the top of the memory column. In the
laboratory frame the data pulses leave the medium downward, and new results appear
in the medium at positions above the old data, at the positions of newly entering *z*
pulses.

Figure 12.21 shows a concrete example of a composite logical operation, an XOR
gate—the SUM bit of a half adder—implemented in the conventional way with
NAND gates [54] and COPY operations.

Fig. 12.21 Implementation of an XOR gate with NAND gates and COPY operations. The results are deposited above the inputs in the data column. Two speeds are necessary to achieve the fanout

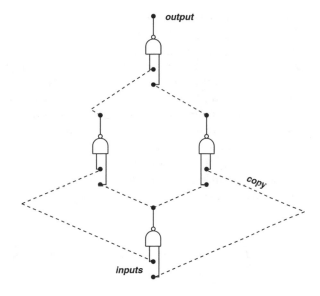

12.6.8 Some Comments on the Universality Result

We note that the result described here differs from the universality results for the ideal billiard ball model [27], the Game of Life [31], and Lattice Gasses [55], for example, in that no internal mirrors or structures of any kind are used inside the medium. To the authors' knowledge, to what extent internal structure is necessary in these other models is open.

Finally, we remark that the model used is reversible and dissipationless. The fact that some of the gate operations realized are not in themselves reversible is not a contradiction, since extra, "garbage" solitons [27] are produced that save enough state to run the computation backwards.

12.7 Multistable Collision Cycles

The computation scheme described up to this point, coding information in the polarization state of vector solitons, is still very far from practical realization. Many practical problems lie mainly in the realization of systems that are close to the ideal Manakov. But even if such systems could be engineered—and they do use well established physics—there would still remain a critical difficulty, which we now address. This is the problem of preventing the accumulation of small errors due to noise over what may well be millions or billions of steps. The way this problem is solved in today's digital computers, and what makes modern computers possible, in fact, is to use bistable systems that restore voltage levels after every step. We will next describe equivalent bistable configurations for Manakov state: bistable cycles of collisions that act like embedded flip-flops. We will then discuss the ways in which such bistable cycles might be used to implement collision-based logic. The results described in this section are reported in more detail in [56, 57].

It is important to realize that the multistability described next occurs in the polarization states of the beams; the solitons themselves do not change shape and remain the sech-shaped solutions of the 3-NLS and Manakov equations. This is in contrast to multistability in the modes of scalar solitons (see, for example, the review [58]). The phenomenon also differs from other examples of polarization multistability in specially engineered devices, such as the vertical-cavity surface-emitting laser (VCSEL) [59], in being dependent only on simple soliton collisions in a completely homogeneous medium.

12.7.1 The Basic Three-Cycle and Computational Experiments

Figure 12.22 shows the simplest example of the basic scheme, a cycle of three beams, entering in states A, B, and C, with intermediate beams a, b, and c (see Fig. 12.22). For convenience, we will refer to the beams themselves, as well as their states, as A, B, C, etc. Suppose we start with beam C initially turned off, so that $A = a$. Beam a then hits B, thereby transforming it to state b. If beam C is then turned on, it will hit A, closing the cycle. Beam a is then changed, changing b, etc., and the cycle of state changes propagates clockwise. The question we ask is whether this cycle converges, and if so, whether it will converge with any particular choice of complex parameters to exactly zero, one, two, or more foci. We answer the question with numerical simulations of this cycle.

A typical computational experiment was designed by fixing the input beams A, B, C, and the parameters k_1 and k_2, and then choosing points a randomly and independently with real and imaginary coordinates uniformly distributed in squares of a given size in the complex plane. The cycle described above was then carried out until convergence in the complex numbers a, b, and c was obtained to within 10^{-12} in norm. Distinct foci of convergence were stored and the initial starting points a were

Fig. 12.22 The basic cycle
of three collisions

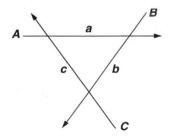

Fig. 12.23 The two foci and their corresponding basins of attraction in the first example, which uses a cycle of three collisions. The states of the input beams are $A = -0.8 - i \cdot 0.13$, $B = 0.4 - i \cdot 0.13$, $C = 0.5 + i \cdot 1.6$; and $k = 4 \pm i$

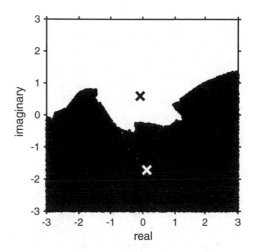

categorized by which focus they converged to, thus generating the usual picture of basins of attraction for the parameter a. Typically this was done for 50,000 random initial values of a, effectively filling in the square, for a variety of parameter choices A, B, and C. The following results were observed:

- In cases with one or two clear foci, convergence was obtained in every experiment, almost always within one or two hundred iterations.
- Each experiment yielded exactly one or two foci.
- The bistable cases (two foci) are somewhat less common than the cases with a unique focus, and are characterized by values of k_R between about 3 and 5 when the velocity difference Δ was fixed at 2.

Figure 12.23 shows a bistable example, with the two foci and their corresponding basins of attraction. Numerical results suggest that the three-collision cycle can have no more than two stable foci, but that n-collision cycles can have up to $n - 1$ foci. The reader is referred to [56] for further examples.

12.7.2   Proposed Physical Arrangement

Our computations assume that the angles of collisions, which for spatial solitons are determined by the unnormalized velocities in laboratory units, are equal. In situations with strong interactions the angles are small, on the order of a few degrees, at the most. We can arrange that all three collisions take place at the same angle by feeding back one of the beams using mirrors, using an arrangement like that shown in Fig. 12.24. Whether such an arrangement is experimentally practical is left open for future study, but it does not appear to raise insurmountable problems. Note that it is also necessary to divert the continuation of some beams to avoid unwanted collisions.

Fig. 12.24 One way to
control the collision angles

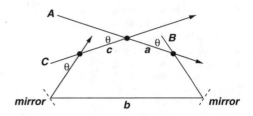

12.7.3 State Restoration

As mentioned, we aim at combating the effects of noise by using bistable collision
cycles to restore state, thus making it feasible to think of cascading a large number
of operations. The basic idea of state-restoration for digital computing has been well
understood for more than half a century; see [60], for example, for an excellent and
early discussion.

12.7.4 Controlling a Bistable Cycle

In order to use these bistable collision cycles for data storage and logic, we need to
develop a method by which we can individually address these devices. In other words,
given a bistable configuration of Manakov solitons with certain constant inputs, we
must be able to switch between binary states of the cycle reliably.

   We accomplish this by temporarily disrupting the bistability of the cycle. For
example, colliding a control beam, or beams, with A (as shown by the dashed lines
in Fig. 12.25) changes the input state A to D. Through careful design of the control beams, we can ensure that A changes in such a way that the cycle (cycle (3)
in Fig. 12.25), which demonstrated bistability without the control beams, becomes
monostable, yielding only one possible steady-state value for the intermediate and
output solitons of cycle (3). Subsequently, when the control beams are turned off,
A equals D^1 and cycle (3) recovers its bistable configuration, but now the initial
state of the cycle is known. This initial condition will lie in one of the two basins
of attraction, causing the cycle to settle to the focus corresponding to that basin. In
this fashion, we control the output state of a bistable soliton collision cycle, where
the value of the monostable focus is controlled by changing the state of the control
beam.

---

[1]We assume here that there is sufficient separation between collisions to ensure that this equality is
true.

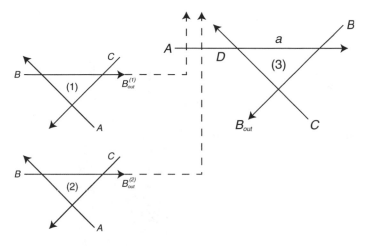

Fig. 12.25 Schematic of NAND gate using bistable collision cycles

12.7.5 NAND and FANOUT Gates

The schematic of a NAND gate is shown in Fig. 12.25. It consists of three cycles: cycles (1) and (2) are the inputs to cycle (3), which represents the actual gate. All three cycles have identical bistable configurations, with input solitons $A = -0.2 + 0.2i$, $B = 0.9 + 1.5i$, $C = -0.5 - 1.5i$ and $k = 4 \pm i$. The output of any cycle is B_{out}, and an input is described by a collision with A. Using the method described in Sect. 12.7.4, cycles (1) and (2) can be set in either binary state 0 or 1. When the inputs from cycles (1) and (2) are active, cycle (3) will become monostable and depending on the values of the inputs, there are four possible monostable foci for cycle (3). Turning off the inputs will place cycle (3) in the state corresponding to the NAND operation. By using identical bistable collision cycles, we ensure that the output is standardized and can serve as input for the next level of logic.

The bistable configuration of all three cycles, along with the values of the monostable foci which correspond to the four inputs, are shown in Fig. 12.26. Only when the inputs are both in state 1 will the cycle be put into state 0. A variability on the inputs will change the position of the monostable foci slightly. We can see from Fig. 12.26 that this change will not affect the position of the output state, unless the change is greater than a specified amount. Quantifying the noise margins of this system remains a topic for future work.

Figure 12.27 shows the schematic of a FANOUT gate, where solitons y and z are chosen in such a way that a copy of soliton in is created at the output, as indicated by out'.

Explicitly, we define the transformations in Eqs. 12.8 and 12.10 as $\rho_2 \equiv L(\rho_1, \rho_L)$ and $\rho_R \equiv R(\rho_1, \rho_L)$, respectively. The value of out' is then $R(y, L(in, z))$. The original input soliton is recovered using the inverse property of Manakov collisions, as described above and in [48]. When viewed as an operator, each polarization state

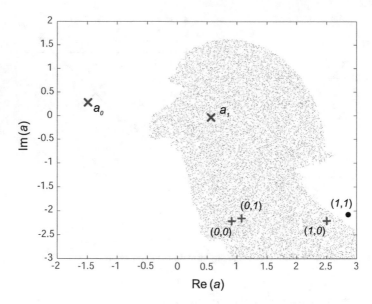

Fig. 12.26 Map of beam a in the complex plane showing NAND gate operation. The two foci, a_0 and a_1, are shown with their corresponding basins of attraction. The + signs are monostable foci which indicate inputs where the cycle reaches state 1, the • is the monostable focus acquired with a $(1, 1)$ input

Fig. 12.27 Schematic of FANOUT gate, where each • indicates a collision

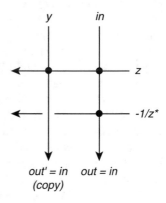

ρ has an inverse defined as $-1/\rho^*$. As such, an arbitrary soliton ρ_1 which collides with another soliton ρ_2, followed by a collision with its inverse $-1/\rho_2^*$, restores the original state ρ_1. Thus the original input soliton in is restored by a collision with the inverse of z, $-1/z^*$.

As a useful example, we design a FANOUT gate for the case of input soliton $in = B_{out}$, where B_{out} is taken from the output of a NAND gate. The bistable configuration of the NAND gate provides for two possible outputs, B_{out0} and B_{out1}, corresponding to binary states 0 and 1, respectively. The FANOUT design stipulates that B_{out0}

maps to B_{out0} and B_{out1} maps to B_{out1}, which can be expressed by the following two simultaneous complex equations in two complex unknowns:

$$R(y, L(B_{out0}, z)) = B_{out0},$$
$$R(y, L(B_{out1}, z)) = B_{out1}. \tag{12.16}$$

Solving Eqs. (12.16) numerically yields $y = 0.6240 - 0.4043i$ and $z = -1.1286 + 0.7313i$. This example thus demonstrates that the output from a NAND gate can be used to drive two similar NAND gates.

12.8 Conclusion

The line (or perhaps tree) of work traced in this chapter suggests many open questions, some theoretical, some experimental, and some a mixture of the two—and even some of interest in their own right without regard to embedded computation. We conclude by mentioning some of these.

In the theoretical area:

- What is a complete mathematical characterization of the state LFTs obtainable by composing either a finite number—or an infinite number—of the Manakov collisions?
- How can we "program" Manakov solitons? Is the Manakov collision system universal without the device of time-gating? Is the temporal system universal?
- Is the complex-valued polarization state used here for the Manakov system also useful in other multi-component systems, especially those that are near-integrable and support spatial solitons?
- What is the theoretical computational power of systems other than Manakov? In particular, which systems in $1 + 1$ or $2 + 1$ dimensions, integrable or noninte-grable, are Turing-equivalent?
- Can $2 + 1$ and higher-dimensional soliton systems be used for efficient computation in uniform media? For example, can a $2 + 1$ integrable system simulate the billiard-ball model of computation, and can such a system be useful without fixed barriers off which balls bounce?
- What is the dynamic behavior of a collision cycle in reaching its steady state? In particular, how fast does the state settle down?
- Do multistable collision cycles occur in other vector soliton systems, such as the nonintegrable saturable systems in photorefractives [42, 61–64]? Can such multistable systems be coupled to implement logical operations like shift registers and arithmetic?
- Is it true that the number of stable foci in a collision cycle of n Manakov solitons is bounded by $n - 1$? Is the $n - 1$ always achievable? What is the dynamic behavior of more complicated collision topologies, can we characterize their stable foci, and can they be used to do useful computation?

- Can scalar or other systems of solitons be used for computation? In this regard we mention the recent interesting work of Bakaoukas and Edwards [65], where they describe a scheme that uses scalar 3-NLS solitons and additional hardware to detect the nature of the collisions to launch additional solitons; and [66], where they use second-order, as well as first-order, scalar 3-NLS solitons, and examine outputs in various time slots.
- What is the connection between discrete (CA) solitons and continuous (PDE) solitons? Why does the same phenomenon manifest itself in two such widely different mathematical frameworks?

On the experimental side of things:

- Can the Manakov system be implemented in a simple, accurate, and practical way?
- Can saturable materials like photorefractive crystals be made that are highly trans-active with acceptable radiation?
- What new physical systems might be found that support solitons which can be easily used to compute?

As we've seen there are many fascinating questions of interest—to both computer scientists and physicists—about soliton information processing. The very notion that nonlinear waves/particles can encode and process information remains largely unexplored.

The work we've discussed here reflects only one aspect of what is called "unconventional" or "nonstandard" computation, and which comprises alternatives to the lithographed silicon-chip-based paradigm as a physical basis for computation. See the *International Journal of Unconventional Computing* for reports of progress in this growing and fascinating field.

Acknowledgments We owe a debt of gratitude to colleagues and students, too many to enumerate, for useful comments and discussions over the years. Most notably, Stephen Wolfram provided an important spark three decades ago, and Mordechai Segev two decades ago. The description of state-restoring logic is based on work with Darren Rand and Paul Prucnal.

References

1. Wolfram, S. (ed.): Theory and Application of Cellular Automata. World Scientific, Singapore (1986)
2. Wolfram, S.: Universality and complexity in cellular automata. Phys. D **10D**, 1–35 (1984)
3. Park, J.K., Steiglitz, K., Thurston, W.P.: Soliton-like behavior in automata. Phys. D **19D**(3), 423–432 (1986)
4. Parks, T.W., Burrus, C.S.: Digital Filter Design. Wiley, New York (1987)
5. Steiglitz, K., Kamal, I., Watson, A.: Embedding computation in one-dimensional automata by phase coding solitons. IEEE Trans. Comput. **37**(2), 138–145 (1988)
6. Goldberg, C.H.: Parity filter automata. Complex Syst. **2**, 91–141 (1988)
7. Fokas, A.S., Papadopoulou, E., Saridakis, Y.: Particles in soliton cellular automata. Complex Syst. **3**, 615–633 (1989)

8. Fokas, A.S., Papadopoulou, E., Saridakis, Y.: Coherent structures in cellular automata. Phys. Lett. **147A**(7), 369–379 (1990)
9. Ablowitz, M.J., Keiser, J.M., Takhtajan, L.A.: Class of stable multistate time-reversible cellular automata with rich particle content. Phys. Rev. A **44A**(10), 6909–6912 (1991)
10. Bruschi, M., Santini, P.M., Ragnisco, O.: Integrable cellular automata. Phys. Lett. A **169**, 151–160 (1992)
11. Siwak, P.: On automata of some discrete recursive filters that support filtrons. In: Domek, S., Kaszynski, R., Tarasiejski, L. (eds.) Proceedings of Fifth International Symposium on Methods and Models in Automation and Robotics, vol. 3 (Discrete Processes), pp. 1069–1074, Międzyzdroje, Poland, Aug. 25–29. Wydawnictwo Politechniki Szczecińskiej (1998)
12. Siwak, P.: Filtrons and their associated ring computations. Int. J. Gen. Syst. **27**(1–3), 181–229 (1998)
13. Siwak, P.: Iterons, fractals and computations of automata. In: Dubois, D.M. (ed.) Second International Conference on Computing Anticipatory Systems, CASYS '98, Conference Proceedings 465, pp. 45–63. Amer. Inst. Phys, Woodbury, New York (1999)
14. Squier, R.K., Steiglitz, K.: Programmable parallel arithmetic in cellular automata using a particle model. Complex Syst. **8**, 311–323 (1994)
15. Kung, H.T.: Why systolic architectures? IEEE Comput. **15**(1), 37–46 (1982). January
16. Frisch, U., d'Humie'res, D., Hasslacher, B., Lallemand, P., Pomeau, Y., Rivet, J.P.: Lattice gas hydrodynamics in two and three dimensions. Complex Syst. **1**, 649–707 (1987)
17. Jakubowski, M.H., Steiglitz, K., Squier, R.K.: Implementation of parallel arithmetic in a cellular automaton. In: Cappello, P., et al. (eds.) 1995 International Conference on Application Specific Array Processors, Strasbourg, France, Los Alamitos, CA, July 24–26. IEEE Computer Society Press (1995)
18. Jakubowski, M.H.:: Computing with Solitons in Bulk Media (Ph.D. Thesis). Princeton University, Princeton, NJ (1998)
19. Leighton, F.T.: Introduction to Parallel Algorithms and Architectures. Morgan Kaufman Publishers, San Mateo (1992)
20. Liu, H.-H., Fu, K.-S.: VLSI arrays for minimum-distance classifications. In: Fu, K.S. (ed.) VLSI for Pattern Recognition and Image Processing. Springer, Berlin (1984)
21. Hordijk, W., Crutchfield, J.P., Mitchell, M.: Embedded-particle computation in evolved cellular automata. In: Toffoli, T., Biafore, M., Leão, J. (eds.) Proceedings of Fourth Workshop on Physics and Computation (PhysComp96), Boston, Mass., Nov. 22–24, pp. 153–158. New England Complex Systems Institute (1996)
22. Boccara, N., Nasser, J., Roger, M.: Particlelike structures and their interactions in spatiotemporal patterns generated by one-dimensional deterministic cellular-automaton rules. Phys. Rev. A **44**(2), 866–875 (1991)
23. Takahashi, D.: On a fully discrete soliton system. In: Boiti, M., Martina, L., Pempinelli, P. (eds.) Proceedings of 7th Workshop on Nonlinear Evolution Equations and Dynamical Systems (NEEDS'91), pp. 245–249. World Scientific, Singapore (1991)
24. Santini, P.M.: Integrability for algebraic equations, functional equations and cellular automata. In: Makhankov, V., Puzynin, I., Pashev, O. (eds.) Proceedings of8th Workshop on Nonlinear Evolution Equations and Dynamical Systems (NEEDS '92), pp. 214–221. World Scientific, Singapore (1992)
25. Adamatzky, A.I.: On the particle-like waves in the discrete model of excitable medium. Neural Network World, pp. 3–10 (1996)
26. Adamatzky, A.I.: Computing in Nonlinear Media & Automata Collectives. Taylor & Francis (2001)
27. Fredkin, E., Toffoli, T.: Conservative logic. Int. J. Theor. Phys. **21**(3/4), 219–253 (1982)
28. Margolus, N.: Physics-like models of computation. Phys. D **10D**, 81–95 (1984)
29. Adamatzky, A.I.: Universal dynamical computation in multidimensional excitable lattices. Int. J. Theor. Phys. **37**(12), 3069–3108 (1988)
30. Atrubin, A.J.: An iterative one-dimensional real-time multiplier. IEEE Trans. Electron. Comput., **EC-14**:394–399 (1965)

31. Berlekamp, E.R., Conway, J.H., Guy, R.K.: Winning Ways for Your Mathematical Plays, vol. 2. Academic Press Inc. [Harcourt Brace Jovanovich Publishers], London (1982)
32. Serizawa, T.: Three-state Neumann neighbor cellular automata capable of constructing self-reproducing machines. Syst. Comput. Jpn 18(4), 33–40 (1987)
33. Wolfram, S.: A New Kind of Science. Wolfram Media (2002)
34. Scott, A.C., Chu, F.Y.F., McLaughlin, D.W.: The soliton: a new concept in applied science. Proc. IEEE 61(10), 1443–1483 (1973)
35. Drazin, P.G., Johnson, R.S.: Solitons: An Introduction. Cambridge University Press, Cambridge (1989)
36. Ablowitz, M.J., Clarkson, P.A.: Solitons, Nonlinear Evolution Equations, and Inverse Scattering. Cambridge University Press, Cambridge (1991)
37. Rebbi, C., Soliani, G.: Solitons and Particles. World Scientific, Singapore (1984)
38. Makhankov, V.G.: Soliton Phenomenology. Kluwer Academic Publishers, Norwell (1990)
39. Jakubowski, M.H., Steiglitz, K., Squier, R.K.: When can solitons compute? Complex Syst. 10(1), 1–21 (1996)
40. Segev, M., Valley, G.C., Crosignani, B., DiPorto, P., Yariv, A.: Steady-state spatial screening solitons in photorefractive materials with external applied field. Phys. Rev. Lett. 73(24), 3211–3214 (1994)
41. Shih, M., Segev, M., Valley, G.C., Salamo, G., Crosignani, B., DiPorto, P.: Observation of two-dimensional steady-state photorefractive screening solitons. Electron. Lett. 31(10), 826–827 (1995)
42. Shih, M.F., Segev, M.: Incoherent collisions between two-dimensional bright steady-state photorefractive spatial screening solitons. Opt. Lett. 21(19), 1538–1540 (1996)
43. Tikhonenko, V., Christou, J., Luther-Davies, B.: Three-dimensional bright spatial soliton collision and fusion in a saturable nonlinear medium. Phys. Rev. Lett. 76, 2698–2702 (1996)
44. Jakubowski, M.H., Steiglitz, K., Squier, R.K.: Information transfer between solitary waves in the saturable Schrödinger equation. Phys. Rev. E 56, 7267–7273 (1997)
45. Radhakrishnan, R., Lakshmanan, M., Hietarinta, J.: Inelastic collision and switching of coupled bright solitons in optical fibers. Phys. Rev. E 56(2), 2213–2216 (1997)
46. Manakov, S.V.: On the theory of two-dimensional stationary self-focusing of electromagnetic waves. Sov. Phys. JETP 38(2), 248–253 (1974)
47. Kang, J.U., Stegeman, G.I., Aitchison, J.S., Akhmediev, N.: Observation of Manakov spatial solitons in AlGaAs planar waveguides. Phys. Rev. Lett. 76(20), 3699–3702 (1996)
48. Jakubowski, M.H., Steiglitz, K., Squier, R.: State transformations of colliding optical solitons and possible application to computation in bulk media. Phys. Rev. E 58(5), 6752–6758 (1998)
49. Yariv, A., Yeh, P.: Optical Waves in Crystals. Wiley, New York (1984)
50. Bennett, C.H.: Logical reversibility of computation. IBM J. Res. Dev. 17(6), 525–532 (1973)
51. Steiglitz, K.: Time-gated Manakov spatial solitons are computationally universal. Phys. Rev. E 63(1), 016608 (2000)
52. Stegeman, G.I., Segev, M.: Optical spatial solitons and their interactions: Universality and diversity. Science 286(5444), 1518–1523 (1999)
53. Mano, M.M.: Computer Logic Design. Prentice-Hall, Englewood Cliffs (1972)
54. Mowle, F.J.: A Systematic Approach to Digital Logic Design. Addison-Wesley, Reading (1976)
55. Squier, R.K., Steiglitz, K.: 2-d FHP lattice gasses are computation universal. Complex Syst. 7, 297–307 (1993)
56. Steiglitz, K.: Multistable collision cycles of Manakov spatial solitons. Phys. Rev. E 63(4), 046607 (2001)
57. Rand, D., Steiglitz, K., Prucnal, P.R.: Noise-immune universal computation using Manakov soliton collision cycles. In: Proceedings of Nonlinear Guided Waves and Their Applications. Optical Society of America, x (2004)
58. Enns, R.H., Edmundson, D.E., Rangnekar, S.S., Kaplan, A.E.: Optical switching between bistable soliton states: a theoretical review. Opt. Quantum Electron. 24, S1295–S1314 (1992)
59. Kawaguchi, H.: Polarization bistability in vertical-cavity surface-emitting lasers. In: Osinski, M., Chow, W.W. (eds.) SPIE Proceedings. Physics and Simulation of Optoelectronic Devices V, volume 2994, pp. 230–241. National Labs, Sandia Park, NM, USA (1997)

60. Lo, A.W.: Some thoughts on digital components and circuit techniques. IRE Trans. Elect. Comp., **EC-10**:416–425 (1961)
61. Christodoulides, D.N., Singh, S.R., Carvalho, M.I., Segev, M.: Incoherently coupled soliton pairs in biased photorefractive crystals. Appl. Phys. Lett. **68**(13), 1763–1765 (1996)
62. Chen, Z., Segev, M., Coskun, T., Christodoulides, D.N.: Observation of incoherently coupled photorefractive spatial soliton pairs. Opt. Lett. **21**, 1436–1439 (1996)
63. Anastassiou, C., Segev, M., Steiglitz, K., Giordmaine, J.A., Mitchell, M., Shih, M., Lan, S., Martin, J.: Energy switching interactions between colliding vector solitons. Phys. Rev. Lett. **83**, 2332–2335 (1999)
64. Anastassiou, C., Steiglitz, K., Lewis, D., Segev, M., Giordmaine, J.A.: Bouncing of vector solitons. In *Conference on Lasers and Electro-Optics*, San Francisco, CA, May 8–12 (2000)
65. Bakaoukas, A.G., Edwards, J.: Computing in the 3NLS domain using first order solitons. Int. J. Unconv. Comput. **5**, 489–522 (2009)
66. Bakaoukas, A.G., Edwards, J.: Computation in the 3NLS domain using first and second order solitons. Int. J. Unconv. Comput. **5**, 523–545 (2009)

Chapter 13
Soliton-Guided Quantum Information Processing

Ken Steiglitz

Abstract We describe applications of solitons and soliton collisions to the transport, transfer, and beam-splitting of qubits carried by optical photons. The transport and transfer realize the "flying qubits" necessary for quantum information processing, and the beam-splitting leads, in theory, to an implementation of quantum computing using linear optics. These proposed applications are embedded in a uniform optical fiber and require no special device fabrication, no cooling, and no vacuum.

The pioneering papers of Feynman [1] and Deutsch [2] in the 1980s sparked the rapid development of the field of quantum information processing. The theoretical and experimental progress has been remarkable, with the development, for example, of quantum error correction and a fast algorithm for factoring, and the exploration of a wide variety of physical implementations. In the latter category, the optical photon as the carrier of a qubit has played an important role in the experimental demonstration of quantum cryptography and other important applications to communications and information processing. At the same time there has been tremendous progress in our understanding of classical nonlinear waves, and, in particular, solitons in optical fibers. In this chapter we will explore what role solitons and soliton collisions might play in the development of quantum information processing with optical photons. For a more detailed account, the reader is referred to [3–5], from which the material in this chapter was drawn.

13.1 Photon Trapping

A pulse traveling down a fiber forms a soliton when the dispersion, which tends to widen the pulse, is counterbalanced by the nonlinear Kerr effect, whereby the electric field changes the index of refraction of the material. Such a soliton is called a

K. Steiglitz (✉)
Computer Science Department, Princeton University, Princeton, NJ 08544, USA
e-mail: ken@cs.princeton.edu

© Springer International Publishing Switzerland 2017   297
A. Adamatzky (ed.), *Advances in Unconventional Computing*,
Emergence, Complexity and Computation 23,
DOI 10.1007/978-3-319-33921-4_13

temporal soliton, as opposed to a spatial soliton, where a beam is confined spatially when diffraction is counterbalanced by a nonlinear effect in the material. In this chapter we will restrict our attention to temporal solitons. Hasegawa and Tappert predicted that stable optical solitons will form in a fiber in 1973 [6], and they were observed experimentally in 1980 by Mollenauer et al. [7]. Since then, because of important potential applications to communications, there has been intense activity in both the theoretical and experimental aspects of solitons in optical fibers.

The induced waveguide What is important to us here is the fact that the soliton creates a local distortion of the index of refraction that travels with it down the fiber. This moving distortion can act as a waveguide that can trap and shepherd another, much weaker, light pulse that can differ from the soliton in both frequency and polarization. The strong soliton pulse is called the *pump*, denoted by P, and the weak, shepherded, pulse is called the *probe*, denoted by u. We will follow the model of such a pump/probe system that was laid out by Manassah [8]. It consists of two coupled equations: the first is the standard, integrable cubic nonlinear Schrödinger equation (3-NLS) that describes the formation of the pump; the second, which describes the propagation of the probe, is, in fact, precisely the *linear* Schrödinger wave equation with a potential determined by the pump.

The solution for the pump is the well known soliton solution, a complex wave with a carrier and sech-shaped envelope. The relative phase of two of these solitons on collision determines the nature of the collision. In particular, when the relative phase is π, the collision is repulsive, and the induced waveguide will look like a smoothly bent waveguide. We will be using collisions of this type throughout.

The solution for the probe is, as we might expect, an eigenvalue problem, which we solve by separation of variables, using as ansatz the complex wave

$$u(z, t) = u(t)e^{-iEz}, \tag{13.1}$$

where z is distance along the fiber, and t is time in the frame moving with the pump soliton, which we refer to as *local time*. In the z direction it is simply a phasor of constant intensity. In the t direction, which we can think of in the z-t plane as the lateral direction in the induced waveguide, the probe is more interesting. The reduced equation with independent variable t is the associated Legendre equation, with solutions $u_{\ell m}$ of degree ℓ and order m that are non-singular, physically acceptable, and zero at infinity for integers $\ell \geq m > 0$. Letting $\xi = \tanh(k_R t)$, where k_R is a parameter that determines the energy of the soliton, each $u_{\ell m}$ is the product of $(1 - \xi^2)^{m/2}$ and a polynomial in ξ of degree $(\ell - m)$ and parity $(-)^{\ell-m}$, with $(\ell - m)$ zeros in the interval $-1 \leq \xi \leq +1$ [9, 10]. As functions of t the solutions of the reduced equation take the form $\text{sech}^m(k_R t)$ times a polynomial in $\tanh(k_R t)$ of degree $(\ell - m)$. The degree ℓ of the wave functions supported in the induced waveguide is determined solely by the ratio of corresponding parameters in the pump and probe equations, and is therefore fixed for any given physical fiber implementation.

Assume, then, that the given fiber implementation is such that ℓ and m are integers, and denote the probe solution of degree ℓ and order m by $|\ell m\rangle$. Then there are exactly ℓ eigenfunctions supported by the induced waveguide, corresponding to $m = 1, \ldots, \ell$, with corresponding energy eigenvalues $E_1, \ldots, E_\ell$. When the superposition of more than one of these co-propagate (the reduced equation is linear) the difference in these energy levels causes beating in the z-direction (see Eq. 13.1), as discussed in [8].

The quantum limit Up to now we have described the formation of an electromagnetic probe wave trapped in the waveguide induced by a soliton, where this probe is weak compared to the soliton. If we let the probe get weaker and weaker we reach the point at which the probe can no longer behave like a wave, but must behave like a particle—a photon.

We next must consider the critical question of whether it is possible to detect a probe photon in the presence of the (much larger) pump. There are two ways in which we can separate the probe and pump to make this detection feasible: First, they can be orthogonally polarized in a polarization-maintaining fiber. Second, they can be separated in wavelength. As discussed in more detail in [3], it is reasonable to expect the detection of single probe photons in the collisions described here to be possible at a wavelength of 1550 nm within about one or two kilometers of fiber. This experiment would be the next step in pursuing a physical demonstration of the ideas discussed in this chapter.

13.2 Photon Transfer

We look next at the simple situation where a faster soliton which is not carrying a photon overtakes a slower soliton that is, as sketched in Fig. 13.1. What happens, with appropriate choice of parameters, is that the photon will be transferred to the faster soliton, as shown by the numerical simulation illustrated in Fig. 13.2. The repulsive collision of the two pump solitons is shown at the top. At the bottom we see the deflection of the probe wave (now a photon) to the faster soliton. The probe in this case is in the single-peaked ground state $|11\rangle$. This setup and the ones described in the following sections correspond exactly to what is known for classical waves as a *directional coupler*, and such couplers induced by spatial solitons have been studied since the 1990s [11–13].

Fig. 13.1 Sketch illustrating the conditions for photon transfer when a fast soliton overtakes a slower one

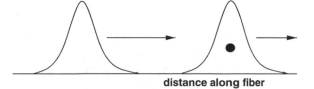

distance along fiber

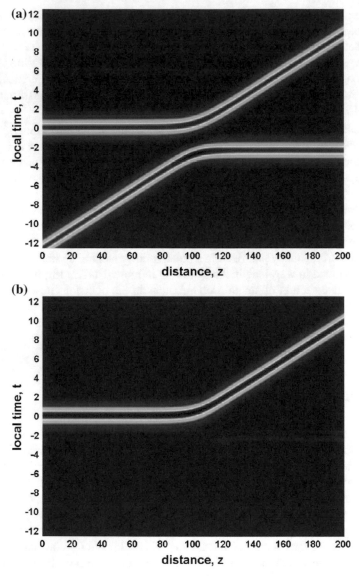

Fig. 13.2 An example of photon transfer. **a** The pump solitons in a repulsive collision. **b** The probe when launched in the state |11⟩. The photon is transferred to the overtaking soliton (See [3] for details.)

In some sense the experiment we have just described is analogous to a photon bouncing off a mirror: the photon is simply diverted. If the photon carries a qubit of quantum information in a photon/no-photon representation, it is a flying qubit in the sense described by DiVincenzo [14] in his well known paper outlining the requirements for a physical implementation of quantum computing and communication. The present fiber scheme would then provide a means for routing flying qubits with solitons.

13.3   Beam Splitters

If we change the system parameters we can arrange an experiment that corresponds, not to an ordinary mirror, but to a half-silvered mirror—a beam-splitter. For this we use a collision between probe solitons of order 2 instead of 1, and a greater relative velocity, as shown in Fig. 13.3.

This is analogous to a non-polarizing beam-splitter. The photon is deflected or transmitted with certain probabilities, in the case shown, both 1/2 for a 50/50 beam-splitter. For a system that functions as a polarizing beam-splitter, we can use a probe in a state that is the superposition of states $|22\rangle$ and $|21\rangle$. It turns out that with the proper choice of parameters, the photon in state $|21\rangle$ is not deflected, while the photon in state $|22\rangle$ is transferred. When the superposition of the two photon states is used as the input probe, the modes are separated, just as an ordinary polarizing filter will separate the horizontal and vertical components of a light wave of mixed polarization. An example of a polarizing beam-splitter is shown in Fig. 13.4.

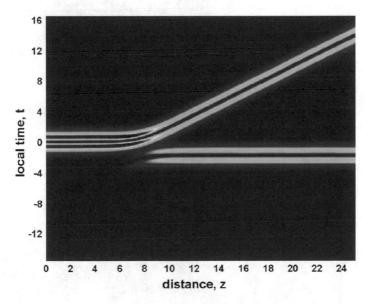

Fig. 13.3   The probe when launched in the state $|21\rangle$. In this case the soliton collision takes place at a greater relative velocity and the system acts as an ordinary non-polarizing beam-splitter (See [3] for details.)

Fig. 13.4 Illustrating a soliton-guided polarizing (mode-separating) beam-splitter. **a** The probe when launched in the excited state $|21\rangle$. The photon in this case stays in large part with its original captor soliton. **b** The probe when launched in the state $|22\rangle$. The photon is transferred to the faster soliton, as in the $|11\rangle$ case shown in Fig. 13.2. **c** The probe when an equal linear combination of ground and excited states, $|22\rangle + |21\rangle$, is launched. The system is analogous to a polarizing beam-splitter (See [3] for details.)

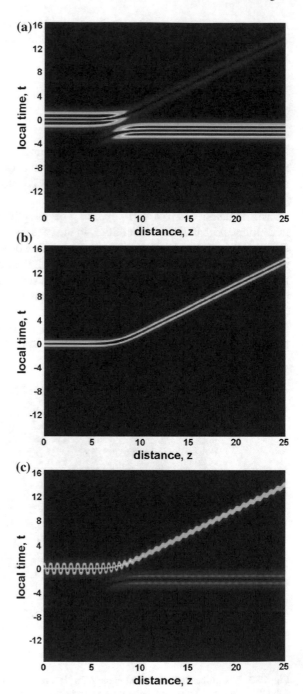

Note that in this soliton-induced beam-splitter, the modes of the probe play the role of polarization axes, and these axes should not be confused with the polarization modes of the fiber medium itself. It thus might be more proper to call this system a "mode-separating" beam-splitter.

13.4 Manipulating Photon Phase

There is a missing piece if we want to accomplish general quantum computing, as we shall see in the next section—we must be able to shift the phase of a single photon. But the same ideas used in the previous sections can be used for this purpose. A phase shifter can work as follows: Two solitons, A and B, are launched at the same velocity in the z direction, first B, then A. Initially, soliton A carries a photon. Soliton C, a third soliton, is then launched at a greater velocity. When C overtakes A, the photon is captured by C; and C carries the photon with it until it overtakes B, at which point the photon is transferred from C to B. The net effect is that soliton C ferries the photon from A to B. The photon accumulates extra phase during the time it travels at an altered velocity, and the amount of the phase shift can be controlled by adjusting that time. Figure 13.5 shows the probe in an example. The reader is referred to [4] for details about how the solitons and probe are designed to accomplish this phase shifting.

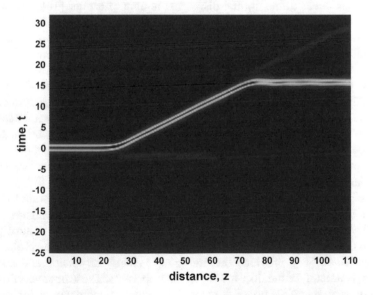

Fig. 13.5 The probe signal in a phase shifter. A photon is being carried by soliton A; The faster soliton C overtakes soliton A, picks up the photon, and ferries it to soliton B, where it is deposited. In this example from [4] the overall phase shift achieved is π (See [4] for details.)

13.5   General Quantum Computing

In 2001 Knill et al. [15] published a surprising and important paper: they showed that general quantum computing can be implemented using only components we have described—beam-splitters, phase shifters—plus single-photon sources and photo-detectors. The latter plays a crucial role in providing the necessary nonlinear aspect to the system: feedback from photo-detectors. The reader is referred to [15] for the details of this clever scheme, and to [16] and its references for recent improvements.

We have thus described a way in which soliton-guided photons can be used to implement general quantum computing, at least in theory. We should quickly point out, however, that the scheme proposed by Knill et al. carries with it an overhead that may be, although polynomial in the problem size as required by the theory, prohibitively large. This is balanced, however, by the relative simplicity of the physical components, most of which are either off-the-shelf or close to it. In addition, the use of soliton-guided flying qubits provides a natural and uniform way to implement the required routing and switching.

13.6   Using Dark Solitons

Dark solitons occur in the normal-dispersion regime of a fiber [18], and occur as dips in a uniform background, in contrast with the bright solitons we have so far been considering. They offer some real advantages over bright soliton collisions in controlling light waves: First, dark solitons are known to be more stable in the presence of noise and are generally more robust than bright solitons [18, 19]. Second, the probe, which is of much lower intensity, peaks at the dip in the intensity of its host soliton, thus increasing the signal-to-noise ratio and making it easier, in principle, to detect. Third, the characteristics of the dark soliton beam splitter do not depend on the relative phase or relative speed of the colliding solitons, whereas bright solitons need to have their phases and speeds carefully controlled to produce a given result. The improvement in signal-to-noise ratio for detecting single photons may prove to be especially important in any practical implementation.

Figure 13.6 shows collisions of two dark solitons, and also illustrates the case when the probe photon is unaffected by the collision, but simply remains with its captor. The waveguides induced by dark solitons can be used to control photon probes in the same way that bright solitons can, in contrast with the zero-crosstalk case reported in [17], provided that the group velocity dispersion and nonlinear coupling parameter for the fiber are chosen appropriately. These degrees of freedom are readily available if we use different wavelengths and polarizations for the pump and probe. For the probes corresponding to the degree-1 associated Legendre modal functions, the dark soliton junctions behave in a way that is very closely analogous to a beam-splitter made of crossed optical polarizers, with a single parameter playing

Fig. 13.6 a The pump signal, a dark soliton collision. **b** The corresponding probe signal illustrating the case when the probe is not affected, the so-called "zero-crosstalk" case [11, 17] (See [5] for details.)

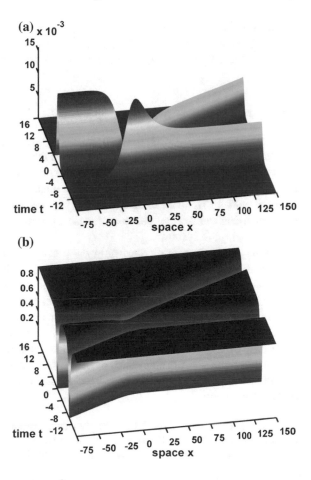

Fig. 13.7 The probe signal in the same induced waveguide with the probe parameters adjusted so that probe is split equally (See [5] for details.)

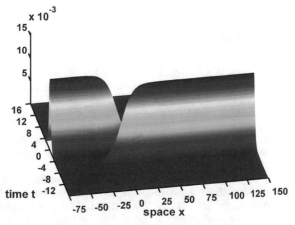

the role of angle between polarizing filters. For the probes corresponding to the degree-2 associated Legendre functions, the junction can act as a mode-separating beam-splitter. Figure 13.7 illustrates such a dark-soliton-guided beam-splitter.

13.7 Conclusion and Open Problems

We have seen how, in theory (and that is a big qualification), solitons in optical fibers can be used to provide a kind of "substrate" for manipulating qubits. Transport, transfer (and therefore routing), and even general quantum computing, using the scheme of Knill et al. [15], all fit naturally in this picture. At the least, this way of implementing flying qubits may prove of practical use in many quantum communication and cryptographic systems.

Open questions remain concerning the practicality of physical implementation: are fibers, photon sources, and photon detectors available that have the required physical characteristics? Perhaps the most logical next step, and a project of interest in itself, would be the experimental verification of photon capture and transport in an optical fiber, by both bright and dark solitons.

Also of interest are the questions, both theoretical and experimental, of the susceptibility of trapped photons to decoherence, as compared with that of ordinary photons in fibers—a problem that, to the author's knowledge, not been studied. It would also be very interesting if soliton-guided photons could be used to realize quantum repeaters.

Acknowledgments I owe a special debt of gratitude to Darren Rand, coauthor of [3], the springboard for this line of work. He shares any credit for this work, but not any blame. I've benefited also from discussions with Sanjeev Arora, Andrew Houck, Steve Lyon, and Herschel Rabitz.

References

1. Feynman, R.P.: Simulating physics with computers. Int. J. Theor. Phys. **21**(6/7), 467–488 (1982)
2. Deutsch, D.: Quantum theory, the Church–Turing principle and the universal quantum computer. Proc. R. Soc. Lond. **A400**, 97–117 (1985)
3. Steiglitz, K., Rand, D.: Photon trapping and transfer with solitons. Phys. Rev. A **79**, 021802(R) (2009)
4. Steiglitz, K.: Soliton-guided phase shifter and beam splitter. Phys. Rev. A **81**, 033835 (2010)
5. Steiglitz, K.: Making beam splitters with dark soliton collisions. Phys. Rev. A **82**, 043831 (2010)
6. Hasegawa, A., Tappert, F.: Transmission of stationary nonlinear optical physics in dispersive dielectric fibers I: anomalous dispersion. Appl. Phys. Lett. **23**(3), 142–144 (1973)
7. Mollenauer, L.F., Stolen, R.H., Gordon, J.P.: Experimental observation of picosecond pulse narrowing and solitons in optical fibers. Phys. Rev. Lett. **45**, (1980)
8. Manassah, J.T.: Ultrafast solitary waves sustained through induced phase modulation by a copropagating pump. Op. Lett. **15**(12), 670–672 (1990)
9. Messiah, A.: Quantum Mechanics, 1st edn. North-Holland, Amsterdam (1961)

10. Schiff, L.I.: Quantum Mechanics, 3rd edn. McGraw-Hill, New York (1968)
11. Akhmediev, N., Ankiewicz, A.: Spatial soliton X-junctions and couplers. Op. Commun. **100**, 186–192 (1993)
12. Lan, S., DelRe, E., Chen, Z., Shih, M.-F., Segev, M.: Directional coupler with soliton-induced waveguides. Op. Lett. **24**, 475–477 (1999)
13. Guo, A., Henry, M., Salamo, G.J., Segev, M., Wood, G.L.: Fixing multiple waveguides induced by photorefractive solitons: directional couplers and beam splitters. Op. Lett. **26**, 1274–1276 (2001)
14. DiVincenzo, D.P.: The physical implementation of quantum computation. Fort. der Phys. **48**, 771–783 (2000). http://arxiv.org/pdf/quant-ph/0002077v3.pdf
15. Knill, E., Laflamme, R., Milburn, G.: A scheme for efficient quantum computation with linear optics. Nature **409**(46), 46–52 (2001)
16. Marinescu, D.C., Marinescu, G.M.: Classical and quantum information. Academic Press, New York (2012)
17. Miller, P.D.: Zero-crosstalk junctions made from dark solitons. Phys. Rev. E **53**(4), 4137–4142 (1996)
18. Agrawal, G.P.: Nonlinear Fiber Optics, 4th edn. Academic Press, London (2006)
19. Luther-Davies, B., Xiaoping, Y.: Waveguides and Y junctions formed in bulk media by using dark spatial solitons. Op. Lett. **17**(7), 496–498 (1992)

Chapter 14
Models of Computing on Actin Filaments

Stefano Siccardi and Andrew Adamatzky

Abstract Actin is a filament-forming protein forming cytoskeleton and information processing network of eukaryotic cells. To speculate about a range of computing operation that could be implemented in actin filaments we design quantum automata, non-linear electrical circuits and one-dimensional latices of nodes with Morse interaction models of the actin filaments. In numerical experiments we implement quantum gates, one-bit binary adder, multi-valued logic gates, gates based on (un)forced pulses, and collision-based soliton circuits.

14.1 Introduction

Two key components of the cytoskeleton—actin filaments and tubule microtubules—are responsible for cell's motility and transduction, and possibly processing, of signals inside the cell. These networks of proteins are a key physical substrate of cell-level learning [19, 22, 37, 46, 61, 77–79, 94]. By modulating dendritic ion channel activity, actin filaments determine rule of neural information processing and facilitate computational abilities of dendritic trees via facilitation of ionic condensation and ion cloud propagation [76]. Psychiatric and neurological disorders are caused by disfunction in actin assembly or the actin association with other proteins and intracellular components [28, 54, 74, 84, 90, 104]. Computational models of tubulin microtubules have been developed in 1990s and used to demonstrate that computation could be implemented in tubulin protofilaments by classical and quantum means [36, 43, 47, 79]. Less attention was paid to actin double helix filaments, despite importance of the actin in learning and information pre-processing as might be hinted by predominant presence of actin networks in synapses [16, 25, 29, 50]. In [4] we proposed a model of actin filaments as two chains of one-dimensional

S. Siccardi (✉) · A. Adamatzky
Unconventional Computing Centre, University of West of England, Bristol, UK
e-mail: stefano.siccardi@uwe.ac.uk

A. Adamatzky
e-mail: andrew.adamatzky@uwe.ac.uk

© Springer International Publishing Switzerland 2017   309
A. Adamatzky (ed.), *Advances in Unconventional Computing*,
Emergence, Complexity and Computation 23,
DOI 10.1007/978-3-319-33921-4_14

binary-state semi-totalistic automaton arrays. We analysed the complete rule space of actin automata using integral characteristics of space-time configurations generated by these rules and discovered state transition rules that support travelling and mobile localizations. We found that some properties of actin automata rules may be predicted using Shannon entropy, activity and incoherence of excitation between the polymer chains. We also showed that it is possible to infer whether a given rule supports travelling or stationary localizations by looking at ratios of excited neighbours that are essential for generations of the localizations. We speculate that a computation in actin filaments is implemented when localisations (defects, conformational changes, ionic clouds, solitons), which represent data, collide with each other and change their velocity vectors or states. Parameters of the localisations between the collision are interpreted as values of input variables. Parameters of the localisation after the collision are values of output variables. The computation is implemented at the collision site [2, 5, 93, 105]. The electrical potential or ionic waves [95] or quantum protein transitions [36, 41] might play the role of signal passing through the proteins and branches of the cytoskeletal network, colliding, changing their velocities or states, and thus performing Boolean logical operations [2]. We discuss several families of actin models, from quantum automata to lattice with Morse potential and show how logical gates and circuits can be implemented on a single actin filament or an ensemble of the filaments.

14.2 Quantum Actin Automata

14.2.1 Quantum Cellular Automata

In quantum cellular automata (QCA) cell states are represented by qubits or cubits, instead of classical bits. Several formal definitions have been proposed, see e.g. [6, 24, 33]. Methods to translate classical automata into quantum ones can be found in [56, 83]. We start our discussion with two classical definitions of one-dimensional QCA (1QCA) as proposed in [33, 55, 99].

Definition I A 1QCA $\mathcal{A} = \langle Q, \lambda, N, \delta \rangle$ is determined by a finite set Q of states, a *quiescent state* λ, a neighbourhood $N = \{n_1, \ldots, n_r\} \subseteq Z$, with $n_1 < n_2 < \cdots < n_r$ and a local transition function

$$\delta : Q^{r+1} \rightarrow C_{[0,1]}$$

satisfying the following three conditions:

1. local probability: for any $(q_1, \ldots, q_r) \in Q^r$,

$$\sum_{q \in Q} |\delta(q_1, \ldots, q_r, q)|^2 = 1$$

2. stability of the quiescent state: if $q \in Q$, then

$$\delta(\lambda, \ldots, \lambda, q) = 1 \text{ if } q = \lambda, \ 0 \text{ otherwise}$$

3. To state the third condition we introduce the automaton's mapping as follows. A configuration $c : Z \to Q$ is a mapping such that $c(i) \neq \lambda$ only for finitely many i. Let $C(\mathcal{A})$ denote the set of all configurations. Computation of $\mathcal{A}$ is done in the space $H_{\mathcal{A}} = l_2(C(\mathcal{A}))$ with the basis $\{|c\rangle \mid c \in C(\mathcal{A})\}$. In one step, $\mathcal{A}$ transfers from one basis state $|c_1\rangle$ to another $|c_2\rangle$ with the amplitude

$$\alpha(c_1, c_2) = \prod_{i \in Z} \delta(c_1(i + n_1), \ldots, c_1(i + n_r), c_2(i))$$

A state in $H_{\mathcal{A}}$, in general, has the form

$$|\phi\rangle = \sum_{c \in C(\mathcal{A})} \alpha_c |c\rangle$$

with normalised α_c. The evolution operator $E_{\mathcal{A}}$ of $\mathcal{A}$ maps any state $|\phi\rangle$ into $|\psi\rangle = E_{\mathcal{A}}|\phi\rangle$ such that

$$|\psi\rangle = E_{\mathcal{A}}|\phi\rangle = \sum_{c \in C(\mathcal{A})} \beta_c |c\rangle$$

where

$$\beta_c = \sum_{c' \in C(\mathcal{A})} \alpha_{c'} \alpha(c', c)$$

Now we can state the third condition: unitarity. The mapping $E_{\mathcal{A}}$ must be unitary.

In general, to decide whether the unitarity condition is satisfied is a nontrivial problem [1, 27, 34]. Partitioned 1QCA, that are well suited to represent cells and their states, are easier to deal with.

Definition II A partitioned 1QCA (P1QCA) is a 1QCA $\mathcal{A} = \langle Q, \lambda, N, \delta \rangle$ that satisfies the following conditions:

1. The set of states Q is the cartesian product $Q = Q_1 \times \cdots \times Q_r$ of $r = |N|$ nonempty sets.
2. The local transition function δ is the composition of two functions:

$$\delta_c : Q^r \to Q \text{ a classical mapping}$$

and

$$\delta_q : Q \to C^Q \text{ a quantum mapping}$$

where $\delta_c((q_{1,1}, \ldots, q_{1,r}), \ldots, (q_{r,1}, \ldots, q_{r,r})) = (q_{r,1}, q_{r-1,2}, \ldots, q_{1,r})$.

3. The function δ_q defines a 1QCA with operator

$$U_{\mathcal{A}_q}(q_2, q_1) = [\delta_q(q_1)](q_2)$$

$U_{\mathcal{A}_q}$ is also called a local transition matrix of $\mathcal{A}$. It can be shown that the evolution of a P1QCA is unitary if and only if this operator is unitary.

14.2.2 Application to Actin-Like Structures

Our next step is to translate the actin automaton defined in [4] to a quantum automaton based on the P1QCA definition. The original actin automaton is composed of two layers of cells, labelled x_k and y_k. In our model we use only one type of cells, q_i, identifying $q_{2k+1} = x_k$ and $q_{2k} = y_k$ like in Fig. 14.1. With this convention, the neighborhood of any q_i is $(q_{i-2}, q_{i-1}, q_i, q_{i+1}, q_{i+2})$, that corresponds both to the neighbours of cells x_k and y_k in the original notation. Moreover, the state of the generic cell q_i is a vector $(q_{i1}, q_{i2}, q_{i3}, q_{i4}, q_{i5})$ and the first step of the evolution includes building of the intermediate state

$$\tilde{q}_i = (q_{i+2,1}, q_{i+1,2}, q_{i,3}, q_{i-1,4}, q_{i-2,5})$$

In the classical actin automaton [4], the node/cell transition rules has the following form $((a_{00}, a_{01}, a_{02}, a_{03}, a_{04}), (a_{10}, a_{11}, a_{12}, a_{13}, a_{14},)) = (X_0, X_1)$
with $a_{ik} \in \{0, 1\}$ and X_i is a 5-elements array. To compute the state of cell q_i at time $t + 1$, the sum

$$S = q_{i-2} + q_{i-1} + q_{i+1} + q_{i+2}$$

is computed at time t, and the new state of q_i takes the value $a_{q_i(t),S}$ where $q_i(t)$ is the state of q_i at time t.

In the quantum case, we must specify the matrix (of dimension 32) $U(q_1, q_2) = [u_{a_1a_2a_3a_4a_5,b_1b_2b_3b_4b_5}]$ whose elements give the probability that a state $\tilde{q} = (a_1a_2 a_3a_4a_5)$ evolves into another state $q = (b_1b_2b_3b_4b_5)$.

For the generic element of U we have: $u_{a_1a_2a_3a_4a_5,b_1b_2b_3b_4b_5} = 0$ if $b_3 \neq X_{a_3,S}$ and
$u_{a_1a_2a_3a_4a_5,b_1b_2b_3b_4b_5} =$

$$1 \text{ if } b_3 = X_{a_3,S} \text{ and } b_{i\neq3} = B_i(S) \tag{14.1}$$

Fig. 14.1 A scheme of quantum actin automata

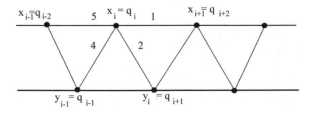

where $B_i(S)$ is an array or a rule, possibly dependent on S, that must be specified. In the P1QCA case we have to specify also how the four 'extra' states evolve. Several choices are possible, so that this kind of automata is quite flexible and suitable to model different situations. In other words, when using quantum automata we have three choices to make: (1) the specific definition of the automaton, e.g. a partitioned quantum automaton; (2) the transition matrix type; (3) the rule for the evolution, like in the classical case.

Note that:

1. The value $\lambda = (0, 0, 0, 0, 0)$ must be stable, so the rules $(1, x, y, z, w), (any))$ are excluded
2. We have to check if the matrix U is unitary for specific rules
3. The above automaton can be very similar to a classical one: we can use a mechanism that generates superposition of the states or does not; in the latter case, we can start with cells in superposition or without. If we start with classical—not superposed—states, the evolution is classical
4. We used the following definition to manage the states transition:

$u_{a_1 a_2 a_3 a_4 a_5, b_1 b_2 b_3 b_4 b_5} =$

$$1 \text{ if } b_3 = X_{a_3,S} \text{ and } b_{i \neq 3} = a_i \tag{14.2}$$

that is, the four "extra" substates of a node/cell are copied from the neighbours of the node/cell.

14.2.3 Implementation

14.2.3.1 Unitarity

Unitarity check must be performed, as explained in the above discussion, for the 512 rules considered in the classical actin automata (excluding rules 1xxxx because of quiescent states stability). Using definition (14.2), where a cell inherits the substates of its neighbours, one obtains 16 unitary matrices, for rules $(i, 31 - i)$ and $i = 0, \ldots, 15$. It is worth noting that for all the rules, conditions that excite a quiescent cell will deactivate an excited one.

14.2.3.2 Evolution

In the first runs, we used seeds like those shown in our previous paper [4], five x_i and five y_i cells, all the others being in the quiescent state. The automata had a total of 60 cells and the automaton was evolved for 60 steps. As our aim at this stage was a qualitative comparison of the classical and quantum models, we tried to obtain comparable results, so we prepared pictures in terms of x and y (that is q_{2n+1} and q_{2n}) cells.

Initial state (aa1aa), matrix [17]

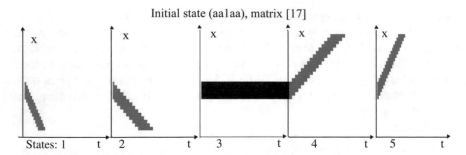

States: 1   t   2   t   3   t   4   t   5   t

Fig. 14.2 Rule $(0, 31)$, initial state is $(aa1aa)$ as explained in the text, matrix (14.2)

Figure 14.2 shows the evolution of the 5 substates of the x cells; time is the horizontal axis and x the vertical axis. We recall that the states are actually qubits, that is they are a linear combination of the two basis elements $|0\rangle$ and $|1\rangle$. Accordingly, the picture represents the module of the coefficient of $|1\rangle$ in gray, black meaning 1 and white 0. Of course, it is expected that the identity rule $(0, 31)$ will leave the cells in their initial condition, and it is indeed the case starting with five x_i and five y_i cells in the state (00100).

It is worth to note that, if we start with cells in a superposition state, e.g. $1/4 \sum (a_j a_k 1 a_i a_l)$, with all the combinations of $a = 0$ and $a = 1$, we get an evolution for the 'lateral' substates, even with the rule $(0, 31)$, as shown in Fig. 14.2.

We did not aim at a systematic analysis of seeds, but looked for seeds that are particles (gliders, localisations) moving to the right or to the left and that could be used as building blocks for some logical functions. Examples of evolutions starting with one seed generating a particle moving to the right and another seed generating a particle moving to the left are shown in Fig. 14.3. In this and in the subsequent pictures, we make no distinctions between x_i and y_i, and the pictures show the evolution of all the cells together, that will be labelled generically x.

Seeds are just two sets of three adjacent cells each, at positions (6, 7, 8) and (12, 13, 14). Cell 7 is initialised at 00010, cell 8 and 12 at 00100, cell 13 at 01000. Cells 6 and 14 carry in their rightmost and leftmost states respectively the bits that we want to use in an operation. In the four runs we have the four combinations 00, 01, 10 and 11.

When the seeds meet at the third step, the resulting central state of cell 10 (circled in red in the picture) is 1 unless both cells 8 and 12 are initialised at 0. This means that this automaton is acting like a classical gate OR: if it could be 'prepared' in the initial states containing the two seeds at specified positions, and if it could be 'read' at the middle position after the specified number of steps, it could be used as a building block for a classical computer.

The same automaton—when initialised in a superposition of states—propagates superposition. For instance one could initialise cell 7 and 13 at a superposition of excited and idle state, e.g. $1/\sqrt{2} \cdot 00010 + 1/\sqrt{2} \cdot 00000$ and $1/\sqrt{2} \cdot 01000 + 1/\sqrt{2} \cdot 00000$ respectively and let cell 8 and 12 start at 00100 like in the previous

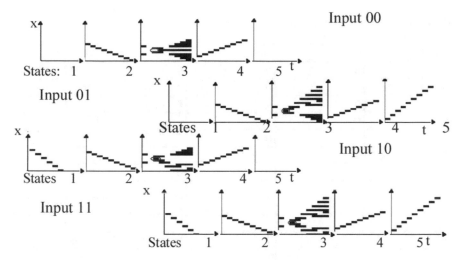

Fig. 14.3   Rule (14, 17), matrix (14.2) of two cells at positions 6 and 14 in initial *right* and *left* states 00, 01, 10, 11. The 'output' of the logical OR gate is the *red circles*

case. Cells 6 and 14 that carry the qubits used in the operation are initialised at $1/2 \cdot 00000 + \sqrt{3}/2 \cdot 00001$ and $1/2 \cdot 00000 + \sqrt{3}/2 \cdot 10000$. What one obtains is that the output state (state 3 of cell 10 at step 3) is 0 if both inputs are 0, it is $1/2|0\rangle + \sqrt{3}/2|1\rangle$ if one of the inputs is in the active state $1/2 \cdot 00000 + \sqrt{3}/2 \cdot 00001$, and it is $1/4|0\rangle + \sqrt{15}/4|1\rangle$ if both input cells are active.

So we see that although the automaton is perfectly working with quantum states, the translation of classical to quantum logical operations is not completely straightforward. A systematic study of the behaviour of matrix (14.2) with classical input has been done in [85] in terms of moves and collisions of two and five particles. In this way it is found that a number of logical gates can be built.

14.2.4   Gates

We now discuss all the logical gates we can build in the evolution of given consecutive cells.

We recall that, considering two input qubits, we have 16 logical gates as in Table 14.1. With five cells, each carrying an active substate, that is a qubit, we can choose the input qubits in 10 different ways, and for each choice we can choose each of the given qubits as the output. Moreover, for each choice of the input qubits we can choose in eight different ways the initial state of the other three 'ancillary' qubits.

Table 14.1 Logical gates definition, columns contain the output value, given the input values of the headings

(0, 0)	(0, 1)	(1, 0)	(1, 1)	Gate
0	0	0	0	Constant 0
0	0	0	1	$x \wedge y$
0	0	1	0	$x \wedge \neg y$
0	0	1	1	$x \forall y$
0	1	0	0	$\neg x \wedge y$
0	1	0	1	$y \forall x$
0	1	1	0	$x \oplus y$
0	1	1	1	$x \vee y$
1	0	0	0	$\neg x \wedge \neg y$
1	0	0	1	$x \oplus \neg y$
1	0	1	0	$\neg y \forall x$
1	0	1	1	$x \vee \neg y$
1	1	0	0	$\neg x \forall y$
1	1	0	1	$\neg x \vee y$
1	1	1	0	$\neg (x \wedge y))$
1	1	1	1	Constant 1

So, for each rule, we have $10 \times 5 \times 8 = 400$ possible gates, and we run the evolution for the values (0, 0), (0, 1), (1, 0), (1, 1) to check which logical gates it is possible to obtain. In experiments, we considered three time steps, to check if the results are stable, or if we could have different gates, just waiting one or two more time steps.

As expected, the first rules like (1, 30) give less interesting results than others, that can excite states starting with a lower number of cells.

Results for rule (4, 27) are reported in Tables 14.2 and 14.3; the first time step of the collision is reported, similar results have been found for second and third steps later.

14.2.5 Adders

Using the gates shown in Tables 14.2 and 14.3 for rule (4, 27), we can build a simple two-bit half-adder.

To solve this problem we use a slightly different matrix, that makes active substates 1 and 2 to be swapped to substates 5 and 4, when they meet in a cell that has an active substate 3. In this way it is possible, e.g. to get a half adder whose output state for the carrier is substate 2 of cell 3 at step 6 (Fig. 14.4). Cells are represented along the vertical axe and time evolves on the horizontal one from left to right. In general 3 or more time steps are shown in the pictures. Each cell is composed by 5 squares

Table 14.2 Logical gates depending on the choice of 2 input qubits, 1 output qubit and the initial state of the qubits other than the 2 input ones

inp1	inp2	out	000	001	010	011
1	2	3	$x \wedge y$	$x \oplus y$	$x \oplus y$	$\neg x \wedge \neg y$
1	3	3	$y \forall x$	$x \oplus y$	$x \oplus y$	$x \oplus \neg y$
1	4	3	$x \wedge y$	$x \oplus y$	$\neg(x \wedge y))$	$x \oplus \neg y$
1	5	3	$x \wedge y$	$x \oplus y$	$\neg(x \wedge y))$	$x \oplus \neg y$
2	3	3	$y \forall x$	$x \oplus y$	$x \oplus y$	$x \oplus \neg y$
2	4	3	$x \wedge y$	$x \oplus y$	$\neg(x \wedge y))$	$x \oplus \neg y$
2	5	3	$x \wedge y$	$x \oplus y$	$\neg(x \wedge y))$	$x \oplus \neg y$
3	4	3	$x \forall y$	$x \oplus y$	$x \oplus y$	$x \oplus \neg y$
3	5	3	$x \forall y$	$x \oplus y$	$x \oplus y$	$x \oplus \neg y$
4	5	3	$x \wedge y$	$\neg(x \wedge y))$	$x \oplus y$	$x \oplus \neg y$

Rule (4, 27), time step 1, ancillary states 000–011

Table 14.3 Logical gates depending on the choice of two input qubits, one output qubit and the initial state of the qubits other than the two input ones

inp1	inp2	out	100	101	110	111
1	2	3	$\neg(x \wedge y))$	$x \oplus \neg y$	$x \oplus \neg y$	$x \vee y$
1	3	3	$x \oplus y$	$x \oplus \neg y$	$x \oplus \neg y$	$y \forall x$
1	4	3	$x \oplus y$	$\neg x \wedge \neg y$	$x \oplus \neg y$	$x \vee y$
1	5	3	$x \oplus y$	$\neg x \wedge \neg y$	$x \oplus \neg y$	$x \vee y$
2	3	3	$x \oplus y$	$x \oplus \neg y$	$x \oplus \neg y$	$y \forall x$
2	4	3	$x \oplus y$	$\neg x \wedge \neg y$	$x \oplus \neg y$	$x \vee y$
2	5	3	$x \oplus y$	$\neg x \wedge \neg y$	$x \oplus \neg y$	$x \vee y$
3	4	3	$x \oplus y$	$x \oplus \neg y$	$x \oplus \neg y$	$x \forall y$
3	5	3	$x \oplus y$	$x \oplus \neg y$	$x \oplus \neg y$	$x \forall y$
4	5	3	$x \oplus y$	$x \oplus \neg y$	$\neg x \wedge \neg y$	$x \vee y$

Rule (4, 27), time step 1, ancillary states 100–111

representing its substates and a dark bar divides a cell from its neighbours. Note also that cells are drawn in a single column to make figures simpler to read, but they are actually arranged in two layers, like in Fig. 14.1, so that e.g. cell 3 interacts with its neighbours 1, 2, 4 and 5. Empty squares represent 0 values for the states, a black dot represents 1.

In such a way it is possible, in principle, to send the output of a gate to another one in an arbitrarily long chain. However, as states go on moving without the possibility of being turned off, it is challenging to program even slightly more complex circuits, like e.g. a full adder, because particles interact in a lot of unwanted ways.

To get usable devices it is necessary to implement separate automata, each one performing a specific task. In such a way we can program for instance a full adder. Example of the full adder states for $0+1+0 = 1$, $1+0+1 = 10$ and $1+1+1 = 11$ are shown in Figs. 14.5, 14.6 and 14.7.

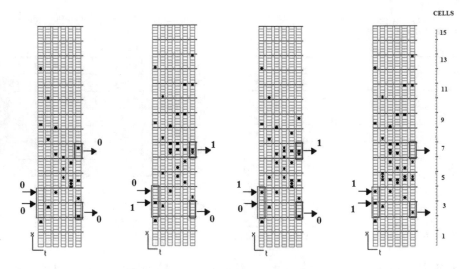

Fig. 14.4 A two-bit half-adder with gates built using rule (4, 27). *Blue squares*: input cells, *red square*: output cells, substate 3 for the sum and substate 2 for the carrier

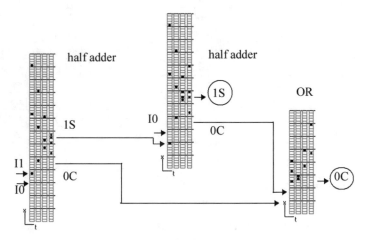

Fig. 14.5 A three-bit full-adder with gates built using rule (4, 27): $0 + 1 + 0 = 1$. The final sum and carrier are encircled

14.3 Multi-valued Logic on Quantum Actin Automata

The following results have been already obtained on relationships between the multiple-valued logic and the quantum information. Multiple-valued logic gates for quantum computing are designed in [71]. Extensions of QCA dealing with three-valued logics are proposed in [9, 10, 58]. In particular, [9] implemented Łukasiewicz NOT, AND and OR functions with a QCA. What these approaches have in common

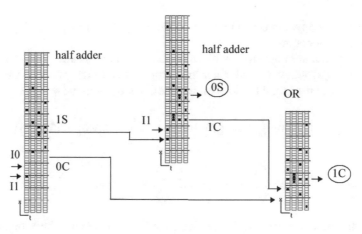

Fig. 14.6 A three-bit full-adder with gates built using rule (4, 27): $1 + 0 + 1 = 10$. The final sum and carrier are encircled

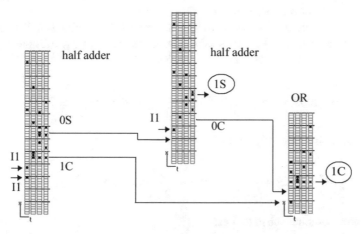

Fig. 14.7 A three-bit full-adder with gates built using rule (4, 27): $1 + 1 + 1 = 11$. The final sum and carrier are encircled

is that they propose a version of the QCA, sometimes called an Extended QCA, where cells can take three states—qutrits—instead of two states as in the standard QCA models. This is achieved by using cells consisting of eight quantum dots with a specific structure. We adopt a different approach. We use a standard QCA, the one whose states can be expressed in qubits, but we *interpret* superposed states as the third logical value ($\frac{1}{2}$). This approach is inspired by [17, 18], where the intermediate logical value is interpreted in an epistemic way as 'unknown' or 'controversial'. In [86] we explicitly implemented logical operations of several three-valued logics: Łukasiewicz, Gödel, Jaśkowski, Sobociński and Sette. There we studied how the automata can be programmed and coupled together, even if it is straightforward that all three-valued operations can be represented using e.g. Łukasiewicz logic and the augmentation operator.

As we are mainly concerned with substates and their evolution and movements, we now exemplify how the substates are computed and interpreted in the spirit of [18]. We start at the initial time t with two particles x_1, x_2 in the states:

$$x_1 \equiv (|0\rangle|0\rangle|0\rangle|0\rangle \frac{1}{\sqrt{2}}(|0\rangle + |1\rangle)) = \frac{1}{\sqrt{2}}(|00000\rangle + |00001\rangle)$$

$$x_2 \equiv (|0\rangle|0\rangle|0\rangle \left(\frac{1}{2}|0\rangle + \frac{\sqrt{3}}{2}|1\rangle)|0\rangle\right) = \frac{1}{2}|00000\rangle + \frac{\sqrt{3}}{2}|00010\rangle$$

all the other particles being in the resting state $|00000\rangle$. We use the actin automata rule (01100, 10011) [4] whose decimal form is (12, 19). The rule (12, 19) defines the following state transition: the state 3 is activated when 1 or 2 other states, coming from the neighbouring particles, are active.

We now compute the state of particle x_3 at time $t + 1$. The first step is to compute the intermediate state, that inherits substates: substate 5 from x_1, substate 4 from x_2, substate 3 = substate 2 = substate 1 = $|0\rangle$. We get:

$$\frac{1}{2\sqrt{2}}|00000\rangle + \frac{1}{2\sqrt{2}}|00001\rangle + \frac{\sqrt{3}}{2\sqrt{2}}|00010\rangle + \frac{\sqrt{3}}{2\sqrt{2}}|00011\rangle$$

Now we apply the rule (12, 19) and obtain:

$$\frac{1}{2\sqrt{2}}|00000\rangle + \frac{1}{2\sqrt{2}}|00101\rangle + \frac{\sqrt{3}}{2\sqrt{2}}|00110\rangle + \frac{\sqrt{3}}{2\sqrt{2}}|00111\rangle$$

We can now compute substate 3 as:

$$\frac{\sqrt{1}}{2\sqrt{2}}|0\rangle + \frac{\sqrt{7}}{2\sqrt{2}}|1\rangle \qquad (14.3)$$

14.3.1 Reading Automata Output, Concatenating Automata and Interpreting Superposition

We note that the proposed automaton, without other hypotheses or transformations, is not, in general, suitable for dealing with possibility distributions and the logic that can be associated with them [14, 26]. For example, for the possibility measures Π it is generally required that $\Pi(\alpha \vee \beta) = \max(\Pi(\alpha), \Pi(\beta))$, a condition that is not satisfied by our automaton. In any practical implementations, reading a state means to run the automaton and to perform a state measure, with result necessarily be either $|0\rangle$

or $|1\rangle$. The $|0\rangle$ and $|1\rangle$ coefficients are square roots of state probabilities. When these coefficients are obtained the interpretation in terms of $\{0, \frac{1}{2}, 1\}$ comes in. The first interpretation is obtained by dividing the interval $[0, 1]$ of probabilities of finding an excited output state $|1\rangle$ in three parts, and identifying $0 = [0, \frac{1}{3})$, $\frac{1}{2} = [\frac{1}{3}, \frac{2}{3}]$, $1 = (\frac{2}{3}, 1]$. The second interpretation is in identifying *any* superposed states with the $\frac{1}{2}$ truth value.

In scenarios of both interpretations the automaton is not capable, however, of performing all the operations, e.g. all the gates of Łukasiewicz or Sobociński logics. We need one more assumption to deal with these constructions. We suppose to have also the possibility of interpreting any superposed states as 1. This interpretation, that is not a part of the automaton but of the system that reads and transmits the signals, is identified with the augmentation operator ∇: $\nabla 0 = 0$; $\nabla \frac{1}{2} = \nabla 1 = 1$ The practical implementation could be a little tricky, and depending on the exact characteristics of physical devices it could be necessary to reassign this. Below we describe a possible abstract device.

Suppose we want to apply the ∇ operator to an input qubit. As the automaton must be run a number N of times to collect statistics of the output states, we must have the possibility to repeatedly send these input qubits to the operator. The operator must measure the input to decide if it must pass a $|0\rangle$ or $|1\rangle$ state. Each measure will give either $|0\rangle$ or $|1\rangle$ and it is possible that $|1\rangle$ is read before N trials, so the operator must pass $|1\rangle$ for all the N runs. It is also possible that $|0\rangle$ is read for all the N trials, so the operator must pass $|0\rangle$ for all the runs. In both cases the input qubits are lost in the measures, and the ∇ operator must initialise a set of N new qubits of the type that has detected reading the input. This means that the automaton must wait, or just discard the final output, until the ∇ operator starts sending qubits. In an artificial setting, it would be sufficient for the ∇ operator to send a (classical) signal to the detectors of the final output, to switch them on; in a natural setting a suitable threshold mechanism has to be found. With the aid of this operator we will be able to manage many other logic operators.

14.3.2   Realizations of Logic Operators

14.3.2.1   Łukasiewicz Operators

Using the interpretation with three equal intervals, the automaton can perform the Łukasiewicz conjunction $\wedge_L$ and disjunction $\vee_L$, see Table 14.4a, b. The Łukasiewicz implication can be obtained as $x \rightarrow_L y = \neg x \vee_L y$, see Table 14.4c.

The automaton evolution for operators involutive NOT, $\vee_L$, and $\wedge_L$ is shown in Fig. 14.8a, b, c. The rule (00100, 11011) [4], or (4, 27), has been used in these computations and in the following ones. In the figures for three-valued logic, empty squares represent 0 values for the states, a black dot represents 1 and a gray one represents 1/2. Each separate group of cells represents an automaton; input and output

Table 14.4 Operators of Łukasiewicz logic

(a) $\vee_L$

	0	1	$\frac{1}{2}$
0	0	1	$\frac{1}{2}$
1	1	1	1
$\frac{1}{2}$	$\frac{1}{2}$	1	1

(b) $\wedge_L$

	0	1	$\frac{1}{2}$
0	0	0	0
1	0	1	$\frac{1}{2}$
$\frac{1}{2}$	0	$\frac{1}{2}$	0

(c) $\rightarrow_L$

	0	1	$\frac{1}{2}$
0	1	1	1
1	0	1	$\frac{1}{2}$
$\frac{1}{2}$	$\frac{1}{2}$	1	1

states are highlighted with arrows and values; in some figures several automata are linked by arrows, meaning that the output of one of them is sent as input to another. When an automaton in such a schema is used to perform a specific logic operation, e.g. $\vee$, a label has been added on top of its cells.

14.3.2.2 The Gödel Operators

From now on we identify *any* superposed states with the $\frac{1}{2}$ truth value. In this hypothesis the automaton can perform the Gödel conjunction $\alpha \wedge_G \beta = \min(\alpha, \beta)$ and the disjunction $\alpha \vee_G \beta = \max(\alpha, \beta)$, see Table 14.5 and automaton simulations in Fig. 14.9. The involutive negation $\neg$ is the same as in the previous case, see Fig. 14.8a. An implication operator can again be obtained by composing $\neg$ and $\vee_G$. Note that here operators max and min work well only because we identify $\frac{1}{2}$ with any superposed states, disregarding their actual probability.

14.3.2.3 Łukasiewicz Operators Again

Łukasiewicz implication, Table 14.4c, can be written as $x \rightarrow_L y = (\nabla \neg x \vee y) \wedge (\nabla y \vee \neg x)$, where $\vee = \text{Max}$ and $\wedge = \text{Min}$. We show in Fig. 14.10a the automata, their connections and the evolution for the case $\frac{1}{2} \rightarrow_L \frac{1}{2}$, whose truth value is 1. It is important to stress that we cannot just send the input qubits in parallel to two gates, as we would do if they were classical bits, because the no cloning theorem [12]

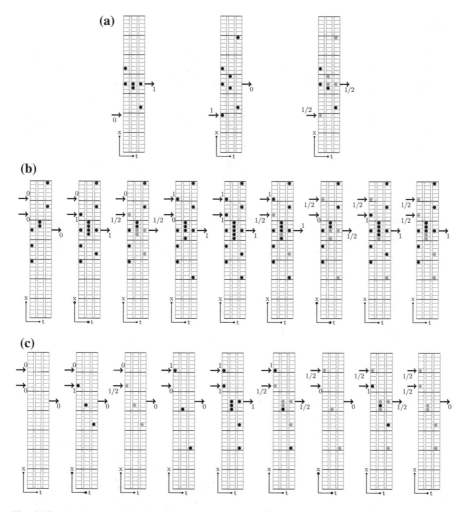

Fig. 14.8   Automaton evolution for **a** involutive NOT, **b** $\vee_L$, **c** $\wedge_L$

affirms that quantum states cannot be copied. What actually happens, instead, is that the qubits pass through the first gate and can be used again as input for the second gate. This happens for instance for the x qubit passing the first NOT gate and for y passing the first $\vee$ in Fig. 14.10a. Jaśkowski implication and Sobociński and Sette operators can be implemented in similar ways, see details in [86].

We have shown that a simple partitioned quantum automaton model, inspired by actin double-helix protein polymer, can perform Boolean and three-valued logical operations. The realizability of such an automaton depends on the physical substrate characteristics. Namely, the automaton's states must be quantum states and they must influence each other accordingly to a summation rule: a particle (node, G-actin unit) must switch its state if a fixed number of its neighbours are active. It would be

(a)

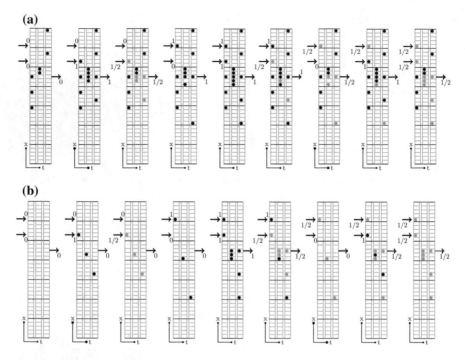

Fig. 14.9 Automaton evolution for **a** $\vee_G$ and **b** $\wedge_G$

Table 14.5 Operators max and min

(a) $\vee_G$

	0	1	$\frac{1}{2}$
0	0	1	$\frac{1}{2}$
1	1	1	1
$\frac{1}{2}$	$\frac{1}{2}$	1	$\frac{1}{2}$

(b) $\wedge_G$

	0	1	$\frac{1}{2}$
0	0	0	0
1	0	1	$\frac{1}{2}$
$\frac{1}{2}$	0	$\frac{1}{2}$	$\frac{1}{2}$

necessary to compute a mechanism through which signals or data travelling along actin filaments can interfere constructively to activate a state, and to engage in a destructive interference to deactivate the state. There can be even more effects in a physical system that must be taken into account, e.g. errors, that can arise when the signals travel or are composed. Moreover, the interactions might fade with time, something that might have both positive and negative effects on the computations,

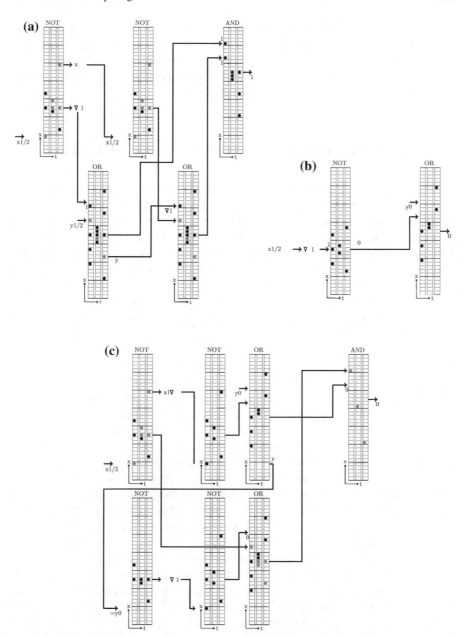

Fig. 14.10 Automaton evolution for **a** $\frac{1}{2} \to_L \frac{1}{2}$, **b** $\frac{1}{2} \to_J 0$, **c** $\frac{1}{2} \to_S 0$

and different sections of the system could follow different rules. To make a real computer, based on the described automata, it is necessary to find an efficient way to prepare the input states and to read the output states and routes for storing information have to be defined. In the following sections we consider few simple mechanisms of implementing computation in actin filaments.

14.4 Actin Filaments as Nonlinear RLC Transmission Lines

In [87] we presented a model of actin in terms of RLC (resistance, capacitance, inductance) non-linear electrical transmission lines that can implement logical gates via interacting voltage impulses. These models take inspiration from the actin automata, but do not fit exactly in its definition, as cells do not possess discrete states, there are no intrinsic time steps, and no rules are defined for state transitions. However, the states can be digitised by defining a suitable threshold, a time step can be defined by convenience and transitions are governed by Kirchhoff's circuit laws so that a deterministic evolution takes place even in absence of explicit transition rules.

Our starting point is a model of electrical lines as chains of circuits composed of resistors, capacitors and inductors. Our model is based on Tuszynski et al. [95] model of actin monomers in terms of electrical components. This latter model was developed in order to explain experimental observations of ionic conductivity along actin filaments [59]. This model exhibits, in a continuous limit, standing waves and travelling impulses. We use the same formalism as in [95] but without invoking the continuous limit approximation and study the behaviour of several types of solutions for tens of monomers. We show how an excitation moves along the actin filament and how it collides with another excitation coming from elsewhere. We demonstrate that these collisions can be used to implement logical operations.

The underlying mechanism is the following. There is a potential difference between the filament core and the ions surrounding the filament. A time-dependent current is generated by the ions' moving along helical paths of the actin filament, responsible for the inductance L. There is a resistive component R_1 to the ionic currents, due to viscosity, in series with L. There is also a resistance in parallel to these components, R_2, between the ions and the surface of the filament; and a capacitance C_0. More details can be found in the original paper [95]. Referring to Fig. 14.11, where an actin monomer unit in a filament is shown by the dotted lines, we assume that capacitors are nonlinear, see discussion in [63, 103] where formulas of electrochemical capacitance for a quantum version of a leaky capacitor are derived. In the model the two plates have a DC coupling, and tunneling effects between the two plates:

$$Q_n = C_0(V_n - bV_n^2) \qquad (14.4)$$

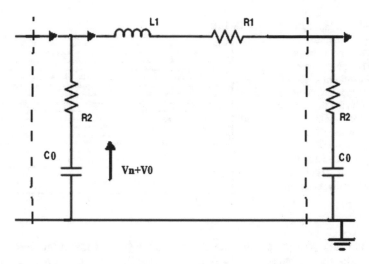

Fig. 14.11   A circuit diagram for the nth unit of an actin filament (from [95])

The main equation is:

$$LC_0 \frac{d^2}{dt^2}(V_n - bV_n^2) = V_{n+1} + V_{n-1} - 2V_n - R_1 C_0 \frac{d}{dt}(V_n - bV_n^2)$$

$$- R_2 C_0 \{2\frac{d}{dt}(V_n - bV_n^2) - \frac{d}{dt}(V_{n+1} - bV_{n+1}^2)$$

$$- \frac{d}{dt}(V_{n-1} - bV_{n-1}^2)\} \qquad (14.5)$$

The above is the basic equation, that is derived using Kirchhoff's law and that will be used to compute the cells' behaviour. In [95] formulas are derived to obtain values for the relevant parameters, that could be tuned by changing the temperature and the ion concentrations. We use the following parameters:

$$L = 1.7 \text{ pH}, \; C_0 = 96 \times 10^{-6} \text{ pF},$$
$$R_1 = 6.11 \times 10^6 \Omega, \; R_2 = 0.9 \times 10^6 \Omega \qquad (14.6)$$

With these parameters we measure time in nanoseconds and voltage in arbitrary units. Let us consider the case when an initial voltage is set at one end of the filament for some time as, for example, could take place when an actin filament is in close proximity of an ion channel. We use for voltage the form:

$$V_0 = \frac{1}{2} - \frac{e^{t-t_0} - e^{-t+t_0}}{2(e^{t-t_0} + e^{-t+t_0})} \qquad (14.7)$$

In all the following computations we consider $t_0 = 3$ ns.

Fig. 14.12 An input pulse travelling along actin filament with parameters (14.6). The figure represents the cell evolutions, each cell voltage is represented by a diagram with the time on the horizontal axis and the voltage V on the vertical one. The diagrams are tiled vertically with cell 1 at the *bottom* and cell 20 at the *top*, with a unitary displacement

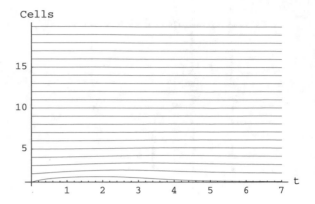

We find that pulses actually travel, but after a few nanoseconds they become damped (Fig. 14.12). From the diagram we see that the 12th unit is reached in about 4 ns, so that the speed is approximately 5.4 nm × 12/4 ns, that is about 16 m/sec. That is consistent with the estimates given in [95]. If we decrease the resistance R_1 by an order of magnitude, the speed doubles and we obtain a much longer lasting pulse.

In the next two sections we consider two types of inputs: forced and unforced. An unforced input is an impulse applied just once at input sites of actin-based gate. A forced input is an impulse applied continuously during evolution of the gate.

14.4.1 Gates with Unforced Pulses

Let us consider interactions and possible gates implemented using only initial conditions of cells, without any forced pulses applied. They are based on the assumption that we can initialize any cells with a pre-determined electrical potential value and reliably measure output potentials at given units of actin filament.

Using parameters (14.6) and $b = 0.1$, and initial condition $V = V_0 \neq 0$, we look at cells having voltages greater than a fixed percentage of initial potential V_0. We find that some logical operations can be implemented if we consider signals above $\approx 0.1V_0$ as output corresponding to logical TRUE, to read outputs of the logical gates. The threshold is a heuristic value, that has been chosen because on one hand the signals, in the simulation conditions, stay above this threshold for a time long enough to interact with other signals; on the other hand it seems high enough to allow us differentiate between signal and noise. In groups of monomers we can allow for higher thresholds because the signals there fade slowly, see Fig. 14.18.

Figure 14.13 shows the AND operation obtained with two input cells at positions 8 and 15. The output is in cells 11 and 12, that are excited at the end of the evolution if both the input cells are excited initially. The OR operation is obtained with two input cells at positions 9 and 11. The output is in cell 10 that is excited at the end of the evolution if one of the input cells or both of them are excited initially.

(a)

(b)

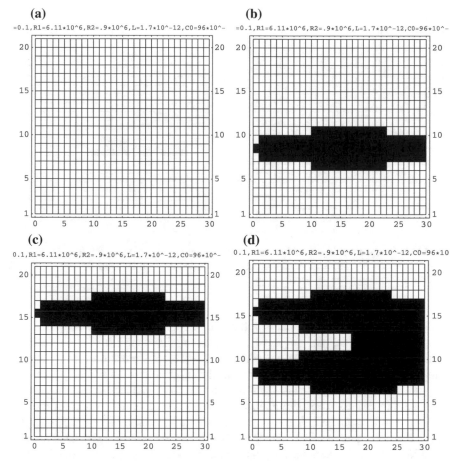

Fig. 14.13 Unforced pulses, AND operation, with parameters (14.6). Inputs are **a** 00, **b** 10, **c** 01 and **d** 11 in cells 8 and 15; output in cells 11 and 12. Cells excited above $0.1V_0$ are *black*; time evolves along the horizontal axis and cells are aligned along the vertical axis

As all cells are considered identical, what matters in the choice of input positions is just their mutual distance, as signals must travel for a suitable number of time steps to meet with the appropriate phases and intensities. The intensities are summed or subtracted from each other depending on the signal phases, and the results must be higher than the chosen threshold. The proposed positions were found after running several simulations and varying the input cells' positions.

To obtain a NOT gate, we initialize an auxiliary cell at position 9 at $-V_0$ and place the input cell at position 11. The output is found in cell 10 as shown in Fig. 14.14.

To build an XOR gate we use the formula

$$a \oplus b = (\neg(a \wedge b)) \wedge (a \vee b) \tag{14.8}$$

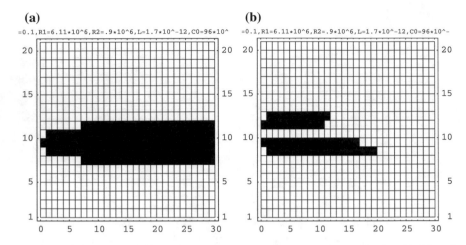

(a)

=0.1,R1=6.11*10^6,R2=.9*10^6,L=1.7*10^-12,C0=96*10^

(b)

=0.1,R1=6.11*10^6,R2=.9*10^6,L=1.7*10^-12,C0=96*10^-

Fig. 14.14 Unforced pulses, NOT operation, with parameters (14.6). Input is **a** 0 and **b** 1 in cell 11; output in cell 10. Cells excited above $0.1V_0$ are *black*; time evolves along the horizontal axis and cells are aligned along the vertical axis

as shown in Fig. 14.15. Each run is composed of four gates, and it is supposed that some mechanism reads the output of each gate, amplifies the signal and passes the signal to the next gate. In the figure the arrays represent signals travelling between gates. The first from the left is an AND gate with input cells 8, 15 and output cell 12. The gate's output is cascaded to the next gate: the NOT gate having input in cell 11 and output in cell 10. The original inputs, coming out at cells 8 and 15 of the first gate, are sent to input cells 9 and 11 of the OR gate in the third section. Finally, the output of the OR gate in cell 10 and the output of the NOT gate are sent to input cells 8 and 15 of the final AND gate. The final output of the XOR operation is found in cells 11 and 12 at the right end.

14.4.2   Gates with Forced Pulses

Let us consider interactions and possible gates implemented by applying input pulses continuously at input sites. We experimented with forcing pulses with both step function (14.7), function $\approx 1/(1 + exp(\alpha t))^2$ and sinusoidal pulses. We can obtain a rich set of gates by sending a sinusoidal pulse to inputs during the whole evolution of the system; the intensity of the input pulse remains constant.

In numerical model we use slightly different values for resistances: $R_1 = 9.23 \times 10^6\Omega$, $R_2 = 1.32 \times 10^6\Omega$, that can be obtained changing concentrations of K^+ and Na^+; excitations are sinusoidal with a 1 ns period.

Figure 14.16 shows the XOR operation obtained with two input cells at positions 10 and 14. The output is found in cells 12 that is excited at the end of the evolution

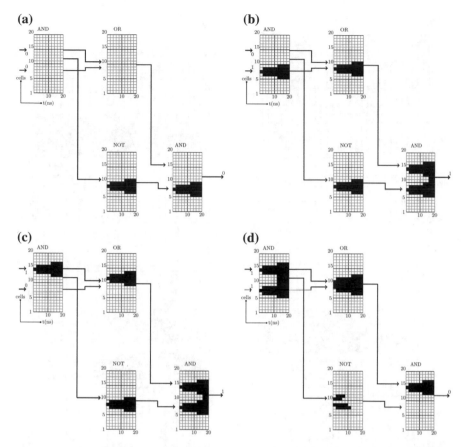

Fig. 14.15 Unforced pulses, XOR operation. Inputs are **a** 00, **b** 10; **c** 01 and **d** 11 in cells 8 and 15 of the *leftmost* gate; output in cells 11 and 12 of the *rightmost*. Cells excited above $0.1V_0$ are *black*; time evolves along the horizontal axis and cells are aligned along the vertical axis

if only one the input cells is. The input pulses are sinusoidal with opposite phases. The other logical gates can be implemented in a similar way.

A one-bit half-adder is shown in Fig. 14.17. The circuit is composed of AND gate followed by an XOR gate. The AND gate has input cells 8 and 14 and outputs the most significant bit (the carry) in cell 11. The XOR gate has input cells 10 and 14. The gate outputs the less significant bit in cell 12. Input in cell 14 is common to both gates, while input in cell 8 of the AND gate travels to cell 10 of the XOR changing its phase.

One could argue that it could be difficult in practice to send an input pulse to exactly one monomer (or to initialize it in an excited state), which requires a localization precision of under 5 nm. We therefore compute some evolutions supposing that the pulse hits a number N of monomers. Under these conditions, the values of resistance,

(a)

(b)

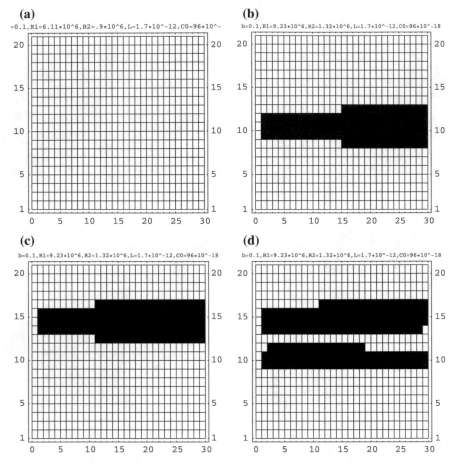

(c)

(d)

Fig. 14.16 Forced pulses, XOR operation, with parameters (14.6) with $R_1 = 9.23 \times 10^6 \Omega$, $R_2 = 1.32 \times 10^6 \Omega$. Inputs are **a** 00, **b** 10, **c** 01 and **d** 11 in cells 10 and 14; output is in cell 12

inductance and capacitance must be computed with appropriate addition rules. For resistance we have:

$$R_{2\,tot} = \left(\sum_{i=1}^{N} R_{2,i}^{-1} \right)^{-1}, \quad R_{1\,tot} = \sum_{i=1}^{N} R_{1,i} \tag{14.9}$$

Total inductance and capacitance are obtained by summing those of individual monomers. Figure 14.18 shows the XOR gate with input sent to groups of ten monomers in positions 8–17 and 23–32. As in Fig. 14.16, the input pulses are sinusoidal with opposite phases. Output is in cell 20. In our approximation, we do not consider the detailed evolution inside the groups of monomers receiving the input, they are treated as a single cell, with R, L, C parameters computed by summing

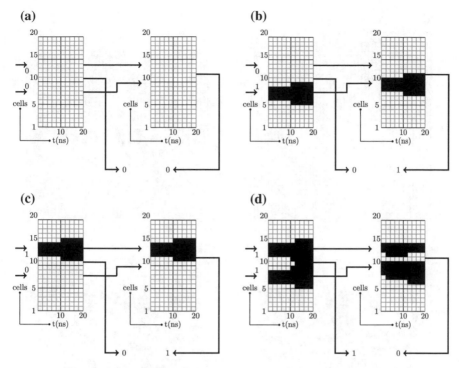

Fig. 14.17 Forced pulses, half-adder. Inputs are **a** 00, **b** 10, **c** 01 and **d** 11 in cells 8, 10 and 14; outputs in cell 11 and 12

them on ten monomers. We note that the results are robust: pulses fade slowly and a threshold of 0.3 V of the input signal can be used, instead of 0.1 V as in the previous computations.

14.4.3   RLC Networks

Actin filaments are usually found in networks, forming cross links that can have a parallel or orthogonal shape. These links can be made by several kinds of binding proteins that originate different network topologies. We consider for instance filamin A, see e.g. [72], that builds orthogonal networks, where two actin filaments cross link. We suppose that electric pulses generated as output of a gate in a filament can reach the other filament using the filamin molecule as a connecting wire. Branching of actin networks has been observed in cells and it requires the presence of ARP2/3 protein as a branch point from which two actin filaments propagate at acute angles [45]. Assuming that auxiliary proteins have the same electrical characteristics as actin, we can compute R_1, R_2, C_0, and L for filamin, taking into account just its dimensions. In

(a)

(b)

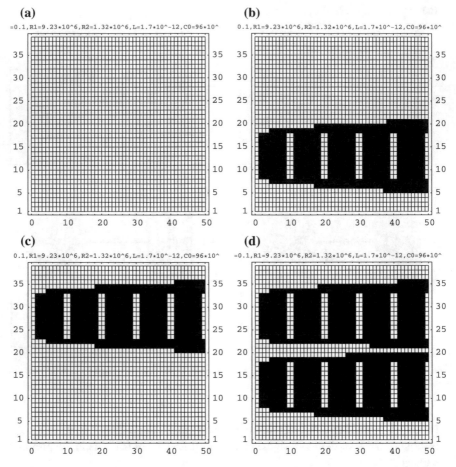

Fig. 14.18 Forced pulses, XOR operation, with parameters (14.6), $R_1 = 9.23 \times 10^6 \Omega$, $R_2 = 1.32 \times 10^6 \Omega$. Inputs are **a** 00, **b** 10, **c** 01 and **d** 11 in cells 8–17 and 23–32; output in cell 20

such way, in principle, any number of actin gates could be connected. However there is a problem: pulses intensity will fade away rather quickly, unless some amplifying method is found. Some studies, e.g. [81], have found additional roles of filamin A in signal transduction, and others like [82] have found that actin filaments can regulate some voltage-gates ion channels. We can therefore think of a system where the output voltage of a gate opens a ion channel that in turn activates the next gate and so on. Of course this subject must be treated in more depth to draw any practical conclusions, as e.g. the spatial distribution of the ion channels interfaced the actin network, their dimensions, the necessary electrochemical gradients, the type of ions that can be used, their diffusion rates etc. must be carefully evaluated to check the feasibility of a logical device.

14.5   Solitons in Actin Networks

An idea of computing with solitons has been under radar of engineers, computer scientists and physicists since late 1980s [91] and got an additional boost with the rise of the collision-based computing paradigm [3]. A logical gate can be implemented with solitons in two ways. In the first approach we encode values of logical variables into the states—shape, phase, speed, energy level—of solitons. When two solitons collide they might change their states. States of the solitons before collision represent input logical values. States of the solitons after collision represent output logical variables. In the second approach we encode values of logical variables in presence or absence of a soliton in a given loci of space. Some loci of spaces are assigned to be inputs, others outputs. If there is a soliton in the input loci the input is considered TRUE, if no soliton in the loci then input is FALSE. When two solitons collide they can change their velocity vectors, reflect, annihilate or merge. Possible post-collision output trajectories of the solitons are interpreted in terms of logical operations.

To evaluate the feasibility of developing computing circuits using programmable polymerisation of branching actin networks we simplify the actin filaments as one-dimensional lattices of nodes with nearest neighbour interactions and study propagation of the localised excitations, or solitons, and their interaction at the junctions, or branching sites, of the lattices (see [88]). We adopt Morse interaction potential between nodes as proposed by Velarde et al. in [96–98]. Advantages of this approach is that one can consider movement of both the electrons and the lattice excitation, or soliton, together. For example, it was discovered in [96, 98] that when two solitons with opposite vector vectors collide they exchange electron probability density between each other. However we will not consider electrons in the present work.

We consider one-dimensional lattices of nodes. Each node has a potential energy. The potential energy of a node depends on the positions of the neighbouring nodes and their potential energies. Nodes oscillate around their equilibrium positions, a minimum of the potential energy. When a few nodes are excited the oscillation travels along the lattice as a solitary wave.

The Hamiltonian of the lattice nodes is as follows [97]:

$$H_{Mo} = \sum_{i=1}^{N} \left[\frac{1}{2} \frac{p_i^2}{M} + D(1 - exp[-b(q_i - q_{i-1})])^2 \right] \qquad (14.10)$$

where N is the number of nodes, M is the mass of a lattice node (all nodes have the same mass), q_i is the displacement of node i from its equilibrium position, p_i is the momentum of node i, D is the depth of the well defined by the potential (dissociation energy), b is the stiffness of the spring constant in the Morse potential.

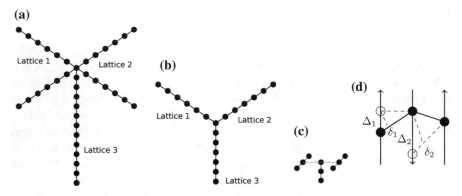

Fig. 14.19 Junctions of lattices. Nodes are *black discs*. Interactions between nodes along the same lattice is shown by *solid lines*. **a** Geometry 1. **b** Geometry 2. **c** Perspective view of details of a generic junction of geometry 1. **d** Neighbourhood structure at the junction, *filled circles* refer to equilibrium positions, *empty circles* and *dotted lines* to displaced positions

The dynamics of the motion of nodes is as follows

$$\frac{d^2 q_i}{dt^2} = (1 - exp[-(q_{i+1} - q_i)])exp[-(q_{i+1} - q_i)]$$
$$- (1 - exp[-(q_i - q_{i-1})])exp[-(q_i - q_{i-1})] \qquad (14.11)$$

Following [97] and [98] we take $(2Db^2/M)^{-1/2}$ as a unit of time, $(2D)$ as unit of energy, b^{-1} as displacement. The values used in [97] and [98] are $b = 4.45\,\text{Å}^{-1}$, $D = 0.1\,\text{eV}$, and the time unit is $3.29 \times 10^{-13}\,\text{s}$. The values correspond to, e.g., electron transport along hydrogen bonded polypeptide chains such as α-helices. For actin monomers, we can use $b = 4.89\,\text{nm}^{-1}$, $D = 7.67\,\text{eV}$ (see [7]) and $M = 6.7 \times 10^{-23}\,\text{kg}$ corresponding to a time unit of $2.5 \times 10^{-11}\,\text{s}$.

The one-dimensional lattices cross each other and form junctions (Fig. 14.19). Geometry 1 (Fig. 14.19a) consists of two chains forming a junction at an internal point and of a third chain starting at the junction. Geometry 2 (Fig. 14.19b) consists of three lattices forming an Y-shape, the junction is formed by the extreme points of the lattices.

At the junction a node of a lattice interacts also with its nearest neighbours in complementary lattices. This interaction is shown by grey lines in Fig. 14.19c. We use the same potential for the junctions as for the lattices, but we introduce an attenuation parameter to account for a weaker interaction between nodes of different lattice. When nodes in the junction oscillate around their equilibrium positions they move along their lattices. Their distances from their nearest neighbours in the complementary filaments changes and their interactions are affected. This is shown in Fig. 14.19c, where the filled circles represent nodes in their equilibrium positions, empty circles are nodes in displaced positions and dotted lines are the corresponding distance changes. We approximate the variations of the distances between the three nodes of a junction with the variations of their positions in their respective lattices.

We note that the coordinate q_i of the node i is relative to the internal reference of its filament. If the node is part of a junction, an increase of its q_i may result in an increase or a decrease of its distance from the other nodes in the junction, depending on the positions and directions of the filaments. For example, given the orientations and positions as shown in Fig. 14.19d, an increase Δ_1 of the coordinate of the left-most node will result in a decrease δ_1 of its distance from the central node (discordant variations). An increase Δ_2 of the coordinate of the central node will result in an increase δ_2 of the central node's distance from the rightmost node (concordant varia-tions). We assume that a displacement q_i of the ith node along its lattice corresponds to a fraction of the displacement along the junction. We incorporate this effect in the exploratory parameter α in the expression of the potential acting between the nodes of the junction. We assume all structures have concordant variations of the q_i unless discordant variation is explicitly stated.

The Hamiltonian for the complex of three lattices is

$$H_{lattice} = \sum_{i=1}^{N_1} [\frac{1}{2}\frac{p_i^2}{M} + D(1 - exp[-b(q_i - q_{i-1})])^2]$$

$$+ \sum_{i=N_1+1}^{N_1+N_2} [\frac{1}{2}\frac{p_i^2}{M} + D(1 - exp[-b(q_i - q_{i-1})])^2]$$

$$+ \sum_{i=N_1+N_2+1}^{N_1+N_2+N_3} [\frac{1}{2}\frac{p_i^2}{M} + D(1 - exp[-b(q_i - q_{i-1})])^2]$$

$$+ \alpha D(1 - exp[-b(q_{i_1} - q_{i_2})])^2] + \alpha D(1 - exp[-b(q_{i_2} - q_{i_3})])^2]$$

$$(14.12)$$

A set of equations is written for each lattice. The first and last equation of each set lack the terms with q_{i-1} and q_{i+1}, respectively, and a term is added for the nodes in the junctions. where N_1, N_2, N_3 are the numbers of nodes in the lattices 1, 2 and 3, $q_{i_1}, q_{i_2}, q_{i_3}$ are the nodes in the junction, α parametrizes the strength of the interaction in the junction. This Hamiltonian refers to structures with concordant variations of the coordinates; if discordant variations were considered, some of the q_{i_k} would have the opposite sign.

We derive equations of motion analogous to the equations of motion in (14.11). They have been numerically solved for geometries 1 and 2. The conservation of energy has been checked at every integration step.

14.5.1 AND *gate*

Inputs are lattices 1 and 2. Output is lattice 3. Presence of a soliton at the top end of lattice 1 or 2 means logical input TRUE. Presence of a soliton at the bottom end of

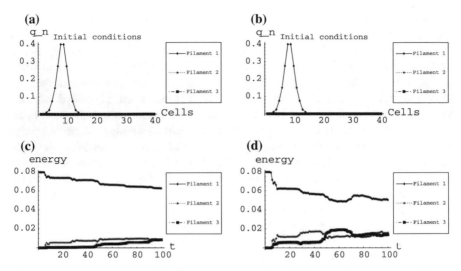

Fig. 14.20 Gate AND in geometry 1 (Fig. 14.19a): only one input is TRUE, a soliton is excited either in lattice 1 or in lattice 2. **a, b** Initial energy distribution along lattices. **c, d** Energy evolution with time. **a, c** $\alpha = 0.1$. **b, c** $\alpha = 0.4$. Energy for lattice 1 is shown by *solid line* with rhomb markers, lattice 2 *dashed line* with *square* markers, lattice 3 *dotted line* with start markers. Values q_n are measured in b^{-1}, energy in $2D$ and time in $(2Db^2/M)^{-1/2}$ units

lattice 3 means logical output TRUE. Absences of the solitons corresponds to logical values FALSE.

In all conditions, the excitation spreads in part to all the lattices, see e.g. Fig. 14.22 that shows an example of the evolution of the structure. Therefore by presence of a soliton we mean that the lattice energy exceeds a specified threshold. Absence of a soliton corresponds to the energy falling below the threshold.

The AND gate is implemented in geometry 1 with three lattices of 40 nodes each. Propagation and interaction of solitons is tested for $\alpha = 0.1$ and 0.4. Runs with initial conditions with no solitons, one at the beginning of lattice 1 or lattice 2 show that a single soliton's energy stays in its lattice, with just a small amount passing to others lattices (Fig. 14.20). If there is a soliton only in one of input lattices no soliton will appear in the output lattice.

When both inputs of the gate are TRUE two solitons generated in input lattices 1 and 2 transfer a considerable amount of energy to output lattice 3, if their motions along the filaments are in phase (that is they start at the same positions or very near). The amount of the energy transferred from input lattices to output lattice depends on the strength of the junction (Fig. 14.21). Thus, when solitons in input lattices are in phase the geometry 1 implement AND gate.

As an illustration, in Fig. 14.22 we show a three-dimensional plot of the displacement of nodes around their equilibrium positions of the lattices for the case $\alpha = 0.1$ for a single input TRUE: a single initial soliton in lattice 1. The excitation is

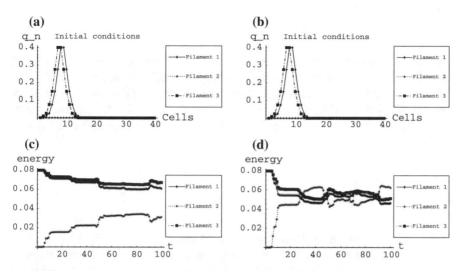

Fig. 14.21 Gate AND in geometry 1 (Fig. 14.19a): both inputs are TRUE, solitons are excited in lattice 1 and lattice 2. **a, b** Initial energy distribution along lattices. **c, d** Energy evolution with time. **a, c** $\alpha = 0.1$. **b, c** $\alpha = 0.4$. Values q_n are measured in b^{-1}, energy in $2D$ and time in $(2Db^2/M)^{-1/2}$ units. Energy for lattice 1 is shown by *solid line* with rhomb markers, lattice 2 *dashed line* with *square* markers, lattice 3 *dotted line* with start markers

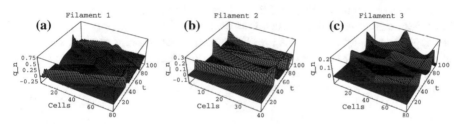

Fig. 14.22 Evolution of a single soliton starting at the beginning of lattice 1. **a** Evolution lattice 1; **b** Evolution lattice 3; **c** Evolution lattice 2; $\alpha = 0.1$

transmitted to lattices 2 and 3. Two solitons with opposite directions of motion are excited in lattice 2, even if most of energy stays in its original lattice.

The excitations are reflected at the ends of the filaments and they meet again and again. As they are soliton-like but not perfect solitons, they interfere and may change or lose their shapes. Changing the geometry of the filament structures, that is making them cross at different points, may change the relative phases of the excitation periodic motions along the filaments, so that they meet at different points with different consequences on their changes. In case of segments of infinite length this phenomenon would not occur.

14.5.2   OR *and* NOT *gates*

The OR gate can be implemented in geometry 2 and the NOT gate in geometry 2 with discordance of nodes' coordinates in the junction. The lattices 2 and 3 are in such a position that the variations of their coordinates is so that an increase of the internal coordinate q_i (Fig. 14.19d) of the node of the lattice 2 in the junction results in a decrease of the node's distance from the central node in the junction (for details see [88]).

14.5.3   *Cascading Gates*

To make a NOR gate, we connect NOT gate to OR gate as shown in Fig. 14.23a. There five filaments are connected in two junctions. The coordinates of first junction (black circle) are concordant. The coordinates of the second junction (empty circle) are discordant. All lattices but the bottom right lattice have 40 nodes. This bottom right lattice has 80 nodes. It carries the auxiliary signal necessary to implement negation. The input signals, starting at the top lattices, travel along the same number of nodes as the auxiliary signal before they collide with each other.

The energy diagrams (Fig. 14.23b, c, d, e) show that the circuit works correctly. If no input signals are applied, the output energy is high. If one or two input signals are applied, the output energy is low. From a quantitative point of view, the values of input energies are 0.0797 (in units 2D). We consider an output represents TRUE when energy is 0.031 (39 % of the input energy), see Fig. 14.23b. Logical FALSE corresponds to energy 0.01 (15 % of the input), see Fig. 14.23c, d, e.

14.6   Discussion

All results presented are theoretical. There is a very wide gap between theoretical designs of logical circuits in actin and their prototyping in experimental laboratory conditions. For example, we have shown that a simple partitioned quantum automaton model, inspired by actin double-helix protein polymer, is powerful enough to perform many three-valued logical operations. The realizability of such an automaton depends on the physical substrate characteristics. The substrate's states must be quantum. The states must influence each other by a summation rule. We can encode such states in the dynamics of actin molecule via the interplay between electrical dipoles [53] and magnetic dipoles [92] in the actin filaments. Another option would be to encode values of input and output logical variables in the electrostatic surface charge on actin monomers, as configurations of intermixed pockets of negative charge and positive charges. The configurations of pockets encoding the variables can be updated by charge density waves (formed by clouds of counter ions) propagating from one actin

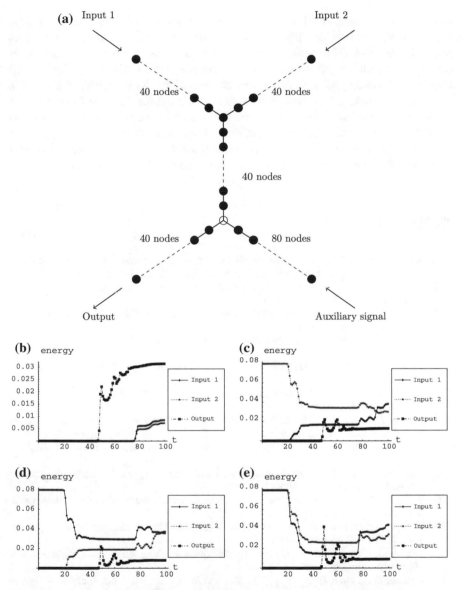

Fig. 14.23 A NOR gate. **a** Structure. **b** Energy evolution with input (FALSE, FALSE). **c** Energy evolution with input (FALSE, TRUE). **d** Energy evolution with input (TRUE, FALSE). **e** Energy evolution with input (TRUE, TRUE). Energy for input 1 is shown by *solid line* with rhomb markers, input 2 *dotted line* with start markers, input 3 *dashed line* with *square* markers

unit to its neighbours [100]. Cascading logical gates is yet another open question. It is necessary to have at disposal a way to transmit states from one automaton, actin filament, to another automaton, actin filament. Cross-linking of actin filaments could be a candidate for such a transmission mechanism. The cross-linking will require additional hypothesis on how phase transitions, ionic clouds, or dispersive solitary wave propagate via cross-linking proteins: α-actinin, spectrin, filamin, fibrin, dystrophin. Actin based computing circuits based on interactions between solutions sounds more feasible. At least we can initiate solutions and detect them using existing measurement equipment. Achieving programmable polymerisation of actin networks with desired geometry would be the most significant challenge in this case.

References

1. Aaronson, S., Watrous, J.: Closed timelike curves make quantum and classical computing equivalent. Proc. R. Soc. A Math. Phys. Eng. Sci. **465**(2102), 631–647 (2009)
2. Adamatzky, A.: Collision-based computing in biopolymers and their automata models. I J. Mod. Phys. C **11**, 1321–1346 (2000)
3. Adamatzky, A.: New media for collision-based computing. In: Collision-Based Computing. Springer, Heidelberg (2002)
4. Adamatzky, A., Mayne, R.: Actin Automata: Phenomenology and Localizations (2014). arXiv:1408.3676
5. Adamatzky, A.: Topics in reaction-diffusion computers. J. Comput. Theor. Nanosci. **8**, 295–303 (2011)
6. Aoun, B., Tarifi, M.: Introduction to Quantum Cellular Automata (2004). arXiv:quant-ph/0401123
7. Aprodu, I., Redaelli, A., Soncini, M.: Actomyosin interaction: mechanical and energetic properties in different nucleotide binding states. Int. J. Mol. Sci. **9**(10), 1927–1943 (2008)
8. Arruda, A.I.: On the imaginary logic of N.A. Vasil'ev. Stud. Logic Found. Math. **89**, 3–24 (1977)
9. Bajec, I.L., Mraz, M.: Towards multi-state based computing using quantum-dot cellular automata. Unconv. Comput. 105–116 (2005)
10. Bajec, I.L., Mraz, M.: Multi-valued logic based on quantum-dot cellular automata. Int. J. Unconv. Comput. **3**(4), 311 (2007)
11. Belnap, N.D. Jr.: A useful four-valued logic. In: Modern Uses of Multiple-valued Logic, pp. 5–37. Springer, Heidelberg (1977)
12. Benenti, G., Casati, G., Strini, G.: Principles of Quantum Computation and Information, vol. I-II. World Scientific, Singapore (2007)
13. Bochvar, D.A.: On a Three Valued Calculus and Its Application to the Analysis of Contradictories (1939)
14. Boutilier, C.: Modal logics for qualitative possibility and beliefs. In: Proceedings of the Eighth international conference on Uncertainty in artificial intelligence, pp. 17–24. Morgan Kaufmann Publishers Inc., Burlington (1992)
15. Cantiello, H.F., Patenaude, C., Zaner, K.: Osmotically induced electrical signals from actin filaments. Biophys. J. **59**(6), 1284–1289 (1991)
16. Cingolani, L.A., Goda, Y.: Actin in action: the interplay between the actin cytoskeleton and synaptic efficacy. Nat. Rev. Neurosci. **9**(5), 344–356 (2008)
17. Ciucci, D., Dubois, D.: Three-valued logics for incomplete information and epistemic logic. In: Logics in Artificial Intelligence, pp. 147–159. Springer, Heidelberg (2012)

18. Ciucci, D., Dubois, D.: From paraconsistent three-valued logics to multiple-source epistemic logic. In: 8th Conference European Society for Fuzzy Logic and Technology (EUSFLAT-13). Atlantis Press, The Netherlands (2013)
19. Conrad, M.: Cross-scale information processing in evolution, development and intelligence. BioSyst. **38**(2), 97–109 (1996)
20. Craddock, T.J.A., Tuszynski, J.A., Hameroff, S.: Cytoskeletal signaling: is memory encoded in microtubule lattices by CaMKII phosphorylation. PLoS Comput. Biol. **8**(3), e1002421 (2012)
21. Damsky, C.H., Werb, Z.: Signal transduction by integrin receptors for extracellular matrix: cooperative processing of extracellular information. Curr. Opin. Cell Biol. **4**(5), 772–781 (1992)
22. Debanne, D.: Information processing in the axon. Nat. Rev. Neurosci. **5**(4), 304–316 (2004)
23. Dehmelt, L., Halpain, S.: Actin and microtubules in neurite initiation: are MAPs the missing link? J. Neurobiol. **58**(1), 18–33 (2004)
24. Delgado, P.: Quantum Cellular Automata: Theory and Applications. Ph.D. thesis, University of Waterloo (2007)
25. Dillon, C., Goda, Y.: The actin cytoskeleton: integrating form and function at the synapse. Ann. Rev. Neurosci. **28**, 25–55 (2005)
26. Dubois, D., Prade, H.: Possibilistic logic: a retrospective and prospective view. Fuzzy Sets Syst. **144**(1), 3–23 (2004)
27. Durr, C., LêThanh, H., Santha, M.: A Decision Procedure for Well-formed Linear Quantum Cellular Automata (1999). arXiv:cs:9906024
28. Fiala, J.C., Spacek, J., Harris, K.M.: Dendritic spine pathology: cause or consequence of neurological disorders? Brain Res. Rev. **39**(1), 29–54 (2002)
29. Fifková, E., Delay, R.J.: Cytoplasmic actin in neuronal processes as a possible mediator of synaptic plasticity. J. Cell Biol. **95**(1), 345–350 (1982)
30. Fisch, M., Turquette, A.: Peirce's triadic logic. Trans of the Charles S. Peirce Society, pp. 71–85 (1966)
31. Gartzke, J., Lange, K.: Cellular target of weak magnetic fields: ionic conduction along actin filaments of microvilli. Am. J. Physiol. Cell Physiol. **283**(5), C1333–C1346 (2002)
32. Griffith, L.M., Pollard, T.D.: Evidence for actin filament-microtubule interaction mediated by microtubule-associated proteins. J. Cell Biol. **78**(3), 958–965 (1978)
33. Gruska, J.: Quantum Computing. Springer, Heidelberg (1999)
34. Gruska, J.: Descriptional complexity issues in quantum computing. J. Automata Lang. Comb. **5**(3), 191–218 (2000)
35. Gupta, S., Bisht, S.S., Kukreti, R., Jain, S., Brahmachari, S.K.: Boolean network analysis of a neurotransmitter signaling pathway. J. Theor. Biol. **244**(3), 463–469 (2007)
36. Hameroff, S.: Quantum computation in brain microtubules? the penrose-hameroff 'orch or' model of consciousness. Phil. Trans. R. Soc. Lond. A **356**, 1869–1896 (1998)
37. Hameroff, S.: Coherence in the cytoskeleton: implications for biological information processing. In: Biological Coherence and Response to External Stimuli, pp. 242–265. Springer, Heidelberg (1988)
38. Hameroff, S.: Quantum walks in brain microtubules: biomolecular basis for quantum cognition? Topics Cogn. Sci. **6**(1), 91–97 (2014)
39. Hameroff, S., Penrose, R.: Orchestrated reduction of quantum coherence in brain microtubules: A model for consciousness. Math. Comput. Simul. **40**(3), 453–480 (1996)
40. Hameroff, S.: Quantum coherence in microtubules: a neural basis for emergent consciousness? J. Consc. Stud. **1**(1), 91–118 (1994)
41. Hameroff, S.: The brain is both neurocomputer and quantum computer. Cogn. Sci. **31**(6), 1035–1045 (2007)
42. Hameroff, S., Craddock, T.J.A., Tuszynski, J.A.: Quantum effects in the understanding of consciousness. J. Integr. Neurosci. **13**(02), 229–252 (2014)
43. Hameroff, S., Dayhoff, J.E., Lahoz-Beltra, R., Samsonovich, A.V., Rasmussen, S.: Models for molecular computation: conformational automata in the cytoskeleton. Computer **25**(11), 30–39 (1992)

44. Hameroff, S., Watt, R.C.: Information processing in microtubules. J. Theor. Biol. **98**(4), 549–561 (1982)
45. Higgs, H.N., Pollard, T.D.: Regulation of Actin Filament Network Formation Through ARP2/3 Complex: Activation by a Diverse Array of Proteins 1. Ann. Rev. Biochem. **70**(1), 649–676 (2001)
46. Jaeken, L.: A new list of functions of the cytoskeleton. IUBMB life **59**(3), 127–133 (2007)
47. Janmey, P.: The cytoskeleton and cell signalling: component localization and mechanical coupling. Phys. Rev. **78**, 763–781 (1998)
48. Joachim, C., Gimzewski, J.K., Aviram, A.: Electronics using hybrid-molecular and mono-molecular devices. Nature **408**(6812), 541–548 (2000)
49. Kauffman, S.A.: Metabolic stability and epigenesis in randomly constructed genetic nets. J. Theor. Biol. **22**(3), 437–467 (1969)
50. Kim, C.-H., Lisman, J.E.: A role of actin filament in synaptic transmission and long-term potentiation. J. Neurosci. **19**(11), 4314–4324 (1999)
51. Kinsner, W.: Complexity and its measures in cognitive and other complex systems. 7th IEEE International Conference on Cognitive Informatics, 2008. ICCI 2008, pp. 13–29. IEEE, New York (2008)
52. Kleene, S.C.: Introduction to Metamathematics (1952)
53. Kobayashi, S., Asai, H., Oosawa, F.: Electric birefringence of actin. Biochimica et Biophysica Acta (BBA)-Specialized Section on Biophysical Subjects, **88**(3), 528–540 (1964)
54. Kojima, N., Shirao, T.: Synaptic dysfunction and disruption of postsynaptic drebrin-actin complex: a study of neurological disorders accompanied by cognitive deficits. Neurosci. Res. **58**(1), 1–5 (2007)
55. Kondacs, A., Watrous, J.: On the power of quantum finite state automata. In: Proceedings of the 38th Annual Symposium on Foundations of Computer Science, 1997, pp. 66–75. IEEE, New York (1997)
56. Inokuchi, S., Mizoguchi, Y.: Generalized Partitioned Quantum Cellular Automata and Quantization of Classical CA. Int. J. Unconv. Comput. **1**, 149–160 (2005)
57. Landa, H., Marcovitch, S., Retzker, A., Plenio, M.B., Reznik, B.: Quantum coherence of discrete Kink solitons in ion traps. Phys. Rev. Lett. **104**, 043004 (2010)
58. Bajec, I.L., Zimic, N., Mraz, M.: Towards the bottom-up concept: Extended quantum-dot cellular automata. Microelectr. Eng. **83**(4), 1826–1829 (2006)
59. Lin, E.C., Cantiello, H.F.: A novel method to study the electrodynamic behavior of actin filaments, evidence for cable-like properties of actin. Biophys. J. **65**(4), 1371–1378 (1993)
60. Lodish, H., Zipursky, S.L.: Molecular cell biology. Biochem. Mol. Biol. Educ. **29**, 126–133 (2001)
61. Ludin, B., Matus, A.: The neuronal cytoskeleton and its role in axonal and dendritic plasticity. Hippocampus **3**(S1), 61–71 (1993)
62. Łukasiewicz, J.: On three-valued logic. In: The Polish Review, pp. 43–44. (1968)
63. Ma, Z., Wang, J., Guo, H.: Weakly nonlinear ac response: theory and application Phys. Rev. B **59**(11), 7575 (1999)
64. Martinez, G., Adamatzky, A., Mayne, R., Chen, F., He, Q.: Majority gates and circular computation in slime mould. 2015 European Conference on Artificial Life, in press (2015)
65. Matus, A.: Actin-based plasticity in dendritic spines. Science **290**(5492), 754–758 (2000)
66. Mayne, R., Adamatzky, A.: The physarum polycephalum actin network: formalisation, topology and morphological correlates with computational ability. In: Proceedings of the 8th International Conference on Bioinspired Information and Communications Technologies, pp. 87–94 (2014)
67. Mayne, R., Adamatzky, A., Jones, J.: On the role of the plasmodial cytoskeleton in facilitating intelligent behaviour in slime mould Physarum polycephalum. arXiv preprint arXiv:1503.03012 (2015)
68. McCulloch, W.S., Pitts, W.: A logical calculus of the ideas immanent in nervous activity. Bull. Math. Biophys. **5**(4), 115–133 (1943)
69. Mitchell, M.: Complex systems: network thinking. Artif. Intell. **170**(18), 1194–1212 (2006)

70. Motoike, I.N., Adamatzky, A.: Three-valued logic gates in reaction-diffusion excitable media. Chaos, Solitons & Fractals **24**(1), 107–114 (2005)
71. Muthukrishnan, A., Stroud Jr., C.R.: Multivalued logic gates for quantum computation. Phys. Rev. A **62**(5), 052309 (2000)
72. Nakamura, F., Osborn, T.M., Hartemink, C.A., Hartwig, J.H., Stossel, T.P.: Structural basis of filamin a functions. J. Cell Biol. **179**(5), 1011–1025 (2007)
73. Oda, T., Mitsusada, I., Tomoki, A., Yuichiro, M., Akihiro, N.: The nature of the globular-to fibrous-actin transition. Nature **457**(7228), 441–445 (2009)
74. Persico, A.M., Bourgeron, T.: Searching for ways out of the autism maze: genetic, epigenetic and environmental clues. Trends Neurosci. **29**(7), 349–358 (2006)
75. Priel, A., Ramos, A.J., Tuszynski, J.A., Cantiello, H.F.: A biopolymer transistor: electrical amplification by microtubules. Biophys. J. **90**(12), 4639–4643 (2006)
76. Priel, A., Tuszynski, J.A., Cantiello, H.F.: Electrodynamic signaling by the dendritic cytoskeleton: toward an intracellular information processing model. Electromagn. Biol. Med. **24**(3), 221–231 (2005)
77. Priel, A., Tuszynski, J.A., Cantiello. H.F.: The dendritic cytoskeleton as a computational device: an hypothesis. In: The Emerging Physics of Consciousness, pp. 293–325. Springer, Heidelberg (2006)
78. Priel, A., Tuszynski, J.A., Woolf, N.J.: Neural cytoskeleton capabilities for learning and memory. J. Biol. Phys. **36**(1), 3–21 (2010)
79. Rasmussen, S., Karampurwala, H., Vaidyanath, R., Jensen, K.S., Hamcroff, S.: Computational connectionism within neurons: a model of cytoskeletal automata subserving neural networks. Physica D: Nonlinear Phenomena **42**(1), 428 449 (1990)
80. Russell, B., MacColl, H.: The existential import of propositions. In: Mind, pp. 398–402 (1905)
81. Sampson, L.J., Leyland, M.L., Dart, C.: Direct interaction between the actin-binding protein filamin-A and the inwardly rectifying potassium channel, Kir2. 1. J. Biol. Chem. **278**(43), 41988–41997 (2003)
82. Schubert, T., Akopian, A.: Actin filaments regulate voltage-gated ion channels in salamander retinal ganglion cells. Neuroscience **125**(3), 583–590 (2004)
83. Schumacher, B., Werner, R.F.: Reversible Quantum Cellular Automata (2004). arXiv:quant-ph/0405174
84. Sekino, Y., Kojima, N., Shirao, T.: Role of actin cytoskeleton in dendritic spine morphogenesis. Neurochem. Int. **51**(2), 92–104 (2007)
85. Siccardi, S., Adamatzky, A.: Actin quantum automata: communication and computation in molecular networks. Nano Commun. Netw. **6**(1), 15–27 (2015)
86. Siccardi, S., Adamatzky, A.: A Quantum actin automata and three-valued logics, To be found (2015)
87. Siccardi, S., Tuszynski, J.A., Adamatzky, A.: Boolean gates on actin filaments. Phys. Lett. A **380**(1), 88–97 (2016)
88. Siccardi, S., Adamatzky, A.: Logical gates implemented by solitons at the junctions between one-dimensional lattices Submitted to International Journal of Bifurcation and Chaos (2015)
89. Solé, R.V., Luque, B.: Phase transitions and antichaos in generalized Kauffman networks. Phys. Lett. A **196**(5), 331–334 (1995)
90. van Spronsen, M., Hoogenraad, C.C.: Synapse pathology in psychiatric and neurologic disease. Curr. Neurol. Neurosci. Rep. **10**(3), 207–214 (2010)
91. Steiglitz, K., Kamal, I., Watson, A.: Embedding computation in one-dimensional automata by phase coding solitons. IEEE Trans. Comput. **37**(2), 138–145 (1988)
92. Torbet, J., Dickens, M.J.: Orientation of skeletal muscle actin in strong magnetic fields. FEBS Lett. **173**(2), 403–406 (1984)
93. Simple collision-based chemical logic gates with adaptive computing: Toth, R., Stone, C., de Lacy Costello, B., Adamatzky, A., Bull, L. Int. J. Nanotechnol. Mol. Comput. (IJNMC) **1**, 1–16 (2009)

94. Tuszynski, J.A., Brown, J.A., Hawrylak, P.: Dielectric polarization, electrical conduction, information processing and quantum computation in microtubules. are they plausible? In:Philosophical Transaction Royal Society of London. Series A: Mathematical, Physical and Engineering Sciences, pp. 1897–1925 (1998)
95. Tuszyński, J.A., Portet, S., Dixon, J.M., Luxford, C., Cantiello, H.F.: Ionic wave propagation along actin filaments. Biophys. J. **86**(4), 1890–1903 (2004)
96. Velarde, M.G., Ebeling, W., Chetverikov, A.P.: On the possibility of electric conduction mediated by dissipative solitons. Int. J. Bifurc. Chaos **15**(1), 245–251 (2005)
97. Velarde, M.G., Ebeling, W., Hennig, D., Neißner, C.: On soliton-mediated fast electric conduction in a nonlinear lattice with Morse interactions. Int. J. Bifurc. Chaos **16**(4), 1035–1039 (2006)
98. Velarde, M.G., Ebeling, W., Chetverikov, A.P., Hennig, D.: Electron trapping by solutions. Classical versus quantum mechanical approach. Int. J. Bifurc Chaos **18**(2), 521–526 (2008)
99. Watrous, J.: On one-dimensional quantum cellular automata. In: Proceedings of the 36th Annual Symposium on Foundations of Computer Science, 1995, pp. 528–537. IEEE, New York (1995)
100. Woolf, N.J., Priel, A., Tuszynski, J.A.: The cytoskeleton as a nanoscale information processor: electrical properties and an actin-microtubule network model. In: Nanoneuroscience, pp. 85–127. Springer, Heidelberg (2010)
101. Wuensche, A.: The ghost in the machine. In: Santa Fe Institute Studies in the Sciences of Complexity, vol. 17, pp. 465–465. Addison-Wesley Publishing (1994)
102. Wuensche, A.: Attractor basins of discrete networks. Cognitive Science Research Paper, vol. 461 (1997)
103. Wang, B., Zhao, X., Wang, J., Guo, H.: Nonlinear quantum capacitance. Appl. Phys. Lett. **74**(19), 2887–2889 (1999)
104. Van Woerko, A.E.: The major hallucinogens and the central cytoskeleton: an association beyond coincidence? Towards sub-cellular mechanisms in schizophrenia. Med. Hypotheses **31**(1), 7–15 (1990)
105. Zhang, L., Adamatzky, A.: Collision-based implementation of a two-bit adder in excitable cellular automaton. Chaos, Solitons and Fractals **41**, 1191–1200 (2009)

Chapter 15
Modeling DNA Nanodevices Using Graph Rewrite Systems

Reem Mokhtar, Sudhanshu Garg, Harish Chandran,
Hieu Bui, Tianqi Song and John Reif

Abstract DNA based nanostructures and devices are becoming ubiquitous in
nanotechnology with rapid advancements in theory and experiments in DNA self-
assembly which have led to a myriad of DNA nanodevices. However, the modeling
methods used by researchers in the field for design and analysis of DNA nanostructures and nanodevices have not progressed at the same rate. Specifically, there
does not exist a formal system that can capture the spectrum of the most frequently
intended chemical reactions on DNA nanostructures and nanodevices which have
branched and pseudo-knotted structures. In this paper we introduce a graph rewriting system for modeling DNA nanodevices. We define pseudo-DNA nanostructures
(**PDN**s), which describe the sequence information and secondary structure of DNA
nanostructures, but exclude modeling of tertiary structures. We define a class of
labeled graphs called DNA graphs, that provide a graph theoretic representation
of PDNs. We introduce a set of graph rewrite rules that operate on DNA graphs.
Our DNA graphs and graph rewrite rules provide a powerful and expressive way
to model DNA nanostructures and their reactions. These rewrite rules model most
conventional reactions on DNA nanostructures, which include hybridization, dehybridization, base-stacking, and a large family of enzymatic reactions. A subset of
these rewrite rules would likely be used for a basic graph rewrite system modeling
most DNA devices, which use just DNA hybridization reactions, whereas other of
our rewrite rules could be incorporated as needed for DNA devices for example
enzymic reactions. To ensure consistency of our systems, we define a subset of DNA
graphs which we call *well-formed DNA graphs*, whose strands have consistent $5'$ to
$3'$ polarity. We show that if we start with an input set of well-formed DNA graphs, our
rewrite rules produce only well-formed DNA graphs. We give four detailed example
applications of our graph rewriting system on (1) Yurke et al. [82] DNA tweezer
system, (2) Yurke et al. [77] catalytic hairpin-based triggered branched junctions, (3)
Dirks and Pierce [17] HCR, and (4) Qian and Winfree [59] scalable circuit of seesaw

R. Mokhtar (✉) · S. Garg · H. Bui · T. Song · J. Reif
Department of Computer Science, Duke University, Durham, NC, USA
e-mail: reem@cs.duke.edu

H. Chandran
Google, Mountain View, CA, USA

© Springer International Publishing Switzerland 2017   347
A. Adamatzky (ed.), *Advances in Unconventional Computing*,
Emergence, Complexity and Computation 23,
DOI 10.1007/978-3-319-33921-4_15

gates. Finally, we have a working software prototype (DAGRS) that we have used to generate automatically well-formed DNA graphs using a basic rewriting rule set for some of the examples mentioned.

15.1 Introduction

15.1.1 Motivation

A *DNA nanostructure* is a macromolecule composed of DNA strands that are hybridized together and have a predetermined target structure [62]. A *DNA nanodevice* is a biochemical system composed of DNA nanostructures that undergo dynamic reactions which modify their structure [38]. DNA nanodevices often act as nanoscale machines and are designed to perform specific tasks, including DNA walkers [2, 25, 27, 28, 45, 61, 66–68, 70, 74, 78, 79], tweezers [8, 82], gears [71], and other nanodevices [6, 44, 48, 50].

DNA nanostructures are known to obey structural constraints, which are difficult to model in detail [56]. These include geometric and conformational constraints such as the persistence length, helical pitch, major and minor grooves, supercoiling, conformations under the effect of different hydration levels, and other physical properties [69]. Most commonly intended reactions in structural and dynamic DNA nanotechnology research can be treated on a fundamental level while temporarily discounting these constraints.

We propose a method of coarse-grained modeling that provides a high-level abstraction of DNA nanostructures. Our coarse-grained modeling of DNA nanostructures includes a class of structures which we call pseudo-DNA nanostructures, similar to the conventional cartoon descriptions of DNA nanostructures, and an equivalent class of labeled graphs which we call well-formed DNA graphs.

A general approach to modeling a biophysical structure is to develop a coarse-grained model. The term "coarse-grained modeling" has conventionally been used to refer to models which more concerned with biophysical models [18, 46, 52]. By *coarse-grained modeling*, we refer to the modeling of the primary and secondary structure, and some other properties that affect hybridization and enzymatic reactions, or transformations, but do not attempt to model or provide specification of tertiary structure. Tertiary structure is not generally of key importance to these reactions on DNA nanostructures with relatively smaller scales of complexity (where steric hindrance does not greatly influence reactions).

15.1.2 Prior and Related Work

The visual representation of DNA nanostructures often termed *DNA cartoons* is widely used (such as in Fig. 15.1b as a representation of Fig. 15.1a). They provide a visual representation of the secondary structure of DNA nanostructures, including 5′

to 3′ directionality and hybridizations between single strands of DNA. Their goal is to abstract away individual base-pairings and helical twists to provide a domain-based representation. Though there are many variations on this, most cartoon renderings represent a strand with a line that terminates in an arrow at the 3′-end. When two complementary antiparallel strands hybridize to each other through base pairing, a series of lines between the strands represent the hybridization bonds between each base pair. DNA cartoons are frequently used in practice for describing DNA nanostructures and may also be used by visual GUI software, e.g., for graphical specification and rendering of DNA nanostructures. Alternatives to string-based methods, such as explicit graph structures, are needed in order to model certain DNA nanostructures, especially those with branched or pseudo-knotted regions. Many earlier examples of such graph-based approaches that provide abstractions of DNA nanostructures and their reactions have been developed. Jonoska made early use of graphs for representing DNA-based computations [31]. Reif [60] developed a graph model for representing DNA nanostructures with representation of 5′ to 3′ directionality and hybridizations between single strands of DNA [60]. In addition, Birac et al. [3] have made use of graph-based data structures for the design and modeling of DNA nanostructures [3]. Other more recent graph models for DNA nanostructures include those of Kawamata et al. [30, 32]. None of these model base-stacking which can be essential for some DNA reactions [76]. There are various prior works on coarse-grained modeling of dynamic DNA nanodevices that transition between distinct DNA nanostructures. Cardelli [5], Phillips and Cardelli [54] and Lakin et al. [40, 42] developed an algebra for representing a restricted set of DNA-based reactions, such as see-saw gates [58]. Yin et al. [77] made use of a graphical representation of biomolecular self-assembly pathways, in which DNA was rendered into an abstract nodal representation, then used to manipulate hybridization self-assembly reaction in constructing nanostructures [77]. Kumara et al. [39] presented analysis

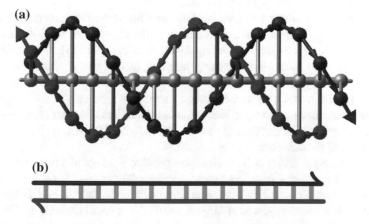

Fig. 15.1 **a** A rendering of a double-stranded DNA duplex (nanoengineer). **b** A cartoon rendering of the duplex

of assembly pathways for different types of DX tiles that previously refused to assemble into parallel structures [39]. Ibuki et al. [30] also developed a graph-structure to model basic DNA reactions [30]. Finally, McCaskill and Niemann [49] presented the most relevant example of a DNA graph replacement scheme, which provides a framework for computing intermediary and target DNA and RNA structures, and more specifically enzymatic reactions [49]. Klavins developed graph grammar models for robotic self-assembly, and also adapted this in DNA self-assembly [33–36].

Graph grammars and graph rewriting systems (defined in Sect. 15.2.4) have been described at length in [11, 12, 19, 65]. Graph grammars were introduced as a generalization of formal grammars on strings (Chomsky grammars [9, 10]) to those on graphs [20]. Its use to describe distributed assembly started with [35]. Here, we are attempting to use this same formalism to model commonly intended DNA hybridization reactions. Graph grammars were used by Flamm et al. [22, 22], Andersen et al. [1] and Mann et al. [47] to model chemical reactions, and a Graph Grammar Library (GGL) software developed to that effect.

Related work in the field of DNA nanotechnology includes [26] domain-based DNA reaction enumerator, in which strands and nanostructures are represented as formal structures which produce the most relevant reaction pathway. The method is based on condensing large reaction graphs by excluding a number of transient (fleeting or fast reaction) states and combining strongly connected components in the graph. In Sect. 15.9, we provide an example of how our graph rewriting system can utilize this concept to decrease the number of states. Finally, oxDNA [18, 46, 52] is currently the most advanced coarse-grained modeling system at the time of writing. However, oxDNA uses a biophysical model that treats each nucleotide as a rigid body, with a plane perpendicular to the base so as to capture its planar orientation, and collinear interacting sites. This paper disregards these physical properties and attempts to abstract them away, since one advantage of this approach is being able to demonstrate lesser-likely reaction pathways.

The behavior of many complex systems can be explicitly rendered into graph transformations. Graph rewriting systems provide a powerful, intuitive and flexible method of analyzing and modeling DNA reactions [15, 55]. Our DNA graph model allows us to model base-stacking interactions between strands in addition to the conventional hybridization bonds in DNA nanostructures. We introduce a graph-theoretic approach that uses graph transformations to represent reactions on DNA nano structures; these graph transformations correspond to DNA reactions, and transform well-formed DNA graphs into modified well-formed DNA graphs that represent the products of these corresponding reactions.

Our DNA graphs and graph rewrite rules provide a powerful and expressive way to model DNA nanostructures and their reactions. These rewrite rules model most conventional reactions on DNA nanostructures, which include hybridization, dehybridization, base-stacking, and a large family of enzymatic reactions. To ensure consistency of our systems, we define a subset of DNA graphs which we call *well-formed DNA graphs*, whose strands have consistent $5'$ to $3'$ polarity. We show that if we start with an input set of well-formed DNA graphs, our rewrite rules produce only well-formed DNA graphs. Initially it was thought that one of the advantages of our approach is that existing rewrite systems which automate the graph rewriting process can be used, but have since realized that they require some modification, which is why we chose to develop our own rewriting systems. Some rule-based systems that were explored included Kappa and KaSim [13, 21], Porgy and Tulip [55], amongst others, but found that they could not easily perform the domain-level modeling that were intended using graph rewriting within these systems. As an example using Kappa, separating out the domains as distinct but connected "agents" while respecting the order in which they occur on a strand, and simultaneous attempting to incorporate branched and pseudo-knotted structures, while accommodating a rule-set that is intended to be generalizable did not seem readily feasible.

15.1.3 Overall Organization

In Sect. 15.2, we define the class of pseudo-DNA nanostructures, with an example, define the class of well-formed DNA graphs, and finally present our graph rewriting systems. Section 15.3 introduces a set of non-enzymatic graph rewrite rules. In Sect. 15.4, we show that every graph generated by our rewriting systems belongs to the class of well-formed DNA graphs, and we show that this class of graphs is equivalent to the class of pseudo-DNA nanostructures. Algorithms 1 and 2 contain algorithms for the construction of members of the classes **PDN** and **WFDG** as part of the proof in Sect. 15.4. In Sect. 15.5, we show an example application of the graph rewriting systems to Yurke et al. [82] DNA tweezer system. Section 15.6 briefly describes our software system for automating our graph rewriting systems. Section 15.7 displays further examples of the application of graph rewriting systems to existing DNA reaction systems, including the catalytic hairpin-based branched junctions [17, 77] HCR [17] and the [59] scalable circuit comprised of seesaw gates [59]. Section 15.8 has an extended (enzymatic) rewrite system that is used in conjunction with the basic system. Section 15.7 Provides further example applications of our graph rewrite rules. Section 15.8 Provides enzymic graph rewrite rules. Sections 15.9, 15.10 and 15.11 discuss further details of the graph rewrite rules and how they can be combined with existing work. Section 15.12 contains a summary of our results as well as a critical analysis of the limitations and advantages of our approach.

15.2 Definitions

15.2.1 Pseudo-DNA Nanostructures

In order to accommodate certain physically feasible structures which may be generated by our rewriting systems, but at the same time maintain a general method of characterizing the most frequently intended DNA reactions, we define a class of structures, *pseudo-DNA nanostructures* (**PDN**). The class describes only the sequence information and secondary structure of DNA nanostructures, and excludes tertiary structure information. Any pseudo-DNA nanostructure $p \in$ **PDN** is a set of DNA strands in the nanostructure, each described by domains (which can be one nucleobase or a sequence of nucleobases) and the hybridizations between domains.

Formally, we define $p \in$ **PDN** as a quintuple $(\mathcal{D}, \mathcal{S}, \delta, \mathcal{H}, \mathcal{B})$, where:

1. $\mathcal{D}$ is a set of domains $\{d_1, d_2, \ldots\}$, and their complements $\{\bar{d}_1, \bar{d}_2, \ldots\}$ (if they exist in the structure). This is necessary because all domains involved in the PDN dictate the relationships in the set of hybridizations. If the complementary domains do not exist, then the set of hybridizations is empty.

2. $\mathcal{S}$ is a set of strands $\{s_1, s_2, \ldots\}$

3. $\delta : \mathcal{S} \to \mathcal{D}^*$ maps each strand in $\mathcal{S}$ to a string over the set of domains in $\mathcal{D}$, and each string's domains are read in the 5' to 3' direction.

4. $\mathcal{H}$ is the set that maps the hybridization relationship between domains of strands in $\mathcal{S}$. $\mathcal{H}$ is a set of 4-tuples. Each $h \in \mathcal{H}$ is of the form (s_1, i, s_2, j), indicating that the ith domain of strand s_1 is hybridized to the jth domain of strand s_2. Each domain of any strand can hybridize to no more than one complementary domain of another (or the same) strand.

5. $\mathcal{B}$ is the set that maps base-stacking interactions, specifically those that involve domains on different strands in $\mathcal{S}$. That is, the members in $\mathcal{B}$ define blunt-end, overhang, and frayed-end base-stacking interactions between domains which are located on different strands (whether on the 3' or 5' in the case of blunt-end, or at an unpaired location in the case of frayed-end base-stacking, or both in the case of overhang base-stacking). $\mathcal{B}$ is a set of 12-tuples. Each $b \in \mathcal{B}$ is $(s_1, i, e_1, s_2, j, e_2, s_3, k, e'_1, s_4, \ell, e'_2)$. This 12-tuple describes the hybridization of two pairs of strands (or two duplexes), (s_1, i, s_3, k) and (s_2, j, s_4, ℓ). The base stacking exists between s_1's ith domain's e_1th end with s_3's kth domain's e'_1th end, and simultaneously between s_2's jth domain's e_2th end with s_4's lth domain's e'_2th end. In this case, e'_1 and e'_2 refer to the corresponding domain ends of s_1 and s_2, respectively.

An example is shown in Fig. 15.2, the three-arm branched structure may be represented as a member of **PDN** as follows:

1. $\mathcal{D} = \{a, b, c, \bar{a}, \bar{b}, \bar{c}\}$
2. $\mathcal{S} = \{s_1, s_2, s_3\}$.
3. $\delta(s_1) = ab, \delta(s_2) = \bar{b}c, \delta(s_3) = \bar{c}\bar{a}$.
4. $\mathcal{H} = \{(s_1, 1, s_2, 2), (s_1, 2, s_3, 1), (s_3, 2, s_2, 1)\}$

Fig. 15.2 A three-arm junction

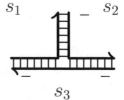

5. $\mathcal{B}$ may be any one of the following sets, depending on the junction's conformation [43]:

   a. $\emptyset$

   b. $(s_1, 1, 3', s_3, 2, 5', s_2, 2, 5', s_3, 1, 3')$

   c. $(s_1, 1, 3', s_3, 2, 5', s_1, 2, 5', s_2, 1, 3')$

   d. $(s_2, 1, 3', s_1, 2, 5', s_3, 1, 3', s_1, 2, 5')$

15.2.2 DNA Graph Notation

We will define the following notation for DNA graphs in order to facilitate the introduction of our DNA graph rewriting systems (note that we will subsequently show that a subset of DNA graphs, called well-formed DNA graphs, are equivalent to pseudo-DNA nanostructures). A *DNA graph* is a vertex-labeled and edge-labeled graph formally defined as an 8-tuple $G = (V, E, L_V, L_E, \Sigma, vl, el, \delta)$ where:

1. V is a finite set of vertices. Each vertex represents an unhybridized DNA strand domain, a hybridized DNA strand domain, a $3'$-end or a $5'$-end of a DNA strand.
2. E is a finite set of edges. Each edge represents a relationship between two vertices.
3. $L_V = \{\bigcirc, \bullet, 3', 5'\}$ is the finite set of vertex labels, where:

   a. $\bigcirc$: unhybridized vertex, where every base in the domain that this vertex is mapped to is unhybridized.

   b. $\bullet$: hybridized vertex, where every base in the domain that this vertex is mapped to is hybridized.

   c. $3'$: 3 prime end.

   d. $5'$: 5 prime end.

4. $L_E = \{\;\longmapsto,\;\cdots\cdots,\;\longrightarrow,\;\dashv\}$ is the finite set of edge labels, where:

   a. $\longmapsto$: base-stacking, where the arrow direction indicates $5'$ to $3'$ directionality. Here, we only explicitly represent base-stacking interactions between adjacent bases on different strands, and not the base-stacking interactions between adjacent bases on a single strand.

   b. $\cdots\cdots$: hybridization.

   c. $\longrightarrow$: covalent bond linking base pairs between two domains, where the arrow direction indicates strand directionality.

   d. $\dashv$ is an edge label that signifies a strand-end, where the vertical bar location indicates strand directionality.

5. Σ is the set of all possible domain names. For each $d \in \Sigma$, its complement is denoted as $\bar{d}$, and it represents the reverse-complement base sequence. It is assumed that each $d \in \Sigma$ has a unique complement $\bar{d} \in \Sigma$ s.t. no other symbol in $\Sigma - \{d\}$ can have a complement $\bar{d}$. Each domain d is mapped to a base sequence, which is a string over the set $\{A, C, T, G\}$, and the complement of each member is $\bar{A} = T, \bar{T} = A, \bar{C} = G, \bar{G} = C$.

6. $vl : V \mapsto L_V$, is a function that assigns a label to each vertex.
7. $el : E \mapsto L_E$, is a function that assigns a label to each edge.
8. $m : V \mapsto \Sigma$ is a mapping between vertices and domain names.

15.2.3   Well-Formed DNA Graphs

To be concise, we will define a *DNA strand graph* here as a maximal subgraph that consists of one or more vertices with a label in the vertex label subset $\{\bigcirc, \bullet, 3', 5'\}$ connected only via edges labeled ⟶. In other words, it is a graph representing only one strand, with no hybridization or base-stacking labeled edges (as shown in Fig. 15.3).

There are two types of strands: *circular* (see Sect. 15.4) and *non-circular*, where a non-circular strand has an explicit start and end, as opposed to a circular strand, which maintains directionality without having ends.

A *well-formed DNA graph* (referred to as **WFDG**), is a DNA graph that adheres to the six constraints below. Some of the constraints apply differently to circular strands, which will be made clear as we define the constraints in this section.

The *in-degree* of a vertex is the number of edges which are directed into that vertex. The *out-degree* of a vertex is the number of edges which are directed out of that vertex. The *degree* of a vertex is the sum of its in-degree and out-degree. The direction of an edge is indicated by the edge label's end marker location. For example, in Fig. 15.3a, the direction of the edge with label ⊣ between vertices v_1, u_1 (where label(u_1) = 3', and v_1 is the hybridized vertex, labeled ● and marked with the domain name "d_1" for ease of reference) is towards u_1.

Constraint 1: Strand Ends In circular strands, no ends exist. That is, by definition, $\nexists u, v \in V$ s.t. their labels are 5' and 3', respectively.

In a non-circular strand, $\exists u, v \in V$ s.t. their labels are 5' and 3', respectively, subject to:

1. in-degree(u) = 0 and out-degree(u) = 1, and a directed edge labeled ⊣ from u (i.e. the 5'-end) to some vertex w s.t. label(w)$\in \{\bigcirc, \bullet\}$.
2. in-degree(v) = 1 and out-degree(v) = 0, and a directed edge labeled ⊣ to v (i.e. the 3'-end) from some vertex w s.t. label(w)$\in \{\bigcirc, \bullet\}$.

Fig. 15.3 a An example well-formed DNA graph. **b** An example of a DNA strand graph

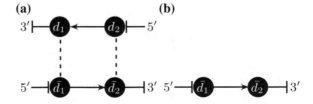

Constraint 2: Unhybridized Vertex Degree $\forall$ vertices $v \in V$ s.t. $\text{label}(v) = \bigcirc$ then in-degree$(v) = 1$ and out-degree$(v) = 1$.

Constraint 3: Hybridized Vertex Degree $\forall$ vertices $v \in V$ s.t. $\text{label}(v) = \bullet$ then $3 \leq \deg(v) \leq 5$. A discussion of the reasoning behind this can be found in Sect. 15.10 (on P. XXX).

Constraint 4: Vertex Edge Labels For each vertex $v \in V$ the following conditions apply:

1. If the vertex v has $\text{label}(v) = \bullet$, then it has exactly one hybridization-labeled edge $\text{label}(e) = \text{-----}$. This edge is treated as two edges going in both directions, but is represented as one undirected edge. Which means it counts both an an in-degree and out-degree for each participating vertex.
2. If the vertex v has $\text{label}(v) = \bullet$, then $\exists i$ base-stacking-labeled edges $e \in E$ connected to v s.t. $\text{label}(e) = \longmapsto$ and i is an integer s.t. $0 \leq i \leq 2$
3. $\exists j$ covalently-labeled edges $e \in E$ connected to v s.t. $\text{label}(e) = \longrightarrow$ and j is an integer s.t. $0 \leq j \leq 2$, where there is at most one edge directed towards v and at most one edge directed away from v.
4. $\exists k$ edges labeled with $\longmapsto$ connected to v, s.t. $k = |2 - j|$, at most one edge is directed towards v and at most one edge directed away from v.

Constraint 5: Complementarity For every vertex v where $\text{label}(v) = \bullet$, and $m(v) = d \in \Sigma^+$ there is a corresponding vertex $\bar{v}$ s.t.: $e(v, \bar{v}) \in E$, $\text{label}(e) = \text{-----}$, $\text{label}(\bar{v}) = \bullet$ and $m(\bar{v}) = \bar{d} \in \Sigma^+$, the complement of d.

Constraint 6: Directionality If $\exists w, x, y \in V$ s.t. w is directly connected to x, and x is directly connected to y through edges labelled with $\longrightarrow$, then exactly one of the following is true:

1. Either the edge $e(w, x)$ is directed towards x iff the edge $e(x, y)$ is directed towards y, or the edge $e(w, x)$ is directed towards w iff the edge $e(x, y)$ is directed towards x.
2. In non-circular strands (recall **Constraint 1**), if $\exists u, v \in V$ s.t. their labels are $5'$ and $3'$, then exactly one of the following statements has to be true:

   a. There exists a directed path of covalently-labeled edges from u to v that passes through w, x, y, in that order.
   b. There exists a directed path of covalently-labeled edges from u to v that passes through y, x, w, in that order.

3. In circular strands, this is not applicable (recall **Constraint 1**). A rendering of the repercussions on circular strands is made in Fig. 15.4.

15.2.4 Graph Rewriting Systems

A *graph rewriting system* is a set of rewrite rules that transforms a graph instance to another, both belonging to the same class of graphs. Each rule in our graph rewriting

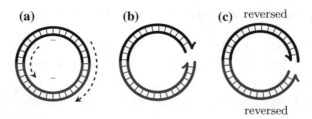

Fig. 15.4 Directionality is necessary for well-formed DNA graphs to ensure context-free nature of rewriting systems. **a** A circular duplex structure, where the domain sequences are ordered in the direction of the dotted arrow. **b** Directionality of the same duplex in a if cut by an enzyme. **c** Directionality of the same duplex in a if cut by an enzyme, but had the reverse direction of the domain sequences in **b**

systems has the form (also known as single pushout) [51] of a triple $p = \langle L \xrightarrow{r} R \rangle$, where the left-hand side in each rule, L, undergoes a partial graph morphism r, to a right-hand side R. The rule is applied to a *host* graph, or the graph to be transformed via the application of a rewrite rule. An occurrence of L in a graph G makes the rule applicable. This occurrence is called a *match* [64]. In other words, a host graph graph G can be transformed into another graph H, first by finding a matching between the left-hand side L of a rule r within the host graph, and transforming it to the right-hand side R in H. In DAGRS (DNA Graph Rewriting System), a matching is found by applying the subgraph isomorphism algorithm by Ullmann [73] (explained in Sect. 15.6).

15.2.5   Our DNA Graph Rewriting Systems

For the purpose of brevity, we have arranged the set of rewrite rules that define our graph rewriting systems between Sect. 15.3 (non-enzymatic) and Sect. 15.8 (enzymatic): Sect. 15.3 includes only hybridization and base-stacking rules; and Sect. 15.8 lists the extended rewriting rules that accompany the basic set.

Each side of the rewrite rule is a graph that follows the DNA graph notation defined in Sect. 15.2.2; and the well-formedness constraints defined in Sect. 15.2.3. As explained previously, the left-hand side of each rule is a graph that matches a subgraph of a larger, host graph G. This subgraph encodes: the vertices, edges, vertex-labels and edge-labels, and directionality. The host graph G in this case is a *DNA state graph*, which is some DNA graph G containing one or more DNA *strands* (as defined in Sect. 15.2.3), which comprise a well-formed DNA graph. These rules are intended for coarse-grained modeling, and are independent of any external parameters like salt concentration and the pH value of the buffer (An extension of this work may involve applying these rules within the confines of reaction rates, which would dramatically decrease the number of possible rules that may be applied at each transformation junction starting from the input species).

15.3   Non-enzymatic DNA Graph Rewriting Rules

The rules are structured as follows:

1. Cartoon Rendering: a DNA cartoon depiction of the DNA strands or complex.
2. Rewrite Rule: the corresponding graph rewrite rule
3. r and l: the direction of rule application. r indicates that the rule transforms the left-hand side L to the right-hand side R. If the rule also has the direction l, then it can be used to transform R to L.
4. L refers to the left-hand side, and R is the right-hand side.

Rule #1: Domain Binding

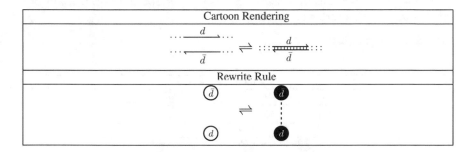

Rewrite Rule: Two complementary DNA domains d and $\bar{d}$ hybridize together to form a duplex, or double-stranded DNA. Note, the cartoon represents two strands each made up of one domain, but the strands may be surrounded by other domains (vertices) in either direction, while maintaining their original directionality.

L: The unhybridized domains d and $\bar{d}$ are each represented by a circle, with no edge between them.

R: They are each represented as solid circles (hybridized). The dashed edge represents the hydrogen bond(s) between d and $\bar{d}$.

Rule #2: Strand Displacement

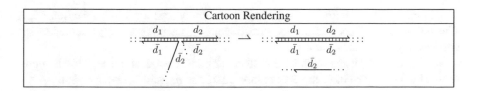

Rewrite Rule: Five domains are involved in this transformation. The vertex with domain label $\bar{d_2}$, which corresponds to the domain covalently bound to the complex on the left-hand side, displaces a vertex labeled with the same domain $\bar{d_2}$, which corresponds to the domain hybridized to d_2.

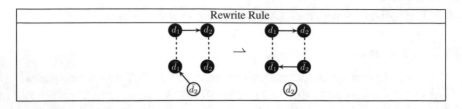

L: The upper strand region is composed of domains d_1, d_2. d_1 is hybridized to $\bar{d}_1$, and d_2 is hybridized to its complement $\bar{d}_2$.

R: The hybridization bond between vertex $\bar{d}_2$ and d_2 has been removed, and $\bar{d}_2$ has now been relabeled with $\bigcirc$. The vertex with domain $\bar{d}_2$, which is covalently bound to $\bar{d}_1$, is relabeled as hybridized ($\bullet$), and a hybridization edge has now been added to connect this vertex to vertex $\bar{d}_2$.

Rule #3: Base stacking

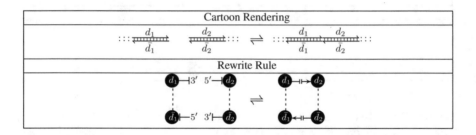

Rewrite Rule: Two double stranded DNA complexes, each comprised of two vertices with domains d_1, $\bar{d}_1$ and d_2, $\bar{d}_2$, respectively, are present. All the vertices are attached to 5′ or 3′-labeled vertices, where the vertex labeled d_1 is connected to a vertex labeled with a 3′, and d_2 to a vertex labeled 5′, etc. A base-stacking bond between them is formed. Since base-stacking may be assumed to be directionally independent [76], the directionality of the domains on each duplex relative to their neighbors do not really matter, so this rule can be extended to include all configurations of antiparallel duplex directionality.

L: Solid vertices with labels d_1 and $\bar{d}_1$ are connected by a hybridization edge. The vertex labels 3′ and 5′ show the directionality of each domain represented by the vertices. Similarly for d_2 and $\bar{d}_2$.

R: The two duplexes form a base-stacking bond with each other. This is represented by the edges between vertices labeled d_1 and d_2, and between $\bar{d}_2$ and $\bar{d}_1$.

Rule #4: Base stacking with overhangs

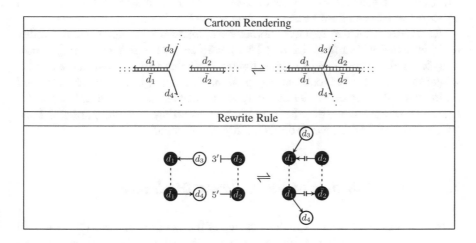

Rewrite Rule: Similar to rule #3, however, here one of the double stranded DNA complexes actually has overhangs (domains d_3, d_4) on both ends.

  L: Vertices with labels d_1 and $\bar{d}_1$ have covalently-labeled edges with d_3 and d_4, and the edge directions show the strands' 5' to 3' directionality. The vertices labeled d_2 and $\bar{d}_2$ explicitly show directionality, through the 5' and 3' labeled vertices connected to them via strand-end edges.

  R: The vertices with 5' and 3' labels are destroyed. Base stacking bonds are formed between d_2 and d_1. Likewise between $\bar{d}_1$ and $\bar{d}_2$.

Rule #5: Remote Toehold [23] mediated Strand Displacement

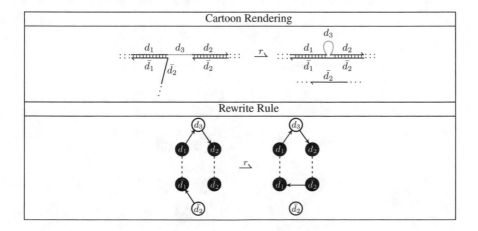

Rewrite Rule: Similar to rule #2, but the double-stranded DNA complex now has an extra unhybridized domain d_3, which separates the domains d_1 and d_2. The overhanging domain $\bar{d}_2$ has to locate the hybridized domain $\bar{d}_2$, and then displace it.

L: Vertex with domain d_1 is hybridized to $\bar{d}_1$, and d_2 to $\bar{d}_2$. The vertices labeled d_1, d_2, d_3 are connected via covalent bonds, and the d_3 domain is hollow-labeled to indicate that it is unhybridized.

R: A vertex labeled $\bar{d}_2$ replaces another vertex mapped to the same domain. Vertex $\bar{d}_2$ is relabeled with a hybridization label, and the other vertex labeled $\bar{d}_2$ is now unhybridized, hence it is relabeled as hollow. Note that variations of rule # 5 exist, where the separator between the two parts of the duplex is not a single-stranded region, but different structures that may act to separate both sides (a hairpin, for example), without preventing the reaction from occurring. In addition, rules # 2 and # 5 are also valid, with reverse directionality of the strands.

15.3.1 Distal Toehold Mediated Strand Displacement

In order to describe reactions in the last step of the example in Sect. 15.7, we need to use conditional graph rewriting rules. A conditional check is performed before the application of this rule, whether the two distal parts participating in this reaction are connected via hybridization and covalent bonds. In other words, the structure connecting the two parts is irrelevant, so long as it connects them (Fig. 15.5).

15.4 Correctness of Our DNA Graph Rewriting Systems

To prove the correctness of our DNA graph rewriting systems, or that every graph generated by the systems is a pseudo-DNA nanostructure, we need to prove two things: (1) the class of structures generated by our graph rewriting systems and the

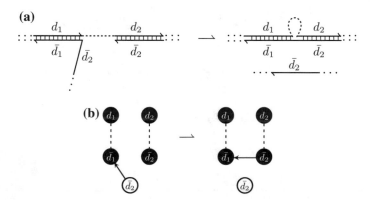

Fig. 15.5 a Cartoon rendering of distal toehold mediated strand displacement. **b** Distal toehold mediated strand displacement graph rewriting rule

class of well-formed DNA graphs **WFDG** are equivalent and (2) that **WFDG** and the class of pseudo-DNA nanostructures are equivalent.

The first part involves first proving a) that every DNA graph obtained by applying a rewrite rule to a well-formed DNA graph is also well-formed (Theorem 1), and (b) that any member of the class of well-formed DNA graphs can be generated by our graph rewriting systems, by demonstrating that every well-formed DNA graph can be obtained, given a well-formed input set of DNA strand graphs (Lemma 1), thus showing that our graph rewriting systems does not produce DNA graphs that do not belong to the class of well-formed DNA graphs.

Together, these two propositions (Theorem 1 and Lemma 1) show that, given a well-formed DNA graph, any DNA graph produced by our graph rewriting systems is also well-formed, and that every well-formed DNA graph may be produced by our graph rewriting systems, hence, the class of graphs produced by our graph rewriting systems is equivalent to the class of well-formed DNA graphs.

In the second part (Theorem 2), we show that the class of well-formed DNA graphs is equivalent to that of the class of pseudo-DNA nanostructures. This is done by proving (a) every pseudo-DNA nanostructure has a corresponding well-formed DNA graph (Lemma 2) and (b) every well-formed DNA graph represents a pseudo-DNA nanostructure (Lemma 3).

These two theorems demonstrate that the class of pseudo-DNA nanostructures and the class of graphs produced by our graph rewriting systems are equivalent.

Theorem 1 *For the rules in Sect. 15.3, any application of a DNA graph rewriting system rule on a well-formed DNA graph produces a well-formed DNA graph.*

Proof Using induction, we consider the following cases:

Single application
Let the initial graph G_0 denote an arbitrary, well-formed, DNA graph. For every rule, given a subgraph $L \subseteq G_0$ that matches the left-hand side, $r : L \to R$ transforms the subgraph to $R \subseteq G_1$.

- **Rule #1: Domain Binding** If rule #1 is applicable, then a single application of the rule to one pair of vertices transforms the subgraph in G_0 that matches the left-hand side of rule #1 to the right-hand side, which rewrites the labels of u, v to ●, and adds an edge between them with the label ...., resulting in a new graph G_1. G_1, which does not violate any well-formedness constraints:

  Constraint 1—Strand Ends: No such vertices are affected, because the rule does not apply to them.
  Constraint 2—Unhybridized Vertex Degree: Since these unhybridized vertices have not been transformed, then they remain unhybridized, where their in-degree is 1, and out-degree is 1.
  Constraint 3—Hybridized Vertex Degree: For vertices u, v, their degree has changed by adding one edge. Either it has increased from 2 to 3, 3 to 4 (in the case of a pre-existing base-stacking-labeled edge), or 4 to 5 (in the case

of two pre-existing base-stacking-labeled edges). For all other hybridized vertices, since they have have not been involved in the transformation, then their degrees remain the same.

Constraint 4—Vertex Edge Labels: For vertices u, v, one new edge has been added which has the label .... Since prior to the transformation no such edge existed, then exactly one edge label($e(u,v)$) = ··········. No other edges nor labels have been affected.

Constraint 5—Complementarity: The rule would only be applied to u, v if $m(u) = d \in \Sigma^+$ and $m(v) = \overline{d} \in \Sigma^+$, so this rule has not been violated.

Constraint 6—Directionality: If the nanostructure does not allow for circular strands, then the hybridization respects anti-parallel directionality.

- **Rule #2: Strand Displacement** If rule #2 is applicable, then the subgraph L consists of the vertices $v_1, v_2, v_3, v_4, v_5 \in V$ (starting from the unhybridized (hollow) vertex v_1 with domain d_2), and edges such that label($e(v_4, v_5)$) = ..., and label($e(v_1, v_2)$) = $\longrightarrow$. The rewriting rule destroys the edge $e(v_4, v_5)$, changes the label of v_5 to $\bigcirc$ and v_1 to $\bullet$, and adds a hybridization edge $e(v_1, v_4)$, which does not violate any of the well-formedness constraints.

  Constraint 1—Strand Ends: No such vertices are affected, because the rule does not apply to them.

  Constraint 2—Unhybridized Vertex Degree: Except for v_1 (displacing domain d_2), unhybridized vertices have not been transformed and remain unhybridized, where their in-degree is 1, and out-degree is 1. v_5 has become unhybridized. An edge between v_4, v_5 has been removed, so the vertex degree has been lessened by 1, which means that its degree has changed from 3 to 2, 4 to 3 (if there is a pre-existing base-stacking-labeled edge), or 5 to 4 (in case of two pre-existing base-stacking labeled edges). In addition, v_5 has been relabeled to $\bigcirc$.

  Constraint 3—Hybridized Vertex Degree: For vertex v_1, which has become hybridized, its degree has changed by adding one edge. Either it has increased from 2 to 3, 3 to 4 (in the case of a pre-existing base-stacking-labeled edge), or 4 to 5 (in the case of two pre-existing base-stacking-labeled edges). For all other hybridized vertices, since they have have not been involved in the transformation, then their degrees remain the same.

  Constraint 4—Vertex Edge Labels: For vertices v_1, v_4, one new edge has been added which has the label .... Since prior to the transformation no such edge existed, then exactly one edge label($e(v_1,v_4)$) = ··········. One edge $e(v_4, v_5)$ with label .... has been removed.

  Constraint 5—Complementarity: The rule would only be applied if $m(v_1) = m(v_5) = d_2 \in \Sigma^+$ and $m(v_4) = \overline{d_2} \in \Sigma^+$, so this rule has not been violated.

  Constraint 6—Directionality: If the nanostructure does not allow for circular strands, then the hybridization respects anti-parallel directionality.

- **Rule #3: Base Stacking** If rule #3 is applicable, then the subgraph $L \subseteq G_0$ consists of the vertices $v_1, v_2, v_3, v_4 \in V$ labeled ● with domains $d_1, \bar{d}_1, d_2, \bar{d}_2$, respectively. Each of these vertices are connected to u_1, u_2, u_3, u_4 in the same order, with vertex labels $3', 5', 5', 3'$, respectively. There are no conditions on the directionality of the strands. The subgraph L undergoes the removal of the edges between each pair $e(u_i, v_i)$, and the vertices u_i, which violates no well-formedness constraints. Then, v_1 and v_2 are connected via a base-stacking edge, labeled ⟶⊢⟶, and likewise for v_3 and v_4. No well-formedness constraints are violated in G_1.

  Constraint 1—Strand Ends: The vertices u_1, u_2, u_3, u_4 are destroyed.
  Constraint 2—Unhybridized Vertex Degree: These vertices are not affected.
  Constraint 3—Hybridized Vertex Degree: Vertices v_1, v_6 have had an edge added with label ⟶⊢⟶.
  Constraint 4—Vertex Edge Labels: No vertices are relabeled.
  Constraint 5—Complementarity: No hybridization reactions have occurred.
  Constraint 6—Directionality: The directionality of the strands are maintained in the direction of the base-stacking arrows.

- **Rule #4: Base Stacking with Overhangs** If rule #4 is applicable, then the subgraph $L \subseteq G_0$ consists of the vertices v_1 through v_6 and u_1, u_2. v_1, v_6 (labeled d_1 and $\bar{d}_1$) are labeled ●, which are connected to v_2, v_5 (labeled d_3 and d_4), respectively. In addition, vertices v_3, v_4 (labeled $d_2, \bar{d}_2$) are connected to u_1, u_2, which are $5'$ and $3'$ labeled vertices, respectively. It is assumed that there are no conditions on the directionality of the strands. The subgraph L undergoes the destruction of the vertices u_1, u_2, and the edges between each pair $e(u_i, v_j)$. Then, v_1 and v_3 are connected via a base-stacking edge, labeled ⟶⊢⟶, and likewise for v_4 and v_6, which does not violate any well-formedness in G_1.

  Constraint 1—Strand Ends: Two strand-end vertices (u_1, u_2) are destroyed.
  Constraint 2—Unhybridized Vertex Degree: Since these unhybridized vertices have not been transformed, then they remain unhybridized, where their in-degree is 1, and out-degree is 1.
  Constraint 3—Hybridized Vertex Degree: Since these hybridized vertices have not been transformed, then they remain hybridized.
  Constraint 4—Vertex Edge Labels: For vertices v_1, v_3 (and v_4, v_6), one new edge has been added which has the label ⟶⊢⟶. Since prior to the transformation no such edge existed, then exactly one edge ⟶⊢⟶.
  Constraint 5—Complementarity: The rule does not affect the hybridization state of any domain, so this constraint remains unviolated.
  Constraint 6—Directionality: The rule does not affect the hybridization state of any domain, so directionality is not violated. Directionality of the duplexes relative to each other does not matter [76].

- **Rule #5: Remote Toehold mediated Strand Displacement** If rule #5 is applicable, then the subgraph $L \subseteq G_0$ consists of the vertices $v_1, v_3, v_4, v_6 \in V$ labeled ● (with domains $d_1, d_2, \bar{d_2}, \bar{d_1}$), and v_2, v_5 are labeled ○ (with domains $d_3, \bar{d_2}$, respectively). v_5 is connected covalently to v_6, and v_5 is labeled as unhybridized. Between v_1, v_3, the unhybridized vertex v_2 resides. As the rule is applied, the edge $e(v_3, v_4)$ is destroyed, and vertex v_4 is relabeled as an unhybridized vertex, and v_5 is then connected to v_3 via a hybridization-labeled edge; these operations violate no well-formedness constraints in G_1.

  Constraint 1—Strand Ends: No such vertices are affected, because the rule does not apply to them.

  Constraint 2—Unhybridized Vertex Degree: Vertex v_6 (domain $\bar{d_2}$) has changed from hybridized to unhybridized (and relabeled to ○). An edge $e(v_5, v_6)$ was removed, so the degree of v_6 has decreased by 1. All other unhybridized vertices have not been transformed and remain unhybridized, where their in-degree is 1, and out-degree is 1.

  Constraint 3—Hybridized Vertex Degree: Vertex v_1 (domain $\bar{d_2}$) has changed from unhybridized to hybridized. An edge $e(v_1, v_5)$ has been added. For vertices u, v, their degree has changed by adding one edge. Either it has increased from 2 to 3, 3 to 4 (in the case of a pre-existing base-stacking-labeled edge), or 4 to 5 (in the case of two pre-existing base-stacking-labeled edges). For all other hybridized vertices, since they have have not been involved in the transformation, then their degrees remain the same.

  Constraint 4—Vertex Edge Labels: For vertices u, v, one new edge has been added which has the label .... Since prior to the transformation no such edge existed, then exactly one edge $\mathsf{label}(e(e,v)) = $ ----------. No other edges nor labels have been affected.

  Constraint 5—Complementarity: The rule would only be applied to u, v if $m(u) = d \in \Sigma^+$ and $m(v) = \bar{d} \in \Sigma^+$, so this rule has not been violated.

  Constraint 6—Directionality: If the nanostructure does not allow for circular strands (which in the current framework we have decided to exclude to ensure this rule's consistency), then the hybridization respects anti-parallel directionality.

Using induction, we have proven that each rewrite rule of our graph rewriting systems generates a well-formed DNA graph, given a well-formed DNA graph as input. We next show that the set of well-formed DNA graphs, has a corresponding member in the set of (tertiary structure-abstracted) DNA nanostructures, or pseudo-DNA Nanostructures. This will be accomplished through Lemmas 1–3, which follow. In Lemma 1, we start by proving that every member W in the class **WFDG** can be generated using the basic set of graph rewriting rules in Sect. 15.3, by starting from the simplest version of W, a DNA strand graph which corresponds to this well-formed DNA strand graph W.

Lemma 1 *Every* $W \in$ **WFDG** *can be produced by a sequence of non-enzymatic rule applications on its corresponding input set of well-formed DNA strand graphs.*

Proof Consider any well-formed DNA graph w. A well-formed DNA graph consists of four types of edge labels, as mentioned in Sect. 15.2.2: base-stacking, hybridization, covalent and strand-ends. Of these, consider only the base-stacking edge set b and the hybridization edge set h. For this $w \in$ **WFDG**, construct a graph w' such that $w' = w \setminus \{b, h\}$. This results in a set of DNA strands represented by disconnected subgraph components within w'. We consider w' as the input set of DNA strand graphs. Clearly, $w' \in$ **WFDG**. Applying rule # 1 on w' yields a new graph sg_1. To keep track of states, we can store the graphs in a state graph (called SG), where the root is represented by w'. A child node represents the result of applying a rule to a parent strand graph $sg_i \in SG$. Subsequent applications of rule # 1, # 3, # 4, which have matching subgraphs in the current sg_i, yields up to two child strand graphs each, until there are no more possible applications. We ignore rules # 2 and # 5, because they result in graphs which are isomorphic to the ones made by the other rules. Let $w'_{i'} \in SG$ represent a leaf node of SG. Assuming that there are no isomorphic strand graphs, we claim that at least one leaf node w is equivalent to either a $w'_{i'}$, or any sg_i in SG. That is because there exists at least one graph $sg_i \in SG$ which is isomorphic to w. Hence, by some sequence of non-enzymatic rule applications on a set of input DNA strand graphs, we have constructed a well-formed DNA graph. The same holds for any well-formed DNA graph **WFDG**.

Lemma 2 *Every* **PDN** *can be represented by a well-formed DNA graph. The algorithm in Algorithm 1 does not violate any* **WFDG** *constraints.*

Proof Given an input **PDN**, $p = (\mathscr{D}, \mathscr{S}, \mathscr{H}, \mathscr{B})$, we first start by constructing all the strand graphs. For each strand starting from the $5'$ vertices, an $\bigcirc$-labeled vertex is constructed for each domain until no more are encountered, and a $3'$ vertex is created to terminate the strand. Once all the strand graphs are created, incorporate the hybridization and base-stacking edges ($\cdots\cdots$, $\longrightarrow\!\mapsto$), and relabel the hybridized vertices with $\bullet$.

We give an algorithm (Algorithm 1) that constructs a graph W. For each strand, only one $5'$ and one $3'$ vertex is created. This ensures constraint 1 is valid. Lines 2–15 ensure constraints 2 and 6 are valid. Lines 19–26 add a single hybridization edge to all hybridized vertices, while lines 28–35 can add at most two base-stacking edges to each vertex. This ensures constraints 3 and 4 are valid. Line 23 ensures constraint 5 is valid.

Lemma 3 *Every well-formed DNA graph in* **WFDG** *has a corresponding* **PDN**.

Proof Given an input well-formed DNA graph $W = (V, E, L_v, L_e, \Sigma, vl, el, m) \in$ **WFDG**. We give an algorithm that constructs a **PDN** $p = (\mathscr{D}, \mathscr{S}, \mathscr{H}, \mathscr{B})$ in Algorithm 2. The pseudocode in Algorithm 2 constructs a valid **PDN** given a well-formed DNA graph. A valid **PDN** contains a set of DNA strands ($\mathscr{S}$), the set of domains ($\mathscr{D}$) that hybridize ($\mathscr{H}$), and the set of domains that base-stack ($\mathscr{B}$). Lines 2–15 create a set of domains $\mathscr{D}$ and strands $\mathscr{S}$. Each strand can have only one 5' and one 3'-end. Lines 16–20 create a set of hybridizations $\mathscr{H}$. By constraint 4, there can be at most 1 hybridization per vertex. Lines 21–30 create a set $\mathscr{B}$ of base stacking tuples. By constraint 4, there can be at most two base-stackings per vertex. The ends of the vertex signify whether the 5' or the 3' end of the vertex base stacks with another edge. Each of the well-formedness constraints ensures that a **PDN** is created.

Algorithm 1 Construct $W \in$ **WFDG** from input $p = (\mathscr{D}, \mathscr{S}, \mathscr{H}, \mathscr{B})$

```
 1: procedure CONSTRUCTW(p = (𝒟,𝒮,ℋ,ℬ))
 2:     for s ∈ S do
 3:     ▷ Convert strands into graphs
 4:         construct a 5' vertex u
 5:         construct a vertex v for the first domain in strand s
 6:         construct edge e(u,v) labeled ⊣
 7:         previous_vertex ← v
 8:         for the next domain d in strand s (5' to 3') do
 9:             if ( then∄ a domain after d)
10:                 Break
11:             end if
12:             construct a vertex w for d with label ◯.
13:             create an edge (previous_vertex,w) labeled ⟶
14:             previous_vertex ← w
15:             MAP(s,i) = w ▷ create a map, given a strand s and location i that returns vertex w
16:         end for
17:         construct a 3' vertex x
18:         create an edge (previous_vertex,x) labelled ⊣
19:     end for
20:     for each tuple h ∈ H do
21:     ▷ Create hybridization edge, where h of the form (s₁,i,s₂,j)
22:         if v₁ and v₂ have complementary sequences then
23:             v₁ ← LOOKUP(s₁,i)
24:     ▷ Lookup vertex given a strand and location
25:             v₂ ← LOOKUP(s₂,j)
26:             create a hybridization edge e(v₁,v₂);
27:             relabel v₁ and v₂ to ●
28:         end if
29:     end for
30: ▷ Create base-stacking edges b of the form ((s₁,i₁,e₁,s₂,i₂,e₂),(s₃,i₃,e₃,s₄,i₄,e₄))
31:     for each b ∈ B do
32:         v₁ ← LOOKUP(s₁,i₁)
33:         v₂ ← LOOKUP(s₂,i₂)
34:         v₃ ← LOOKUP(s₃,i₃)
35:         v₄ ← LOOKUP(s₄,i₄)
36:         create a base-stacking edge e(v₁,v₂) (with label ⊣↦)
37:         create a base-stacking edge e(v₃,v₄) (with label ⊣↦)
38:     ▷ The edge directionality depends on the order of the tuples of b
39:     end for
40: end procedure
```

Algorithm 2 Construct a $p \in$ **PDN**, given a well-formed DNA graph $W \in$ **WFDG**

1: **procedure** CONSTRUCTPDN(W)
2:    **for** each vertex $v \in V$ s.t. $vl(v) = 5'$ **do**
3:
4:       ▷ construct strand s **DFS**(vertex, el) and returns all vertices linked to vertex v transitively via edge label el, and forms a strand s
5:          $s \leftarrow$ DFS(v, covalent) CREATEMAP(s)
6:
7:       ▷ create a map that maps all vertices v in s to their relative positions in s. Hence, **map**(v_1) returns strand s and its position i.
8:       CREATECOMPLEMENT(s)
9:
10:      ▷ create complement each vertex v on the same strand in s, and maps its position its complement. Hence COMPLEMENT(s_1, i_1) gives (s_2, i_2). This function uses the map created on line 6
11:         insert s into $\mathscr{S}$
12:         $d \leftarrow$ domain(v)
13:         **if** $d \notin \mathscr{D}$ **then**
14:            add d to $\mathscr{D}$;
15:         **end if**
16:    **end for**
17:    DFS_CIRCULAR(V)
18:    ▷ We may simply visit the rest of the vertices to identify and construct circular strands
19:    **for** each edge $e(v_1, v_2) \in E$ such that $el(e)$ is a hybridization edge **do**
20:    ▷ constructing $\mathscr{H}$
21:         $s_1, i_1 \leftarrow$ LOOKUP(v_1)
22:         $s_2, i_2 \leftarrow$ COMPLEMENT(s_1, i_1)
23:         add tuple (s_1, i_1, s_2, i_2) to $\mathscr{H}$
24:    **end for**
25:    **for** For each edge $e \in E$ such that el is a base stacking edge **do**
26:    ▷ constructing $\mathscr{B}$
27:         e is ($v_1, end1, v_2, end2$)
28:         $s_1, i_1 \leftarrow$ LOOKUP(v_1)
29:         $s_2, i_2 \leftarrow$ LOOKUP(v_2)
30:         COMPLEMENT(s_1, i_1) = s_3, i_3
31:         COMPLEMENT(s_2, i_2) = s_4, i_4
32:         **if** ($s3, i3, s4, i4$) have a base stacking edge between them **then**
33:
34:         ▷ end'_1 is the opposite of end_1 e.g. if end_1 is $5'$, then end'_1 is $3'$
35:            add tuple (($s_1, i_1, end_1, s_2, i_2, end_2$), ($s_3, i_3, end'_1, s_4, i_4, end'_2$)) to B
36:         **end if**
37:    **end for**
38: **end procedure**

Theorem 2 *Both classes* **WFDG** *and* **PDN** *are equivalent: A member of* **PDN** *exists if and only if it has one corresponding member of* **WFDG***.*

In Lemma 2 we have demonstrated that a pseudo-DNA nanostructure can be transformed into a well-formed DNA graph by constructing vertices and edges for each strand out of the strand domains, and hybridization edges between vertices out of the set of hybridizations, and finally base-stacking edges out of the set of base-stacking interactions, while maintaining the constraints that define the class of well-formed DNA graphs. In Lemma 3, we have demonstrated that every member of **WFDG** has a respective member belonging to **PDN**. These three propositions prove that the classes **WFDG** and **PDN** are equivalent. Following that, **PDN** and the class of structures generated by our grammar are equivalent.

15.5   Example: Yurke et al. DNA Tweezer

Table 15.1 illustrates one possible reaction sequence of the DNA tweezer nanomachine, developed by Yurke et al. [82], using DNA graph rewrite rules. In the beginning, the tweezer consists of two duplex DNA arms with toehold domains at both
ends (i.e., d_2 and d_3). $\bar{d}_3$ and $\bar{d}_2$ domains from the DNA fuel strand undergo toehold
binding (rule # 1) with d_3 and d_2 of the tweezer, respectively. As a result of this
process, the set strand pinches two arms of the tweezer together and results in the
"closed" state with a remaining exposed toehold domain d_5. To open the tweezer, $\bar{d}_5$
domain of the unset strand $\bar{d}_5, \bar{d}_3, \bar{d}_2$ binds to domain d_5 of the closed-state tweezers.
To completely displace the strand $\bar{d}_2, \bar{d}_3, d_5$ the strand displacement rewrite rule (rule
2) is applied and results in waste duplex. As a result of this process the original
tweezer state is restored. Note this is one path of many possible rule-application
sequences. This is also shown in the corresponding states of this path generated by
DAGRS in Fig. 15.21.

15.6   Brief Description of Our Prototype Software System

We implemented a prototype that generates all the states possible depending on a
subset of the basic rules defined in Sect. 15.3. Choosing a subset of the states, we
display one specific sequence of reactions in Table 15.1. The DNA Graph Rewriting
System (DAGRS) is a simple prototype implemented in Python 2.7 [57] that uses
the graph-tool library [53] to parse, store and apply graph rewriting rules, and uses
the [73] subgraph isomorphism algorithm to match the set of left-hand side of the
graph rewriting rules to some subgraph in a given host DNA graph G. PyQt5 [63]
was used to render the states, which were explored in a breadth-first search manner.
This is implemented as follows: a list is maintained of states (E) that have been fully
explored (where the given set of rules may no longer be applied to these states), and a
queue of states currently being explored Q. The program attempts to find a matching
for the first rule supplied by the user on a state removed from the queue (which
is some well-formed DNA graph s_i). If no such matching exists, this is repeated
with the next rule in the set. If no matching can be found, the program terminates.
If a matching is found, a new state is generated, where the matched subgraph is
transformed into the right-hand side of the rule applied. This state is then added to
Q, and the next rule is applied to the current state being explored (s_i). The prototype
is able to generate the landscape of states depending on the input species, and the
subset of rules supplied. We demonstrated that it is able to accommodate simple
systems, including the previous example in Sect. 15.5, as well as further examples
mentioned in Sect. 15.7. Depending on the rewrite rules used, all possible applications

Table 15.1 One sequence of rule-applications representing one possible reaction pathway

Cartoon rendering	Intermediate graph G_i	Matching rule
		Initial input species (G_0) has matching subgraph consisting of vertices with domains d_3 and $\bar{d}_3$, which matches the LHS of Rule #1.
		Rule #1
		Rule #1
		Rule #2
		Rule #2

(continued)

Table 15.1 (continued)

Cartoon rendering	Intermediate graph G_i	Matching rule
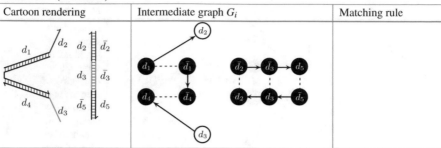		

are considered by using a basic breadth-first exploration of the state-space, starting from an input state. To select the next rule to apply, a subgraph isomorphism test is performed on the current state graph G_i to check for the existence of some subgraph sg_j matching the left-hand side of any rewrite rule in a given rule-set. There is the possibility of not generating a large, even exponential, number of structures, if graph rewriting were the only method applied. However, one can restrain the number of generated states depending on other restrictions that are not currently handled, such as domain length (as discussed in the Conclusion section). In essence, the current system does not handle constraining the number of states. However, to show that this graph-centered paradigm can be utilized with at least one alternative way of collapsing the number of states, we applied the results from the work by Grun et al. [26] to an example in Sect. 15.9. Reactions are divided into transient (fast) and non-transient (slow), which has the advantage of decreasing the state-space size.

15.7 Examples of Non-enzymic Devices

15.7.1 Catalytic Hairpin-Based Triggered Branched Junction

Figure 15.7 shows a detailed sequence of graph rewrite rule applications, which can be applied in succession to produce a 3-arm junction. From the cartoon representation of the 3-arm junction in Fig. 15.6, we see the sequence of steps that are involved in the catalytic creation of the DNA complex.

We start with a set of 4 DNA complexes, 3 hairpins A, B, C and an initiator I. Initiator I and hairpin A hybridize via a toehold a, followed by subsequent strand displacement to give complex $A + I$ (Figs. 15.6 and 15.7). On strand displacement of hairpin A, the stem loop is opened, exposing domain $\bar{b}$, and allowing it to react

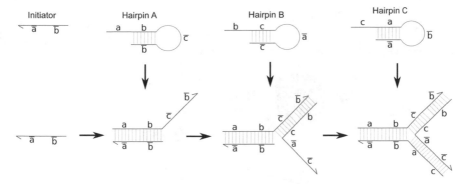

Fig. 15.6 Catalytic hairpin-based trigger branched junction

with hairpin B. Again, this is followed by stem loop opening of hairpin B, forming complex $A + B + I$. Domain $\bar{c}$ is now exposed, which hybridizes with hairpin C and subsequently opens the stem loop.

Note that a region of hairpin C, namely $\bar{b}\bar{a}$ is complementary to the arm ab of the original hairpin A. By a process of remote-toehold mediated strand displacement [23], the domain $\bar{b}\bar{a}$ displaces the attached initiator, creating the final 3-arm junction $A + B + C$, and releasing initiator I, which is free to catalyze the formation of another structure $A + B + C$ (Figs. 15.9 and 15.10).

15.7.2 Qian and Winfree's Seesaw Gate

Figure 15.11 shows a sequence of rules that have been applied for the see-saw system, via which [59] developed a circuit that can compute the square root of a fixed integer. We use two graph rewrite rules in the simulation of this complex: domain binding and strand displacement. The rules can also be used to show the leaks, both fuel-gate and gate-gate, if the domains are segmented further to represent shorter sequences.

The Qian and Winfree [59] system shown above contains four single-stranded DNA: input (S_2TS_1), output (S_3TS_2), fuel (S_4TS_2), and gate ($\bar{T}\bar{S_2}\bar{T}$). We start with a set of three DNA complexes, two single-stranded DNA: input and fuel, and one double-stranded DNA: gate-output complex as shown above. The input and gate-complex hybridize via a toehold (rule #1), followed by strand displacement of the domain S_2 (rule #2). Following this, toehold T dehybridizes, releasing the output strand. A fuel strand reacts with the new input-gate complex, and the same reactions as above take place, only in the reverse order. The input strand is released, and is free to catalyze the release of another output strand.

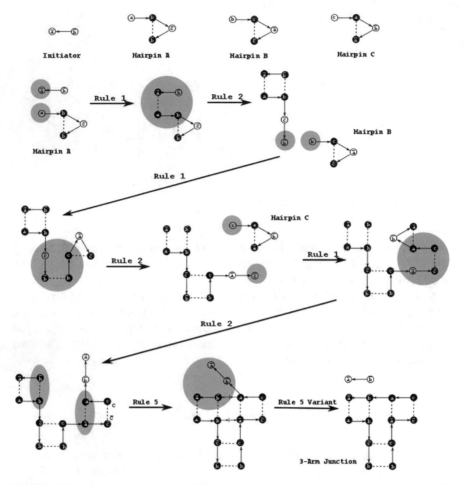

Fig. 15.7 The figure shows a sequence of graph rewrite rules that can be applied in succession. We obtain the 3-arm junction in the above design by Yin et al. [77]. Note that the application of rule #5 here includes the variation of the original (see Sect. 15.3.1)

15.7.3 Hybridization Chain Reaction

Figure 15.12 shows how to apply the rewrite rules to basic HCR system designed by Dirks and Pierce [17]. We use two graph rewrite rules: toehold binding and strand displacement.

A description of basic HCR system as shown in Fig. 15.13: there are two types of hairpins ($H1$ and $H2$) and one type of single strand (I) in this system. The reaction starts with hybridization between $\bar{a}$ domain of I and a domain of $H1$ (rule #1).

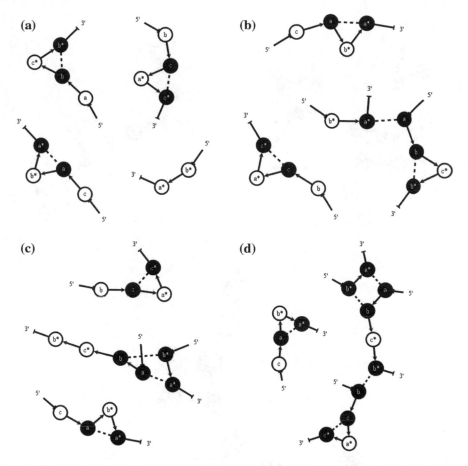

Fig. 15.8 a–g A subset of the sequence of states generated by our DNA Graph Rewriting System (DAGRS)

Hairpin $H1$ is opened by initiator I after toehold binding (rule #1). c domain of $I * H1$ hybridizes with $\bar{c}$ domain of hairpin $H2$ (rule #1). Hairpin $H2$ is opened by $I * H1$ after toehold binding (rule #2). The $\overline{ba}$ domain of $I * H1 * H2$ will be the initiator in the next cycle (Fig. 15.15).

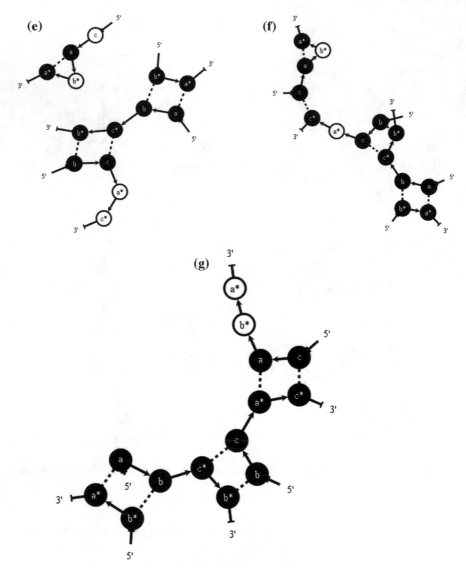

Fig. 15.8 (continued)

15.8 Enzymatic DNA Graph Rewriting Rules

These rules differ from the non-enzymatic ones defined in Sect. 15.3 in that they require a regular expression, which represents a restriction site, to be matched on one or both strands (depending on the enzyme). This is why each rule is separate than its perceived reverse reaction. These rules have not been implemented in DAGRS.

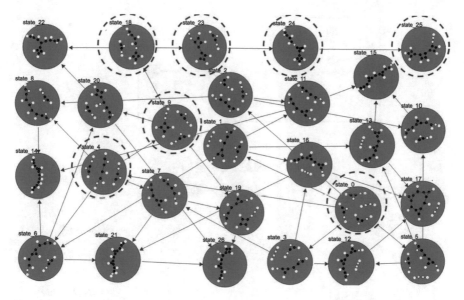

Fig. 15.9 The set of states generated by the graph rewriting rules given the input species. There are 26 states. Those circled in red (state_0) and blue mark the chosen subset of states in Fig. 15.8

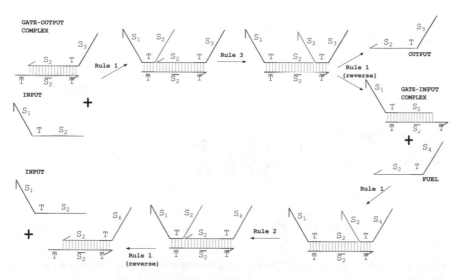

Fig. 15.10 Cartoon rendering of seesaw gate system

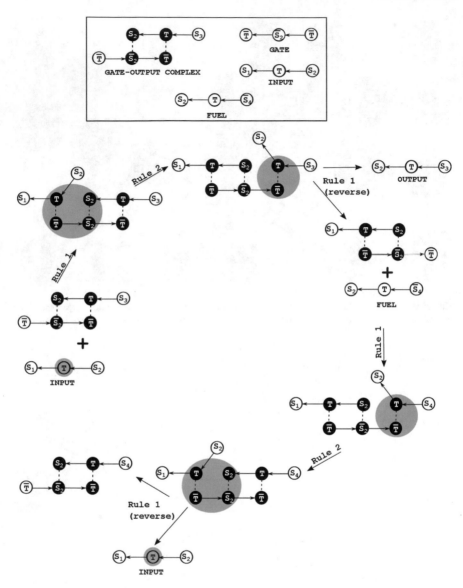

Fig. 15.11 This shows a sequence of graph rewriting rules that can be applied in succession, for the DNA circuit system based on see-saw gates. The input strand in the above system acts as a catalyst, and helps in releasing the output strand, via help from a fuel strand [59]

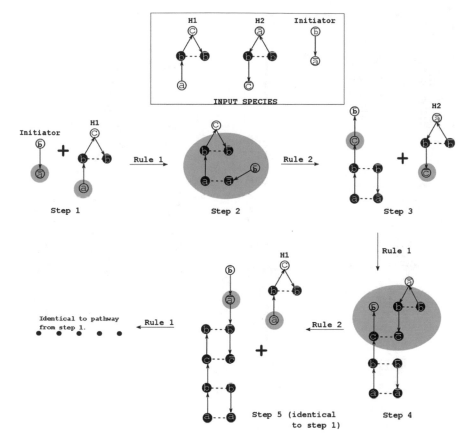

Fig. 15.12 The figure shows how to apply our graph rewriting systems to the basic HCR system designed by Dirks and Pierce [17]

Rule #6: *Restriction Enzyme Cutting (Overhang formation)*

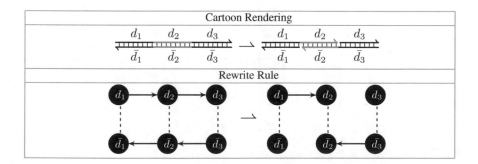

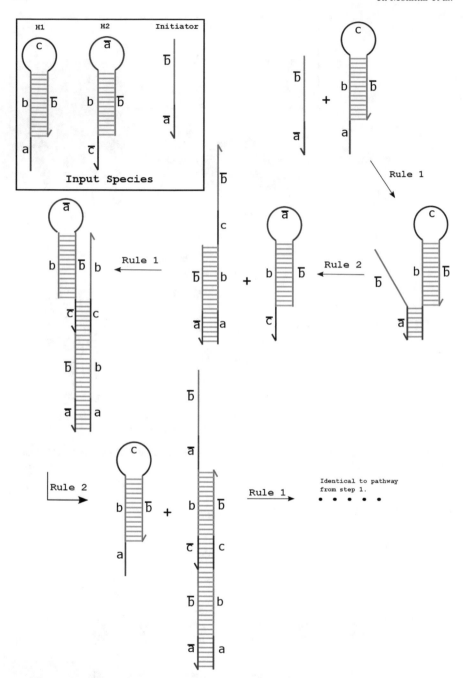

Fig. 15.13 Depiction of HCR system reaction

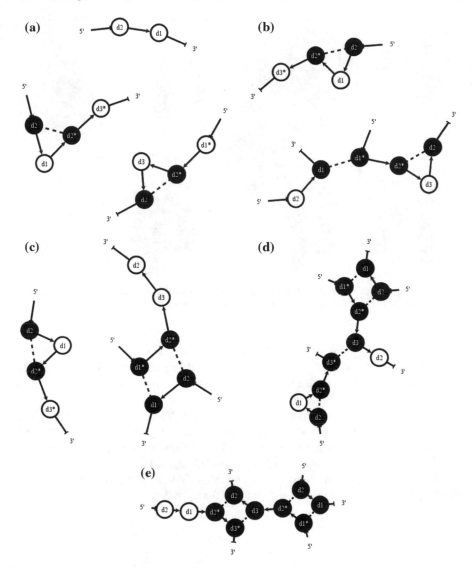

Fig. 15.14 A subset of the states generated by DAGRS, in sequence

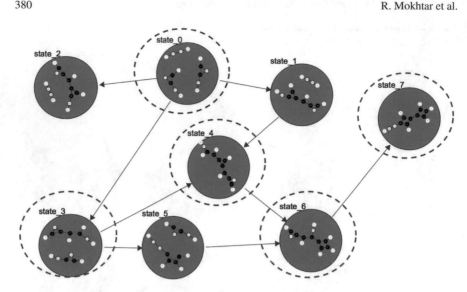

Fig. 15.15 States generated by DAGRS given the subset of rules supplied. Those circled in *red* (state_0) and *blue* mark the chosen subset of states in Fig. 15.14

Rewrite Rule: A double-stranded DNA complex is made up of two strands with domains d_1, d_2, d_3 and $\bar{d}_1, \bar{d}_2, \bar{d}_3$. A restriction enzyme recognizes the restriction site d_2, and cuts the strands at different base pair locations. Strand d_1, d_2, d_3 is cut after d_2, and strand $\bar{d}_1, \bar{d}_2, \bar{d}_3$ after $\bar{d}_2$.

L: Each domain is represented by a hybridization-labeled vertex. The hybridization-labeled edges show hydrogen bonding. The edges between the vertices labeled d_1, d_2, d_3, which have covalent labels, show directionality and covalent bonding between the domains (and likewise for $\bar{d}_3, \bar{d}_2, \bar{d}_1$).

R: The edge between vertices with domains d_2 and d_3 are replaced by base-stacking labeled edges, and likewise for vertices $\bar{d}_2$ and $\bar{d}_1$.

Rule #7: *Restriction Enzyme Cutting (Blunt-end formation)*

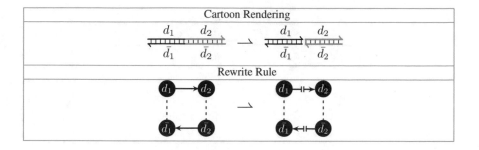

Rewrite Rule: A double-stranded DNA complex is made up of two strands with domains d_1, d_2 and $\bar{d}_1, \bar{d}_2$. A restriction enzyme recognizes the restriction site d_1, and cuts the strands at the same base pair location.

L: Each domain d_1, d_2 is represented by solid vertices. Solid edges show directionality and covalent bonding, while the dotted edges show hydrogen bonding.

R: The covalently-labeled edges are relabeled to base-stacking edges.

Rule #8: *Polymerization*

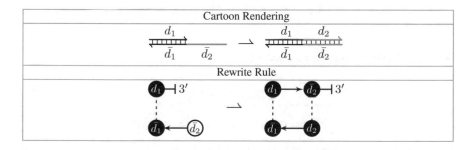

Rewrite Rule: A double-stranded DNA made up of a vertex with domain d_1 hybridized to $\bar{d}_1$, which is covalently bound to vertex $\bar{d}_2$ (which is unhybridized). The $3'$ vertex is connected to d_1 and indicates the end of the strand. A DNA polymerase extends the $3'$ end that is connected to d_1, and forms a complementary domain to vertex $\bar{d}_2$.

L: Hybridized vertices d_1 and $\bar{d}_1$, and hollow vertex $\bar{d}_2$. A $3'$ vertex is connected to d_1.

R: A new vertex with domain d_2 is created, with domain complementary to $\bar{d}_2$. It is connected via a covalent edge with d_1, with the same directionality (the edge is pointing towards vertex d_1). The edge between vertex d_1 and the $3'$ vertex is removed, and replaced with an edge from vertex d_1.

Rule #9: *Strand-displacement via polymerization*

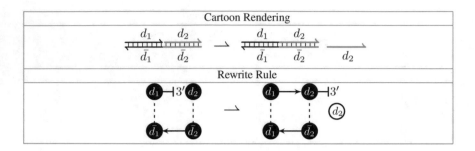

Rewrite Rule: A region of double-stranded DNA made up of three strands, comprised of four vertices with domains d_1, d_2 and on the third strand, $\bar{d}_2\bar{d}_1$. The domains d_1 and d_2 are both hybridized to $\bar{d}_2\bar{d}_1$. A DNA polymerase extends the $3'$; end of vertex v_1, resulting in the domain d_2 which is complementary to $\bar{d}_2$. In the process, it displaces the existing vertex v_2 (with domain d_2).

L: Hybridized v_1 and v_4, each connected via hybridization edges to v_5 and v_3. In addition, vertices v_1 and v_2 have a base-stacking edge between them.

R: A new vertex v_5 with domain complementary to vertex v_3 (d_2) is added. It is connected via a covalently-labeled edge with v_1, and it is connected to vertex v_4 via a hybridization-labeled edge. The existing vertex u now becomes a hollow vertex outside the complex.

Rule #10: *Restriction Enzyme Nicking*

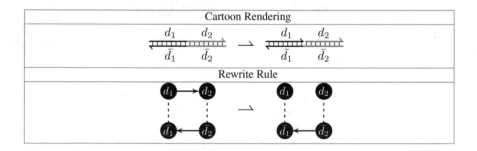

Rewrite Rule: A double-stranded DNA region made of two strands, with domains d_1, d_2 and $\bar{d}_2, \bar{d}_1$. A restriction enzyme creates a "nick" at the base pair location between domains d_1 and d_2, which is represented by the removal of the edge between domains d_1 and d_2.

    L: The edge between domains d_1, d_2 is a covalently-labeled edge.

    R: The edge between the vertices labeled d_1, d_2 is removed.

Rewrite Rule: A double stranded DNA region made of three strands. These are comprised of four vertices, with domains d_1, d_2 and $\bar{d}_2, \bar{d}_1$. A ligase joins the domains d_1 and d_2, by adding a covalently-labeled edge between the vertices with labels d_1 and d_2.

    L: The edge between vertices labeled d_1, d_2 is removed.

    R: The edge between vertices labeled d_1, d_2 is a covalently-labeled edge.

Rule #11: *Single-stranded Ligation*

Cartoon Rendering
(diagram)
Rewrite Rule
(diagram)

Rule #12: *Blunt end Ligation*

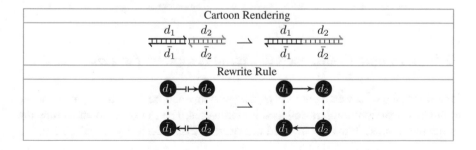

Cartoon Rendering
(diagram)
Rewrite Rule
(diagram)

Rewrite Rule: Two double-stranded DNA regions, made up of four vertices, with domains $d_1, d_2, \bar{d}_2, \bar{d}_1$. A base-stacking labeled edge exists between the two vertices with domains d_1 and d_2. A ligase creates two covalent bonds at the same location on both strands, connecting the vertices with domains d_1 and d_2 together, and likewise for $\bar{d}_2$ and $\bar{d}_1$. This results in a single double-stranded DNA complex.

L: The edges with domains d_1 and d_2, as well as $\bar{d}_2$ and $\bar{d}_1$ are base-stacking labeled edges.

R: The edges are relabeled as covalent.

Rewrite Rule: A double-stranded DNA region made of 4 strands, which are represented by 6 vertices, with domains d_1, d_2, d_3 and $\bar{d}_3, \bar{d}_2, \bar{d}_1$. A ligase creates two covalent bonds, at different locations on both strands. This relabels the edges between vertices with domains d_2 and d_3 together, as well as $\bar{d}_2$ and $\bar{d}_1$. This results in a double-stranded DNA region.

L: The edges between the vertices with domains d_2 and d_3, as well as $\bar{d}_2$ and $\bar{d}_1$ are base-stacking labeled edges.

R: The edges are relabeled to be covalent.

Rule #13: *Sticky-end Ligation*

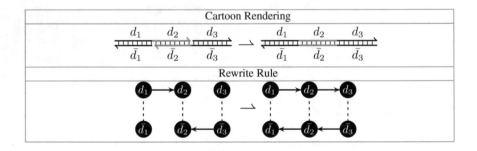

15.9   Example Condensed Reaction Graph of Tweezer

Below we show how our graph rewriting systems can be used to supplement the condensed reaction graph introduced by Grun et al. [26], in order to show how the graph rewriting systems can be used to extend it (Figs. 15.16, 15.17 and 15.18).

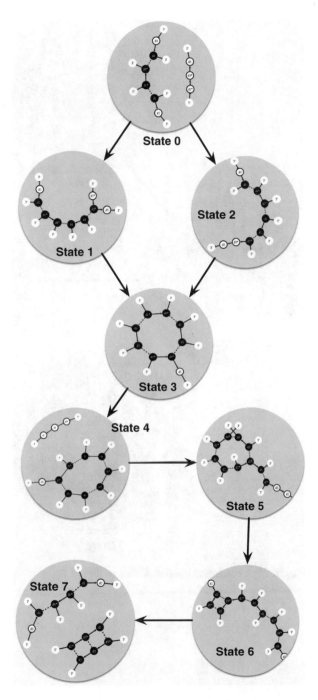

Fig. 15.16 Under the definitions in [26] States 0, 4, and 7 would be considered resting states and states 1, 2, 3, 5, and 6 would be considered transient

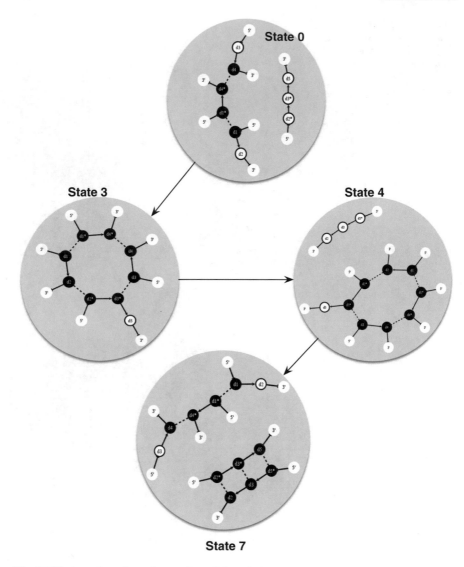

Fig. 15.17 A condensed reaction version of Fig. 15.16

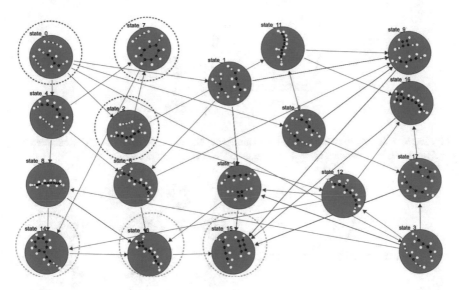

Fig. 15.18 The states circled in blue lead up to the closed state of the tweezer. Those marked in green lead up to the opened state

15.10   An Explanation of Constraint 3

We required that $3 \leq deg(v) \leq 5$ (on P. XXX) because there is a limited number of possible combinations of edges around a hybridized vertex that are accepted in a member of **WFDG**. Specifically:

1. v_i will be connected to another vertex via a hybridization edge (1)
2. If v_i is a domain on the 3′ or 5′ ends, then it will connect to either a 5′ or 3′ vertex via a ⊣ edge. ($j = 0, j = 1$ or $j = 2$)
3. If v_i is a domain that is on the 3′ or 5′ ends, but base-stacked against another strand's 3′ or 5′ end, then instead of the previous edge, it will be connected via a ⊣↦ edge to this domain ($k = |2 - j|$, so $k = 0, k = 1$ or $k = 2$)
4. If v_i is adjacent to another domain on its strand, then it will be connected via ⟶ ($i = 0, i = 1$, or $i = 2$).

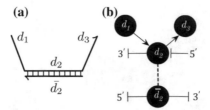

Fig. 15.19 Example showing the minimum number of edges around vertex d_2. **a** Cartoon rendering of 3 duplexes, base-stacked against each other. **b** DNA graph rewriting notation

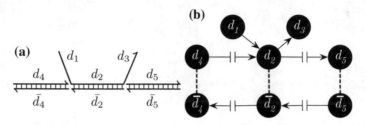

Fig. 15.20 Example showing the maximum number of edges around vertex d_2. **a** Cartoon rendering of 3 duplexes, base-stacked against each other. **b** DNA graph rewriting notation

In Fig. 15.19a, b, we provide an example in vertex d_2, which is connected to other domains using the maximal result of combining all edges while simultaneously maintaining the vertex's compliance with well-formedness constraints (disregarding the rest of the structure, which does not comply in this case. The example provided is meant to convey how a vertex can be surrounded by 5 edges). Here we see that $j = 2$, $k = 0$ and $i = 2$. A corresponding example is provided in Fig. 15.20a, b for the case of d_2 being unhybridized.

15.11   Software Output for Yurke et. al Tweezer Example

See Fig. 15.21.

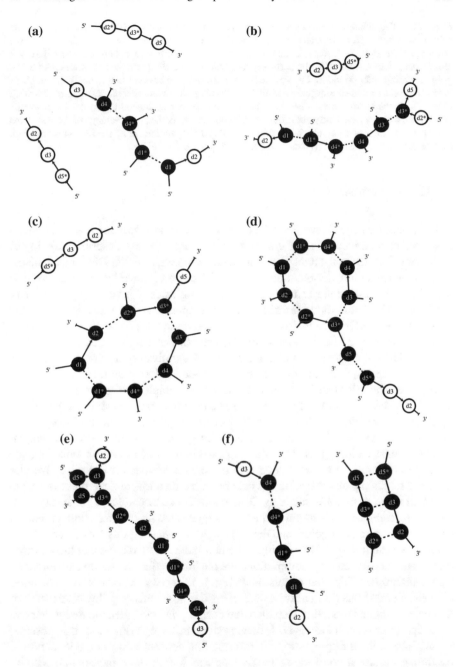

Fig. 15.21 The subfigures are a subset of the states generated by DAGRS for the tweezer. This subset was chosen from 18 states, to correspond to the pathway represented in Table 15.1 (the sequence: state_0, state_3, state_8, state_11, state_12, state_15) from the open state (state_0) to the closed. This demonstrates that the software does indeed generate graphs which correspond to the graph rewriting rules provided. The systems do not discriminate between energetically favorable or unfavorable states, and simply generates all the states possible given a specific set of graph rewriting rules. There are states which arise that may or may not occur, depending on the experiment's conditions, which are currently not taken into account. We think the key advantage of these systems is that it does display all states, intended or not, depending on the set of rewrite rules supplied. **a** State 0. **b** State 2. **c** State 7. **d** State 14. **e** State 13. **f** State 15

15.12 Conclusion

In this paper, we have shown that DNA hybridization and enzymatic reactions can be described using a specific set of graph-rewriting rules. We have described a family of DNA graph rewriting systems that can model some well-known DNA mechanisms such as hybridization, dissociation, strand displacement, as well as various enzymic reactions. This technique of using graph rewriting rules is interesting since it can form the basis of a computational tool that can enumerate the transition pathways between complex DNA nanostructures. However, these rules do not take into account tertiary structures, concentrations, or external conditions that can affect their application. In addition, the inclusion of some rules that significantly increases the number of states, however, they do not need to be included unless they must be. A subset of these rewrite rules can likely be used for a basic graph rewrite system modeling most DNA devices, some of which have been demonstrated, which use just DNA hybridization reactions, whereas other of our rewrite rules could be incorporated as needed for DNA devices for example using enzymic reactions. In practice, we would suggest including as few rules as possible to limit to combinatorial explosion of states that can occur if certain of these rewrite rules are included. We aim to model reaction pathways using domain based verification, and we make the same assumptions as commonly made by DNA nanoscience researchers. For instance, in strand displacement, it is assumed that (1) sequences with different domain names are designed to be orthogonal, and that (2) single stranded regions are designed to be secondary-structure free. By keeping these in mind, we can assume that the intermediate states as generated by our grammar are the intermediate states that are intended by the researcher. The goal of this modeling, is to assist researchers in verification of their intended reaction pathway, and to ensure no incorrect pathways exist. If at this coarse-grained modeling step there are design problems, then the researcher can re-design their model. However, following domain level verification, it is possible to take into account tertiary structure information, concentrations, kinetic rates etc., by further modification of the software. One way to take these factors into account is to provide additional information alongside the graph-rewriting rules, that can be mapped onto the vertex, edge and even subgraph levels during the process of applying these graph-rewriting rules. For example, each vertex (domain) can be assigned a length, and each edge (reaction) can be assigned a transition probability. Though non-trivial, these reaction rates can be approximated based on experimental data [7, 83]. In addition, structural checks (steric hindrance check, persistence length check) can be performed after each transition (at intermediate states) to verify no tertiary rules are violated. These graph rewrite rules originate from the commonly known dynamic mechanisms of interaction between DNA strands (strand displacement, base stacking). With accurate kinetic rates for each of these mechanisms being determined [83], the next step would involve the association of kinetic rates with each graph rewrite rule. The rates would be parameterized depending on the properties (e.g., domain lengths, sequence G/C content) of the DNA graph which the rules are applied to.

Current language-based computational tools [40–42, 54, 81] simulate strand displacement between DNA structures, but they lack in one major areas. The language does not allow the modeling of complex DNA nanostructures involving branched or pseudo-knotted structures, even when enzymes are present [80]. Our graph rewrite rule systems attempt to bridge this gap, by using a simple abstract model for each DNA mechanism.

We have implemented a prototype that performs graph rewriting to generate any sequence of states based on a given input species and the set or subset of rules that can be applied. For future work, we envision allowing a user to input a set of DNA nanostructures and their concentrations, and visualize the various transitions that these nanostructures can undergo. These systems would be useful for both the design and verification of large scale DNA computing systems. Verification would allow the user to avoid spurious pathways that the nanostructures could undertake. Additionally, the user would be allowed to selectively choose a set of rules that can operate on an input set. The rules would be plugin based, so that users can design and add in their own rules and their corresponding kinetic rates.

A *conditional graph rewrite system* is a set of rewrite rules, where each rule is of the form $p = e|\langle L \xrightarrow{r} R \rangle$, such that the rule is applicable if and only if all conditions $c \in e$ hold, and there is a match $m : L \to G$ [16, 51]. The application of rules can be driven by the satisfying of conditions such as domain lengths, depending on the kinetic model used [83]. Using existing kinetic models, we can use this property of graph rewriting systems to conditionally apply rules based on the length and sequence information of domains. To model systems such as that of the abstract tile assembly model, one can apply certain restrictions on the application of the rule-set. The results from [24] show that there exists a rule-set that can be consistent with the geometry of a plane, by using planarity constraints. In our case, this can be taken into consideration by only allowing applications of a rule-set that maintain the planarity of the resulting graph, using standard planarity tests [4, 29, 75]. In addition, multiple copies of a tile can be modeled by using a multiset of DNA complexes as input to the system. Furthermore, there would be a combinatorial explosion in the worst case in the number of nanostructures and re-write rules that can be applied as the reaction proceeds. We aim to leverage graph isomorphism heuristics and probabilistic graph rewriting [14, 37, 72] to help reduce this complexity.

Acknowledgments This work was supported by the National Science Foundation under NSF CCF 1217457 and NSF CCF 1320360.

References

1. Andersen, J.L., Flamm, C., Merkle, D.: Inferring chemical reaction patterns using rule composition in graph grammars. J. Syst. Chem. **4**(1), 4 (2013)
2. Bath, J., Green, S., Turberfield, A.: A free-running DNA motor powered by a nicking enzyme. Angew. Chem. Int. Edit. **44**(28), 4358–4361 (2005)
3. Birac, J.J., Sherman, W.B., Kopatsch, J., Constantinou, P.E., Seeman, N.C.: Architecture with GIDEON, a program for design in structural DNA nanotechnology. J. Mol. Gr. Model. **25**(4), 470–480 (2006)
4. Booth, K.S., Lueker, G.S.: Testing for the consecutive ones property, interval graphs, and graph planarity using PQ-tree algorithms. J. Comput. Syst. Sci. **13**(3), 335–379 (1976)
5. Cardelli, L.: Strand algebras for DNA computing. Nat. Comput. **10**(1), 407–428 (2011)
6. Chen, Y., Wang, M., Mao, C.: An autonomous DNA nanomotor powered by a DNA enzyme. Angew. Chem. Int. Edit. **43**(27), 3554–3557 (2004)
7. Chen, Y.-J., Dalchau, N., Srinivas, N., Phillips, A., Cardelli, L., Soloveichik, D., Seelig, G.: Programmable chemical controllers made from DNA. Nat. Nanotechnol. **8**(10), 755–762 (2013)
8. Chhabra, R., Sharma, J., Liu, Y., Yan, H.: Addressable molecular tweezers for DNA-templated coupling reactions. Nano Lett. **6**(5), 978–983 (2006)
9. Chomsky, N.: Three models for the description of language. IRE Trans. Inf. Theory **2**(3), 113–124 (1956)
10. Chomsky, N.: Syntactic Structures, The Hague (1971)
11. Claus, V., Ehrig, H., Rozenberg, G. (eds.): Graph-Grammars and Their Application to Computer Science and Biology. Lecture Notes in Computer Science, vol. 73. Springer, Berlin (1979)
12. Courcelle, B.: Graph rewriting: an algebraic and logic approach. Handbook of Theoretical Computer Science, pp. 194–242. Elsevier, Amsterdamm (1990)
13. Danos, V., Feret, J., Fontana, W., Harmer, R.: Graphs, rewriting and causality in rule-based models (2012)
14. Danos, V., Harmer, R., Honorato-Zimmer, R.: Thermodynamic Graph-Rewriting. Springer, Berlin (2013)
15. Danos, V., Laneve, C.: Graphs for Core Molecular Biology. Springer, Berlin (2003)
16. Dershowitz, N., Jouannaud, J.-P.: Rewrite systems. Handbook of Theoretical Computer Science, vol. B. North-Holland, Amsterdam (1991)
17. Dirks, R., Pierce, N.: Triggered amplification by hybridization chain reaction. Proc. Natl. Acad. Sci. USA **101**(43), 15275–15278 (2004)
18. Doye, J.P.K., Ouldridge, T.E., Louis, A.A., Romano, F., Šulc, P., Matek, C., Snodin, B.E.K., Rovigatti, L., Schreck, J.S., Harrison, R.M., Smith, W.P.J.: Coarse-graining DNA for simulations of DNA nanotechnology. Phys. Chem. Chem. Phys. **15**(47), 20395–20414 (2013)
19. Ehrig, H.: Introduction to the algebraic theory of graph grammars (a survey). Proceedings of the International Workshop on Graph-Grammars and Their Application to Computer Science and Biology, pp. 1–69. Springer, London (1979)
20. Ehrig, H., Pfender, M., Schneider, H.J.: Graph-grammars: an algebraic approach. In: Automata Theory, pp. 167–180
21. Feret, J., Krivine, J.: Kasim: a simulator for kappa (2008–2013)
22. Flamm, C., Andersen, J.L., Merkle, D., Stadler, P.F.: Inferring chemical reaction patterns using rule composition in graph grammars. arXiv.org (2012)
23. Genot, A., Zhang, D., Bath, J., Turberfield, A.: Remote toehold: a mechanism for flexible control of DNA hybridization kinetics. J. Am. Chem. Soc. **133**(7), 2177–2182 (2011)

24. Ghrist, R., Lipsky, D.: Grammatical self assembly for planar tiles. In: 2004 International Conference on MEMS, NANO and Smart Systems (ICMENS'04), pp. 205–211. IEEE (2004)
25. Green, S., Bath, J., Turberfield, A.: Coordinated chemomechanical cycles: a mechanism for autonomous molecular motion. Phys. Rev. Lett. **101**, 238101 (2008)
26. Grun, C., Sarma, K., Wolfe, B., Shin, S.W., Winfree, E.: A domain-level DNA strand displacement reaction enumerator allowing arbitrary non-pseudoknotted secondary structures. http://dna.caltech.edu/Papers/Peppercorn2014-VEMDP.pdf (2014). Accessed 4 Nov 2014
27. Gu, H., Chao, J., Xiao, S.-J., Seeman, N.: A proximity-based programmable DNA nanoscale assembly line. Nature **465**(7295), 202–205 (2010)
28. He, Y., Liu, D.: Autonomous multistep organic synthesis in a single isothermal solution mediated by a DNA walker. Nat. Nanotechnol. **5**(11), 778–782 (2010)
29. Hopcroft, J.E., Tarjan, R.E.: Efficient planarity testing. J. ACM **21**(4), 549–568 (1974)
30. Ibuki, K., Fumiaki, T., Masami, H.: MPS. Abstraction of DNA graph structures for efficient enumeration and simulation. **2011**(12), 1–6 (2011)
31. Jonoska, N., Karl, S.A., Saito, M.: Graph structures in DNA computing. Computing with Bio-Molecules, Theory and Experiments, pp. 93–110. Springer, Berlin (1998)
32. Kawamata, I., Aubert, N., Hamano, M., Hagiya, M.: Abstraction of graph-based models of bio-molecular reaction systems for efficient simulation. Computational Methods in Systems Biology, pp. 187–206. Springer, Berlin (2012)
33. Klavins, E.: Universal self-replication using graph grammars. In: 2004 International Conference on MEMS, NANO and Smart Systems (ICMENS'04), pp. 198–204. IEEE (2004)
34. Klavins, E.: Programmable self-assembly. IEEE Control Syst. Mag. **27**(4), 43–56 (2007)
35. Klavins, E., Ghrist, R., Lipsky, D.: Graph grammars for self assembling robotic systems. In: Proceedings. ICRA'04. 2004 IEEE International Conference on Robotics and Automation, 2004, vol. 5, pp. 5293–5300 (2004)
36. Klavins, E., Ghrist, R., Lipsky, D.: A grammatical approach to self-organizing robotic systems. IEEE Trans. Autom. Control **51**(6), 949–962 (2006)
37. Krause, C., Giese, H.: Probabilistic Graph Transformation Systems. New Trends in Image Analysis and Processing—ICIAP 2013, pp. 311–325. Springer, Berlin (2012)
38. Krishnan, Y., Simmel, F.C.: Nucleic acid based molecular devices. Angew. Chem. Int. Edit. **50**(14), 3124–3156 (2011)
39. Kumara, M.T., Nykypanchuk, D., Sherman, W.B.: Assembly pathway analysis of DNA nanostructures and the construction of parallel motifs. Nano Lett. **8**(7), 1971–1977 (2008)
40. Lakin, M.R., Cardelli, L., Youssef, S., Phillips, A.: Abstractions for DNA circuit design. J. R. Soc. Interface **9**(68), 470–486 (2012)
41. Lakin, M.R., Parker, D., Cardelli, L., Kwiatkowska, M., Phillips, A.: Design and analysis of DNA strand displacement devices using probabilistic model checking. J. R. Soc. Interface R. Soc. **9**(72), 1470–1485 (2012)
42. Lakin, M.R., Youssef, S., Polo, F., Emmott, S., Phillips, A.: Visual DSD: a design and analysis tool for DNA strand displacement systems. Bioinform. (Oxf. Engl.) **27**(22), 3211–3213 (2011)
43. Lilley, D.M.J.: Structures of helical junctions in nucleic acids. Q. Rev. Biophys. **33**(02), 109–159 (2000)
44. Liu, D., Balasubramanian, S.: A proton-fuelled DNA nanomachine. Angew. Chem. Int. Edit. **42**(46), 5734–5736 (2003)
45. Lund, K., Manzo, A.J., Dabby, N., Michelotti, N., Johnson-Buck, A., Nangreave, J., Taylor, S., Pei, R., Stojanovic, M.N., Walter, N.G., Winfree, E., Yan, H.: Molecular robots guided by prescriptive landscapes. Nature **465**(7295), 206–210 (2010)
46. Machinek, R.R.F., Ouldridge, T.E., Haley, N.E.C., Bath, J., Turberfield, A.J.: Programmable energy landscapes for kinetic control of DNA strand displacement. Nat. Communi. **5**, 5324 (2014)
47. Mann, M., Ekker, H., Flamm, C.: The graph grammar library-a generic framework for chemical graph rewrite systems. arXiv.org (2013)
48. Mao, C., Sun, W., Shen, Z., Seeman, N.: A Nanomechanical device based on the B-Z transition of DNA. Nature **397**, 144–146 (1999)

49. McCaskill, J.S., Niemann, U.: Graph replacement chemistry for DNA processing. DNA Comput. **2054**, 103–116 (2001) (Chapter 8)
50. Modi, S., Krishnan, Y.: A method to map spatiotemporal pH changes inside living cells using a pH-triggered DNA nanoswitch, pp. 61–77 (2011)
51. Nupponen, K.: The design and implementation of a graph rewrite engine for model transformations. Master's thesis (2005)
52. Ouldridge, T.E., Louis, A.A., Šulc, P., Romano, F., Doye, J.P.K.: DNA hybridization kinetics: zippering, internal displacement and sequence dependence. Nucleic Acids Res. **41**, 8886–8895 (2013)
53. Peixoto, T.P.: Graph-tool: efficient network analysis (Version 2.2.31) [Software]. http://graph-tool.skewed.de/ (2014). Accessed 23 June 2014
54. Phillips, A., Cardelli, L.: A programming language for composable DNA circuits. J. R. Soc. Interface **6**(11), 419–436 (2009)
55. Pinaud, B., Melançon, G., Dubois, J.: PORGY: a visual graph rewriting environment for complex systems. Comput. Gr. Forum **31**(3), 1265–1274 (2012)
56. Potoyan, D.A., Savelyev, A., Papoian, G.A.: Recent successes in coarse-grained modeling of DNA. Wiley Interdiscip. Rev.: Comput. Mol. Sci. **3**(1), 69–83 (2012)
57. Python Software Foundation: Python™. https://www.python.org/download/releases/2.7/ (2001–2014)
58. Qian, L., Winfree, E.: A simple DNA gate motif for synthesizing large-scale circuits. DNA Computing, pp. 70–89. Springer, Berlin (2009)
59. Qian, L., Winfree, E.: Scaling up digital circuit computation with DNA strand displacement cascades. Science **332**(6034), 1196–1201 (2011)
60. Reif, J.: Parallel biomolecular computation: models and simulations. Algorithmica **25**(2–3), 142–175 (1999)
61. Reif, J.: The design of autonomous DNA nano-mechanical devices: walking and rolling DNA. DNA Computing, pp. 439–461. Springer, Berlin (2003)
62. Reif, J., Chandran, H., Gopalkrishnan, N., LaBean, T.: Self-assembled DNA nanostructures and DNA devices, pp. 299–328. Nanofabrication Handbook. CRC Press, Taylor and Francis Group, New York (2012)
63. Riverbank Computing Limited: PyQt5 (version 5.3.2) [Software]. http://www.riverbankcomputing.com/software/pyqt/download5 (2014)
64. Rozenberg, G.: Handbook of Graph Grammars and Computing by Graph Transformation: Volume I. Foundations. World Scientific, Singapore (1997)
65. Rozenberg, G., Ehrig, H.: Handbook of Graph Grammars and Computing by Graph Transformation, vol. 1. World Scientific, Singapore (1999)
66. Sekiguchi, H., Komiya, K., Kiga, D., Yamamura, M.: A design and feasibility study of reactions comprising DNA molecular machine that walks autonomously by using a restriction enzyme. Nat. Comput. **7**(3), 303–315 (2008)
67. Sherman, W., Seeman, N.: A precisely controlled DNA biped walking device. Nano Lett. **4**, 1203–1207 (2004)
68. Shin, J.-S., Pierce, N.: A synthetic DNA walker for molecular transport. J. Am. Chem. Soc. **126**(35), 10834–10835 (2004)
69. Sinden, R.R.: DNA Structure and Function. Gulf Professional Publishing (1994)
70. Tian, Y., He, Y., Chen, Y., Yin, P., Mao, C.: A DNAzyme that walks processively and autonomously along a one-dimensional track. Angew. Chem. Int. Edit. **44**(28), 4355–4358 (2005)
71. Tian, Y., Mao, C.: Molecular gears: a pair of DNA circles continuously rolls against each other. J. Am. Chem. Soc. **126**(37), 11410–11411 (2004)
72. Torrini, P., Heckel, R., Ráth, I.: Stochastic simulation of graph transformation systems. Fundamental Approaches to Software Engineering, pp. 154–157. Springer, Berlin (2010)
73. Ullmann, J.R.: An algorithm for subgraph isomorphism. J. ACM **23**(1), 31–42 (1976)
74. Wang, Z.-G., Elbaz, J., Willner, I.: DNA machines: bipedal walker and stepper. Nano Lett. **11**(1), 304–309 (2011)

75. Wei-Kuan, S., Wen-Lian, H.: A new planarity test. Theor. Comput. Sci. **223**(1–2), 179–191 (1999)
76. Woo, S., Rothemund, P.W.K.: Programmable molecular recognition based on the geometry of DNA nanostructures. Nat. Chem. **3**(8), 620–627 (2011)
77. Yin, P., Choi, H., Calvert, C., Pierce, N.: Programming biomolecular self-assembly pathways. Nature **451**(7176), 318–322 (2008)
78. Yin, P., Turberfield, A., Sahu, S., Reif, J.: Designs for autonomous unidirectional walking DNA devices. DNA Comput. pp. 410–425. Springer, Berlin (2004)
79. Yin, P., Yan, H., Daniell, X., Turberfield, A., Reif, J.: A unidirectional DNA walker moving autonomously along a linear track. Angew. Chem. Int. Edit. **116**(37), 5014–5019 (2004b)
80. Yordanov, B., Kim, J., Petersen, R.L., Shudy, A., Kulkarni, V.V., Phillips, A.: Computational design of nucleic acid feedback control circuits. ACS Synth. Biol. **3**(8), 600–616 (2014)
81. Yordanov, B., Wintersteiger, C.M., Hamadi, Y., Phillips, A., Kugler, H.: Functional Analysis of Large-Scale DNA Strand Displacement Circuits. Springer International Publishing, Cham (2013)
82. Yurke, B., Turberfield, A., Mills, A., Simmel, F., Neumann, J.: A DNA-fuelled molecular machine made of DNA. Nature **406**(6796), 605–608 (2000)
83. Zhang, D.Y., Winfree, E.: Control of DNA strand displacement kinetics using toehold exchange. J. Am. Chem. Soc. **131**(48), 17303–17314 (2009)

Chapter 16
Computational Matter: Evolving Computational Functions in Nanoscale Materials

Hajo Broersma, Julian F. Miller and Stefano Nichele

Abstract Natural evolution has been manipulating the properties of proteins for billions of years. This 'design process' is completely different to conventional human design which assembles well-understood smaller parts in a highly principled way. In evolution-in-materio (EIM), researchers use evolutionary algorithms to define configurations and magnitudes of physical variables (e.g. voltages) which are applied to material systems so that they carry out useful computation. One of the advantages of this is that artificial evolution can exploit physical effects that are either too complex to understand or hitherto unknown. An EU funded project in Unconventional Computation called NASCENCE: Nanoscale Engineering of Novel Computation using Evolution, has the aim to model, understand and exploit the behaviour of evolved configurations of nanosystems (e.g. networks of nanoparticles, carbon nanotubes, liquid crystals) to solve computational problems. The project showed that it is possible to use materials to help find solutions to a number of well-known computational problems (e.g. TSP, Bin-packing, Logic gates, etc.).

16.1 Introduction

Conventional or classical computation is based on an abstract model of a machine called a Turing Machine [69]. Such a machine can write or erase symbols on a possibly infinite one dimensional tape. Its actions are determined by a table of instructions that determine what the machine will write on the tape (by moving one square left

H. Broersma (✉)
University of Twente, Enschede, The Netherlands
e-mail: h.j.broersma@utwente.nl

J.F. Miller
University of York, YO10 5DD Heslington, York, England
e-mail: julian.miller@york.ac.uk

S. Nichele
Norwegian University of Science and Technology, Trondheim, Norway
e-mail: nichele@idi.ntnu.no

© Springer International Publishing Switzerland 2017
A. Adamatzky (ed.), *Advances in Unconventional Computing*,
Emergence, Complexity and Computation 23,
DOI 10.1007/978-3-319-33921-4_16

or right) given its state (stored in a state register) and the symbol on the tape. Turing showed that the calculations that could be performed on such a machine accord with the notion of computation in mathematics. Von Neumann proposed a design for a computer architecture based on the ideas of Turing that formed the foundation of modern stored programs computers [52]. These are digital in operation. Although they are made of physical devices (i.e. transistors), computations are made on the basis of whether a voltage is above or below some threshold. Classical computers are based on a *symbolic* notion of computation.

Unconventional computing looks at systems that carry out computation which do not conform to the Turing model of computation. An obvious and highly plentiful source of such unconventional computing systems are living organisms. These carry out prodigious amounts of computation in their everyday tasks of survival and reproduction. Unlike classical programs the computational instructions that underlie living organisms have not been designed but rather have been evolved. Symbolic notions of computation have a severe drawback compared with evolving physical computational systems. The former has no obvious way of making use of the natural computational power of physical systems. According to Conrad this leads us to pay "The Price of Programmability" [12], whereby in conventional programming and design we proceed by excluding many of the processes that may lead to us solving the problem at hand. The question then emerges: Is there a way to use classical computation to *exploit* rather than exclude, the properties of physical systems to solve computational problems? This chapter describes work that answers this question by attempting to use computer controlled evolution to 'program' useful devices. This allows artificial evolution to directly manipulate a physical system. In this way it is hoped to create novel and useful devices in physical systems whose operational principles are not necessarily understood or are hitherto unknown.

Computer-controlled evolution is referred to by a variety of names: evolutionary algorithms [14], genetic algorithms [23], genetic programming [27, 56]. It is part of a wide research area known as bio-inspired computation. The main elements of an evolutionary algorithm are:

> Generate initial population of size p. Set number of generations, $g = 0$
> REPEAT
>> Calculate the fitness of each member of the population
>> Select a number of parents according to quality of fitness
>> Recombine some, if not all, parents to create offspring genotypes
>> Mutate some parents and offspring
>> Form a new population from mutated parents and offspring
>> Optional: promote a number of unaltered parents from step 4 to the new population
>> Increment the number of generations $g \leftarrow g + 1$
>> UNTIL(g equals the number of generations required) **OR** (the fitness is acceptable)

In evolutionary computing, the term *genotype* (or chromosome) is used to refer to the string of numbers that defines a solution to a search problem. The individual elements of the genotype are commonly referred to as *genes*. To solve a computational problem requires an assessment of how well a particular genotype represents a solution to the computational search problem. This is called a *fitness function*. The "survival-of-the-fittest" principle of Darwinian evolution is implemented by using a form of fitness-based selection that is more likely to choose solutions for the next generation that are fitter rather than poorer. *Mutation* is an operation that changes a genotype by making random alterations to some genes, with a certain probability. *Recombination* is a process of generating one or more new genotypes by recombining genes from two or more genotypes. Sometimes, genotypes from one generation are promoted directly to the next generation, this is referred to as *elitism* (see the optional step in the above Algorithm).

Evolution-in-materio (EIM) is a term coined by Miller and Downing [38] that refers to the manipulation of physical systems using computer controlled evolution (CCE) [19–21, 38, 41]. It is a type of unconstrained evolution in which, through the application of physical signals, various intrinsic properties of a material can be heightened or configured so that a useful computational function is achieved. Yoshihito discussed a closely related concept of "material processors" which he describes as material systems that can process information by using the properties of the material [75]. Zauner describes a related term which he refers to as "informed matter" [76]. It is interesting that inspection of much earlier research publications reveals that ideas similar to evolution-in-materio, albeit without computers, were conceived in the late 1950s (particularly by Gordon Pask, see [6, 55]).

The concept of EIM grew out of work that arose in a sub-field of evolutionary computation called evolvable hardware [16, 22, 61, 77], particularly through the work of Adrian Thompson [66, 68]. In 1996, Thompson famously demonstrated that unconstrained evolution on a silicon chip called a Field Programmable Gate Array (FPGA) could utilise the physical properties of the chip to solve computational problems [65].

In 2002, Miller and Downing discussed the concept of evolution-in-materio and suggested that liquid crystal might be a suitable material for attempting to evolve computation in materials [38]. They also discussed many other materials whose properties can be affected reversibly via physical signals. Utilising the evolvable motherboard concept of Layzell [32], Harding constructed a liquid crystal analogue processor (LCAP) that utilises the physical properties of liquid crystal for computation [18]. The experimental setup used by Harding was similar in concept to that used by Thompson [65, 67].

The work described in this chapter was carried out as part of an EU funded research project called NASCENCE (Nanoscale Engineering of Novel Computation Using Evolution) [5] in which various computational problems were investigated using evolution-in-materio using micro-electrode arrays.

The plan of the chapter is as follows. Section 16.2 explains the central concept of evolution-in-materio. Section 16.3 describes the used configurable nano-materials, and in Sect. 16.4 an overview of the EIM hardware control system is given. A

brief introduction of the computational problems under investigation is presented
in Sect. 16.5, and Sect. 16.6 gives a detailed summary of the experimental results for
the solved computational problems, together with an explanation of possible emergent behaviours of carbon nanotube materials and gold nanoparticle materials. In
Sect. 16.7 different models and simulation tools for nano-materials are introduced,
each at a different abstraction level. Finally, Sect. 16.8 concludes the chapter and
outlines directions for further work.

16.2 Conceptual Overview

The central idea of evolution-in-materio is that the application of some physical
signals to a material (configuration variables) can cause it to alter how it responds to
an applied physical input signal and how it generates a measurable physical output
(see Fig. 16.1) [38]. Physical outputs from the material are converted to output data
and a numerical fitness score is assigned based on how close the output is to a desired
response. This fitness is assigned to the member of the population that supplied the
configuration variables. Ideally, the material would be able to be reset before the
application of new configuration instructions. This is likely to be important as without
the ability to reset the material, it may retain a memory from past configurations.
This could lead to the same configuration having different fitness values depending
on the history of interactions with the material.

Mappings need to be devised which convert problem domain data into suitable
signals to apply to the material. An *input-mapping* needs to be devised to map problem
domain inputs (if any) to physical input signals. An *output-mapping* is required to
convert measured variables from the material into a numerical value which can be
used to solve a computational problem. Finally, a *configuration-mapping* is required
to convert numerical values held on a computer into physical variables that are used
to "program or configure" the material. In the course of this chapter we will see
examples of a number of mappings.

A difficult question which to some extent can only be answered by experiments,
concerns which types of physical variables should be manipulated to obtain the best
response from the material. As will be seen, generally in the NASCENCE project,
only electrical stimuli to the materials have been investigated. This was largely chosen
because it is relatively straightforward to manipulate such signals.

There are two main ways that computational problems can be solved using EIM.
In the first, a material is used in the mapping of genotype to a fitness value. In this
approach the material is seen as an assistant in an evolutionary search process. It
provides a "black-box" mapping from genotype to output data (from which fitness
is assessed). The thinking behind this is that the material may provide a more evolvable genotype-to-phenotype mapping, since physical variables can be exploited that
could not be exploited if a purely algorithmic mapping was used (as is standard in
evolutionary computation). In this type of *hybrid* system, much of the data required
for solving a particular problem would remain on a computer. The role of the material

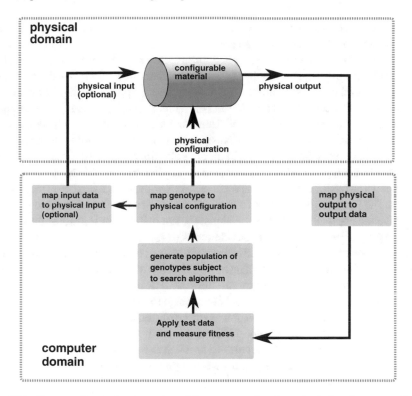

Fig. 16.1 Concept of evolution-in-materio. There are two domains: physical and computer. In the physical domain there is a material to which physical signals can be applied or measured. These signals are either input signals, output signals or configuration signals. A computer controls the application of physical inputs applied to the material, the reading of physical signals from the material and the application to the material of other physical inputs known as configurations. A genotype of numerical data is held on the computer and is transformed into configuration variables that physically affect the material. The genotypes are subject to an evolutionary algorithm. Physical output signals are read from the material and converted to output data in the computer. A fitness value is obtained from the output data and supplied as a fitness of a genotype to the evolutionary algorithm

would be to improve the search process itself. Thus in this case the material does not necessarily require any input data. Examples of computational problems that can be tackled using this approach are: Travelling Salesman Problem (TSP). Function Optimisation and Bin-packing. The TSP is the well known problem of determining the shortest tour through a number of cities. Function optimisation is the problem of determining a vector of numbers which minimises a complex function. Bin-packing is the problem of packing a number of items into as few bins as possible, assuming that each bin has a fixed weight capacity. To obtain solutions to such problems using EIM requires that a set of configuration signals are determined that cause the material to output a suitable vector of measured values.

In the second approach, the evolutionary algorithm determines a configuration which allows the material to act as a *stand alone* computational device. This is a device which provided with the evolved configuration signals, carries out the desired computational mapping. A number of such problems have been considered: digital logic gates, data classification, robot control and graph colouring. For example, suppose that one desired to carry out data classification using a material. Assuming that a stand alone device could be built that used the material and some circuitry to provide the evolved set of configuration signals, one could potentially feed data into the material at a very fast rate and obtain data classification at very high speed and low power consumption. Similarly, evolved logic gates may be able to operate at high speeds and low power etc. We will see examples of both approaches being used in this chapter. The term *configuration* of a material can have a number of meanings. It can merely be the application of physical signals to the material so that some underlying physical properties change, e.g. conductance or resistance. As a result, the material is put into a state that allows the desired computation to take place. Or alternatively, it may be that when the physical signals are applied the material physically changes in some way. For instance, the underlying (electrical) network might be rearranged, or the molecules at the nano-scale could self-organise to a desired state so that the target computational function takes place. An example of the latter is provided by the work with liquid crystal of Harding and Miller [18]. In this case, applied configuration signals caused liquid crystal molecules to twist, thus there was a physical change in the material when configuration signals were applied. Finally, both these effects may happen at the same time.

16.3   Configurable Materials and Micro-electrode Arrays

Although computational materials may be configured by different kinds of stimuli, e.g. electrical signals, magnetic fields, temperature variations, light, etc., it was decided to only manipulate electrical signals within the NASCENCE project. Two types of evolvable material systems were constructed. Both are based on electrode arrays. A material is deposited in the vicinity of the electrodes. Some of the electrodes are chosen as inputs (if the computational problem demands inputs), some are chosen as outputs, and a number of electrodes are chosen as configuration electrodes. In one system, the material deposited was a mixture of single-walled carbon nanotubes (SWCNT) randomly mixed in an insulating material. This is shown in Fig. 16.2. The insulating material was either PMMA/PBMA (Polymethy/butyl methacralate) [49]. Carbon nanotubes are conducting or semi-conducting and the role of the PMMA/PBMA is to introduce insulating regions within the nanotube network, to create non-linear current versus voltage characteristics.

In the other system, shown in Fig. 16.3, a disordered network of gold nanoparticles interconnected by insulating molecules (1-octanethiols) is trapped in a small region surrounded by electrodes [4].

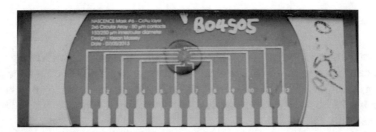

Fig. 16.2 Circular twelve electrode array. The material in the centre is a mixture of SWCNT and PMMA. The concentration of SWCNT is 0.05 % by weight. SWCNTs are mixed with PMMA or PBMA and dissolved in anisole (methoxybenzene). 20 μL of material is drop dispensed onto the electrode array. This is dried at 100°C for 30 min to leave a film over the electrodes [49]

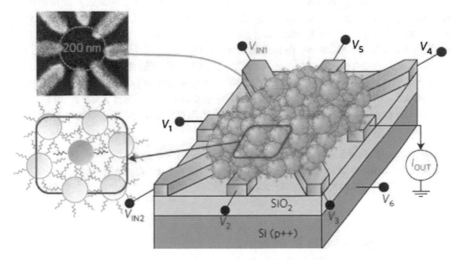

Fig. 16.3 Circular eight electrode array. The material is a disordered network of 20 nm Au NPs interconnected by insulating molecules (1-octanethiols). The NPs are trapped in a circular region (200 nm in diameter) between radial metal (Ti/Au) electrodes on top of a highly doped Si/SiO₂ substrate, which functions as a back gate. The device operates at temperatures below 1°K [4]

At low temperatures, a nanoparticle with capacitance C has a charging energy $E = e^2/C$ which is larger than the thermal energy.[1] In this case nanoparticles exhibit Coulomb blockade and act as a single electron transistor (SET). One electron at a time can tunnel when sufficient energy is available (ON state), either by applying a voltage across the SET or by electrostatically shifting its potential. Otherwise, the transport is blocked because of the Coulomb blockade (OFF state).

---

[1] e is the charge on an electron.

16.4   EIM Hardware Control Systems

In order to be able to apply an evolutionary algorithm to determine a set of signals that should be applied to the electrode arrays, one requires a hardware interface system between a computer and the material. The hardware system needs to allow a variety of signals to be applied to the electrodes. In the NASCENCE project the signals used were one of the following:

- Digital voltages, 0 and 3.5 V
- Analogue voltages in some range
- Square-wave signals

One also needs to be able to sample and record voltages detected on electrodes, since from these measurements a fitness value is determined. Thus one needs equipment that allows the user to choose a sampling frequency and store the values (in a buffer). Since it is not known in advance to which electrodes input signals should be applied, generally one needs a way of allowing the evolutionary algorithm to choose which electrodes will receive inputs (if the computational problem requires inputs) and which electrodes will be designated as outputs, and finally which electrodes will be the configuration inputs. A variety of different hardware systems have been explored for doing this.

- Digital acquisition cards together with programmable switch arrays [9]
- Mbed microcontrollers with digital to analogue converters [37]
- Purpose built platforms [4, 34]

16.5   Computational Problems

The NASCENCE consortium investigated a diverse range of computational problems. The list of problems is given below.

1. Logic gates

   a. Two-input single output Boolean functions (e.g. (N)AND, (N)OR, XOR) [4, 26, 34]
   b. Three/Four input single output Boolean functions (e.g. even-3 and 4 parity) [43]
   c. Two-input two-output Boolean functions (e.g. half adder) [4, 26, 37]
   d. Three-input, two-output Boolean functions (e.g. full-adder)

2. Travelling Salesman

   This has no inputs and as many outputs as there are cities [9]

3. Classification

   a. Standard machine learning benchmarks (Iris, Lens, banknote): number of inputs equals the number of attributes, number of outputs is equal to the number of classes [8, 44]

   b. Frequency classification: this requires one input for carrying the source signal whose frequency is to be classified and two outputs which are used to decide the class of the frequency (high or low) [45, 49]

   c. Tone discriminator: this has the same number of inputs and outputs as the frequency classifier [49]

4. Function Optimisation

   a. This has no inputs and as many outputs as there are dimensions in the function to be optimised [47, 49]

5. Bin-Packing

   a. This has no inputs and as many outputs as there are items to be placed into bins [46]

6. Robot control

   a. This has as many inputs as robot sensors and as many outputs as robot actuators (e.g. motors) [42]

7. Graph colouring

   a. This has been looked at with a single input (graph select) and as many outputs as there are nodes to be coloured. Each output selects the colour of the node [33]

The seven classes of problems cover many types of problems involving markedly different numbers of inputs, outputs and number of instances. Some problems like TSP, Function Optimisation and Bin-packing have no inputs. The material acts like a form of genetic programming and via evolved configurations generates a solution from its outputs. This is standard practice in genetic programming. In this type of approach a material is used in the genotype-phenotype mapping. However, one must be careful that in problems that have no inputs, evolution is not merely evolving configuration signals to produce outputs that are desired. In other words, that it is not directly wiring configuration signals to outputs, thus effectively ignoring the material.

In the work on evolving logic gates various functions present much greater difficulty due to their inherent non-linearity. It is well known that parity functions are difficult to evolve non-linear functions. Indeed, they have been used as benchmark problems in genetic programming for some time.

16.6   Experimental Investigations

Below we give more details on some of the computational tasks that have been used in our experimental work.

16.6.1   Travelling Salesman Problem

Solutions to TSP problems were evolved using twelve electrodes (3 × 4) and sixteen electrodes (4 × 4) [9]. Figure 16.4 shows results using a 3 × 4 electrode array for

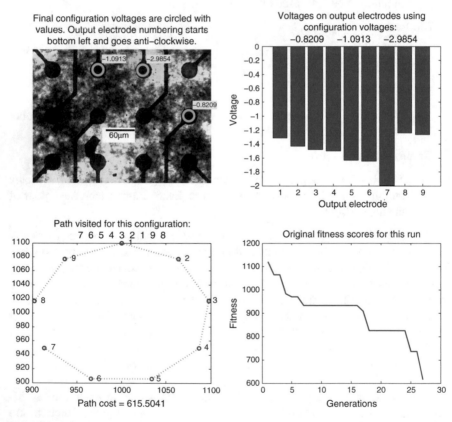

Fig. 16.4 *Top left* shows the CNT dispersal over an earlier prototype 3 × 4 grid electrodes array (×200). Note unevenness of material over electrodes and the mask fault on the third electrode does not appear to affect the evolutionary search when finding the shortest tour of the TSP (the evolutionary history is shown by the best performing genotype for each generation, *bottom right*). In this final configuration, voltages are applied to the circled electrodes and the remaining electrodes provide the floating point values into the TSP. *Top right* recorded voltages which when sorted determine the order to visit cities. *Bottom left* Optimum tour solution of the TSP [9]

a nine-city TSP problem. The particular TSP instances were generated by placing cities on a circle so that they were equidistant from one another. The genotype defined a number of real-values voltages and to which electrodes these configuration voltages would be applied. The latter was accomplished by using digitally configurable analogue cross point switches. A DAQ card first digitally configures the switch connections and then inputs analogue configuration voltages to the material and records the corresponding analogue outputs. The number of configuration voltages deployed depends on the problem being tackled and the availability of spare electrodes on the array. The configuration voltages and electrodes to which they connect were decided by a 1+4 evolutionary algorithm. The range of voltages values was restricted to ±3 V and all connections are one to one (i.e. one configuration voltage can only go to one electrode). Configuration voltages were applied for 1 s and a mean value of sampled voltages from the output electrodes was calculated from the last 0.2 s of sampled values. This was done to exclude any "settling periods" within the material. The time required to configure the analogue switch and set up channels on the DAQ card means that testing a configuration takes several seconds. Actually further investigations revealed that signals from the SWCNT-PMMA materials have negligible noise levels after the initial 50ms so that sampling times could be substantially reduced.

The method of obtaining a tour of cities (i.e. a permutation) is as follows. A vector of voltage values with as many elements as cities is read from the electrode array. The ith element represents city i. The vector is sorted and the city indexes form a permutation, thus defining a tour (see [9] for details). The graphic at the top right of Fig. 16.4 shows a set of voltages read from a 3×4 array. Choosing the lowest voltages (y-axis) in sequence and observing which electrode corresponds to that, one obtains the permutation: 7 6 5 4 3 2 1 9 8.

To assess the effectiveness of the technique, results using the electrode array were compared with a software-based evolutionary technique called Cartesian Genetic Programming [40] in which a graph of mathematical operations is evolved that takes a number of random real-valued inputs to produce a real-valued vector with as many elements as cities. It was found that results using the electrode array were comparable with the CGP method. It remains for future experiments to determine whether the material system scales well as the number of cities increases.

16.6.2 Classification

Using the purpose-built evolutionary platform Mecobo 3.0 and subsequently Mecobo 3.5, experiments were carried out to investigate whether well-known classification problems could be solved [44, 48]. Two relatively small problems were selected from the UCI Machine Learning repository [2]. The problems are known as Lens and Iris. We only report here on the results with the Iris dataset (see [48] for results with the Lens dataset). The Iris dataset is a list of measurements taken from one of three types of Iris flowers. It has four attributes which are classified into one of the three classes. The dataset contains 150 instances with real-valued attributes. The first fifty

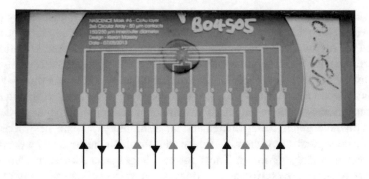

Fig. 16.5 Organisation of inputs, outputs and configuration inputs for a randomly chosen genotype example for Iris data classification

instances are class 1, the second fifty class two and the third set of fifty are class 3. The dataset was divided into two groups (training and test set) of 75 instances each. Each set contained exactly 25 instances of each class.

All of these experiments were performed with electrode arrays having 12 electrodes. The data inputs, outputs and configuration signals applied to the electrode array are shown in Fig. 16.5. Four electrodes were used to input data mapped from the attribute data, three electrodes were used as outputs (i.e. to define the class) and five electrodes were used as configuration inputs (shown in grey). Each output electrode was used for each output class. Each genotype defined which electrodes were outputs, inputs or received the configuration inputs. Class was decided by the leftmost output with the largest value (e.g. if the leftmost output was the largest value, the data would be designated as class one).

In the case of Mecobo 3.0, attribute information was converted by an input-mapping to the frequency of a square wave input signal. This was done by creating a linear mapping from two defined limiting square wave frequencies to the maximum and minimum attribute values [44, 48]. The output-mapping looked at the numbers of bits in the output buffers between transitions from zero to one. The length in bits between these transitions was measured and the average transition gap was used to determine the output classes. The configuration mapping took the allowed gene values and using the Mecobo 3.0 hardware, these were converted into various configuration signals. The genes defined:

- Which electrode would the signal be applied to (0–11)
- Signal type (0 or 1) indicating either a constant voltage (0 or 3.5 V) or square wave configuration signal
- Amplitude (0 or 1) deciding whether the constant applied is 0 or 3.5 V
- Frequency of square wave (500 Hz–10 KHz)
- Phase of square wave
- Duty cycle of square wave (0–100)

Thus a genotype for a twelve electrode array requires 72 genes (6 for each electrode). The first 24 genes were reserved for inputs, that is to say the first gene of each group of six would decide to which electrode the input signals would be applied to, the remaining five genes in each group are then redundant. The next 30 genes decided what kind of configuration signal would be applied to which electrode. Of the remaining three groups of six genes the first genes defined which electrode would be an output. The remaining five in each group were redundant. Thus in this way, a genotype of 72 genes could define which electrodes would receive input signals, which would supply outputs and which would receive configuration signals (of various types).

In the case of Mecobo 3.5, the amplitude of the static analogue input signal was used for input mapping by creating a linear mapping between maximum and minimum attribute values and the maximum and minimum voltages that could be applied to the electrodes. In this case, the output-mapping was simply the average of the sample values of the output buffers.

Here, the genotype is much simpler and consists of 24 genes. The first gene in each of the first four pairs defines where the inputs will be applied. The first gene in each of the last three pairs of genes defines which electrode would be chosen as an output. The remaining genes define the configuration electrodes and the voltage that is applied to those electrodes.

Twenty 1+4-evolutionary algorithms were executed over 50 generations and the two methods compared. The results are shown in Table 16.1.

The fitness calculation is based on the confusion matrix. This required counts to be made of the number of true positives TP, true negatives TN, false positives FP, and false negatives FN. The way this was done is as follows. If the predicted p is correct, then it is a true positive, so TP should be incremented. It is also a true negative for the other two classes, hence TN should be incremented by two. If the predicted class is incorrect, then it is a false positive for the class predicted, so FP should be incremented. It is also a false negative for the actual class of the instance, so FN should be incremented. Finally, the remaining class is a true negative, so TN should be incremented.

Once all instances had been classified, the fitness of a genotype was calculated using Eq. 16.1.

$$ fitness = \frac{TP.TN}{(TP + FP)(TN + FN)} \tag{16.1} $$

Table 16.1 Performance results for Mecobo 3.0 and Mecobo 3.5

Accuracy	Mecobo 3.5 analogue (%)	Mecobo 3.0 digital (%)
Training	91.33	66.93
Test	86.6	60.73

Twenty evolutionary runs of 50 generations of 1+4-evolutionary algorithm were carried out using the Iris data set. Accuracy is the percentage of the training or test set correctly predicted [44]

Table 16.2 Comparative results of different percentages of SWCNT in PMMA

%SWCNT (%)	Average training accuracy (%)	Average test accuracy (%)
1.0	82.13	72.27
0.71	80.67	71.07
0.50	81.73	71.6
0.10	81.73	71.6
0.05	85.07	72.27
0.02	80.93	69.47

Ten 1+4 evolutionary runs of 500 generations were performed on the Iris dataset with Mecobo 3.0. The first column shows the weight percent fraction of SWCNT in PMMA. The second and third columns show the average training accuracy and average test accuracy found. Accuracy is the percentage of the training or test set correctly predicted. Note that no evolution was possible using percentages (by weight) of SWCNT to PMMA polymer lower than 0.02 % since output buffers contained only zeroes

So, if all instances are correctly predicted, the fitness is 1, since in this case $FP = 0$ and $FN = 0$. In the case that all instances are incorrectly predicted, $TP = 0$ and $TN = 0$, so the fitness is zero.

Statistical tests were carried out and they supported the hypothesis that solving the Iris classification is easier when evolution manipulates analogue voltages rather than more complex digital signals [44]. This also has relevance for creating a stand alone system. It would be relatively straightforward to build a circuit that could supply fixed configuration voltages to an electrode array. It would also be straightforward to build a system to automatically map numerical attribute information into applied voltages to be input to the device.

Experiments were performed to investigate how the classification results depended on the concentration of carbon nanotubes. The findings are detailed in Table 16.2.

It was found that when the percentage by weight of SWCNT in the polymer was above 0.02 % no statistical difference could be found in the results.

The classification results using Mecobo 3.5 were also compared with an implementation of CGP for classification under the same evolutionary conditions [44]. The implementation of CGP for this problem was similar to that used for the TSP problem. That is to say, the function set was a set of arithmetic and mathematical operators. The inputs were a fixed set of randomly chosen constants. There were as many outputs as classes. The class was decided by the leftmost largest output and fitness was calculated using (16.1). The results with the material turned out to be not statistically significantly different from those obtained with CGP. This again shows that evolving classifiers in materials is promising.

16.6.3 Logic Gates

A number of Boolean logic functions have been realised by applying evolution-in-materio using various evolvable platforms and materials.

Using a twelve electrode array similar to that shown in Fig. 16.2 and the Mecobo 3.0 evolution-in-materio hardware platform [34] an exhaustive search was carried out over the twelve electrodes. All possible combinations of choosing two input electrodes, one output electrode and nine configuration electrodes were examined. This experiment revealed that with the right configuration inputs any of the sixteen possible two-input Boolean functions can be obtained.

In other experiments the MBed experimental platform was used [26, 37]. Experiments were carried out to see if materials could implement threshold logic gates. In these logic gates one assumes continuous variables are divided into ranges and values in these ranges are designated logical one or zero based on whether the value of the variable is above or below certain real-valued thresholds. To obtain logical OR or AND requires just one threshold, however more complex gates such as XOR may require two. Using the derivative-free search methods Nelder-Mead [51] and Differential Evolution [63] it was shown to be possible to find thresholds that allowed a number of logic functions to be implemented, for example XOR, half and full one bit adders. On the more complex functions, Nelder-Mead proved to be less effective. In later work it was found that using concentrations of SWCNT to polymer (by weight) greater than 0.11 % consistently produced poorer results which worsened with increasing concentrations [37].

Using Mecobo 3.0 it was found to be possible to implement both even-3 and even-4 parity functions [43]. Even parity functions output one when an even number of inputs are one, and zero otherwise. It is well-known that such functions are extremely hard to find using random search and as a consequence have been frequently used as a benchmark for genetic programming methods [27]. A sixteen electrode array was used and a genetic representation similar to that used in the classification work (see Sect. 16.6.2). However, the phase of the applied square waves was not used in the study, as previous work had shown that manipulating the phase of applied square waves had no utility in problem solving. The parity problems were formulated as classification problems, so that two outputs were assumed, the largest leftmost value deciding whether the output was zero or one. Binary inputs were represented as one of two different frequency square waves (500 Hz represented logical zero and 10 KHz represented logical one). As with previous work on classification an average transition gap was calculated from the output buffers. In these experiments fitness was measured by computing a confusion matrix and using the Matthews Correlation Coefficient. Other experiments were performed where inputs and configuration signals were either 0 or 3.3V. It was found that using square waves inputs and evolving configurations that choose different square wave frequencies performed better under evolutionary search than using amplitudes [43].

We close this subsection with a brief description of recently published work [4] in which the nanoparticle networks from Fig. 16.3 were used to evolve all Boolean logic gates. In Fig. 16.6a we see an atomic force micrograph (AFM) image of a real nanoparticle network, where the two input electrodes and the output electrode are denoted by V_{IN1}, V_{IN2} and I_{OUT}, respectively. Time dependent signals in the order of a hundred mV are applied to the input electrodes as illustrated in Fig. 16.6b, and a time dependent current in the order of a hundred pA is read from the output

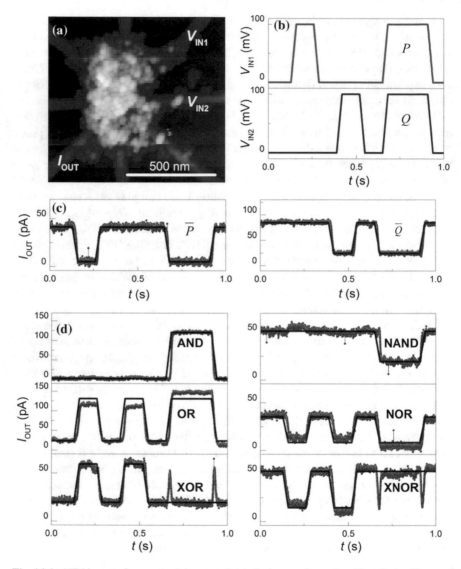

Fig. 16.6 AFM image of a nanoparticle network (**a**), the input voltages in mV applied to V_{IN1} and V_{IN2} (**b**) and the different logic outputs in pA read from I_{OUT} (**c** and **d**)

electrode. The other five electrodes and the back gate can be used to apply different sets of static configuration voltages. Using a genetic algorithm, suitable sets of configuration voltages have been found to produce the output functions of Fig. 16.6c, d. Red symbols are experimental data, solid black curves are expected output signals (matched to amplitude of experimental data). We observe two clear negators (inverters) for the input functions P and Q in Fig. 16.6c, and we observe a variety of

Boolean logic gates in Fig. 16.6d, including the universal NAND and NOR gate. All these gates show a great stability and reproducibility. For the exclusive gates (XOR, XNOR) spike-like features are observed at the rising and falling edges of the (1,1) input, as expected for a finite slope in the input signals. More details can be found in [4]. The remarkable thing here is not that we can produce logic gates using the electrical and physical properties of charge transport in neighbouring nanoparticles. It is remarkable that we can do this with one and the same sample of a disordered network of nanoparticles in a circular region of about 200 nm in diameter, and by using only six configuration voltages. A similar designed reconfigurable device based on current transistor technology would require about the same space. This shows the great potential for our approach.

16.6.4 Other Computational Problems

Below we briefly describe the results of other experimental investigations with the carbon nanotube composites.

16.6.4.1 Function Optimisation

The evolution-in-materio technique was also used to help find the minima of complex multi-dimensional mathematical functions [47, 49]. These kinds of problems are well-known in the research field of evolutionary computation and indeed there are extensive benchmark suites containing highly nonlinear multi-modal optimisation functions. Using the Mecobo 2.0 platform experiments were conducted on a suite of 23 of these functions. A similar genotype representation to that used for classification was employed in this work. However, function optimisation like TSP requires no inputs. Instead one wants to generate a vector which optimises a function. Unlike the classification work, outputs from the electrode arrays were calculated as a scaled average number of ones in the output buffers (i.e. we did not use a calculation of average transition gap). Many of the suite of benchmark functions are thirty dimensional, meaning that the vector that optimises the function in question has thirty elements. This raises an immediate problem when using an electrode array with only sixteen electrodes. In order to evolve solutions that could provide a large number of outputs a multi-chromosomal genotype was devised in which each chromosome applied configuration signals to the electrode array and an output was read. Then the next chromosome was loaded, another evolved set of configuration signals applied and the next output value was computed. Once again the results using the material were compared with the CGP technique. As before the implementation of CGP uses a small number of inputs which are random constants and the genotype represents a network of mathematical operations. It generates as many real-valued outputs as the dimension of the optimisation function. Inputs and internal node operations are defined in the interval $[-1, 1]$. The outputs are then linearly mapped into the defined ranges of the dimensions of the optimisation functions. Using this technique CGP has been compared with Differential Evolution (DE), Particle Swarm Optimization (PSO), and a Standard Evolutionary Algorithm (SEA). Comparisons showed that in 15/20 benchmarks CGP is the same or better than DE, in 19/20 cases CGP is the same or better than PSO or SEA [39].

The experiments show that in 7/23 functions the best results with the experimental material are equal to optimum results and in case of 11/23 functions the best results are very close to optimum results. In four cases the average results with the experimental material are equal to optimum results and in thirteen cases average results are very close to optimum results. In 10/23 functions the best results of experimental material are better than or equal to the best results of CGP. Given the competitiveness of the CGP technique, the results for the material are very encouraging.

16.6.4.2 Bin Packing

Bin-packing is a well-studied NP-hard problem [11]. In the bin packing problem, a total number of n_0 items, consisting of items with different weights (or sizes), have to be placed in bins. Each bin however has a maximum weight (size) capacity c_j. The objective is to place all the items in the least number of bins such that no bin has its weight (size) limit exceeded.

Scholl and Klein have collected bin-packing benchmarks [59]. The datasets are divided into three classes, according to difficulty. The best result for each dataset has been obtained by Scholl et al. [60] using an algorithm called BISON which combines a successful heuristic meta-strategy tabu search and a branch and bound procedure.

Experiments were performed with an electrode array having twelve electrodes [46]. The material consisted of PMMA and with a concentration of SWCNT equal to 0.71 % (expressed as a weight % fraction of the PMMA).

Like TSP and function optimisation bin-packing problems require no inputs. The total number of outputs required is equal to the number of items that need to be packed into bins. Since bin-packing problems typically have 50 or more items multiple chromosomes must be used. Each chromosome defines a number of configuration signals to the electrode array and the remaining outputs supply recorded values in output buffers. The number of chromosomes required is given by the number of items to be packed into bins divided by the number of outputs chosen. For instance for a problem with 50 items, if two outputs are chosen the genotype requires 25 chromosomes. Thus in this case, there will be 10 configuration signals applied for each chromosome processed.

The genotype representation was similar to that used for classification experiments. A series of output values (0 or 1) were read from a buffer of samples taken from output electrode(s). The output values read from electrode(s) were linearly mapped between values -1.0 and 1.0. These values were then used to define the index of the bin (bin_i) in which the object i of the bin packing problem would be placed. So, the total number of outputs must be equal to the total number of objects (n_o).

The linearly mapped output values, x_i corresponding to each chromosome were used to decide the bin index, bin_i which denotes which bin item, i will be placed in. Assuming the number of items is n_o, the bin index is given by Eq. 16.2.

$$bin_i = \left\lfloor n_o \frac{(x_i + 1.0)}{2 + \varepsilon} \right\rfloor \tag{16.2}$$

The floor function $\lfloor z \rfloor$ returns the nearest integer less than or equal to its argument, z. Here *epsilon* is a very small positive quantity. Essentially, Eq. 16.2 divides the interval $[-1, 1]$ into n_o equal intervals corresponding to bins, so that the mapped output values decide which bin an item will be placed in. For instance, assuming the number of items, $n_o = 50$ if x_i is -1.0, bin_i is 0, and if x_i is 1.0, bin_i is 49.

The fitness was assessed in two stages. First the total bin overflow was subtracted from the bin capacity. This results in a negative value when some bins overflow. However, as soon as a solution is reached in which no bin is overfull, the fitness is the number of unused bins.

Twenty evolutionary runs of 5000 generations were carried out using a 1+4 evolutionary algorithm. This took more than two days to complete. On a suite of bin packing benchmarks a range of results were obtained. In some cases (on easier benchmarks) it was possible to find solutions that were near to the known minimum number of bins. Some experiments were done over more generations (25,000) and much better results were obtained. This indicates that the evolution was not stuck in a local optimum and continued to improve with more generations.

16.6.4.3 Robot Control

Experiments were also undertaken to investigate both simulated and actual robot control using a micro-electrode array [42]. Electrode arrays were used with either 12 or 16 electrodes and PMBA (polybutyl methacralate). The sample used in the experiments consisted of 1.0 % carbon nanotubes by weight (99 % by PBMA).

Robots used either six or eight IR distance sensors. The distance values were linearly mapped to determine the duty cycle of a square wave with frequency 5KHz. The duty cycle varies from 0 to 100, where a value 0 means a constant 0 V signal is applied and 100 means a constant 3.3 V signal is applied. A duty cycle of 50 means that a regular square wave of 5 KHz is applied. High values of duty cycle produce a more separated series of 3.3 V pulses.

Two outputs were taken from the electrode array corresponding to two robot motors. The fraction of ones in two output buffers sampled from the electrode array were linearly mapped to determine motor speeds.

The genotype representation was similar to classification except that applied signal phase was not used.

Robot controllers were evolved that allowed both simulated and actual robots to continuously explore mazes with as little obstacle collision as possible. In some cases, the robot controllers showed good generalisation ability in that the evolved controller could control a robot placed in a maze not seen during training (evolution). However, robot controllers transferred to real robots did not work as well as their simulated counterparts. This is not surprising as real robots differ in a number of ways. Direct evolution of control hardware with real robots was too computationally expensive to be practicable.

16.6.4.4 Graph Colouring

Using the purpose-built Mecobo platform, solutions were evolved to the well known graph colouring problem [33]. Graph colouring in our experiments was formulated as follows: given 3 different colours, colour a graph such that no two neighbouring nodes of the graph are assigned the same colour. The target device was the one capable of producing two valid 3-colourings of a simple graph with 4 nodes. Two such colourings are shown in Fig. 16.8. A Genetic Algorithm was used to construct (evolve) a device that produced such colourings in the material connected to the Mecobo board. A conceptual overview of the sought devices is depicted in Fig. 16.7. Graph select was used as input, in order to define which instance ought to be evolved. Eight configuration signals were used to configure the material and 4 outputs signals were mapped to the four graph nodes. The output voltage range was divided in 3 intervals and the highest output voltage observed on each pin was defined as the chosen colour.

The chosen problem instances are fairly simple. The goal of the experiments described herein was to investigate which type of configuration signals produced best results on the chosen problem rather than comparing the performances against known benchmarks. As such, three different groups of configuration signals were tested: only static analogue voltages as configuration signals, only square waves

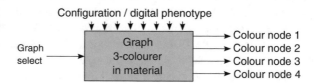

Fig. 16.7 The target device: the graph select input is used to select which of the two instances of the problem, i.e. two different colourings in Fig. 16.8, should be solved in output [33]

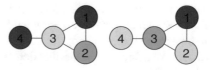

Fig. 16.8 Two valid 3-colourings of the same graph. Both are valid solutions because there are no two neighbours with the same colour. The numbers are node labels, which are mapped to outputs from the device [33]

as configuration signals, mixed configuration signals, i.e. both the previous were allowed. Each of the different signal forms produced working devices. However, the most successful signal representation was square waves, both in terms of successful working devices and number of generations required to obtain a working device.

16.6.5   Electrical Behaviour of Carbon Nanotubes

As mentioned earlier, carbon nanotube-based material may be configured by different kinds of stimuli, e.g. electrical signals, magnetic fields, temperature variations, light, etc. CNT materials have been investigated within the NASCENCE project by means of different electric signals. In particular, the following electrical signals have been considered and explored:

- Static voltages;
- Square waves;
- A mixture of the above.

In order to be able to produce any kind of computation within the underlying nanotubes network, the input data and the configuration data must allow the exploitation and manipulation of underlying physical properties in the CNT-polymer material. Moreover, such manipulated properties must be observable and give a measurable response, i.e. output response. As such, the choice of the input signal and configuration types play an important role and define which physical properties are available and utilised. In [33], a comparison of different signals have been investigated for the evolution of solutions to a well known computational problem, i.e. graph colouring (see Sect. 16.6.4.4).

In the case of static voltages, the parameters under evolutionary control are typically the physical pin to which the signal is applied to, the starting time, the ending time, and the voltage amplitude. In the experiments in [33], the range of amplitude was limited to 0–3.3 V. In the case of square wave signals, the evolved parameters are the physical pin to which the signal is applied to, the starting time, the ending time, the frequency, and the duty cycle. In the experiments in [33], the square wave amplitude was fixed at 0 and 3.3 V. The results show that it was possible to evolve working solutions using all three types of signals (static voltages only, square waves only, a mixture of static voltages and square waves). However, the choice of signal types influenced the evolutionary results. Square wave signals showed promising potential and the ability to produce rich dynamics [53].

One model of the CNT material suggested that if only static DC voltages are applied it behaves as a network of resistors [35]. It was shown that TSP problems [9] could be solved using a SPICE model of the material as a 'cloud' of resistors [50]. In case of applied square wave signals, the CNT material could be seen as an RC circuit, i.e. the CNT material holds capacitance. It is then possible to create macroscopic pin-to-pin models for every pin couple and thus model the material as a whole. Inspection of evolved solutions in [54] showed that the exploited physical properties are often unanticipated. In particular, it was observed that evolution was able to create and exploit signal delays, signal inversions and signal canceling. All the mentioned properties may provide a source of non-linearity and rich dynamics that may be potentially exploited for physical implementations of reservoir computing [24, 36] in CNT materials.

16.6.6 Behaviour of Gold Nanoparticles

Unlike the electrical properties and behaviour of the CNT materials we have been using within the NASCENCE project, the electrical properties and physical effect of Coulomb blockade behind the charge transport in the used gold nanoparticle networks

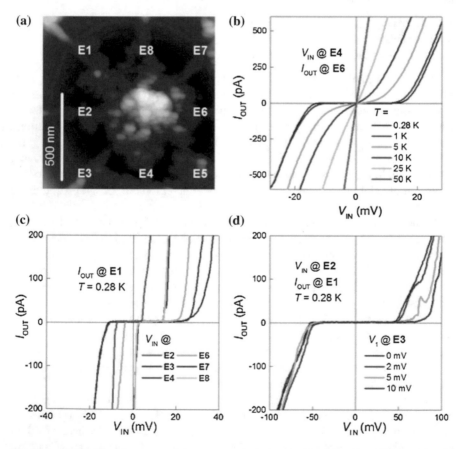

Fig. 16.9 AFM image of a nanoparticle network (**a**), the input voltages in mV applied to electrode E_4 and the output current in pA read at electrode E_6 at different temperatures (**b**) the outputs in pA read from E_1 for different input electrodes at 0.28 K (**c**), and the effect of a static voltage applied at E_3 on the I-V curves of a fixed pair at 0.28 K (**d**)

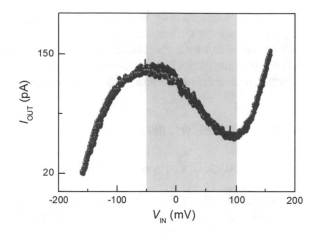

Fig. 16.10 NDR behaviour of a gold nanoparticle network: the output current I_{OUT} increases with an increasing input voltage V_{IN} in the interval $-150\,\text{mV} < V_{IN} < -50\,\text{mV}$, but the I-V curve bends down in the region of $-50\,\text{mV} < V_{IN} < 100\,\text{mV}$

are pretty well understood. As described in Sect. 16.3, under suitable energy conditions and restrictions, the charge transport is governed by the Coulomb blockade effect [25, 71]. The particles act as single electron transistors (SETs) with a high ON/OFF ratio and strong non-linear behaviour. This makes them potentially good candidates for interesting nontrivial functionalities. As an illustration, in Fig. 16.9 we included some I-V characteristics of one of the nanoparticle network we used in [4]. The nonlinear behaviour is very clear from the figures. We also observed a special form of nonlinearity usually referred to as negative differential resistance (NDR), as shown in Fig. 16.10: a gate that was evolved to be a negator (inverter) for $0\,\text{mV} < V_{IN} < 100\,\text{mV}$ exhibits NDR within the considerably larger range $-50\,\text{mV} < V_{IN} < 100\,\text{mV}$.

This behaviour is interesting and plays a key role in the evolvability of more complex functions. For instance, if we compare an XOR with an OR, then it can be observed that the OR could in principle be based on simple linear behaviour, where a high input signal gives rise to a high output signal, no matter whether both input signals are high or just one of the input signals. In case of an XOR this is different: we should only have a high output signal if precisely one of the input signals is high and the other is low; two high input signals should yield a low output signal. This is clearly possible if the evolvable system exhibits NDR behaviour.

16.7 Simulations

Apart from the experimental work and results, the NASCENCE consortium has also worked on the theoretical underpinning of the experimental work and on simulations. The latter are based on physical or mathematical models of the material systems. In case of the nanoparticle networks the physics is pretty well understood, but for the composites of nanotubes the situation is quite different and much more complex.

We start this subsection by describing four different approaches to modeling the nanotube composites, reflecting four different levels of abstraction. The subsection will be completed by a short description of the physical model that underlies the behaviour of jumping electrons in nanoparticle networks, as well as an alternative approach to simulating these networks using neural networks.

16.7.1 Physical Models

16.7.1.1 Models of Carbon Nanotube Materials

Modeling of the physical nanoscale structures may be performed at several abstraction levels, ranging from the low level local interactions between neighboring particles to the high level "black box" behavioral models. Other intermediate levels are also important, particularly for describing emergence of properties at different intermediate scales. Four different models are described here:

1. Model of computation based on collective property of wave functions which describe electrons moving within the material;
2. Model of electrical properties based on DC or AC circuits;
3. Model of conductivity based on dynamical hierarchies and cellular structure;
4. Model of abstract behavior and computational classes using cellular automata.

Model of Computation Based on Collective Electrodynamics in Aggregates of Carbon Nanotubes

Information processing performed by the CNT material is described in [29] within the framework of Ashby's systems theory [1], as introduced in classical cybernetics. Electrical properties exploited for computation arise as an emergent property of the stimulated material. At the lower level of the hierarchy, the wave functions of electrons are manifested as an electromagnetic field, which is one of the main physical phenomena manipulated for computation in an EIM setting. Even if the computation happens at the nanoscale and quantum level, what is captured by the measuring instrumentation is an approximation of the true physicality of computations in the material. As such, the electric field in the nanocomposite can be considered as a manifestation of a collective emergent property of electrons in the computing substrate. In the proposed framework, a future research direction is proposed that may consider the manipulation of quantum properties of electrons in the material so that the emerging electromagnetic properties can be used for computation. Another aspect of the proposed framework is to allow the manipulation of parameters, e.g. temperature, in the description of the system state. This may allow control of parameters during the computation and the system may be described with a bigger choice of variables. This is close to polymorphic electronics [62], where there may be different functionalities for different operating temperatures.

Model of Electrical Properties Based on DC/AC Circuits

Observing the behavior of CNT materials under varying inputs, e.g. static voltages or square wave signals, allows macroscopic modeling of pin-to-pin characteristics with simple RC circuits. In [33], two SPICE models [50] are presented, one for describing the electrical behavior of carbon nanotube materials when stimulated with static voltages applied to input pins, and one for capturing the behavior when square waves are used as manipulation signals. A simple SPICE model, consisting of a 'cloud' of resistors and connectors between them, has been successfully used to replicate results by Clegg et al. [9] for solving an instance of the traveling salesman problem. In this case, using only DC voltages as configuration parameters, the material behaves as a network of resistors. It must be noted that each sample of nanotube material contains a wide variety of such networks and the different configuration signals allow the selection of suitable networks to solve a wide variety of problems [9, 42, 46, 47]. Another model that captures the behavior of CNT materials under the influence of square waves has been proposed in [33]. This model consists of simple circuit elements such as capacitors and resistors. As such, each pin pair can be modeled by a simple RC circuit and by using one such model for each pin pair, a complete model of the material slide can be constructed.

Model of Conductivity Based on Dynamical Hierarchies and Cellular Structure

The approach in [30] aims at modeling conductivity dependence on the concentration of carbon nanotubes and varying electric potential in the material. The approach is based on two main paradigms: dynamical hierarchies [57] and cellular computation [10]. Each material sample is divided in a grid where each cell can represent a sub area of the sample, with relative content, i.e. polymer molecules, nanotubes bundles, electrodes. Each cell behaves according to the physics of the material it contains and interactions with neighboring cells. Results show that higher concentration of CNTs lead to more percolation paths and consequently more current flow. Different cell shapes may be considered for future works, e.g. dodecahedron.

Model of Abstract Behavior Using Cellular Automata

The wide variety of problems solved in CNT materials does not give any direct indication of the computational properties and computational power of the materials used. However it is clear that the materials can be exploited at the computational level required to solve the given task. Cellular automata (CA) offer a broader knowledge of different complexity levels and computational classes, e.g. Wolfram classes [74]. As such, CA models of the material may allow a framework to be established that relates measurable physical properties to abstract CA behaviors. In [15], cellular automata transition tables of different complexities have been evolved in-materio.

An interesting future direction is the possibility to evolve universal cellular automata [3, 13] in the CNT material. In addition, ongoing work attempts to relate the evolved in-materio cellular automata with CA parameters, e.g. lambda [31], and connect material computation with the notion of edge of chaos.

16.7.1.2 Models of Nanoparticle Materials

As we explained earlier, the charge transport in the nanoparticle networks we have been using in the experiments that have led to [4] is based on a physical phenomenon that is known as the Coulomb blockade effect [25, 71]. The individual gold nanoparticles act as single electron transistors (SETs). Electrons can jump between neighbouring particles when the energy conditions are favourable. One electron at a time can tunnel between two particles if sufficient energy is available (ON state), either by applying a voltage across the particle or by electrostatically shifting its potential; otherwise, the transport is blocked due to Coulomb blockade (OFF state). These disordered assemblies of nanoparticles therefore provide an almost random network of interconnected robust, non-linear, periodic switches, as a result of the Coulomb oscillations of the individual nanoparticles. We have observed experimentally that electron transport below 5 K is dominated by Coulomb blockade, and strongly depends on the used input and output electrodes, as well as on the static voltages applied to the remaining electrodes.

Due to the high costs and time consuming experiments involved in the experimental work, it was highly desirable to develop a simulation tool to explore the potential functionalities of such nanoparticle networks without the burden of spending many hours in the lab and wasting expensive resources to look for such functionalities experimentally. In addition, the simulations can also inform us on the minimum requirements that are needed for obtaining the targeted functionality if we were able to produce these nanoparticle networks according to a predetermined design. This could lead to new devices for the digital industry, possibly replacing purpose-built assemblies of transistors. Moreover, simulations can provide us with evidence concerning the scalability of our approach. Simulations can also give us new insights into the dynamics of the charge transport that might lead to a better understanding as to why and how the networks reveal the functionalities we observe. Furthermore, there are many questions on the use of these networks that are difficult to answer experimentally, because there are serious challenges in fabricating examples with smaller central gaps or with more control electrodes using the same area.

The simulation tool we developed in [70] is an extension of existing tools for simulating nanoparticle interactions, like SPICE [50] or SIMON [73]. Since the dynamics of our nanoparticle networks is governed by stochastic processes: electrons on particles can tunnel through junctions with a certain probability, there are basically two simulation methods to our disposal: Monte-Carlo Methods and the Master Equation Method [71, 72]. Since the number of particles is large, this rules out the second approach, hence the Monte-Carlo Method is the only suitable candidate. This method simulates the tunnelling times of electrons stochastically. To get meaningful results, one needs to run the algorithm in the order of a million times. Doing so, the stochastic process gives averaged values of the charges, currents, voltages, etc. More details on the simulation tool can be found in [70].

We have validated our tool for designed systems with small numbers of particles that are experimentally known from literature, and that have also been simulated before [72]. We have also used our tool to examine other structures of nanoparticle networks. Interestingly, we have shown through simulations that all Boolean logic gates that we evolved experimentally in [4] can be evolved in a regular 4×4 grid consisting of only 16 nanoparticles. We refer to [70] for more details. Currently, we are not aware of any production techniques for constructing these regular grids of nanoparticles.

Although our simulation tool can in principle handle arbitrary systems of any size, scalability is a serious issue if we consider the computation time. Even a parallellized CUDA code we have developed for a GPU does not really solve the problem if we want to simulate networks consisting of hundreds of particles. Moreover, as the networks in [4] cannot be produced according to a predefined specific design, it is not possible to use an accurate physical model for such systems.

With these drawbacks in mind, in the next subsection we present an alternative approach. This novel approach is based on training artificial neural networks in order to model and investigate the nanoparticle networks.

16.7.1.3 Neural Network Simulation Model

To support future experimental work on our evolvable systems, we developed simulation tools for predicting candidate functionalities. One of these tools that we have briefly described in the previous subsection is based on a physical model, but the one we present here is based on a neural network model.

Neural networks have proven to be powerful function approximators and have been successfully applied in a wide variety of domains [7, 28, 58, 64]. Being essentially black-boxes themselves, neural networks do not facilitate a better understanding of the underlying quantum-mechanical processes. For that purpose the physical models we described before are more appropriate. But in contrast to physical models, neural networks provide differentiable models and thus offer interesting possibilities to explore the computational capabilities of the nano-material.

Before this exploration can take place, a neural network must first be trained, using data collected from the material. In our case, since we already have a physical model and an associated validated simulation tool for the nanoparticle networks, to show that this approach is useful we can restrict ourselves in the first instance to training data obtained from the simulated material. This gives us the opportunity to predict functionalities in small nanoparticle networks, also networks that have not been fabricated yet, like the 4×4 grid structure we mentioned above. This in turn can inform electrical engineers on the minimum requirements necessary for obtaining such functionalities without the burden of costly and time-consuming fabrication and experimentation.

One of the advantages of the neural network approach is that we do not need to have any detailed information on the structure or physical properties of the material. We only need as many input-output data combinations as we can get from the simulation tool or from measurements on a particular material sample, in order to train a neural network that models this specific sample. The more independent data we use, the more accurate the trained neural network is expected to model the sample.

Another advantage of the neural network approach is that one can optimise the input configuration through gradient descent instead of performing a black-box optimisation. In other words, as soon as we have trained the neural network with sufficiently many input-output combinations, searching for arbitrary functions is very fast and can happen independently of the material or the physical model.

To show that this approach is worthwhile, in [17] we used data obtained from the physical-model-based simulations of the previous subsection to train a neural network. We show there that the neural network can model the simulated nanomaterial quite accurately. The differentiable neural network model of the evolvable nanoparticle network is then used to find logic gates, as a proof of principle.

This shows that the new approach has great potential for partly replacing costly and time-consuming experiments. We are currently using the neural network approach on real data collected from samples of the nanoparticle networks and the carbon nanotube composites. It is too early to report on the results of this approach here.

16.8   Conclusions

Evolution-in-materio is a bottom-up approach where the intrinsic underlying physics of materials is exploited as a computational medium. In contrast to a traditional design process where a computational substrate, e.g. silicon, is precisely engineered, EIM uses a bottom-up approach to manipulate materials with the aim of producing computation. This idea is rather old. Gordon Pask pioneered this work in the late 1950s, by growing neural structures (dendritic wires) in ferrous sulphate materials by electrical stimulation without computers. The EIM ideas became popular again with the work of Adrian Thompson in 1996. Thompson demonstrated that artificial evolution could utilize physical properties of an FPGA chip to solve computational problems. Miller and Downing suggested that many materials could be exploited and coined the term "evolution-in-materio" [38]. The work described in this chapter was carried out within the EU funded project NASCENCE. The goal of the project was to demonstrate that computer-controlled evolution could exploit the physical properties of carbon nanotubes / polymer nano-composites and networks of gold nanoparticles for solving difficult computational problems. Experimental results have shown that EIM is a plausible, competitive and efficient method for solving computational functions. Proof of concept has been given on several instances of problems within various complexities, and different number of inputs and outputs. In particular, solutions have been evolved in-materio for logic gates, travelling salesman problem, machine learning classification, frequency classification, tone discrimination, function optimization, bin-packing, robot control, and graph colouring. The results outlined herein are very promising and lay the foundation for further work. Future work includes the investigation of novel materials and bigger instances of the solved problems. Being able to scale-up the instances of problems tackled may allow real world applications to be targeted. The long term goal of the EIM research community is to build information processing devices by exploiting bottom-up architectures without reproducing individual components. We envision that such devices will be potentially very fast, energy efficient and rather cheap compared to traditional von Neumann-based computers.

Acknowledgments The research leading to these results has received funding from the European Community's Seventh Framework Programme (FP7/2007-2013) under grant agreement number 317662.

References

1. Ashby, W.R.: Design for a Brain, the origin of adaptive behaviour. Chapman & Hall Ltd., New York (1960)
2. Bache, K., Lichman, M.: UCI machine learning repository (2013). http://archive.ics.uci.edu/ml
3. Berlekamp, E.R., Conway, J.H., Guy, R.K.: Winning ways for your mathematical plays, vol. 4. AMC 10, p. 12 (2003)
4. Bose, S.K., Lawrence, C.P., Liu, Z., Makarenko, K.S., van Damme, R.M.J., Broersma, H.J., van der Wiel, W.G.: Evolution of a designless nanoparticle network into reconfigurable boolean logic. Nat. Nanotechnol. (2015). doi:10.1038/NNANO.2015.207
5. Broersma, H., Gomez, F., Miller, J.F., Petty, M., Tufte, G.: Nascence project: nanoscale engineering for novel computation using evolution. Int. J. Unconv. Comput. **8**(4), 313–317 (2012)
6. Cariani, P.: To evolve an ear: epistemological implications of Gordon Pask's electrochemical devices. Syst. Res. **3**, 19–33 (1993)
7. Ciresan, D.C., Meier, U., Masci, J., Schmidhuber, J.: A committee of neural networks for traffic sign classification. In: International Joint Conference on Neural Networks (IJCNN), pp. 1918–1921 (2011)
8. Clegg, K., Miller, J., Massey, M., Petty, M.: Practical issues for configuring carbon nanotube composite materials for computation. In: Proceedings of the 2014 IEEE International Conference on Evolvable Systems (ICES), pp. 61–68 (2014)
9. Clegg, K.D., Miller, J.F., Massey, M.K., Petty, M.C.: Travelling salesman problem solved 'in materio' by evolved carbon nanotube device. In: Proceedings of bthe 13th International Conference on Parallel Problem Solving from Nature - PPSN XIII. LNCS, vol. 8672, pp. 692–701. Springer (2014)
10. Codd, E.F.: Cellular Automata. Academic Press, New York (1968)
11. Coffman Jr., E.G., Garey, M.R., Johnson, D.S.: Approximation algorithms for bin packing: a survey. In: Hochbaum, D.S. (ed.) Approximation Algorithms for NP-hard Problems, pp. 46–93. PWS Publishing Co., Boston (1997)
12. Conrad, M.: The price of programmability. In: Herken, R. (ed.) The Universal Turing Machine A Half-Century Survey, pp. 285–307. Oxford University Press, Oxford (1988)
13. Cook, M.: Universality in elementary cellular automata. Complex Syst. **15**(1), 1–40 (2004)
14. Eiben, A.E., Smith, J.E.: Introduction to Evolutionary Computing. Springer, New York (2003)
15. Farstad, S.: Evolving cellular automata in-materio. In: Master Thesis Semester Project, Norwegian University of Science and Technology, Supervisor: Stefano Nichele, Gunnar Tufte. NTNU (2015)

16. Greenwood, G., Tyrrell, A.M.: Introduction to Evolvable Hardware. IEEE Press, New Jersy (2007)
17. Greff, K., van Damme, R., Koutník, J., Broersma, H., Mikhal, J., Lawrence, C., van der Wiel, W., Schmidhuber, J.: Unconventional computing using evolution-in-nanomaterio: neural networks meet nanoparticle networks. Preprint (2015)
18. Harding, S., Miller, J.F.: Evolution in materio: a tone discriminator in liquid crystal. In: Proceedings of the Congress on Evolutionary Computation 2004 (CEC'2004), vol. 2, pp. 1800–1807 (2004)
19. Harding, S.L., Miller, J.F.: Evolution in materio: evolving logic gates in liquid crystal. Int. J. Unconv. Comput. **3**(4), 243–257 (2007)
20. Harding, S.L., Miller, J.F., Rietman, E.A.: Evolution in materio: exploiting the physics of materials for computation. Int. J. Unconv. Comput. **4**(2), 155–194 (2008)
21. Harding, S., Miller, J.F.: Evolution in materio. In: Meyers, R.A. (ed.) Encyclopedia of Complexity and Systems Science, pp. 3220–3233. Springer, Berlin (2009)
22. Higuchi, T., Liu, Y., Yao, X.: Evolvable hardware. Springer, New York (2006)
23. Holland, J.H.: Adaptation in Natural and Artificial Systems: An Introductory Analysis with Applications to Biology, Control and Artificial Intelligence. MIT Press, Cambridge (1992)
24. Jaeger, H.: The "echo state" approach to analysing and training recurrent neural networks-with an erratum note. Bonn, Germany: German National Research Center for Information Technology GMD. Technical Report 148, 34 (2001)
25. Korotkov, A.: Coulomb Blockade and Digital Single-Electron Devices, pp. 157–189. Blackwell, Oxford (1997)
26. Kotsialos, A., Massey, M.K., Qaiser, F., Zeze, D.A., Pearson, C., Petty, M.C.: Logic gate and circuit training on randomly dispersed carbon nanotubes. Int. J. Unconv. Comput. **10**, 473–497 (2014)
27. Koza, J.: Genetic Programming: On the Programming of Computers by Natural Selection. MIT Press, Cambridge (1992)
28. Krizhevsky, A., Sutskever, I., Hinton, G.E.: Imagenet classification with deep convolutional neural networks. In: Advances in Neural Information Processing Systems (NIPS 2012), p. 4 (2012)
29. Laketić, D., Tufte, G., Lykkebø, O.R., Nichele, S.: An explanation of computation - collective electrodynamics in blobs of carbon nanotubes. In: Proceedings of 9th EAI International Conference on Bio-inspired Information and Communications Technologies (BIONETICS), IN PRESS. ACM (2015)
30. Laketić, D., Tufte, G., Nichele, S., Lykkebø, O.R.: Bringing colours to the black box - a novel approach to explaining materials for evolution-in-materio. In: Proceedings of 7th International Conference on Future Computational Technologies and Applications. XPS Press (2015)
31. Langton, C.G.: Computation at the edge of chaos: phase transitions and emergent computation. Phys. D: Nonlinear Phenom. **42**(1), 12–37 (1990)
32. Layzell, P.: A new research tool for intrinsic hardware evolution. In: Proceedings of The Second International Conference on Evolvable Systems: From Biology to Hardware. LNCS, vol. 1478, pp. 47–56 (1998)
33. Lykkebø, O., Tufte, G.: Comparison and evaluation of signal representations for a carbon nanotube computational device. In: Proceedings 2014 IEEE International Conference on Evolvable Systems (ICES), pp. 54–60 (2014)
34. Lykkebø, O.R., Harding, S., Tufte, G., Miller, J.F.: Mecobo: A hardware and software platform for in materio evolution. In: Ibarra, O.H., Kari, L., Kopecki, S. (eds.) Unconventional Computation and Natural Computation. LNCS, pp. 267–279. Springer International Publishing, Cham (2014)
35. Lykkebø, O., Nichele, S., Tufte, G.: An investigation of square waves for evolution in carbon nanotubes material. In: Proceedings of the 13th European Conference on Artificial Life (ECAL2015), pp. 503–510. MIT Press (2015)
36. Maass, W., Natschläger, T., Markram, H.: Real-time computing without stable states: a new framework for neural computation based on perturbations. Neural Comput. **14**(11), 2531–2560 (2002)

37. Massey, M.K., Kotsialos, A., Qaiser, F., Zeze, D.A., Pearson, C., Volpati, D., Bowen, L., Petty, M.C.: Computing with carbon nanotubes: optimization of threshold logic gates using disordered nanotube/polymer composites. J. Appl. Phys. **117**(13), 134903 (2015)
38. Miller, J.F., Downing, K.: Evolution in materio: looking beyond the silicon box. In: Proceedings of NASA/DoD Evolvable Hardware Workshop, pp. 167–176 (2002)
39. Miller, J.F., Mohid, M.: Function optimization using Cartesian genetic programming. In: Genetic and Evolutionary Computation Conference(GECCO) Companion, pp. 147–148 (2013)
40. Miller, J.F., Thomson, P.: Cartesian genetic programming. In: Poli, R., Langdon W.B., et al. (eds.) Proceedings of EuroGP 2000. LNCS, vol. 1802, pp. 121–132. Springer (2000)
41. Miller, J.F., Harding, S.L., Tufte, G.: Evolution-in-materio: evolving computation in materials. Evol. Intell. **7**, 49–67 (2014)
42. Mohid, M., Miller, J.: Evolving robot controllers using carbon nanotubes. In: Proceedings of the 13th European Conference on Artificial Life (ECAL2015), pp. 106–113. MIT Press (2015)
43. Mohid, M., Miller, J.: Solving even parity problems using carbon nanotubes. In: 2015 15th UK Workshop on Computational Intelligence (UKCI). IEEE Press (2015, in press)
44. Mohid, M., Miller, J.: Evolving solution to computational problems using carbon nanotubes. Int. J. Unconv. Comput. (2016, in press)
45. Mohid, M., Miller, J., Harding, S., Tufte, G., Lykkebø, O., Massey, M., Petty, M.: Evolution-in-materio: a frequency classifier using materials. In: Proceedings of the 2014 IEEE International Conference on Evolvable Systems (ICES): From Biology to Hardware, pp. 46–53. IEEE Press (2014)
46. Mohid, M., Miller, J., Harding, S., Tufte, G., Lykkebø, O., Massey, M., Petty, M.: Evolution-in-materio: solving bin packing problems using materials. In: Proceedings of the 2014 IEEE International Conference on Evolvable Systems (ICES): From Biology to Hardware, pp. 38–45. IEEE Press (2014)
47. Mohid, M., Miller, J., Harding, S., Tufte, G., Lykkebø, O., Massey, M., Petty, M.: Evolution-in-materio: solving function optimization problems using materials. In: 2014 14th UK Workshop on Computational Intelligence (UKCI), pp. 1–8. IEEE Press (2014)
48. Mohid, M., Miller, J.F., Harding, S.L., Tufte, G., Lykkebø, O.R., Massey, M.K., Petty, M.C.: Evolution-in-materio: solving machine learning classification problems using materials. In: Proceedings of the 13th International Conference on Parallel Problem Solving from Nature - PPSN XIII. LNCS, vol. 8672, pp. 721–730. Springer (2014)
49. Mohid, M., Miller, J., Harding, S., Tufte, G., Massey, M., Petty, M.: Evolution-in-materio: Solving computational problems using carbon nanotube-polymer composites. Soft Comput. (2016, in press)
50. Nagel, L., Pederson, D.: Simulation program with integrated circuit emphasis. Memorandum ERL-M382, University of California, Berkeley (1973)
51. Nelder, A., Mead, R.: A simplex method for function minimization. Comput. J. **7**, 308–313 (1965)
52. Neumann, J.v.: First draft of a report on the EDVAC. Technical report, University of Pennsylvania (1945)
53. Nichele, S., Laketić, D., Lykkebø, O.R., Tufte, G.: Is there chaos in blobs of carbon nanotubes used to perform computation? In: Proceedings of 7th International Conference on Future Computational Technologies and Applications. XPS Press (2015)
54. Nichele, S., Lykkebø, O.R., Tufte, G.: An investigation of underlying physical properties exploited by evolution in nanotubes materials. In: Proceedings of 2015 IEEE International Conference on Evolvable Systems. IEEE Symposium Series on Computational Intelligence, IN PRESS. IEEE (2015)
55. Pask, G.: Physical analogues to the growth of a concept. Mechanisation of Thought Processes, no. 10 in National Physical Laboratory Symposium, pp. 877–922. Her Majesty's Stationery Office, London, UK (1958)
56. Poli, R., Langdon, W.B., McPhee, N.F.: A Field Guide to Genetic Programming. Lulu Enterprises, UK Ltd (2008)

57. Rasmussen, S., Baas, N.A., Mayer, B., Nilsson, M., Olesen, M.W.: Ansatz for dynamical hierarchies. Artif. Life **7**(4), 329–353 (2001)
58. Sak, H., Senior, A.W., Beaufays, F.: Long short-term memory based recurrent neural network architectures for large vocabulary speech recognition. CoRR **abs/1402.1128** (2014). http://arxiv.org/abs/1402.1128
59. Scholl, A., Klein, R.: Bin packing. http://www.wiwi.uni-jena.de/Entscheidung/binpp/index.htm
60. Scholl, A., Klein, R., Jürgens, C.: Bison: a fast hybrid procedure for exactly solving the one-dimensional bin packing problem. Comput. Oper. Res. **24**(7), 627–645 (1997)
61. Sekanina, L.: Evolvable components: From Theory to Hardware Implementations. Natural Computing. Springer (2004)
62. Sekanina, L.: Design methods for polymorphic digital circuits. In: Proc. of the 8th IEEE Design and Diagnostics of Electronic Circuits and Systems Workshop DDECS, pp. 145–150 (2005)
63. Storn, R., Price, K.: Differential evolution — a simple and efficient heuristic for global optimization over continuous spaces. J. Glob. Opt. **11**(4), 341–359 (1997)
64. Sutskever, I., Vinyals, O., Le, Q.V.: Sequence to sequence learning with neural networks. In: Ghahramani, Z., Welling, M., Cortes, C., Lawrence, N.D., Weinberger, K.Q. (eds.) Advances in Neural Information Processing Systems 27: Annual Conference on Neural Information Processing Systems 2014, 8–13 December 2014, Montreal, Quebec, Canada, pp. 3104–3112 (2014). http://papers.nips.cc/paper/5346-sequence-to-sequence-learning-with-neural-networks
65. Thompson, A.: An evolved circuit, intrinsic in silicon, entwined with physics. In: T. Higuchi, M. Iwata, L. Weixin (eds.) Proceedings of the 1st International Conference on Evolvable Systems (ICES'96). LNCS, vol. 1259, pp. 390–405. Springer (1997)
66. Thompson, A.: Hardware evolution: automatic design of electronic circuits in reconfigurable hardware by artificial evolution. Distinguished dissertation series. Springer (1998)
67. Thompson, A., Harvey, I., Husbands, P.: Unconstrained evolution and hard consequences. In: Sanchez, E., Tomassini, M. (eds.) Towards Evolvable Hardware: The Evolutionary Engineering Approach. LNCS, vol. 1062, pp. 136–165. Springer (1996)
68. Thompson, A., Layzell, P., Zebulum, R.S.: Explorations in design space: unconventional electronics design through artificial evolution. IEEE Trans. Evol. Comput. **3**(3), 167–196 (1999)
69. Turing, A.M.: On computable numbers, with an application to the entscheidungsproblem. Proc. Lond. Math. Soc. **42**(2), 230–265 (1936)
70. van Damme, R., Broersma, H., Mikhal, J., Lawrence, C., van der Wiel, W.: A simulation tool for evolving functionalities in disordered nanoparticle networks. Preprint (2015)
71. Wasshuber, C.: Computational Single-Electronics. Springer, Berlin (2001)
72. Wasshuber, C.: Single-Electronics – How it works. How it's used. How it's simulated. In: Proceedings of the International Symposium on Quality Electronic Design, pp. 502–507 (2012)
73. Wasshuber, C., Kosina, H., Selberherr, S.: A simulator for single-electron tunnel devices and circuits. IEEE Trans. Computer-Aided Des. Integr. Circuits Syst. **16**, 937–944 (1997)
74. Wolfram, S.: Universality and complexity in cellular automata. Phys. D: Nonlinear Phenom. **10**(1), 1–35 (1984)
75. Yoshihito, A.: Information processing using intelligent materials - information-processing architectures for material processors. Intell. Mater. Syst. Struct. **5**, 418–423 (1994)
76. Zauner, K.P.: From prescriptive programming of solid-state devices to orchestrated self-organisation of informed matter. In: Banâtre, J.P., Fradet, P., Giavitto, J.L., Michel, O. (eds.) Unconventional Programming Paradigms: International Workshop UPP 2004, vol. 3566, pp. 47–55. Springer (2004)
77. Zebulum, R., Pacheco, M., Vellasco, M.: Evolutionary Electronics – Automatic Design of Electronic Circuits and Systems by Genetic Algorithms. The CRC Press International Series on Computational Intelligence (2002)

Chapter 17
Unconventional Computing Realized with Hybrid Materials Exhibiting the PhotoElectrochemical Photocurrent Switching (PEPS) Effect

Kacper Pilarczyk, Przemysław Kwolek, Agnieszka Podborska, Sylwia Gawęda, Marek Oszajca and Konrad Szaciłowski

Abstract Increasing demand for high computational power and high density memories enforces rapid development of microelectronic technologies. However, classical, silicon-based electronic elements cannot be miniaturized infinitely. Therefore, in order to sustain rapid development of information processing devices, new approaches towards future computing devices are needed. These approaches encompass either search for new material technologies or new information processing paradigms. In this chapter we present our contribution to the field including both

K. Pilarczyk (✉) · K. Szaciłowski (✉)
Academic Centre for Materials and Nanotechnology, AGH University of Science
and Technology, al. Mickiewicza 30, 30-059 Kraków, Poland
e-mail: kacper.pilarczyk@fis.agh.edu.pl

K. Szaciłowski
e-mail: szacilow@agh.edu.pl

K. Pilarczyk
Faculty of Applied Physics and Computer Science, AGH University of Science
and Technology, al. Mickiewicza 30, 30-059 Kraków, Poland

P. Kwolek
Department of Materials Science, Rzeszow University of Technology,
ul. W. Pola 2, 35-959 Rzeszow, Poland
e-mail: pkwolek@prz.edu.pl

A. Podborska
Faculty of Non-Ferrous Metals, AGH University of Science and Technology,
al. Mickiewicza 30, 30-059 Kraków, Poland

S. Gawęda
Faculty of Chemistry, Jagiellonian University, Ingardena 3, 30-069 Kraków, Poland

M. Oszajca
Nanograde AG, Laubisruetistrasse 50, 8712 Staefa, Switzerland

© Springer International Publishing Switzerland 2017   429
A. Adamatzky (ed.), *Advances in Unconventional Computing*,
Emergence, Complexity and Computation 23,
DOI 10.1007/978-3-319-33921-4_17

approaches. We introduce classical, Boolean logic devices based on different materials and nanoscale implementations of ternary logic, fuzzy logic and neuromimetic computing.

17.1 Introduction

The modern world functions based on the information exchange. A vast part of our everyday lives depend on the increasing computing power, fast data transmission and huge databanks. In every minute a lot of simple computing tasks are unnoticeably performed by so called "smart" products, including tags, labels, smart windows and smart clothes. Numerous aspects of products design are taken into consideration before the introduction onto the market—these involves materials engineering, the development of new technologies and computational paradigms. It includes new, more efficient algorithms, new computer architectures, and novel materials used for the construction of electronic devices. Whereas the classical designs are based mainly on the silicon structures, the new solutions require cooperation of chemists, physicists and information scientists to develop systems, in which the information processing would be faster, more efficient and could be incorporated at the smaller scale, preferably at the molecular level.

The most versatile computing devices are created by nature. The nervous systems of animals are extremely powerful in tasks with which current computers can barely cope, including human voice, handwriting and face expression recognition. At the same time, we perform these processes in the real time without any significant conscious effort, despite the fact that the elementary computing devices in our brains (neurons and synapses) are relatively slow and the brain itself is less efficient in terms of the raw computing power than modern supercomputers.

The other differences between microprocessors and nervous cells lie in the nature of the transmitted signals and the medium in which they propagate. The communication between electronic devices is governed by electrons (and holes) flow in metals and semiconductors, whereas biological structures (including neural systems) use ions which propagate in aqueous media or interact with cellular membranes. This difference makes the direct interfacing of computers and neural systems quite complicated [1, 2]. That is one of the main motivations to search for fully artificial structures capable of operating in a neuron-like and synapse-like manner.

In this chapter we present an attempt to develop the semiconductor nanoparticle-based information processing devices. We start with basic physics of electrolyte-semiconductor interfaces, mechanisms of photoelectrochemical photocurrent switching (the PEPS effect) and the principles of binary logic devices operation based on this phenomenon. We conclude the chapter with the newest works approaching the fields of ternary and fuzzy logic and neuromimetic computing systems. Despite the obvious differences, the nanoparticulate electrodes share a common feature with biological components: the information processing occurs in a distributed matrix of elements, which lead to the conclusion, that an appropriate modification of the former should give us the artificial systems working in a similar manner to the latter.

17.2 Semiconductor-Electrolyte Interfaces

One of the most important features of semiconductors is their ability to separate electron-hole pairs upon illumination within the appropriate wavelength range. This property is extensively utilized in photovoltaics and photocatalysis. It also enables the use of semiconducting materials in optoelectronics and laser technologies. As for the information processing the charge separation is embodied in the PhotoElectrochemical Photocurrent Switching (PEPS) effect, which can be applied in the construction of switches, logic gates and other basic elements used in the computing systems.

As a rule of thumb one can assume that under illumination of n-type semiconductors anodic photocurrent is generated (these materials behave as photoanodes), whereas in the case of p-type semiconductors—cathodic photocurrent will be recorded. However, sometimes, under appropriate polarization with external potential, a transition between anodic and cathodic photocurrent can be observed (Fig. 17.1) [3–6]. In 2006 the name and the catchy abbreviation for this phenomenon was coined—the PEPS effect—and since then it has been extensively investigated. The PEPS effect may be defined as the change in the photocurrent polarity due to the change in either the semiconducting electrode potential or the wavelength of incident light [7].

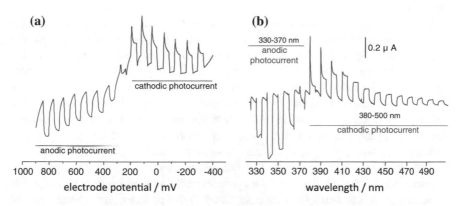

Fig. 17.1 A linear sweep voltammetry performed for n-type semiconductor (TiO$_2$ modified with [Fe(CN)$_6$]$^{4-}$) under pulsed illumination at 350 nm (**a**) and the photocurrent action spectrum of the same material recorded at 200 mV versus Ag/AgCl reference electrode with varying wavelength of the incident light

In order to observe the switching effect, several conditions must be met. The most obvious is the illumination of a studied system (semiconductor, which may be modified with organic or inorganic species) within its absorption range in order to generate an electron-hole pair. The second one, is the presence of efficient electron donors/acceptors dissolved in electrolyte. This aspect influences both thermodynamics and kinetics of the system and should be considered at the design step for each device. From the thermodynamic point of view, the redox potential of an electron donor/acceptor must be located between the conduction and the valence band edges. Only in such a case, ΔG (i.e. the change of Gibbs energy) for the reduction of oxidized species from the electrolyte with an electron from the conduction band as well as the oxidation of reduced species with a hole from the valence band is negative (thus, the process is thermodynamically allowed). We assume that the reduction is fully dependent on the conduction band whereas the oxidation relies on the valence band as it is usually adopted in the photocatalysis [8]. This assumption is usually valid when the band gap of a semiconductor is not too wide and for relatively high reorganization energy value. Popular, artificial redox couples applied in experiments on the switching effect are: I^-/I_3^- (in Ref. [9]), Ce^{4+}/Ce^{3+} (in Ref. [10, 11]), sacrificial electron donors like ethylenediaminetetraacetic acid (EDTA) are also commonly used [12].

It is noteworthy, that the PEPS effect can be also observed in redox-innocent supporting electrolytes (e.g. aqueous solution of KNO_3), when the band gap of a semiconductor is sufficiently wide (more than ca. 2 eV) [13]. In such a case, in equilibrium with air, two redox couples, namely $OH^\cdot/OH^-$ and $O_2/O_2^{\cdot-}$, can be distinguished. Their formal potentials are equal to 1.9 and -0.16 V versus standard hydrogen electrode (SHE) respectively [14, 15]. When the electrolyte is purged with oxygen, its reduction potential shifts anodically which facilitates cathodic photocurrent flow leading to the occurrence of the switching effect at sufficiently negative electrode potential.

The kinetic aspect of these processes is also crucial. The electron donor/acceptor system present in electrolyte should ensure the charge transfer through the semiconductor-liquid junction in a fast enough manner to avoid the recombination events. From the viewpoint of the thermodynamics, appropriate reduction potential values of the redox couples versus the conduction and the valence band edges ensure the electrochemical stability of an investigated material. On the one hand, at sufficiently cathodic polarization of the electrode and in the absence of an efficient electron acceptor, the reduction of metal ions from the crystal lattice may occur. On the other hand, holes photogenerated under irradiation in the valence band may oxidize metal ions from the lattice. Both processes lead to the electrode degradation and may induce drastic changes in its photoelectrochemical properties. For oxide semiconductors it is usually not the case, however this problem has to be considered for sulphides and iodides [8].

Finally, the third factor having a significant impact on the occurrence and the mechanism of the PEPS effect is a potential barrier characteristics at the semiconductor-electrolyte interface. The stronger the electric field at the junction, the higher is

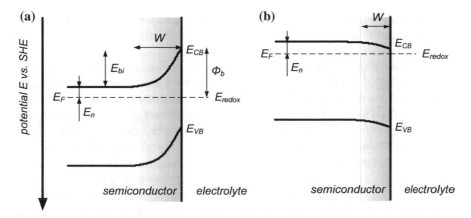

Fig. 17.2 The energy diagram for the n-type semiconductor-electrolyte junction at the equilibrium. The depletion layer (**a**) and the accumulation layer (**b**) are presented. The sign of the charge collected within the semiconductor part of the junction depends on the relative positions of the Fermi level E_F and the reduction potential of the redox couple E_{redox} in the electrolyte. E_{CB}, E_{VB}—the conduction and the valence band edge potentials, E_n—the Fermi level potential versus E_{CB}, E_{bi}—the potential barrier at the junction, Φ_b—the potential difference between E_{CB} and E_{redox}, W—the depletion layer width

the intensity of anodic and lower of cathodic photocurrents. In order to observe the switching effect, one should prepare a material with a moderate height of the potential barrier. Its value for a given system may be tuned using two factors—the concentration of donor states and the space charge layer width according to the Eq. (17.1). The initial difference between Fermi levels of both counterparts in the junction is the driving force of the majority charge carriers flow. For the n-type semiconductor, the resulting depletion layer is usually 10–1000 nm wide (Fig. 17.2) [16]. At the 1 V initial difference between Fermi levels, the strength of the electric field is around 10^5 V·cm^{-1}, which enables 1 μm separation of charge carriers from an electron-hole pair within picosecond timescale [17]. Since the potential distribution in the depletion layer is parabolic, the equation describing the potential barrier at the interface may be formulated as follows [1, 18]

$$E_{bi} = \frac{eN_D}{2\varepsilon_0\varepsilon_r} W^2 \tag{17.1}$$

where N_D—donor atoms density, e—the elementary charge, ε_r and ε_0—dielectric constants of a semiconductor and the vacuum respectively, W—the depletion layer width. The accumulation layer in turn is much thinner (around 0.1 nm, Fig. 17.2) [18]. It should be noted here, that in the described model it is assumed that the band edges are pinned to the interface. It is valid for oxide semiconductors immersed in aqueous electrolyte in the dark [8]. However, the presence of surface states (even at the level of 1 % of a monolayer coverage) complicates such a simple description. The electron flow from these states to the electrolyte hampers the space charge layer and thus the potential barrier development. It causes, as the effect, partial or complete Fermi level pinning [16].

In the case of oxide materials, the concentration of electron donors in the lattice may be controlled during the annealing step in the oxygen-rich atmosphere. Lowering its partial pressure increases oxygen vacancies concentration which are the most prominent electron donors in oxide semiconductors. Contrary, high oxygen partial pressure decreases donor density and the potential barrier height at the solid-liquid interface. However, it should be noted here that it also decreases the electric conductivity of the electrode [19]. Alternatively, one may compensate a semiconductor by the implementation of acceptor levels within the energy gap (e.g. in the form of metal vacancies) [19, 20]. Semiconductor compensation via the introduction of defects into the lattice is believed to be responsible for the photocurrent switching effect observed in the case of bismuth/lanthanum vanadate solid solutions $(Bi_xLa_{1-x}VO_4)$ [13].

However, usually it is easier to control the width of the space charge layer. The thinner it is, the higher is the probability of the electron transfer from the conduction band to electrolyte to occur. The space charge layer width may be easily decreased by the reduction in semiconductor grains size. For the sufficiently fine powder, the space charge layer is only partially developed resulting in a weak electric field at the interface. Yet, one must note that increasing cathodic photocurrent intensity concomitantly decreases anodic photocurrents. In the case of a small potential barrier, the relative photocurrent intensities depend mainly on the kinetics of the electron transfer from the conduction band to electron acceptors in electrolyte and the hole transfer from the valence band to electron donors [21].

So far, the mechanism of the PEPS effect was usually described for the surface-modified, n-type semiconductors (mainly TiO_2, also CdS). To observe a significant signal from the organic or inorganic modifier adsorbed onto the oxide surface it is necessary to ensure as high surface-to-volume ratio as possible. Thus, mainly nanoscopic semiconducting powders (i.e. with the diameter within the range of tens of nanometres) were investigated [7, 22–29]. In such cases, the variations in the space charge layer width can be neglected. The separation of electron-hole pairs depends mainly on the kinetics of the charge transfer through the solid-liquid junction.

17.3   The PEPS Effect in Unmodified Semiconductors

Several different systems have been already described with regard to the PEPS effect occurrence in unmodified semiconductors. Nonetheless, in such cases it is easier to observe the switching effect upon the change in the electrode polarization, rather than the change in the incident light wavelength. The switching following the latter is also possible, but it requires the presence of trap states within the band gap that efficiently exchange charge carriers with electrolyte. The photocurrent switching effect was observed for unmodified oxide semiconductors such as bismuth orthovanadate [30, 31], $Bi_xLa_{1-x}VO_4$ solid solutions (x between 0.23 and 0.93) [13], lead molybdate,

[11] V-VI-VII semiconductors like bismuth oxyiodide, [6, 9] bismuth oxynitrate [30] and antimony sulfoiodide [10] as well as cadmium sulphide (CdS) [26, 31] and lead sulphide [32].

Figure 17.3a, b present the examples of the photocurrent action spectra recorded at different potential values and reveal the switching effect in the presence of oxygen dissolved in the electrolyte. The mechanism of anodic photocurrent generation in the

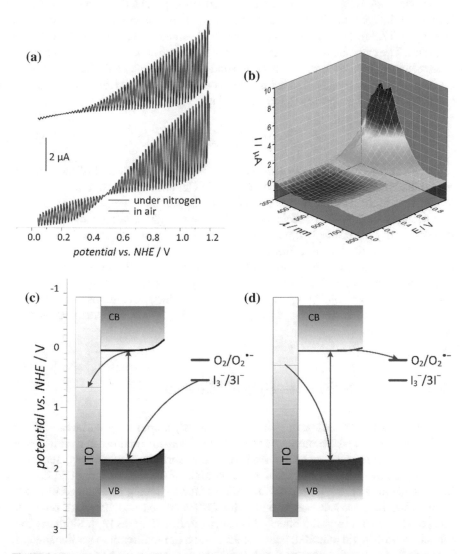

Fig. 17.3 The photoelectrochemistry measurements results for BiOI. The linear sweep voltammetry of BiOI photoelectrode under pulsed 450 nm irradiation (**a**). The photocurrent action spectra at different photoelectrode potentials in 10 mM KI equilibrated with air (**b**). The mechanisms of photocurrent generation: anodic (**c**) and cathodic (**d**). NHE stands for Normal Hydrogen Electrode

case of n-type semiconductor is very simple. Holes photogenerated in the valence band oxidize an electron donor D_{aq} from the solution according to the Eq. (17.2):

$$D_{aq} + h^+ \xrightleftharpoons{\hspace{1.5cm}} D_{aq}^+ \tag{17.2}$$

Hydroxyl ions either from aqueous electrolyte or these adsorbed onto semiconductors surface may behave as sufficient electron donors (provided that the quasi-Fermi level potential for holes is more positive than the oxidation potential of the electron donor, Fig. 17.3c). The anodic photocurrent is observed above the switching potential value. The cathodic photocurrent in turn is generated at more negative potentials, when electrons excited to the conduction band are able to reduce the electron acceptor from electrolyte (e.g. dissolved oxygen or oxidized species from the introduced redox couple A_{aq}) according to the Eq. (17.3):

$$A_{aq} + e^- \xrightleftharpoons{\hspace{1.5cm}} A_{aq}^- \tag{17.3}$$

Such a process may occur when the quasi-Fermi level potential of electrons in the semiconductor under illumination (which in fact is very close to the Fermi level of electrons in the dark for n-type semiconductors) is more negative than the reduction potential of the redox species in the electrolyte. A hole in the valence band is simultaneously neutralized with an electron from the conductive substrate inducing the net electrons flow towards the electrolyte (Fig. 17.3d).

To sum up, in the case of unmodified semiconductors, the observation of the switching effect is possible only for sufficiently low potential barrier E_{bi} at the solid-liquid junction—this may be achieved by the reduction in grains sizes of semiconductor and the decrease in the doping level, [33] or when appropriate surface states are present.

17.4 The PEPS Effect in Hybrid Materials

Another class of materials exhibiting the PEPS effect consist of surface-modified semiconductors, in which modifiers may be both inorganic and organic compounds which become adsorbed onto the semiconductors surface. Alternatively, semiconducting particles may be dispersed in some kind of organic or inorganic matrix. The first detailed description and explanation of the switching mechanism in regard to this class of materials was provided for CdS modified with Prussian blue and TiO_2 modified with Prussian blue and hexacyanoferrate(II) ions [27]. Similarly to a semiconductor-liquid junction, upon the adsorption of modifier onto semiconductor, in the dark, the equilibration of the Fermi levels is realized via the charge transfer through the interface. The current flow direction depends on the relative positions of the Fermi levels for electrons in the n-type semiconductor and the molecule of modifier. Since, usually very fine powder is subjected to the modification, the space charge layer is developed only to a small extent. Thus, the band bending may be neglected. In such a system there is a possibility to excite electrons in semiconductor from the valence to the conduction band, but also another processes involving surface molecules take place. These may include both electrons transfer and holes transfer from the surface molecules, however the latter process is very rare, [34] therefore is not discussed in detail.

When molecular species are adsorbed onto a surface of semiconductor the electronic structures of both counterparts are perturbed by this interaction [35]. In the case of metal oxides the surface metal ions, which are coordinatively unsaturated, form coordination compounds with molecular species. This process always involves a certain level of electronic coupling [36]. The electron transfer in such a system may be described by a semi-classical approach, which in turn may be divided into two diverse scenarios, depending on the strength of electronic coupling. While the strongly-coupled systems are characterized by the Creutz–Brunschwig–Sutin model, [37–40] the Sakata–Hiramoto–Hashimoto model is applied for weakly-coupled systems [41, 42]. Our preliminary computational results indicate that both models can be applied to describe higher catechol homologues adsorbed on titanium dioxide surfaces, depending on the number of fused aromatic rings and the structure of anchoring groups [43–45].

When the electronic coupling between the molecule and the surface is strong enough, new surface coordination species and new energy levels are formed. The origin of which are the molecular levels of the ligands and the surface states of semiconductor counterpart. In the case of TiO_2, which is the n-type semiconductor, the interaction between Highest Occupied Molecular Orbital (HOMO) of the molecule (usually an aromatic π orbital) with empty surface states (titanium d orbitals) results in the formation of a bonding orbital. While the HOMO level of the system is mostly located at the molecule, the Lowest Unoccupied Molecular Orbital (LUMO) belongs to the conduction band (Fig. 17.4a). When the excitation of a surface complex occurs, the optical electron transfer (OET) is observed i.e. an electron is promoted directly to the conduction band. It is possible in extreme situations, that d orbitals of only one Ti^{4+} ion may be involved in the optical electron transfer [46, 47].

In the case of the weak electronic coupling, there is only one way of photosensitization which is called the photoinduced electron transfer (PET). Initially, when the photon with sufficient energy is absorbed, an electron is promoted from the HOMO to the LUMO of the system (Fig. 17.4b). In the case of conjugated aromatic photosensitizers (e.g. perylene, porphyrins, etc.) the ground state of the adsorbed molecule is its π-state, while the photoexcited state consists of π^*-orbitals of the same aromatic system [48]. If the LUMO level is located above the conduction band edge, the electron transfer process is possible and an electron is injected into the conduction band to the energy level that is in resonance with the π^* level [49]. Since non-exponential kinetics of this step was observed, it is possible, that the empty surface states, (probably coordinatively unsaturated Ti^{4+} surface states with oxygen vacancies) may contribute to the photoinduced electron transfer [50].

Depending on the local energetics and/or the electron transfer kinetics, electrons excited to the surface states/the conduction band may either reduce an electron

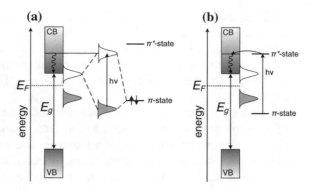

Fig. 17.4 The direct photosensitization of an n-type semiconductor according to the Creutz–Brunschwig–Sutin model (**a**) and the indirect photosensitization according to the Sakata–Hiramoto–Hashimoto model (**b**). The conduction band states may be either fully "compatible" with molecules leading to a strong coupling and consequent formation of new energy levels (**a**) or may overlap only partially with molecules orbitals resulting in a weak coupling, still with a possibility for the electron transfer (**b**). *White* and *grey* bell-shaped envelopes represent empty and occupied surface states respectively, E_F is the Fermi energy, E_g—the band gap, CB and VB stems for conduction and valence band respectively

acceptor in the electrolyte or be transferred to the conducting support. In this way cathodic or anodic photocurrents are generated, respectively. In most of the cases both processes take place simultaneously and the resulting photocurrent polarity is a result of the competition between them and depends on the wavelength of incident light (possibility of the excitation of both or only one counterpart of the hybrid material), the electrode potential and the interaction between a semiconductor surface and an organic modifier. Possible electron pathways for a semiconductor-modifier system are summarized in Fig. 17.5. Since the application of a very fine powder was assumed, the band bending at the semiconductor-molecule interface was neglected.

The simplest case, described in the previous section i.e. the anodic photocurrent generation occurring entirely within the semiconductor is presented in Fig. 17.5a. It may be observed for the wavelength range, in which the semiconductor is excited exclusively. The illumination within the absorption band of a modifier may also produce the anodic photocurrent (Fig. 17.5b), however the mechanism is more complicated and depends on the nature of the interaction between semiconductor surface and molecules.

In the case of a weak electronic coupling, the photoinduced electron transfer is the only possible way of the photosensitization—both HOMO and LUMO levels of such a system are located mainly on the modifier (e.g. π and π^* states respectively for conjugated aromatic modifiers such as perylene and porphyrins) [48]. The LUMO level of the system must be located at the potential lower than the conduction band edge. In such a case the electron injection to the conduction band may occur for the energy level in the resonance with the π^* level [49]. Subsequently, the injected

Fig. 17.5 Tentative mechanisms of the photocurrent switching for surface modified semiconductors, where HOMO and LUMO denotes highest occupied and lowest unoccupied molecular levels associated with the surface species: the photocurrent generation at semiconductor upon fundamental (**a** and **c**) and modifier (**b** and **d**) excitations at gradually decreasing photoelectrode potential. The anodic photocurrent generation at semiconductor upon fundamental (**a**) and modifier (**b**) excitations. The cathodic photocurrent generation at semiconductor upon fundamental (**c**) and modifier (**d**) excitations

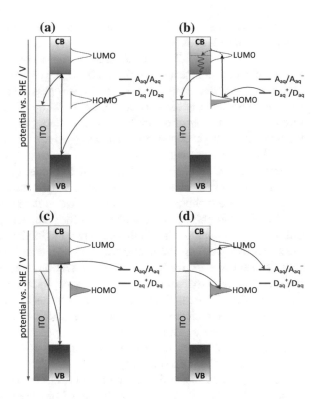

electron moves towards the conductive substrate, whereas an electron donor from the electrolyte neutralizes the adsorbed molecule, provided that the redox potential of the electron donor is located between HOMO and LUMO levels. The most common examples of systems exhibiting PET are: TiO_2—alizarin, [51, 52] TiO_2—carminic acid [25] and TiO_2—phthalic acid [49].

In the case of a strong electronic coupling (like in a CT complex), new energy levels are formed from empty surface states of the semiconductor (e.g. d-states of titanium in the case of TiO_2) and HOMO levels of the ligand. Together they constitute a new LUMO level of the system, which mainly belongs to the conduction band, whereas the HOMO level is entirely located on the molecule. Upon the illumination, an electron is transferred between these two states and further, to the conductive substrate [43]. Subsequently, the oxidized ground state of the molecule is reduced with an electron from the electron donor present in the electrolyte. The OET has been observed for titanium dioxide modified with catechol [40, 48, 51] and salicylic acid [53].

The electron injected into the conduction band should afterwards move inside the bulk material and through the relaxation move towards the band edge. The timescale of these two processes is between one and two hundred femtoseconds. However, the recombination processes (involving both the chromophore and the sacrificial electron acceptor from electrolyte) are also possible, but these occur in the picosecond timescale—they are relatively slow as compared to the electron injection which takes place in hundreds of fs [48].

Both, PET and OET processes may be observed when the redox potential of adsorbed species is located within the band gap of the semiconductor—it prevents modifier molecules from the electrochemical degradation. Moreover, the electron transfer must occur at lower energy than the value of the band gap. Otherwise, the intense interband transitions will mask this effect [40]. In practice, the anodic photocurrent is often generated simultaneously via the molecule and the semiconductor excitation. The contributions of these two components in the net, measured photocurrent depend, among the others, on the number of adsorbed molecules, which in turn, is proportional to the surface area of the semiconductor. For relatively big grains, the anodic photocurrent generated within the semiconductor is dominant.

The cathodic photocurrent is observed for sufficiently negative electrode potential. One may conclude that in this case two distinctive processes exist, namely the mechanisms involving the semiconductor or the molecule which is excited. The former case was described in the previous section, where an electron from the conduction band directly reduces an electron acceptor from the electrolyte (Fig. 17.5c). In the latter case, the modifier excited state behaves as a very efficient electron donor which reduces the electron acceptor in the electrolyte, while an electron from the conductive substrate neutralizes the surface molecule (Fig. 17.5d). Such a behaviour has been observed for instance in the case of TiO_2—Prussian blue [22] and TiO_2—ferrocenoboronic acid [23].

For sufficiently big grains of semiconductors (or the materials which are sufficiently highly-doped), the electron transfer from the conduction band to the electrolyte is very slow resulting in very low cathodic photocurrents intensity. On the other hand, the excitation of a modifier may result in a very intensive cathodic photocurrent generation. In such a way it is possible to induce the switching effect for hybrid materials, whereas it does not occur for the unmodified semiconductor. The presence of a potential barrier at the semiconductor-molecule interface arising from sufficiently coarse grains yields very interesting results, namely the anodic photocurrent is generated mainly in the semiconductor while the cathodic one—in the molecules. This allows us to treat both counterparts separately leading to an enhanced control over systems properties at the level of intermolecular processes.

The described model may be further developed, if one takes into consideration that modifiers can undergo redox reactions within the experimental potential range—as e.g. in the case of TiO_2 modified with folic [24] and carminic acids [25]. In these situations the switching effect occurs only due to the polarization of an electrode. The most important hybrid materials, for which modifiers undergo redox reactions are TiO_2 nanoparticles modified with penta- and hexacyanoferrates [22, 23, 27–29, 54, 55]. For these systems, the photocurrent switching resulting either from the change in electrode potential or the wavelength of incident light was observed thanks to the possibility of oxidation/reduction of cyanoferrates as well as because of the metal-to-band charge transfer complexes (MBCT) of Fe^{2+} with Ti^{4+} ions present at the surface.

The anodic photocurrents in such systems are observed because of the semiconductor excitation provided that adsorbed molecules are oxidized. At sufficiently low potential (the surface species are reduced), the excitation of the surface complex within the MBCT band yields the cathodic photocurrent (an electron injected to the conduction band reduces the electron acceptor from the electrolyte). Still, the switching due to the wavelength change is possible at the potential value, for which the modifier is only partially oxidized. Depending on the light wavelength either the semiconductor or the MBCT complex is excited leading to the generation of the anodic or the cathodic photocurrent respectively.

The last group of inorganic materials exhibiting the PEPS effect are bulk p-n heterojunctions. Also in this case it is possible to observe the switching effect due to the change of either the electrode polarization or the wavelength. Two conditions that must be met are: different band gap values of p- and n-type semiconductors (i) and an appropriate alignment of the band edges in the energetic scale (ii). The former enables the selective excitation of only one counterpart of the junction, the latter prevents the parasitic electron transfer between particles of different conductivity types. If these two prerequisites are not fulfilled, only one photocurrent direction will be preferred and no switching effect will be observed [33]. The most significant examples of the switching effect observed for p-n heterojunctions are: n-TiO_2-N/p-CuI, [12] n-$BiVO_4$/p-Co_3O_4, [56] n-TiO_2/p-Se, [3, 4] n-CdS/p-CdTe, [57] p-$PbMoO_4$/n-Bi_2O_3 [58] and p-CuO/n-$CuWO_4$ [59].

A similar behaviour can be observed also in the case of inorganic n-type semiconductors modified with organic semiconductors. Nonetheless, only few such materials have been reported so far: titanium dioxide/ferrocene, [23] titanium dioxide/hexacyanobutanedienide, [60] titanium dioxide/hexacyanodiazahexadienide [60] and cadmium sulphide/fullerene-oligothiophene [61] systems.

The molecular semiconductors, like fullerene-porphyrin dyads also exhibit similar functionalities. This class of systems has been studied in detail with a good example of zinc porphyrin-fullerene tetramer (Fig. 17.6a). These molecules, deposited in the form of Langmuir–Blodget films on the gold substrate generated photocurrents spectrum which follows the absorption spectrum of the compound within the region characteristic for tetrapyrrolic chromophore (Fig. 17.6b) [62]. The photocurrent polarity depends, however on the photoelectrode potential (Fig. 17.6c), which is a result of the asymmetrical spacial conformation of the molecule and the formation of the Schottky barrier at the molecular layer-ITO interface. Fortunately, this dependence can be utilized to mimic the behaviour of 2:1-demultiplexer (vide infra).

17.5 Boolean Logic Devices Based on the PEPS Effect

Since the PEPS effect allows the straightforward control on the photocurrent polarity via two independent channels (i.e. the wavelength of incident light and the photoelectrode potential), materials exhibiting this effect are suitable for the construction of Boolean logic gates. In most of the cases it is sufficient to assign appropriate Boolean

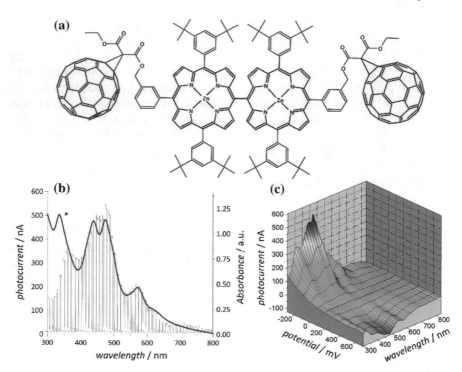

Fig. 17.6 The molecular structure of fullerene-porphyrin tetrad (**a**), its absorption spectrum (*red*) and the photocurrent action spectrum (**b**) together with the photocurrent action spectra recorded in the function of photoelectrode potential (**c**) [62]

values (0 and 1) to the specific ranges of the input stimuli and to the output states. In more complex systems, two different wavelengths can be used as the inputs and the generated photocurrent can be interpreted as the output. The most convenient material for light-only control is titanium dioxide modified with hexacyanoferrate(II) anions. The surface interactions lead to the formation of cyanobridged, two-dimensional assembly with Fe-C≡N-Ti bonding framework.

In an original experiment two LEDs (with wavelengths equal to 400 and 460 nm) were used as the light sources. Boolean 0 and 1 states were assigned to off and on states of each diode acting as two independent inputs. At the exit, the Boolean 0 was assigned to null net photocurrent, while any non-zero photocurrent intensity was interpreted as the Boolean 1 (irrespectively of its polarity). During the pulsed irradiation with the violet diode (400 nm) at the potentials that oxidize completely the surface species (+400 mV vs Ag/AgCl reference electrode), the anodic photocurrent is generated and the irradiation with the blue LED (460 nm) does not give any signal. The synchronized irradiation with both diodes lead to the same effect as the use of the violet diode alone (Fig. 17.7, Table 17.1, output 1) [28, 29].

At the lower potentials the electrochemical reduction of surface species takes place. The irradiation of this material with violet and blue diodes leads to the gen-

(a)

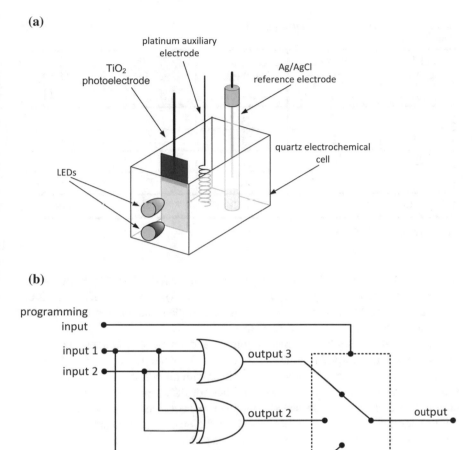

(b)

Fig. 17.7 The experimental setup for the measurement of a reconfigurable logic device prototype (**a**) and an electronic equivalent circuit for the logic system based on a modified titanium dioxide electrode (**b**). Output 1 follows the input 1 signal, output 2 computes the XOR function of input data while output 3 corresponds to the logic sum (OR) of input data. The three-position switch represents the programming functionality of the device through the photoelectrode potential

eration of cathodic photocurrent. Simultaneous irradiation by both LEDs effects in photocurrent of much higher intensity. This kind of behaviour of the photoelectrode at $-200\,$mV versus Ag/AgCl corresponds to the OR logic gate operation (Fig. 17.7, Table 17.1, output 3). The partial oxidation of the surface complex results in yet another observation. The violet light pulses generate the anodic photocurrent, which is consistent with the excitation of an inner part of semiconductor particles. The blue light pulses generate the cathodic photocurrent and the synchronized irradiation

Table 17.1 The truth table for logic gates based on $[Fe(CN)_6]^{4-}$ modified titanium dioxide photoelectrodes

input 1	input 2	output		
400 nm	460 nm	400 mV	250 mV	−200 mV
0	0	0	0	0
0	1	0	1	1
1	0	1	1	1
1	1	1	0	1
Boolean function		YES/ BUFFER	XOR	OR

Adapted from Refs. [28, 63]

with both LEDs gives null net current as the anodic and the cathodic photocurrents compensate effectively (Fig. 17.7, Table 17.1, output 2). It corresponds to the XOR gate function. This system represents the first example of the photoelectrochemical XOR logic gate with two optical inputs.

Furthermore, the system is reconfigurable and its logic characteristics can be changed via an appropriate polarization of the photoelectrode [28, 63]. The rich chemistry of cyanoferrate complexes combined with the reactivity of wide band gap semiconductors creates numerous possibilities for other, even more complex logic circuits based on simple chemical systems. The information is supplied to the device through the pulses of light, is processed by the electrochemical interactions and retrieved in the form of current spikes. This behaviour allows facile communication between various electronic components and chemical logic systems [27–29].

Similar, albeit more simplistic behaviour was observed for titanium dioxide modified with thiamine, [55] folic acid, [24] and carminic acid [25]. Thiamine itself cannot bind to the surface of titanium dioxide and it requires an additional anchoring group. It can be bound to the surface of nanocrystalline titania via a pentacyanoferrate linker. Materials obtained by the immobilization of these chromophores onto the surface of TiO_2 behave similarly to those obtained via the deposition of redox-active cyanoferrate complexes. This is possible because of organic chromophores containing π-electron systems that can act here as electron buffers, donating or accepting electrons if necessary [64]. In principle, any electrochemically active, amphoteric molecule with its energy levels properly aligned with the semiconductor bands [65] should yield an optoelectronic switch [54, 63]. It seems easy, but it is a nontrivial task, as the molecule must be photostable and should be resistant to the photocatalytic degradation in the contact with semiconducting surfaces [66, 67].

Table 17.2 The truth table for the 1:2 demultiplexer

light	input 1	photoelectrode potential	input 2	photocurrent	output 1	output 2
OFF	0	negative	0	NO	0	0
ON	1	negative	0	cathodic	1	0
OFF	0	positive	1	NO	0	0
ON	1	positive	1	anodic	0	1

Adapted from Refs. [24, 63]

The PEPS effect allows the use of such systems as optoelectronic two-channel demultiplexers (Table 17.2). In this case it is also possible to assign the logic values to input and output signals for the logic operation analysis. '0' value can be related to the negative polarization and '1' value to the positive polarization of a photoelectrode. In the case of the optical input the simplest and the most natural signal assignment is achieved when the logical '0' and the logical '1' correspond to the light turned off and on respectively. In the case of photocurrent switching, when both polarities of photocurrent are allowed, it is convenient to split the electric output into two channels, one associated with the anodic photocurrent, and the other with the cathodic one. The photoelectrochemical system defined this way follows the behaviour of the 1:2 demultiplexer (Fig. 17.8).

An interesting example of a single nanoparticle which could be used in the construction of an optoelectronic 1:2 demultiplexer is provided by nanocrystalline S-doped CdS [68]. The photoelectrode made of sulphur-doped CdS generates the anodic and/or the cathodic photocurrents depending on the incident light wavelength or the applied bias. This unexpected behaviour of nonstoichiometric CdS was attributed to the presence of additional sulphur atoms which results in the existence of additional energy levels within the band gap [26]. Under the potential higher than the conduction band edge potential the anodic photocurrent which is characteristic for CdS was observed. However, when then the potential has been lowered beyond the

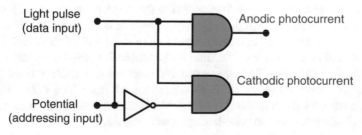

Fig. 17.8 The electronic-equivalent circuit of a surface modified titania photoelectrode working as the two channel optoelectronic demultiplexer

potential of the sulphur-associated doping states only the cathodic photocurrent was recorded. The polarity of the generated photocurrent depends also on the incident light wavelength. The illumination with light of higher energy results in anodic photocurrent—the excitation within fundamental transition occurs—whereas lowering the energy results in the cathodic photocurrent generation associated with the valence band to sulphur-centred gap states transition.

The peculiar photoelectrochemical behaviour of S-doped CdS nanoparticles of 5–7 nm in diameter can be used to build the 1:2 demultiplexer operating on the same principle as in the case of surface-modified titanium dioxide. Interestingly, all the switching phenomena occur at the single nanoparticle level and do not require any cooperative interactions between nanoparticles. Despite the fact, that the operation of these systems has been demonstrated only on a large set of nanoparticles, similar devices have been reported to work at a single nanowire level [69–72].

The photoelectrochemical properties of nanoparticulate CdS deposited on conducting substrates open the way for the construction of nanoscale photoelectrochemical arithmetic systems. The optoelectronic binary half-adder can be built from nanoparticle-based ITO/CdS Schottky junctions working in iodide-loaded, semisolid ionic liquid electrolyte [73]. The single Schottky junction was made by the chemical bath deposition of thin layer of CdS on the ITO surface or by the cast-coating onto the conductive substrate with thiourea-capped CdS nanoparticles (Fig. 17.9a). At no external bias the illumination of the junction results in the anodic photocurrent generation. The combination of two Schottky photodiodes with identical polarities results in the operation resembling AND logic gate as the high intensity photocurrent can be recorded only upon the illumination of both junctions (Fig. 17.9b). The opposite polarization of photodiodes naturally leads to the XOR gate, as the illumination of any single junction generates certain net photocurrent, while the concomitant illumination of both junctions results in the compensation of photocurrents (Fig. 17.9c).

An appropriate concatenation of these two optoelectronic systems (Fig. 17.10) leads to the construction of an optoelectronic binary half-adder. In this device two input signals (light pulses) generate current pulses in two different circuits (corresponding to the AND and the XOR logic gates) and thus yield the binary representation of the sum of the input signals [73]. These devices, similarly to the previously described, do not require any interactions between individual particles and therefore the whole arithmetic unit can be confined within the size of two small nanoparticles.

The described materials give the possibility for the photocurrent direction control through both stimuli: the change in the wavelength or the photoelectrode potential. These structures are not composites comprising of two types of semiconductors but are engineered as uniform materials and the photocurrent switching phenomenon is their intrinsic feature resulting from a specific electronic structure of the system. In this sense the materials described in this chapter are unique. The PEPS effect may serve as a basis for the construction of tunable, light-harvesting antennae, chemical switches, logic gates, and many other optoelectronic devices.

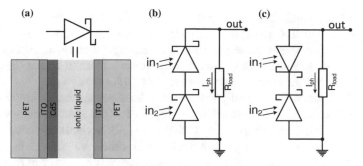

Fig. 17.9 The construction of the Schottky junction (**a**), schematics of the optoelectronic AND (**b**) and XOR (**c**) logic gates based on ITO/CdS Schottky junctions

Fig. 17.10 The connection diagram of the binary half-adder based on ITO/CdS Schottky junctions. R_{load} denotes internal resistance of the potentiostat at 0 V

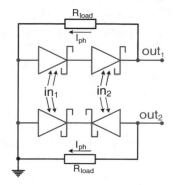

17.6 Application of PEPS Effect in Ternary Logic

When we think about conventional computing devices the obvious association concerns the binary system, which is the most widespread in both hardware and software implementations. It is quite natural that we differentiate between only two distinctive states, which we call TRUE and FALSE, on and off or high and low, as this kind of dualism is present in most of the modern languages and its fundamentals have been paved very early in the history of science (works by Aristotle on so called term logic) [74]. The traditional laws provide deterministic description of statements and the formalism is not only intuitive but also greatly simplifies operations within the two-valued Boolean algebra. However, its use is associated with some limitations, which are easy to overlook at the first glance.

The greatest advantage of the classical two-valued logic is at the same time its main flaw—the simplicity resulting from the use of only two logic values is achieved at the cost of reduction systems ability of describing undefined states (the law of excluded middle and the principle of bivalence) [75]. The significance of such limitation can be presented on the basis of the sea battle paradox in which a statement cannot be interpreted as either true or false, as such interpretation would lead to the restriction of free will and would hinder the possibility to influence the future events at the unacceptable level. Therefore, it is necessary to introduce another state, which will allow both output values to be possible. This example clearly shows that binary logic, despite its apparent intuitiveness, fails in the description of certain situations—nevertheless it has become dominant till the early 20th century.

In 1917 the attempt to formally work around the aforementioned limitations was made by Polish logician Jan Łukasiewicz [76] through the addition of an unknown state (which also has been proposed by C.S. Peirce) [77]. Several other concepts have arisen in following years giving a firm fundament for the multi-valued logic, probability logic, fuzzy logic and other more complex ideas. It is important to note, that shortly after these works it became evident that multi-valued systems (both infinite and finite) are closely related to some fundamental phenomena observed in nature, such as gene regulation pathways [78]—where three different outcomes influencing the production of proteins can be achieved (promotion, inhibition and "no effect")—or cell signalling patterns [79]. Another example of the natural implementation of the ternary logic can be found in the Lake Titicaca basin, in Bolivia and Peru, where the Aymara language is spoken. This, in contradiction to the most of the modern languages, seems to be based on the principles of the three-valued system and the formal analysis of its syntax reveals that it naturally incorporates the third value to express the unknown state [80].

As for fuzzy logic, just after the coinage of the term and the development of the mathematical description supporting the system, it occurred that it is strikingly natural for human comprehension, as we use it unconsciously in our everyday life (when we can't precisely tell if something is cold or hot and we select neither of two limiting values), and as such can be utilized in numerous devices programmable through the linguistic commands easily understandable for people (like thermostats) [81].

On the other hand, at the beginning of the road to a modern computer it occurred reasonable (from the technological point of view) and convenient to reduce the number of possible logic states to the minimum, as the optimisation of electronic circuits has become much simpler for binary system [82]. The straightforward distinguishability of two states was the main reason behind the bivalent system domination—in simple words, due to its fault tolerance in the case of signal distortions and common types of noise. Nonetheless, the fast development of transmission media, particularly the introduction of optical fibres (in which the information can be encoded not only through the on/off states but also in the polarization direction) on the large scale, and the increasing accuracy of information processing devices (leading to a better signal recognition), provoked the discussion on the renaissance of the alternative approaches based on multi-valued logic systems [83].

We used the term "renaissance" to highlight that many applications of the multi-valued logic emerged in the course of the 20th century technological development. The best example is provided by the ternary Setun (Сетунь) built in the Soviet Union in 1958 and several of its successors (e.g. Сетунь-70) [84, 85]. At the moment of construction its performance surpassed—in terms of the power consumption and the computing speed—the specifications of contemporary binary machines. Nevertheless, due to some factors of partially political nature (the negative decision given by the USSR authorities) the mass production of this class of computers has never come true.

The described attempts however have not been the only examples of the ternary logic occurrence in both hardware designs and software implementations. One of the most interesting is the interpretation of circulating superconducting currents present in a Josephson junction in terms of the three-valued logic [86]. Such currents, particularly their on/off state and their direction, could provide a good platform for the storage and processing of information based on the balanced ternary system. On the other hand, some concepts utilized in programming languages use the additional, third logic value (as for example SQL which is used for databases handling). In the case of SQL it is so called NULL flag which indicates that the data is missing, hence the logic value of the comparison with a field containing the missing data is unknown (0 in the balanced ternary logic) [87].

As for the more prototypic applications of the ternary logic system we can mention the use of carbon nanotubes-based pseudo n-Type FETs demonstrated recently by J. Liang et al., [88] the quantum-dot cellular automaton realizing multi-valued logic operations presented by I. Lebar Bajec, N. Zimic and M. Mraz [89] or encoders constructed with the use of carbon nanotubes FETs proposed by P. Viswa Saidutt et al. [90] Nonetheless, among these systems, the lack of devices capable of optical signal processing is evident. As it was mentioned in one of the former paragraphs, the true advantage of three-valued systems over classically used binary components could be achieved in the case of optoelectronic devices and the information encoded in pulses of light. This results directly from the nature of the used carrier—the light can be either turned on or off and polarized in two orthogonal planes giving three easily distinguishable logic values. Another alternative is to combine two different types of inputs (e.g. the electrode potential and the incident light wavelength or the electrode potential and the chemical environment). Such systems however have been presented by only a few authors (Shutian Liu, Uwe Pischel) [91, 92] and have been usually limited to the in-solution devices.

Only recently have we demonstrated [60] the first case of the solid-state system capable of reproducing two ternary logic gates functions—namely *accept anything* and *consensus* operations. By a simple modification carried out for a wide band gap semiconductor (i.e. titanium dioxide) nanoparticulate film we managed to change the photocurrent response of a device based on this hybrid material, in such a way that resembles the outputs of two aforementioned logic gates. The main idea behind this alteration is related to the introduction of additional electronic states related with the presence of cyanocarbon modifiers (hexacyanobutadienide or hexacyanodiaza-hexadienide anions, denoted as X^- in Fig. 17.11) and a new excitation path.

Due to the presence of the additional path on which photocurrents may be generated (Fig. 17.11c) the recorded photocurrent action spectra (Fig. 17.12) have three distinctive areas in which either anodic (Fig. 17.11a) or cathodic (Fig. 17.11c) signals are generated or the net photocurrent is close to zero. The last scenario is realized because of the competitive character of the two processes denoted in Fig. 17.11b. The situation could be compared to two different semiconductors, one of the n-type and the second one of the p-type, merged together and active in two different regions of

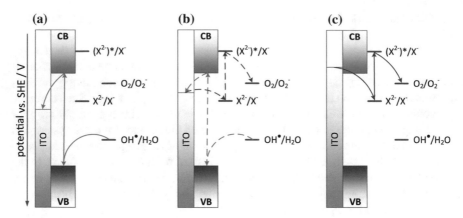

Fig. 17.11 A simplified energy diagram and the mechanistic explanation of three distinguishable modes in which the photocurrent may be generated within the system

the electrode potential and the incident light wavelengths. This remarkable property is utilized to encode the ternary information processing function by an appropriate selection of the electrical and optical inputs.

As in the case of modified wide band-gap semiconductors discussed in the former paragraphs in relation to the binary logic systems we can easily associate three different logic values (we will use balanced ternary) with respect to three different output ranges obtained from the cyanocarbon modified titanium dioxide electrode. In particular, the anodic photocurrents region will be tied to logical "1" (TRUE), the cathodic region will correspond to logical "−1" (FALSE) and obviously the area with no net photocurrents can be interpreted as logical "0" (UNKNOWN). The mapping seems natural if we take into consideration the mechanisms responsible for the signal generation in all three regions. Especially the situation depicted in the Fig. 17.11b is translated directly into the regime of the three-valued logic—the two competitive processes realized by the n-type and the p-type parts of the system lead to the situation in which the state of the device is actually unknown.

The transition between both gates can be achieved by a simple shift of the current threshold. Both of them can find an application in the construction of arithmetic devices and since they can be fairly easily concatenated (by an appropriate connection of the current output through a resistor with the voltage input of another gate) and can be powered by one of the inputs (the energy carried by the light pulses) they present a great platform for computing systems of higher complexity. Moreover, the proposed structure would not suffer from undesired feedback loops as the unidirectional information flow can be easily achieved by the introduction of e.g. Schottky diode. Similar approach with slightly different hybrid materials lead to the construction of ternary-to-binary decoders and other three-valued logic gates, such as *exclusive OR* gate.

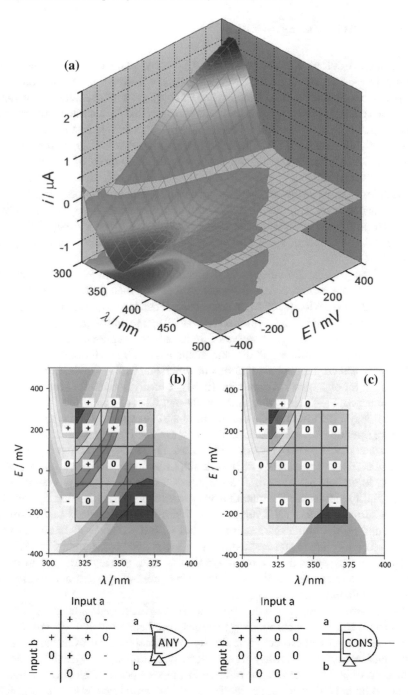

Fig. 17.12 The photocurrent action map (**a**) with its interpretation as two dual ternary logic gates: accept anything (**b**) and consensus (**c**). The transition between them is realised by the change of current intensity threshold

17.7 Towards Fuzzy Logic...

We mentioned before that nature utilizes also some more complex systems than commonly used binary logic and it is quite easy to prove this assumption as our everyday communication with the environment bases to a great extent on the vagueness. The sensory system provides a good example that inaccurate information, if its combined and interpreted in a proper way by our nervous system, can provide us with reliable picture of the situation [93]. It is so because our brains deal with the chunks of information which are blurred by the limitations of our bodies—i.e. the finite number of sensors per area unit, the potential loss of a signal during the transmission, *etc.* Nonetheless, we are capable of information processing at the far more advanced level than the most powerful computers (obviously, when it comes to the raw computational power, the artificial systems surpass our capabilities, but still the variety of information we are prepared to deal with makes our brains one of a kind) [94].

One of the greatest limitations of the binary system which hinders the universal description of our world, comes from the rule of excluded middle [75]. As it was mentioned in the previous section, in some cases it can lead to paradoxes or so called false dichotomies. One of the solutions is provided by the multi-valued logic systems. On the other hand, even if the number of logic states is increased we are still forced to operate with crisp values and the statements have to be well defined. Obviously, in the real world we encounter situations in which no absolute answer exists and the truth is a matter of subjective reasoning (it can be easily noticed for sensory systems of different people—their brains interpret the same set of conditions in a different, individual way). It shows clearly that we need another approach, in which elements of sets we operate with will not be exclusively assigned to a particular set.

That concept was formally developed by Zadeh in 1965 [95] (although some other mathematicians have used it before like Tarski, Łukasiewicz or Russell) [76, 96] and the fuzzy logic was introduced. The major advantage of using this system comes from its ability to translate the quantitative data into the qualitative description given by some human-like linguistic rules. Thus, it allows the operations on information bits resembling the human-like reasoning but with a formal mathematical system attached. The formalism and methodology was expanded by Mamdani in 1975 [97] and has been used since then in many types of systems and algorithms, particularly to provide the interaction with a user in various control systems and to aid some artificial intelligence algorithms [98].

With the growing interest in the application of the fuzzy logic numerous hardware and software solutions incorporating this idea have been provided. One of the most innovative is an attempt to combine it with the concepts taken from the ground of molecular electronics. The fuzzy logic systems operating at the molecular scale have become an object of interest not only from the viewpoint of electronics but also as a potential candidate to model the functions on neural networks found in living organisms.

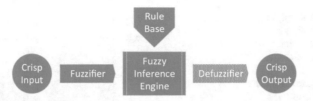

Fig. 17.13   The schematic representation of a fuzzy logic system. The defuzzifier is optional and it can be omitted if Takagi–Sugeno inference is applied or if the expert system is meant to provide a linguistic output

One of the first molecular devices harnessed to work as fuzzy logic systems (FLS) were presented by researchers like Gentili [93, 94, 99–104] and Deaton [105] at the beginning of the century and lately by Zadegan [106]. One of the next steps which should be taken is to transplant this approach onto the field of hybrid materials, which interconnects the molecular world with nanoscale structures of different kind, particularly of the semiconducting nature. One of the possible realizations of this idea is based on the use of systems exhibiting the PEPS effect.

In the case of PEPS-based systems both types of inputs together with the output can be characterized by functions which are smooth in nature (i.e. continuous and differentiable). This feature of the device allows us to divide particular ranges of the variables and label them in the process called fuzzification, then perform some operations according to the set of adopted rules inside a fuzzy inference machine, to finally obtain the output with (in the case of Mamdani method) or without (in the Takagi-Sugeno approach) defuzzification of the result (Fig. 17.13).

In the first step we express crisp values by their membership functions and label them in an association with different ranges (Fig. 17.14). The process we incorporate is very similar to the approach presented by Gentili [99] in his works on fuzzy logic realized in the molecular scale (we used FuzzyLite library with QTFuzzyLite graphic user interface and XFuzzy 3.3 design tool). We use triangular and trapezoidal type membership functions and create normal, convex fuzzy sets, which are presented in the Fig. 17.14.

It is important to note that the plot for the system output (Fig. 17.14c) corresponds directly with the surface presented in Fig. 17.12 and six chosen regions are related to the different photocurrent ranges or the situation in which no net photocurrent is generated. In the next step, in order to create a Mamdani FLS we formulate linguistic rules which will enable mapping of the input onto the output within the inference engine. These are in the form of IF...THEN... statements where the former part is called the antecedent or the premise and the latter is called the consequence or the conclusion. We can also incorporate three basic operators—AND, OR and NOT—to connect multiple antecedents. In the case of a larger base of rules and a rich data set, some implications may lead to contradictions, i.e. the same antecedents will lead to different consequences. To avoid that situation one can introduce weights which will strengthen or weaken certain rules, nevertheless for a chosen system it is not the case.

Fig. 17.14 Partitions in fuzzy sets of both inputs—**a** the wavelength of incident light and **b** the electrode potential—and **c** the generated photocurrent. The ordinates represent the membership functions values and the labels correspond to the symbols used in Table 17.3—Very Low (VL), Low (L), Medium (M), High (H), Very High (VH) and Extremely High (EH)

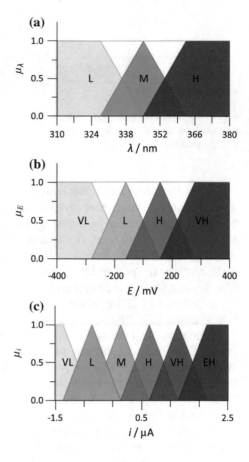

Many different rules can be created in the accordance with Figs. 17.14 and 17.12 some of which could be easily split into more basic implications. One of the examples is provided below where the instruction (17.4a) can be replaced by three simpler implications connected with the AND operator (17.4b). This equivalence is nontrivial as it results from the chosen definition of t-norm and s-norm, nonetheless from the linguistic point of view we understand that this interchangeability is valid. Obviously, even more complex formulations could be found—but usually we want to avoid unnecessary complication of the inference system.

$$\text{IF } (\lambda = \text{L AND } E = \text{VL}) \text{ OR } (\lambda = \text{M AND } E = \text{L}) \text{ OR } (\lambda = \text{H AND } E = \text{H})\textbf{THEN } i = \text{L} \tag{17.4a}$$

$$\begin{aligned}
&\text{IF } \lambda = \text{L AND } E = \text{VL THEN } i = \text{L} \\
&\text{IF } \lambda = \text{M AND } E = \text{L THEN } i = \text{L} \\
&\text{IF } \lambda = \text{H AND } E = \text{H THEN } i = \text{L}
\end{aligned} \tag{17.4b}$$

Although the explicit enumeration of implications may seem to be the clearest representation of rules base, in the case of systems with a large number of labels it

Table 17.3 The rules matrix for the fuzzy logic system based on the photocurrents map presented in Fig. 17.12

		VL	L	H	VH	Potential
	L	L	H	VH	EH	
Wavelength	M	VL	L	M	M	
	H	VL	VL	L	M	

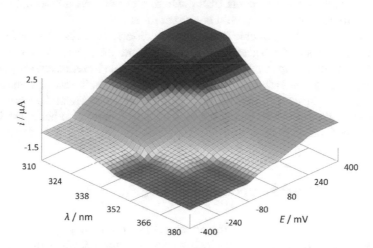

Fig. 17.15 The fuzzy inference system output as the approximation of the experimental result shown in Fig. 17.12

becomes tedious. Sometimes, a better result can be achieved if we chose only one operator which will connect our inputs and the rules base in the form of a matrix. For the device described here we can decide to work with the AND operator and compose the matrix in the form given in Table 17.3. Since we have two independent inputs represented by the fuzzy sets containing three and four labels (for wavelengths and potentials respectively) we will end up with the 3×4 matrix.

Now, having the rules base and the fuzzy sets we can operate on, we can propel our inference engine and obtain the output in a linguistic form. If we are unsatisfied with the answer in this form and would rather get the crisp values from our system we have to perform the defuzzification. By an appropriate selection of the defuzzification method we can use our FLS as a so called approximator which will be able to predict the crisp values of generated photocurrent based on the fuzzified representation of inputs values. In this way we can also try to recreate the photocurrent map presented in Fig. 17.12. Such an attempt is shown in Fig. 17.15. We can easily note that the main features of the experimental result are well reproduced.

In the presented case, the fuzzy logic system is rather simple, i.e. it is based on the relatively small fuzzy sets with a limited number of membership functions (labels) and a simplified set of rules. We also used standard t-norm and s-norm operations (i.e. the minimum for conjunction and the maximum for disjunction), basic accumulation and implication methods (the maximum and the minimum respectively), finally we utilized the defuzzification based on the centroid calculation. Still, we were able to recreate some of the features of the experimental results. All of the abovementioned parameters have an impact on the FLS prediction capabilities and to optimize the FLS as an approximator they should be further tuned.

At the same time it is possible to add some additional features into our FLS such as so called hedges [107]. These are the terms put into the rules base that influence the shape of membership functions in fuzzy sets. They can be represented by adverbs like: very, slightly, somewhat, etc. On the other hand we can introduce some aspects of learning mechanisms found in neural networks and create a neural adaptive fuzzy inference system capable of tuning the membership functions to better predict the output values [108, 109].

17.8 …and Neuromimetic Computing

One of the biggest challenges for the researchers working on unconventional computing systems is the quest for mimicking natural neural networks functionalities. Despite the fact, that with the modern supercomputers we have already reached the computational speed of our brains (as matter of fact we have even surpassed it by the order of magnitude) [110] the complexity of cognitive functions of our nervous system are still beyond the reach of classical architectures—we are definitely better in synthesising big portions of inaccurate data into a well-defined output and our reasoning is well-tailored for tasks such as pattern recognition.

The superiority of structures found in nature—in terms of the learning efficiency—results from their plasticity [111–114]. On the one hand, this feature bases on the capability of establishing new synaptic connections in order to enhance the reasoning processes (for example by the increase in the number of active regions within the neuron, which can connect to the nearby cells). On the other hand, it follows from the difference in used paradigms, as we do not limit ourselves to merely binary type of information and easily interpret all the intermediate states, as it happens, for example in the case of data provided by our sensory system which is burdened with the uncertainty due to the way the receptors gather and process information. That kind of flexibility is an unique feature of our brain and is not present in any artificial system (at least not to that extent) [114].

Nonetheless, some attempts to create similar algorithms have been already made and some of them were at least partially successful. One of the most prominent examples is the development of the neural networks, which—through an appropriate training—are capable of working as approximators, control, pattern and sequence recognition systems. The basic ideas behind the neural networks have been later combined with some biomimetic concepts (e.g. related to the functions of neocortex) giving rise to the another machine learning model, namely the hierarchical temporal memory (HTM), which also found various applications [115, 116].

The HTM networks, are the systems in which an advanced information processing occurs within the large hierarchical network of simple computing nodes. This approach mimics the structure of the cerebral cortex, where information is processed in layered circuits (usually six main layers) which undergo characteristic bottom-up, top-down, and horizontal interactions. Such computational approach does not require a high numerical efficiency of a single node, as it is based on a collective behaviour of a large number of simple, interconnected nodes. This idea somehow resembles the concept of a 'smart dust', first introduced by Kristofer Pister in 1997 [117]. It is based on a large number of small ($\sim$1 mm^3) ubiquitous and autonomous computing devices, equipped with sensors, processing units, memory and communication facilities conglomerated into a dynamic mobile network. This idea was further developed as various sensing and labelling nanostructures became available, [118, 119] and finally achieved the complexity of a 'lab-on-a-particle' via the integration of nanoparticles with functional biomolecules [120–124]. Apart from sensing applications, similar concepts have been applied in the construction of simple logic devices based on micellar and liposomic structures with embedded molecular sensors, switches and logic gates [125–127].

At the same time, human reasoning may be partially characterised in terms of the fuzzy logic—the brain deals with data which is often imprecise and blurry and its interpretation often takes the form of implications gathered in the form of a rules base. Even some aspects of the information processing occurring at the neural and synaptic levels seem to be well explained by this paradigm. Due to that fact, some researchers postulate the possibility to simulate some of the functions of biological neural networks through chemical systems, which characteristics can be related to the concepts of the fuzzy logic [93]. This approach is interesting, as the same systems could be utilised as bio- and chemosensors, providing a good model for the sensory system found in living organisms.

Another approach is more hardware-centred and base on the idea of reconstructing the connections found in biological structures within fully artificial systems, the complex networks of specially designed processors equipped with memory cells or elements exhibiting some memristive properties, for example in the form of multiple interfaces present in hybrid materials. Still, in the most of proposed solutions, information is encoded in the form of chemical or electrical signals and these artificial synapses stay blind. Furthermore, in most of the cases, the reaction times of these systems are incomparably longer than the ones found in biological structures [128–134].

The systems utilizing the PEPS effect, could on the other hand, be programmed with light and enrich the set of possible interactions. Moreover, since the components described in the previous paragraphs could be concatenated and they can be powered through one of the inputs, the task of building a complex network of such devices seams—at least theoretically—to be fairly simple. The preliminary data obtained for several hybrid materials exhibiting the PEPS effect and time-dependant photocurrent

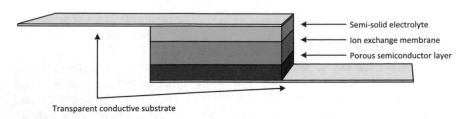

Transparent conductive substrate

Fig. 17.16 The schematic representation of a photoactive electrode exhibiting synapse-like behaviour

generation profiles indicate that the synapse-like characteristics of the output can be achieved fully artificially and with the use of optical input.

Particularly, the system composed of cadmium sulphide nanoparticles combined with thermally activated multi-walled carbon nanotubes (MWCNT) in the presence of ionic liquid present a good candidate for the observation of a Hebbian-like plasticity, [135] characteristic for neural connections. According to the experimental data, it seems that this peculiar behaviour is due to the subtle equilibrium between charge trapping/detrapping mechanisms and processes controlled by the diffusion. This assumption is supported by the results of a numerical experiment, in which the equivalent circuit is proposed for the system and its response to the pulsed stimulus is computed.

The main part of the experiment involves irradiation with the short (approx. $100 \mu s$) pulses of blue light. The electrode was made of CdS/MWCNT hybrid material [26, 136] deposited onto ITO@PET foil and covered with another slide of ITO@PET with semi-solid electrolyte (Fig. 17.16). Upon illumination, the device generates anodic photocurrents, which characteristics strongly depend on the time parameters (number of pulses, their length and the interval between them) of the input.

This behaviour could be compared to the response of a living neuron, which depends on the time correlation between post- and presynaptic spikes. Moreover, in similarity to the biological structures and in accordance with the spike-timing-dependant plasticity (STDP) rule, [137] here we also observe two distinctive modes of the system operation—the potentiation (when the consecutive photocurrent spike is enhanced) and the depression (when the intensity decreases for the consecutive spike)—depending on the electrode potential and the pulses length (Fig. 17.17).

In the case of the information processing realised on the neural level in living organisms, the change of the second spike intensity in series can be modelled mathematically and the details of a particular model will depend on the type of a synapse and its internal parameters. The basic form of such relation within the STDP model can be represented by the combination of exponential functions and is usually called the learning window [138]. Similarly, for the artificial system presented above, it is also possible to fit a function to the plot depicting the increase/decrease of the second photocurrent spike intensity with respect to the first one in series. The relation will take the form given below (Eq. 17.5):

Fig. 17.17 The effect of
applied potential on the
photocurrent response of a
photoactive electrode
composed of CdS/MWCNT
hybrid and ionic liquid—**a** in
the potentiation and **b** in the
depression modes

(a)

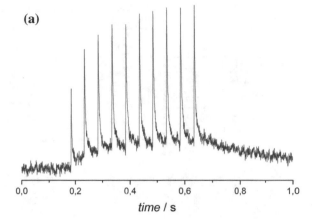

time / s

(b)

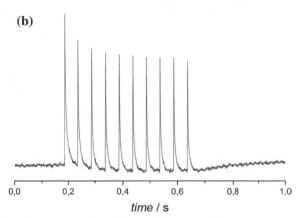

time / s

$$f(\Delta t) = \alpha_1 \exp\left(-\frac{\Delta t}{T_1}\right) + \alpha_2 \exp\left(-\frac{\Delta t}{T_2}\right) + \beta \tag{17.5}$$

where Δt is the time interval between the preceding and the following spikes, $\alpha_{1,2}$
can be interpreted as the parameters determined by synaptic weights and $T_{1,2}$ are
analogues of the characteristic time constants. The fitting parameters are equal to:
$\alpha_1 = 3.014 \pm 0.034$, $\alpha_2 = 4.80 \pm 0.21$, $T_1 = 116.4 \pm 3.7$ ms, $T_2 = 6.88 \pm 0.31$ ms
and $\beta = 1.722 \pm 0.024$. It is remarkable that the time constants are comparable to
the ones found in biological structures, indicating that the presented hybrid system
successfully mimics the behaviour characteristic for natural synapses [139].

In order to better understand the mechanism behind the observed short-term memory effect a numerical experiment was designed. The equivalent circuit was proposed
(Fig. 17.18a)—with three RC loops, two resistors resembling the ohmic resistance
of electrolyte and CdS/MWCNT junctions and two diodes as the representation of
Schottky junctions between the support and the material and between MWCNTs
and CdS nanoparticles [22, 27, 140]—and the simulation was performed. As the

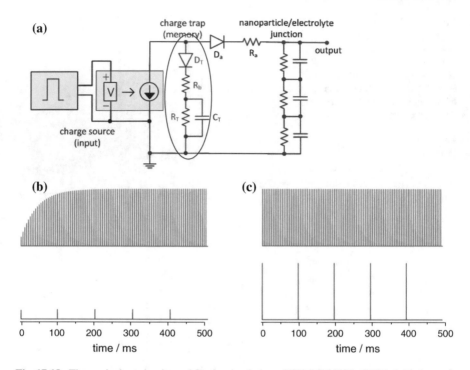

Fig. 17.18 The equivalent circuit used for the simulation of ITO/MWCNTs/CdS hybrid electrode behaviour (**a**) with the results of a numerical experiment performed for the system with (**b**) and without (**c**) a branch responsible for trapping/detrapping acts (marked with the *red line*)

result, we managed to recreate the behaviour of the hybrid material-based device, particularly the potentiation of current signals with the decreasing interval between consecutive pulses (Fig. 17.18b, c).

The most interesting part of the results interpretation is related to the presence of a branch responsible for charge trapping/detrapping events in the circuit (marked with a red line in Fig. 17.18a). This part of the electronic system is essential for the synapse-like behaviour, as its removal results in the lack of any memory effect upon stimulation and the obtained plot agrees with the experimental result for an unmodified semiconductor. It is also remarkable that upon the change of the either potential or the pulses length we can induce the transition between two described modes—i.e. the potentiation and the depression.

To fully describe the interaction between individual constituents of the photo-electrode an energetic diagram can be constructed. We know that the profile of each photocurrent spike does not depend on the stimulus parameters, as only the increase/decrease in their intensity is observed. Moreover, the photocurrent action spectrum profile does not change upon modification with multi-walled carbon nanotubes with respect to neat cadmium sulphide nanoparticles. It allows us to conclude that the photocurrent is generated entirely on CdS sites and MWCNTs are

Fig. 17.19 The energetic diagram with the explanation of the mechanism responsible for the observed synapse-like behaviour

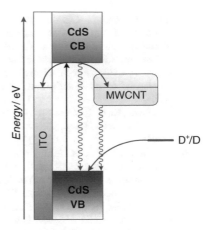

fully responsible for trapping/detrapping of charge carriers. Finally, the redox-active species present in electrolyte control the time domain of the memory effect through the diffusion coefficients, as they take part in closing the charge transfer cycle (Fig. 17.19) [141].

As it was mentioned in the former paragraphs it is possible to concatenate the photoactive electrodes into more complex systems which should exhibit more sophisticated interaction modes—hopefully with some kind of the information processing interference within the superstructure leading to the self-programming which would eventually result in the enhancement of the reasoning and learning capabilities. These experiments are planned to be performed in a near future with semiconductors active in different wavelength ranges and several different optical inputs connected to the matrices of such photoactive cells. The used modifications and the construction of cells could also vary to further tune the properties of the system.

Acknowledgments The financial support from the National Science Centre (grants no. UMO-2012/05/N/ST5/00327, UMO-2013/11/D/ST5/03010 and UMO-2015/18/A/ST4/00058), the Ministry of Science and Higher Education (grant no. IP2012 030772) and the Foundation for Polish Science (grant no. 71/UD/SKILLS/2014 carried-out within the INTER programme, co-financed from the European Union within the European Social Fund) is gratefully acknowledged. P. Kwolek was supported by the Foundation for Polish Science within the START fellowship.

References

1. Abidian, M.R., Martin, D.C.: Multifunctional nanobiomaterials for neural interfaces. Adv. Funct. Mater. **19**, 573–585 (2009)
2. Martino, N., Ghezzi, D., Benfenati, F., Lanzani, G., Antognazza, M.R.: Organic semiconductors for artificial vision. J. Mater. Chem. B **1**, 3768–3780 (2013)

3. Somasundaram, S., Chenthamarakshan, C.R., de Tacconi, N.R., Ming, Y., Rajeshwar, K.: Photoassisted deposition of chalcogenide semiconductors on the titanium dioxide surface: mechanistic and other aspects. Chem. Mater. **16**, 3846–3852 (2004)
4. de Tacconi, N.R., Chenthamarakshan, C.R., Rajeshwar, K., Tacconi, E.J.: Selenium modified titanium dioxide photochemical diode/electrolyte junctions: photocatalytic and electrochemical preparation, characterization and model simulations. J. Phys. Chem. B **109**, 11953–11960 (2005)
5. Rajeshwar, K., de Tacconi, N.R., Chenthamarakshan, C.R.: Spatially directed electrosynthesis of semiconductors for photoelectrochemical applications. Curr. Opin. Solid State **8**, 173–182 (2004)
6. Poznyak, S.K., Kulak, A.I.: Photoelectrochemical properties of bismuth oxyhalide films. Electrochim. Acta **35**, 1941–1947 (1990)
7. Szaciłowski, K., Macyk, W.: Chemical switches and logic gates based on surface modified semiconductors. Comptes Rendus Chim. **9**, 315–324 (2006)
8. Memming, R.: Semiconductor Electrochemistry, 1st edn. Wiley-VCH, Weinheim (2001)
9. Kwolek, P., Szaciłowski, K.: Photoelectrochemistry of n-type bismuth oxyiodide. Electrochim. Acta **104**, 448–453 (2013)
10. Kwolek, P., Pilarczyk, K., Tokarski, T., Mech, J., Irzmański, J., Szaciłowsk, K.: Photoelectrochemistry of n-type antimony sulfoiodide nanowires. Nanotechnology **26**, 105710-1–105710-9 (2015)
11. Kwolek, P., Pilarczyk, K., Tokarski, T., Lapczynska, M., Pacia, M., Szacilowski, K.: Lead molybdate - a promising material for optoelectronics and photocatalysis. J. Mater. Chem. C. **3**, 2614–2623 (2015)
12. Beránek, R., Kisch, H.: A hybrid semiconductor electrode for wavelength-controlled switching of the photocurrent direction. Angew. Chem. Int. Ed. **47**, 1320–1322 (2008)
13. Kwolek, P., Pilarczyk, K., Tokarski, T., Lewandowska, K., Szacilowski, K.: BixLa1-xVO4 solid solutions: tuning of electronic properties via stoichiometry modifications. Nanoscale **6**(4), 2244–2254 (2014)
14. Wardman, P.: Reduction potentials of one-electron couples involving free radicals in aqueous solution. J. Phys. Chem. Ref. Data **18**, 1637–1755 (1989)
15. Sawyer, D.T., Valentine, J.S.: How super is superoxide? Acc. Chem. Res. **14**, 393–400 (1981)
16. Bard, A.J., Stratmann, M., Licht, S.: Encyclopedia of Electrochemistry, Semiconductor Electrodes and Photoelectrochemistry. Wiley, Weinheim (2002)
17. Walter, M.G., Warren, E.L., McKone, J.R., Boettcher, S.W., Mi, Q., Santori, E.A., et al.: Solar water splitting cells. Chem. Rev. **110**, 6446–6473 (2010)
18. Tan, M.X., Laibinis, P.E., Nguyen, S.T., Kesselman, J.M., Stanton, C.E., Lewis, N.S.: Principles and applications of semiconductor photoelectrochemistry. Prog. Inorg. Chem. **41**, 21–144 (1994)
19. van de Krol, R., Grätzel, M.: Photoelectrochemical Hydrogen Production. Springer, New York (2011)
20. Yin, W.J., Wei, S.H., Al-Jassim, M.M., Turner, J., Yan, Y.: Doping properties of monoclinic BiVO4 studied by first-principles density-functional theory. Phys. Rev. B **83**(15), 1–11 (2011)
21. Hodes, G., Howell, I.D.J., Peter, L.M.: Nanocrystalline photoelectrochemical cells a new concept in photovoltaic cells. J. Electrochem. Soc. **139**, 3136–3140 (1992)
22. Szaciłowski, K., Macyk, W., Stochel, G.: Synthesis, structure and photoelectrochemical properties of the TiO_2 - Prussian blue nanocomposite. J. Mater. Chem. **16**, 4603–4611 (2006)
23. Macyk, W., Stochel, G., Szaciłowski, K.: Photosensitization and photocurrent switching effect in nanocrystalline titanium dioxide functionalized with iron(II) complexes: a comparative study. Chem. Eur. J. **13**, 5676–5687 (2007)
24. Gaweda, S., Stochel, G., Szaciłowski, K.: Bioinspired nanodevice based on the folic acid/titanium dioxide system. Chem. Asian J. **2**, 580–590 (2007)
25. Gaweda, S., Stochel, G., Szaciłowski, K.: Photosensitization and photocurrent switching in carminic acid/titanium dioxide hybrid material. J. Phys. Chem. C **112**, 19131–19141 (2008)

26. Podborska, A., Gaweł, B., Pietrzak, Ł., Szymańska, I.B., Jeszka, J.K., Łasocha, W., et al.: Anomalous photocathodic behavior of CdS within the Urbach tail region. J. Phys. Chem. C **113**, 6774–6784 (2009)
27. Hebda, M., Stochel, G., Szaciłowski, K., Macyk, W.: Optoelectronic switches based on wide bandgap semiconductors. J. Phys. Chem. B **110**, 15275–15283 (2006)
28. Szaciłowski, K., Macyk, W., Stochel, G.: Light-driven OR and XOR programmable chemical logic gates. J. Am. Chem. Soc. **128**, 4550–4551 (2006)
29. Szaciłowski, K., Macyk, W.: Working prototype of an optoelectronic XOR/OR/YES reconfigurable logic device based on nanocrystalline semiconductors. Solid State Electron. **50**, 1649–1655 (2006)
30. Lu, B., Zhu, Y.: Synthesis and photocatalysi performances of bismuth oxynitrate with layared structures. Phys. Chem. Chem. Phys. **16**, 16509–16514 (2014)
31. El Harakeh, M., Alawieh, L., Saouma, S., Halaoui, L.I.: Charge separation and photocurrent polarity-switching at CdS quantum dots assembly in polyelectrolyte interfaced with hole scavengers. Phys. Chem. Chem. Phys. **11**, 5962–5973 (2009)
32. Ogawa, S., Hu, K., Fan, F., Bard, A.J.: Photoelectrochemistry of films of quantum size lead sulfide particles incorporated in self-assembled monolayers on gold. J. Phys. Chem. B **101**, 5707–5711 (1997)
33. Gaweda, S., Kowalik, R., Kwolek, P., Macyk, W., Mech, J., Oszajca, M., et al.: Nanoscale digital devices based on the photoelectrochemical photocurrent switching effect: preparation, properties and applications. Isr. J. Chem. **51**, 36–55 (2011)
34. Kuncewicz, J., Ząbek, P., Kruczała, K., Szaciłowski, K., Macyk, W.: Photocatalysis involving a visible light-induced hole injection in a chromate(VI)-TiO$_2$ system. J. Phys. Chem. C **116**, 21762–21770 (2012)
35. Chen, X., Yeganeh, S., Qin, L., Li, S., Xue, C., Braunschweig, A.B., et al.: Chemical fabrication of heterometallic nanogaps for molecular transport junctions. Nano Lett. **12**, 3974–3979 (2009)
36. Prezhdo, O.V., Duncan, W.R., Prezhdo, V.V.: Photoinduced electron dynamics at the chromophore-semiconductor interface: a time-domain ab initio perspective. Prog. Surf. Sci. **84**, 30–68 (2009)
37. Creutz, C., Brunschwig, B.S., Sutin, N.: Interfacial charge transfer absorption: semiclassical treatment. J. Phys. Chem. B **109**, 10251–10260 (2005)
38. Creutz, C., Brunschwig, B.S., Sutin, N.: Interfacial charge transfer absorption: application to metal-molecule assemblies. Chem. Phys. **324**, 244–258 (2006)
39. Adams, D.M., Brus, L., Chidsey, C.E.D., Creager, S., Creutz, C., Kagan, C.R., et al.: Charge transfer on the nanoscale: current status. J. Phys. Chem. B **107**, 6668–6697 (2003)
40. Creutz, C., Brunschwig, B.S., Sutin, N.: Interfacial charge transfer absortion: 3. application to semiconductor-molecule assemblies. J. Phys. Chem. B **110**, 25181–25190 (2006)
41. Sakata, T., Hashimoto, K., Hiramoto, M.: New aspects of electron transfer on semiconductor surface: dye-sensitization system. J. Phys. Chem. **94**, 3040–3045 (1990)
42. Kitao, O.: Photoinduced electron transfer in dye-sensitized solar cells: modified Sakata-Hashimoto-Hiramoto model (MSHH). J. Phys. Chem. C **111**, 15889–15902 (2007)
43. Kwolek, P., Oszajca, M., Szaciłowski, K.: Catecholate and 2,3-acenediolate complexes of d^0 ions as prospective materials for molecular electronics and spintronics. Coord. Chem. Rev. **56**, 1706–1731 (2012)
44. Macyk, W., Szaciłowski, K., Stochel, G., Buchalska, M., Kuncewicz, J., Łabuz, P.: Titanium(IV) complexes as direct TiO$_2$ photosensitizers. Coord. Chem. Rev. **254**, 2687–2701 (2010)
45. Oszajca, M., Kwolek, P., Mech, J., Szaciłowski, K.: Substituted polyacenes as prospective modifiers of TiO$_2$ surface. Curr. Phys. Chem. **1**, 242–260 (2011)
46. Rego, L.G.C., Batista, V.S.: Quantum dynamics simulations of interfacial electron transfer in sensitized TiO$_2$ semiconductors. J. Am. Chem. Soc. **125**, 7989–7997 (2003)
47. Abuabara, S.G., Rego, L.G.C., Batista, V.S.: Influence of thermal fluctuations on interfacial electron transfer in functionalized TiO$_2$ semiconductors. J. Am. Chem. Soc. **127**, 18234–18242 (2005)

48. Prezhdo, O.V., Duncan, W.R., Prezhdo, V.V.: Dynamics of the photoexcited electron at the chromophore-semiconductor interface. Acc. Chem. Res. **41**(2), 339–348 (2008)
49. Duncan, W.R., Prezhdo, O.V.: Theoretical studies of photoinduced electron transfer in dye-sensitized TiO_2. Annu. Rev. Phys. Chem. **58**, 143–184 (2007)
50. Ardo, S., Meyer, G.J.: Photodriven heterogeneous charge transfer with transition-metal compounds anchoder to TiO_2 surfaces. Chem. Soc. Rev. **38**, 115–164 (2009)
51. Duncan, W.R., Stier, W.M., Prezhdo, O.V.: Ab initio nonadiabatic molecular dynamics of the ultrafast electron injection across the alizarin-TiO_2 interface. J. Am. Chem. Soc. **127**, 7941–7951 (2005)
52. Rajh, T., Chen, L.X., Lukas, K., Liu, T., Thurnauer, M.C., Tiede, D.M.: Surface restructuring of nanoparticles: an efficient route for ligand-metal oxide crosstalk. J. Phys. Chem. B **106**, 10543–10552 (2002)
53. Regazzoni, A.E., Mandelbaum, P., Matsuyoshi, M., Schiller, S., Bilmes, S.A., Blesa, M.A.: Adsorption and photooxidation of salicylic acid on titanium dioxide: a surface complexation description. Langmuir **14**, 868–874 (1998)
54. Szaciłowski, K., Macyk, W.: Photoelectrochemical photocurrent switching effect: a new platform for molecular logic devices. Chimia **61**, 831–834 (2007)
55. Szaciłowski, K., Macyk, W., Hebda, M., Stochel, G.: Redox-controlled photosensitization of nanocrystalline titanium dioxide. Chem. Phys. Chem. **7**, 2384–2391 (2006)
56. Long, M., Beránek, R., Cai, W., Kisch, H.: Hybrid semiconductor electrodes for light-driven photoelectrochemical switches. Electrochim. Acta **53**, 4621–4626 (2008)
57. Agostinelli, G., Dunlop, E.D.: Local inversion of photocurrent in cadmium telluride solar cells. Thin Solid Films **431–432**, 448–452 (2003)
58. Nam, K.M., Park, H.S., Lee, H.C., Meekins, B.H., Leonard, K.C., Bard, A.J.: Compositional screening of the Pb-Bi-Mo-O system. Spontaneous formation of a composite of p-$PbMoO_4$ and n-Bi_2O_3 with improved photoelectrochemical efficiency and stability. J. Phys. Chem. Lett. **4**, 2707–2710 (2013)
59. Zheng, J.Y., Song, G., Kim, C.W., Kang, Y.S.: Facile preparation of p-CuO and p-CuO/n-CuWO4 junction thin films and their photoelectrochemical properties. Electrochim. Acta **69**, 340–344 (2012)
60. Warzecha, M., Oszajca, M., Pilarczyk, K., Szaciłowski, K.: A three-valued photoelectrochemical logic device realising accept anything and consensus operations. Chem. Commun. **51**, 3559–3561 (2015)
61. Lewandowska, K., Podborska, A., Kwolek, P., Kim, T.-D., Lee, K.-S., Szaciłowski, K.: Optical signal demultiplexing and conversion in the fullerene-oligothiophene-CdS system. Appl. Surf. Sci. **319**, 285–290 (2014)
62. Lewandowska, K., Szaciłowski, K.: Molecular photodiode and two-channel demultiplexer based on the [60]fullerene-porphyrin tetrad. Aust. J. Chem. **64**, 1409–1413 (2011)
63. Gaweda, S., Podborska, A., Macyk, W., Szaciłowski, K.: Nanoscale optoelectronic switches and logic devices. Nanoscale **1**, 299–316 (2009)
64. Bendikov, M., Wudl, F., Perepichka, D.F.: Tetrathiafulvalenes, oligoacenes and their buckminsterfullerene derivatives: the bricks and mortar of organic electronics. Chem. Rev. **104**, 4891–4945 (2004)
65. Ishii, H., Sugiyama, K., Ito, E., Seki, K.: Energy level alignment and interfacial electronic structures at organic/metal and organic/organic interfaces. Adv. Mater. **11**, 605–625 (1999)
66. Linsebigler, A.L., Lu, G., Yates Jr., J.T.: Photocatalysis on TiO_2 surfaces: principles, mechanisms, and selected results. Chem. Rev. **95**, 735–758 (1995)
67. Fujishima, A., Zhang, X., Tryk, D.A.: TiO_2 photocatalysis and related surface phenomena. Surf. Sci. Rep. **63**, 515–582 (2008)
68. Podborska, A., Szaciłowski, K.: Towards 'computer-on-a-particle' devices: optoelectronic 1:2 demultiplexer based on nanostructured cadmium sulfide. Aust. J. Chem. **63**, 165–168 (2010)
69. Maharjan, A., Pemasiri, K., Kumar, P., Wade, A., Smith, L.M., Jackson, H.E., et al.: Room temperature photocurrent spectroscopy of single zincblende and wurtzite InP nanowires. Appl. Phys. Lett. **94**, 193115 (2009)

70. Yang, F., Huang, K., Ni, S., Wang, Q., He, D.: $W_{18}O_{49}$ nanowires as ultraviolet photodetector. Nanoscale Res. Lett. **5**, 416–419 (2010)
71. Fang, X., Bando, Y., Liao, M., Gautam, U.K., Zhi, C., Dierre, B., et al.: Single-crystalline ZnS nanobelts as ultraviolet-light sensors. Adv. Mater. **21**, 2034–2039 (2009)
72. Zhai, T., Fang, X., Li, L., Bando, Y., Goldberg, D.: One-dimensional CdS nanostructures: synthesis, properties and applications. Nanoscale **2**, 168–187 (2010)
73. Mech, J., Kowalik, R., Podborska, A., Kwolek, P., Szaciłowski, K.: Arithmetic device based on multiple Schottky-like junctions. Aust. J. Chem. **63**, 1330–1333 (2010)
74. Łukasiewicz, J.: Aristtle's Syllogistic from the Standpoint of Modern Formal Logic. Claredon Press, Oxford (1951)
75. Goble, L.: The Blackwell Guide to Philosophical Logic. Wiley-Blackwell, Hoboken (2001)
76. Łukasiewicz, J.: In: Borkowski, L. (ed.) Selected Works. North-Holland, Amsterdam (1970)
77. Fisch, M., Turquette, A.: Peirce's triadic logic. TCS Peirce Soc. **2**, 71–85 (1966)
78. Elati, M., Neuvial, P., Bolotin-Fukuhara, M., Barillot, E., Radvanyi, F., Rouveirol, C.: LICORN: learning cooperative regulation networks from gene expression data. Bioinformatics **23**, 2407–2414 (2007)
79. Morris, M.K., Saez-Rodriguez, J., Sorger, P.K., Lauffenburger, D.A.: Logic-based models for the analysis of cell signaling networks. Biochemistry **49**, 3216–3224 (2010)
80. IGad, R.: Logical and Linguistic Problems of Social Communication with Aymara People. International Development Research Centre (IDRC), Ottawa (1984)
81. Miriel, J., Fermanel, F.: Classic wall gas boiler regulation and a new thermostat using fuzzy logic – Improvements achieved with a fuzzy thermostat. Appl Energ. **68**, 229–47 (2001)
82. Glaser, A.: History of Binary and Other Nondecimal Numeration. Tomash Publishers, Los Angeles (1971)
83. Jin, Y., He, H., Lü, Y.: Ternary optical computer principle. Sci. China Ser. F **46**, 145–150 (2003)
84. Brousentsov, N.P., Maslov, S.P., Ramil Alvarez, J., Zhogolev, E.A.: Development of ternary computers at Moscow State University: Russian virtual computer museum. Available from: http://www.computer-museum.ru/english/setun.htm (2010). Accessed 26 March 2010
85. Stakhov, A.: Brousentsov's ternary principle, Bergman's number system and ternary mirror-symmetrical arithmetic. Comput. J. **45**, 221–236 (2002)
86. Morisue, M., Endo, J., Morooka, T., Shimizu, N., Sakamoto, M.: In: 28th International Symposium on Multi Valued Logic, pp. 19–24 (1998)
87. Natarajan, J., Coles, M., Cebollero, M.: Pro T-SQL Programmer's Guide. Apress, New York (2015)
88. Liang, J., Chen, L., Han, J., Lombardi, F.: Design and evaluation of multiple valued logic gates using pseudo N-type carbon nanotube FETs. IEEE Trans. Nanotechnol. **13**, 695–708 (2014)
89. Bajec, I.L., Zimic, N., Mraz, M.: The ternary quantum-dot cell and ternary logic. Nanotechnology **17**, 1937–1942 (2006)
90. Saidutt, P.V., Srinivas V., Phaneendra, P.S., Muthukrishnan, N.M. (eds.): In: 2012 Asia Pacific Conference on Postgraduate Research in Microelectronics and Electronics (PrimeAsia) (2012)
91. Liu, S., Li, C., Wu, J., Liu, Y.: Optoelectronic multiple-valued logic implementation. Opt. Lett. **14**, 713–715 (1989)
92. Remón, P., Ferreira, R., Montenegro, J.M., Suau, R., Pérez-Inestrosa, E., Pischel, U.: Reversible molecular logic: a photophysical example of a Feynman gate. ChemPhysChem **10**, 2004–2007 (2009)
93. Gentili, P.L.: The human sensory system as a collection of specialized fuzzifiers: a conceptual framework to inspire new artificial intelligent systems computing with words. J. Intell. Fuzzy Syst. **27**, 2137–2151 (2014)
94. Gentili, P.L.: Small steps towards the development of chemical artificial intelligent systems. RSC Adv. **3**, 25523–25549 (2013)
95. Zadeh, L.A.: Fuzzy sets. Inf. Control **8**, 338–353 (1965)

96. Bergmann, M.: An Introduction to Many-Valued and Fuzzy Logic: Semantics, Algebras, and Derivation Systems. Cambridge University Press, Cambridge (2008)
97. Mamdani, E.H., Assilian, S.: An experiment in linguistic synthesis with a fuzzy logic controller. Int. J. Hum. Comput. Stud. **51**, 135–147 (1999)
98. Bryan, L.A., Bryan, E.A.: Programmable Controllers: Theory and Implementation. Industrial Text Company, Marietta (1997)
99. Gentili, P.L., Horvath, V., Vanag, V.K., Epstein, I.R.: Belousov-Zhabotinsky "chemical neuron" as a binary and fuzzy logic processor. Int. J. Uncov. Comput. **8**, 177–192 (2012)
100. Gentili, P.L.: Boolean and fuzzy logic gates based on the interaction of flindersine with bovine serum albumin and tryptophan. J. Phys. Chem. A **112**, 11992–11997 (2008)
101. Gentili, P.L.: Boolean and fuzzy logic implemented at the molecular level. Chem. Phys. **336**, 64–73 (2007)
102. Gentili, P.L.: The fundamental fuzzy logic operators and some complexoolean logic circuits implemented by the chomogenism of a spirooxazine. Phys. Chem. Chem. Phys. **13**, 20335–20344 (2011)
103. Gentili, P.L.: The fuzziness of a chromogenic spiroxazine. Dyes Pigments **110**, 235–248 (2014)
104. Gentili, P.L.: Molecular processors: from qubits to fuzzy logic. ChemPhysChem **12**, 739–745 (2011)
105. Deaton, R., Garzon, M.: Fuzzy logic with biomolecules. Soft Comput. **5**, 2–9 (2001)
106. Zadegan, R.M., Jepsen, M.D.E., Hildebrandt, L.L., Birkedal, V., Kjems, J.: Small **11**, 1811–1817 (2015)
107. Huynh, V.N., Ho, T.B., Nakamori, Y.: A parametric representation of linguistic hedges in Zadeh's fuzzy logic. Int. J. Approx. Reason. **30**, 203–223 (2002)
108. Jang, J.S.R.: ANFIS: adaptive-network-based fuzzy inference system. IEEE Trans. Syst. Man Cybern. **23**, 665–684 (1993)
109. Jang, J.S.R., Sun, C.T.: Neuro-fuzzy modeling and control. Proc. IEEE **83**, 378–405 (1995)
110. Available from http://top500.org/
111. Song, S., Miller, K.D., Abbott, L.F.: Cometitive Hebbian learning through spike-timing-dependent synaptic plasticity. Nat. Neurosci. **3**, 919–926 (2000)
112. Caporale, N., Dan, Y.: Spike timing-dependent plasticity: a Hebbian learning rule. Annu. Rev. Neurosci. **31**, 25–46 (2008)
113. Feldman, D.E.: The spike-timing dependence of plasticity. Neuron **75**(4), 556–571. PubMed PMID: 22920249. Pubmed Central PMCID: 3431193, 23 Aug 2012
114. Martin, S.J., Grimwood, P.D., Morris, R.G.M.: Synaptic plasticity and memory: an evaluation of the hypothesis. Annu. Rev. Neurosci. **23**, 649–711 (2000)
115. George, D., Jaros, B.: The HTM learning algorithms. Numenta Inc Report (2007)
116. Hawkins, J., George, D.: Hierarchical Temporal Memory. Concepts, Theory, and Terminology. Numenta Inc Report (2006)
117. Warneke, B., Last, M., Liebowitz, B., Pister, K.S.J.: Smart dust: communicating with a cubic-millimeter computer. Computer **34**, 2–9 (1997)
118. Link, J.R., Sailor, M.J.: Smart dust: self-assembling, self-orienting photonic crystals of porous Si. Proc. Natl. Acad. Sci. **100**, 10607–10610 (2003)
119. Sailor, M.J., Link, J.R.: Smart dust: nanostructured devices on a grain of sand. Chem. Commun. **11**, 1375–1383 (2005)
120. Katz, E., Willner, I.: Integrated nanoparticle-biomolecule hybrid systems: synthesis, properties and applications. Angew. Chem. Int. Ed. **43**, 6042–6108 (2004)
121. Katz, E., Willner, I., Wang, J.: Electroanalytical and bioelectroanalytical systems based on metal and semiconductor nanoparticles. Electroanalysis **16**, 19–43 (2004)
122. Burns, A., Ow, H., Wiesner, U.: Fluorescent core-shell silica nanoparticles: towards "lab-on-a-paricle" architectures for nanobiotechnology. Chem. Soc. Rev. **35**, 1028–1042 (2006)
123. Burns, A., Sengupta, P., Zedayko, T., Baird, B., Wiesner, U.: Core/shell fluorescent silicananopaticles for chemical sensing: towards single-particle laboratories. Small **2**, 723–726 (2006)

124. Gill, R., Zayats, M., Willner, I.: Semiconductor quantum dots for bioanalysis. Angew. Chem. Int. Ed. **47**, 7602–7625 (2008)
125. Uchiyama, S., McClean, G.D., Iwai, K., de Silva, A.P.: Membrane media create small nanospaces for molecular computation. J Am Chem Soc. **127**, 8920–1 (2005)
126. de Silva, A.P., Dobbin, C.M., Vance, T.P., Wannalerse, B.: Multiply reconfigurable 'plug and play' molecular logic via self-assembly. Chem. Commun. **11**, 1386–1388 (2009)
127. Pallavicini, P., Diaz-Fernandez, Y.A., Pasotti, L.: Micelles as nanosized containers for the self-assembly of multicomponent fluorescent sensors. Coord. Chem. Rev. **253**, 2226–2240 (2009)
128. Erokhin, V., Berzina, T., Smerieri, A., Camorani, P., Erokhina, S., Fontana, M.P.: Bio-inspired adaptive networks based on organic memristors. Nano Commun. Netw. **1**, 108–117 (2010)
129. Erokhin, V., Berzina, T., Camorani, P., Smerieri, A., Vavoulis, D., Feng, J., et al.: Material memristive device circuits with synaptic plasticity: learning and memory. BioNanoScience **1**, 24–30 (2011)
130. Erokhin, V., Berzina, T., Camorani, P., Fontana, M.P.: Non-equlibrium electrical behaviour of polymeric electrochemical junctions. J. Phys. Condens. Matter **19**, 205111 (2007)
131. Berzina, T., Smerieri, A., Barnabó, M., Pucci, A., Ruggieri, G., Erokhin, V., et al.: Optimization of an organic memristor as an adaptive memory element. J. Appl. Phys. **105**, 124515 (2009)
132. Ohno, T., Hasegawa, T., Tsuruoka, T., Terabe, K., Gimzewski, J.K., Aono, M.: Short-term plasticity and long-term potentiation mimicked in single inorganic synapses. Nat. Mater. **10**, 591–595 (2010)
133. Kim, K., Chen, C.-L., Truong, Q., Shen, A.M., Chen, Y.: A carbon nanotube synapse with dynamic logic and learning. Adv. Mater. **25**, 1693–1698 (2013)
134. Bichler, O., Zhao, W., Alibart, F., Pleutin, S., Vuillaume, D., Gamrat, C.: Functional model of a nanoparticle organic memory transistor for use as a spiking synapse. IEEE Trans. Electron Devices **57**, 3115–3122 (2010)
135. Gerstner, W., Kistler, W.M.: Mathematical formulations of Hebbian learning. Biol. Cybern. **87**, 404–415 (2002)
136. Datsyuk, V., Kalyva, M., Papagelis, K., Parthenios, J., Tasis, D., Siokou, A., et al.: Chemical oxidation of multiwalled carbon nanotubes. Carbon **46**, 833–840 (2008)
137. Bi, G.Q., Poo, M.M.: Synaptic modifications in cultured hippocampal neurons: dependence on spike timing, synaptic strength, and postsynaptic cell type. J. Neurosci. **18**, 10464–10472 (1998)
138. Legenstein, R., Pecevski, D., Maass, W.: A Learning Theory for Reward-Modulated Spike-Timing-Dependent Plasticity with Application to Biofeedback. PLoS Comput Biol. **4**, e10000180 (2008)
139. Fiete, I.R., Senn, W., Wang, C.Z.H., Hahnloser, R.H.R.: Spike-time-dependent plasticity and heterosynaptic competition organize networks to produce long scale-free sequences of neural activity. Neuron **65**, 563–576 (2010)
140. Rutherglen, C., Burke, P.: Carbon nanotube radio. Nano Lett. **7**, 3296–3299 (2007)
141. Pilarczyk, K., Podborska, A., Lis, M., Kawa, M., Migdal, D., Szaciłowski, K.: Adv. Electron. Mater. (2016) doi:10.1002/aelm.201500471

Chapter 18
Organic Memristor Based Elements for Bio-inspired Computing

Silvia Battistoni, Alice Dimonte and Victor Erokhin

Abstract Bio-based/bio-inspired systems are attracting the interest of many studies even if we are far from reproducing the simplest living cell property. The concept of memory is particularly well suited for mimicking learning behavior in biosystems and in information processing systems being capable of coupling inherently memory and logic capabilities. Bio-electronics is another challenging platform, mostly if we consider organic devices based on conductive and biocompatible polymers. This chapter deals with several examples of devices developed by joining unconventional computing, organic memristors and living being. Starting from organic memristors we realized logic gates with memory and a single layer perceptron. We developed hybrid systems based on living beings as key elements for the proper device working, in particular with *Phyarum polycephalum* and neurons. These devices enable new and unexplored opportunities in such emerging field of research.

18.1 Introduction

Memristor is a device, predicted theoretically by L. Chua in 1971 [1] that can vary its conductivity as a function of the current that has passed through it. This property well corresponds to the main feature of the synapse, responsible for learning of living beings according to the rule of Hebb [2]. An explosive increase of the works in the field of memristors occurred after the report of the first experimental realization of this device [3]. Even if the ideal memristor hardly exists in nature [4, 5];

S. Battistoni (✉) · A. Dimonte · V. Erokhin
IMEM-CNR, Parco Area delle Scienze 37/A, Parma, Italy
e-mail: silvia.battistoni@imem.cnr.it

A. Dimonte
e-mail: alice.dimonte@imem.cnr.it

V. Erokhin
e-mail: victor.erokhin@fis.unipr.it

S. Battistoni
University of Parma, Parco Area delle Scienze 7/A, Parma, Italy

© Springer International Publishing Switzerland 2017
A. Adamatzky (ed.), *Advances in Unconventional Computing*,
Emergence, Complexity and Computation 23,
DOI 10.1007/978-3-319-33921-4_18

469

resistance-switching electronic compounds are seriously considered as key elements for mimicking synapse properties, and most of them are now called memristors or, more correctly, "memristive devices". Mainly, these devices are based on thin layers of metal oxides, however, there are several realizations of memristors, based on other materials.

Electrochemical Polyaniline (PANI) based adaptive element, called also "organic memristive device", was designed and realized for performing synapse-mimicking properties in electronic circuits, capable to perform Hebbian learning [6]. It has demonstrated non-linear electrical characteristics with hysteresis loop and rectification. Several deterministic [7, 8] and stochastic [9, 10] adaptive networks were realized with these devices, prooving the learning possibilities in DC and pulse modes [11]. Direct demonstration of the synapse-mimicking properties was done by the artificial reconstruction of the part of the nervous system of the pond snail, responsible for learning, where organic memristive devices were placed in the positions of synapses [12].

Nevertheless, the other branch of bio-inspired computational systems can be based on the direct integration of living beings into electronic (or other, for example, optical) systems. In this case it would be possible to use some unique properties of biosystems that are difficult to reproduce with man made devices.

In this chapter we will try to give some ideas of the application of organic memristive devices for both areas of bio-inspired computing.

The structure of the chapter is the following one: initially, we will briefly describe working principles of the PANI based organic memristor without many details, referring to the appropriate published papers. Then, we will describe bio-inspired systems of the first type: memristive devices are used as key elements for the realization of logic gates with memory and perceptron. Next section will be dedicated to the bio-inspired systems of the second type: hybrid devices including living beings: nervous cells and slime mould. In the first case, the integration was done with PANI device, while in the second one the other polymer—poly(3,4-ethylenedioxythiophene)polystyrene sulfonate (PEDOT:PSS)—was used as an active channel. As these results are new, we will give more details on the particularities of the architecture and properties of these elements.

18.2 Organic Memristive Device

The working principle of the device is based on the significant difference in the conductivity of PANI in the reduced and oxidized states [13]. Thus, the device shown in Fig. 18.1 contains a thin PANI layer, deposited on the insulating support [14].

Two electrodes, called as Source (S) and Drain (D) respectively, are connected to the PANI channel. S electrode in maintained at a ground potential level. A narrow layer of solid electrolyte (Li+ salt doped polyethylene oxide (PEO)) is deposited in the center of the channel. The contact of PANI with PEO is an active zone, as all redox reaction occur there according to the potential difference of the PANI in

Fig. 18.1 Schematic representation of the organic memristor with external power supply and measuring devices

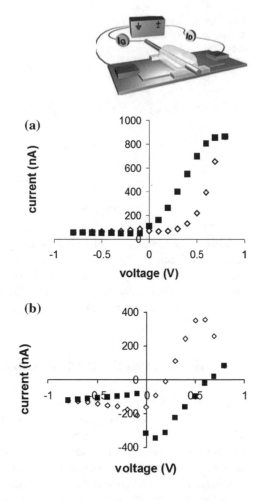

Fig. 18.2 Cyclic voltage-current characteristics for electronic (**a**) and ionic (**b**) currents of organic memristive device (empty rhombuses—increase of the voltage, *filled squares*—decrease of the voltage). Reprinted with permission from [6]

the contact with respect to the reference electrode (G), connected to the PEO layer. Typical cyclic voltage current characteristics for the electronic and ionic currents are presented in Fig. 18.2.

In the case of the application of fixed bias voltage higher than the oxidation potential or lower that the reduction one, we can observe reinforcing or inhibition of the device conductivity, respectively. The experimental dependences are shown in Fig. 18.3 and they establish a basis for unsupervised (a) and supervised (b) learning.

Initially, ON/OFF ratio of these devices was equal to 100. However, optimization of the used materials [15, 16] and the device architecture [17] resulted in the increase of this ratio up to 105 with also a significant improvement of the stability of its properties.

The variation of the conductivity occurs in the active zone of the device, that is accompanied by the motion of Li+ ions between PANI and PEO layers according to the following reaction:

Fig. 18.3 Typical temporal dependences of the drain current at fixed applied voltages: $+0.6$ V (**a**) and -0.2 V (**b**)

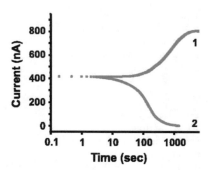

$$PANI^+ : Cl^- + Li^+ + e^- \rightleftharpoons PANI + LiCl, \qquad (18.1)$$

This statement was directly confirmed by microRaman [18] and X-ray fluorescence experiments [19]. Thus, it was established that the conductivity value of the device is proportional to the ionic charge passed through the junction. Some comparisons of the networks' features, composed from organic memristive devices, with elements of nervous system can be found in [20].

18.3 Bioinspired Elements of Unconventional Computers

What distinguishes a brain from a computer? They both are able to memorize and process information but is it possible to find common features between them? A lot of answers could be given[1] but the strongest and the most effective feature is that in a computer architecture processing and memorizing of the information are done in separate times and they correspond to different part of it (processor and memory). In a brain these two process are done concurrently, we are all able to understand, compare and memorize new items in the same time, and this allow us to learn. Moreover, it is not so easy to identify just one part of the brain responsible for this ability but it is absolute out of discussion that the main role should be attributed to the synapsis. By definition, a synapse is an highly specialized junction between two neural cells (neurons) necessary for the transmission of signals. This transmission could be done with the release of chemical compounds (named transmitters) by the first cell (pre-synaptic cell) and the reception of them by the target cell (post synaptic cell). This type of synapse is called the chemical synapse but it is also possible that two neurons form the electrical synapse in which the communication is done through direct electrical signals. Synapsis modifications are deeply related to their activity and the description of this phenomena is given by the Hebb's rule [21]:

---

[1]http://scienceblogs.com/developingintelligence/2007/03/27/why-the-brain-is-not-like-a-co/.

When an axon of cell A is near enough to excite a cell B and repeatedly or persistently takes part in firing it, some growth process or metabolic change takes place in one or both cells such that A's efficiency, as one of the cells firing B, is increased.

This rule and this attitude is another good example of the difference between the brain and a computer: if you ask to a computer to solve the similar task for different times, it will calculate it in the same way from the first to the last approach without performing any structural modification in its architecture. Human brain does not work in this way and in the following chapter we will try to explain how it can be simulated and realized at a hardware level with memristive devices. Thus Hebbian rule learning can be performed by networks of electronic synapsis (organic memristive devices).

18.3.1 Boolean Logic Versus Memristive Logic

Computer function is based on the classical Boolean logic in which, defined the logical operation (AND, OR or NOT) and the values of the variable ("1" or "0"), the solution must be given by the truth table. So the initial configuration of the variables and of the operation is crucial for the final solution because this latter one only depends on that. In this sense living beings do not follow the tighten Boolean logic rules since their decisions are influenced not only by the variables and operations configuration but depend also on the "experiences" that they have accrued in their previous activity. One good example of this is the following: image to be at a crossroads and to have to reach a castle situated far away from your position. The streets in front of you are equivalents, they both reach the castle, but one of them, the shortest one, is dark and gloomy. A computer facing the decision of which street to take, will probably go for the shortest one because it is the only logical choice but a human with experience of horror movies that has to face the same thing, will decide for the safer one.

This concept of "experience" is deeply connected to the concept of learning and memory and its description in terms of logical operations is quite complicated. That is the reason why a new type of logic has been introduced and called: the memristive logic.

As the name suggest, memristors are the basis of this kind of logic since they are able to vary and memorize their physical properties and it was widely proved that they can emulate and reproduce some properties of synapses [12, 22–24] and neurons [25, 26], such as the Hebbian type of learning [2]. In the past, our group demonstrated that a simple circuit of one or more memristors is able to mimic homo- and hetero-synaptic properties of a pond snail [12].

Fig. 18.4 M-AND element. Picture is taken from [27]

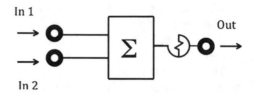

18.3.2 M-AND Gate

The realization of a memristive logical operation with organic and inorganic devices was reported in [27]. The logical operation chosen for that purpose was the M-AND (memorized AND) and it was realized with the circuit shown in Fig. 18.4. In the truth table of the AND operation, 1-1 is the only possible combination of the input variables that results in the 1 logic output. All the other combinations of variables (1-0; 0-1 and 0;0) provide 0. In Fig. 18.4, the memristor is connected to a summator (Σ), which performs the sum of the two input signals (In1 and In2); in our configuration these two inputs are voltage values: in the case of "0" logic the voltage value is 0 V and, in the case of "1" logic, it is 0.3 V. The output signals (OUT) is the current that flowing through the memristive device.

This correspondence between voltage values and logic inputs has been chosen in order to respect the truth table of the AND logic gate. In fact the oxidation of the PANI (and so the switch on of the memristive device conductivity) does not occur until the voltage applied to the electrodes reaches the value of about 0.6 V, that would happen exactly in the case of the combination 1-1 inputs of our currently experiment. In all the other combinations, the potential applied would be lower than this threshold (e.g. 0.3 V or 0 V) and this would not induce any redox process in the device. In Fig. 18.5 the output current (panel (a)) and the sum of voltage inputs (panel (b)) are shown as a function of time and of the combination of input values. For the first 300 s (Step 1), the combination chosen is the 0-0 logic and so a total voltage of 0 V is applied to the device. For the successive 900 s (Step 2) the combination is changed and it is replaced with the 1-0 and the 0-1 logic; both of them correspond to a voltage value of 0.3 V. After 300 s (Step 3) of a new application of the 0-0 combination, the forth step was done corresponding to the application of the 1-1 logical combination (in total 0.6 V). Taking now in consideration the panel (a) of Fig. 18.5, it is possible to notice that in the case of Step 1 and 3 the returned output current is very low and close to 0 μA while in the other two steps the values are positive but with different values (0.7 μA for Step 2 and 2 μA for Step 4). If we now want to classify and interpret these two current results, we must set a current threshold, in the same way that we did for the input voltage. From the comparison between this threshold and the current obtained in the Step, we will be able to distinguishes the 0 and the 1 logic: if the step output is below the threshold, it will be classified as 0 and vice versa if it is above the threshold, it will be identified as 1 logic. In our case, if we set for example this threshold to a value of 1.5 μA, Step 2 would correspond to an 0 output while Step 4 to an 1 logic, confirming the AND truth table.

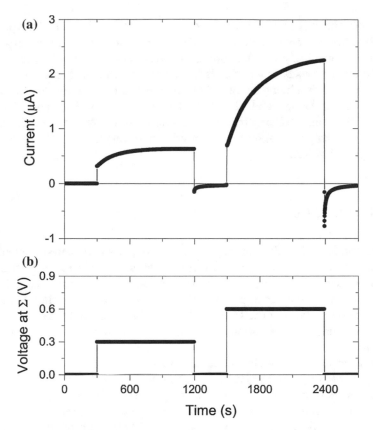

Fig. 18.5 M-AND element: panel **a** current output; panel **b** sum of the voltage inputs. Picture taken from [27]

Thus, we have successfully demonstrate that we can use a simple circuit based on memristive device as a logic gate able to perform an AND operation. Another interpretation of this result must be given thanks to the so called "fuzzy" logic. The fuzzy logic is considered as an extension of the Boolean logic since it admits more than just 2 possible output (0 and 1). Memristors are considerate as good candidates for performing this particular kind of logic since their current output naturally depends on the applied voltage values and time (or frequency) of its application. In this term, we can interpret our 3 current outputs, 0, 0.63 and 2.25 μA as three different logical outputs (in principle, all intermediate values are also possible).

Looking now at the shape of the output current in Step 2 and Step 4 we can conclude that the kinetics of these two curves are different. In step 2, once the potential is applied, the curve saturated in a brief time, around 300 s and the difference between the current value acquired after 1 s and 15 min is pretty small (0.32 μA and 0.63 μA, respectively). In the Step 4, where the oxidation of PANI (conductivity switching) is taking place, this difference is much more accentuated (0.70 and 2.25 μA) and it is clear that the curve depends on the duration of the application of the potential, since the current is not completely saturated. This kind of behavior is very similar to what happened in the learning process of living beings described by the Hebbian rule. Time and intensity of the stimulation are crucial parameters that can affect the learning process of a living being and the same is valid for the memristive devices: the value of the voltage applied and the duration of the step influence the output current. This strong similarity between the behavior of the synaptic learning and the memristor output profile allows us to consider these devices as good candidates for the synapse mimicking.

18.3.3 Perceptron

After the considerations done in the previous section, in which we have demonstrated the ability to realize organic memristive devices that can be used for the realization of AND logic gate with memory, the natural implementation would be to try to perform more difficult and complex logic operations. To do that, a simple circuit as the one used for the AND gate is not sufficient and our ambitious goal requires the realization of a more complex configuration of devices. Let us introduce now the concept of Perceptron.

A Perceptron is an artificial neural network (ANN) able to learn different tasks such as pattern recognition and classification, approximation, prediction, and others. These tasks, especially in the case of incomplete input data, are quite difficult to solve for a computer while for a human brain it is pretty easy to find a solution. The first concept was developed by Rosenblatt in 1957 [28] and it represented a model of a brain Perceptron, but subsequently more complicated models of mathematical learning algorithms were proposed. In our case [29] we realized a single-layer elementary Perceptron made by 3 memristive devices connected in the circuit shown in Fig. 18.6.

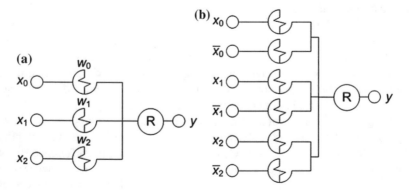

Fig. 18.6 Scheme of an elementary Perceptron based on one (**a**) and two (**b**) organic memristors per every link weight. Picture taken from [29]

As mentioned above, a Perceptron is able to classify input data in different classes. The concept behind the Perceptron classification is very close to what we have already described in the AND logic gate. There are several input data (x_i) that can assume the 0 and the 1 logic values. The output value, y (x), can be interpreted with the following relation:

$$y(x_i) = \sum w_i x_i, \tag{18.2}$$

where the weights of the linear combination are indicated as w_i. Once again, if we want to classify the output data obtained for a fixed set of x_i, we need to compare this value of y with a threshold value fixed before: if y exceeds our "a priori" value (b), it will corresponds to the class No1, and so, for example, to 1 logic, otherwise it is within the class No2, so 0 logic.

$$\sum w_i x_i \begin{cases} > b = 1 \\ < b = 0 \end{cases} \tag{18.3}$$

18.3.3.1   Organic Memristive Devices Based Perceptron and the Learning Procedure

In our case, since we used just 3 memristive devices, our input data (x_0, x_1, x_2) are voltage values and they can be 0.2 V or 0.4 V for "0" and "1" logic respectively; input x_0 is fixed to the value of "1" logic. The output is the sum of the currents that flow in the three devices and our threshold is set to the value of $3.0 \pm 0.5\,\mu A$. From simple considerations it is easy to understand that in a case of memristors, w_i are related to the memresistance. They will play also an important role in the learning process of this Perceptron.

The choice of the input voltage values is not trivial; as we observed in the section of the characterization of a classical memristor, two voltage values are critical for the working principle of the device. The first one is 0.6 V that corresponds to the oxidation of the PANI layer and that brings the entire device in its ON state (high conductivity); the other one is 0.1 V that corresponds to the reduction of PANI and so to the OFF state of the device (high resistance). All the potential values that are in the middle between these two values do not correspond to any redox reaction in the device active zone and would allow just the reading of the conductivity state without its variation. That's why we chose 0.2 and 0.4 V as our "0" and "1" logic values. For the sake of completeness, if we look at the formula describing the Perceptron, we have to notice that we may need to have negative weights w_i. Since they represent the ratio between the voltage and the current, they cannot be naturally negative and so we need to invert the sign of the voltage of the two variable inputs (x_1 and x_2) respect to the sign of x_0. An alternative solution would be using a more complex circuit (shown in Fig. 18.6b) made by the double number of the devices. To perform the learning of the classification of the inputs, we decided to use the logic function of NAND and NOR using the algorithm with the error correction suggested by Rosenblatt. In this method, an error function must be defined:

$$e = y - y_t, \tag{18.4}$$

where y is the logical output (determined by the threshold method) and y_t is the desired output determined by the truth table of the chosen logic function. The sign and the value of the e determine if it is necessary to increase or decrease the memristance of the devices and so if it is necessary to induce the oxidation or the reduction of them; in fact if $e < 0$, it will be necessary to potentiate the network applying the voltage value of 0.7 V for a 600 s. Instead if $e > 0$, the network must be depressed with the application of -0.2 V for 30 s (duration of the training was varied: number of necessary training cycles was found to depend on the duration of each training cycle). In the case of a zero error no action is required. The action of potentiation or de-potentiation of the network has a deep meaning in term of the mathematical formulation of the perceptron: in these processes in fact, the weight of the linear combination will be increased or decreased in order to accord the obtained current value with the expected one. So the iterative procedure necessary for the learning is pretty tricky and we are going to describe it step by step; it starts with the application of the first combination of input to the devices, for example 1,0,0 ($0.4V$, $\|0.2V\|$, $\|0.2V\|$). In the case of a NAND function, the desired output is 1. The output current of the system is 1.8 μA, that, compared with the threshold, is interpreted a 0 output. So in this first case our error value is negative and this means that we need to potentiate the network. Once that we perform this latter step, we can skip to the next input combination are repeating the procedure. The learning procedure stops when 4 consecutive combinations of input result in a zero error. All the steps for the learning of the Perceptron are reported in Table 18.1.

Table 18.1 shows that we successfully trained the system to perform the NAND function in 15 steps and after one hour of the learning procedure, this ability remain memorized. After this first test on the NAND function, we tried to re-train the system to perform the NOR function. Surprisingly just 2 potentiation steps were required for the total re-train of the system. In conclusion, we demonstrated to be able to realize an elementary Perceptron made by organic memristors. This latter one was trained, at the beginning, in order to perform the NAND logic function and consequently re-trained to perform the NOR function.

Table 18.1 Results of the Perceptron learning to perform NAND function

N	Input			Output (μA)	Logic output	Desired output	Error	Correction time		
	x_0	x_1	x_2					0	1	2
1	1	0	0	1.8	0	1	−1	p600	–	–
2	1	0	1	−0.3	0	1	−1	p600	–	d30
3	1	1	0	4.8	1	1	0	–	–	–
4	1	1	1	3.5	1	0	1	d30	p600	p600
5	1	0	0	4.5	1	1	0	–	–	–
6	1	0	1	4.1	1	1	0	–	–	–
7	1	1	0	0	0	1	−1	p600	d30	–
8	1	1	1	3.8	1	0	1	d30	p600	p600
9	1	0	0	5.3	1	1	0	–	–	–
10	1	0	1	5.1	1	1	0	–	–	–
11	1	1	0	0	0	1	−1	p600	d30	–
12	1	1	1	2.5	0	0	0	–	–	–
13	1	0	0	4.6	1	1	0	–	–	
14	1	0	1	4.5	1	1	0	–	–	–
15	1	1	0	4.0	1	1	0	–	–	–

In the correction time the duration of potentiation (p) and depression (d) is reported in second with the relative prefix. Table reprinted with permission from [29]

18.4 Hybrid Systems with Biological Molecules

18.4.1 Interface with Neurons

The possibility of interfacing living systems with artificial devices pave the way to different important applications in Biomedicine. Such challenging field of research, has to deal with the problems of two worlds: the first one is the biological world that has to guarantee the viability of the cells/living been that are part of the interface and the latter one is the organic and inorganic device's world that has to deal with functionality of the integrated device. A large number of groups worked in this direction, using different kind of cells (cardiomyocyte [30], skeletal muscle [31]) to realize bio-integrated devices but the strongest efforts have been addressed to realize what it can be called "neuromorphic devices" so devices able to mimic special functionalities of the natural living tissues. Our group has been working in this direction for years, approaching the problem in two ways: the first one is the realization of memristive devices able to mimic some typical behaviors of synapsis, such as the synaptic plasticity (see Sect. 18.3.1); on the other hand we are now currently working on the direct interface between neuronal cells and organic devices.

In this contest, one of the most critical aspect to take into consideration is the reduction of the interfacial strains caused by the mismatch of stiffness between the tissue and the electrodes. Conductive polymers appear as perfect substrate for the growth and the stimulation of neuron activity since they have mechanical properties between the rigid metal electrode and the softer tissue and conformational properties that can increase the charge transport [32–35].

18.4.1.1 Preparation of the Substrates

In previous papers, Poly(aniline) (PANI) and poly(3,4-ethylenedioxythiophene) polystyrene sulfonate (PEDOT:PSS) were interfaced with different living being and cells [36–38], but, mostly in the case of PANI, the biocompatibility was not considered generalized but restricted to few cell types only [37]. So, with the aim of looking at the interface between PANI and different type of cells, we will provide in this chapter, the results of a novel and multidisciplinary study in favor of the PANI biocompatibility. The preparation of the PANI substrates was reported in several previous papers [27, 39]. Briefly, we chose to use the well-known Langmuir–Schaefer (LS) technique in which a solution of PANI (Sigma–Aldrich Mw = 100,000) in 1-methyl-2- pyrrolidinone (Sigma–Aldrich ACS reagent 99.0 %) with an addition of 10 % in volume of Toluene (AnalaR NORMAPUR®ACS) is spread on the interface between a subphase (in our case Milli-Q water) and air. This process leads to the formation of a polymer monolayer that is called as the "Langmuir Film" whose surface pressure can be controlled. Once that the film is formed and stable, touching it with a substrate (an insulating cover glass with a diameter of 12–15 mm) is possible to make the transfer of the polymer from the water to the sample. The adhesion of the monolayer is provided by the electrostatic forces between the backbone of the PANI and the glass substrate. Repeating this process it is possible to create a stacked structure with a desirable number of monolayers. The morphology of the prepared samples (shown in Fig. 18.7) was analysed by AFM characterization (Cypher AFM-system(AsylumResearch, Santa Barbara, CA)) in AC mode in air. Film thickness measurements were done with a P-6 profilometer (KLA Tencor, USA).

The thickness of the film was estimated to be about 120 ± 13 nm while the roughness was found to be $25,6 \pm 6.0$ nm. Morphological properties of PANI films were quite uniform especially the thickness of the analysed samples. This fact confirms that the chosen deposition technique guarantees a very good control over the morphology and reproducibility of the samples.

18.4.1.2 Testing the Biocompatibility of PANI

Different cells cultures were prepared to test the biocompatibility of the polymeric layers. Human immortalized and tumor cells, HeLa and HEK293T cells, were cultured with the protocol described in [40] on bare and PANI covered glass slides (Fig. 18.8). The uncoated glass was used as control. From the comparison between

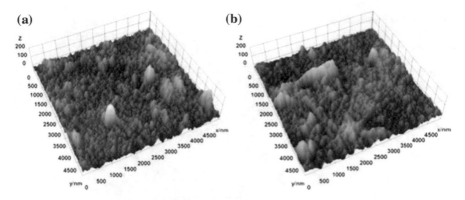

Fig. 18.7 AFM images of different areas of the PANI samples. **a** Zone 1 and **b** Zone 2. Pictures are from [40]

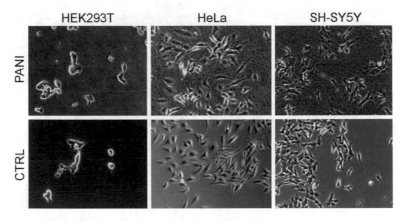

Fig. 18.8 Different cell lines growing on PANI. Picture taken from [40]

the sample with and without PANI layer it is possible to observe that these two types of cells grow on PANI in a comparable way in which they grow on the control sample. After this first test, the human neuroblastoma cell line, SH-SY5Y, was tested. If cultured in normal culturing conditions, this cell line can produce short neurites but they are able to differentiate into a more complex neuronal phenotype if treated with retinoic acid (RA) and brain derived neurotrophic factor (BDNF) following a specific protocol. After the first day after the plating in a complete medium, this latter is changed with DMEM +15 % FBS + 10 nM all-trans-retinoic acid and kept for 5 days. After that, it was changed with DMEM +50 ng/ml BDNF.

Looking at the images obtained with Bright field microscope (Fig. 18.9), it is possible to observe all the modifications of the cells induced by a different composition of the culturing medium. In general we can conclude that also this cell line grows normally on PANI.

Fig. 18.9 Differentiation of SH-SY5Y cells: after 1 day (**a**), 5 days (**b**) of treatment with 10 M retinoic acid and (**c**) after 5 days of treatment with 10 M retinoic acid and 3 days with BDNF 50 ng/ml on PANI and on microscope glass slides. Picture taken from [40]

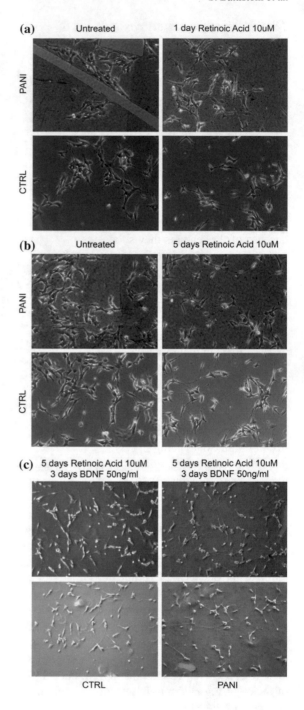

18.4.2   Heterostructures with Slime Mold

Here we report two typologies of hybrid devices based on poly(3,4-ethylenedioxy-thiophene), doped with polystyrene-sulfonate (PEDOT:PSS), and *Physarum poly-cephalum* slime mold (PP). They have both memristive features, resulting from electrochemical variations occurring in the polymer upon application of appropriate potentials across the semiconducting PEDOT:PSS channel. Moreover, we take advantage of the slime mold as a living bio-electrolyte. The ionic flux taking place between PP and PEDOT:PSS induces an over-oxidation and a conductivity switch of the polymer. Indeed, the detection and the records of ion fluxes are basic points to understand life and communication of living organisms. Therefore, these devices can be classified as bioelectronics elements, resulting from the interplay of a biological living organism and electronic components, but also as organic memristive devices, because of the features they showed. Bioelectronics encompasses a wide range of research areas at the interface between biology and electronics.[2] On the one hand, biological problems are approached by electronics, as well as the detection and the characterization of biological materials. On the other hand, biological systems are exploited in electronic applications (e.g. processing novel components from DNA). Moreover, as mentioned before, the devices we built can also be classified as memristor, an emerging family of circuital elements. Memristors can be briefly defined as resistors with memory, whose value depends on the history of the current passed through it. Of particular interest is that, since their prediction, memristive devices changed the computing field and, if based on organic materials, they are considered promising candidates to mimic synapses and realize circuits capable of learning. In [20] Erokhin et al. realized an organic memristor based on the redox properties of polyaniline, a redox electrochromic polymer; indeed, organic materials are preferred to non-organic ones being closer to the biological world. More interestingly, in [12], a particular algorithm based on a memristor net results able to simulate and reproduce the way of learning in a snail.

18.4.2.1   Device Fabrication

PEDOT:PSS channels were realized by a standard method [41], obtaining a 10 mm stripe with a thickness of 100 nm on a glas substrate. 20 % of di-ethylene glycole and 0.5 % of dodecylbenzenesulfonic acid (DBSA) have been added to the PEDOT:PSS, to improve its conductivity and film formation properties [42]. Source and drain electrodes were obtained by applying silver conductive paste at the edges of the PEDOT:PSS channel. In case of first device, we used wire of Silver, Gold and Platinum as gate electrodes, the second device required just a Silver one. Electrical measurements were performed by two Keithley 2400 SourceMeters, controlled by a home-made Matlab software. *Physarum polycephalum* slime mold, in its vegetative form, was cultivated into petri dishes with agar 1,5 % non nutrient gel, fed with oats

---

[2]http://www.nist.gov/pml/div683/upload/bioelectronics_report.pdf.

and kept in a dark and humid environment. The colony was periodically replanted to a new fresh agar to guarantee its safety. By realizing these devices we made a step forward in the field of bioelectronics and memristor creating organic memory elements interfaced with a living organism.

18.4.2.2 *Physarum Polycephalum* (PP)

Physarum polycephalum slime mold is a mass of cytoplasm, a single cell whose shape changes continuously responding to the external environmental stimuli [43–46]. Therefore, in case of food abundance it creates pancake-like structures, while many distributed food sources induce tree-shape like structure. Several protoplasmic tubes, branching towards the food sources, characterize the latter. PP creates optimized networks in order to reach the food by the minimum length path [47] and to optimise geometry of its protoplasmic network [48, 49]. PP's network has been defined by Matsumoto as a system of biochemical oscillators [50, 51] in which waves of excitation or contraction result from external stimuli and perturbations. Moreover, the contraction waves are associated to a variation in the electrical potential; indeed, waves observed in plasmodium [52] can be compared to excitable chemical systems, such as Belousov–Zhabotinsky medium [53–55]. Pershin et al. in 2009 proved that it has memory patterns [56]: if exposed to periodic environmental shocks, PP regulates autonomously its speed, remembering that, after a certain period the environment will income into an improper condition. This behaviour can be associated to learning, generally considered a feature of more complex species.

18.4.2.3 The Electronics Counterpart

PEDOT:PSS is highly used in electrochemistry: OECTs, based on PEDOT:PSS, demonstrated to work efficiently also with more complex solutions as cell culture medium [57] and solid-gels [58]. Therefore, considering all these interesting features we can justify the choice of this polymer, and the results we are presenting here are in good agreement with the positive element from which we started our work.

In both cases we realized elements with memristive features resulting by electrochemical changes occurring into the polymer upon application of anodic potentials across the semiconducting PEDOT:PSS channel. Additionally, we take advantage of the features of the conducting polymer and of the PP as a living bio-electrolyte. An over-oxidation process induces a conductivity switch in the polymer, due to the ionic flux taking place at the interface between the mold and the polymer. This behavior endows a current-depending memory effect to the device.

18.4.2.4 First Device

The first realized device, as in [36], can be defined as a bio-organic sensing/memristive device (BOSMD) thanks to its sensing feature. This structure device is innovative and unique, because it can work both as a transistor (in particular and organic electrochemical Transistor) and as a memristor by simply connecting/disconnecting one of the contact (e.g. sweeping from the three terminal to the two terminal configuration). Therefore, it well suits for monitoring bio-electro-chemical processes, but it works, also, as a bio-hybrid memory device. The BOSMD has been characterized with a complete measurement series and, in addition, three kind of gate electrodes materials have been tested. We used silver, platinum and gold, widely exploited in electrochemistry. The memristor is particularly well suited for mimicking the learning behavior of biosystems, opening novel perspectives in the information processing field [59]. OECT [60] are promising electronic devices being fully biocompatible and consisting in a semiconducting polymer channel in contact with an electrolyte, confined in a well. The gate electrode is immersed into the electrolyte and the overlapping area between polymer and electrolyte defines the channel of the OECT and the site in which ionic interchanges take place [61].

The device working mechanism is based on the reversibility doping/dedoping of the channel. When a voltage is applied between drain-source (Vds), holes drift within the transistor channel, generating a current and the device is switched on. When a positive voltage is applied between gate-source (Vgs), cations from the electrolyte enter into the polymeric channel that is de-doped switching the device in the off state [62]. The exploited electrolytes are typically physiological solutions, e.g. NaCl or phosphate buffered saline [57].

Figure 18.10 shows a scheme of the BOSMD, when source, gate and drain are all connected, it works in the OECT configuration, while, taking out the drain (just lifting up the tip) the device is switched to the memristor configuration.

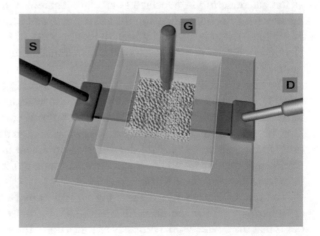

Fig. 18.10 Schematic diagram of the OECT based on PEDOT:PSS (the *black* stripe is the PEDOT:PSS film) where the gate electrode (G) is immersed into the *Physarum polycephalum* cell (the *yellow* area in the figure). S and D are the silver electrodes. Figure taken from [36]

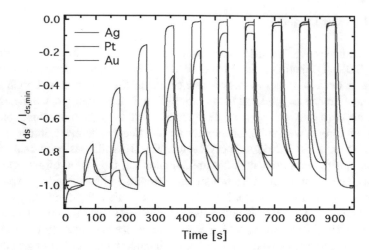

Fig. 18.11 Normalized kinetics curves (Ids/Ids, min vs. time) of the BOSMD. The voltage applied at the gate has been switched in the range 0–1.6 V with a 0.2 step. Picture is taken from [36]

Figure 18.11 shows the kinetic curves trends recorded with the three different gate materials. The current between drain and source has been acquired switching the gate between 0 V and 1.6 V. The comparison among the typical current modulations, observed for the different gates, shows how the mold reacts electrochemically in presence of different materials. To this regard, it is well known that there are two operating regimes for an OECT: Faradaic and non-Faradaic or capacitive, respectively [63]. In the Faradaic mode, a redox reaction occurs at the gate surface, generating a current between gate and source. Thus, ions enters into the polymer film, that is de-doped, with a significant decrease of the channel current and an increased current modulation. On the other side, during the capacitive mode, a double layer at the gate/electrolyte interface is created. Therefore, a significant potential drop reduces the effective voltage applied at the gate and acting on the underneath polymeric film, limiting its de-doping. The main point of these phenomena is that the electrode material determines the electrochemical regime under which operates the mold-OECT.

The silver is redox-active and, in fact, generates the highest current modulation (see Fig. 18.11) over the whole voltage range investigated. On the other hand, platinum and gold, being both inert from a redox point of view, modulate lower currents. The voltage gate shift between silver, gold and platinum depends on the specific reactivity of the gate material with the saline environment inside PP, providing a direct control information about the electrochemical state of the mold and, in particular, of its interaction with the environment.

The graphs Fig. 18.12a–c show the typical output characteristics (Ids vs. Vds at different Vgs) for the three different gate electrodes investigated. What stands out is that the device works properly as transistor. Moreover, the biocompatibility of PEDOT:PSS with the mold was confirmed has evidences reported in literature for this

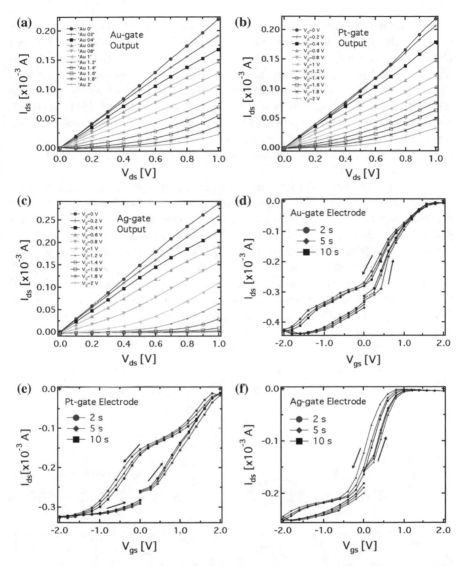

Fig. 18.12 Output characteristics (Ids vs. Vds at different Vgs) of the device OECT-Physarum polycephalum, using Au (**a**), Pt (**b**) and Ag (**c**) wires as gate electrodes inserted into the mold. The transfer characteristics (Ids vs. Vgs) are reported in (**d**), (**e**) and (**f**), respectively [36]

polymer. The graphs in Fig. 18.12d–f, respectively report the transfer characteristics (Ids vs. Vgs, Vds = −0.4 V) of our device. The characterization has been developed by a cyclic sweep mode, first increasing gate voltage and then decreasing. Thus, starting from a Vgs of 0 V with a 0,2 V step, the voltage was first incremented up to 2, then lowered up to −2 V and finally increased back to 0V. In the positive range cations are pushed into the PEDOT:PSS channel, de-doping it. Hence, a current decrease is recorded up to a saturation, depending on the gate material. In the negative range a higher gate voltage is required to restore the doping level of the PEDOT:PSS channel, i.e. for the cations to desorb from the polymer backbone towards PP. The high negative Vgs required to (re-)dope the polymer could be related to the high viscosity of the electrolyte (the mold), thus a higher force is necessary to ions for crossing the cell membrane that works as a "quasi" solid electrolyte [64]. Moreover, also, the hysteresis curves are dependent on the gate material; the lowest was found for the silver and the largest for the platinum. The hysteresis arises from the competition between dynamics of cations the timescale for the doping/de-doping effect of the polymeric channel [65].

As already mentioned the same device was used, also, in a 2-terminals configuration, with the PEDOT:PSS stripe as reference electrode and a bias voltage applied between the gate, acting as working electrode, and one of the channel electrodes (in particular the source). In Fig. 18.13 are the I-V curves for Au, Pt and Ag (panels (a), (b) and (c)) working electrodes, respectively. The Pt-electrode shows an IV curve with a reduction peak at about −3 V and a broad oxidation peak, between +0.1 and +2.1 V. The reduction peaks in case of Au and Ag electrodes are at −2.8 and −1.9, respectively, while narrower oxidation peaks at +1.6 V and +0.7 V, respectively. The redox peaks trend is in good agreement with the measurement reported in Fig. 18.11d–f. The oxidation peak for the Pt-electrode (Fig. 18.13b) seems to be

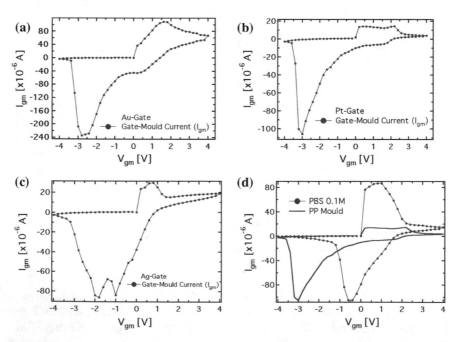

Fig. 18.13 **a, b, c** I-V characteristics for the 3 different working electrodes (Pt, Au and Ag). The voltage was applied between −4 to +4V, with a 0.2 V step. **d** The I-V measurement comparing a Pt-gate electrode with the mold and with a standard PBS 0.1 M electrolyte. Picture taken from [36]

the convolution of those obtained using Ag and Au electrodes. Pt electrodes, even though substantially inert, can sustain faradaic reactions in presence of biomolecules, [66] and the cytoplasm of PP is an environment rich of complex molecular species. As far as the electrodes reduction is concerned, the displacement between peaks of reduction-oxidation leads to a memristive-like behavior for our device. This memristive effect is probably due to a competition between the capacitive coupling at the interface mold/polymer and the choice of the working electrode, that promotes a flux of ionic species towards the underneath polymer. However, of the memristive systems reported in literature, as fas as we know, can be considered as an "ideal memristor", since several processes are always responsible for their properties [4].

The trends in Fig. 18.13, present an hysteresis loop and a rectification, allowing us define the system as a memristive device in a wide sense. If we compare this device, in its memristive configuration, with the polyaniline-based devices, the first visible advantage is switching velocity that is of 1 s in the present case with respect to about 60 s in the case of polyaniline memristive devices. If we consider oxide memristors, the organic nature of our system allows better biocompatibility and could be easily integrated in living organisms. Moreover, the BOSMD work efficiently for more than ten cycles before the natural degradation of the organic polymer. Finally, it has to be underlined that the PPC-OECT is a good candidate for an innovative memristive element satisfying the requirements sought for memristors, defined as electronic elements with memory properties.

18.4.2.5 Second Device

The second device here presented and related to the paper [67] is an hybrid memristive device based on PEDOT:PSS and *Physarum polycephalum* slime mold. As we were building a typical memristor we used a three-terminal configuration exactly as in [68]. However, in this case PP is used as electrolyte with the gate, in silver, placed in it and kept at zero-bias. As in the first device the channel was in PEDOT:PSS for the reasons described above. We recorded the current through the polymeric channel and that at the gate electrode providing information on the ion flux contribution at the interface between the polymer and the mold. As we expected, the device behaves as a memristive device, with meta-stable redox states very close to the write-once read-many memories (WORMs) [69]. In this case, we realized an innovative device with memory properties and based on the association of a living and an organic systems.

Figure 18.14 shows the schematic set-up of the device, it is visible the glass substrate with the polymeric channel with at the edge the source and drain electrodes. Then in the middle of the channel, a stripe of *Physarum* is placed with a silver wire inside. Measurements were carried out by recording the hysteresis loops current between source and drain. The voltage was swept in a symmetrical range ($-Vds$, $+Vds$), with 0.1 V step kept for 1 s. The applied Vds were respectively: 1, 3, 7.5, and 10 V. The measurement trend started from 0 V to $-Vds$, then from $-Vds$ to $+Vds$ and back to 0 V. In the meantime, the gate-source current Igs was monitored. All the Voltages measurements were performed on the same device by increasing the

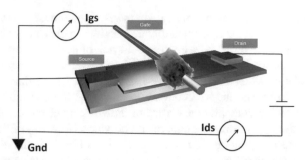

Fig. 18.14 Scheme of the memristor architecture. The *yellow blob* represent the mold that was placed in the middle of the channel and the *silver* wire (gate) is in it. Igs is the ionic current that flows at the interface between the mold and the PEDOT:PSS thin film (the *dark blue* channel) deposited on the glass substrate (*light blue*). Ids is the current that flows in the polymer layer. The silver wire is grounded as one of the electrode the source-drain potential is swept from positive to negative voltages. Picture taken from [67]

ranges. Figure 18.15 presents two graphs of the measurements performed: the (a) panel indicates the hysteresis loop of the drain current between -1 V $/+1$ V (black curve) and -3 V$/+3$ V (red curve). It is only in the second case that a clear rectifying behaviour can be observed. Panel (b) represents the measurement extended over a wider range (-7.5, $+7.5$ V) and, as in other cases, we recorded that at high Vds voltages (e.g. over a threshold between 6 and 10 V) the Ids current drops down of several orders of magnitude. This suggest a transition of the PEDOT:PSS from its conductive to non-conductive state, we hypothesized this effect on the basis of other cases of over-oxidation observed in thiophene-based organic semiconductors upon application of high anodic potentials in a wet environment [70, 71] and, more strictly related to our case in [72, 73]. The latter work attributes this effect to the cations of the electrolyte. In our case we have to take into account that one of the most abundant ions contained in PP's cellular matrix are Ca++ [74]. Therefore, it is reasonable that when PEDOT:PSS is directly interfaced with slime mold and a sufficient high voltage is applied the over-oxidation effect can happen. Moreover, the threshold voltage capable to over-oxidate PEDOT:PSS in our case is much lower than that used in ion pumps ($+25$ V). This could depend from the higher ionic content and the major complexity of the intracellular matrix of PP. Once the switch happened, the device shows, interestingly, a hysteresis behavior in current-voltage characteristic typical of the memory-like effect [75]. Measuring a cyclic current-voltage trend in a wider range (-10, $+10$ V), we obtained an Ids versus Vds loop with a large hysteresis area characterized by two well-defined lobes and two peaks oxidation/reduction (located around $+4.5$ V and -5 V, respectively). Moreover, our device's memory effect presents a non-volatile nature very close to WORMs [69] where information is stored by the resistive switching above a certain voltage level and the actual state of the device is read at a lower voltage. Hence, this hybrid memristor can beseen

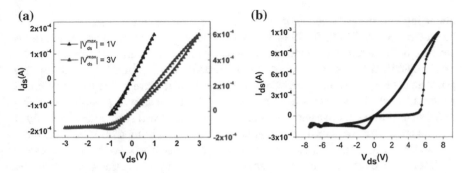

Fig. 18.15 Current voltage characteristic of the memristive device: panel **a** depicts Ids versus Vds profiles in (−1, +1) V *black curve* and (−3, +3) V, *red curve*. In **b** the hysteresis loop measured in the Vds range (−7.5, +7.5) V. The conductivity switch of PEDOT:PSS towards a less conductive form can be observed in the positive voltage branch (the switch can be related to an over-oxidation of the polymer) picture taken from [67]

as a biologically relevant memory device, based on the interaction of an organic semiconductor and a living system. The ionic flux across PP/PEDOT:PSS plays a key role both in the rectifying behavior and in the conductivity switch observed in our hybrid system.

The current in the conducting channel depends on doping/de-doping states, e.g. upon injection of cations from the electrolyte in contact with the polymer channel [76]. Moreover, the current is the result of two contributing flux: the ions injected into the polymer and its intrinsic charge carriers. Therefore, the gate current let us monitor only the ionic species (Fig. 18.15b).

Monitoring the current gate during the voltage sweep of the channel we expected to record also the movement of the ions between PP/PEDOT:PSS (Fig. 18.16).

Our study presents two bio-inspired devices based on a living system, the *Physarum polycephalum*, and the conducting polymer PEDOT:PSS. Most of the observed effects are induced by electrochemical changes triggered by the mold and

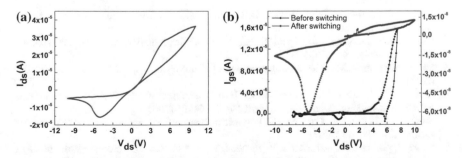

Fig. 18.16 **a** Hysteresis curve measured on one of PP/PEDOT:PSS device soon after the conductivity switch. Vds was swept between (−10, +10) V. **b** Igs versus Vds before (*black curve*) and after (*blue curve*) the occurrence of the conductivity switch. Picture taken from [67]

occurring into the polymer upon application of potentials. PP efficiently works as an electrolyte and PEDOT:PSS is used both as a transistor channel and as the redox element of the memristive devices. The first device, considering the obtained results, can be classified as a bio-organic sensing/memristive device (BOSMD) enabling new possible developments in bioelectronics. The particular architecture of the BOSMD we built allows a direct monitoring of the internal cellular activities. Moreover, being in the same time a sensitive and a memory device, the BOSMD offers the capability of mimicking the behavior of living organisms, studying their interaction with the environment and implementing new neuromorphic systems. The second device presents an interesting over-oxidation mechanism, triggered by the cationic flux between PEDOT:PSS and PP. The device can be accounted as an element with properties similar to those of synapses, useful to implement bio-inspired computation.

18.5 Conclusions

In this chapter we have described several examples of the use of organic memristive devices in bio-inspired information processing systems. In particular, we have demonstrated the possibility of the realization of logic gates with memory and a single layer Perceptron, capable, after adequate training, to perform some logic operations. In the second part it was demonstrated the realization of hybrid systems, including living beings. In particular, we have interfaced memristive devices with slim mold and with nervous cells.

References

1. Chua, L.O.: Memristor-the missing circuit element. IEEE Trans. Circuit Theory **18**(5), 507–519 (1971)
2. Hebb, D.O.: The Organization of Behavior: A Neuropsychological Theory. Psychology Press, Abingdon (2005)
3. Strukov, D.B., Snider, G.S., Stewart, D.R., Williams, R.S.: The missing memristor found. Nature **453**(7191), 80–83 (2008)
4. Pershin, Y.V., Di Ventra, M.: Memory effects in complex materials and nanoscale systems. Adv. Phys. **60**(2), 145–227 (2011)
5. Sascha Vongehr, X.M.: The missing memristor has not been found. Sci. Rep. vol. 5(11657) (2015)
6. Erokhin, V., Berzina, T., Fontana, M.P.: Hybrid electronic device based on polyaniline-polyethyleneoxide junction. J. Appl. Phys. **97**(6), 064501 (2005)
7. Erokhin, V., Berzina, T., Fontana, M.: Polymeric elements for adaptive networks. Crystallogr. Rep. **52**(1), 159–166 (2007)
8. Smerieri, A., Berzina, T., Erokhin, V., Fontana, M.: A functional polymeric material based on hybrid electrochemically controlled junctions. Mater. Sci. Eng.: C **28**(1), 18–22 (2008)
9. Erokhin, V., Berzina, T., Camorani, P., Fontana, M.P.: Conducting polymer - solid electrolyte fibrillar composite material for adaptive networks. Soft Matter **2**(10), 870–874 (2006)

10. Erokhin, V., Berzina, T., Gorshkov, K., Camorani, P., Pucci, A., Ricci, L., Ruggeri, G., Sigala, R., Schüz, A.: Stochastic hybrid 3d matrix: learning and adaptation of electrical properties. J. Mater. Chem. **22**(43), 22881–22887 (2012)

11. Smerieri, A., Berzina, T., Erokhin, V., Fontana, M.: Polymeric electrochemical element for adaptive networks: pulse mode. J. Appl. Phys. **104**(11), 114513 (2008)

12. Erokhin, V., Berzina, T., Camorani, P., Smerieri, A., Vavoulis, D., Feng, J., Fontana, M.P.: Material memristive device circuits with synaptic plasticity: learning and memory. BioNanoScience **1**(1–2), 24–30 (2011)

13. Kang, E.T., Neoh, K.G., Tan, K.L.: Polyaniline: a polymer with many interesting intrinsic redox states. Prog. Polym. Sci. **23**, 277–324 (1998)

14. Erokhin, V., Fontana, M.: Thin film electrochemical memristive systems for bio-inspired computation. J. Comput. Theor. Nanosci. **8**(3), 313–330 (2011)

15. Berzina, T., Smerieri, A., Bernabò, M., Pucci, A., Ruggeri, G., Erokhin, V., Fontana, M.: Optimization of an organic memristor as an adaptive memory element. J. Appl. Phys. **105**(12), 124515 (2009)

16. Berzina, T., Smerieri, A., Ruggeri, G., Erokhin, V., Fontana, M., et al.: Role of the solid electrolyte composition on the performance of a polymeric memristor. Mater. Sci. Eng.: C **30**(3), 407–411 (2010)

17. Erokhin, V., Berzina, T., Camorani, P., Fontana, M.P.: On the stability of polymeric electrochemical elements for adaptive networks. Coll. Surf.: Physicochem. Eng. Asp. **321**(1), 218–221 (2008)

18. Berzina, T., Erokhin, V., Fontana, M.: Spectroscopic investigation of an electrochemically controlled conducting polymer-solid electrolyte junction. J. Appl. Phys. **101**(2), 024501 (2007)

19. Berzina, T., Erokhina, S., Camorani, P., Konovalov, O., Erokhin, V., Fontana, M.: Electrochemical control of the conductivity in an organic memristor: a time-resolved x-ray fluorescence study of ionic drift as a function of the applied voltage. ACS Appl. Mater. Interfaces **1**(10), 2115–2118 (2009)

20. Erokhin, V., Schüz, A., Fontana, M.P.: Organic memristor and bio-inspired information processing. Int. J. Unconv. Comput. **6**(1), 15–32 (2010)

21. Hebb, D.O.: The Organization of Behavior. Wiley, New York (1949)

22. Jo, S.H., Chang, T., Ebong, I., Bhadviya, B.B., Mazumder, P., Lu, W.: Nanoscale memristor device as synapse in neuromorphic systems. Nano Lett. **10**(4), 1297–1301 (2010)

23. Adhikari, S., Yang, C., Kim, H., Chua, L.: Memristor bridge synapse-based neural network and its learning. IEEE Trans. Neural Netw. Learn. Syst. **23**(9), 1426–1435 (2012)

24. Indiveri, G., Linares-Barranco, B., Legenstein, R., Deligeorgis, G., Prodromakis, T.: Integration of nanoscale memristor synapses in neuromorphic computing architectures. Nanotechnology **24**(38), 384010 (2013)

25. Seok Jeong, D., Kim, I., Ziegler, M., Kohlstedt, H.: Towards artificial neurons and synapses: a materials point of view. RSC Adv. **3**, 3169–3183 (2013)

26. Krzysteczko, P., Mönchenberger, J., Schöfers, M., Reiss, G., Thomas, A.: The memristive magnetic tunnel junction as a nanoscopic synapse-neuron system. Adv. Mater. **24**(6), 762–766 (2012)

27. Baldi, G., Battistoni, S., Attolini, G., Bosi, M., Collini, C., Iannotta, S., Lorenzelli, L., Mosca, R., Ponraj, J., Verucchi, R., et al.: Logic with memory: and gates made of organic and inorganic memristive devices. Semicond. Sci. Technol. **29**(10), 104009–104014 (2014)

28. Rosenblatt, F.: The perceptron: a probabilistic model for information storage and organization in the brain. Psychol. Rev. **65**(6), 386 (1958)

29. Demin, V., Erokhin, V., Emelyanov, A., Battistoni, S., Baldi, G., Iannotta, S., Kashkarov, P., Kovalchuk, M.: Hardware elementary perceptron based on polyaniline memristive devices. Org. Electron. **25**, 16–20 (2015)

30. Horiguchi, H., Imagawa, K., Hoshino, T., Akiyama, Y., Morishima, K.: Fabrication and evaluation of reconstructed cardiac tissue and its application to bio-actuated microdevices. IEEE Trans. NanoBiosci. **8**, 349–355 (2009)

31. Akiyama, Y., Iwabuchi, K., Furukawa, Y., Morishima, K.: Long-term and room temperature operable bioactuator powered by insect dorsal vessel tissue. Lab Chip **9**, 140–144 (2009)
32. Povlich, L., Kim, D.H., Richardson-Burns, S.: In: Reichert, W.M. (ed.) Indwelling Neural Implants: Strategies for Contending with the In Vivo Environment. Soft, fuzzy, and bioactive conducting polymers for improving the chronic performance of neural prosthetic devices. CRC Press, Boca Raton (2008)
33. Bonetti, S., Pistone, A., Brucale, M., Karges, S., Favaretto, L., Zambianchi, M., Posati, T., Sagnella, A., Caprini, M., Toffanin, S., Zamboni, R., Camaioni, N., Muccini, M., Melucci, M., Benfenati, V.: A lysinated thiophene-based semiconductor as a multifunctional neural bioorganic interface. Adv. Healthc. Mater. **4**(8), 1190–1202 (2015)
34. Benfenati, V., Toffanin, S., Bonetti, S., Turatti, G., Pistone, A., Chiappalone, M., Sagnella, A., Stefani, A., Generali, G., Ruani, G., et al.: A transparent organic transistor structure for bidirectional stimulation and recording of primary neurons. Nat. Mater. **12**(7), 672–680 (2013)
35. Buchko, C.J., Slattery, M.J., Kozloff, K.M., Martin, D.C.: Mechanical properties of biocompatible protein polymer thin films. J. Mater. Res. **15**, 231–242 (2000)
36. Tarabella, G., D'Angelo, P., Cifarelli, A., Dimonte, A., Romeo, A., Berzina, T., Erokhin, V., Iannotta, S.: A hybrid living/organic electrochemical transistor based on the Physarum polycephalum cell endowed with both sensing and memristive properties. Chem. Sci. **6**, 2859–2868 (2015)
37. Bidez, P.R., Li, S., MacDiarmid, A.G., Venancio, E.C., Wei, Y., Lelkes, P.I.: Polyaniline, an electroactive polymer, supports adhesion and proliferation of cardiac myoblasts. J. Biomater. Sci. Polym. Edit. **17**(1–2), 199–212 (2006)
38. Ghasemi-Mobarakeh, L., Prabhakaran, M.P., Morshed, M., Nasr-Esfahani, M.H., Ramakrishna, S.: Electrical stimulation of nerve cells using conductive nanofibrous scaffolds for nerve tissue engineering. Tissue Eng. Part A **15**(11), 3605–3619 (2009)
39. Troitsky, V., Berzina, T., Fontana, M.: Langmuir–Blodgett assemblies with patterned conductive polyaniline layers. Mater. Sci. Eng.: C **22**(2), 239–244 (2002)
40. Juarez-Hernandez, L.J., Cornella, N., Pasquardini, L., Battistoni, S., Vidalino, L., Vanzetti, L., Caponi, S., Serra, M.D., Lannotta, S., Pederzolli, C., Macchi, P., Musio, C.: Bio-hybrid interfaces to study neuromorphic functionalities: new multidisciplinary evidences of cell viability on poly(anyline) (pani), a semiconductor polymer with memristive properties, Biophys. Chem. (2015)
41. DeFranco, J.A., Schmidt, B.S., Lipson, M., Malliaras, G.G.: Photolithographic patterning of organic electronic materials. Org. Electron. **7**(1), 22–28 (2006)
42. Ouyang, J., Xu, Q., Chu, C.-W., Yang, Y., Li, G., Shinar, J.: On the mechanism of conductivity enhancement in poly(3,4-ethylenedioxythiophene):poly(styrene sulfonate) film through solvent treatment. Polymer **45**(25), 8443–8450 (2004)
43. Stephenson, S.L., Stempen, H., Hall, I.: Myxomycetes: A Handbook of Slime Molds. Timber Press Portland, Oregon (1994)
44. Adamatzky, A.: Physarum Machines: Computers from Slime Mould, vol. 74. World Scientific, Singapore (2010)
45. Adamatzky, A., Jones, J.: Towards physarum robots: computing and manipulating on water surface. J. Bionic Eng. **5**(4), 348–357 (2008)
46. Adamatzky, A., Erokhin, V., Grube, M., Schubert, T., Schumann, A.: Physarum chip project: growing computers from slime mould. IJUC **8**(4), 319–323 (2012)
47. Tero, A., Takagi, S., Saigusa, T., Ito, K., Bebber, D.P., Fricker, M.D., Yumiki, K., Kobayashi, R., Nakagaki, T.: Rules for biologically inspired adaptive network design. Science **327**(5964), 439–442 (2010)
48. Adamatzky, A.: Bioevaluation of World Transport Networks. World Scientific, Singapore (2012)
49. Adamatzky, A., Martínez, G.J., Chapa-Vergara, S.V., Asomoza-Palacio, R., Stephens, C.R.: Approximating mexican highways with slime mould. Nat. Comput. **10**(3), 1195–1214 (2011)
50. Matsumoto, K., Ueda, T., Kobatake, Y.: Reversal of thermotaxis with oscillatory stimulation in the plasmodium of physarum polycephalum. J. Theor. Biol. **131**(2), 175–182 (1988)

51. Nakagaki, T., Yamada, H., Ueda, T.: Modulation of cellular rhythm and photoavoidance by oscillatory irradiation in the physarum plasmodium. Biophys. Chem. **82**(1), 23–28 (1999)
52. Yamada, H., Nakagaki, T., Baker, R.E., Maini, P.K.: Dispersion relation in oscillatory reaction-diffusion systems with self-consistent flow in true slime mold. J. Math. Biol. **54**(6), 745–760 (2007)
53. Adamatzky, A.: Physarum machines: encapsulating reaction-diffusion to compute spanning tree. Naturwissenschaften **94**(12), 975–980 (2007)
54. Costello, B.D.L., Adamatzky, A.: Experimental implementation of collision-based gates in Belousov–Zhabotinsky medium. Chaos, Solitons Fractals **25**(3), 535–544 (2005)
55. Adamatzky, A., Teuscher, C.: From Utopian to Genuine Unconventional Computers. Luniver Press, Europe (2006)
56. Pershin, Y.V., La Fontaine, S., Di Ventra, M.: Memristive model of amoeba learning. Phys. Rev. E **80**(2), 021926 (2009)
57. Lin, P., Yan, F., Yu, J., Chan, H.L., Yang, M.: The application of organic electrochemical transistors in cell-based biosensors. Adv. Mater. **22**(33), 3655–3660 (2010)
58. Khodagholy, D., Curto, V.F., Fraser, K.J., Gurfinkel, M., Byrne, R., Diamond, D., Malliaras, G.G., Benito-Lopez, F., Owens, R.M.: Organic electrochemical transistor incorporating an ionogel as a solid state electrolyte for lactate sensing. J. Mater. Chem. **22**(10), 4440–4443 (2012)
59. Malliaras, G.G.: Organic bioelectronics: a new era for organic electronics. Biochimica et Biophysica Acta (BBA) - General Subject **1830**(9), 4286–4287 (2013). Organic Bioelectronics - Novel Applications in Biomedicine
60. Svennersten, K., Larsson, K.C., Berggren, M., Richter-Dahlfors, A.: Organic bioelectronics in nanomedicine. Biochimica et Biophysica Acta (BBA) - General Subjects **1810**(3), 276–285 (2011). Nanotechnologies - Emerging Applications in Biomedicine
61. Tarabella, G., Mahvash, F.M., Coppede, N., Barbero, F., Lannotta, S., Santato, C., Cicoira, F.: New opportunities for organic electronics and bioelectronics: ions in action. Chem. Sci. **4**, 1395–1409 (2013)
62. Nilsson, D., Chen, M., Kugler, T., Remonen, T., Armgarth, M., Berggren, M.: Bi-stable and dynamic current modulation in electrochemical organic transistors. Adv. Mater. **14**(1), 51–54 (2002)
63. Cicoira, F., Sessolo, M., Yaghmazadeh, O., DeFranco, J.A., Yang, S.Y., Malliaras, G.G.: Influence of device geometry on sensor characteristics of planar organic electrochemical transistors. Adv. Mater. **22**(9), 1012–1016 (2010)
64. Kamiya, N., Kuroda, K.: Studies on the velocity distribution of the protoplasmic streaming in the myxomycete plasmodium. Protoplasma **49**(1), 1–4 (1958)
65. Larsson, O., Laiho, A., Schmickler, W., Berggren, M., Crispin, X.: Controlling the dimensionality of charge transport in an organic electrochemical transistor by capacitive coupling. Adv. Mater. **23**(41), 4764–4769 (2011)
66. Tarabella, G., Pezzella, A., Romeo, A., D'Angelo, P., Coppede, N., Calicchio, M., d'Ischia, M., Mosca, R., Iannotta, S.: Irreversible evolution of eumelanin redox states detected by an organic electrochemical transistor: en route to bioelectronics and biosensing. J. Mater. Chem. B **1**, 3843–3849 (2013)
67. Romeo, A., Dimonte, A., Tarabella, G., D'Angelo, P., Erokhin, V., Lannotta, S.: A bio-inspired memory device based on interfacing physarum polycephalum with an organic semiconductor, APL Mater. **3**(1) (2015)
68. Erokhin, V., Berzina, T., Smerieri, A., Camorani, P., Erokhina, S., Fontana, M.: Bio-inspired adaptive networks based on organic memristors. Nano Commun. Netw. **1**(2), 108–117 (2010)
69. Möller, S., Perlov, C., Jackson, W., Taussig, C., Forrest, S.R.: A polymer/semiconductor write-once read-many-times memory. Nature **426**(6963), 166–169 (2003)
70. Tang, H., Zhu, L., Harima, Y., Yamashita, K.: Chronocoulometric determination of doping levels of polythiophenes: influences of overoxidation and capacitive processes. Synth. Metals **110**(2), 105–113 (2000)

71. Kim, J.-S., Ho, P., Murphy, C., Baynes, N., Friend, R.: Nature of non-emissive black spots in polymer light-emitting diodes by in-situ micro-Raman spectroscopy. Adv. Mater. **14**(3), 206–209 (2002)
72. Tehrani, P., Kanciurzewska, A., Crispin, X., Robinson, N.D., Fahlman, M., Berggren, M.: The effect of ph on the electrochemical over-oxidation in pedot:pss films. Solid State Ion. **177**(39–40), 3521–3527 (2007)
73. Isaksson, J., Kjäll, P., Nilsson, D., Robinson, N., Berggren, M., Richter-Dahlfors, A.: Electronic control of Ca^{2+} signalling in neuronal cells using an organic electronic ion pump. Nat. Mater. **6**(9), 673–679 (2007)
74. Ridgway, E., Durham, A.: Oscillations of calcium ion concentrations in physarum polycephalum. J. Cell Biol. **69**(1), 223–226 (1976)
75. Heremans, P., Gelinck, G.H., Muller, R., Baeg, K.-J., Kim, D.-Y., Noh, Y.-Y.: Polymer and organic nonvolatile memory devices. Chem. Mater. **23**(3), 341–358 (2010)
76. Bernards, D.A., Malliaras, G.G.: Steady-state and transient behavior of organic electrochemical transistors. Adv. Funct. Mater. **17**(17), 3538–3544 (2007)

Chapter 19
Memristors in Unconventional Computing: How a Biomimetic Circuit Element Can be Used to Do Bioinspired Computation

Ella Gale

Abstract Memristors differ from resistors by possessing a memory, and both synapses and neurons have been discussed as biological memristors. The short-term memory of the memristor (or spiking profile) is similar in form to neural spikes. Thus, memristors are obvious candidates for building biomimetic circuits and computers. In this chapter, we review some recent experimental results in the area of memritor-based spike computing. We demonstrate how memristor spikes are a real-world memristor model of an inhibitory neuron, then we demonstrate the complex emergent behaviour from networks of memristors which resembles neural dynamics, we then expose these memristor networks to living neural cells, where the memristor state is altered by cellular action. Further investigation of the memristor spiking process allows us to elucidate design rules for spiking logic gates, and we demonstrate a novel full adder instantiated in a single memristor. Spiking memristor computation might be the best route to truly neuromorphic computers.

19.1 What Are Memristors?

The resistor, capacitor, and inductor are the three well-known fundamental circuit elements, discovered in 1745, 1827 and 1831 respectively. Based on an assumption of completeness (see Fig. 19.1), in 1971 Leon Chua postulated a 4th fundamental circuit element [31] which would relate charge q to magnetic flux φ and would have the distinction of being the first non-linear circuit element.

No physical instantiations of the memristor were generally acknowledged until 2008 when Strukov et al. realised that their molecular electronic switches' behaviour was due to the titanium electrodes and not the organic layer [149] and announced that they had found the memristor [134]. At that time, it was believed that memristors could not have been made contemporary with the other fundamental circuit elements

E. Gale (✉)
Department of Chemistry, University of Bath,
Claverton Down Road, Bath BA2 7AY, UK
e-mail: e.gale@bath.ac.uk

© Springer International Publishing Switzerland 2017
A. Adamatzky (ed.), *Advances in Unconventional Computing*,
Emergence, Complexity and Computation 23,
DOI 10.1007/978-3-319-33921-4_19

497

Fig. 19.1 The four circuit
measurables and the six
relationships between them

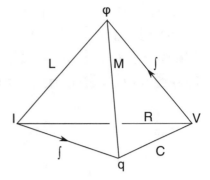

as memristance was a nanoscale phenomenon [134], 'essentially unobservable at the millimetre scale and above', [149] this has been since experimentally disproved by the fabrication of macroscopic memristors [7, 55, 88, 151].

It later became apparent that memristor-like devices had been fabricated before 2008: resistance switching was first observed in oxides in a gold-silicon monooxide-gold sandwich in 1963 [115], the first metal oxide resistance switch was reported in nickel oxide in 1964 [13], memristor-like switching curves were observed in TiO_2 thin films in 1968 [9] and the memistor [148], a 3-terminal memristive system, was fabricated in 1960 (although this is disputed as being an example of a memristive system [5]). These examples inspired some searches for the earliest known memristor [26, 125] with the singing arc [65] in 1880 and the coherer from 1899 [60] being the oldest devices identified so far.

In fact, memristor theory was used to describe Resistive Random Access Memory (ReRAM) devices that were made out of a range of materials before 2008. ReRAM is usually based on transition metal oxides such as TiO_2, $SrTiO_3$ [135, 139], NiO [66], CuO [156], ZnO [24], MnO_x [158], HfO_x [101], Ta_2O_5 [109], Ti_2O_{5-x}/TiO_y [155], TaO_x/TiO_{2-x} [154]; both binary and perovskite oxides are capable of resistance switching. memristors can be made out of TiO_2, chalcogenides [117], polymers [80, 123], atomic switches [69], spintronic systems [118] and quantum systems [120]. Whether ReRAM and memristors are different technologies or the same thing, is a question that has involved some controversy. The term ReRAM refers to a suggested use for a range of materials (namely as memory), whereas memristors are named after a property, and have been investigated not just for RAM, but also for such uses as neuromorphic hardware (e.g. [53, 59, 73, 82, 83, 120, 157]), chaotic circuits (e.g. [22, 76, 113, 114]), processors (e.g. [51, 94, 119, 127, 129, 138]), vision processors (e.g. [62, 77, 102, 111, 157]) and robot control (e.g. [48, 73, 144]) to name a few. Memristor measurements tend to be a.c. measurements, often done at different frequencies. This is partly because initially the memristor was thought to be a purely a.c. device and only recently have the d.c. aspects of the device received attention [18, 46, 82, 128]. ReRAM measurements tend to concentrate on the resistance and persistence of the switching states.

In this chapter we shall primarily concern ourselves with memristors made of TiO_2—the most investigated material to date—although the work presented is expected to be applicable to any memristor/ReRAM device presenting a spiking I-t profile.

19.2 Chemical Mechanisms of Operation

Many suggested mechanisms have been put forward to explain the causes of resistance switching and of memristance, and TiO_2-based devices been described by all of them. The high resistance material is generally believed to be stoichiometric TiO_2 [134], but there is debate about what the low resistance material is, and, given that the memristors/ReRAM are made in several different ways there is no expectation that there is a single explanation or material cause.

It is generally agreed that the ions in the material hold the device's state. The most popular suggested mechanisms can be split into two groups: ionic and thermal, where ionic mechanisms concentrate the motion of ions in the material, and thermal mechanisms investigate the change of material, due to the motion of ions.

19.2.1 Ionic

The ionic mechanism in TiO_2 devices involve the migration of oxygen vacancies (although suggestions of migrating OH^- ions have been made [14, 152]). This movement of vacancies creates auto-doped phases which are metallically conducting for $TiO_{(2-x)}$ for $x > 1.5$ [145]. The oxygen ions may combine at the anode and evolve O_2 gas, this has been seen in ReRAM [137], the Strukov memristor when fabricated with macroscale electrodes [153] and the sol-gel memristor [56]. There is some evidence that the choice of electrode material can hinder the production of O_2, for example, it is thought aluminium supports TiO_2 memristance by acting as a source or sink of oxygen ions [79, 96]. Most memristor-based modelling has concentrated on modelling the flow of these oxygen vacancies (see later). It has been suggested that switching is dominated by ionic motion rather than charge trapping [110]. A mixed mechanism involving oxygen vacancy transport between the tip of a conducting filament (itself formed by thermal mechanism) and an electrode has been suggested [17]. Alternative electrochemical mechanisms that explain memristance via a change in titanium oxidation state, Ti(IV)→Ti(III), have been put forward [152] although not widely adopted.

19.2.2 Thermal

Joule heating is the process where the application of an electric field and flowing current heats the material and changes its structure. TiO_2 atomic deposition thin films can form conducting filaments as extended defects along grain boundaries [136], the ions drift, forming a path which breaks with excess heat (i.e. too high a voltage) and can be reformed via the same mechanism. This mechanism has been credited with causing both high and low resistance states [25]. The conduction channels are usually ohmic and it has been shown that the resistance and reset current bears little relation to the composition of the material or whether it is operated in a unipolar or bipolar mode, which suggests, and has been verified by simulation, that the mechanism is thermally activated dissolution of conducting filaments [75]. This could explain why compliance current choice controls the resistance and reset current and suggests that larger resistance values are more related with the heat sink effects of the metal oxide, rather than its electronic properties.

The form of these conducting filaments/channels is debated. In 2009 A Pt-TiO_2-Pt ReRAM showed a Magnéli phase [96], Ti_nO_{2n-1}, crystallographic shear plane from rutile [21], roughly 10–20 nm in diameter and which shows metallic (i.e. ohmic) conduction. Magnéli phases are considered likely to be the material cause as they are more thermodynamically stable than vacancies spread through-out the material [100]. Single crystal TiO_2 nanorods have shown bipolar switching [159] which have a rectifying effect. Conical Magnéli phase filaments have been observed and shown to be the cause of fusing and anti-fusing BPS [97]. Hourglass-shaped Magnéli phases were seen in [96], which were formed by alternating polarity voltage and proved to be more stable in operation. Fractal conducting filaments, have been suggested by simulation [98], observed [131] and are thought to relate to dielectric breakdown [116]. From spectromicroscopy and TEM, electroforming the Strukov memristor has been found to produce an ordered Ti_4O_7 Magnéli phase [133].

19.2.3 Other Mechanisms

Other mechanisms have been put forward such as: metal-semiconductor transitions [130], crystalline TiO_2-amorphous TiO_2 phase transition via conduction heating and breaking [25], raising and lowering Schottky barriers via bulk (UPS) or interface (BPS) transport of oxygen [139], and conductance heating causing lateral transport of conducting filaments [75]. There is also a debate as to what the structure of the less-conducting thin-film TiO_2 is, with rutile (r-TiO_2) [25] and amorphous titanium dioxide (a-TiO_2) being the most popular suggestions. Note, that a-TiO_2 has also been suggested as the conducting form of TiO_2 and that Magnéli phases are sheer planes from r-TiO_2.

Currently, the Magnéli conducting filaments are the most generally accepted mechanism, but as all the materials are treated differently, measured differently, and titanium dioxide has a wealth of behaviours, it is unlikely that there will emerge only one explanation for all published TiO_2 devices.

19.2.4 Uses of Memristors

The first commercial suggestion for memristors was computer memory because the memristor's small feature size could offer an increase in memory density, wouldn't require power to hold a value (so it is low energy electronics) and could be used for multi-state memory. However, alongside the usual manufacturing faults, memristors suffer some unique issues on which there has been much work and many attempted fixes, both based on memristor theory [68, 70] and experiment [91]. For example, according to the theory, memristor state will change when read, affecting data integrity. This can be solved by using a read pulse below a threshold time [70] or a reading algorithm that reads and then rewrites the memory cells [142] (this is also useful for reducing sneak-path-related read errors). Other work has focused on improving memory tests to identify undefined states [68] and fabrication errors [85] (by making use of sneak paths to test several memristors at the same time).

A big problem for memristor-based cross-bar memory is sneak paths which can result in erroneous memory values being read and are a big problem for shrinking memory size. Attempted solutions include: Hewlett–Packard's reading algorithm [142] described above; unfolding the memory (only one memristor per column/row, which greatly reduces the possible gains from shrinking memory size; adding an active element like a diode [106], which might add delay, or a transistor gate [89], which would remove the advantage of the memristor's small size; using anti-series memristors as the memory element so that the total resistance is always $R_{on} + R_{off}$ [84], this would require differentiating between $R_{on} - R_{off}$ and $R_{on} + R_{off}$ for logical values; using multiple access points [161]; using a.c. sensing [126] at the cost of increased complexity; making use of the memristors own non-linearity (instead of a diode) [91, 154]; making use of a 3-terminal memristor [160] (instead of a transistor) which forces sneak paths to have a higher than R_{off} resistance (which is the best case failure for sneak paths) but requires an additional column line. The authors of [160] concentrated on the memistor [148] as an example device, but intriguingly the idea could be applied to the 3-terminal plastic memristor [38, 39]. For memristor-based memory to be adopted, there is much further work to be done in this area, especially on experimentally testing these approaches.

An area memristors are perhaps uniquely suited to is the design and fabrication of neuromorphic processors using memristors, which will require novel approaches and hardware instantiations. Because memristors can implement IMPLY logic, Bertrand Russell's logical system [147] is getting something of a renaissance. If memristors natively implement IMPLY logic [18], perhaps as spikes [47, 51], then there will be investigations into engineering memristors into current circuit design approaches (which traditionally use AND, OR, NOT, NAND and similar). Several workers have suggested doing digitised stateful logic in CMOS-compatible cross-bar memory arrays (using the IMPLY and FALSE operation set) and have started to look at design methodologies: as in [95], making encoders/decoders out of transistor-memristor arrays [143] and making use of the parallelisability of the architecture for simultaneous bitwise vector operations [99]. Recently, several researchers have started to look at hysteresis [15, 35, 36, 63] and other figures of merit [86, 124, 146] of practical use to engineers.

An interesting and unexpected outcome of memristor research is the discovery that evolution has made use of memristors and memristive mechanisms. Biological material and mechanisms like sweat ducts [67], leaves [93, 141], blood [92], slime mould [49, 121], synapses [103, 157] can be described as memristive and memristor theory has been used to understand learning in the synapses and simple organisms and to update neuron models such the Hodgkin–Huxley [28]. For the blood, leaves and sweat duct results, it is likely that the memristive behaviour arises as a result of the form of the structures and is not used to perform a biological function, however from the results, it can be suggested that synapses and neurons use memristive functionality to perform their biological function. This could suggest that electrophysiology is a field ripe for memristor research and perhaps the most ground-breaking memristor work will come from linking the manufactured devices with living memristors, either directly or via the creation of bio-inspired computers which may usher in an entirely new paradigm of computational approaches. Early results in the field will be described in this chapter.

The Hodgkin–Huxley model of a neuron is an equivalent electronic circuit for a neural cell membrane, which allows the propagation of voltage spikes along a nerve cell. The circuit consists of a capacitor to model the cell membrane bilayer, a battery, to model energy input into the system and two 'time-varying' resistors to model to action of the sodium and potassium protein-pumps. Neurons use energy to pump ions outside the cell so there is a chemical potential gradient, and when a neural signal needs to be sent, the protein pumps open and the ions flood back across the membrane, causing a voltage spike. The time-varying resistors in the model were required to model the switching action and refractivity period of the protein pumps. Chua's recent work [27–29] has highligthed that a charge-dependent memristor explains the protein dynamics more naturally (in that a membrane protein is more sensitive to the charges on either side of it, than some time-varying clock). All the components of the model could be bought in a hardware store, but the time-varying resistors, used to model protein pumps, did not really have a hardware analogue. If the ion pumps in neural cell membranes are modelled as time-varying resistors, then biological brains should have huge impedances (and a large resulting power draw) which we know they do not. By replacing the time-varying resistor with time-invariant memristors responding to the time-varying ion concentration, the model no longer requires unnatural impedances [27].

19.3 Memristor Theory

A resistor, of resistance, R, relates current, I, and voltage, V, thus:

$$V = RI. \tag{19.1}$$

The memristor relates the time-integrals of current and voltage [31], namely the charge, q and magnetic flux, φ, thus:

$$\varphi = Mq, \tag{19.2}$$

where M is a function of q, i.e. the memristance of the device depends on the amount of charge that has passed through it, and this value is referred to as the state of the memristor. The charge and magnetic flux are not measured however, we measure the voltage and current as:

$$V(t) = M(q(t))I(t) \tag{19.3}$$

where the resistance sampled by the circuit is the memristance at that point in time, thus, a memristor can be viewed as a type of time-varying resistor (in fact, many theoretical descriptions start from this point), because the resistance can vary, the memristor gives non-linear behaviour. The memristance is not strictly a time-varying quantity, because the memristance depends only on the charge. However, the charge is often a function of time, essentially the memristance in an implicit function of time through its explicit dependence on charge. Note that, Eq. 19.3 also has a term $\frac{dM(q(t))}{dt}\frac{dq(t)}{dt}$, which is often excluded in the simple model.

As Eq. 19.3 is currently understood, the charge, q, stands for the time-integral of the current rather than a charge on the device. The model in [58], used to model memristors in general rather than neurons, is based on a two-level description, assigns the charge q to the ions present in the system, which associates the magnetic flux with the magnetic flux of the ions (which is many, many times smaller than that associated with the electrons). The current in Eq. 19.3 is then the current associated with the electrons (or total current, the ionic current is so small as to be and irrelevant component for the TiO_2 memristor). This approach has the ions responding to voltage changes slower than the electrons, and thus holding the memory (or state) of the device. This state, which is $M(q)$, is then sampled by the electrons. This idea, although not widely accepted in the field, does explain why so many memristors have an ionic 'memory' property and suggests that when modelling neurons we should start with the ionic behaviour and then build up to an 'electronic' description, i.e. memristors and neurons can both be thought of as ionic-electronic 'devices'.

Fig. 19.2 Single memristor
dynamics. At τ_0 a voltage
step is applied, the
short-term memory of the
memristor lasts until the
memristor achieves $I(\tau_\infty)$.
Note that the voltage step is
applied at τ_0 and not turned
off until after 200 s

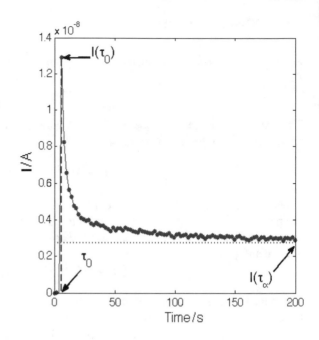

19.3.1 Short Term Memory

The input of a sudden 'spike' in ionic concentration across a neural membrane
leads to an associated voltage spike. The memristors, if given a changing voltage,
exhibit a quick current spike (starting a τ_0), with a slowly decaying tail, see Fig. 19.2.
This is different to neurons as they exhibit voltage spikes in response to a current,
but the similarity in shape suggests that building neural architecture based on the
complement to neuron's current measurables is possible. When there is a change in
voltage, ΔV, across a memristor, the device exhibits a current spike (see Fig. 19.2,
the physical cause of which is discussed at length in [46, 59]). This spike is highly
reproducible and repeatable and is related to the size of the voltage change [46]. The
spike's size is highly reproducible within our measurement ability (future quicker,
high-accuracy measurements may find some variance), the current then relaxes to a
stable long-term value taking approximately 2–3 s to get to this value.

This slow relaxation is thought to be the d.c. response of the memristor [46, 59][1]
or short term memory, and if a second voltage change happens within this time frame,
its resulting current spike is different to that expected from the ΔV alone. The size

---

[1]Note that there has been some discussion as to whether the memristors have a d.c. response, within
which is has been generally agreed that the perfect memristor may not, but non-ideal memristors
(which all real devices are) may. This is not a settled issue and generated much discussion at a
recent conference (CASFEST 2014).

Fig. 19.3 The effect of adding spikes close in time. The response spikes are the negative current spikes. When a positive spike it included but not allowed to relax the corresponding negative spike is smaller

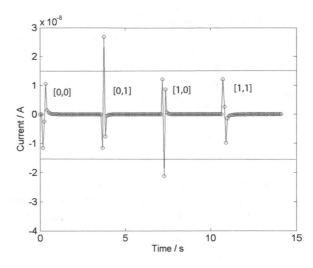

Fig. 19.4 The input voltage for Fig. 19.3

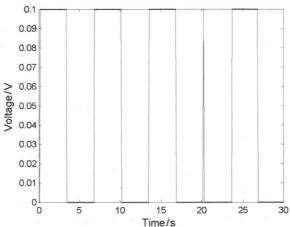

and direction of this current spike depends on the direction and magnitude of ΔV and the short-term memory of the memristor.

As an example, consider a memristor pulsed with a positive 1V voltage square wave as in Fig. 19.4 (where the pulses are repeated to demonstrate the repeatability) with a timestep of ≈ 0.02 s. The current response is shown in Fig. 19.3 and we can see there is a positive current spike associated with the $+\Delta V$ and a perhaps less obvious negative current spike associated with the $-\Delta V$ transition from $+1V \rightarrow 0V$. At approximately 20 s, we shortened the square wave to a single time step, and the memory of the system has caused the response spike (responding to the $-\Delta V$) to be smaller (and as it is smaller, it suggests that there is some physical property of the device which has not adjusted to its $+V$ value). Thus the response is subtractive in current and additive in resistance state.

Table 19.1 Voltage steps for voltage input combinations

t_1	t_2	Direction of energy decrease	ΔV
V_A	V_A	$V_A \longrightarrow V_A$	$+\Delta V$
V_A	V_B	$V_A \longrightarrow V_B$	$+\Delta V$
V_B	V_A	$V_B \longleftarrow V_A$	$-\Delta V$
V_B	V_B	$V_B \longrightarrow V_B$	$+\Delta V$

To try and understand the subtleties of this apparent 'addition', consider the following system: two voltages, V_A and V_B are sent to the memristor, one after the other separated by one time-step (i.e. before the memristor has equilibrated), where $V_B > V_A$ and $V_B = 0.12$ V. We look at two situations:

1. $V_A(t) \rightarrow V_B(t+1)$;
2. $V_B(t) \rightarrow V_A(t+1)$.

These two situations are drastically different if we look at the transitions, ΔV, as situation 2 has a negative $\Delta V_{B \rightarrow A}$, all the other transitions are positive see Table 19.1. As the addition takes place against a 'background' of decreasing current response, a second spike at the same voltage level is an addition of energy into the system, and thus the transition from $V_x \longrightarrow V_x$ is taken as being positive. Situation 1 shows that if the smaller voltage is sent first ($V_A \rightarrow V_B$), the current of the first transition $\Delta i_{0 \rightarrow A}$ increases with the size of V_A, and the second transition $\Delta i_{A \rightarrow B}$ decreases with the size of ΔV_A, due to the decrease in the effective $\Delta V_{A \rightarrow B}$. However, the sum of these two effects is non-linear, so that the total current transferred (approximated as the sum of the spikes here, but actually the area under the two current transients) is not the same as that shown for situation 2 (until $V_B = V_A$). This shows that more current is being transferred and demonstrates that the spikes are dependent on ΔV. Furthermore, it makes it clear that $\Delta i_{0 \rightarrow A} + \Delta i_{A \rightarrow B} \neq \Delta i_{0 \rightarrow B} + \Delta i_{B \rightarrow A}$, (except in the trivial case where $V_B = V_A$) and that spike based 'addition' is non-commutative and therefore the order in which the spikes are sent is relevant to the output of the calculation. More details and a discussion of the use of this effect are given in Sect. 19.7.

Thus, memristors possess a short-term memory effect, and voltage spikes input within this window (τ_∞) cause a non-linear but definite 'subtractive summation' of the current responses. It is interesting that memristors possess a short-term memory, as the brain operates using short-term memory effects in the neural spikes. Memristor current spikes are similar to neural spikes: what are the ramifications of this?

19.4 Memristors Acting Like Neurons

Of interest to us here is another property of neurons habituation:. Habituation [132] is the the learned ability to ignore unnecessary stimuli, such as an author not attending to the surrounding background noises of a coffee shop when writing (incidentally, as this learning is unconscious, it takes a great deal of mental training to turn it off, although long-time meditators have been shown to be able to do this [8]). One of the first and most famous habituation experiments involved the 'sea snail' (confusingly also known as 'sea hare') or *aplysia*. This invertebrate has a sensitive feeding tube which it doesn't want to get damaged and so will withdraw if the tube is stimulated (like a child will stop sticking its tongue out if you threaten to grab it). In an experiment the feeding tube was lightly brushed with a delicate paintbrush (a harmless intervention), and, over time, the *aplysia* learned not to attend to this stimulus and not to withdraw its tongue. *Aplysia* is a simple creature with only 20,000 neurons, and the neural pathways for this response has been mapped out (and requires only 4 neurons). The response at the sensory neuron does not diminish, but with training the corresponding motor response signals (recorded at motor neuron $L7_g$) decrease in size and this is the physiological cause of habituation.

Chua demonstrated theoretically that an ideal memristor stimulated by repeated voltage spikes would alter its memristive state and present a decreasing current response spike (see Fig. 7 in [27]) and this result strengthened previous theoretical work demonstrating that the neural protein pumps in the Hodgkin–Huxley model of the neuron [71] was best described theoretically as a memristor [28, 29]. We shall replicate this theoretical result with a real memristor in Sect. 19.4.1.

The advantages of using spike interactions are many-fold. The memristor switching itself can be slow [56] but the spikes can interact much faster, the output of which is 'held' in the short-term memory of the memristor, which gives rise to, if not faster processing, more complex operations within a given time-frame than is usually the case in standard electronics.

19.4.1 Re-Creation of Biological Experiment Using a Memristor

A famous study of an amnesiac patient H.M. who, as a result of ill-advised brain surgery, had his medial temporal lobe removed to treat his epilepsy. As a result he was not able to convert short term memories into long-term memories (a mental deficiency made famous by the film Memento), however, he was able to learn and improve at certain tasks without ever being aware that he had learnt it, this is the type of 'automatic' skills learnt through practice, like martial arts or driving a car [10, 140]. This type of memory is called non-declarative memory and includes 'how-to' types of memory, and elementary reflexive (as in it becomes a reflex) learning such as habituation, sensitisation (ability to attend to dangerous stimuli) and classical (association of two stimuli) and operant (association of a subject's actions and a stimuli) conditioning (trained ingrained responses), all of which is stored in the unconscious mind.

Fig. 19.5 Experimental replication of the *Apylsia* experiment: **a** The applied voltage input: 0.1 V spikes lasting 1 time step, with 0 voltage rests of 1 timestep (around 0.08 s). N.B. error bars not shown as the electrometer can measure to fA and μV. **b** *Bars* current response to positive spikes applied sequentially; *line* current response to positive spikes applied after the short-term memory evaporates (for comparison); **c** *Bars* negative 'bounceback' response to zeroing voltage, note the first bar is the response to zero after zeroing. If the spike inputs are separated by more than the short term memory (lines in (**b**) and (**c**)) the response to the spikes are the same each time indicating no learning has taken place. If the spikes a input within the time frame of the memristor's short-term memory, we see habituation, as more spikes are input, the response is decreased

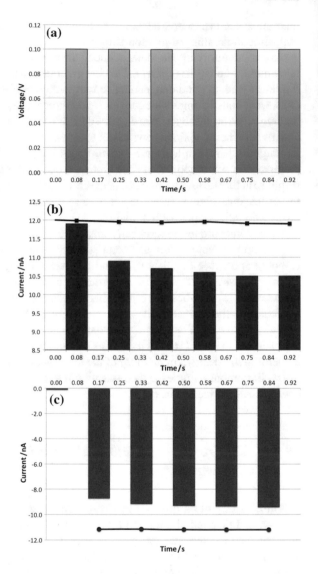

Figure 19.5 demonstrates the same response in our memristors, Fig. 19.5b compares the positive current non-habituation response, the line, with the habituation response, the blue bars, to the voltage input in Fig. 19.5a. We see a clear decrease in the level of the current response with increasing numbers of stimuli demonstrating electronic habituation with a single memristor. Unlike [27] this response is itself a non-linear decrease which follows the same shape as the decay curves in [54, 59]. This is (to my knowledge) the first time habituation has been demonstrated in such a manner in a single memristor circuit (although these results were hinted at in [37] which is very interesting as this is a different type of memristor and suggests that

this is a general property of the devices rather than being specific to a material). Unlike neurons, this habituation learning is plastic and only persists as long as the short-term memory of the memristor, allowing for plastic habituation and learning in memristor networks (again, see Erokhin's et al work's for interesting work involving the plastic (rewrite-able) and permanent learning in memristor networks [41]). Our devices also express the state of their memory when the device is returned to zero voltage (bounceback, see Sect. 19.7.1.2 for a description) and this is shown in Fig. 19.5c, where the line shows the response of the device to switching to zero after the memory has been lost (non-habituation) and the red bars show the response to zeroing after continued stimuli (habituation), the first bar is the response to the first zero, i.e. after the device has been fully zeroed, and this small value serves as a control for the effect. We see that the negative 'bounceback' response (Fig. 19.5c) shows a similar learning effect which is smaller than the positive effect, unlike the positive spike response this effect moves towards the non-habituation response rather than away from it.

19.5 Emergent Spike Dynamics in a Network

Another aspect of spiking neurons is the emergence of long-range correlated dynamics (brainwaves) and complexity which is the concept of being on the edge of chaos. This area of phase-space, has very rich dynamics, and it is known that most of life occurs on the edge of chaos. In this section, I will show that a network containing only memristors can produce dynamics of a similar level of complexity (later experiments [52] demonstrated that these memristors are capable of actual chaotic behaviour).

Memristors have been compared to both neurons [28, 29, 32] and synapses [40, 103], and as such, have been enthusiastically received by the neuromorphic computing community. The first experimental memristor paper [134] suggested that because memristors combined processing and memory in the same component, they were similar to neurons and could be the basis of a brain-like computer. This same group later presented the 'neuristor' [122]—a combination of memristors and capacitors that presented repetitive brain-like dynamics.

Brain dynamics are known to be chaotic and it has been suggested that neurons are poised at the edge of chaos [29], so the search for chaos using memristors (a search also inspired by the similarity of the memristor's operation to the Chua diode in Chua's circuit [105, 107]—the simplest chaotic circuit) is relevant to us here. From simulations, it has been shown that a Chua circuit can be built including a memristor [11, 12, 16, 22, 108]. Experimentally, chaos-like dynamics have been observed in memristor circuits [43] and chaos has been demonstrated as arising from a single memristor [52].

We decided to look for complex, emergent dynamics in simple networks consisting of just memristors (although it should be understood that because they are non-ideal memristors, the devices likely contain some capacitance, resistance and inductance, like all real circuit elements). The networks are shown in Table 19.2. We take a measure of circuit complexity, whereby increasing the number of memristors in the circuit, number and type of interactions (i.e. out of series and anti-parallel) increases the circuit complexity. For example, experiment 7 is the most complex circuit, experiment 1 is the least.

All networks were probed with the same voltage step, which is shown in Fig. 19.6. To try an understand the dynamics, we used techniques from statistics. The periodogram of the input presents the periods of Fourier transform frequencies, the

Table 19.2 Constructed experiments. Anti-parallel memristor interactions are represented by 'p', anti-polarity series memristor interactions by 's'

Experiment number	Wiring diagram	Number and type of anti-polarity interactions
1		0
2		1s
3		0
4		1p
5		0
6		0
7		1s, 1p
8		2p

Fig. 19.6 Standard voltage step input

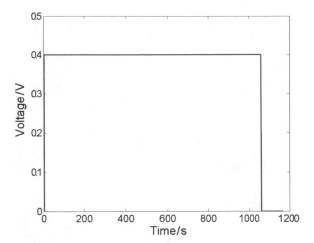

autocorrelation function demonstrates how dependent a point is on the one directly proceeding it, and demonstrates whether there is a trend in the data. The partial autocorrelation function shows the persistence of the data, i.e. the number of autoregression functions required to fit it, this can be thought of as a way of looking at the simplicity of the dynamics. Examples of network dynamics follows.

19.5.1   'Ideal' Dynamics

The simplest dynamics were those which were single memristor-like, and showed a current transient decaying with time, see Fig. 19.7 and compare to the single memristor response in Fig. 19.2. This type of response is given the classification '1' in Table 19.2, and accounted for 2/22 experiments. Single memristor-like $I–t$ curves with a few single spikes was designated '1s' (4/22). Graphs which looked like a single memristor switching state were designated 'w' (2/22), see Fig. 19.8. '1', '1s' and 'w' were all variations on the single memristor $I–t$ in that they have a switching spike when the voltage was turned on or off and decay with time either without spikes ('1'), with a few spikes ('1s') or with several switching spikes ('w'). These dynamics are characterised by regular and increasing frequency packets in the periodograms, a long-tail (indicating a trend) in the autocorrelation function and a small number of autoregression functions required to fit it (the number of functions is the number of points above the lines of statistical significance, plotted in blue on the graphs). By comparing the statistical data of 'w' type dynamics with those shown in Fig. 19.7 we see that the 'w' are most similar to the ideal dynamics and result from simple superposition of several small versions of the 'ideal dynamics'.

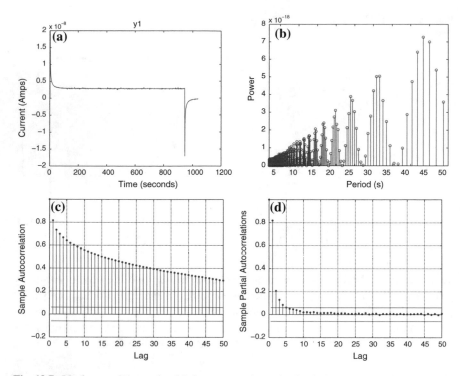

Fig. 19.7 Ideal network example of 3 A-type memristors in circuit 6: **a** I–t curve in response to constant voltage; **b** a periodogram; **c** autocorrelation function; **d** partial autocorrelation function. This demonstrates that using similar memristors with low circuit complexity gives the expected result. Voltage applied is shown in Fig. 19.6

19.5.2 Emergent Spiking Dynamics

The more complex behaviour is the spiking behaviour and this is broken down into two types: 'SpO' which is oscillations with a few single spikes overlaid (5/22) and 'Sp' which is oscillations with bursting spikes (6/22). Examples of 'SpO' responses are shown in Fig. 19.9. An example of 'Sp' type dynamics are shown in Fig. 19.10. Neither the 'Sp' or 'SpO' circuits showed switching spikes when the voltage was turned on or off, which we think is because that spike is the impulse to the circuit and the energy emerges at a later time in the observed spikes.

The remaining type is designated as 'fi', where the current is several orders of magnitude higher and the response more linear, both facts suggest that a conducting filament in one (or more) of the memristors has nearly bridged the electrodes (the current is linear when it finally connects). These circuits are not interesting from the dynamical point of view, however the memristors in this state are excellently suited for the task of holding a state and are useful for resistive Random Access Memory (ReRAM).

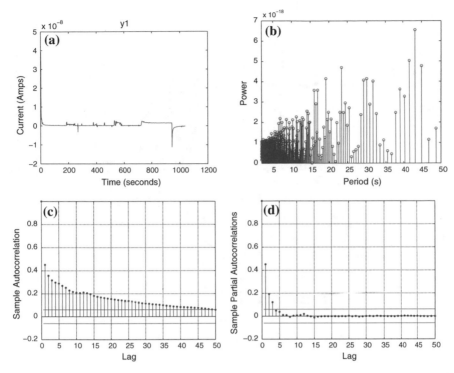

Fig. 19.8 Switching network type ('w') network example using 3 A-type memristors in circuit 7: **a** *I–t* curve in response to constant voltage; **b** a periodogram; **c** autocorrelation function; **d** partial autocorrelation function. This demonstrates that increasing compositional complexity can move an ideal network away from ideality. Voltage applied is shown in Fig. 19.6

The periodograms for the emergent spiking are different to those seen for the ideal dynamics, with the greatest power frequencies being the fast frequencies (small period). The autocorrelation functions shows no background trend and a partial autocorrelation functions show that autoregression functions needed to fit the data have to include both long-time-scale and short-time-scale interactions.

We have shown that anti-parallel interactions give rise to more complex behaviour, specifically an increased likelihood of spiking and more complex spiking dynamics. This result demonstrates that the chaotic dynamics seen in simulations [122] is likely due to the sub-circuit of two anti-parallel memristors rather than numerical effects. Our results could also suggest that the repeating spiking oscillations seen in the neuristor [22] may arise from the anti-parallel sub-circuit rather than the interaction between a memristor and a capacitor. We know from the data presented here that the spikes are not random, but further work will be required to demonstrate if they are chaotic in nature—the fact that 3-memristor networks possess a richer, more persistent and quantitatively different dynamics to 2-memristor networks is suggestive (because chaotic systems need at least 3 state variables).

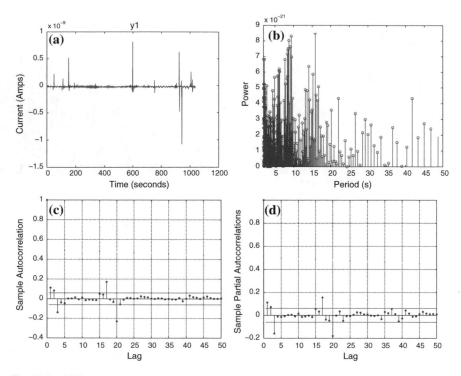

Fig. 19.9 Spiking network type 'SpO' example from B type memristors in circuit 7: **a** *I*–*t* curve in response to constant voltage; **b** a periodogram; **c** autocorrelation function; **d** partial autocorrelation function. The ACF clearly shows the interaction of several oscillations and a dampening directly after the single peak, the PACF shows we need an AR of order 7 to fit the data. Voltage applied is shown in Fig. 19.6

We have classified our results into two types: The first are ideal networks which show voltage-switch-related spikes followed by smooth exponential decay. The second are switching networks, which lack the voltage-switch-related spikes, have a flat baseline and possess both oscillatory dynamics and bursting spikes. The ideal networks combine like resistors, in that a network of only memristors addressed by their joint 1-port entry is indistinguishable from a single memristor. The spiking networks add up differently and emergent dynamics would make it easy to tell that there was more than one memristor present if we were given a 'black-box' circuit and asked to identify whether it contained a single memristor or a network (this is the sort of thought experiment the phrase 'addressed by their joint 1-port entry' describes [30]).

What is the underlying cause of the spikes? These spikes are widely observed in memristors and referred to as current transients—which is a description of the dynamics rather than a cause—the cause is often ascribed to capacitance in the circuit. These networks were tested without a capacitor in the system, but there is of course parasitic capacitance in the circuit and all real devices include some measure of capacitance, resistance and inductance. Furthermore our devices have aluminium

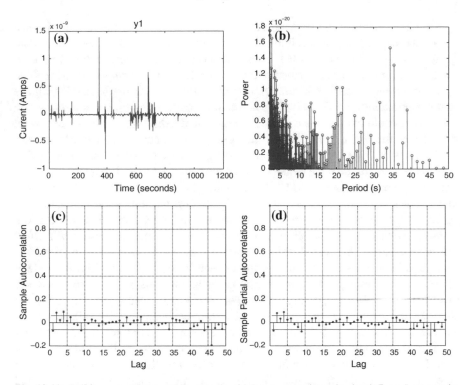

Fig. 19.10 Spiking network type 'Sp' example of 3 B type memristors in circuit 7. : **a** I–t curve in response to constant voltage; **b** a periodogram; **c** autocorrelation function; **d** partial autocorrelation function. We see groups of bursting spikes and underlying oscillations. Voltage applied is shown in Fig. 19.6

electrodes which are known to act as sources and sinks of oxygen ions [14, 33, 81, 87], which can stabilise switching [25, 150], and can be thought of as a slow, ionic capacitor.

However, the spikes can be fitted by memristor theories (both ours [44] and others [134], see [46]), which suggests that they may arise from the memristance in the devices. When it first appeared, the memristor was thought of as an a.c. device and the question of the d.c. response was not addressed. But obviously, the ions within the memristor will drift to voltage regardless of whether it is varying or constant (with the proviso that if the voltage varies too fast the ions won't have time to drift one way significantly before changing direction; this is covered by the part of the memristor definition which described the hysteresis as the frequency tends to infinity, see [6, 31]). Thus, we expect there to be a time-varying response to a d.c. current, and the observed current spike is a compelling candidate. We have experimentally demonstrated that this spike shape is related to frequency-related hysteresis changes [54], a fingerprint of the memristor. This fact is suggests that the spikes are memristance, but the question is not conclusively answered yet.

We shall now take the position that the spikes are mainly due to memristance rather than capacitance. A question of interest is then why does the current equilibrate (which we know it does, see [46]). Due to this behaviour, we have called the spikes the short-term memory of the memristor and have shown that single memristor-like spikes can interact within this time window and can be used to implement logic gates and perform addition [47, 51]. If the memristors in the circuit possess a short-term memory it could explain the cause of the emergent behaviour. As these tests were done across the whole circuit, we do not know how the state of the 3 memristors vary individually. If there is a time delay in a voltage induced current spike propagating across the network, then the response of a memristor to it would change the voltage across another memristor by ΔV. If these changes are not synchronised then these small ΔV's can move around the network causing the individual memristors to spike and propagate a different ΔV (remember the entire network is subject to a constant voltage so any small change in resistance on one memristor will affect the voltage across the others): this situation is called the 'roving ΔV in [50]. This idea explains the loss of the voltage spikes seen in ideal networks and the sudden emergence of bursting spikes from an almost flat baseline that has been observed (see the control in [61]). Thus, the oscillations can then be explained as the 'ringing' of the network that results from constructive and destructive interactions of spikes as a ΔV is passed around and the bursting spikes would occur when the spikes constructively interact. If the bursting spikes were caused by this sort of system, it would explain the correlation between spikes that has been observed. Finally, the ΔV would be more likely to become unsynchronised if the network contained more sources of delay and difference, such as those which would be caused by introducing more memristors, more types of memristors, mismatched polarity, increased numbers of junctions, and so on. These sources of delay match the conditions under which spiking networks were observed as reported in this paper. This is an explanation, however it is not the only one we have entertained. Another explanation is to think of the boundary between on and off resistance material as continually oscillating slightly around equilibrium as ions diffuse around: it could be the interaction of these individual oscillators that causes the oscillations in the network (as is seen in other systems our group studies such as *Physarum* [2] and BZ reactions [3, 4]) and the interaction of oscillations could cause the bursting spikes.

The future directions for this work have been described above, but what are the future uses for this technology? If memristor spiking networks are operating under similar principles to those in the brain, it is possible that spiking memristor circuitry would be a good choice for brain-machine interfaces for use in neural prosthetics, disease treatment (e.g. a memristor-based circuit breaker for epileptic fits or a spike generator for dealing with the symptoms of Parkinson's disease) and functionalised prosthetics (i.e. artificial limbs that interface directly with the nervous system). We have started to investigate this by combining spiking memristor networks with spiking neural cell culture to see if the two spiking networks can influence each other electrically [61], see Sect. 19.6. Another area of interest is biomimetic robotics where spiking memristor networks could be used to process sensory inputs and control a robot. Finally, a spiking memristor computer may be capable of a more bio-inspired approach to computing and could prove better at task at which biology excels, such as pattern recognition, fuzzy processing, learning and so on.

19.6 Spikes and Cells

In this section we describe results from linking the networks in Table 19.2 to neural cell cultures which exhibit spikes as the cells grow and attempt to find each other. The set-up is shown in Fig. 19.12, cells are isolated in a multielectrode array (MEA), which is a dish that contains micro-electrodes. Figure 19.11 is a micrograph showing the cells at an early period of growth as they are attempting to grow towards each other. See Sect. 19.10 for full experimental details.

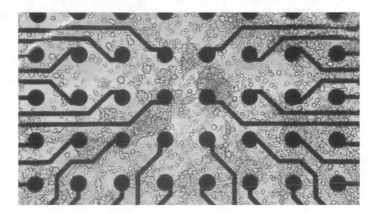

Fig. 19.11 IMR-32 cells in an MEA dish

Fig. 19.12 Scheme

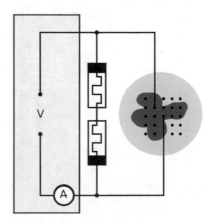

19.6.1 Memristor Network Connected to MEA Dish Without Cells

There were 3 control experiments, the network by itself, as shown in Sect. 19.5, and the memristor network connected to the MEA dish containing medium or water. Cell culture medium is an aqueous solution containing a mix of ions, sugars, colloids, fats and proteins, therefore any observed effect needs to be separated from that of connecting a memristor network to an aqueous sink of charged particles. Thus, we compared the results for an MEA dish filled with RPMI and distilled water. To completely remove the effects of ions on any conductivity, ultra-pure deionised water was also tested.

Results for the cell culture medium control are shown in Fig. 19.13 and this response is the expected response for a single memristor [46] or a memristor network with low compositional complexity and 'ideal' memristors. The other two controls (distilled water and deionised water) are identical and all start with the current of $60\,\mu A$ (a measured resistance of $8333\,\Omega$). The memristor spiking is suppressed and it appears that the current spikes are removed from the memristor network or otherwise damped by connection to a large sink of ions. The control experiments show that this effect is due to the loss of current from the networks to the aqueous medium, and not to cellular activity. Thus, it seems that connection to an external current sink removes the non-idealities of a memristor network (not permanently, because the network still exhibits standard memristor network dynamics after the experiments). We believe that the memristor network spiking is due to spike-response-caused voltage changes across individual memristors being out of sync (as postulated in [50]). If this is hypothesis is correct, then the connection to a sink of current would likely prevent this mechanism occurring by damping the system dynamics.

Fig. 19.13 Control experiment 2: A 2-memristor circuit connected to cell culture set-up containing cell medium but no cells. Voltage applied is shown in Fig. 19.6

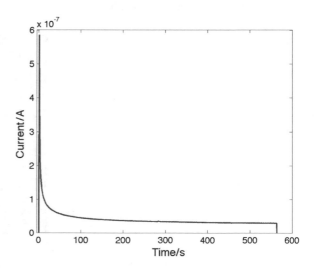

19.6.2   The Effect of Coupling to an External Pool of Spiking Neurons

Figure 19.14 shows the effect on the memristor I–t profile of coupling the 2-memristor circuit to an MEA dish full of living, spiking neuronal cells. The memristor spiking is again suppressed, as with the control experiment involving medium. However, by comparing Fig. 19.14 with the control in Fig. 19.13, we can see that there is a difference due to the action of the cells: there is a discontinuity at 30 s and a spike seen at 410 s, as marked by arrows in Fig. 19.14. The discontinuity is similar to that seen when a voltage changes across the memristor and could be the result of a voltage spike changing the 'state' of the memristor, the spike may be caused by the cells themselves or a memristor spike retained in the network.

Figure 19.15 shows the same experiment for a different dish that contained more cells, had been incubated for an extra day and where the electrodes were closer (neighbours along the edges of the rectilinear grid rather than on the diagonal). This data clearly shows an effect of coupling to the cells at 300 and 400 s. The spike at 400 s is consistent to a memristor being switched to a higher voltage, and is suggestive of a cellular change altering the resistance of the memristor. The drop and spike at around 300 s is not consistent with this, and could be the memristor network reacting to neuronal dynamics.

A full set of control experiments for electrode separation (not shown) have been run comprising of 1, 2 and 3 diagonals, equivalent to 283, 566 and 849 μm, which demonstrate that the amount of perturbation of the memristor's I–t curve is increased for shorter distances, i.e. the closer the electrodes, the shorter distance any signal has to travel to affect the exterior memristor circuit, thus the more likely it is that cellular processes will affect the memristors.

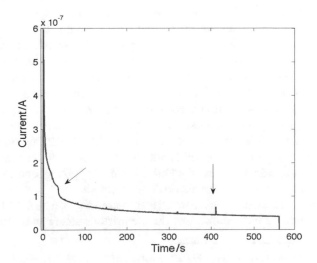

Fig. 19.14 The effect of cells. Measured I–t curve for the 2-memristor circuit. *Arrows* indicate the possible effect of cellular processes. Voltage applied is shown in Fig. 19.6

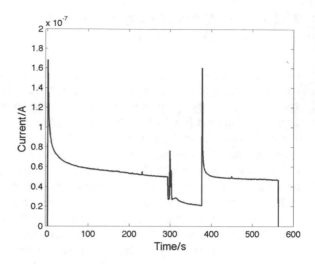

Fig. 19.15 Repeat of experiment, using a dish containing more cells and after an extra day's growth. Voltage applied is shown in Fig. 19.6

19.6.3   The Effect of Extra Spiking

Figure 19.16 show the effects of adding capsaicin to chemically stimulate cell spiking. The cells were subject to $10\,\mu L$ at approximately 10 min before starting the experiment and a second amount at 200 s into this run. The main result is that increased cell activity has greatly changed the shape to a sloping background[2] with many spikes superimposed on it. This shows that an external memristor circuit can 'record' spiking activity from an electrically connected living neuronal network. The second result is that the effect of the increased capsaicin dose can be seen 110 s after it was added in the form of an increased number and magnitude of the spikes.

There are 16 individual data sets compared, which is enough to give an indication of an effect, but too few for statistical analysis. Bearing this in mind, we can see that the step superimposed with a curve is likely in differentiated cells (with a capsaicin-induced increase in spiking rate); conversely, the addition of capsaicin to undifferentiated cells is more likely to lead to a line with spikes. A standard curve with spikes overlaid was only seen under capsaicin, implying that these spikes come from the cells and not the memristors themselves (a capsaicin control was run and gave the same results for water). It is possible for some experiments to exhibit no spikes in the memristor circuit at all, even though the memristors were only connected to areas that were seen to be actively spiking, so in those cases cell spiking activity either stopped or was not transmitted to the memristor circuit. Finally, the cells with no chemical additives to induce spiking seem to have a range of behaviours as measured by the memristor sub-circuit.

This work demonstrates that memristors can be switched by neuronal cell action and we have shown that the memristor's state can be changed by cellular activity.

---

[2] Similar to that seen in Fig. 19.15 between 300 and 400 s, and this is a shape not seen in the memristor networks without cells.

Fig. 19.16 Capsaicin added whilst memristors were connected. Voltage applied is shown in Fig. 19.6

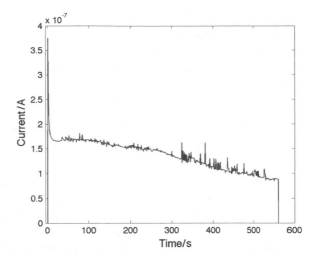

This is an exciting result as this cellular input could be put into an artificial neural net (either one built from memristors or other components) allowing the A.I. to react to cellular processes.

Furthermore, it shows that connecting spiking neuronal cells to a spiking memristor changes the memristor network dynamics in a qualifiable way, and related some of these changes to extra spiking activity. Thus memristors can be used as a biosensor for spiking activity and these sort of nanoscale networks placed in different parts of the brain could be inputs into wearable expert systems for disease monitoring or diagnosis.

19.7   Computing with Spikes

Having shown that memristor s can have complex, emergent dynamics like neural cells, and can interact with neural cells, we now demonstrate how a spiking memristor computer might be built. Memory allows the association of events which happen (and have happened) at different points in time. In this paper, I will present how this allows the device to store data temporarily and allow input bits to interact through time. This idea is essentially the opposite of parallelisation, when building parallel circuitry we are essentially replacing the time (as in time for a computation) with space (extra processors). Here we are replacing space (components on a circuit board) with time (increased number of steps) using the memory to store the computation through time. The gates made using this motivates the question of what is the circuit and logical complexity of such a mode of operation; essentially what is our conversion rate between space and time? We need to know the trade-offs to allow us to get the maximum efficiency for a circuit. For example, we are reaching the limits of how far we can continue shrinking computers in accordance with Moore's law [112]. To overcome this, the International Technology Roadmap for Semiconductors suggested the approach of 'More-than-Moore' where the components are capable of performing more functions [78]. To plan a circuit with these components we would need to precisely compare and contrast this extra functionality.

19.7.1 Experimental Schemes for Computing with Memristors

The result presented in Sect. 19.4.1 demonstrate that the memristor is capable of subtraction and we can use this effect to design logic gates.

We are free to choose arbitrary properties of the physical world to represent our logic, however, whatever we choose must make sense within the device and how it actually acts. From experiments and what we know of physical reality we can expect the forward flow of time (which causes directionality, see Sect. 19.7.1.2) and energy conservation to arise from the material properties of the memristor. Directionality suggests the use of a sequential logic approach. By choosing which material property is associated with our computation. Energy conservation in this section refers to the idea that energy is not created or destroyed, and thus higher energy inputs into an electronic device will cause a higher energy state. This is the route of most of the difficulty in designing novel logic gates in nanomaterials, an obvious approach is to take 1 and 0 as having different energies. If this is done, then to do Turing complete computation, at some point, one must 'create' energy (or rather take it from somewhere else) to do the computation $[1, 1] \longrightarrow 0$.

19.7.1.1 Sequential Logic

We shall make use of sequential logic (as first implemented in [57]), which works with the spike interactions seen in the memristor. Memristor sequential logic allows the computation through time by storing a state and allowing it to interact with the input; thus a one terminal device can do two-input (or higher) logical operations, if we are willing to wait for the output.

As shown in Fig. 19.17 the memristor's state is stored in its short-term memory and the current output to a voltage change is actually a function of its zeroed/null state and the input. Sequential logic makes use of the memristor's short-term memory to store the first bit, A, of an operation before the transmission of the second bit B. The output at time, t_A, is a function of A and the memristor's starting state (which is $\emptyset$ if the device has been properly zeroed), given by $f(A, \emptyset)$. At time t_B (where t_B is one measurement step after t_A) the output would be $f(B, A)$. The response step, t_1 is measured one measurement step after t_b. Thus, this voltage data is input at t_A and t_B and measured at t_1, t_2 and so on where $t_a < t_b < t_1 < t_2$.

Fig. 19.17 Sequential logic. The output of the memristor is a function of its state, as shown in the box, and the input. As the state is stored for the duration of the short-term memory logical values can be combined if they are input sequentially

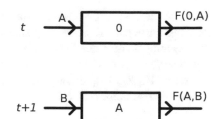

19.7.1.2 Elucidated Rules

Essentially, memristors operate a sequence-sensitive 'subtractive summation': where inputs add additional energy to the system, but where that additional energy is subject to subtraction due to energy loss and the whole process operates under nonlinear dynamics. The rules required to design a memristor logic gate follow.

19.7.1.3 Directionality

The memristor naturally implements IMP. The memristor is directional: e.g. The response at t_1, for $A \rightarrow B$ does not equal the output (t_1) for $B \rightarrow A$. The cause of this is that the memristor responds to the difference in voltage. This naturally allows memristor-based sequential logic to compute implication logic as Implication, IMP or $\rightarrow$, requires that $0 \rightarrow 1 \neq 1 \rightarrow 0$ and thus the order in which the two values are input has a meaning. Naturally, sequential logic, as it separates the values in time, implements this ordering.

19.7.1.4 'Summation' via Energy Conservation

If the logical 1 is taken as being a high voltage then more energy is imparted to the system from the logical combination [1, 1] compared to [0, 0]. This approach can allow the creation of memristor based time-limited summators of use in leaky integrate and fire neurons. Note that by the phrase 'energy conservation' we are talking about the fact if we put more energy into the memristor system, we will have a higher energy state. As the system is non-conservative, this higher energy state is not just the sum of the two input energies.

19.7.1.5 'Bounceback'

The application of a voltage spike produces a resultant current spike in the direction of the difference between the starting voltage and the ending voltage, e.g. the first voltage change $V_0 \rightarrow V_A$ causes a positive current response, $+i_A$, if V_A is positive, and negative, $-i_A$, if V_A is negative. If the system is then returned to zero there is a smaller current spike of the opposite polarity, i.e. $-i_0$ and $+i_0$ respectively for the two examples mentioned above. If several spikes are input before returning to zero, i.e. a sequence of $[V_0, V_A, V_A, V_0]$ the current spike, i_0 is larger than would be the case for $[V_0, V_A, V_0]$, although not twice as large due to losses in the system (as the system is non-conservative).

19.7.1.6 'Diminishing Returns'

As discussed in the example of $[V_0, V_A, V_A, V_A]$ above each addition of each consecutive spike has a reduced effect compared with the first. This property is seen with spikes of the same polarity and with changing polarity i.e. the response spike to $[V_0, +V_A, -V_A, +V_A, -V_A]$ is smaller than $[V_0, +V_A, -V_A]$. This only happens to spikes input into the memristor's short-term memory as waiting for the device to return to a blank state refreshes the property that reacts to the voltage step. We expect that the physical property in question is related to the ions in the device, as it they that 'hold' the memory, but this has not been experimentally verified. The diminishing returns effect is behind the habituation shown in Sect. 19.4.1.

19.7.2 Examples of Logical Systems

Knowledge of these rules and effects allows us to design logical computation systems which perform a surprising amount of computation with only a single memristor. We have found that, in these schemes that the summation effect is important in magnitude logic, the 'bounceback' effect is more relevant in polarity logic (although both affect the outcome). As these can be balanced and set in opposition to each other, the richest effects came from using the mixed logics (as presented in Table 19.3: we will now present a few examples.

There are two variables we can utilise when assigning logical values: the magnitude, as represented by M for a high magnitude and m for a low magnitude; and the sign, as represented by a $+$ for positive and $-$ for negative. The 4 different logical assignations that can be applied using these values is shown in Table 19.3. To implement logical operations, voltage spikes are applied for one time-step and the response recorded at the same frequency. In between logical operations, the devices were left for longer than the equilibration time (τ_∞ in [46] which is around 3.5 s) to zero the memristor by removing its short term memory.

Changing the values of M and m can allow the results to be tuned or balanced against the effect of polarity, but in this paper we shall just deal with qualitative examples.

Table 19.3 Four different methods of implementing logical 1 and 0 with memristor spikes: M refers to a high magnitude voltage, m to a low magnitude voltage and '+' and '−' refer to its polarity

Logical value	Magnitude logic	Polarity logic	Mixed logic 1	Mixed logic 2
One	M	+	+M	−M
Zero	m	−	−m	+m

19.7.3　Logic Gates

We can do Boolean logic with the spike interactions by sending the second bit of information one time-step (0.02 s) after the first. We take the input as the current spikes caused by the voltage spike inputs. The output is the response current as measured after the 2nd bit of information. at that timestep, i.e. V_B. After a logic operation the device is zeroed by being taken to 0 V for approximately 4 s, and this removes the memristor's memory.

We have some freedom in how we assign the 1 and 0 states to device properties and these give different logic. The following examples will demonstrate some approaches and build an OR gate or an XOR gate: i.e. 1 is any positive current output, 0 is any negative current output, inputs are positive (+1) and negative (0) voltages.

The XOR truth table is shown in column 9 in Table 19.4. If we take logical 1 to be the current resulting from a positive voltage and a logical 0 to be the current resulting from a negative voltage, then, the response is the current when the 2nd bit is input (not after, although it could be designed that way but it is slower). We get a high absolute value of current if and only if the two inputs are of different signs, i.e. we have {1 0} or {0 1} which gives us an exclusive OR operation. For this logical system, we used the same voltage level and allowed a change in sign to indicate logical zero or logical one:

- $0, 0 = -0.1\,\text{V}, -0.1\,\text{V}$
- $0, 1 = -0.1\,\text{V}, +0.1\,\text{V}$
- $1, 0 = +0.1\,\text{V}, -0.1\,\text{V}$
- $1, 1 = +0.1\,\text{V}, +0.1\,\text{V}.$

Table 19.4 XOR truth table (exclusive OR)

Input 1	Input 2	Output
0	0	0
0	1	1
1	0	1
1	1	0

Fig. 19.18 The input
voltage for the XOR gate

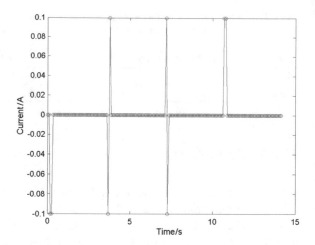

With a pause between operations to allow the memristor to lose its memory, the XOR operation is reproducible. Note, as our devices are slightly asymmetric, $F[0, 1]$ does not exactly equal $F[1, 0]$ and $F[0, 0]$ does not exactly equal $F[1, 1]$. This arises from the material and the method of synthesis (see [56]) and has nothing to do with the voltage being positive per se, the devices are always wired up the same way round before measurement. Larger voltage ranges highlight the difference between positive and negative, demonstrating how different logical combinations can be split out or alternatively balanced by sensitive voltage choice (Fig. 19.18).

As XOR A = NOT A, if we always take the 2nd point after the first (and only bit in this case) as being the response bit (as we did above for the XOR gate), we have a NOT gate.

19.7.4 Full-Adder

It is possible to compute an unconventional instantiation of a full-adder, as shown in Fig. 19.19 (admitting that we require a voltage spike to current spike conversion). The two input and carry bits are input as a series of spikes using mixed logic 2 with input 1 represented by -0.5 V and input 0 represented by $+0.001$ V. The input sequence is $[A, B, C, 1, 2, 3, 4]$, with the logic input at $t_A - t_C$, the response spike recorded at t_1, an extra read voltage of -0.15 V input at t_2. This gate requires a clock to operate. Figure 19.19 shows the response of the memristor to this scheme, for the three inputs of a full adder, the read spike at t_2 is marked with an * to make it easier to understand, and the data of the memristor losing its short-term memory is not shown.

From this set-up the following things can be deduced from knowing the maximum positive and negative current spikes within 4 time-steps of an input (although this requirement need not be too stringent if we have a way of recording the maximum

Fig. 19.19 A full adder using mixed logic 2. The first three input bits are the logical inputs, the system has one timestep to respond ($\tilde{1}$s) before a read spike is sent in, as marked by an *. The numbers of the ranges correspond to the list

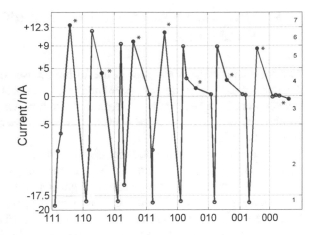

current within the ranges in between zeroing the system, which we can do with knowledge of the read pulse clock).

The resulting information from the current is thus:

1. if a negative current is recorded in the range -17.5 to -20 nA: we have had a 1 input into the system
2. if a negative current is recorded in the range -5 to -17.5 nA: we have a carry bit from the operation
3. if a negative current is recorded in the range 0 to -5 nA: we have had a zero in the system (this is redundant information)
4. if the maximum positive current is recorded in the range 0 to $+5$ nA: the result of the calculation is 0
5. if the maximum positive current is recorded in the range $+5$ to $+9$ nA: the result of the calculation is 1
6. if the maximum positive current is recorded in the range $+9$ to $+12.3$: the result is '2' (or 1 for the carry bit, 0 for the summation bit)
7. if the maximum positive current is recorded over 12.5 nA: the result is '3' (or 1 for both the carry and summation bit in binary logic).

The output in the negative is purely a result of the input voltages to the system. The positive system includes the 'bounceback', and the summation effect as probed by the read voltage which gives thresholded values of the memristor's state.

With switches, it would be possible to send on the logical result as binary. Region two of the plot encodes the carry bit for the operation, because only if there are two $-M$ spikes (which encode 1) within 3 time-steps of each other we will see a current response in that range. The summation bit is not encoded in as direct a manner, the maximum of the positive currents encodes the numerical sum, and so the summation bit for the value 3 is in a different place to that for the value 1. If we only require knowledge of the carry and summation bit, we can do without the read voltage and corresponding spikes. Changing the values of M and m can tune the effect and might allow us to change the relative values of the output spikes.

19.8 Memristor Networks as Used in Artificial Intelligence Simulations

The memristor concept has proved useful in neuroscience. It has been shown that synapse action [72, 83] is memristive and the ion channels involved in neuronal signal transport are biological memristors, in fact, the time-varying resistors in Hodgkin–Huxley's electrical model of the neuron have been more properly identified as memristors [28, 29].

In a computer neural net, the memristors can act as synapses where the conductivity of a memristor connection is taken as the synaptic weight of that connection. During the training period, the synaptic weight increases as the resistance of the memristor decreases. Some examples of memristive synapses in neural nets are [72, 74, 83]. As memristors are non-linear devices and the resistance-spike number curve is non-linear, the synapses are inherently non-linear, which introduces richer behaviour to the system. In [74] it was demonstrated that depending on the initial orientation of the curve, memristors can be included that are trained easily and quickly after few spikes, but slow their rate of change after many spikes (reducing the chance of network saturation) or memristors can be included to be slow to respond to low numbers of spikes and to change more drastically after larger numbers.

Memristor networks have only recently started to be investigated for A.I. Other than neural nets, memristor networks have been shown (through simulation) to be useful to adding in memory to network dynamics [45, 50, 62].

An example [50] is a trained network that composed music where each possible pair of notes was connected by a pair of memristors (one each for each direction) which gave a fully connected directed graph. The network was trained by 'playing' music of a specific genre and altering connection weights between pairs of notes based on the amount that note combination was used. Tempo was included from a fully connected directed graph of note lengths. Once trained the network could compose music deterministically using Markov chain processes (Markov chains give the next note with only the knowledge of the previous note) where the memristors allowed network to maintain a memory beyond that allowed by standard Markov chain processes which is necessary for recognisable music complete with recurring themes and genre-specific style. This design could also be used with the spiking networks shown in Sect. 19.5.

In another simple application [62], a matrix of light sensors attached to memristors acting under cellular automata rules was used to find the edges in an image, see Fig. 19.20. The memristor's short-term memory was used as a real-world cellular automata (where the automata's states were associated with the memristor losing it's short-term memory). This was then simulated for low quality, wide-angle distorted real world images of horizons to demonstrate that the system was capable of finding the horizon. This system is cheap, light-weight, low-power and composed of all-analogue electronics in hardware, which is suitable for robust tilt correction in Unmanned Aerial Vehicles, UAVs.

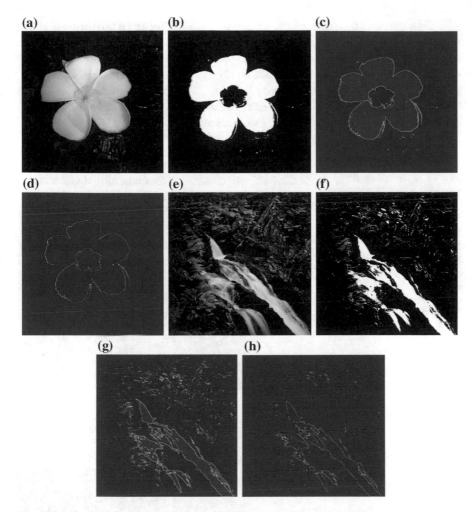

Fig. 19.20 Images of a flower and a waterfall with an edge detection algorithm run on memristor modelling cellular automata. Taken from [62]. **a** Flower—Original, **b** Flower—B & W, **c** Flower—Gener. 2 CA, **d** Flower—Gener. 3 CA, **e** Waterfall—Original, **f** Waterfall—B & W, **g** Waterfall—Gener. 2 CA, **h** Waterfall—Gener. 3 CA

Models of memristor short-term memory have been used in simulation designs of scheduling algorithms inspired by ant behaviour [45]. In these algorithms, the non-linear response of the memristor is used to model the decreasing freshness of data over time or the importance of sorted data over transmission. This was modelled for the collection of data from a large area network of sensors, but is applicable to sorted big data applications—important mission critical data is transferred first, allowing for quick decision making in an emergency, and nice-to-know useful data is transferred after allowing for slower, more considered data analysis when time allows.

The algorithm was then applied to compressing images for quick transfer, where the most important part of the image (or data) could be sent first. Robot control systems that allowed the robot processor to switch on a continuum between fast, emergency decisions with only very relevant data (such as, is the robot's heat sensor too hot, indicating it is on fire) and considered learning and decisions (such as, how should the new information gained that day change the robots operation in the future) [48]. And example of the algorithms, using singular vectors of a picture to stand in for the information and importance of the information, is shown in Fig. 19.21.

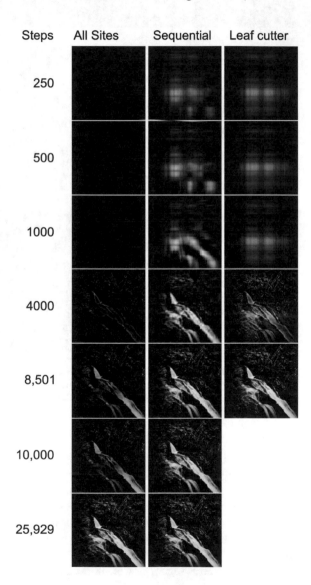

Fig. 19.21 Information transfer example using the singular value decomposition vectors to represent the information chunks taken from the environment via a robot's sensors. As time goes on the robot gets a more complete picture of the environment. Taken from [45]

19.9 Conclusions

I have shown that memristors possess spiking behaviour, similar in form to neural spikes, with current and voltage switched. This means that we can copy neural architectures with memristors. Because of the 'subtractive summation' interaction of memristor current responses, we can do habituation with memristors, directly simulating a neuron process in hardware. These interactions can be used to perform standard logic operations, including the creation of a full adder in a single memristor.

The short-term memory allows interactions of inputs through time (as in events that happen at different time effect the output). This is the source of the full adder's high degree of processing power for a single circuit element. It also leads to emergent spiking behaviour in networks (containing only memristors) similar to brain dynamics and these networks can interact with neural cell cultures.

Memristors do seem to be an excellent analogue for neurons and this area has the exciting possibilities of creating novel, neuromorphic, in fact, neuromimetic computers, and even interlinking with living neurons.

19.10 Methodology and Experimental Details

19.10.1 Making the Memristors

The standard memristors were prepared by first sputtering aluminium onto a plastic substrate via a mask. The sol-gel preparation is described in detail elsewhere [55]. The final preparation was diluted 1:10 in dried methanol to make the spin-coat solution. 3 ml of spin-coat solution was spun at 33 r/s for 60 s and left for at least an hour (the time taken for the sol to convert to the gel [90]) before the second set of electrodes were sputtered on top. The electrodes were 4 mm wide and crossed at 90° giving a 16 mm^2 active area. Profilometer tests revealed that the TiO$_2$ gel layer was approximately 44 nm ($\pm$2.5 nm) thick.

6 batches of devices were fabricated and in total 130 devices were tested, and example device behavior is given in Fig. 19.22. To test the fabrication methods devices were left in clean room air to dehydrate (R series) or put into 5–10 mTorr vacuum (V series) or the top electrode size was changed to 1, 2, 3, 4 or 5 mm, giving active areas of 4, 8, 12, 16 or 20 mm^2 (D series). The D series memristors were dehydrated under vacuum. Comparison devices were made with: A. gold electrodes top and bottom (Au-TiO$_2$-Au memristors); B. aluminium bottom electrodes and gold top electrode (Au-TiO$_2$-Al memristors); and C. the standard aluminium top and bottom electrode (Al-TiO$_2$-Al), all these devices were made with 4 mm wide electrodes. From the memristor literature it is expected that the Au-TiO$_2$-Au devices should switch similarly to the Al-TiO$_2$-Al memristors, from the ReRAM literature it is expected that the aluminium oxide has a role to play in the switching.

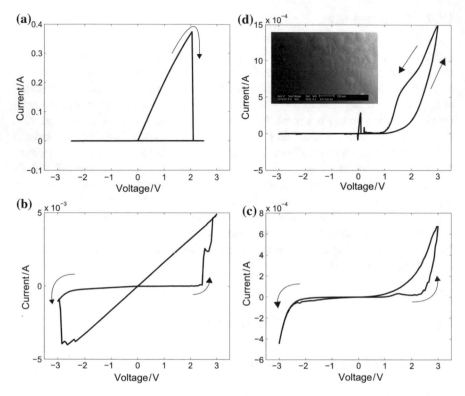

Fig. 19.22 Examples of the types of behaviour found in TiO_2 sol-gel memristors. **a** Au-TiO_2-Au devices show fuse-breaking behaviour, with an LRS of 10^{-1} A and no repeatability over subsequent runs. Al-TiO_2-Al memristors show two types of behaviour, triangular switching with an ohmic LRS on the order of 10^{-3}, (**b**), and curved switching with a LRS most often on the order of 10^{-4} A, see **c**. Mixed devices, Al-TiO_2-Au show memristor like behaviour only when aluminium is negative with respect to the gold electrode, an example of curved half-memristance is shown in **d** (half triangular memristance was also observed). When the gold electrode is negative with respect to the aluminium, the memristance is suppressed and oxygen evolution from the TiO_2 distends the gold electrode as shown in the insert. The *sharp dividing line* shows the location of the *bottom* electrode and allows the comparison between un-distended and distended electrode surface

Virgin memristor devices were used for all comparisons and were measured with linear $I-V$ sweeps on a Keithley 2400 sourcemeter between a range of ± 3.5 V which is the same range as [64] or ± 2 V, with a voltage step size of 0.05 V and a dwell time of 2 s: this is the D.C. equivalent of an A.C. voltage of 1 mHz, a frequency that it too slow to source any other way from the sourcemeter. Twelve memristors were tested in each case, the source was always connected to the bottom electrode to ensure comparability. For the Al-TiO_2-Au memristor virgin runs, six devices were connected one way (Earth to Au, source to Al) and six the other way. For these devices a second run was done connecting the wires the opposite way round (to reverse the asymmetry). The compliance current was 2 mA unless stated.

As in [64] our electrodes were sputter coated, the $TiO_4(sol)$ is then spun on-top and dehydrates to the gel [34]:

$$Ti(OH)_4(sol) \rightarrow H_2O + TiO_2(gel)$$

before the next set of electrodes are sputtered on top. As aluminium oxide forms very quickly and the TiO_2 may oxidise the bottom electrode, and we expected vacuum prepared memristors to have an Al_2O_3 layer, however it may be a different thickness or the TiO_2 layer may have a different form under the two different preparation methods.

Photographs were taken with a Canon Powershot G6 using frontal and side-light from a tungsten desk lamp using a sheet of white paper as a soft-box. Micrographs were taken using on a 3D Hirox digital microscope using both frontal and side-lighting.

19.10.2 Emergent Network Dynamics Experiment

The circuits were wired up according to the Table 19.2 and connected to a Keithley 2400 Sourcemeter in current-sensing, voltage sourcing mode. To get the I–t curves, the memristor circuits were taken to $+0.4$ V for 1000 timesteps (1 timestep $= 1.6$ s) the voltage source was then switched to 0 V and data was gathered for a further 100 timesteps. 22 different experiments were analysed. To investigate whether a slow changing voltage had an effect, a sinusoidal voltage of 1600 timesteps of 2 s was used. In all experiments voltages were kept very low to avoid the creation of filaments via Joule heating which would lead to filamentary memristors switching into lower resistance states.

The I–t time-scries plots were analysed using MatLab. The periodograms were calculated using discrete Fourier transforms using a fast-Fourier transform algorithm [42]: all periodograms were calculated with the same sample frequency of 0.942 Hz. Time-series auto-correlation function (ACF) plots were plotted to test for significant persistence as measured by the number of timesteps (lag) required to predict the next step. Partial auto-correlation function (PACF) gives a measure of the signals 'complexity' by measuring the minimum order of auto-regression (AR) function required to fit it, where an AR(0) is a memoryless system, AR(1) requires 1 preceding step and so on. Auto-correlation was performed using MatLab's 'autocorr' function [19, 104] and the partial auto-correlation using MatLab's 'parcorr' function [19]. Output I–t plots were tested for a departure from randomness based on the ACF data using the Ljung-Box-Pierce Q-test [19], implemented in MatLab as 'lbqtest', using a threshold p-value of 0.05 (below which we reject the null hypothesis of the data being random). The test was performed lags from 1 to 50 based on the observation that bursting spikes did not tend to persist for more than 100 s.

19.10.3 Cells

The memristors used in this study have two different behaviours [56]. Previously we reported that using a mixture of both types connected either in series with opposite polarity or in an anti-parallel configuration was most likely to lead to spiking dynamics [57]. For this study, a circuit of two memristors of different types in anti-polarity series was used, see Fig. 19.12. The current was measured at 0.56 s intervals for 560 s using a Keithley sourcemeter (controlled via MatLab) that provides a voltage of +0.5 V.

The cells used in this study are IMR-32 neuroblastoma cells. The cells were cultured in RPMI medium (RPMI-1640, Sigma-R0883) with 10 % heat-inactivated fetal bovine serum (Sigma-F0804), 2 mM L-glutamine (Sigma-G7513), essential amino acids (RPMI 1640 amino acids solution, Sigma-R7131) and 0.01 % penicillin-streptomycin (Sigma-P1333). The cells (25,000 cells/dish) were transferred to a multi-electrode array (MEA) dish (Avenda Biosystems) coated with PEI [20] and incubated at physiological temperatures for between 24 and 48 h. The MEA comprises 64 microelectrodes 30 μM diameter and 200 μM apart which allows the measurement of physiological voltages (typically ≤200 mV). The two electrodes in close proximity of each other with the largest amount of spiking activity were selected to be connected to the external memristor circuit.

Two different ways of promoting cell spiking were used: 1. A subset of the IMR-32 cells were placed in a medium with added [dibutyryl cAMP, 0.5 mg/ml, Sigma-D0627], which causes differentiation and promotes spiking [23]; 2. 10 μL of 130 mM capsaicin (Sigma-M2028) in phosphate buffered saline was added 200 s into experiments with either undifferentiated, or differentiated cells—capsaicin is known to promote TRPV1-dependent calcium signalling, and differentiated IMR-32 cells have been shown to express TRPV1 [1]. The first dose of capsaicin was administered when the cells were not connected to the memristors to check that capsaicin caused increased spiking, then the memristors were connected for the second and third doses, further doses caused a negligible effect due to cell desensitisation.

Sixteen experiments will be discussed: 7 experiments with undifferentiated, unstimulated cells (with 3 control experiments comparing electrode separation) and 9 with stimulation of spiking by adding capsaicin, 3 of which were differentiated cells. For the memristor network control, the cables to the MEA dish were disconnected and left as open-circuit. For the MEA dish controls, a clean MEA dish was filled with (in turn) distilled water, deionised water or RPMI medium as made up for the cells (and warmed to body temperature).

References

1. Abdullah, H., Cosby, S., Heaney, L., McGarvey, L.: S123 the effect of rhinovirus infection on cough receptors on human sensory nerve and human primary bronchial epithelial cells. Thorax **66**(Suppl 4), A57–A57 (2011)

2. Adamatzky, A.: Physarum Machines: Computers from Slime Mould. World Scientific Series on Nonlinear Science Series A, vol. 74. Prentice-Hall, Upper Saddle River (1994)
3. Adamatzky, A., Costello, B.L.: Reaction-diffusion computing. In: Rozenberg, G., Bäck, T., Kok, J. (eds.) Handbook of Natural Computing, pp. 1897–1920. Springer, Berlin (2012)
4. Adamatzky, A., Holley, J., Bull, L., de Lacy Costello, B.: On computing in fine-grained compartmentalised Belousov-Zhabotinsky medium. Chaos, Solitons Fractals **44**, 779–790 (2011)
5. Adhikari, S.P., Kim, H.: Why are memristor and memistor different devices?. Memristor Networks. Springer International Publishing, Switzerland (2014)
6. Adhikari, S.P., Sah, M.P., Kim, H., Chua, L.O.: Three fingerprints of memristor. IEEE Trans. Circuits Syst. **60**, 3008–3021 (2013)
7. Alan Doolittle, W., Calley, W., Henderson, W.: Complementary oxide memristor technology facilitating both inhibitory and excitatory synapses for potential neuromorphic computing applications. In: International Semiconductor Device Research Symposium, ISDRS'09, pp. 1–2 (2009). doi:10.1109/ISDRS.2009.5378162
8. Antonova, E., Chadwick, P., Kumari, V.: More meditation, less habituation? the effect of mindfulness practice on the acoustic startle reflex. PLoS One **10**, e0123512 (2015)
9. Argall, F.: Switching phenomena in titanium oxide thin films. Solid State Electron. **11**, 535–541 (1968)
10. Bailey, C., Chen, M.: Morphological basis of long-term habituation and sensitisation in Aplysia. Science **220**, 91–93 (1983)
11. Bao, B.C., Liu, Z., Xu, J.P.: Steady periodic memristor oscillator with transient chaotic behaviours. Electron. Lett. **46**, 237–238 (2010)
12. Bao, B.C., Xu, J.P., Liu, Z.: Initial state dependent dynamical behaviors in a memristor based chaotic circuit. Chin. Phys. Lett. **27**, 070504 (2010)
13. Beadle, W.: Switching properties of thin NiO films. Solid State Electron. **7**, 785–797 (1964)
14. Bernede, J.C.: Polarized memory switching in MIS thin films. Thin Solid Films **81**, 155–160 (1981)
15. Biolek, D., Biolek, Z., Biolkova, V.: Interpreting area of pinched hysteresis loop. Electron. Lett. **50**, 74–75 (2014)
16. Bo-Cheng, B., Ping, X.J., Guo-Hua, Z., Zheng-Hua, M., Ling, Z.: Chaotic memristive circuit: equivalent circuit realization and dynamical analysis. Chin. Phys. B **20**, 120502, 7pp (2011)
17. Borghetti, J., Strukov, D.B., Pickett, M.D., Yang, J.J., Stewart, D.R.: Electrical transport and thermometry of electroformed titanium dioxide memristive switches. J. Appl. Phys. **106**, 125504, 5pp (2009)
18. Borghetti, J., Snider, G.D., Kuekes, P.J., Yang, J.J., Stewart, D.R., Williams, R.S.: Memristive switches enable 'stateful' logic operations via material implication. Nature **464**, 873–876 (2010)
19. Box, G., Jenkins, G., Reinsel, G.: Time Series Analysis: Forecasting and Control, 3rd edn. Prentice-Hall, Upper Saddle River (1994)
20. Bull, L., Uroukov, I.S.: Initial results from the use of learning classifier systems to control in vitro neuronal networks. In: Proceedings of the 9th annual conference on genetic and evolutionary computation, pp. 369–376. ACM (2007)
21. Bursill, L., Hyde, B.: Crystallographic shear in the higher titanium oxides: structure, texture, mechanisms and thermodynamics. Prog. Solid State Chem. **7**, 177–253 (1972). doi:10.1016/0079-6786(72)90008-8. http://www.sciencedirect.com/science/article/pii/0079678672900088
22. Buscarino, A., Fortuna, L., Frasca, M., Gambuzza, L.V.: A chaotic circuit based on Hewlett-Packard memristor. Chaos **22**, 023136 (2012)
23. Carbone, E., Sher, E., Clementi, F.: Ca currents in human neuroblastoma IMR32 cells: kinetics, permeability and pharmacology. Pflügers Arch. **416**(1–2), 170–179 (1990)
24. Chang, W.Y., Lai, Y.C., Wu, T.B., Wang, S.F., Chen, F., Tsai, M.J.: Unipolar resistive switching characteristics of ZnO thin films for nonvolatile memory applications. Appl. Phys. Lett. **92**, 022110 (2008)

25. Choi, B., Jeong, D., Kim, S., Rohde, C., Choi, S., Oh, J., Kim, H., Hwang, C., Szot, K., Waser, R., Reichenberg, B., Tiedke, S.: Resistive switching mechanism of TiO_2 thin films grown by atomic-layer deposition. J. Appl. Phys. **98**, 033715 (2005)
26. Chua, L.: Resistance switching memories are memristors. Appl. Phys. A: Mater. Sci. Process. **102**, 765–782 (2011)
27. Chua, L.: Memristor, Hodgkin-Huxley, and edge of chaos. Nanotechnology **24**(38), 383001 (2013). http://stacks.iop.org/0957-4484/24/i=38/a=383001
28. Chua, L., Sbitnev, V., Kim, H.: Hodgkin-Huxley axon is made of memristors. Int. J. Bifurc. Chaos **22**, 1230011, 48pp (2012)
29. Chua, L., Sbitnev, V., Kim, H.: Neurons are poised near the edge of chaos. Int. J. Bifurc. Chaos **11**, 1250098, 49pp (2012)
30. Chua, L.O.: Introduction to Nonlinear Network Theory, 1st edn. McGraw-Hill, New York (1969)
31. Chua, L.O.: Memristor - the missing circuit element. IEEE Trans. Circuit Theory **18**, 507–519 (1971)
32. Chua, L.O., Kang, S.M.: Memristive devices and systems. Proc. IEEE **64**, 209–223 (1976)
33. Colle, M., Buchel, M., de Leeuw, D.M.: Switching and filamentary conduction in non-volatile organic memories. Org. Electron. **7**, 305–312 (2006)
34. Drakakis, E., Yaliraki, S., Barahona, M.: Memristors and bernoulli dynamics. In: 12th International Workshop on Cellular Nanoscale Networks and their Applications (CNNA) (2010)
35. Duarte, J.C., Martins, E.V., Alves, L.N.: Amplitude characterization of memristive devices. In: International Conference on Electronics, Circuits and Systems (ICECS), pp. 45–49. IEEE, Abu Dhabi, UAE (2013)
36. Duarte, J.C., Martins, E.V., Alves, L.N.: Frequency characterisation of memristive devices. In: European Conference on Circuit Theory and Design (ECCTD), pp. 1–4. IEEE, Dresden, Germany (2013)
37. Erokhin, V.: Organic memristive devices: architecture, properties and applications in neuromorphic networks. In: IEEE 20th International Conference on Electronics, Circuits, and Systems (ICECS), pp. 305–308 (2013). doi:10.1109/ICECS.2013.6815415
38. Erokhin, V., Berzina, T., Fontana, M.P.: Hybrid electronic device based on polyaniline-polyethylenoxide junction. J. Appl. Phys. **97**, 064501 (2005)
39. Erokhin, V., Schuz, A., Fontana, M.: Organic memristor and bio-inspired information processing. Int. J. Unconv. Comput. **6**, 15–32 (2009)
40. Erokhin, V., Berzina, T., Camorani, P., Smerieri, A., Vavoulis, D., Feng, J., Fontana, M.P.: Material memristive device circuits with synaptic plasticity: learning and memory. BioNanoSci (2011). doi:10.1007/s12668-011-0004-7
41. Erokhin, V., Berzina, T., Gorshkov, K., Camorani, P., Pucci, A., Ricci, L., Ruggeri, G., Sigala, R., Schuz, A.: Stochastic hybrid 3d matrix: learning and adaptation of electrical properties. J. Mater. Chem. **22**, 22881–22887 (2012). doi:10.1039/C2JM35064E
42. Frigo, M., Johnson, S.: FFTW: an adaptive software architecture for the FFT. In: Proceedings of the International Conference on Acoustics, Speech, and Signal Processing
43. Gabuzza, L.V., Fortuna, L., Frasca, M., Gale, E.: Experimental evidence of chaos from memristors. Int. J. Bifurc. Chaos (forthcoming) (2015)
44. Gale, E.: The memory-conservation theory of memristance. In: Proceedings of the 16th UKSim-AMSS International Conference of Modelling and Simulation (UKSIM2014), pp. 598–603 (2014)
45. Gale, E., de Lacy Costello, B., Adamatzky, A.: Memristor-based information gathering approaches, both ant-inspired and hypothetical. Nano. Commun. Netw. **3**, 203–216 (2012)
46. Gale, E., de Lacy Costello, B., Adamatzky, A.: Observation, characterization and modeling of memristor current spikes. Appl. Math. Inf. Sci. **7**, 1395–1403 (2013)
47. Gale, E., de Lacy Costello, B., Adamatzky, A.: Is spiking logic the route to memristor-based computers? In: International Conference on Electronics. Circuits and Systems (ICECS), pp. 297–300. IEEE, Abu Dhabi, UAE (2013)

48. Gale, E., de Lacy Costello, B., Adamatzky, A.: Design of a hybrid robot control system using memristor-model and ant-inspired based information transfer protocols. In: Proceedings of Workshop Fr-Ws-09 on 'Unconventional Approaches to Robotics and Automation Inspired by Nature' (UARACIN) at International Conference on Robotics and Automation (ICRA), pp. 34–36 (2013)
49. Gale, E., Adamatzky, A., de Lacy Costello, B.: Slime mould memristors (2013). arXiv:1306.3414v1
50. Gale, E., Matthews, O., de Lacy Costello, B., Adamatzky, A.: Beyond Markov chains, towards adaptive memristors network-based music generation. In: 1st AISB Symposium on Music and Unconventional Computing, Annual Convention of the Society for the Study of Artifical Intelligence and the Simulation of Behaviour (2013)
51. Gale, E., de Lacy Costello, B., Adamatzky, A.: Boolean logic gates from a single memristor via low-level sequential logic. In: Mauri, G., Dennunzio, A., Manzoni, L., Porreca, A.E. (eds.) Unconventional Computation and Natural Computation. Lecture Notes in Computer Science, vol. 7956, pp. 79–89. Springer, Berlin (2013)
52. Gale, E., de Lacy Costello, B., Adamatzky, A.: Emergent spiking in non-ideal memristor networks. Microelectron. J. **45**, 1401–1415 (2014)
53. Gale, E., de Lacy Costello, B., Adamatzky, A.: Spiking in memristor networks. Memristor Networks. Springer International Publishing, Switzerland (2014)
54. Gale, E., de Lacy Costello, B., Erokhin, V., Adamatzky, A.: The short-term memory (D.C. response) of the memristor demonstrates the causes of the memristor frequency effect. In: Proceedings of CASFEST (2014)
55. Gale, E., Mayne, R., Adamatzky, A., de Lacy Costello, B.: Drop-coated titanium dioxide memristors. Mater. Chem. Phys. **143**, 524–529 (2014)
56. Gale, E., Pearson, D., Kitson, S., Adamatzky, A., de Lacy Costello, B.: The effect of changing electrode metal on solution-processed flexible titanium dioxide memristors. Mater. Chem. Phys. **162**, 20–30 (2015). doi:10.1016/j.matchemphys.2015.03.037. http://www.sciencedirect.com/science/article/pii/S0254058415002072
57. Gale, E., de Lacy Costello, B., Adamatzky, A.: Boolean logic gates from a single memristor via low-level sequential logic. In: Submitted
58. Gale, E.M., de Lacy Costello, B., Adamatzky, A.: Filamentary extension of the Mem-Con theory of memristance and its application to titanium dioxide Sol-Gel memristors. In: IEEE International Conference on Electronics Design, Systems and Applications (ICEDSA 2012). Kuala Lumpur, Malaysia (2012)
59. Gale, E.M., de Lacy Costello, B., Adamatzky, A.: Observation and characterization of memristor current spikes and their application to neuromorphic computation. In: International Conference on Numerical Analysis and Applied Mathematics (ICNAAM 2012). Kos, Greece (2012)
60. Gandhi, G., Aggarwal, V., Chua, L.O.: The detectors used in the first radios were memristors. Memristor Networks. Springer International Publishing, Switzerland (2014)
61. Gater, D., Iqbal, A., Davey, J., Gale, E.: Connecting spiking neurons to a spiking memristor network changes the memristor dynamics. In: International Conference on Electronics. Circuits and Systems (ICECS), pp. 534–537. IEEE, Abu Dhabi, UAE (2013)
62. Georgilas, I., Gale, E., Adamatzky, A., Melhuish, C.: UAV horizon tracking using memristors and cellular automata visual processing. Advances in Autonomous Robotics. Lecture Notes in Computer Science, vol. 8069, pp. 64–75. Springer, Berlin (2013)
63. Georgiou, P., Yaliraki, S., Drakakis, E., Barahona, M.: Quantitative measure of hysteresis for memristors through explicit dynamics. Proc. R. Soc. A **468**, 2210–2229 (2012)
64. Gergel-Hackett, N., Hamadani, B., Dunlap, B., Suehle, J., Richer, C., Hacker, C., Gundlach, D.: A flexible solution-processed memrister. IEEE Electron Device Lett. **30**, 706–708 (2009)
65. Ginoux, J.M., Rossetto, B.: The singing arc: the oldest memristor? Chaos, CNN, Memristors and Beyond. World Scientific, Singapore (2013)
66. Goux, L., Lisoni, J., Jruczak, M., Wouters, D., Courtade, L., Muller, C.: Coexistence of bipolar and unipolar resistive-switching modes in NiO cells made by thermal oxidation of Ni layers. J. Appl. Phys. **107**, 024512 (2010)

67. Grimes, S., Lutken, C., Martinsen, O.: Memristive properties of human sweat ducts. World Congr. Med. Phys. Biomed. Eng. **25**(7), 696–698 (2009)
68. Hamdioui, S., Taouil, M., Haron, N.: Testing open defects in memristor-based memories. IEEE Trans. Comput. **99**, TC.2013.206 (2013)
69. Hasegawa, T., Nayak, A., Ohno, T., Terabe, K., Tsuruoka, T., Gimzewski, J.K., Aono, M.: Memristive operations demonstrated by gap-type atomic switches. Appl. Phys. A **102**, 811–815 (2011)
70. Ho, Y., Huang, G., Li, P.: Dynamical properties and design analysis for nonvolatile memristor memories. IEEE Trans. Circuits Syst. **I**(58), 724–736 (2011)
71. Hodgkin, A.L., Huxley, A.F.: A quantitative description of membrane current and its application to conduction in nerve. J. Physiol. **117**, 500–544 (1952)
72. Howard, G., Gale, E., Bull, L., de Lacy Costello, B., Adamatzky, A.: Towards evolving spiking networks with memristive synapses. In: IEEE Symposium on Artificial Life (ALIFE), pp. 14–21 (2011). doi:10.1109/ALIFE.2011.5954655
73. Howard, G.D., Gale, E., Bull, L., de Lacy Costello, B., Adamatzky, A.: Evolution of plastic learning in spiking networks via memristive connections. IEEE Trans. Evol. Comput. **16**, 711–719 (2012)
74. Howard, G.D., Bull, L., de Lacy Costello, B., Gale, E., Adamatzky, A.: Evolving spiking networks with variable resistive memories. Evol. Comput. **22**, 79–103 (2014)
75. Ielmini, D., Nardi, F., Cagli, C.: Universal reset characteristics of unipolar and bipolar metal-oxide RRAM. IEEE Trans. Electron Devices **58**, 3246–3253 (2011)
76. Itoh, M., Chua, L.: Memristor oscillators. Int. J. Bifurc. Chaos **18**, 3183–3206 (2008)
77. Itoh, M., Chua, L.O.: Memristor cellular automata and memristor discrete-time cellular neural networks. Int. J. Bifurc. Chaos **19**(11), 3605–3656 (2009). doi:10.1142/S0218127409025031. http://www.worldscientific.com/doi/abs/10.1142/S0218127409025031
78. ITRS: International technology roadmap for semiconductors 2012 update overview. Technical report, International Technology Roadmap for Semiconductors (2012). http://www.itrs.net/Links/2012ITRS/Home2012.htm
79. Jeong, H.Y., Lee, J.Y., Choi, S.Y., Kim, J.W.: Microscopic origin of bipolar resistive switching of nanocale titanium oxide thin films. Appl. Phys. Lett. **95**, 162108 (2009)
80. Jeong, H.Y., Kim, J.Y., Kim, J.W., Hwang, J.O., Kim, J.E., Lee, J.Y., Yoon, T.H., Cho, B.J., Kim, S.O., Ruoff, R.S., Choi, S.Y.: Graphene oxide thin films for flexible nonvolatile memory applications. Nano Lett. **10**(11), 4381–4386 (2010). doi:10.1021/nl101902k. http://pubs.acs.org/doi/abs/10.1021/nl101902k
81. Jeong, H.Y., Lee, J.Y., Choi, S.Y.: Interface-engineered amorphous TiO_2 based resistive memory devices. Adv. Funct. Mater. **20**, 3912–3917 (2010)
82. Jin, X., Rast, A., Galluppi, F., Davies, S., Furber, S.: Implimenting spike-timing dependent plasticity on SpiNNiker neuromorphic hardware (2010)
83. Jo, S.H., Chang, T., Ebong, I., Bhadviya, B.B., Mazumder, P., Lu, W.: Nanoscale memristor device as a synapse in neuromorphic systems. Nanoletters **10**, 1297–1301 (2010)
84. Jung, C.M., Choi, J.M., Min, K.S.: Two-step write scheme for reducing sneak-path leakage in complementary memristor array. IEEE Trans. Nanotechnol. **11**(3), 611–618 (2012). doi:10.1109/TNANO.2012.2188302
85. Kannan, S., Rajendran, J., Sinanoglu, O., Karri, R.: Sneak path testing of memristor-based memories. In: Proceedings of the 2013 26th International Conference on VLSI Design and the 12th International Conference on Embedded Systems, pp. 386–391 (2013)
86. Kavehei, O., Al-Sarawi, S., Cho, K.R., Eshraghian, K., Abbott, D.: An analytical approach for memristive nanoarchitectures. IEEE Trans. Nanotechnol. **11**, 374–385 (2012)
87. Kever, T., Bottger, U., Schlindler, C., Waser, R.: On the origin if bistable resistive switching in metal organic charge transfer complex memory cells. Appl. Phys. Lett. **91**, 083506 (2007)
88. Kim, D.J., Fisk, Z.: A Kondo insulating memristor. Appl. Phys. Lett. **101**(1), 013505 (2012). doi:10.1063/1.4733328. http://scitation.aip.org/content/aip/journal/apl/101/1/10.1063/1.4733328

89. Kim, S., Jeong, H.Y., Kim, S.K., Choi, S.Y., Lee, K.J.: Flexible memristive memory array on plastic substrates. Nano Lett. **11**(12), 5438–5442 (2011). doi:10.1021/nl203206h. http://pubs.acs.org/doi/abs/10.1021/nl203206h

90. Kim, J.Y., Kim, S.H., Lee, H.H., Lee, K., Ma, W., Gon, X., Heeger, A.J.: New architecture for high-efficiency polymer photovoltaic cells using solution-based titanium oxide as an optical spacer. Adv. Mater. **18**, 572–576 (2006)

91. Kim, K.H., Gaba, S., Wheeler, D., Cruz-Albrecht, J., Hussain, T., Srinivasa, N., Lu, W.: A functional hybrid memristor crossbar-array/CMOS system for data storage and neuromorphic applications. Nano Lett. **12**, 389–395 (2012)

92. Kosta, S., Kosta, Y., Bhatele, M., Dubey, Y., Gaur, A., Kosta, S., Gupta, J., Patel, A., Patel, B.: Human blood liquid memristor. Int. J. Med. Eng. Inf. **3**, 16–29 (2011)

93. Kosta, S.P., Kosta, Y., Gaur, A., Dube, Y.M., Chuadhari, J.P., Patoliya, J., Kosta, S., Panchal, P., Vaghela, P., Patel, K., Patel, B., Bhatt, R., Patel, V.: New vistas of electronics towards biological (biomass) sensors. Int. J. Acad. Res. 511–526 (2011)

94. Kvatinsky, S., Nacson, Y., Etsion, Y., Friedman, E., Kolodny, A., Weiser, U.: Memristor-based multithreading. Comput. Arch. Lett. **99**, 1–1 (2013). doi:10.1109/L-CA.2013.3

95. Kvatinsky, S., Wald, N., Satat, G., Friedman, E.G., Kolodny, A., Weiser, U.C.: Memristor-based material implication (IMPLY) logic: design principles and methodologies. IEEE Trans. Very Large Scale Integr. (VLSI) **22**, 1–13 (2013)

96. Kwon, D., Jeon, J.M., Jang, J., Kim, K., Hwang, C., Kim, M.: Direct observation of conducting paths in TiO$_2$ thin film by transmission electron microscopy. Microsc. Microanal. **15**, 996–997 (2009)

97. Kwon, D.H., Kim, K.M., Jang, J.H., Jeon, J.M., Lee, M.H., Kim, G.H., Li, X.S., Park, G.S., Lee, B., Han, S., Kim, M., Hwang, C.S.: Atomic structure of conducting nanofilaments in TiO$_2$ resistive switching memory. Nat. Nanotechnol. **5**, 148–153 (2010)

98. Lee, J., Lee, S., Chang, S., Gao, L., Kang, B., Lee, M.L., Kim, C., Noh, T., Kahng, B.: Scaling theory for unipolar resistance switching. Phys. Rev. Lett. **105**, 205701-1–205701-4 (2010)

99. Lehtonen, E., Poikonen, J., Laiho, M.: Memristive stateful logic. In: Adamatzky, A., Chua, L. (eds.) Memristor Networks, pp. 603–623. Springer International Publishing, Cham (2014)

100. Liborio, L., Harrison, N.: Thermodynamics of oxygen defective Magneli phases in rutile: a first principles study. Phys. Rev. B **77**, 104104 (2008)

101. Lien, C., Chen, Y., Lee, H., Chen, P., Chen, F., Tsai, M.J.: The highly scalable and reliable hafnium oxide ReRAM and its future challenges. In: Proceedings of 10th IEEE International Conference on Solid-State and Integrated Circuit Technology (ICSICT), pp. 1084–1087 (2010)

102. Lim, C.K.K., Prodromakis, T.: Computing motion with 3D memristive grids (2013). arXiv:1303.3067v1

103. Linares-Barranco, B., Serrano-Gotarredona, T.: Memristance can explain spike-time dependent plasticity in neural synapses. Nat, Preced (2009)

104. Ljung, G.M., Box, G.: On a measure of lack of fit in time series models. Biometrika **65**, 297–303 (1978)

105. Madan, R.N.: Chua's Circuit: A Paradigm for Chaos. World Scientific, Singapore (1993)

106. Manem, H., Rose, G.S., He, X., Wang, W.: Design considerations for variation tolerant multilevel CMOS/Nano memristor memory. In: Proceedings of the 20th Symposium on Great Lakes Symposium on VLSI, GLSVLSI'10, pp. 287–292. ACM, New York (2010). doi:10.1145/1785481.1785548. http://doi.acm.org/10.1145/1785481.1785548

107. Masumoto, T.: A chaotic attractor from Chua's circuit. IEEE Trans. Circuits Syst. **CAS–31**, 1055–1058 (1984)

108. Messias, M., Nespoli, C., Botta, V.A.: Hopf bifurcation from lines of equilibria without parameters in memristor oscillators. Int. J. Bifurc. Chaos **20**, 437–450 (2010)

109. Miao, F., Strachan, J.P., Yang, J.J., Zhang, M.X., Goldfarb, I., Torrezan, A.C., Eschbach, P., Kelley, R.D., Medeiros-Ribeiro, G., Williams, R.S.: Anatomy of a nanoscale conduction channel reveals the mechanism of a high-performance memristor. Adv. Mater. **23**(47), 5633–5640 (2011). doi:10.1002/adma.201103379

110. Miao, F., Yang, J.J., Borghetti, J., Medeiros, G., Williams, R.S.: Observation of two resistance switching modes in TiO$_2$ memristive devices electroformed at low current. Nanotechnology **22**, 254007, 7pp (2011)
111. Mittal, A., Swaminathan, S.: Image stabilization using memristors. In: 2nd International Conference on Mechanical and Electrical Technology (ICMET), pp. 789–792 (2010). doi:10.1109/ICMET.2010.5598474
112. Moore, G.E.: Cramming more components onto integrated circuits. Electron. Mag. **38**, 4 (1965)
113. Muthuswamy, B.: Memristor based circuit chaos. IETE Tech. Rev. **26**, 1–15 (2009)
114. Muthuswamy, B., Chua, L.: Simplest chaotic circuit. Int. J. Bifurc. Chaos **20**, 1567 (2009)
115. Nielsen, P.H., Bashara, N.M.: The reversible voltage-induced initial resistance in the negative resistance sandwich structure. IEEE Trans. Electron Devices **11**, 243–244 (1964)
116. Niemeyer, L., Pietronero, L., Wiesmass, H.: Fractal dimension of dielectric breakdown. Phys. Rev. Lett. **52**, 1033–1036 (1983)
117. Oblea, A., Timilsina, A., Moore, D., Campbell, K.: Silver chalcogenide based memristor devices. In: The 2010 International Joint Conference on Neural Networks (IJCNN), pp. 1–3 (2010). doi:10.1109/IJCNN.2010.5596775
118. Pershin, Y.V., Ventra, M.D.: Spin memristive systems: spin memory effects in semiconductor spintronics. Phys. Rev. B **78**, 113309-1–113309-4 (2008)
119. Pershin, Y.V., di Ventra, M.: Solving mazes with memristors: a massively parallel approach. Phys. Rev. E **84**, 046703 (2011). doi:10.1103/PhysRevE.84.046703. http://link.aps.org/doi/10.1103/PhysRevE.84.046703
120. Pershin, Y.V., Ventra, M.D.: Neuromorphic, digital and quantum computation with memory circuit elements. Proc. IEEE **100**, 2071–2081 (2012)
121. Pershin, Y.V., Fontaine, S.L., di Ventra, M.: Memristive model of amoeba's learning. Phys. Rev. E **80**, 021926, 6p (2009)
122. Pickett, M.D., Medeiros-Ribeiro, G., Williams, R.S.: A scalable neuristor built with Mott memristors. Nat. Mater. **16**, 114–117 (2012)
123. Pincella, F., Camorani, P., Erokhin, V.: Electrical properties of an organic memristive system. Appl. Phys. A **104**, 1039–1046 (2011)
124. Prodromakis, T., Peh, B.P., Papavassiliou, C., Toumazou, C.: A versatile memristor model with non-linear dopant kinetics. IEEE Trans. Electron Devices **30**, 3099–3105 (2011)
125. Prodromakis, T., Toumazou, C., Chua, L.: Two centuries of memristors. Nat. Mater. **11**, 478–481 (2012)
126. Qureshi, M., Yi, W., Medeiros-Ribeiro, G., Williams, R.: AC sense technique for memristor crossbar. Electron. Lett. **48**, 757–758(1) (2012). http://digital-library.theiet.org/content/journals/10.1049/el.2012.1017
127. Rast, A.D., Galluppi, F., Jin, X., Furber, S.B.: The leaky integrate and fire neuron: a platform for synaptic model exploration on the SpiNNaker chip (2010)
128. Rivas-Pérez, M., Linares-Barranco, A., Cerdá, J., Ferrando, N., Jiménez, G., Civit, A.: Visual spike-based convolution processing with a cellular automata architecture (2010)
129. Robinett, W., Pickett, M., Borghetti, J., Xia, Q., Snider, G.S., Medeiros-Ribeiro, G., Williams, R.S.: A memristor-based nonvolatile latch circuit. Nanotechnology **21**(23), 235203 (2010). http://stacks.iop.org/0957-4484/21/i=23/a=235203
130. Rozenberg, M.J., Inoue, I.H., Snchez, M.J.: Strong electron correlation effects in nonvolatile electronic memory devices. Appl. Phys. Lett. **88**(3), 033510 (2006). doi:10.1063/1.2164917. http://scitation.aip.org/content/aip/journal/apl/88/3/10.1063/1.2164917
131. Song, S.J., Seok, J.Y., Yoon, J.H., Kim, K.M., Kim, G.H., Lee, M.H., Hwang, C.S.: Real-time identification of the evolution of conducting nano-filaments in TiO$_2$ thin film ReRAM. Sci. Rep. **3**, 3443 (2013)
132. Squire, L., Kandel, E.: Memory: From Mind to Molecules. Owl Book. Henry Holt and Company (2000). http://books.google.ae/books?id=pqoYubI2GhsC
133. Strachan, J.P., Pickett, M.D., Yang, J.J., Aloni, S., Kilcoyne, A.L.D., Medeiros-Ribeiro, G., Williams, R.S.: Direct identification of the conducting channels in a functioning memristive device. Adv. Mater. **22**, 3573–3577 (2012)

134. Strukov, D.B., Snider, G.S., Stewart, D.R., Williams, R.S.: The missing memristor found. Nature **453**, 80–83 (2008)
135. Sun, X., Li, G., Zhang, X., Ding, L., Zhang, W.: Coexistence of the bipolar and unipolar resistive switching behaviours in Au/StTiO3/Pt cells. J. Phys. D: Appl. Phys. **44**, 125404, 5pp (2011)
136. Szot, K., Speier, W., Carius, R., Zastrow, U., Beyer, W.: Localized metallic conductivity and self-healing during thermal reduction of SrTiO. Phys. Rev. Lett. **88**, 075508 (2002). doi:10.1103/PhysRevLett.88.075508. http://link.aps.org/doi/10.1103/PhysRevLett.88.075508
137. Szot, K., Speier, W., Bihlmayer, G., Waser, R.: Switching the electrical resistance of individual dislocations in single-crystalline $SrTiO_3$. Nat. Mater. **5**, 312–320 (2006)
138. Taha, T., Hasan, R., Yakopcic, C., McLean, M.: Exploring the design space of specialized multicore neural processors. In: The 2013 International Joint Conference on Neural Networks (IJCNN), pp. 1–8 (2013). doi:10.1109/IJCNN.2013.6707074
139. Tang, M., Wang, Z., Li, J., Zeng, Z., Xu, X., Wang, G., Zhang, L., Xiao, Y., Yang, S., Jiang, B., He, J.: Bipolar and unipolar resistive switching behaviors of sol-gel-derived SrTiO3 thin films with different compliance currents. Semicond. Sci. Technol. **26**, 075019, 4pp (2011)
140. Thompson, R., Spencer, W.: Habituation: a model phenomena for the study of neural substrates of behaviour. Psychol. Rev. **173**, 16–43 (1966)
141. Volkov, A.G., Tucket, C., Reedus, J., Volkova, M.I., Markin, V.S., Chua, L.: Memristors in plants. Plant Signal. Behav. **9**, e28152, 8pp (2014)
142. Vontobel, P.O., Robinett, W., Kuekes, P.J., Stewart, D.R., Straznicky, J., Williams, R.S.: Writing to and reading from a nano-scale crossbar memory based on memristors. Nanotechnology **20**(42), 425204 (2009). http://stacks.iop.org/0957-4484/20/i=42/a=425204
143. Vourkas, I., Sirakoulis, G.C.: Memristor-based combinational circuits: a design methodology for encoders/decoders. Microelectron. J. **45**, 59–70 (2014)
144. Wang, L., Fang, X., Duan, S., Liao, X.: PID controller based on memristive CMAC network. Abstr. Appl. Anal. **2013**, 510238, 6 pp (2013). doi:10.1155/2013/510238
145. Waser, R., Baiatu, T., Härdtl, K.H.: DC electrical degradation of perovskite-type titanates: I, ceramics. J. Am. Ceram. Soc. **73**(6), 1645–1653 (1990). doi:10.1111/j.1151-2916.1990.tb09809.x
146. Wey, T., Jemison, W.: Variable gain amplifier circuit using titanium dioxide memristors. IET Circuits Syst. **5**, 59–65 (2011)
147. Whitehead, A.N., Russell, B.: Principia Mathematica, vol. 1, pp. 394–508. Marchant Books (1910)
148. Widrow, B.: An adaptive 'adaline' neuron using chemical 'memistors'. Technical report, Stanford University (1960)
149. Williams, R.: How we found the missing memristor. IEEE Spectr. **45**(12), 28–35 (2008). doi:10.1109/MSPEC.2008.4687366
150. Won, S., Go, S., Lee, K., Lee, J.: Resistive switching properties of $Pt/TiO_2/n+$-Si ReRAM for non-volatile memory application. Electron. Mater. Lett. **4**, 29–33 (2008)
151. Wu, H., Cai, K., Zhou, J., Li, B., Li, L.: Unipolar memristive switching in bulk negative temperature coefficient thermosensitive ceramics. PLoS One **8**, e79832 (2013). doi:10.1371/journal.pone.0079832
152. Wu, J., McCreery, R.L.: Solid-state electrochemistry in molecule/TiO_2 molecular heterojunctions as the basis of the TiO_2 "memristor". J. Electrochem. Soc. **156**, P29–P37 (2009)
153. Yang, J.J., Miao, F., Pickett, M.D., Ohlberg, D.A.A., Stewart, D.R., Lao, C.N., Williams, R.S.: The mechanism of electroforming of metal oxide memristive switches. Nanotechnology **20**, 215201-1–215201-9 (2009)
154. Yang, J.J., Zhang, M., Pickett, M., Miao, F., Strachan, J., Li, W., Ohlberg, D.A.A., Yi, W., Choi, B., Wu, W., Nickel, J.H., Medeiros-Ribeiro, G., Williams, R.: Engineering nonlinearity into memristors for passive crossbar applications. Appl. Phys. Lett. **100**, 113501, 4pp (2012)
155. Yang, Y., Sheridan, P., Lu, W.: Complementary resistive switching in tantalum oxide-based resistive memory devices. Appl. Phys. Lett. **100**, 203112, 4pp (2012)

156. Yasuhara, R., Fujiwara, K., Horiba, K., Kumigashira, H., Kotsugi, M.: Inhomogeneous chemical states in resistance-switching devices with a planar-type Pt/CuO/Pt structure. Appl. Phys. Lett. **95**, 012110 (2009)
157. Zamarreno-Ramos, C., Camuñas, L.A.C., Pérez-Carrasco, J.A., Masquelier, T., Serrano-Gotarredona, T., Linares-Barranco, B.: On spike-timing dependent plasticity, memristive devices and building a self-learning visual cortex. Front. Neuromorphic Eng. **5**, 26(1)–26(20) (2011)
158. Zhang, S., Long, S., Guan, W., Liu, Q., Wang, Q., Liu, M.: Resistive switching characteristics of MnOx-based ReRAM. J. Phys. D: Appl. Phys. **42**, 055112 (2009)
159. Zhang, F., Gan, X., Li, X., Wu, L., Gao, X., Zheng, R., He, Y., Liu, X., Yang, R.: Realization of recitifying and resistive switching behaviors of TiO$_2$ nanorod arrays for nonvolatile memory. Electrochem. Solid-State Lett. **14**, H422–H425 (2011)
160. Zidan, M.A., Fahmy, H., Hossein, M., Salama, K.N.: Memristor based memory: the sneak paths problem and solutions. Microelectron. J. **44**, 176–183 (2013)
161. Zidan, M.A., Fahmy, H., Eltawil, A., Kurdahi, F., Salama, K.N.: Memristor multi-port readout: a closed-form solution for sneak-paths. IEEE Trans. Nanotechnol. **13**, 274–282 (2014)

Chapter 20
Nature-Inspired Computation: An Unconventional Approach to Optimization

Xin-She Yang

Abstract Nature-inspired computation plays an increasingly important role in many areas such as computational intelligence, optimization and data mining. From the perspective of traditional algorithms, such nature-inspired, iterative problem-solving methods are an unconventional approach to optimization. Both the number of algorithms and the popularity have increased significantly in recent years. This chapter provides a critical analysis of some nature-inspired algorithms and strives to identify the most essential characteristics among these algorithms. We also look at different algorithmic structures and ways of generating new solutions in a mathematical framework, which will provide some insight into these algorithms. We also discuss some key open problems concerning nature-inspired metaheuristics.

20.1 Introduction

Nature-inspired computation is now playing an increasingly important role in many areas, including artificial intelligence, computational intelligence, data mining, machine learning and optimization. In recent years, nature-inspired optimization algorithms, especially those based on swarm intelligence, have gained a huge popularity [40]. Good examples of such algorithms are ant colony optimization, particle swarm optimization, cuckoo search, firefly algorithm, bat algorithm, bee algorithms and others [7, 18, 28, 40]. Obviously, such popularity can be attributed to many reasons and one key thing is that these algorithms are simple, flexible, efficient and highly adaptable. In addition, from the implementation point of view, they are very simple to be implemented in any programming language. As a result, these algorithms have been applied in a wide spectrum of problems in real-world applications.

Among nature-inspired computation, another class of important algorithms is the slime mould approach for network modelling and design [1, 42]. Such biomimicry

X.-S. Yang (✉)
School of Science and Technology, Middlesex University London,
London NW4 4BT, UK
e-mail: x.yang@mdx.ac.uk

© Springer International Publishing Switzerland 2017
A. Adamatzky (ed.), *Advances in Unconventional Computing*,
Emergence, Complexity and Computation 23,
DOI 10.1007/978-3-319-33921-4_20

543

of using *Physarum polycephalum* can provide in-depth insight of the working mechanisms of complex systems and may help to design better algorithms for modelling networks and solving optimization problems [2].

The main purpose of this chapter is to review some of the recent nature-inspired algorithms for optimization, to analyze their key characteristics and to highlight the essence of algorithms from a mathematical perspective. Therefore, the chapter is organized as follows. Section 20.2 discusses the key features among nature-inspired algorithms such as exploration and exploitation, followed by the introduction of some widely used algorithms in Sect. 20.3, including cuckoo search, firefly algorithm, particle swarm optimization and others. Then, Sect. 20.4 highlights the similarities and differences in local search and global search. Section 20.5 analyzes the iterative generation of new solutions from a mathematical point of view. Finally, some conclusions and some issues are outlined in Sect. 20.6.

20.2 Key Features of Nature-Inspired Algorithms

The key features of nature-inspired optimization algorithms can be analyzed from many different perspectives. From the algorithmic development point of view, we can analyze algorithms in terms of algorithm search behaviour, algorithm components and mathematical equations for iterations.

20.2.1 Search Behaviour: Exploration and Exploitation

The essence of an algorithm is to generate new solutions that are better than previous solutions. For an ideal algorithm, new solutions should be always better than existing solutions and it can be expected that the most efficient algorithm is to find the best solution in the least minimum efforts (ideally in one move). However, such ideal algorithms may not exist at all. For stochastic algorithms, solutions do not always get better. In fact, it can be advantageous to select not-so-good solutions, which can help the search process escape from being trapped at any local optima. Though this may be counter-intuitive, such stochastic nature now forms the essential component of modern metaheuristic algorithms [28].

From the search behaviour point of view, the search moves can be explorative or exploitative. Exploration is the process in which search moves tend to explore the search space more effectively, often far from existing solutions. Exploitation usually uses the information obtained from existing solutions about the objective landscape so as to generate better solutions, often close to the optimal solutions in the neighbourhood of current solutions. Exploration is also called diversification and exploitation is also called intensification [5].

For example, the well-known Newton's method uses the gradient information to generate new solutions by main exploitation. In Newton's root-finding algorithm, the iterative or updating equation can be written as

$$x_{k+1} = x_k - \frac{f(x_k)}{f'(x_k)} = G(x_k),$$ (20.1)

where k is the iteration counter. This algorithm heavily uses local derivative information and thus can typically have a quadratical convergence rate. For a sequence of iterations, the solutions can form a sequence: $x_0, x_1, x_2, \ldots, x_k, \ldots$, which can converge towards a fixed point $x_* = \lim_{k \to \infty} x_k$ (or the true root) under proper conditions. In this case,

$$\lim_{k \to \infty} \frac{|x_{k+1} - x_*|}{|x_k - x_*|^p} = \Lambda \leq \infty,$$ (20.2)

where p is the order of convergence and Λ is the asymptotic error constant. Then, the iterative procedure $x_{k+1} = G(x_k)$ is said to be of order p. In the case of $p = 1$ and $\Lambda < 1$, the iterative sequence is linearly convergent, while the sequence is quadratically convergent if $p = 2$. In the case of $0 < p < 1$, the convergence is sublinear, while the case of $1 < p < 2$ corresponds to the superlinear convergence. Therefore, higher p is desired in algorithm designs. Fixed point theorems and convergence theory show that the sequence can have quadratic convergence if $G' = 0$ and $|G''| < B \leq \infty$. In the case of Newton's method, we have

$$G'(x) = \frac{f(x)f''(x)}{[f'(x)]^2}, \quad G''(x) = \frac{f'(x)f''(x) + ff'''(x)}{[f'(x)]^2} - \frac{2f(x)f''(x)}{[f'(x)]^3}.$$ (20.3)

Since at the fixed point x_*, we have $f'(x_*) = 0$, which gives $G'(x_*) = 0$ and $G''(x_*) = f''(x_*)/f'(x_*)$. Thus, we obtain

$$\lim_{k \to \infty} \left| \frac{x_{k+1} - x_*}{(x_k - x_*)^2} \right| = \left| \frac{G''(x_*)}{2} \right| = \left| \frac{f''(x_*)}{2f'(x_*)} \right| = \Lambda < \frac{B}{2}.$$ (20.4)

This clearly shows that Newton's method converges quadratically.

This is almost an extreme case of exploitation because any step or iteration uses the first derivative and the current solution to determine the right amount of moves. On the other hand, purely random search is a good example of exploration because it considers the problem as a black box, and thus no information about the landscape is used for determining new moves.

Obviously, most algorithms nowadays use both components, and an efficient algorithm should use enough exploration and good exploitation. Too much exploration can explore the search space sufficiently and thus increases the probability of finding the global optimality—the best solution in the whole search space, but this can slow down the search process and reduce the converge rate. On the other hand, too much exploitation may focus on the search in a local region and thus reduce the probability of finding the true global optimality if the problem is nonlinear and multimodal, though it does lead to a good rate of convergence. Therefore, a proper balance of exploration and exploitation is needed. However, the balance of exploration and exploitation is itself an optimization problem and the actual balance may depend on the type of problem to be solved and the algorithm to be used. Maybe, the optimal balance may not exist at all [27]. At the moment, how to achieve the optimal balance is still an open, challenging problem.

20.2.2 Algorithm Components

Apart from analyzing an algorithm from the search behaviour, algorithms can also be analyzed by looking in detail the key algorithmic operators used in the construction of the algorithms. For example, in the well-established class of genetic algorithms [17], genetic operators such as crossover (or recombination), mutation and selection are used [4].

Crossover is the operation of generating two new solutions (offsprings) from two existing solutions (parents) by swapping relevant/corresponding parts of their solutions. This is similar to the main crossover feature in the biological systems. Crossover usually provides good mixing of solution characteristics and can usually generate completely new solutions if the two parents are different. Obviously, when the two parents are identical, offspring solutions will also be identical, and thus provides a good mechanism to maintain good convergence.

Mutation is a mechanism to generate a new solution from a single solution by changing a single site or multiple sites. As in the evolution in nature, mutation often generates new characteristics that can adapt to new environments, and thus new solutions during the search process can also be generated in this way.

Both crossover and mutation are ways of generating new solutions, while selection provides a pressure for evolution. In other words, selection provides a measure or mechanism to determine what is better and selects the fittest. This mimics the key feature of the Darwinian evolution in terms of the survival of the fittest. Without selection, new solutions and new characteristics will not be selected properly, which may lead to a diverse and less convergent system. With too much selection pressure, many new characteristics will die away quickly. If the environment also changes and if some solutions/chacteristics dominate the population, the system may lead to premature convergence. Thus, a proper selection pressure is also important to ensure that only good characteristics or solutions that happen to fit into the new environment can survive.

In terms of exploration and exploitation, crossover can provide both exploration and exploitation capabilities, while mutation mainly provides exploration. On the other hand, selection provides a good way of exploitation by selecting good solutions. Thus, in order to provide a good exploration ability to an algorithm, a higher crossover rate is needed, and that is why the crossover probability is typically over 0.9 in genetic algorithms. However, new solutions should not be too far from existing solutions in many cases, and to avoid too much exploration, the mutation rate should be usually low. For example, in genetic algorithms, the mutation rate is typically under 0.05 [28]. It is worth pointing out that these values are based on empirical observations (both numerically and biologically). In different algorithms, it is usually quite difficult to decide what values are most appropriate, and such choices may need detailed parametric studies and numerical experiments.

20.2.3   Mathematical Updating Equations

The Newton's method for finding roots can be modified to solve optimization problems because the main task of finding optimal solutions can be converted into finding the roots of the first derivative of the objective function $f(x)$. In general, the optimal solution for a d-dimensional function $f(x)$ can be obtained by the following iterative formula:

$$x^{t+1} = x^t - \frac{f'(x^t)}{f''(x^t)}, \tag{20.5}$$

where t is an iteration counter or (pseudo)time. Here, we use the superscript to be consistent with the standard notations for iterative algorithms. In most cases, especially for high-dimensional problems, the calculations of the second derivatives can be computationally expensive, and thus some approximations are often used in various variants. For example, if we replace $f''(x^t)$ by the identity matrix and add a step size factor p, we have the quasi-Newton method

$$x^{t+1} = x^t - pf'(x^t) = g(x^t, p). \tag{20.6}$$

As the iterations continue for a long time or $t \to \infty$, it is expected that the iteration should approach the true solution x^* (that we are searching for), and this is only true under certain conditions. In this case, we have

$$\lim_{t \to \infty} x^{t+1} = x^* = g(x^*, p). \tag{20.7}$$

This is essentially the fixed point theorem for iterations. This requires conditions such as $|g'(x)| < 1$. However, in many cases, p is not constant and can also depend on t.

Mathematically speaking, an algorithm A such as the above is an iterative process, which aims to generate a new or better solution x^{t+1} to a given problem from the current solution x^t at iteration t. Therefore, an algorithm can in general be written as

$$x^{t+1} = A(x^t, p), \tag{20.8}$$

where p is an algorithm-dependent parameter. The path of the iterations will form a zig-zag trajectory in the search space. For population-based algorithms with a swarm of n solutions $(x_1, x_2, \ldots, x_n)$, we can extend the above iterative formula to a more general form

$$
\begin{pmatrix} x_1 \\ x_2 \\ \vdots \\ x_n \end{pmatrix}^{t+1} = A\Big((x_1^t, x_2^t, \ldots, x_n^t); (p_1, p_2, \ldots \ldots, p_m); (\varepsilon_1, \varepsilon_2, \ldots, \varepsilon_r) \Big) \begin{pmatrix} x_1 \\ x_2 \\ \vdots \\ x_n \end{pmatrix}^t,
$$

(20.9)

where $p_1, \ldots, p_m$ are m algorithm-dependent parameters and $\varepsilon_1, \ldots, \varepsilon_r$ are r random variables. In addition, these parameters $p_1, \ldots, p_m$ can either be constants that do not change with iterations or be varying with iterations t. In this latter case, these parameters become $p_1(t), \ldots, p_m(t)$ and thus the choice of parameters values requires proper parameter tuning and even optimal control of parameters. An algorithm can be viewed as a dynamical system, Markov chains and an iterative map [40], and it can also be viewed as a self-organization system [3].

It is worth pointing out that A in the above equation can be a matrix, and thus the iterative (or updating) equations can be a set of coupled equations, which form a complicated, dynamic, iterative system. Obviously, the system can be either linear or nonlinear. For example, the updating equations in particle swarm optimization has two linear equations in terms of position x_i and its corresponding velocity v_i for particle i. On the other hand, the firefly algorithm uses a single, nonlinear updating equation, which can lead to richer characteristics and potentially higher efficiency.

Linear systems are easier to analyze, while nonlinear systems can be more challenging to analyze. At the moment, it still lacks in-depth understanding how different algorithms work. Before we proceed to further analysis, we will introduce some nature-inspired algorithms in the next section.

20.3 Some Swarm-Based Algorithms

The number of nature-inspired algorithms has increased dramatically in recent years. To review a good subset of all these algorithms can take the whole length of a long chapter. Here, we only select a few algorithms that are among most recent and most popular algorithms.

A vast majority of nature-inspired algorithms are usually based on swarm intelligence. However, their updating equations can be either linear or nonlinear, though most algorithms have linear updating equations.

Linear systems may be easier to implement, but the diversity and richness of the system behaviour may be limited. For the algorithms with nonlinear updating systems, the characteristics of the algorithm can be richer, which may lead to some advantages over algorithms with linear updating equations. For example, studies show that firefly algorithm can automatically subdivide the whole population into multiple sub-swarms due to its nonlinear distance-dependent attraction mechanism. At the moment, it is still not clear if nonlinear systems are always potentially better.

In the rest of this section, when we describe the key formulations of each algorithm, we will mainly focus on the key characteristics of its updating equations as an iterative system.

20.3.1   Particle Swarm Optimization

Particle swarm optimization (PSO) was developed by Kennedy and Eberhart in 1995 [18] and PSO was among the first algorithms that were based on swarming behaviour of fish or bird schooling in nature. The updating equations are a set of two linear equations for the position x_i and velocity v_i of a particle system:

$$v_i^{t+1} = v_i^t + \alpha \varepsilon_1 \left[g^* - x_i^t \right] + \beta \varepsilon_2 \left[x_i^* - x_i^t \right], \tag{20.10}$$

$$x_i^{t+1} = x_i^t + v_i^{t+1}. \tag{20.11}$$

Here, the behaviour is essentially controlled by two learning parameters α and β with typical values of $\alpha \approx \beta \approx 2$. In addition, two random vectors ε_1 and ε_2 are uniformly distributed in [0,1].

This system is linear and thus it is quite straightforward to carry out stability analysis [6]. The PSO system usually converges very quickly, but it can have premature convergence. Therefore, over 20 different variants have been proposed to remedy the potential premature convergence. One of key improvements is the use of an inertia function by Shi and Eberhart [23]. Other developments include the accelerated PSO by Yang et al. [31] and some reasoning techniques with PSO by Fister Jr. et al. [14].

20.3.2   Firefly Algorithm

The firefly algorithm (FA) was developed by Xin-She Yang in 2008 [28], which was based on the flashing patterns and behaviour of tropical fireflies. The equation of FA is a single nonlinear equation for updating the locations (or solutions) of the fireflies:

$$x_i^{t+1} = x_i^t + \beta_0 e^{-\gamma r_{ij}^2} \left(x_j^t - x_i^t \right) + \alpha \, \varepsilon_i^t, \tag{20.12}$$

where the second term between any two fireflies (i and j) due to the attraction is highly nonlinear because the attraction is distance-dependent. Here, β_0 is the attractiveness at $r = 0$, while α is the randomization parameter. In addition, $\boldsymbol{\varepsilon}_i^t$ is a vector of random numbers drawn from a Gaussian distribution at time t.

The nonlinear nature of the updating equation means that the local short-distance attraction is much stronger than the long-range attraction, and consequently, the whole population can automatically subdivide into multiple subgroups (or multi-swarms). Under the right conditions, each subgroup can swarm around a local mode, thus, FA can naturally deal with multimodal problems effectively. For example, firefly algorithm can be very efficient in solving classifications and clustering problems [22] as well as scheduling problems [41]. The extensive studies of the firefly algorithm and its variants were reviewed by Fister et al. [9–11].

20.3.3 Cuckoo Search

Another nonlinear system of updating equations is the cuckoo search (CS), developed in 2009 by Xin-She Yang and Suash Deb [34], based on the brood parasitism of some cuckoo species [35, 36]. The enhancement of CS via Lévy flights makes CS more efficient [20] than PSO and genetic algorithms.

The nonlinear equations are also controlled by a switching parameter or probability p_a. The nonlinear system can be written as

$$x_i^{t+1} = x_i^t + \alpha s \otimes H(p_a - \varepsilon) \otimes \left(x_j^t - x_k^t\right), \tag{20.13}$$

$$x_i^{t+1} = x_i^t + \alpha L(s, \lambda). \tag{20.14}$$

Here, two different solutions x_j^t and x_k^t are randomly selected. In addition, $H(u)$ is a Heaviside step function, ε is a random number drawn from a uniform distribution, and s is the step size. The so-called Lévy flights are realized by drawing random step sizes $L(s, \lambda)$ from a Lévy distribution, which can be written as

$$L(s, \lambda) \sim \frac{\lambda \Gamma(\lambda) \sin(\pi \lambda / 2)}{\pi} \frac{1}{s^{1+\lambda}}, \quad (s > 0), \tag{20.15}$$

where $\sim$ denotes the drawing of samples from a probability distribution. In addition, $\alpha > 0$ is the step size scaling factor, which should be related to the scales of the problem of interest.

Recent studies suggests that CS can have autozooming capabilities so that the search process can automatically focus on the promising areas due to the combination of self-similar, fractal-like and multiscale search capabilities [40]. By analyzing the algorithmic system, it can be seen that simulated annealing, differential evolution and APSO are special cases of CS, and that is one of the reasons why CS is so efficient. Some recent reviews have been carried out by Yang and Deb [37] and Fister Jr. et al. [12]. Mathematical analysis also suggested that CS can have global convergence [26].

20.3.4 Bat Algorithm

Another example of linear systems is the bat algorithm that was developed by Xin-She Yang in 2010 [29], inspired by the echolocation behavior of microbats. The updating equations for the velocity v_i^t and location x_i^t of a bat at any at iteration t in a d-dimensional search or solution space can be written as

$$f_i = f_{\min} + (f_{\max} - f_{\min})\beta, \qquad (20.16)$$

$$v_i^t = v_i^{t-1} + \left(x_i^{t-1} - x_*\right)f_i, \qquad (20.17)$$

$$x_i^t = x_i^{t-1} + v_i^t, \qquad (20.18)$$

where x_* is the current best solution and $\beta \in [0, 1]$ is a random vector drawn from a uniform distribution. To control exploration and exploitation, variations of the loudness and pulse emission rates are regulated in the following way:

$$A_i^{t+1} - \alpha A_i^t, \quad r_i^{t+1} - r_i^0[1 - \exp(-\gamma t)], \qquad (20.19)$$

where $0 < \alpha < 1$ and $\gamma > 0$ are constants. BA has attracted a lot of interest and thus the literature is expanding. For example, Yang extended it to multiobjective optimization [30, 32] and Fister et al. formulated a hybrid bat algorithm [13, 15].

20.3.4.1 Differential Evolution

Differential evolution (DE), developed by R. Storn and K. Price in 1996 and 1997 [24, 25], uses a linear mutation equation

$$u_i^t = x_r^t + F\left(x_p^t - x_q^t\right), \qquad (20.20)$$

where F is the differential weight in the range of $[0, 2]$. Here, r, p, q, i are four different integers generated by random permutation. However, the crossover operation can be tricky to analyze, depending on the perspective. The crossover operator in DE is controlled by a crossover probability $C_r \in [0, 1]$ and the actual crossover can be carried out in two ways: binomial and exponential. Crossover can typically be carried out along each dimension $j = 1, 1, \ldots, d$ where d is the number of dimensions. Thus, we have

$$x_{j,i}^{t+1} = \begin{cases} u_{j,i}^t & \text{if rand} < C_r \text{ or } j = I, \\ x_{j,i}^t & \text{if rand} > C_r \text{ and } j \neq I, \end{cases} \qquad (20.21)$$

where I is a random integer from 1 to d, so that $u_i^{t+1} \neq x_i^t$.

Selection is essentially the same as that used in genetic algorithms. It is to select the most fittest, and for the minimization problem, the minimum objective value. Therefore, we have

$$
x_i^{t+1} = \begin{cases} u_i^{t+1} & \text{if } f\left(u_i^{t+1}\right) \leq f\left(x_i^t\right), \\ x_i^t & \text{otherwise.} \end{cases} \tag{20.22}
$$

There are more than 10 different variants [21] and many more studies in the literature.

20.3.5   Flower Pollination Algorithm

Flower pollination algorithm (FPA) was developed by Xin-She Yang in 2012 [33], inspired by the flower pollination process of flowering plants. It has been extended to multiobjective optimization problems and found to be very efficient [39]. The main updating equation for a solution/position x_i at any iteration t is

$$
x_i^{t+1} = x_i^t + \gamma L(\lambda)\left(g_* - x_i^t\right), \tag{20.23}
$$

which is essentially linear. Here, g_* is the current best solution found among all solutions and γ is a scaling factor to control the step size. The step sizes as Lévy flights are drawn from a Levy distribution, that is

$$
L(\lambda) \sim \frac{\lambda \Gamma(\lambda) \sin(\pi \lambda/2)}{\pi} \frac{1}{s^{1+\lambda}}, \quad (s \gg s_0 > 0). \tag{20.24}
$$

Here $\Gamma(\lambda)$ is the standard gamma function, and this distribution is valid for large steps $s > 0$. The local search steps are carried out by

$$
x_i^{t+1} = x_i^t + \varepsilon\left(x_j^t - x_k^t\right), \tag{20.25}
$$

which is also linear. Here, x_j^t and x_k^t are pollen from different flowers of the same plant species. Here, ε from a uniform distribution in [0,1]. In essence, this is a local mutation and mixing step, which can help to converge in a subspace.

However, it is worth pointing out that the generation of new moves using Lévy flights can be tricky. There are efficient ways to draw random steps correctly from a Lévy distribution [20, 40].

The switch between two search branches are controlled by a probability p_s. Recent studies suggested that flower pollination algorithm is very efficient for multiobjective optimization [39].

20.3.6   Other Algorithms

Obviously, there are other nature-inspired algorithms such as harmony search [16] and bacterial foraging optimization [19]. However, the main aim of the rest of this chapter is to continue to analyze algorithms in terms of search capabilities, topology, structure and mathematical aspects, so we will not introduce more algorithms.

20.4   Local Search or Global Search?

When we discussed the algorithms from the search behaviour, we focused on two key components: exploration and exploitation. They are also related to local and global search moves. Loosely speaking, exploration tends to be global, while exploitation are mostly local. However, the differences and distinctions can be subtle, and all may depend on the step sizes and actual moves.

In terms of the neighbourhood and search space, the ways of generating new solutions can be analyzed. From the locality point of view, they can be divided into the following subcategories:

- Local modifications.
- Global modifications.
- Mixed (both local and global) or multiscale modifications.

20.4.1   Local Modifications

Local modifications focus on the local region or the neighbourhood of an existing solution. Once a good feasible solution (in many cases, the best solution) is selected, new solutions are generated around this solution with the radius Δ of the neighbourhood, so that the distance or 2-norm from the new solution x_i^{t+1} to the old solution x_i^t satisfies

$$||x_i^{t+1} - x_i^t||_2 \leq \Delta. \tag{20.26}$$

This is usually realized by perturbing the existing solution with a random term w such that

$$x_i^{t+1} = x_i^t + w. \tag{20.27}$$

If w is drawn from a probability distribution such as uniform or Gaussian, then such local modifications become random walks. For example, the pitch adjustment in the harmony search [16] is a local random walk, and the random perturbation term in the firefly algorithm also plays a role of random walks.

Sometimes, it may be useful to do exploration in combination with exploitation. In this case, a very simple way is to carry out local search around the current best solution; that is

$$x_i^{t+1} = x_*^t (\text{best solution}) + \text{perturbation}. \tag{20.28}$$

If we revisit the genetic operators such as crossover and mutation, there is no clear cut to decide if they are local or global. In the case when the mutation rate is very low and the changes are small, mutation can be primarily local. However, a high mutation rate at multiple sites can lead to solutions that are far away from existing solutions, and in this case, mutation can become global. On the other hand, crossover usually occurs in a subset of the solution space if the diversity is limited. For example, with two solutions $u = [101 \dots 1]$ and $v = [100 \dots 1]$ as parents, whatever the crossover operations may be, the first component will always be 1. This means that crossover can be limited to a subspace in the vast solution space. In this case, only the mutation to the first component can make it jump out of this subspace. Even with this, it is difficult to say crossover is a local or global operator, and all may depend on the properties of the existing solutions in the population.

20.4.2 Global Modifications

For more efficient exploration, new solutions should be generated at the global scale. One of the simplest ways to achieve the global modifications is to use

$$x = a + (b - a)\, \text{rand}, \quad x \in [a, b], \tag{20.29}$$

where the random number (rand) is drawn from a uniform distribution $\text{Unif}(0, 1)$. For a d-dimensional vector, a new solution x^t can be generated between a lower bound (vector L) and an upper bound (vector U) using

$$x^t = L + (U - L)\, R_d, \tag{20.30}$$

where R_d is the random vector (the same size as U and L) with each of its components drawn from the uniform distribution in $[0,1]$.

Obviously, for local search, if the size of the neighborhood is increased so that it is the same order of the scale of the problem [that is, $\Delta = O(||U - L||_2)$], then local modifications essentially become global modifications. Alternatively, in the local modification (20.27), if w is drawn from a scaled Guassian distribution or Gaussian with a large variance $N(0, \sigma^2)$ with σ being sufficiently large, then (20.27) becomes a random walk on the global scale.

Therefore, it all depends on the step size and the scale of the modifications. Along the same line of argument, crossover can also become global if the two parents are drawn from a diverse population with a sufficient distance apart, and mutation can also be global when the mutation is high and frequent.

20.4.3   Mixed or Multiscale Modifications

As we have seen earlier, the difference between global or local modifications can be subtle. When the step sizes are large enough, local modifications can become global. Furthermore, these mechanisms for generating new solutions do not always belong to a single mechanism, and they can be a mixture of two or more components. In fact, in Eq. (20.27), when the random perturbation vector w is drawn from a distribution with a fat tail such as Cauchy's distribution, exponential distribution or Lévy distribution, most modifications are local, but quite a fraction of the new modifications can be global. Such local modifications perpetuated by occasional long jumps can produce solutions with both local and global search capabilities. Under the right conditions, such a mixture of local and global search can be also multiscale.

As a good example, Lévy flights in the cuckoo search are essentially multiscale because the modifications are a mixture of both local and global modifications [40]. Both mathematically and numerically, it can be demonstrated that the characteristics of small steps and long-range jumps occur at all scales. By using an approximation to a Lévy distribution with the exponent λ, we have

$$
\begin{aligned}
L_D &= \frac{1}{\pi} \int_0^\infty \cos(ks) \exp[-|k|^\lambda] dk \\
&\approx \frac{\lambda \Gamma(\lambda) \sin(\pi \lambda / 2)}{\pi |s|^{1+\lambda}}, \quad s \to \infty,
\end{aligned} \tag{20.31}
$$

where s is the step size and $\Gamma(n)$ is the standard Gamma function. With a simple set of calculations for $\lambda = 1.5$, we can draw 1000 samples easily. About 66.2 % of the samples have step size less than 1, with 33.8 % of the steps larger than 1 [or $s \geq 1$ (33.8 %)]. Among the larger steps, we have $s \geq 5$ (4 %), $s \geq 10$ (1.1 %), $s \geq 20$ (0.3 %) and $s \geq 40$ (0.1 %). Clearly, this demonstrates that step sizes drawn from a Lévy distribution can be multiscale.

In most algorithms, a good combination of local search with global search is needed. If the global optimality is near, new modifications should be mostly local, while new solutions should be globally explorative if solutions are far from optimality. As theoretical results regarding algorithm analysis are limited, it is quite challenging to decide what types of modifications should be used at what stages. It is likely that such a combination may be problem-dependent as well as algorithm-dependent.

20.5   Mathematical Aspects of Algorithmic Compoments

As discussed in the previous section, the generation of new solutions depends on the structure and topology of the search neighbourhood. New moves can be in either a local neighborhood (e.g., local random walks), the global domain (e.g., Lévy or uniform initialization), or cross-scale (both local and global, e.g., Lévy flights, exponential, power-law, heavy-tailed).

The crossover operator can be written mathematically as

$$
\begin{pmatrix} x_i^{t+1} \\ x_j^{t+1} \end{pmatrix} = C(x_i^t, x_j^t, p_c) = C(p_c) \begin{pmatrix} x_i^t \\ x_j^t \end{pmatrix}, \tag{20.32}
$$

where p_c is the crossover probability, though the exact form of $C(\ )$ depends on the actual crossover manipulations such as at one site or at multiple sites simultaneously. The selection of i and j can be by random permutation. On the other hand, the choice of parents can more often be fitness-dependent, which may implicitly depend on the relative fitness of the parents in the population. In this case, the functional form for the crossover function can be even more complex. For example, $C(\ )$ can depend on all the individuals in the population, which may lead to $C(x_1^t, x_2^t, \ldots, x_n^t, p_c)$ where n is the population size.

Mutation can be written schematically as

$$
x_i^{t+1} = M(x_i^t, p_m), \tag{20.33}
$$

where p_m is the mutation rate. However, the form $M(\ )$ depends on the coding and the number of mutation sites. This can be written in most cases as a random walk

$$
x_i^{t+1}(\text{new solution}) = x_i^t(\text{old solution}) + \alpha(\text{randomization}), \tag{20.34}
$$

where α is a scaling factor controlling how far the random walks can go [28, 29]. However, mutation can also be carried out over a subset of the population or the mutation operator can also be affected by more than one solution. For example, the mutation operator in differential evolution takes the form $x_r^t + F(x_p^t - x_q^t)$, which involves three different solutions. It is worth pointing out that randomization here increases the diversity of the population, but too much randomness may slow down the rate of convergence. However, it is still an open question concerning what amount of randomization is the right amount.

In general, the solutions can be generated in parallel by random permutation, and thus we may have a more generic form for modifications

$$
\begin{pmatrix} x_1^{t+1} \\ x_2^{t+1} \\ \vdots \\ x_n^{t+1} \end{pmatrix} = G(x_1^t, x_2^t, \ldots, x_n^t, \varepsilon, \beta), \tag{20.35}
$$

where $G(\ )$ can be very complex, which also depends on the random variable ε and parameter β. Ideally, we should have some mathematical foundations or analysis such as Eq. (20.4) or something like the stability analysis [6]. However, it still lacks a unified mathematical framework for such analysis for general algorithms as discussed above.

It is worth pointing out that the steps in multidimensional problems may need to adjust differently along different dimensions. When there are some huge differences in the search ranges of different dimensions (for example, in 2D, one may vary from -10^9 to 10^9 in one dimension, the other may vary from 0 to 0.001), proper scalings in steps are needed. Not all local or global modifications can scale properly. Interestingly, modifications using pattern-search-type moves or the differences of solutions can have automatic scaling properties. A good example is that the mutation in differential evolution always uses the differential changes such as $(x_p^t - x_q^t)$, which can scale automatically at any scale.

20.6 Conclusions and Future Topics

Nature-inspired computation has become a powerful unconventional approach to optimization and computational intelligence. The number of nature-inspired algorithms for optimization has increased significantly, along with the increase of their popularity. In this chapter, we have summarized some of the key characteristics of algorithms, especially those based on swarm intelligence. By briefly outlining some recent and widely used algorithms, we have analyzed the search behaviour in terms of exploration and exploitation, local or global search moves. We also discussed such components using a relatively generic mathematical framework.

In addition, we have also analyzed and highlighted the relationships and differences of genetic operators such as crossover and mutation with local and global search. The above analysis can help us to gain deeper understanding into the main mechanisms of nature-inspired metaheuristic algorithms.

However, there are some important issues that still need to be addressed in future research. In order to encourage further research in the area, we would like to list the following open problems:

- Mathematical framework: a systematical framework is needed to study the convergence and stability of existing and new algorithms. From the current literature, we would like to hint that Markov chain theory, dynamical system, fixed-point iteration theory, Bayesian theory, hypothesis testing and self-organization theory can all be useful to do mathematical analysis.

- Parameter tuning and parameter control: all algorithms have algorithm-dependent parameters. The values of such parameters can affect the search behaviour of an algorithm. To carry out parameter tuning can be a time-consuming task [8]. In addition, there is no need to fix the parameter values during iterations. Proper control and variations of these algorithmic parameters can be challenging, but may be advantageous to fine-tune the search dynamics of any algorithm [38].
- Rise of swarm intelligence: most nature-inspired algorithms use some sort of swarm intelligence, based on the inspiration from nature. One of the key questions is how the collective intelligence emerges under local interaction rules? The understanding of the origin and rise of swarm intelligence may be useful to the understanding of nature-inspired algorithms.
- Large-scale applications: from the current literature, almost all algorithms including the particle swarm optimization can only solve problems of small or moderate scales. It is not clear if these algorithms can directly be applied to solve large-scale problems with thousands or even millions of design variables. Such scalability is yet to be validated by complex problems in real-world applications.

Obviously, there are other challenging issues such as how to balance exploration and exploitation. Here, we have only highlighted a subset of such problems. It can be expected that these key problems can inspire more research in these areas in the near future.

References

1. Adamatzky, A., Yang, X.S., Zhao, Y.X.: Slime mould imitates transport networks in China. Int. J. Intell. Comput. Cybern. **6**(3), 232–251 (2013)
2. Adamatzky, A.: Bioevoluation of World Transport Networks. World Scientific Publishing, Singapore (2012)
3. Ashby, W.R.: Princinples of the self-organizing sysem. In: Von Foerster, H., Zopf Jr., G.W. Pricinples of Self-Organization: Transactions of the University of Illinois Symposium. Pergamon Press, London, UK. pp. 255–278 (1962)
4. Booker, L., Forrest, S., Mitchell, M., Riolo, R.: Perspectives on Adaptation in Natural and Artificial Systems. Oxford University Press, Oxford (2005)
5. Blum, C., Roli, A.: Metaheuristics in combinatorial optimisation: Overview and conceptual comparision. ACM Comput. Surv. **35**, 268–308 (2003)
6. Clerc, M., Kennedy, J.: The particle swarm – explosion, stability, and convergence in a multi-dimensional complex space. IEEE Trans. Evol. Comput. **6**(1), 58–73 (2002)
7. Dorigo, M., Di Caro, G., Gambardella, L.M.: Ant algorithms for discrite optimization. Artif. Life **5**(2), 137–172 (1999)
8. Eiben, A.E., Smit, S.K.: Parameter tuning for configuring and analyzing evolutionary algorithms. Swarm and Evol. Comput. **1**(1), 19–31 (2011)
9. Fister, I., Fister Jr., I., Yang, X.S., Brest, J.: A comprehensive review of firefly algorithms. Swarm and Evol. Comput. **13**(1), 34–46 (2013)
10. Fister, I., Yang, X.-S., Brest, J., Fister Jr., I.: Modified firefly algorithm using quaternion representation. Expert Syst. Appl. **40**(18), 7220–7230 (2013)
11. Fister, I., Yang, X.S., Fister, D., Fister Jr., I.: Firefly algorithm: A brief review of the expanding literature. Cuckoo Search and Firefly Algorithm: Theory and Applications. Studies in Computational Intelligence, pp. 347–360. Springer, Heidelberg (2014)

12. Fister Jr., I., Yang, X.S., Fister, D., Fister, I.: Cuckoo search: a brief literature review. Cuckoo Search and Firefly Algorithm: Theory and Applications, vol. 516, pp. 49–62. Springer, Heidelber (2014)
13. Fister Jr., I., Fister, D., Yang, X.S.: A hybrid bat algorithm. Elektrotehniski Vestnik **80**(1–2), 1–7 (2013)
14. Fister Jr., I., Yang, X.S., Ljubič, K., Fister, D., Brest, J., Fister, I.: Towards the novel reasoning among particles in PSO by the use of RDF and SPARQL. Sci. World J. 2014, article ID. 121782, (2014). http://dx.doi.org/10.1155/2014/121782
15. Fister Jr., I., Fong, S., Brest, J., Fister, I.: A novel hybrid self-adaptive bat algorithm. Sci. World J. **2014**, article ID 709738, (2014). http://dx.doi.org/10.1155/2014/709738
16. Geem, Z.W., Kim, J.H., Loganathan, G.V.: A new heuristic optimization: harmony search. Simulation **76**(2), 60–68 (2001)
17. Holland, J.: Adaptation in Natural and Artificial Systems. University of Michigan Press, Ann Anbor (1975)
18. Kennedy, J., Eberhart, R.C.: Particle swarm optimization. In: Proceedings of IEEE International Conference on Neural Networks, Piscataway, NJ, pp. 1942–1948 (1995)
19. Passino, K.M.: Bactorial foraging optimization. Int. J. Swarm Intell. Res. **1**(1), 1–16 (2010)
20. Pavlyukevich, I.: Lévy flights, non-local search and simulated annealing. J. Comput. Phys. **226**(12), 1830–1844 (2007)
21. Price, K., Storn, R., Lampinen, J.: Differential Evolution: A Practical Approach to Global Optimization. Springer, Berlin (2005)
22. Senthilnath, J., Omkar, S.N., Mani, V.: Clustering using firely algorithm: performance study. Swarm Evol. Comput. **1**(3), 164–171 (2011)
23. Shi, Y.H., Eberhart, R.: A modified particle swarm optimizer. In: Proceedings of the 1998 IEEE World Congress on Computational Intelligence, 4–9 May 1998, Anchorage, AK, IEEE Press, USA, pp. 69-73 (1998)
24. Storn, R.: On the usage of differential evolution for function optimization. In: Proceedings of the Biennial Conference of the North American Fuzzy Information Processing Society (NAFIPS). Berkeley, CA **1996**, pp. 519–523 (1996)
25. Storn, R., Price, K.: Differential evolution - a simple and efficient heuristic for global optimization over continuous spaces. J. Glob. Optim. **11**(4), 341–359 (1997)
26. Wang, F., He, X.S., Wang, Y., Yang, S.M.: Markov model and convergence analysis based on cuckoo search algorithm. Comput. Eng. **38**(11), 180–185 (2012). (in Chinese)
27. Wolpert, D.H., Macready, W.G.: No free lunch theorems for optimization. IEEE Trans. Evol. Comput. **1**(1), 67–82 (1997)
28. Yang, X.S.: Nature-Inspired Metaheuristic Algorithms. Luniver Press, Bristol (2008)
29. Yang, X.S.: A new metaheuristic bat-inspired algorithm. In: Nature Inspired Cooperative Strategies for Optimisation (NICSO 2010). Studies in Computational Intelligence, vol. 284, pp. 65-74. Springer, Heidelberg (2010)
30. Yang, X.S.: Bat algorithm for multi-objective optimisation. Int. J. Bio-Inspired Comput. **3**(5), 267–274 (2011)
31. Yang, X.S., Deb, S., Fong, S.: Accelerated particle swarm optimization and support vector machine for business optimization and applications. In: Networked Digital Technologies 2011, Communications in Computer and Information Science, vol. 136, pp. 53–66. Springer, Heidelberg (2011)
32. Yang, X.S., Gandomi, A.H.: Bat algorithm: a novel approach for global engineering optimization. Eng. Comput. **29**(5), 1–18 (2012)
33. Yang, X.S.: Flower pollination algorithm for global optimization. In: Unconventional Computation and Natural Computation, pp. 240–249. Springer, Heidelberg (2012)
34. Yang, X.S., Deb, S.: Cuckoo search via Lévy flights. In: Proceedings of World Congress on Nature & Biologically Inspired Computing (NaBIC 2009), pp. 210–214. IEEE Publications, USA (2009)
35. Yang, X.S., Deb, S.: Engineering optimization by cuckoo search. Int. J. Math. Modelling Numer. Optim. **1**(4), 330–343 (2010)

36. Yang, X.S., Deb, S.: Multiobjective cuckoo search for design optimization. In: Computers and Operations Research, **40**(6), pp. 1616-1624 (2013)
37. Yang, X.S., Deb, S.: Cuckoo search: recent advances and applications. Neural Comput. Appl. **24**(1), 169–174 (2014)
38. Yang, X.S., Deb, S., Loomes, M., Karamanoglu, M.: A framework for self-tuning optimization algorithm. Neural Comput. Appl. **23**(7–8), 2051–2057 (2013)
39. Yang, X.S., Karamanoglu, M., He, X.S.: Flower pollination algorithm: a novel approach for multiobjective optimization. Eng. Optim. **46**(9), 1222–1237 (2014)
40. Yang, X.S.: Nature-Inspired Optimization Algorithms. Elsevier, London (2014)
41. Yousif, A., Abdullah, A.H., Nor, S.M., Abdelaziz, A.A.: Scheduling jobs on grid computing using firefly algorithm. J. Theoret. Appl. Inf. Technol. **33**(2), 155–164 (2011)
42. Zhang, X.G., Adamatzky, A., Chan, F.T., Deng, Y., Yang, H., Yang, X.S., Tsompanas, M.I., Sirakoulis, G.C., Mahadevan, S.: A biologically inspired network design model, Scientific Reports, vol. 5, Article number 10794, June (2015). http://www.nature.com/srep/2015/150604/srep10794/full/srep10794.html

Chapter 21
On Hybrid Classical and Unconventional Computing for Guiding Collective Movement

Jeff Jones

Abstract Collective movement in living systems typically displays complex dynamics which cannot be described by the component parts themselves. Plasmodium of slime mould *Physarum polycephalum* exhibits complex amoeboid movement during its foraging and hazard avoidance which may be influenced by the local placement of attractants, repellents and light irradiation stimuli. Slime mould is a useful inspiration to soft-robotics due to its simple component parts and the distributed nature of its control and locomotion mechanisms. However, it is challenging to interface classical computing devices to a distributed system which utilises self-organised and emergent properties. In this chapter we investigate potential *hybrid* approaches to the task of automatically guiding collective robotics devices, using a multi-agent model of slime mould. We demonstrate a variety of simple open-loop guidance methods. We then describe a hybrid classical/unconventional computing approach using a closed-loop feedback mechanism with attractant and repellent stimuli. Both stimulus types were capable of successful automatic guidance, but we found that repellent stimuli (a light illumination mask) provided faster and more accurate guidance than attractant sources, which were found to exhibit overshooting phenomena at path turns. The method allows traversal of convoluted arenas with challenging obstacles such as narrow channels and complex gratings, and provides an insight into how unconventional computing substrates may be hybridised with classical computing methods to take advantage of the mutual benefits of both approaches.

21.1 Introduction: Collective Movement

Collective movement is a directed movement of multiple individuals which are coupled (directly or indirectly) by some aspect of their environment or special senses. The phenomenon is observed in natural systems which span huge variations in spatial

J. Jones (✉)
Centre for Unconventional Computing, University of the West of England,
Bristol BS16 1QY, UK
e-mail: jeff.jones@uwe.ac.uk

© Springer International Publishing Switzerland 2017
A. Adamatzky (ed.), *Advances in Unconventional Computing*,
Emergence, Complexity and Computation 23,
DOI 10.1007/978-3-319-33921-4_21

scale, temporal scale, and in their environmental medium. At very small spatial scales this includes self-organised movement or aggregation of collectives composed of individual cells to form regular patterns. In bacteria this can be observed, for example, in the formation of regular patterns in *Bacillus subtilis* [36] and *Proteus mirabilis* [43]. Cells of the cellular slime mould *Dictyostelium discoideum* are well known to aggregate under the influence of $_cAMP$ [16, 34]. At slightly larger scales human cells are known to move collectively during embryogenesis [46], wound repair [9] and tumorigenesis [13] in response to a wide range of chemotactic, bioelectric and mechanical stimuli [17, 35].

Collective movement at the population level can result in dynamic and dramatic patterning phenomena, such as swarming [10], flocking [44], herding [18], and shoaling and schooling [22, 42]. The specific biological and generalised coupling mechanisms which generate these emergent patterns from the low-level individual interactions have been studied [15, 51, 61]. In human environments collective movement is seen in walking trail patterns [21], crowd dynamics [20, 67], and car traffic systems [19, 37].

Non-living systems may also exhibit collective movement, for example the phenomenon of sorted patterned ground [31] or the evolution of dune structures [33, 62]. But in these non-living cases the movement of the 'individuals' is passive and undirected, guided only by environmental forces (freeze-thaw cycles and wind transport respectively). In living systems collective movement not only responds to environmental forces but also adds an element of responding to an external stimulus. This directed response enables the dynamical cohesion of a population in space (for example flocking in response to predatory threats), the efficient ordering or movement of a mobile population (trails or raiding fronts of ant colonies), the aggregation of a population towards a single location (for example prior to the assembly of the grex structure in *Dictyostelium*), and the arrangement of complex patterns (the concentric patterns in *B. subtilis* or the arrangement of endothelial precursor cells to form vascular structures). The type of stimulus and the location at which it is presented are an important consideration in the response of the mobile collective.

Collective movement is of interest for computing and robotics applications because the mechanisms which enable collective movement in natural systems are relatively simple, use local communication cues, exploit self-organised patterning and exhibit distributed control. There is typically no central orchestrator of movement in such systems, and the contribution of all entities may be of equal importance. These emergent behaviours result in efficient collectives which contain redundant parts and are resilient to damage or interruption. Simple identical components would reduce the cost of robotic devices. Furthermore, communication between robotic entities and evaluation of current position and future goal position imply a significant computational cost which would be multiplied when the collective contains a large robotic population. The exploitation of strategies exploited by living collective systems in artificial robotic collectives may result in useful physical and computational cost savings, with the benefit of innate autonomy and resilience of the collective. As previously noted, collective movement is observed in a diverse number of target organisms operating at very different scales. If we are to take inspiration from living examples of collective movement for robotics purposes which example system should be chosen?

21.2 Collective Movement in Slime Mould *Physarum Polycephalum*

An ideal candidate for a biological organism to explore collective movement would be capable of the complex sensory integration, movement and adaptation, yet be composed of a relatively simple component parts that are amenable to simple understanding and external influence. The acellular slime mould *Physarum polycephalum* meets these criteria. *P. polycephalum* is a giant single-celled amoeboid organism, visible to the naked eye. During the plasmodium stage of its complex life-cycle [50] it takes the form of a constantly adapting protoplasmic network. This network is comprised of a sponge-like material which exhibits self-organised oscillatory and contractile activity which is harnessed in the transport and distribution of nutrients within this internal transport network. The organism is remarkable in that the control of the oscillatory behaviour is distributed throughout the almost homogeneous medium and is highly redundant, having no critical or unique components.

The plasmodium is amorphous in shape and ranges from the microscopic scale to up to many square metres in size. It is a single cell syncytium formed by repeated nuclear division, comprised of a sponge-like actomyosin complex co-occurring in two physical phases. The gel phase is a dense matrix subject to spontaneous contraction and relaxation, under the influence of changing concentrations of intracellular chemicals. The protoplasmic sol phase is transported through the plasmodium by the force generated by the oscillatory contractions within the gel matrix. Protoplasmic flux, and thus the behaviour of the organism, is affected by changes in pressure, temperature, space availability, chemoattractant stimuli and illumination [11, 12, 32, 38, 41, 53, 59]. The *P. polycephalum* plasmodium can thus be regarded as a complex functional material capable of both sensory and motor behaviour. Indeed it has been described as a membrane bound reaction-diffusion system in reference to both the complex interactions within the plasmodium and the rich computational potential afforded by its material properties [7]. The study of the computational potential of the *P. polycephalum* plasmodium was initiated by Nakagaki et al. [39] who found that the plasmodium could solve simple maze puzzles. This research has been extended and the plasmodium has demonstrated its performance in, for example, path planning and plane division problems [47, 48], spanning trees and proximity graphs [1, 2], simple memory effects [14, 45], the implementation of logic gates and adding circuits [55, 63].

Robotics use of *P. polycephalum* is possible by exploiting its response to changing conditions within its environment. The migration of the plasmodium is influenced by a wide number of external stimuli including chemoattractants and chemorepellents

[64], light irradiation [40], thermal gradients [65], substrate hardness [54], tactile stimulation [5], geotaxis [66] and magnetotaxis [49]. By careful manipulation of such external stimuli the plasmodium may be considered as a prototype robotic micro-mechanical manipulation system, capable of simple and programmable robotic actions including the manipulation (pushing and pulling) of small scale objects [8, 57], transport and mixing of substances [4]. A *Physarum*-inspired approach to amoeboid robotics was demonstrated by Umedachi et al. [60] in which an external ring of coupled oscillators, each connected to passive and tune-able springs was coupled to a fluid filled inner bladder. The compression of the peripheral springs mimicked the gel contractile phase and the flux of sol within the plasmodium was approximated by the coupled transmission of water pressure to inactive (softer) springs, thus deflecting the peripheral shape of the robot. The resulting movement exhibited flexible behaviour and amoeboid movement.

Physarum has been shown to be a useful model organism in the study of distributed robotics. In this article we explore the problem of collective guidance, i.e. how to move and guide a population of independent mobile entities along a pre-determined path. This task represents a only small subset of general robotics challenges which also include the problem of how to survey and map an unknown environment, and how to plan paths between two or more locations in an environment. Approaches to robotics guidance and planning problems directly inspired by *Physarum* include the simultaneous localisation and mapping problem [30], the generation and manual guidance of collective transport [29], and amoeboid movement [28]. Hybrids of unconventional computing and classical computing substrates are relatively uncommon. In the work of [6] a hybrid path planning system was implemented by using waves from a chemical reaction-diffusion processor to represent start points, end points and obstacles. These waves were used to generate a repulsive field which was used to guide a robot along the arena. The *Physarum* plasmodium itself was used as a guidance mechanism in a biological mechanical hybrid approach where the response of the plasmodium to light irradiation stimuli provided by extended sensors from a classical robot device was then used to provide feedback control to the robot's movement actuators [58]. More recently the problem of generating a path between two points in an arena was tackled with a *Physarum*-inspired morphological adaptation approach [27].

In this article we take the next logical step in these robotics challenges by tackling the problem of dynamically guiding a collective of mobile entities along the path whilst avoiding obstacles. We examine the multi-agent virtual plasmodium and its response to stimuli in Sect. 21.3. Section 21.4 demonstrates simple open-loop examples of guidance. Closed-loop approaches involving a hybrid of classical and unconventional substrates are presented in Sect. 21.5 with assessments of both attractant and repellent guidance methods, novel properties seen during path traversal, and a recovery mechanism for any collectives which may become detached from the target path. We conclude in Sect. 21.6 with a summary of the approach, its main properties and contribution to the field.

21.3   A Virtual Collective Inspired by Slime Mould

To explore guided collective movement we use a simple modification to the particle approximation of *P. polycephalum* introduced in [24] which generated dynamical adaptive transport networks. The approach is based on the concept of simple
component parts which exhibit collective emergent behaviour and has shown to be
successful in reproducing a wide range of behaviour seen in *P. polycephalum*. Presentation of external environmental stimuli (both attractant and repellent) has been
shown to be a critical factor in the evolution of patterning and complexity of computational behaviour within this model [26] (for more information, see [25]). In this
approach the plasmodium is represented by a population of mobile particles with
very simple behaviours, within a 2D diffusive environment. A discrete 2D lattice
stores particle positions and also the concentration of a local generic chemoattractant. The chemoattractant concentration represents the hypothetical flux of sol within
the plasmodium. Free particle movement represents the sol phase of the plasmodium and particle positions represent the fixed gel structure (i.e. global pattern) of
the plasmodium. Particles act independently and iteration of the particle population
is performed randomly to avoid introducing any artifacts from sequential ordering.
Particle behaviour is divided into two distinct stages, the sensory stage and the motor
stage. In the sensory stage, the particles sample their local environment using three
forward biased sensors whose angle from the forward position (the sensor angle
parameter, *SA*, set to 90°), and distance (sensor offset, *SO*, set to 15 pixels) may
be parametrically adjusted (Fig. 21.1). The offset sensors represent the overlapping
filaments within the plasmodium, generating local coupling of sensory inputs and
movement to form networks of particles. The *SO* distance is measured in pixels and
the coupling effect increases as *SO* increases.

During the sensory stage each particle changes its orientation to rotate (via the
parameter rotation angle, *RA*, set to 45°) towards the strongest local source of
chemoattractant. After the sensory stage, each particle executes the motor stage

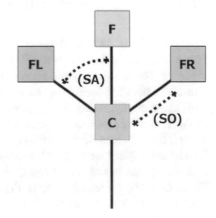

Fig. 21.1 Schematic
illustration of a single agent
particle. Position on the 2D
lattice is indicated by 'C',
three offset sensors 'FL', 'F'
and 'FR' sample the local
chemoattractant
concentration, Sensor Offset
SO parameter sets distance
from 'C' to sensors, Sensor
Angle *SA* indicates angle
between sensors

and attempts to move forwards in its current orientation (an angle from 0–360°) by a single pixel. Each lattice site may only store a single particle and particles deposit chemoattractant (5 arbitrary units) into the lattice only in the event of a successful forwards movement. If the next chosen site is already occupied by another particle the default (non-oscillatory) motor behaviour is to abandon the move, remain in the current position, and select a new random direction.

For experiments with varying population sizes, adaptation of the population size was implemented via tests at regular intervals. The frequency at which the growth and shrinkage of the population was executed determined the turnover rate for the population. The frequency of testing for growth was given by the G_f parameter and the frequency for testing for shrinkage is given by the S_f parameter (both set to 15). Growth of the population was implemented as follows: If there were between G_{min} (0) and G_{max} (10) particles in a local neighbourhood (window size given by G_w, 9 pixels) of a particle, and the particle had moved forward successfully, a new particle was created if there was a space available at a randomly selected empty location in the immediate 3×3 neighbourhood surrounding the particle.

Shrinkage of the population was implemented as follows: If there were between S_{min} (0) and S_{max} (24) particles in a local neighbourhood (window size given by S_w, 5 pixels) of a particle the particle survived, otherwise it was deleted. Deletion of a particle left a vacant space at this location which was filled by nearby particles (due to the emergent cohesion effects), thus causing the blob to shrink slightly.

Diffusion of the collective chemoattractant signal is achieved via a simple 3×3 mean filter kernel with a damping parameter (set to 0.1) to limit the diffusion distance of the chemoattractant. The low level particle interactions result in complex pattern formation. The population spontaneously forms dynamic transport networks showing complex evolution and quasi-physical emergent properties, including closure of network lacunae, apparent surface tension effects and network minimisation. An exploration of the possible patterning parameters was presented in [23].

21.3.1 Generation of Multi-agent Cohesion

Condensation of the multi-agent networks forms uniform sheet-like structures. Figure 21.2 shows the evolution of the stable SA 45°, RA 45° network within a circular arena. The agents coalesce into network trails and the contraction behaviour condenses the network until all interior space is removed and a sheet-like mass remains. This sheet configuration also exhibits unusual properties: the sheet itself forms a minimal surface shape and ripple-like activity can be seen to propagate through the sheet. The sheet also shows relatively stable dissipative 'islands' of greater trail flow. The islands reflect areas where a temporary vacancy of agents exists. The number and size of the islands is related to the sensor offset distance (SO) of the agents. When the SO parameter increases, the number of vacancy islands decreases and the spacing between them increases (Fig. 21.2). This suggests that the vacancy islands

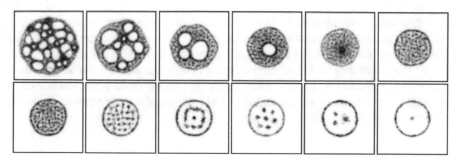

Fig. 21.2 Formation of sheet-like structures and the emergence of dissipative vacancy 'islands'. *Top row (left to right)* Network evolution over time: %*p* = 20 agent trails, *SA* 45°, *RA* 45°. *Bottom row (left to right)* Dissipative vacancy island patterns at *SO*: 9, 13, 19, 23, 28, 38

self-assemble under the influence of the *SO* parameter and represent transient regions of free movement within the mass of particles.

21.3.2 Generation of Oscillatory Dynamics

Although the particle model is able to reproduce many of the network based behaviours seen in the *P. polycephalum* plasmodium such as spontaneous network formation, shuttle streaming and network minimisation, the default motor behaviour does not exhibit oscillatory phenomena and inertial surging movement, as seen in the organism. This is because the default action when a particle is blocked (i.e. when the chosen site is already occupied) is to randomly select a new orientation, resulting in very fluid network evolution. The oscillatory phenomena seen in the plasmodium are thought to be linked to the spontaneous assembly/disassembly of actomyosin and cytoskeletal filament structures within the plasmodium which generate contractile forces on the protoplasm within the plasmodium. The resulting shifts between gel and sol phases prevent (gel phase) and promote (sol phase) cytoplasmic streaming within the plasmodium. To mimic this behaviour in the particle model requires only a simple change to the motor stage. Instead of randomly selecting a new direction if a move forward is blocked, the particle increments separate internal co-ordinates until the nearest cell directly in front of the particle is free. When a cell becomes free, the particle occupies this new cell and deposits chemoattractant into the lattice.

The effect of this behaviour is to remove the fluidity of the default movement of the population. The result is a surging, inertial pattern of movement dependent on population density (the population density specifies the initial amount of free movement within the population). The strength of the momentum effect can be adjusted by a parameter (*pID*) which determines the probability of a particle resetting its internal position coordinates, lower values providing stronger inertial movement. When this simple change in motor behaviour is initiated surging movements are seen and oscillatory domains of chemoattractant flux spontaneously appear within the virtual plasmodium showing characteristic behaviours: temporary blockages of particles (gel phase) collapse into sudden localised movement (solation) and vice versa. The oscillatory domains themselves undergo complex evolution including competition, phase changes and entrainment. We utilise these dynamics below to investigate the possibility of generating useful patterns of regular oscillations which may be coupled to provide motive force.

21.3.3   Generation of Collective Amoeboid Movement

In the absence of stimuli the oscillatory waves propagating through the particle population (a large cohesive 'blob') simply deform the boundary of the blob. This distortion may cause random movement of the blob through the lattice but this movement is not predictable and is subject to changes in direction. To move the blob in any meaningful way it is necessary to distort the blob with regular stimulus inputs in order to shift its position in a chosen direction. These input stimuli are inspired by stimuli which have been shown to influence the movement of slime mould, attractant stimuli and repellent stimuli. Slime mould is known to migrate towards diffusing attractant stimuli, such as nutrient chemoattractant gradients or increasing thermal gradients. Conversely the organism is known to be repelled (moving away from) certain hazardous chemical stimuli and exposure to light irradiation.

Attractant stimuli are represented in the multi-agent model by the projection of spatial values into the diffusive lattice. Since the particles also deposit and sense values from this lattice they will be attracted towards locations which present the same stimulus. An example of the migration of a self-oscillating blob of multi-agent particles towards attractant stimuli can be seen in Fig. 21.3 in which a blob comprising 4000 particles is exposed to the attractant field generated by projection of four discrete attractant stimuli into the diffusive lattice. The initially random distortion of the blob (Fig. 21.3a) is followed by migration of particles at the leading edge of the blob (i.e. closest to the nearest stimulus, Fig. 21.3b). This changes the shape of the blob and causes travelling waves to emerge, moving forwards in the direction of the nearest nutrient, shifting the position of the blob and moving it towards the nutrient. As each nutrient is engulfed it is 'consumed' simply by decrementing the amount projected into the lattice (Fig. 21.3c, d). Removal of the first stimulus point then exposes the blob to the next point, and so on, causing the blob to migrate along the nutrient locations (Fig. 21.3e–h).

Repellent chemical stimuli may be approximated in the model by projecting negative values into the lattice at spatial sites corresponding to repellents, causing the blob to move away from the stimuli. Alternatively, the light irradiation response may be represented by reducing values within the lattice at exposed regions (the shaded region in Fig. 21.4d) whilst reducing the values sampled by agents' sensors within exposed regions. This reduction of stimuli at exposed regions renders them less attractive to individual particles and particles migrate away from these regions.

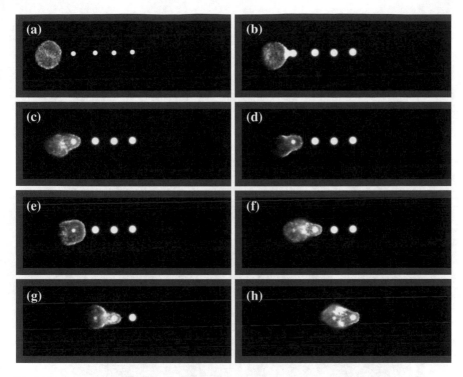

Fig. 21.3 Propagation of self-oscillating blob in the direction of nutrient attractants. **a** Blob comprising of 4000 particles is inoculated on the *left* side of a diffusive lattice containing four nutrient attractants, **b–c** blob shape is deformed as particles move towards the attractants, generating travelling waves within the blob, **d–h** consumption of nutrients exposes the blob to nearby attractants causing the collective to move to the *right* of the arena

This migration is initiated at the interface between exposed and unexposed regions, causing an efflux of particles from exposed regions (Fig. 21.4e). The inherent cohesion of the agent population results in a distortion of the blob shape and a collective movement away from the exposed area 21.4f–h).

In this section we have demonstrated the innate cohesion of the multi-agent population in both oscillatory and non-oscillatory movement modes. The non-oscillatory mode is relatively predictable and attempts to retain a minimal shape profile whilst the oscillatory mode has the advantage of generating self-organised amoeboid movement, at the expense of more unpredictable movement patterns. We have demonstrated how the blob can be influenced by both attractant and repellent stimuli. In the next sections we explore mechanisms to automatically guide the multi-agent population along a pre-selected path.

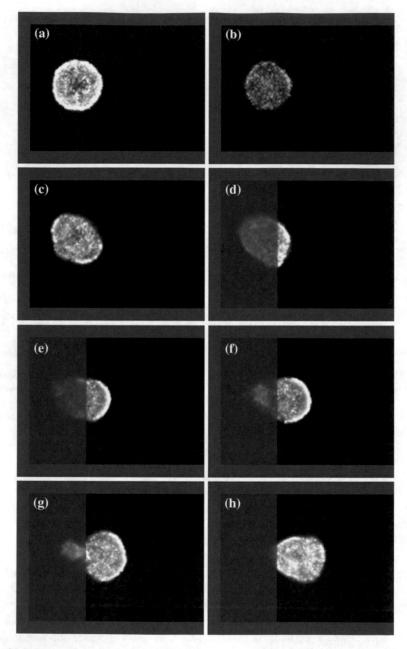

Fig. 21.4 Propagation of self-oscillating blob away from repellent stimulus. **a–c** blob stays in approximately the same position when no stimulus is present, **d** projection of simulated light irradiation to *left* side of the blob (*shaded*) reduces flux and particle sensitivity in the diffusive lattice at these exposed regions, **e–h** particles migrate away from exposed regions towards unexposed regions of the blob, shifting the mass of the blob away from the illuminated region

21.4 Automatically Guided Movement: Open-Loop Methods

In this section we describe relatively simple open-loop approaches to guidance of the virtual plasmodium through an arena. These approaches do not require any input from conventional computing methods and can be implemented completely by projection of spatial patterns to the unconventional computing substrate.

21.4.1 Fuse Method

In this method the entire path is presented as a continuous line of attractant stimuli and a small population is inoculated at the region indicated by the cross-hair (Fig. 21.5a). The population is attracted by the path, growing in size and moving towards it. The stimulus comprising the path is consumed by the population and the population adapts to the consumption of the path by migrating along the path until the end point is reached (Fig. 21.5b–f).

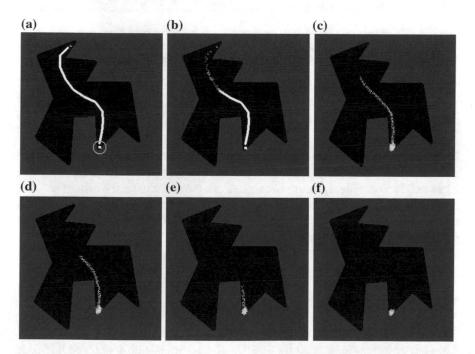

Fig. 21.5 Fuse method. **a** A small population was inoculated at the start point (*cross*) of a path represented by a continuous line of stimuli, **b–c** the population consumes the line, growing and moving along the stimulus path, **d–f** the movement and consumption continues until the blob arrives at the final destination point

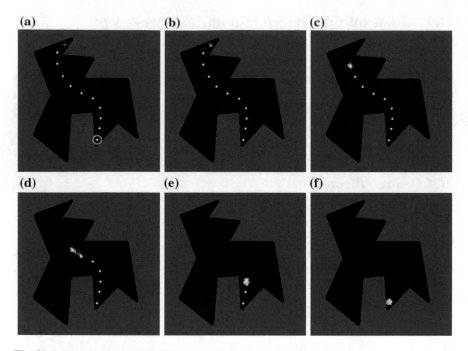

Fig. 21.6 Stepping-stone method. **a** A small population was inoculated at the start point (*cross*) of a discrete array of points, **b–c** the blob consumes each point and is then attracted by the next stimulus point along the path, **d–f** the movement and consumption continues until the blob arrives at the final destination point

21.4.2 Stepping-Stone Method

In the stepping-stone method the path is presented as a sequence of discrete stimulus points (Fig. 21.6a). A small population is again inoculated at the cross-hair location. The small blob of virtual plasmodium consumes the stimulus on the current point on the path and is then attracted to the next point in the path (Fig. 21.6c, d). The blob traverses the path in a series of hops, moving from point to point, until the final goal point is reached (Fig. 21.6f).

21.4.3 Elastic Method

This method utilises the morphological adaptation inherent within the population. The path is again represented by a series of discrete stimuli and a small blob is inoculated at the cross-hair location (Fig. 21.7a). Unlike the fuse and stepping-stone methods, each stimulus point is not immediately consumed, instead the population grows in size to span all points on the path from beginning to end (Fig. 21.7a–c).

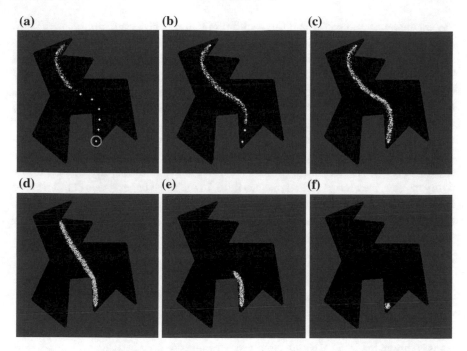

Fig. 21.7 Elastic method. **a–c** A small population was inoculated at the start point (*cross*) and, attracted by the stimulus points of the path, grew to extend across all points, **d–f** all stimulus points except the end point were then removed and the virtual plasmodium adapted to the change of stimulus profile by retracting towards the end point

After reaching the final path point all previous path points are deleted and the blob, which is now only 'anchored' to the attractant of the final point, adapts its shape, shrinking and moving its body plan to the final point, completing the movement.

21.5   Automatically Guided Movement: Closed-Loop Feedback Methods

The open-loop methods, although effective in these examples, do not account for the possibility that the virtual plasmodium may become detached from the path. Once each particular method is set in action, there is no guarantee of the success, or failure, of the motion along the path. We require feedback about the position of the model population within its environment, and mechanisms to dynamically influence the position within the arena to reach the goal. This requires a more complex closed-loop approach.

To provide a real challenge for this approach we utilise the self-oscillatory behaviour of the model plasmodium. This behaviour was introduced in [56] to reproduce the spontaneous and self-organised oscillation patterns observed within small samples of *Physarum* plasmodia [52]. In the model plasmodium these oscillations emerge from interruptions of individual particle movement, as described in Sects. 21.3.2 and 21.3.3. We adjust the momentum of the self-oscillatory amoeboid movement time using the *pID* parameter. Higher values of *pID* (for example 0.05) result in less persistence of direction of individual particles, whereas lower values (for example 0.001) result in much stronger persistence of direction. The accumulation of interruptions in movement of individual particles results in travelling waves of flux forming within the mass of particles as particles occupy vacant spaces within the collective. It was shown in [28] that these travelling waves could shift the mass of particles, effectively moving the blob of virtual plasmodium. In the same paper it was demonstrated how the self-oscillatory dynamics could be influenced by the manual placement of attractant stimuli and simulated light irradiation stimuli, causing the blobs to move towards attractants and away from light hazards.

21.5.1 Momentum Parameter: Effect on Blob Migration

The effect of initiating self-oscillatory behaviour and its *pID* parameter on blob positions is shown in Fig. 21.8 which details experiments with decreasing *pID* parameters and the effect this has on the random migration of an unstimulated self-oscillating blob from its inoculation position (circle) at the centre of the experimental arena. The position of the blob is recorded as the centroid of all the particles comprising the blob and its path is indicated by the line extending from the central point (Fig. 21.8a, c and e).

For the control condition, with no self-oscillatory behaviour initiated, there was very little displacement of the blob from the inoculation site (maximum of 1.41 pixels displacement over 5000 scheduler steps). Any displacement of the blob was caused by the stochastic influences on individual particles within their sensory method. When oscillatory behaviour was initiated, the build-up of momentary interruptions of particle movement caused the position of the blob to be displaced significantly by the emergent travelling waves within the blob. At *pID* 0.1 the distance from the inoculation site increased gradually (Fig. 21.8b). Decreasing *pID* resulted in greater displacement of the blob (Fig. 21.8d and f). Lower *pID* parameter not only resulted in greater movement from the inoculation site but also more persistence in the direction of movement. This was often followed by sudden random changes in direction (for example, Fig. 21.8e). These changes in direction are the cause of the apparent decrease in error from the original position (Fig. 21.8d and f) at decreased *pID*.

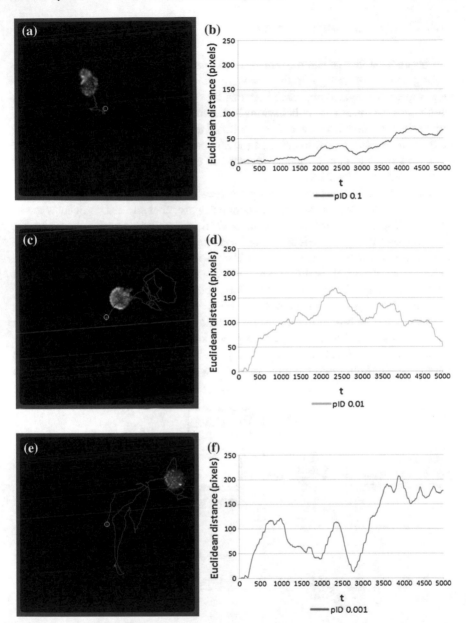

Fig. 21.8 Comparison of spontaneous migration error from inoculation position at different *pID* settings, blob position and error tracked over 5000 steps. **a** record of blob at *pID* 0.1, **b** plot of migration error at *pID* 0.1, **c–d** *pID* 0.01, **e–f** *pID* 0.001

21.5.2 Hybrid Control System

The unpredictability of the movement in the self-oscillating blobs renders it challenging to control and guide their movement. A method for automatic guidance must represent a hybrid approach between the unconventional computing methods which generate the emergent behaviour in the virtual plasmodium (the generation of self-oscillatory travelling waves and amoeboid movement), and classical computing methods to detect the position of the blob and provide the feedback stimuli to guide the blob along the chosen path. A schematic overview of the closed-loop hybrid system is given in Fig. 21.9.

Note that Fig. 21.9 is partitioned by a vertical dashed line. This line indicates the separation of conventional and unconventional approaches and also indicates regions where both approaches interact. The unconventional part of the system generates the emergent oscillatory behaviour of the blob from local and self-organised particle interactions. Information about the blob's collective state is then extracted by the conventional (classical) part of the method which calculates the centroid (centre of mass) of the blob. The position of the blob is compared at every 50 scheduler steps to the points comprising the path in the arena. When the blob is closer to the next point along the path than to the current stimulus location, the *next* point along the path is then selected to provide the new location stimulus for the blob. The target stimulus is then projected into the spatially implemented unconventional part of the method. This stimulus acts to guide the blob towards this new location. Two possible

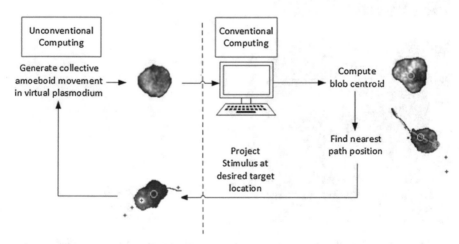

Fig. 21.9 Schematic overview of closed-loop guidance method indicating interface between unconventional computing and conventional computing approaches (*dashed line*). *Left* side of *dashed line* represents the contribution of the unconventional approach, characterised by bottom-up generation of self-organised emergent phenomena, *right* side of *line* represents the contribution of classical computing approaches in generating the control method. Both sides contribute inputs to each other

stimulus types can be used, both of which are seen in the *Physarum* plasmodium and are described in the following sections.

21.5.3   Automatic Guidance with Attractant Stimuli

The first stimulus method used attractant stimuli. In the wild, slime mould is attracted to certain stimuli such as nutrients or localised warm areas. In the model plasmodium attractant stimuli are represented by projecting localised stimuli into the diffusive lattice. These stimuli, when projected within sensory range of the blob, attract the particles comprising the blob. Particles at the outer periphery of the blob move towards the stimulus and the cohesion of the blob (caused by the indirect coupling of the particles' offset sensors) generates travelling waves inside the blob which move directly towards the stimulus. As demonstrated by manual placement in [28], this can be used to guide (or 'pull') the blob *towards* the chosen direction. In the closed-loop method described in this article we can replace the manual placement of stimuli with automated transient placement of point stimuli to guide the blob along the chosen path.

Figure 21.10 shows the results of automated closed-loop guidance of the self-oscillatory blob along a pre-defined path by the attractant method. The path starts at the large cross marker and ends at the circle marker, and individual guidance points on the path are denoted by small crosses (Fig. 21.10a). The path is composed of multiple links between start and end points in an arena populated by solid obstacles (grey shapes) which the agent particles cannot cross. The particle population, comprising 2000 particles, was inoculated at the initial cross marker position and for the first 1000 scheduler steps the blob was allowed to form and stabilise. After 1000 steps the automated guidance mechanism described in the previous sub-section was initiated. The short straight lines connecting the small crosses indicates the pre-defined path (green, online) and the path taken by the oscillatory blob is indicated by the blue (online) markers. The example results include four different momentum (*pID*) parameter settings.

Although the blob follows the path in all examples, at low *pID* (i.e. high momentum) settings there is considerable deviation from the desired path, particularly when the path changes direction (see, for example, Fig. 21.10a and the supplementary video recordings at http://uncomp.uwe.ac.uk/jeff/automatedguidance.htm).

This overshooting of the desired path is caused by the blob position being influenced by the strong oscillatory waves within the blob. At low *pID* values this momentum is particularly strong, causing the blob to overshoot the corners after the momentum of oscillatory waves has been established during straighter sections of the course. Under the strongest momentum condition the blob 'crashes' into the circular obstacle at the lower-right of the arena and the blob has to re-form before its progress can continue. An indication of the strength of the momentum can be seen at the end of each course in Fig. 21.10 where the blob continues to receive attractant input from the final position on the path. Although the position of this stimulus is static, the

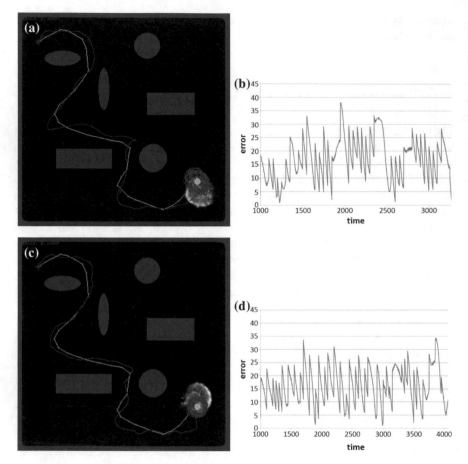

Fig. 21.10 Automated guidance of amorphous amoeboid robot via attractants. **a, c, e, g** Image showing trajectory of blob (*blue online*) as it is guided along path (*green online*) from start (*cross*) to finish (*circle*) with *pID* parameter values 0.001, 0.01, 0.05, 0.1 respectively, **b, d, f, h** plot showing error of blob position (in pixels) compared to path over time

path of the blob (blue, online) shows that the blob continues to traverse around the periphery of the final position. At low *pID* (high momentum) values, the radius of this circular movement is much larger than at high *pID* (low momentum) values.

As the blob traverses the points on the path, there is a repetitive sequence of error minimisation which occurs, as shown in Fig. 21.11 (which shows an enlarged portion of the migration plot of Fig. 21.10f between 2000–3000 steps). This is indicated by the 'sawtooth' profile of the plot. As the blob moves forward (attracted by the stimulus point presented at the next path node) the distance between the current centroid of the blob and the target node is minimised (the diminishing diagonal lines of the plot). When this distance is less than the distance between the current node, the new node is selected. The selection of the next node changes the stimulus point location and

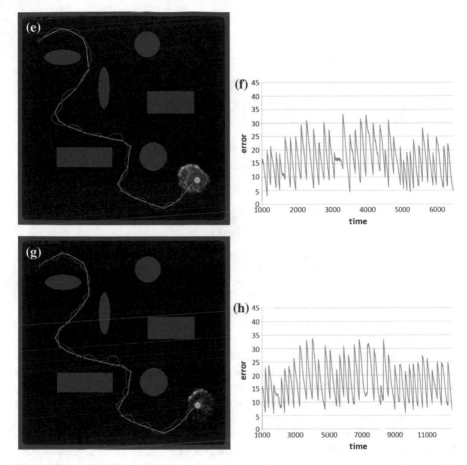

Fig. 21.10 (continued)

also causes a sudden jump in migration error (the vertical lines in the plot). This pattern of minimisation and new target selection occurs until the end node of the path is reached, at which point the blob will circle the final node on the path.

21.5.4 Automatic Guidance with Repellent Stimuli

As an alternative to pulling the blob towards the stimulus, it is also possible to 'push' the blob. This can be achieved by mimicking the response of slime mould to hazardous stimuli, for example exposure to light irradiation. In the face of such stimuli slime mould withdraws parts of its body plan away from exposed regions [40] and can thus be guided away from simple obstacles comprised of light-exposed areas [3]. In the model plasmodium we can reproduce the effect of exposure to light by altering

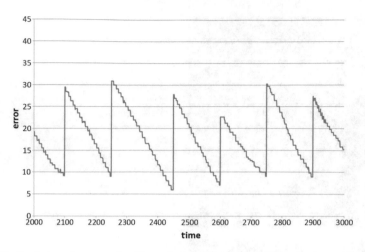

Fig. 21.11 Evolution of automated guidance during path traversal. Chart showing enlarged portion of Fig. 21.10f at time 2000–3000 steps. Sawtooth profile plot shows changes in migration error as the blob is guided from node to node along the path, minimising the distance error at the current node (*downward diagonal*), before the next node is selected, generating a new error distance (*vertical line*)

the sensitivity of particles exposed to illuminated regions of the lattice. This reduces flux within that region of the blob and also attraction to attractant stimuli located in these exposed regions. Due to the cohesion of the blob, the travelling waves moving within the blob are stronger in unilluminated regions and this propels the blob from exposed regions.

We can implement automated guidance by repellent stimuli by having the stimulus point represent an *absence* of illumination, for example a square masked region. *Outside* this region all other areas are temporarily exposed to simulated light exposure. This tends to maintain the blob within the confines of the masked region and also move peripheral parts of the blob that are outside of the protective masked region back inside the mask.

We tested the repellent method on the same obstacle arena as used in the attractant stimulus condition. Figure 21.12 shows examples of blob guidance through this arena at different *pID* values. As indicated by the chart plots showing distance from the chosen path, the general 'sawtooth' pattern of movement along the path is the same as in the attractant guidance method. However, the trajectory of the blob under light irradiation guidance shows much closer adherence to the original path and there is significantly less 'overshoot' than in the attractant guidance method when sudden changes in path direction occur. Across multiple runs of both attractant and repellent conditions the time taken to traverse the path was also shorter in the light-irradiation condition at all *pID* values (Fig. 21.13a). Furthermore, the mean error from the desired path was also lower for the light-irradiation condition, compared to the attractant guided method (Fig. 21.13b).

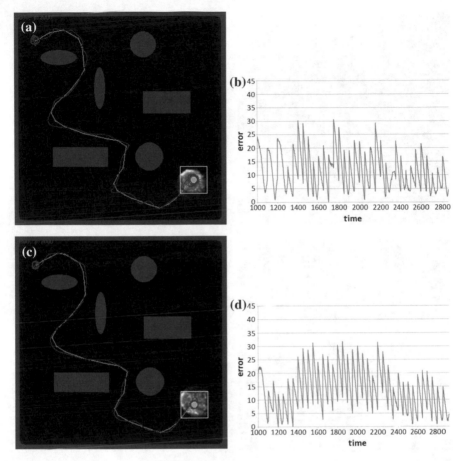

Fig. 21.12 Automated guidance of amorphous amoeboid robot via repellent stimulus of simulated light irradiation. **a, c, e, g** Image showing trajectory of blob (*blue online*) as it is guided along path (*green online*) from start (*cross*) to finish (*circle*) with *pID* parameter values 0.001, 0.01, 0.05, 0.1 respectively. Masked area is indicated by the *square* region surrounding the blob, **b, d, f, h** plot showing error of blob position (in pixels) compared to path over time

Why does the light irradiation guidance method track the path more accurately than the attractant method? The square masked region surrounding the blob (for example, Fig. 21.12a) illuminates all regions outside the mask (i.e. outside of the main blob region), providing simultaneous stimuli at different parts of the blob, compared to the single point stimulus in the attractant condition. Any particles in the illuminated region are subject to the reduction in flux and thus try to return to the unexposed region within the square. This suppression of flux outside the mask square has the effect of damping travelling waves outside the square, causing less momentum to build up, and more accurate traversal of the path.

It should be noted that although the high-momentum travelling waves in the attractant condition were responsible for the increased error from the desired path,

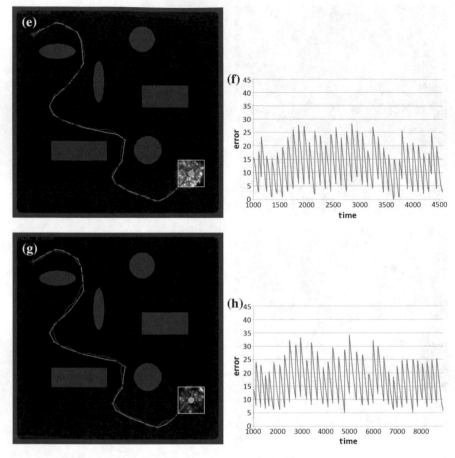

Fig. 21.12 (continued)

the oscillatory travelling waves are of critical importance for the blob movement. Indeed in the attractant stimulus method, the migration of the blob along the entire course could not be completed without oscillatory movement. In the light irradiation stimulus condition, the light mask was sufficient to move the non-oscillatory collective along the path, but the penalty was a greatly increased time of traversal — over 80,000 scheduler steps — compared to a range of 1900–8000 steps under oscillatory conditions.

21.5.5   Novel Properties of Guided Amoeboid Movement

In addition to the tracking abilities of the hybrid unconventional/conventional computing guidance methods, the amorphous and adaptive properties of the collective

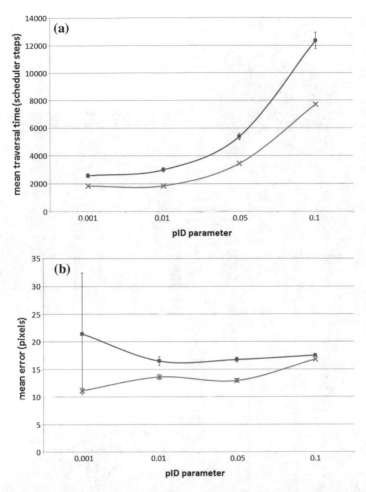

Fig. 21.13 Comparison of attractant and repellent guidance methods in terms of time and guidance errors. **a** comparison of path traversal time for different *pID* values for attractant (*squares*) and repellent (*crosses*) stimuli, **b** mean guidance error from path at different *pID* values for attractant (*squares*) and repellent (*crosses*) stimuli (mean of ten runs per *pID* value, standard deviation indicated). Guidance by repellent stimuli is faster and more accurate

result in some interesting properties during its movement. Figure 21.14 shows the guidance of the blob along a vertical arena (in this example by repellent light irradiation stimuli). The arena is composed of a narrow channel, some horizontal blocks and finally a very narrow grating, before the destination site (Fig. 21.14a). As the blob passes through the narrow channel, the blob elongates, adapting its shape automatically in order to fit through the narrow channel (Fig. 21.14b) before restoring its approximately circular shape once the channel has been crossed.

In the case of the grating at the bottom of the arena, the space between the grating obstacles is so narrow that the blob shape deforms dramatically in regions outside the

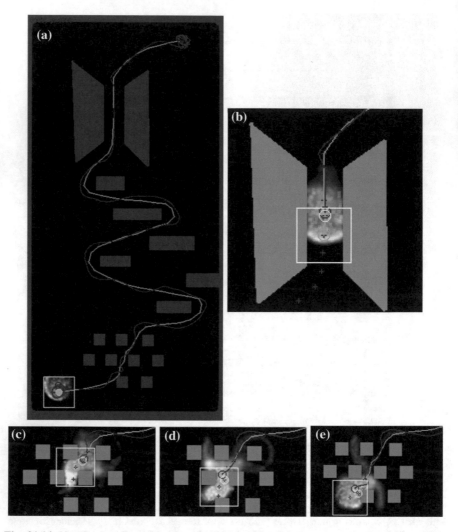

Fig. 21.14 Novel properties of the amoeboid blob as it navigates a complex vertical arena. **a** overview of traversed arena showing obstacles (*grey*), path (*green, online*), and blob trajectory (*blue, online*), **b** blob elongates as it passes through a narrow tunnel, **c–e** blob is distorted as it is forced through a narrow grating by the stimulus mask before re-forming its shape when the obstacle is passed

mask, forming writhing pseudopodium-like tendrils (Fig. 21.14c–e). Again, once the grating has been crossed the blob reforms its shape as it moves to the goal site. Despite this significant distortion of blob shape the path taken by the blob is fairly close to the target path (Fig. 21.14e). These properties are a function of the unconventional computing part of the system, in that they arise as an emergent property of the low-level particle interactions.

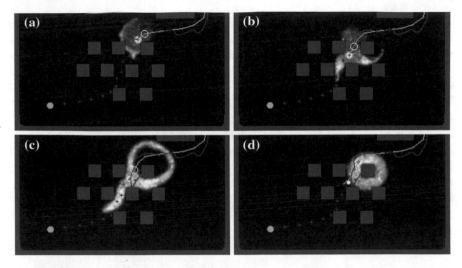

Fig. 21.15 Guided blob becomes stuck on grating using attractant method. **a** blob enters grating area, **b** distortion of blob pattern occurs, **c** blob becomes entwined on grating obstacle, **d** blob minimises around grating and continues to cycle in this position

The distortion and re-formation of the blob shape at the narrow grating does not occur reliably with the attractant based guidance method, however. Figure 21.15 shows a blob guided by a single attractant source entering the grating region (Fig. 21.15a) where its body plan is distorted on contact with the obstacles (Fig. 21.15b). The blob becomes entwined on a single obstacle in the grating (Fig. 21.15c) and minimises its shape to wrap around the obstacle (Fig. 21.15d). The blob remains in this position indefinitely, cycling around this obstacle. Corruption of the X and Y stimulus values with Gaussian noise (to try to present multiple stimulus sites) does not detach the blob from the obstacle. This behaviour again demonstrates the effectiveness of guidance by illumination masked regions compared to the attractant guidance method. Why does the blob not become stuck at this obstacle when guided by the repellent mask? Again, this is because the illumination mask presents multiple stimulus points to the blob (at the interface of the square mask edges which contact the blob), whereas the attractant guidance method only presents a single guidance stimulus.

21.5.6   Emergency Recovery Mode for Lost Collectives

In over 80 experiments with the guidance mechanisms we only encountered one instance (attractant stimulus condition, with low *pID* 0.001) where the blob migrated far from the desired path and was not influenced by the presented attractant stimulus (incidentally, this was the cause for the large deviation bars in the first data point of the attractant series in Fig. 21.13b). In this instance the 'lost' blob spontaneously moved (via random migration) back near to the path where it once again was influenced by the presented stimuli. However, this led us to devise a 'recovery mode' to cater for instances when the amoeboid blobs may lose sight of the path. This mechanism was implemented in the following way (Fig. 21.16). At each scheduled comparison of blob centroid position and the closest nearest path position, the Euclidean distance (in pixels) was compared to a threshold value θ. If θ (set to 100 pixels) was exceeded a binary *lost* flag was set and an attractant stimulus (or light illumination mask for the repellent condition) was projected at random locations along an imaginary line from the blob centroid to the nearest path point (Fig. 21.16b). These additional stimuli attract the blob and continue to be presented until the distance from the blob to the path point was $< \theta$, at which time the *lost* flag was reset and normal guidance resumed. This mechanism is sufficient to guide the errant blob back onto the correct path.

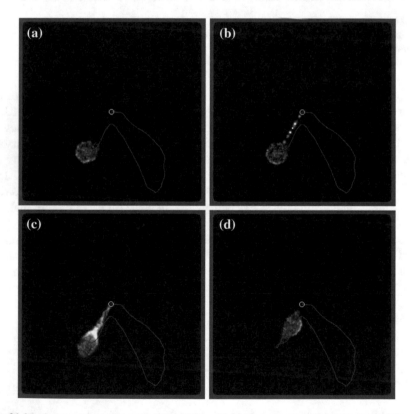

Fig. 21.16 Automated recovery of 'lost' blob. **a** blob migrates away from centre position and exceeds θ, triggering recovery mode, **b** a stream of attractant stimuli is presented between the blob and the target point, **c** the blob is attracted to the stimuli and starts to migrate back to the target, **d** recovery stimuli are halted when the distance $< \theta$

21.6   Discussion

Control and guidance of collective soft-robotics devices is a very challenging problem for classical computing when the robotic devices themselves harness emergent behaviours arising from the local component interactions of the collective itself. In this article we have examined the problem of generating and controlling collective amoeboid movement from an unconventional computing perspective using a multi-agent model of slime mould *Physarum polycephalum*. Collective movement was generated in a morphologically adaptive 'blob' comprising a population of simple particles on a diffusive lattice. Taking inspiration from slime mould, the position of this blob could be altered by the spatial placement of attractant and repellent stimuli. Simple open-loop mechanisms were demonstrated using attractant stimuli which enabled automatic, but uncontrolled, movement along a pre-defined path. In order to automatically guide the movement of the multi-agent collective, however, a hybrid approach utilising features of unconventional and classical computation was required.

The hybrid approach utilised the self-organised generation of blob cohesion and its movement by oscillatory travelling waves (the unconventional computing substrate) in a blob of fixed population size. This was combined with a classically implemented closed-loop mechanism to ascertain the current position of the blob in relation to a pre-defined path. By comparing the current blob position with the closest point on the path, a stimulus (the next available path point) was then presented to the unconventional computing substrate, causing the blob to migrate along the path. Of the two stimulus types investigated (attractant and repellent) to guide the blob, the repellent stimulus (masking the blob from simulated light irradiation) resulted in a faster path traversal with fewer errors (in terms of distance from the pre-defined path) and allowed the blob to pass automatically through very narrow gratings. Passage through a narrow grating could not be achieved using the attractant stimulus condition, due to the lack of simultaneous stimuli to the blob when compared to the repellent mask method. The momentum of the blob could be controlled by adjusting a parameter of the model. Stronger momentum resulted in faster path traversal in both stimulus types, but resulted in a characteristic overshooting of path corners in the attractant stimulus condition.

This hybrid approach successfully combines classical computing (tracking and stimulus location) with unconventional computing (generation of self-organised travelling waves to generate amoeboid movement). The unconventional computing part of the system accounts for novel properties of the amoeboid robot such as cohesion, spontaneous oscillatory movement and automatic deformation of shape (and subsequent re-formation) in the presence of obstacles. The result is a parsimonious combination of classical computing and its benefits (for example, rapid and efficient arithmetic calculations for tracking and guidance), combined with the desirable features provided by unconventional computing (self-organised movement, resilience to deformation) that are a natural fit for unconventional computing substrates and which would be difficult to implement in a classical system. The result of this hybrid approach is a system which combines the best features of both approaches. We hope that this work will provide a useful contribution towards future implementations of soft-bodied robotic systems which utilise hybrid unconventional and classical computing methods.

Acknowledgments This research was supported by the EU research project "Physarum Chip: Growing Computers from Slime Mould" (FP7 ICT Ref 316366).

References

1. Adamatzky, A.: Physarum machines: encapsulating reaction-diffusion to compute spanning tree. Naturwissenschaften **94**(12), 975–980 (2007)
2. Adamatzky, A.: Developing proximity graphs by Physarum polycephalum: does the plasmodium follow the toussaint hierarchy. Parallel Processing Letters **19**, 105–127 (2008)
3. Adamatzky, A.: Steering Todium with Light: Dynamical Programming of *Physarum* Machine (2009). arXiv:0908.0850
4. Adamatzky, A.: Manipulating substances with Physarum polycephalum. Mater. Sci. Eng. C **38**(8), 1211–1220 (2010)
5. Adamatzky, A.: Slime mould tactile sensor. Sens. actuators B: Chem. **188**, 38–44 (2013)
6. Adamatzky, A., de Lacy Costello, B., Melhuish, C., Ratcliffe, N.: Experimental reaction-diffusion chemical processors for robot path planning. J. Intell. Robot. Syst. **37**(3), 233–249 (2003)
7. Adamatzky, A., de Lacy, Costello, B., Shirakawa, T.: Universal computation with limited resources: Belousov-Zhabotinsky and Physarum computers. Int. J. Bifurc. Chaos **18**(8), 2373–2389 (2008)
8. Adamatzky, A., Jones, J.: Towards Physarum robots: computing and manipulating on water surface. J. Bionic Eng. **5**(4), 348–357 (2008)
9. Brugués, A., Anon, E., Conte, V., Veldhuis, J.H., Gupta, M., Colombelli, J., Muñoz, J.J., Brodland, G.W., Ladoux, B., Trepat, X.: Forces driving epithelial wound healing. Nat. Phys. (2014)
10. Buhl, J., Sumpter, D.J.T., Couzin, I.D., Hale, J.J., Despland, E., Miller, E.R., Simpson, S.J.: From disorder to order in marching locusts. Science **312**(5778), 1402 (2006)
11. Carlile, M.J.: Nutrition and chemotaxis in the myxomycete Physarum polycephalum: the effect of carbohydrates on the plasmodium. J. Gen. Microbiol. **63**(2), 221–226 (1970)
12. Durham, A.C.H., Ridgway, E.B.: Control of chemotaxis in Physarum polycephalum. J. Cell Biol. **69**, 218–223 (1976)
13. Friedl, P., Locker, J., Sahai, E., Segall, J.E.: Classifying collective cancer cell invasion. Nat. Cell Biol. **14**(8), 777–783 (2012)
14. Gale, E., Adamatzky, A., de Lacy Costello, B.: Are slime moulds living memristors? Technical report (2013)
15. Garnier, S., Gautrais, J., Theraulaz, G.: The biological principles of swarm intelligence. Swarm Intell. **1**(1), 3–31 (2007)
16. Gholami, A., Steinbock, O., Zykov, V., Bodenschatz, E.: Flow-driven waves during pattern formation of Dictyostelium discoideum. Bull. Am. Phys. Soc. **60** (2015)
17. Gov, N.S.: Collective cell migration. Cell Matrix Mech. 219 (2014)
18. Gueron, S., Levin, S.A.: Self-organization of front patterns in large wildebeest herds. J. Theor. Biol. **165**(4), 541–552 (1993)
19. Helbing, D.: Traffic and related self-driven many-particle systems. Rev. Mod. Phys. **73**(4), 1067 (2001)

20. Helbing, D., Johansson, A.: Eidgenössische Technische Hochschule. Pedestrian, crowd and evacuation dynamics. Swiss Federal Institute of Technology (2009)
21. Helbing, D., Molnar, P., Farkas, I.J., Bolay, K.: Self-organizing pedestrian movement. Environ. Plan. B **28**(3), 361–384 (2001)
22. Herbert-Read, J.E., Perna, A., Mann, R.P., Schaerf, T.M., Sumpter, D.J.T., Ward, A.J.W.: Inferring the rules of interaction of shoaling fish. Proc. Natl. Acad. Sci. **108**(46), 18726–18731 (2011)
23. Jones, J.: Characteristics of pattern formation and evolution in approximations of Physarum transport networks. Artif. Life **16**(2), 127–153 (2010)
24. Jones, J.: The emergence and dynamical evolution of complex transport networks from simple low-level behaviours. Int. J. Unconv. Comput. **6**, 125–144 (2010)
25. Jones, J.: From Pattern Formation to Material Computation: Multi-agent Modelling of Physarum Polycephalum. Springer, Heidelberg (2015)
26. Jones, J.: Mechanisms inducing parallel computation in a model of Physarum polycephalum transport networks. Parallel Process. Lett. **25**(01), 1540004 (2015)
27. Jones, J.: A morphological adaptation approach to path planning inspired by slime mould. Int. J. Gen. Syst. **44**(3), 279–291 (2015)
28. Jones, J., Adamatzky, A.: Emergence of self-organized amoeboid movement in a multi-agent approximation of Physarum polycephalum. Bioinspiration Biomim. **7**(1), 016009 (2012)
29. Jones, J., Tsuda, S., Adamatzky, A.: Towards *Physarum* robots. In: Bio-Inspired Self-Organizing Robotic Systems, pp. 215–251 (2011)
30. Kalogeiton, V.S., Papadopoulous, D.P., Sirakoulis, G.Ch., Vazquez-Otero, A., Faigl, J., Duro, N., Dormido, R., Henderson, T.C., Joshi, A., Rashkeev, K. et al.: Hey *Physarum*! can you perform slam? Int. J. Unconv. Comput. **10**(4) (2014)
31. Kessler, M.A., Werner, B.T.: Self-organization of sorted patterned ground. Science **299**(5605), 380–383 (2003)
32. Kishimoto, U.: Rhythmicity in the protoplasmic streaming of a slime mould, Physarum polycephalum. J. Gen. Physiol. **41**(6), 1223–1244 (1958)
33. Kocurek, G., Ewing, R.C.: Aeolian dune field self-organization-implications for the formation of simple versus complex dune-field patterns. Geomorphology **72**(1), 94–105 (2005)
34. Lee, K.J., Cox, E.C., Goldstein, R.E.: Competing patterns of signaling activity in Dictyostelium discoideum. Phys. Rev. Lett. **76**(7), 1174 (1996)
35. Levin, M.: Morphogenetic fields in embryogenesis, regeneration, and cancer: non-local control of complex patterning. Biosystems **109**(3), 243–261 (2012)
36. Matsushita, M., Wakita, J., Itoh, H., Watanabe, K., Arai, T., Matsuyama, T., Sakaguchi, H., Mimura, M.: Formation of colony patterns by a bacterial cell population. Physica A: Stat. Mech. Appl. **274**(1), 190–199 (1999)
37. Nagatani, T.: The physics of traffic jams. Rep. Progr. Phys. **65**(9), 1331 (2002)
38. Nakagaki, T., Ueda, T.: Phase switching of rhythmic contraction in relation to regulation of amoeboid behavior by the plasmodium of Physarum polycephalum. J. Theor. Biol. **179**, 261–267 (1996)
39. Nakagaki, T., Yamada, H., Toth, A.: Intelligence: Maze-solving by an amoeboid organism. Nature **407**, 470 (2000)
40. Nakagaki, T., Yamada, H., Ueda, T.: Modulation of cellular rhythm and photoavoidance by oscillatory irradiation in the Physarum plasmodium. Biophys. Chem. **82**(1), 23–28 (1999)
41. Nakagaki, T., Yamadaa, H., Ueda, T.: Interaction between cell shape and contraction pattern in the physarum plasmodium. Biophys. Chem. **84**, 195–204 (2000)
42. Parrish, J.K., Turchin, P.: Individual decisions, traffic rules, and emergent pattern in schooling fish. In: Animal Groups in Three Dimensions, pp. 126–142 (1997)
43. Rauprich, O., Matsushita, M., Weijer, C.J., Siegert, F., Esipov, S.E., Shapiro, J.A.: Periodic phenomena in Proteus mirabilis swarm colony development. J. Bacteriol. **178**(22), 6525–6538 (1996)
44. Reynolds, C.W.: Flocks, herds and schools: A distributed behavioral model. In: ACM SIGGRAPH Computer Graphics, vol. 21, pp. 25–34. ACM, New York (1987)

45. Saigusa, T., Tero, A., Nakagaki, T., Kuramoto, Y.: Amoebae anticipate periodic events. Phys. Rev. Lett. **100**(1) (2008)
46. Serini, G., Ambrosi, D., Giraudo, E., Gamba, A., Preziosi, L., Bussolino, F.: Modeling the early stages of vascular network assembly. EMBO J. **22**(8), 1771–1779 (2003)
47. Shirakawa, T., Adamatzky, A., Gunji, Y.-P., Miyake, Y.: On simultaneous construction of Voronoi diagram and Delaunay triangulation by Physarum polycephalum. Int. J. Bifurc. Chaos **19**(9), 3109–3117 (2009)
48. Shirakawa, T., Gunji, Y.-P.: Computation of Voronoi diagram and collision-free path using the Plasmodium of Physarum polycephalum. Int. J. Unconv. Comput. **6**(2), 79–88 (2010)
49. Shirakawa, T., Konagano, R., Inoue, K.: Novel taxis of the *Physarum* plasmodium and a taxis-based simulation of *Physarum* swarm. In: 2012 Joint 6th International Conference on Soft Computing and Intelligent Systems (SCIS) and 13th International Symposium on Advanced Intelligent Systems (ISIS), pp. 296–300. IEEE, New York (2012)
50. Stephenson, S.L., Stempen, H., Hall, I.: Myxomycetes: A Handbook of Slime Molds. Timber Press Portland, Oregon (1994)
51. Sumpter, D.J.T.: Collective Animal Behavior. Princeton University Press, Princeton (2010)
52. Takagi, S., Ueda, T.: Emergence and transitions of dynamic patterns of thickness oscillation of the plasmodium of the true slime mold Physarum polycephalum. Physica D **237**, 420–427 (2008)
53. Takamatsu, A., Fujii, T.: Time delay effect in a living coupled oscillator system with the plasmodium of Physarum polycephalum. Phys. Rev. Lett. **85**, 2026–2029 (2000)
54. Takamatsu, A., Takaba, E., Takizawa, G.: Environment-dependent morphology in plasmodium of true slime mold Physarum polycephalum and a network growth model. J. Theor. Biol. **256**(1), 29–44 (2009)
55. Tsuda, S., Aono, M., Gunji, Y.-P.: Robust and emergent Physarum logical-computing. BioSystems **73**, 45–55 (2004)
56. Tsuda, S., Jones, J.: The emergence of synchronization behavior in Physarum polycephalum and its particle approximation. Biosystems **103**, 331–341 (2010)
57. Tsuda, S., Jones, J., Adamatzky, A.: Towards Physarum engines. Appl. Bionics Biomech. **9**(3), 221–240 (2012)
58. Tsuda, S., Zauner, K.-P., Gunji, Y.-P.: Robot control with biological cells. BioSystems **87**, 215–223 (2007)
59. Ueda, T., Terayama, K., Kurihara, K., Kobatake, Y.: Threshold phenomena in chemoreception and taxis in slime mold Physarum polycephalum. J. Gen. Physiol. **65**(2), 223–234 (1975). February
60. Umedachi, T., Kitamura, T., Takeda, K., Nakagaki, T., Kobayashi, R., Ishiguro, A.: A modular robot driven by protoplasmic streaming. Distrib. Autonom. Robot. Syst. **8**, 193–202 (2009)
61. Vicsek, T., Zafeiris, A.: Collective motion. Phys. Rep. **517**, 71–140 (2012)
62. Werner, B.T.: Eolian dunes: computer simulations and attractor interpretation. Geology **23**(12), 1107–1110 (1995)
63. Whiting, J.G.H., de Lacy Costello, B.P.J., Adamatzky, A.: Slime mould logic gates based on frequency changes of electrical potential oscillation. Biosystems **124**, 21–25 (2014)
64. Whiting, J.G.H., de Lacy Costello, B.P.J., Adamatzky, A.: Towards slime mould chemical sensor: Mapping chemical inputs onto electrical potential dynamics of Physarum Polycephalum. Sens. Actuators B: Chem. **191**, 844–853 (2014)
65. Wolf, R., Niemuth, J., Sauer, H.: Thermotaxis and protoplasmic oscillations in Physarum plasmodia analysed in a novel device generating stable linear temperature gradients. Protoplasma **197**(1–2), 121–131 (1997)
66. Wolke, A., Niemeyer, F., Achenbach, F.: Geotactic behavior of the acellular myxomycete Physarum polycephalum. Cell Biol. Int. Rep. **11**(7), 525–528 (1987)
67. Zheng, X., Zhong, T., Liu, M.: Modeling crowd evacuation of a building based on seven methodological approaches. Build. Environ. **44**(3), 437–445 (2009)

Chapter 22
Cellular Automata Ants

Nikolaos P. Bitsakidis, Nikolaos I. Dourvas, Savvas A. Chatzichristofis and Georgios Ch. Sirakoulis

Abstract During the last decades much attention was given to bio-inspired techniques able to successfully handle really complex algorithmic problems. As such Ant Colony Optimization (ACO) algorithms have been introduced as a metaheuristic optimization technique arriving from the swarm intelligence methods family and applied to several computational and combinatorial optimization problems. However, long before ACO, Cellular Automata (CA) have been proposed as a powerful parallel computational tool where space and time are discrete and interactions are local. It has been proven that CA are ubiquitous: they are mathematical models of computation and computer models of natural systems and their research in interdisciplinary topics leads to new theoretical constructs, novel computational solutions and elegant powerful models. As a result, in this chapter we step forward presenting a combination of CA with ant colonies aiming at the introduction of an unconventional computational model, namely "Cellular Automata Ants". This rather theoretical approach is stressed in rather competitive field, namely clustering. It is well known that the spread of data for almost all areas of life has rapidly increased during the last decades. Nevertheless, the overall process of discovering true knowledge from data demands more powerful clustering techniques to ensure that some of those data are useful and some are not. In this chapter it is presented that Cellular Automata Ants can provide efficient, robust and low cost solutions to data clustering problems using quite small amount of computational resources.

N.P. Bitsakidis (✉) · N.I. Dourvas · G.Ch. Sirakoulis
Democritus University of Thrace, Xanthis 67100, Greece
e-mail: nbitsakidis@gmail.com

N.I. Dourvas
e-mail: ndourvas@ee.duth.gr

G.Ch. Sirakoulis
e-mail: gsirak@ee.duth.gr

S.A. Chatzichristofis
Information Technologies Institute Centre of Research and Technology,
Thessaloniki, Greece
e-mail: schatzic@iti.gr

© Springer International Publishing Switzerland 2017
A. Adamatzky (ed.), *Advances in Unconventional Computing*,
Emergence, Complexity and Computation 23,
DOI 10.1007/978-3-319-33921-4_22

22.1 Introduction

Cellular Automata (CA) are an idealization of a physical system in which space and time are discrete, and the physical quantities take only a finite set of values. CA originally proposed by John von Neumann, who by following a suggestion of Stanislaw Ulam [46] presented them as formal models of self reproducing organisms that can capture the essential features of systems where global behaviour arises from the collective effect of simple components which interact locally [34]. Non-trivial CA are obtained whenever the dependence on the values at each site is non-linear. As a result, any physical system satisfying differential equations may be approximated by a CA, by introducing finite differences and discrete variables [44]. CA forms theoretical background and, at the same time simulation tools and implementation substrates, of mathematical machines with unbounded memory, discrete theoretical structures, digital physics and modelling of spatially extended non-linear systems; massive-parallel computing, language acceptance, and computability; reversibility of computation, graph-theoretic analysis and logic; chaos and undecidability; evolution, learning and cryptography [3, 42, 47]. It is almost impossible to find a field of natural and technical sciences, where CA are not used.

On the other hand, the Ant Colony Optimization (ACO) algorithms are basically a colony of artificial ants or cooperative agents, designed to solve combinatorial optimization problems. These algorithms are probabilistic in nature because they avoid the local minima entrapment and provide very good solutions close to the natural solution [8]. ACO algorithms are extensively used to a variety of applications such as the travel salesman problem [18], image retrieval [28, 29], classification [31], electrical load pattern grouping [11], video games [38], seismic methods [13], communications networks [15], etc. Moreover, ACO algorithms were also used for solving the path planning in a team of robots and most of the effort was to implement the algorithm in real and virtual systems [4, 19, 20, 25–27, 40, 41].

In the last decade the amount of the stored data related to almost all areas of life has rapidly increased. However, the overall process of discovering knowledge from data demands more powerful clustering techniques to ensure that this knowledge is useful. To give a definition, data clustering is an assignment of a set of observations into subsets, namely clusters, so that observations in the same cluster are similar in some trait-often proximity according to a predefined distance. It is an unsupervised learning method, used mainly from statistics and database communities in fields like machine learning, information retrieval, image analysis and pattern recognition. There are several conventional algorithms trying to optimize the clustering techniques in the literature [2, 39]. But there are also some unconventional algorithms inspired from biology that are increasingly considered in clustering because of the low computational cost that they provide. There are efficient approaches in studies based on the ant colonies that try to address the clustering problems [18, 24] as well as modified methods like the Ant Colony Optimization with Different Favor (ACODF) algorithm [45], Constrained Ant Colony Optimization (CACO) algorithm [12] and the modified ACO algorithm proposed by Tiwari et al. [43].

Based on the fact that ants' social behavior, originates from a feature that is common to CA, namely self-organization, i.e. a set of dynamical mechanisms ensuring that the global aim of the system could be achieved through low level interactions between its elements, some new computational intelligence techniques, namely cellular ants have been proposed [10, 27, 47] for application in different scientific fields. In this chapter a cellular automata ants clustering model inspired by the cellular ants algorithm of Vande Moere and his colleagues [32, 33] is presented providing new features in order to overcome some of the previous model limitations. Original rules were modified and extended (long interaction rules) to successfully simulate more complex situations found in different size databases. Furthermore, some limitations like the excessive mobility of cellular ants as well as the discrete data tolerance proposed method limits and the practical flaws of the small size of the grid have been successfully addressed. More over, the idea of proportional distance measurement, the prediction for weights on data dimensions have been effectively applied and tested in different databases. On the other hand, the presented CA model is characterized by as much as low complexity as possible so that the computational recourses are kept low while its computation speed is kept high.

22.2 Cellular Automata and Ant Colony Optimization Principles

Cellular Automata (CA) were originally introduced by John Von Neumann [34] and his colleague Stanislaw Ulam [46]. CA can be considered as dynamical systems in which space and time are discrete and interactions are local. In general, a CA is consisted of a large number of identical entities with local connectivity arranged on a regular array. A finite Cellular Automaton could be defined by the quadruple:

$$\{d, q, N, F\} \tag{22.1}$$

From Eq. 22.1, variable d is a vector of two elements, m and n, denoting the vertical and horizontal CA dimensions, respectively. Both of these variables are expressed in number of cells. At each time step, the state of each cell is updated using a value from the set $Q = 1, 2, \ldots, q - 1$, called set of states. The neighborhood of each cell is defined by the variable N. For a 2-D CA, two neighborhoods are often considered, Von Neumann and Moore neighborhood . Von Neumann neighborhood is a diamond shaped neighborhood and can be used to define a set of cells surrounding a given cell $(x0, y0)$. Equation 22.2 defines the Von Neumann neighborhood of range r.

$$N^v_{(x0, y0)} = (x, y) : |x - x0| + |y - y0| \leq r \tag{22.2}$$

For a given cell $(x0, y0)$ and range r, Moore neighborhood can be defined by the following equation:

$$N^M_{(x0,y0)} = (x, y) : |x - x0| \le r, |y - y0| \le r \qquad (22.3)$$

The transition rule f determines the way in which each cell of the CA is updated. The state of each cell is affected by the cell values in its neighborhood and its value on the previous time step, according to the transition rule or a set of rules. The state of every cell is updated simultaneously in the CA, thus, providing an inherent parallel system.

CA have sufficient expressive dynamics to represent phenomena of arbitrary complexity and at the same time can be simulated exactly by digital computers, because of their intrinsic discreteness, i.e. the topology of the simulated object is reproduced in the simulating device. The CA approach is consistent with the modern notion of unified spacetime. In computer science, space corresponds to memory and time to processing unit. In CA, memory (CA cell state) and processing unit (CA local rule) are inseparably related to a CA cell. Furthermore, CA are an alternative to partial differential equations [36, 44] and they can easily handle complicated boundary and initial conditions, inhomogeneities and anisotropies [21, 37].

The basic element of ACO algorithms is "ants" that is, agents with very simple capabilities which, to some extent, mimic the behavior of real ants [17]. Real ants are in some ways much unsophisticated insects. Their memory is very limited and they exhibit individual behavior that appears to have a large random component. However, acting as a collective, ants collaborate to achieve a variety of complicated tasks with great reliability and consistency [14], such as defining the shortest pathway, among a set of alternative paths, from their nests to a food source [5]. This type of social behavior is based on a common feature with CA, called self-organization, a set of dynamical mechanisms ensuring that the global aim of the system could be achieved through low level interactions between its elements [22]. The most vital feature of this interaction is that only local information is required. There are two ways of information transfer between ants: a direct communication (mandibular, antennation, chemical or visual contact, etc.) and an indirect communication, which is called stigmergy (as defined by Grassé [23]) and is biologically realized through pheromones, a special secretory chemical that is deposited, in many ant species, as trail by individual ants when they move [7]. More specifically, due to the fact that ants can detect pheromone, when choosing their way, they tend to choose paths marked by strong pheromone concentrations. As soon as an ant finds a food source, it evaluates the quantity and the quality of the food and carries some of it back to the nest. During the return trip, the quantity of pheromone that an ant leaves on the ground may depend on the quantity and quality of the food. The pheromone trails will guide other ants to the food source. This behavior is known as "auto catalytic" behaviour or the positive feedback mechanism in which reinforcement of the previously most followed route, is more desirable for future search. In ACO algorithms, an ant will move from point i to point j with probability:

$$\rho_{i,j} = \frac{(\tau_{i,j}^{\alpha})(\eta_{i,j}^{\beta})}{\sum (\tau_{i,j}^{\alpha})(\eta_{i,j}^{\beta})} \tag{22.4}$$

where, $\tau_{i,j}^{\alpha}$ and $\eta_{i,j}^{\beta}$ are the pheromone value and the heuristic value associated with an available solution route, respectively. Furthermore, α and β are positive real parameters whose values determine the relative importance of pheromone versus heuristic information.

During their search for food, all ants deposit on the ground a small quantity of specific pheromone type. As soon as an ant discovers a food source, it evaluates the quantity and the quality of the food and carries some to the nest on their back. During the return trip, every ant with food leaves on the ground a different type of pheromone of specific quantity, according to the quality and quantity of the food. In ACO algorithms, pheromone is updated according to the equation:

$$\tau_{i,j} = (1 - \rho)\tau_{i,j} + \Delta\tau_{i,j} \tag{22.5}$$

where, $\tau_{i,j}$ is the amount of pheromone on a given position (i, j), ρ is the rate of the pheromone evaporation and $\Delta\tau_{i,j}$ is the amount of pheromone deposited, typically given by:

$$\Delta\tau_{i,j} = \begin{cases} 1/L_k & \text{if ant } k \text{ travels on edge } i, j \\ 0, & \text{otherwise} \end{cases} \tag{22.6}$$

where L_k is the cost of the kth tour of an ant (typically is measured as length). Finally, the created pheromone trails will guide other ants to the food source.

Consider for example the experimental setting shown in Fig. 22.1. The ants move along the path from food source F to the nest N. At point B, all ants walking to the nest must decide whether to continue their path from point C or from point H (Fig. 22.1a). A higher quantity of pheromone on the path through point C provides

Fig. 22.1 An example of real ants colony: a An ant follows BHD path by chance. b Both paths are followed with same probability and c Larger number of ants follow the shorter path

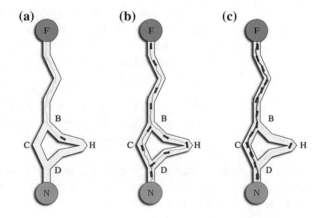

an ant a stronger motivation and thus a higher probability to follow this path. As no pheromone was deposited previously at point B, the first ant reaching point B has the same probability to go through either point C or point H. The first ant following the path BCD will reach point D earlier than the first ant which followed path BHD, due to its shorter length. The result is that an ant returning from N to D will trace a stronger trail on path DCB, caused by the half of all the ants that by chance followed path DCBF and by the already arrived ones coming via BCD. Therefore, they will prefer path DCB to path DHB. Consequently, the number of ants following this path will be increased during time than the number of ants following BHD. This causes the quantity of pheromone on the shorter path to grow faster than the corresponding longer one. Consequently, the probability with which any single ant chooses the path to follow is quickly biased towards the shorter one.

22.3 Cellular Automata Ants for Clustering Problems

22.3.1 Initial Algorithm

In the initial algorithm [32], ants are positioned within a regular, finite grid of cells, similar to the cells inside a CA. Each ant is controlled by a discrete timeline and has a limited perception of the surrounding Moore neighbourhood and is also determined by the same set of rules in order to update its position from one discrete time step to the next. An ant's behavior is solely based on the presence of other ants or pheromone trails in its local neighbourhood and their according data values. At each iteration, the rules are simultaneously applied for all ants, causing them to move their position to one of the eight fields in their neighbourhood or to stay put. As mentioned in [32], the cellular ant algorithm is derived from considering two, intrinsically contradicting concepts: *Edge Repulsion* and *Surface Tension*.

In more details, for each time step, the actions of each cellular ant is determined by the following inducements:

Discrete Data Tolerance. Ants only consider (and thus "count") other ants that are "similar", namely when the distance between their data values in parameter space are below a discrete, predefined similarity tolerance threshold value t. Data similarity between pairs of ants is calculated as follows (with p as the dimensionality of the dataset):

$$
\begin{aligned}
data_i &= \left(z_{i1}, z_{i2}, \ldots, z_{ip}\right) \in R^p, p \in Z^+ \\
d_{ij} &= d\left(data_i - data_j\right) = \left\| data_i - data_j \right\|_p \\
d_{ij} &< t \Rightarrow similar\left(ant_i, ant_j\right) = true
\end{aligned}
\tag{22.7}
$$

where t seems to be similar to the object distance measure variable a in normal ant-based clustering approaches, but results in a simple Boolean parameter ("similar", "not similar"), instead of a continuous similarity value.

Pheromone Trailing. An ant will follow the trail of (a) the most similar ant, that (b) is the freshest, so that ants find similar ants rapidly. Each ant leaves a pheromone trail, consisting of the following attributes: (a) data value(s) from that ant, (b) an ant ID, and (c) the time that has passed since the ant was occupying that cell. The data values allows an ant to follow the "most similar" ant, the ID assures that an ant does not follow its own trail, while the time value enables evaporation.

Surface Tension. Pheromone-following ants tend to generate small, separate clusters that have unstable cohesiveness. Therefore, CA algorithms determine ant actions depending on the discrete amount of similar neighbouring ants. An ant with less than 4 similar neighbours should move to a non-empty cell in its neighbourhood that (a) has no nonsimilar neighbours, and (b) is next to the most similar ant. This rule will cause ants to form large, stable clusters.

Edge Repulsion. Ants in a favorable setting, thus with 6 or more similar neighbours, should still attempt to move away when there are one or more non-similar ants in its neighbourhood. This rule typically will cause large clusters to repulse each other at their outer edges, generating "empty" cells around their perimeter.

Positional Swapping orders ants internally within clusters in relation to data similarity, enabling ants to jump "over" each other to reach more ideal positions within a cluster. In addition, swapping will cause ants that are trapped or positioned in "wrong" clusters to be rapidly "pushed" out to the outer cluster borders. This concept is made possible because the cellular ants method considers ants as CA cells that are able to "sense" data values of neighbouring ants. Ants should organize their positions in the grid relatively to each other according to relative data value gradients (of neighbours in all grid directions) that are as monotonic as possible. At each iteration, each ant picks a random direction (horizontal, vertical, or one of both diagonals) in its neighbourhood, with itself as the middle ant. It then reads the data values of the corresponding neighbours, and calculates the (one-dimensional) *data value distances* d_{ij} between all ants in the multi-dimensional parameter space. Based on these three pair-wise values, an ant is able to determine if it needs to swap its position with one of both its outer neighbours, or if the current constellation is ideal, even for multi-dimensional datasets. If the distance in parameter space between the middle agent and an outer ant is larger than between the outer ants themselves, the middle ant has to swap. Subsequently, the swapping rule will linearly order ants in the chosen grid direction by data similarity, so that "more similar" ants are positioned closer to each other and dissimilar ones are put further apart in the grid. Although this rule organizes ants recursively in randomly chosen directions, an ordered structure will emerge due to the multitude of simultaneous local interactions. The swapping rule ensures that for any three ants that are linearly neighbouring each other, the pair of ants with the largest distance in parameter space will be positioned at the outer grid positions. This data similarity swapping rule still respects the concept of ant decentralization, as ant *B* only considers the data values of its immediate neighbours *A* and *C*. It is generally applicable for multi-dimensional datasets, as it applies to the one-dimensional distance measure d_{ij} in parameter space, calculated between pairs

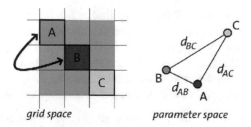

grid space   parameter space

Fig. 22.2 For a random direction of ant B (e.g. *left*-to-*right diagonal*), the largest data value distance in parameter space is d_{BC}, and ant A's data values lie between ant B and C. These dependencies are represented in grid space by swapping ants A and B, so that A lies between B and C in the grid

of ants. For a setting of ants A, B and C, as shown in Fig. 22.2, following swapping rule is valid:

$$d_{AC} > d_{AB}, d_{AC} > d_{BC} \Rightarrow ok$$
$$d_{AC} < d_{AB}, d_{BC} < d_{AB} \Rightarrow swap\,(ant_B, ant_C) \qquad (22.8)$$
$$d_{AC} < d_{AB}, d_{BC} > d_{AB} \Rightarrow swap\,(ant_A, ant_B)$$

The proposed algorithm succeeds, at least in theoretical level, to overcome the clustering problem based on its algorithmic simplicity and efficiency. However, when applied to Iris Data Set, used here for comparison reasons, the results although better, if measured by required iterations, than the standard ant clustering method [30] and the CA inspired ASM approach [10], are not so efficient as expected. This was more pronouncing in large data sets, while in small data sets the already proposed algorithm responded quite well. As mentioned above, the algorithm was originally implemented for the IRIS data set, (using the same parameters, i.e. 150 data items, 4 data dimensions, 3 data types), retrieved from the online Machine Learning Repository [1] and the same algorithmic properties, i.e. threshold t equals to 0.51 and grid sizes 15×15. Some simulation results for different number of iterations are depicted in Fig. 22.9. Taking into consideration the presented algorithm, we tried to enrich its features in order to overcome some of the previous model limitations and to present, if possible, better clustering results.

For example, one of the possible causes for the non-expected results, is the excessive mobility of cellular ants which jump from one side of the grid to its opposite very easily. This problem was tried to be addressed by abandoning the concept of wrapping and replacing the toroidal grid with a rectangular one, an idea that was tested on experimental level with satisfying results.

On the other hand, the presented algorithmic improvements are characterized by as low complexity as possible so that the computational recourses are kept low while its computation speed is kept high. In the next subsections some possible different solutions to probable limitations and/or insufficiencies are examined and presented in details.

22.3.2 The Discrete Data Tolerance Proposed Method Limits

The concept of discrete data tolerance, faces two problems at its implication. The first one, is the diverse (not equally measured) weight of differences between the values of two data dimensions, when the first dimension has a wide allocation of values and the latter a short one. In order to calculate the similarity between two agents that carry data objects, a relatively small difference of values on the first data dimension is valued very important, while, on the contrary, a major difference on the second dimension is valued insignificant. This is a common problem when the similarity of two data objects is tried to be calculated using the absolute value of their dimensions' difference (the Euclidean distance between the two data objects). The common way to solve it is the normalization of data before their further processing, but on this case, a supervised preprocessing of data is not desirable. The second problem is the difficulty, on experimental level, of reaching and finding the ideal and appropriate tolerance threshold. Each data set has its own (that can vary from a very small threshold like 0.5 to a very large one like 3,000), which can be found by continuous try-false experiments, but this process slows down significantly the whole assignment of clustering on large data sets. This delay can make the method nearly useless in practice. In order to solve these problems, a new measurement called proportional distance is proposed.

The idea of Proportional Distance measurement, is the examination of the proportion between the values of two data objects. Two data objects that are identical should have a proportional distance of 0 %, two non-identical should have a larger one. The similarity threshold should distinct now two agents as similar or non-similars, accordingly to their percentage proportional distance.

$$dist\,(AB) = 2 \left| \frac{100 \times \sum_{i=j}^{i=1} \frac{b_i}{a_i + b_i}}{j} - 50 \right| \tag{22.9}$$

where j is the number of data dimensions, b_i is the second agent's value on data dimension i and a_i is the first agent's value on data dimension i.

Another interesting idea could be the necessity of imposing different pre-defined weights for each data dimension in order to calculate the data distance between two data objects or the case of not taking on account one or more data dimensions in order to use the same data set for various kinds of clustering.

Taking into account different weights for each data dimension, is made possible by applying minor changes to the proposed proportional distance measurement rule, as follows:

$$dist\,(AB) = 2 \left| \frac{100 \times \sum\limits_{i=j}^{i=1} \dfrac{b_i}{a_i + b_i}}{\sum\limits_{i=j}^{i=1} weight_i} - 50 \right| \tag{22.10}$$

22.3.3 The Practical Flaws of the Small Size of the Grid

The small size of the grid (only 15–20 % of the cells should be empty, according to the initial method) helps cellular ants to find similar ants on their neighbourhood in order to form clusters, or pheromone trails of similar ants to follow. Trying an initial random allocation of ants inside a large grid leads to the fact that the clustering couldn't even start. But the proposed small size of the grid also causes problems. First of all, it doesn't help the cellular ants to move freely. The cellular ants have a few empty cells to move on, this can slow down the process of forming appropriate clusters. Furthermore, the emerging clusters can't really completely separated. The edge repulsion concept works fine in order to push dissimilar cellular ants away from a cluster to find their similar, but they can't really diverge because of the lack of empty space, which implies that the emerging clusters are stuck to each other. In addition, the smallness of empty space results to the fact that cellular ants are very often positioned on the sides of the grid. A direct Moore neighbourhood consideration is then impossible—one (if the cellular ant is on a side), two (if the cellular ants is on an edge) sides of Moore neighbourhood are outside the grid. Some rules can't be implied in such a small neighbourhood.

Our second approach was to leave the initial setting untouched and allow cellular ants to do an initial forming of clusters for a number of iterations, then expanding the size of grid by adding empty cells on its sides. This concept worked very well, the already emerged clusters had space to move on and expand. Additional expansions could also be tested. So we proposed a new concept, called Progressively Expansive Grid. This concept works fine if it isn't misused, because after one level it doesn't help the process clustering, it simply adds empty cells that restrict the valuable screen space.

22.3.4 The Long Interaction Rule

In order to overcome the problem of trapped cellular ants, which, with no trails to follow, cannot escape from a hostile neighborhood, the concept of Long Interaction Rule is proposed using Hyper-Cellular Automata Ants [6]. Hyper-Cellular Automata Ants have extended visibility and interaction capability, beyond their direct Moore neighbourhood, radius 1; their neighbourhood for interaction is by this way extended.

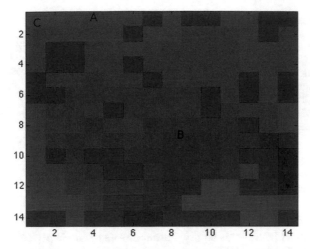

Fig. 22.3 *A* The lone cellular ant can't move freely to find its similar. *B* Neighborhooding clusters can't be separated. *C* The rules can't be applied correctly for cellular ants at the sides of the grid

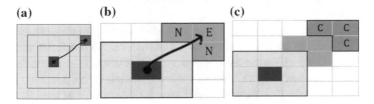

Fig. 22.4 a The hyper-cellular ant's neighborhood (on this image a Moore neighborhood, radius 3). Ant uses "fly-mode" in order to move to a cell far beyond its base. **b** The neighborhood of cellular ant is expanded on up-right direction. Two cells contain ants that are similar, the middle cell is empty—the cellular ant "flies" to the empty cell. **c** If the first extension doesn't provide results, the neighborhood is further extended. The chosen cells are again examined

Furthermore, Hyper-Cellular Automata Ants have the capability of using "fly mode" in order to move to a cell which is far from their base. There should be noted that on "fly mode", hyper-cellular automata ants don't drop pheromone trails (Figs. 22.3 and 22.4).

The Long Interaction Rule results from the restriction that the cellular ants neighbourhood is limited to the Moore neighbourhood, radius 1. The originators have considered this option, we tried to explore it. The two proposed variations of the Long Interaction Rule are the directional expanding long interaction rule and the extended neighbourhood long interaction rule.

More specifically, the directional expanding long interaction rule tries to extend the neighbourhood of an examined cellular ant to one strict direction, vertical, horizontal or diagonal, that faces the center of the grid. The direction to the center is universal, in order that all cellular ants that are trapped should try to form clusters at a single territory, and not being dispersed at all sides of the grid. Examining the three cells

Fig. 22.5 The extended neighborhood long interaction rule. **a** The neighborhood of cellular ant is extended at all sides. **b** If the first extension doesn't provide results, the neighborhood of cellular ant is further extended at all sides

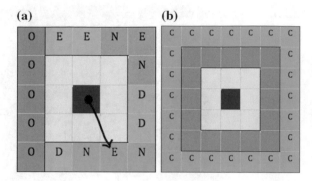

that compose the extended neighbourhood of the examined ant, if one cell is occupied by a similar ant and another is empty, the ant moves to the empty cell. Elsewhere we continue expanding the neighbourhood at this direction until the previous statement is met. The previous rule is overridden and the search is stopped without results, if the whole of extended neighbourhood is outside of the grid or a predefined extension limit is reached.

In the case of the extended neighbourhood long interaction rule, we are trying to extend all the sides of Moore neighbourhood of the examined cellular ant. The new neighbourhood is placed on the outside limits of this square. If there are any cells outside the grid on the new neighbourhood, these are not taken into account. Thereafter a consideration of every sequence of three—continuous on the neighbourhood— cells is examined, until a sequence of a "similar ant—empty cell—similar ant" is taking place. This sequence is checked. At the end of neighbourhood examination, the cellular ant moves to the empty cell of the sequence that has the minimum average data distance difference from it—considering both similar ants. If this statement can't be met (no sequence has been found), a further extension is taking place, until one or more desired sequences are found. The search for a desired movement to a better neighbourhood is stopped, if the whole neighbourhood is outside grid or if an extension limit is reached. As shown in Fig. 22.5a the neighborhood of cellular ant is extended at all sides. Some cells are outside the grid and as a result, they are not examined. Two series of neighborhooding ant-empty cell-neighborhooding cell are found. The cellular ants "flies" to the empty cell of the series that has the two neighborhooding ants with the smallest average data distance from examined cellular ant. In second thought (Fig. 22.5b) if the first extension doesn't provide results, the neighborhood of cellular ant is further extended at all sides. The chosen cells are again examined.

22.3.5 *Progressively Expansive Grid with Inline Gaps*

The inline gaps at the neighbourhoods of the grid, can serve two important goals. The first one, is giving cellular automata ants that are still alone or on groups of 2–3, far from their ideal neighborhood, the chance to move to it. The Long Interaction Rule causes the rapid forming of clustering neighborhoods but their closing too, so some cellular automata ants that should belong to them are excluded. In order to solve this problem, an expansion of the grid combined with insertion of random inline gaps inside the neighborhoods and an exclusive implication of Long Interaction Rule—and no other rule—for alone ants, or ones with no more than 3 similar ants on their neighborhood for this iterations only, is proposed. The exclusive implication of Long Interaction Rule preserves that the empty inline cells won't be captured by neighboring cellular automata ants, but from those ants that should be on those neighborhoods but are until then excluded. The second problem that inline gaps are solving, is the clustering of dynamic data sets. 'Dynamic' here has the meaning either of data objects that change their values during time or of new data objects that enter the grid in order to be clustered. Without inline gaps, the first ones could stay on their previous neighborhoods unable to escape to a neighborhood that pairs with their new values, and the latter could not find their way to their similar data objects that are already clustered as cellular automata ants on tight, unreachable neighborhoods.

As shown in Fig. 22.6a the IRIS data set is clustered after 150 iterations using long interaction rule with proportional distance measurement. Some data objects are lone, far beyond their appropriate clusters. Afterwards, the grid is extended at all sides Fig. 22.6b and the inline gaps are created for each line of the grid Fig. 22.6c. At the already changed grid, inline gaps are created for each column (Fig. 22.6d). The grid is now filled with inline gaps. Practicing long interacting rule only for lone ants, three of them now found their way to their appropriate cluster. Only one is left alone. The clustering is continued on the altered grid, this is the image of grid after further 50 iterations (Fig. 22.6e). The progressive expansion with inline gaps can be further practiced in order to finally cluster all the lone ants (Fig. 22.6f).

22.4 Application of Image Processing on Cellular Automata Ants Clustering

Cellular Automata Ants method had the intention to visualize the clustering of data objects, and it manages well on this purpose. But if a way to identify and separate the emerged clusters was found, it would signal a huge improvement in relation to algorithm's value, because the separated clusters could form new data sets that can be used for further clustering or for statistical analysis. We tried to face this problem, by approaching the visualization as an image. On the scientific field of Digital Image Processing, there are a plenty of ways to unify pixels with nearly common gray level, in order to form large territories and reduce the number of color levels. One common way, described by [35] is by finding a threshold of gray level on a local neighborhood of image—usually the average of gray level on this neighborhood—and unifying the local pixels that have a local variability to the threshold value that is below one predefined level, by altering their previous value on gray level attribute with the threshold value. A second level of unification of similar pixels on gray level, can be done by using Region Growing Method [9], which unifies areas that have weak boundaries (meaning pixels in the boundaries of two areas that have similar values on gray-scale level), in order to form larger areas. This solution can't be really applied on Cellular Automata Ants Method, because finding thresholds (local or global) requires a prior knowledge of the data set, which is a supervised static process—we cannot for example "feed" the grid with dynamic continuous data sets, because they will alter these threshold levels. Instead of using pre-defined or calculated thresholds, we considered a different approach.

The identification algorithm selects random cells that are captured by cellular automata ants, and then extends the area of them, by adding cells that are connected to initial cell and are being captured by similar ants (See [16] for explanation of this process on Digital Image Processing). The continuation of area is preserved by using von Neumann neighborhoods in order to compare the examined cells to the initial. In specific, the whole process is depicted in the following Figs. 22.7. At the beginning, a random cell containing a cellular automata ant is found (Fig. 22.7a). The von Neumann neighborhood of chosen ant is examined. If similar ants are found, the area is extended with their cells as shown in Fig. 22.7b. Then, the von Neumann neighborhood of each area's cell is again examined. If there are ants that are similar to the initial ant, then the area is extended with the cells they occupy (Fig. 22.7c). As a result, the final area which consists of cells that contain the initial ant and ants

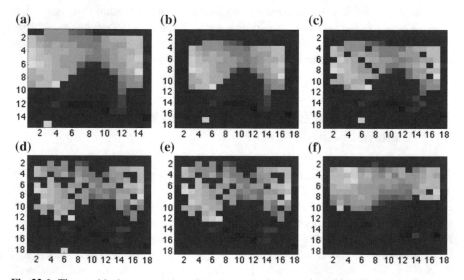

Fig. 22.6 The graphical representation of the progressively expansive grid method with inline gaps

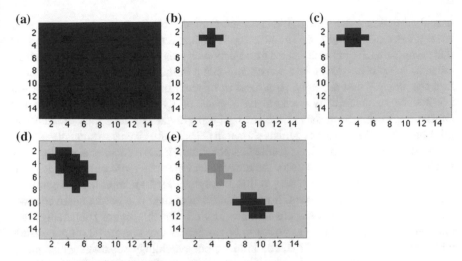

Fig. 22.7 The different steps of the identification algorithm

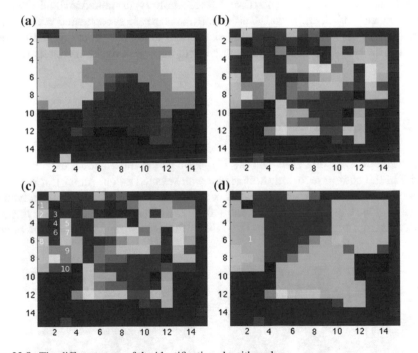

Fig. 22.8 The different steps of the identification algorithm plus

that are similar and presented in Fig. 22.7d. Finally, as given in Fig. 22.7e, using the same technique, another area is found and so on.

The unification of recognized areas is made possible, by merging the areas that have weak boundaries (See [9]). If the cellular automata ants from each side of neighboring areas, are similar, the areas are unified. This process continues until no other areas can't be merged. The last step is recognizing the cellular automata ants that compose each final area—the data objects each area contains form the recognized clusters. In specific, taking into consideration the initial map (Fig. 22.8a) in order to find the clusters that are only visualized in original method, we use Identification Method. More specifically, each cluster consists of agents that are very near ($\pm 1\%$ similarity threshold) on data space (using Proportional Distance measurement) in relation to a random chosen agent as shown in (Fig. 22.8b). We use a small threshold, so the randomness of results is limited as possible. Figure 22.8c visualizes the next step of the method that is to unify the clusters that are very close on data space. We use a bigger similarity threshold (on this example $\pm 3.9\%$) comparing the common borders of neighboring clusters. If more that 90 % of common borders between two clusters are similar, then the clusters are unified. On this figure (Fig. 22.8c), we number the clusters which are going to be unified as one cluster. The final cluster takes the number of one (1) of the similar, but previously separated clusters, as depictured in Fig. 22.8d. Consequently, the result of unifying all similar clusters— each cluster's agents are recognized, and the clustering is presented in Fig. 22.8d. The whole concept of Identification, Unification and Separation concluded with a result that not only visualized the clusters—as Cellular Automata Ants original method tried to achieve—but also recognized, without supervision, the members of each cluster.

22.5 Simulation Results

For readability reasons, Iris database [1], perhaps the most known database to be found in the pattern recognition literature was selected as our initial test bed. The data set contains 3 classes of 50 instances each, where each class refers to a type of iris plant. One class is linearly separable from the other 2; the latter are not linearly separable from each other. In Fig. 22.9a the initial pseudo-random allocation of cel-

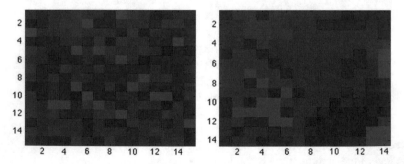

Fig. 22.9 Simulation results for (*left*) the initial grid, (*right*) after 1,500 iterations for the initial algorithm without wrapping

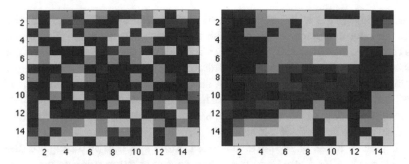

Fig. 22.10 Simulation results for (*left*) initial grid for proportional distance and (*right*) after 1,500 iterations for proportional distance measurement without wrapping

lular automata ants on grid, which has been used on all tests, is presented. The shades represent the difference of data values between each cellular automata ant-agent in relation to a hypothetical cellular automata ant that "carries" a data object with zero data values at each dimension. Two cellular automata ants with similar shade "carry" two similar data objects. The initial algorithm using the aforementioned threshold ($t = 0.51$) and the grid size (15×15) suggested by Vande Moore and Clayden, is not performing as well as expected. The clustering can't provide good results, the cellular automata ants are left on their starting neighbourhoods or switching sides with wrapping, unable to find their similar. The problem is answered by leaving the concept of wrapping, adopting a rectangular grid instead of toroid one. The results were much better but even now not completely satisfactory. In Fig. 22.9b, the grid, after using initial algorithm without wrapping for 1,500 iterations, is presented.

For these reasons we have proceeded with the proportional distance. The same initial pseudo-random allocation has been used on all tests with methods that use proportional distance measurement as depictured in Fig. 22.10a. The difference is that now, a sorting of cellular automata ants is made in prior, in order to determine with which shade each cellular automata ant will be represented at the grid. In Fig. 22.10b, the grid after 1,500 iterations, using proportional distance measurement and setting the threshold to 10 ($\pm 10\%$), is presented. It is clear that the results are good, but a little worse in relation to authentic method after abandoning the concept of wrapping.

According to the presented algorithm improvements, the idea of long interactions has also tested in order to improve the aforementioned results. More specifically, in Fig. 22.11a the grid after 1,500 iterations, using long interaction rule is presented. This time the threshold was set to $t = 0.71$, and the long interaction range was set to the half of the size of the grid. The results were much better, on average the 95.44 % of the first class of IRIS Data Set is clustered together. However, still some lone ants, are not clustered where they should belong. As a result, the combination of long interaction in accordance with proportional distance was applied. Figure 22.11b presents the classification results of the initial grid after only 150 iterations, using long interaction rule and proportional distance measurement. The threshold was set to $t = 10$ ($\pm 10\%$), and the long interaction range was set to the half of the size of the grid. The results were satisfactory, on average the 74.48 % of the first class of IRIS Data Set is clustered together. The most important issue was that algorithm was coming to an end at 150 iterations—only minor movements of cellular automata ants as a result of swapping rule are performed from then on. This corresponds to an improvement of 1066.66 % in relation to all competing algorithms, combined with satisfactory results. The remaining issue is that the 74.48 % of successful clustering is having a devation of 20.3 %, meaning that the least successful clustering attempts could have only about 55 % success.

The idea of inline gaps was also applied. More specifically, in Fig. 22.12a it is shown the grid after 1,500 iterations, using long interaction rule and progressively expansive grid with inline gaps. The results were almost perfect, the clustering is very good and only a small amount of cellular automata ants are positioned outside of their appropriate clusters. Using long interaction rule, proportional distance measurement and progressively expansive grid with inline gaps, the results were almost as good (even better) than when no proportional distance measurement is used. In Fig. 22.12b these results were depictured for the initial grid after 1,500 iterations.

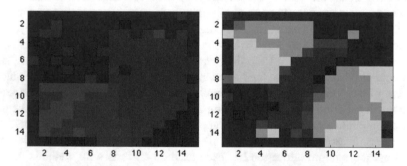

Fig. 22.11 Simulation results (*left*) after applying the long interaction rule at the initial grid for 1,500 iterations and (*right*) after applying the long interaction rule combined with proportional distance for only 150 iterations

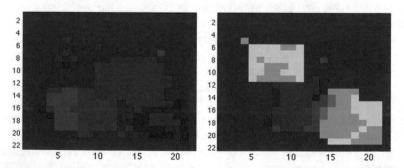

Fig. 22.12 Simulation results (*left*) ater applying long interaction rule and progressively expansive grid with inline gaps at the initial grid for 1,500 iterations and (*right*) after applying the long interaction rule, proportional distance measurement and progressively expansive grid with inline gaps for 1,500 iterations

The last trial was to use a progressively expansive grid with inline gaps. Using both Euclidean measurement and Proportional Distance measurement, the results were completely satisfactory. With classic data distance measurement, setting the step for each expansion of the grid on 600 iterations and ending clustering on 1,500 iterations, the results were almost perfect. The first class of IRIS data set is recognized on average at 96.4 %, but also the deviation from that number is minimized to 2.45 %, meaning that the least successful clustering attempts have a success of 93.5 %. With Proportional Distance measurement, there was a further improvement. Setting the step to each expansion to 150 iterations, and ending the clustering on 400, on average the 88.48 % of the first class of IRIS data set is correctly recognized as a cluster, but also the deviation from that numbered is minimized to 15.46 %, meaning that the least successful clustering attempts have a success of 72 %. Some comparison results can be found in Table 22.1.

Table 22.1 Simulation results for the IRIS dataset when applied different improvements

Method	Time Av.	Success Av.
Initial	126.7303625	60.28
Prop. Dist.	167.4780554	49.92
Long Inter.	220.2520988	95.44
Long Inter. and Prop. Dist.	98.4326502	74.48

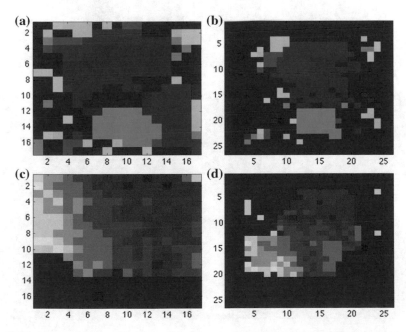

Fig. 22.13 The application of hybrid cellular automata ants to computer hardware dataset

Furthermore, in order to test the functionality of the proposed hybrid cellular automata ants algorithm another dataset was considered, namely the Computer Hardware Dataset, created by Phillip Ein-Dor and Jacob Feldmesser. This database concerns multivariate data set characteristics with 209 number of instances and 9 number of attributes. As it can be found out in Fig. 22.13a, when the computer hardware data set was clustered with long interaction rule the results were satisfactory. In addition, when the computer hardware data set was clustered with long interaction rule with progressively expansive grid and inline gaps the results were even better, as presented in Fig. 22.13b. Moreover, when long interaction rule and proportional distance measurement were applied the results were further improved (Fig. 22.13c) and with all the above, as well as progressively expansive grid with inline gaps, the results were almost perfect. In conclusion, we tried the algorithm using a non-standard data set. The results were satisfactory judging from the view, but we couldn't appreciate quantitatively the success of the clustering because of the lack of class information about this data set.

Finally, the ideas of identification and separation when applied to the hybrid cellular automata ants, were tested exhaustively on IRIS dataset. More specifically, in

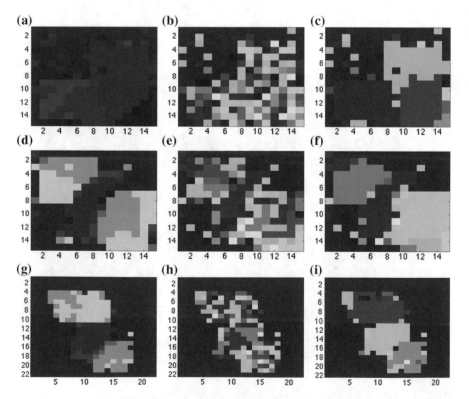

Fig. 22.14 Graphical results of the identification and separation methods when applied to hybrid cellular automata ants in combination with all the previous methods

Fig. 22.14a the initial clustering using long interaction rule is presented. The first class of IRIS data set is on left bottom, waiting to be recognized. During the procedures of identification and separation the identification of small clusters is succeeded (Fig. 22.14b). When the unification process is finished as shown in Fig. 22.14c, the presented results were very good, especially at the recognition of the first class of IRIS data set (which is shown here on red). Some more tests were performed with the application of long interaction rule and proportional distance measurement. As it can be shown in corresponding Figs. 22.14d–f, the results were quite good. Finally, the results were slightly better when expansive grid with inline gaps is used combined with the use of long interaction rule and proportional distance measurement (Figs. 22.14g–i).

From a quantitative point of view, we tried the identification and separation process on three different improvements on algorithm. On long interaction rule, setting the identification threshold to 0.1 and unification threshold to the same as the threshold for clustering (0.71), the results were almost perfect. The first class of IRIS data set, is recognized and separated at 91.92 %, with a deviation of 3.1 %, meaning that even the least successful attempts on identification and separation had 88.8 % success. The large number of clusters (24.88 % on average) remain a problem although. Using long interaction rule and proportional distance measurement, setting the identification threshold to 1 %, and unification threshold to 3.9 %, the results weren't so good. On average the 52 % of first class of IRIS data set is recognized as a cluster. The results were slightly better when expansive grid with inline gaps is used combined with the use of long interaction rule and proportional distance measurement. Setting the identification threshold to 1 %, and the unification threshold to 4.5 %, on average the 61.12 % of the first class of IRIS data set is recognized as a cluster. The explanation on latter—not quite good—results is that we haven't tried to find the ideal threshold for clustering when we used proportional distance measurement, we picked up one (±10%) that simply worked well. As the success of attempt to identificate and separate using long interaction rule prior for clustering indicates, the ideal threshold for clustering should be very near to the one used for identification and separation.

22.6 Conclusions

In this chapter, a cellular automata ants clustering model inspired by the cellular ants algorithm of [32, 33] that provides new features in order to overcome some of the previous model limitations was presented. The original rules were modified and extended (long interaction rules) by introducing the Hyper-Cellular Ants in order to successfully simulate more complex situations found in different size databases. Furthermore, some limitations like the excessive mobility of cellular ants as well as the discrete data tolerance proposed method limits and the practical flaws of the small size of the grid have been successfully addressed. More over, the idea of proportional distance measurement, the prediction for weights on data dimensions have been effectively applied and tested in different databases. On the other hand, the presented CA model is characterized by as much as low complexity as possible so that the computational resources are kept low while its computation speed is kept high. As future work concerns, the expansion of the cellular ants model for handling more complicated clustering situations should be considered. More specifically, at this point, some ideas like the progressively expansive grid with inline gaps in order to help the long interaction rules to boost the model's performance should be further investigated although some preliminary results are rather encouraging. Furthermore, the application of image processing methods on cellular automata ants clustering for cluster identification and separation could be rather beneficial. Finally, the application of the proposed model to different databases with different properties should be also taken under consideration and the preliminary results, for example computer hardware database are also quite promising.

References

1. Asuncion, A.D.N.: UCI machine learning repository (2007). http://www.ics.uci.edu/~mlearn/MLRepository.html
2. Abonyi, J., Feil, B.: Cluster Analysis for Data Mining and System Identification. Birkhäuser, Basel (2007)
3. Adamatzky, A.: Reaction-Diffusion Automata: Phenomenology, Localisations Computation. Springer Publishing Company Incorporated, Berlin (2014)
4. Akst, J.: Send in the bots. Scientist 27(10) (2013). Cited By (since 1996)
5. Beckers, R., Deneubourg, J.L., Goss, S.: Trails and u-turns in the selection of a path by the ant lasius niger. J. Theor. Biol. pp. 397–415 (1992)
6. Bitsakidis, N.P., Chatzichristofis, S.A., Sirakoulis, G.C.: Hybrid cellular ants for clustering problems. Int. J. Ubiquitous Comput. 11(2), 103–130 (2015)
7. Blum, C.: Ant colony optimization: introduction and recent trends. Phys. Rev. 2(4), 353–373 (2005)
8. Bonabeau, E., Dorigo, M., Theraulaz, G.: Inspiration for optimization from social insect behaviour. Nature 406, 39–42 (2000)
9. Brice, C.R., Fennema, C.L.: Scene analysis using regions. Artif. Intell. 1(3–4), 205–226 (1970)
10. Chen, L., Xu, X., Chen, Y., He, P.: A novel ant clustering algorithm based on cellular automata. In: Proceedings of the IEEE/WIC/ACM International Conference on Intelligent Agent Technology. IAT '04, pp. 148–154. IEEE Computer Society, Washington, DC, USA (2004)
11. Chicco, G., Ionel, O.M., Porumb, R.: Electrical load pattern grouping based on centroid model with ant colony clustering. IEEE Trans. Power Syst. 28(2), 1706–1715 (2013)
12. Chu, S.C., Roddick, J.F., Su, C.J., Pan, J.S.: Constrained ant colony optimization for data clustering. In: C. Zhang, H.W. Guesgen, W.K. Yeap (eds.) In: Proceedings of the PRICAI 2004: Trends in Artificial Intelligence, 8th Pacific Rim International Conference on Artificial Intelligence, Auckland, New Zealand, 9–13 August 2004. Lecture Notes in Computer Science, vol. 3157, pp. 534–543. Springer (2004)
13. Conti, C., Roisenberg, M., Neto, G., Porsani, M.: Fast seismic inversion methods using ant colony optimization algorithm. IEEE Geosci. Remote Sens. Lett. 10(5), 1119–1123 (2013)
14. Deneubourg, J., Goss, S.: Collective patterns and decision-making. Ethol. Ecol. Evol. 1(4), 295–311 (1989)
15. Di Caro, G., Dorigo, M.: Antnet: distributed stigmergetic control for communications networks. J. Artif. Int. Res. 9(1), 317–365 (1998)

16. Dillencourt, M.B., Samet, H., Tamminen, M.: A general approach to connected-component labeling for arbitrary image representations. J. ACM **39**(2), 253–280 (1992)
17. Dorigo, M.: Optimization, learning and natural algorithms. Ph.D. thesis, Politecnico di Milano, Italy (1992)
18. Dorigo, M., Gambardella, L.M.: Ant colony system: a cooperative learning approach to the traveling salesman problem. IEEE Trans. Evol. Comput. **1**(1), 53–66 (1997)
19. Garnier, S., Tache, F., Combe, M., Grimal, A., Theraulaz, G.: Alice in pheromone land: an experimental setup for the study of ant-like robots. In: Proceedings of the Swarm Intelligence Symposium, SIS 2007. IEEE, pp. 37–44 (2007)
20. Garnier, S., Combe, M., Jost, C., Theraulaz, G.: Do ants need to estimate the geometrical properties of trail bifurcations to find an efficient route? A swarm robotics test bed. PLoS Comput Biol **9**(3), e1002,903+ (2013)
21. Georgoudas, I., Sirakoulis, G., Scordilis, E., Andreadis, I.: A cellular automaton simulation tool for modelling seismicity in the region of Xanthi. Environ. Modell. Softw. **22**(10), 1455–1464 (2007)
22. Goss, S., Beckers, R., Deneubourg, J., Aron, S., Pasteels, J.: How trail laying and trail following can solve foraging problems for ant colonies. In: Hughes, R. (ed.) Behavioural Mechanisms of Food Selection. NATO ASI Series, vol. 20, pp. 661–678. Springer, Berlin (1990)
23. Grassé, P.P.: La reconstruction du nid et les coordinations interindividuelles chezBellicositermes natalensis etCubitermes sp. la théorie de la stigmergie: Essai d'interprétation du comportement des termites constructeurs. Insectes Sociaux **6**(1), 41–80 (1959)
24. Handl, J., Knowles, J., Dorigo, M.: Ant-based clustering and topographic mapping. Artif. Life **12**, 35–61 (2006)
25. Herianto., Kurabayashi, D.: Realization of an artificial pheromone system in random data carriers using rfid tags for autonomous navigation. In: Proceedings of the International Conference on Robotics and Automation, ICRA '09, pp. 2288–2293. IEEE (2009)
26. Herianto., Sakakibara, T., Kurabayashi, D.: Artificial pheromone system using RFID for navigation of autonomous robots. J. Bionic Eng. **4**(4), 245–253 (2007)
27. Ioannidis, K., Sirakoulis, G.C., Andreadis, I.: Cellular ants: a method to create collision free trajectories for a cooperative robot team. Robot. Auton. Syst. **59**(2), 113–127 (2011)
28. Konstantinidis, K., Sirakoulis, G., Andreadis, I.: Design and implementation of a fuzzy-modified ant colony hardware structure for image retrieval. IEEE Trans. Syst., Man, Cybernetics, Part C: Appl. Rev. **39**(5), 520–533 (2009)
29. Konstantinidis, K., Andreadis, I., Sirakoulis, G.C.: Chapter 3 - application of artificial intelligence methods to content-based image retrieval. In: P.W. Hawkes (ed.) Advances in Imaging and Electron Physics. Advances in Imaging and Electron Physics, vol. 169, pp. 99–145. Elsevier (2011)
30. Lumer, E., Faieta, B.: Diversity and adaptation in populations of clustering ants, from animals to animats. In: Conference on Simulation of Adaptative Behaviour, pp. 501–508 (1994)
31. Martens, D., De Backer, M., Haesen, R., Vanthienen, J., Snoeck, M., Baesens, B.: Classification with ant colony optimization. IEEE Trans. Evol. Comput. **11**(5), 651–665 (2007)
32. Moere, A.V., Clayden, J.J.: Cellular ants: combining ant-based clustering with cellular automata. In: Proceedings of the 17th IEEE International Conference on Tools with Artificial Intelligence, ICTAI '05, pp. 177–184. IEEE Computer Society, Washington, DC, USA (2005). http://dl.acm.org/citation.cfm?id=1105924.1106055
33. Moere, A.V., Clayden, J.J., Dong, A.: Data clustering and visualization using cellular automata ants. In: Australian Conference on Artificial Intelligence, pp. 826–836 (2006)
34. Neumann, J.V.: Theory of Self-Reproducing Automata. University of Illinois Press, Champaign (1966)
35. Niblack, W.: An Introduction to Digital Image Processing. Strandberg Publishing Company, Birkeroed (1985)
36. Omohundro, S.: Modelling cellular automata with partial differential equations. Physica D: Nonlinear Phenom. **10**(1–2), 128–134 (1984)

37. Progias, P., Sirakoulis, G.C.: An FPGA processor for modelling wildfire spreading. Math. Comput. Modell. **57**, pp. 1436–1452 (2013)
38. Recio, G., Martin, E., Estebanez, C., Saez, Y.: Antbot: ant colonies for video games. IEEE Trans. Comput. Intell. AI Games **4**(4), 295–308 (2012)
39. Runkler, T.A.: Ant colony optimization of clustering models. Int. J. Intell. Syst. **20**(12), 1233–1251 (2005)
40. Russell, R.A.: Heat trails as short-lived navigational markers for mobile robots. In: Proceedings of the IEEE International Conference on Robotics and Automation, vol. 4, pp. 3534–3539 (1997)
41. Sirakoulis, G.C., Adamatzky, A.: Robots and Lattice Automata. Springer International Publishing, Cham (2015)
42. Sirakoulis, G.C., Bandini, S. (eds.): Cellular Automata. In: 10th International Conference on Cellular Automata for Research and Industry, ACRI 2012, Santorini Island, Greece, 24–27 September 2012. Lecture Notes in Computer Science, vol. 7495. Springer (2012)
43. Tiwari, R., Husain, M., Gupta, S., Srivastava, A.: Improving ant colony optimization algorithm for data clustering. In: Proceedings of the International Conference and Workshop on Emerging Trends in Technology, ICWET '10, pp. 529–534. ACM, New York, NY, USA (2010)
44. Toffoli, T.: Cellular automata as an alternative to (rather than an approximation of) differential equations in modeling physics. Phys. D: Nonlinear Phenom. **10**(1–2), 117–127 (1984)
45. Tsai, C.F., Wu, H.C., Tsai, C.W.: A new data clustering approach for data mining in large databases. In: Proceedings of the ISPAN, pp. 315–320 (2002)
46. Ulam, S.: Random processes and transformations. Int. Congr. Math. **2**, 264–275 (1952)
47. Was, J., Sirakoulis, G.C., Bandini, S. (eds.): Cellular Automata. In: Proceedings of the 11th International Conference on Cellular Automata for Research and Industry, ACRI 2014, Krakow, Poland, 22–25 September 2014. Lecture Notes in Computer Science, vol. 8751. Springer (2014)

Chapter 23
Rough Set Description of Strategy Games on Physarum Machines

Krzysztof Pancerz and Andrew Schumann

Abstract A *Physarum* machine is a biological computing device implemented in the plasmodium of *Physarum polycephalum* or *Badhamia utricularis* which arc one-cell organisms able to build complex networks for solving different computational tasks. The plasmodial stage of such organisms is a natural transition system which can be considered the medium for bio-inspired strategy games. In the paper, we describe a rough set approach for description of a strategy game created on the *Physarum* machine. The strategies of such a game are approximated on the basis of a rough set model, describing behavior of the *Physarum* machine, created according to the VPRSM (Variable Precision Rough Set Model) approach. Theoretical foundations given in the paper are supplemented with description of a specialized programming language for *Physarum* machines, as well as a software tool developed, among others, for simulation of *Physarum* games.

23.1 Introduction

A *Physarum* machine [3] is a programmable amorphous biological computing device, experimentally implemented in the plasmodium of *Physarum polycephalum* (as well as in the plasmodium of *Badhamia utricularis*). *Physarum polycephalum* is a single cell organism belonging to the species of order *Physarales*, subclass *Myxogastromycetidae*, class *Myxomycetes*, and division *Myxostelida*. In the presented research, the term of *Physarum* machine covers a hybrid device implemented in two plasmodia (cf. [27]), the plasmodium of *Physarum polycephalum*, as well as the plasmodium of *Badhamia utricularis*. *Badhamia utricularis* is another species of order *Physarales*. The plasmodium of *Physarum polycephalum* or *Badhamia utricularis*,

K. Pancerz (✉)
University of Rzeszow, Rzeszow, Poland
e-mail: kpancerz@ur.edu.pl

K. Pancerz · A. Schumann
University of Information Technology and Management, Rzeszow, Poland
e-mail: andrew.schumann@gmail.com

© Springer International Publishing Switzerland 2017   615
A. Adamatzky (ed.), *Advances in Unconventional Computing*,
Emergence, Complexity and Computation 23,
DOI 10.1007/978-3-319-33921-4_23

spread by networks, can be programmable by adding and removing attractants and repellents. In propagating and foraging behavior of the plasmodium, we can perform useful computational tasks (cf. [3–5, 11]). In *Physarum Chip Project: Growing Computers from Slime Mould* (PhyChip) [6] funded by the Seventh Framework Programme (FP7), we are going to construct an unconventional computer on programmable behavior of *Physarum polycephalum*. One of computational tasks which can be performed in *Physarum* machine environments concerns strategy games. Fundamental topics of the research area related to bio-inspired games on *Physarum* machines were earlier considered, e.g. in [26, 27].

As payoffs for the created bio-inspired games on *Physarum* machines, we may define a variety of complex tasks, including:

- achievement of as many as possible attractants occupied by plasmodia of organisms,
- construction of the longest path consisting of attractants occupied by plasmodia.

Determining different payoffs for *Physarum* games appears as an interesting field of research due to a huge number of different methodologies and paradigms which can be applied. In our earlier paper, see [13], we considered rough set models of *Physarum* machines. This model can become a basis for defining strategy approximations in *Physarum* games.

In the presented approach, we use transition system models of behavior of *Physarum* machines in terms of rough set theory. In the behavior of *Physarum* machines, one can notice some ambiguity in plasmodium motions that influences exact anticipation of states of machines in time. In [13], rough set models, created over transition systems, were proposed to model this ambiguity. Rough sets are an appropriate tool to deal with rough (ambiguous, imprecise) concepts in the universe of discourse (cf. [16]). The model described in [13] is based on the original definition of rough sets proposed by Z. Pawlak [16]. In fact, this definition is rigorous in terms of set inclusion. In Sect. 23.3, we extend the model described in [13] to a new one based on a more relaxed and generalized rough set approach, called the Variable Precision Rough Set Model (VPRSM), proposed by W. Ziarko in [29]. The analogous approach was proposed by us for timed transition systems (see [25]). The VPRSM approach was defined on the basis of the notion of majority set inclusion instead of standard set inclusion. The majority set inclusion is parameterized. Therefore, a new model is more flexible. In case of *Physarum* games, the rough set model of *Physarum* machines based on the VPRSM approach enables us to set different levels of difficulty of games.

Thus, this paper is a continuation of our work on the bio-inspired game theory [26, 27], where, on the one hand, we use the notion of concurrent games [1, 2, 18], because the plasmodia move concurrently, and, on the other hand, we use the notion of reflexive games [10, 19], because it is impossible to make a partition of player's strategies into disjoint sets. So, the main features of bio-inspired games are as follows:

- Transitions of the plasmodium of *Physarum polycephalum* and the plasmodium of *Badhamia utricularis* on the same set of states can be regarded as a zero-sum game, because the interests of the parties are strictly contradictory here, i.e., the plasmodium of *Physarum polycephalum* and the plasmodium of *Badhamia utricularis* cannot occupy the same attractants.

- The plasmodium of *Physarum polycephalum* and the plasmodium of *Badhamia utricularis* perform a concurrent game, i.e., they move in parallel.
- Strategies in the meaning of conventional game theory are uncertain and imprecise for plasmodia of *Physarum polycephalum* and *Badhamia utricularis*, i.e., they do not give definite payoffs, but these strategies as well as payoffs they give can be approximated by rough sets.

These features seem to be important for further applications of bio-inspired game theory in fields such as biology, economy, or cognitive science, where we can tune the harshness and uncertainty of behavior.

The third feature can be considered the most important among others. The point is that the plasmodium of *Physarum polycephalum* is a one-cell organism that can behave as an individual and a collective under the same conditions. This ability was defined by us as an individual-collective duality [20]. This means that the plasmodium can follow just one strategy for just one payoff, but it can follow many strategies at once also. So, strategies involved into the plasmodium behavior cannot be atomic in all cases, because sometimes the plasmodium performs a set of strategies as its one strategy. In other words, we can face the situation of emergency when a strategy of the plasmodium is a complex phenomenon that is not atomic in fact. In particular, there is a possibility that strategies for plasmodia have a non-empty intersection and we cannot make a partition of all strategies used in a game. This situation can be observed in many real games such as stock markets. In the same measure, it is impossible to define all trader's strategies as atoms. In this paper, we appeal to the rough set approximation as a tool to deal with non-atomic strategies.

23.2   The Rudiments of Rough Sets

In this section, we recall necessary definitions, notions and notation concerning rough sets.

The idea of rough sets (see [16]) consists of the approximation of a given set by a pair of sets, called the lower and the upper approximation of this set. Some sets cannot be exactly defined. If a given set X is not exactly defined, then we employ two exact sets (the lower and the upper approximation of X) that define X roughly (approximately).

Let $U \neq \emptyset$ be a finite set of objects we are interested in. U is called the universe. Any subset $X \subseteq U$ of the universe is called a concept in U. Let R be any equivalence relation over U. We denote an equivalence class of any $u \in U$ by $[u]_R$. With each subset $X \subseteq U$ and any equivalence relation R over U, we associate two subsets:

- $R_*(X) = \{u \in U : [u]_R \subseteq X\}$,
- $R^*(X) = \{u \in U : [u]_R \cap X \neq \emptyset\}$,

called the R-lower and R-upper approximation of X, respectively. A set $BN_R(X) = R^*(X) - R_*(X)$ is called the R-boundary region of X. If $BN_R(X) = \emptyset$, then X is sharp (exact) with respect to R. Otherwise, X is rough (inexact).

Roughness of a set can be characterized numerically. To this end, the accuracy of approximation of X with respect to R is defined as:

$$\alpha_R(X) = \frac{card(R_*(X))}{card(R^*(X))},$$

where $card$ denotes the cardinality of the set and $X \neq \emptyset$.

The definitions given earlier are based on the standard definition of set inclusion. Let U be the universe and $A, B \subseteq U$. The standard set inclusion is defined as

$$A \subseteq B \text{ if and only if } \forall_{u \in A} u \in B.$$

In some situations, the application of this definition seems to be too restrictive and rigorous. W. Ziarko proposed in [29] some relaxation of the original rough set approach. His proposition was called the Variable Precision Rough Set Model (VPRSM). The VPRSM approach is based on the notion of majority set inclusion. Let U be the universe, $A, B \subseteq U$, and $0 \leq \beta < 0.5$. The majority set inclusion is defined as

$$A \overset{\beta}{\subseteq} B \text{ if and only if } 1 - \frac{card(A \cap B)}{card(A)} \leq \beta,$$

where $card$ denotes the cardinality of the set. $A \overset{\beta}{\subseteq} B$ means that a specified majority of elements belonging to A belongs also to B. One can see that if $\beta = 0$, then the majority set inclusion becomes a standard set inclusion.

By replacing the standard set inclusion with the majority set inclusion in definitions of approximations, we obtain the following two subsets:

- $R_*^\beta(X) = \{u \in U : [u]_R \overset{\beta}{\subseteq} X\}$,
- $R^{*\beta}(X) = \{u \in U : \frac{card([u]_R \cap X)}{card([u]_R)} > \beta\}$,

called the R_β-lower and R_β-upper approximation of X, respectively.

For more details on rough sets, we refer the readers to [16, 17].

23.3 Rough Set Based Descriptions of Behavior of Physarum Machines

The *Physarum* machine comprises an amorphous yellowish mass with networks of protoplasmic veins, programmed by spatial configurations of attracting and repelling stimuli. When several attractants are scattered in the plasmodium range, the plasmodium forms a network of protoplasmic veins connecting those attractants and original points of the plasmodium. Each original point of the plasmodium and each attractant occupied by the plasmodium will be called an active point in *Physarum* machines. As a result, a transition system is built up. Therefore, transition systems become a natural model used to describe behavior of *Physarum* machines.

Formally, in the present context, a transition system is a quadruple

$$TS = (S, E, T, I),$$

where:

- S is the non-empty set of states;
- E is the set of events;
- $T \subseteq S \times E \times S$ is the transition relation;
- $I \subseteq S$ is the set of initial states.

Any transition system $TS = (S, E, T, I)$ can be presented in the form of a labeled graph with nodes corresponding to states from S, edges representing the transition relation T, and labels of edges corresponding to events from E.

In case of *Physarum* machines, the event may correspond to activation of the attractant causing the movement of the plasmodium from a given active point, in fact neighboring, to this attractant and occupation of it.

Formally, a structure of the *Physarum* machine can be described as a triple $\mathcal{PM} = (P, A, R)$ (cf. [13]), where:

- $P = \{p_1, p_2, \ldots, p_k\}$ is a set of original points of plasmodia (*Physarum poly-cephalum* or *Badhamia utricularis*);
- $A = \{a_1, a_2, \ldots, a_m\}$ is a set of attractants;
- $R = \{r_1, r_2, \ldots, r_n\}$ is a set of repellents.

A dynamics (behavior) of the *Physarum* machine $\mathcal{PM}$ can be described by the set $V = \{v_1, v_2, \ldots, v_r\}$ of protoplasmic veins formed by the plasmodia during its action.

To build a model, in the form of a transition system $TS(\mathcal{PM}) = (S, E, T, I)$, of behavior of the *Physarum* machine $\mathcal{PM} = (P, A, R)$ described by the set $V = \{v_1, v_2, \ldots, v_r\}$ of protoplasmic veins, we can use the following set of bijective functions:

- $\sigma : P \cup A \rightarrow S$ assigning a state to each original point of the plasmodium as well as to each attractant;
- $\tau : V \rightarrow T$ assigning a transition to each protoplasmic vein;
- $\iota : P \rightarrow I$ assigning an initial state to each original point of the plasmodium.

Each event of the set of events E is assigned to *Physarum* motions in accordance with the following types of *Physarum* behavior:

- *direction* (the plasmodium moves from one state/attractant/initial point to another state/attractant),
- *fusion* (the plasmodium moves from different states/attractants/initial points to the same one state/attractant),

- *splitting* (the plasmodium moves from one state/attractant/initial point to different states/attractants),
- *repelling* (the plasmodium stops to move in one direction).

The model of behavior of the *Physarum* machine $\mathcal{PM}$ in the form of a transition system $TS(\mathcal{PM}) = (S, E, T, I)$ is the basis for creating the rough set description of behavior of this machine (cf. [13]).

For each state $s \in S$ in the transition system $TS(\mathcal{PM})$, we can determine its direct successors and predecessors. Let:

- $Post(s, e) = \{s' \in S : (s, e, s') \in T\}$;
- $Pre(s, e) = \{s' \in S : (s', e, s) \in T\}$,

then the set $Post(s)$ of all direct successors of the state $s \in S$ is given by

$$Post(s) = \bigcup_{e \in E} Post(s, e)$$

and the set $Pre(s)$ of all direct predecessors of the state $s \in S$ is given by

$$Pre(s) = \bigcup_{e \in E} Pre(s, e).$$

If there exists the state $s \in S$ in the transition system $TS(\mathcal{PM})$ such that $card(Post(s)) > 1$, where $card$ is the cardinality of the set, then $TS(\mathcal{PM})$ is called a non-deterministic transition system. In this case, we observe a splitting of the plasmodium. One can see that, in non-deterministic transition systems, we deal with ambiguity of direct successors of some states, i.e., there exist states having no uniquely determined direct successors. This observation underlies the approach presented in [13], where we managed modeling of this ambiguity using rough set theory. Because of splitting, the plasmodium can move in many directions simultaneously and, as a result, in one plasmodium act we observe many acts in fact. So, we have one complex act consisting of $card(Post(s))$ elements.

Analogously to rough approximation of sets defined in rough set theory, we defined rough anticipation of states over transition systems called predecessor anticipation.

Let $TS(\mathcal{PM}) = (S, E, T, I)$ be a transition system and $X \subseteq S$. The lower predecessor anticipation $Pre_*(X)$ of X is given by

$$Pre_*(X) = \{s \in S : Post(s) \neq \emptyset \text{ and } Post(s) \subseteq X\}.$$

The lower predecessor anticipation consists of all states from which $TS(\mathcal{PM})$ surely goes to the states in X as results of any events occurring at these states.

The upper predecessor anticipation $Pre^*(X)$ of X is given by

$$Pre^*(X) = \{s \in S : Post(s) \cap X \neq \emptyset\}.$$

The upper predecessor anticipation consists of all states from which $TS(\mathcal{PM})$ possibly goes to the states in X as results of some events occurring at these states. It means that $TS(\mathcal{PM})$ can also go to the states from outside X.

If $BN_{Pre}(X) = \emptyset$, then the anticipation of the set X of states on the basis of their direct predecessors is exact. This means that we observe a fusion of the plasmodium. In the opposite case (i.e., $BN_{Pre}(X) \neq \emptyset$), the anticipation of X is rough (inexact). The accuracy of anticipation can be defined analogously to the accuracy of approximation in rough set theory, i.e.:

$$\alpha(X) = \frac{card(Pre_*(X))}{card(Pre^*(X))}.$$

Let us consider an exemplary transition system $TS(\mathcal{PM}) = (S, E, T, I)$ shown in Fig. 23.1 and the set $X_1 = \{s_6, s_8\}$.

For X_1, we obtain:

- $Pre_*(X_1) = \{s_3, s_4\}$,
- $Pre^*(X_1) = \{s_2, s_3, s_4\}$.

Hence, $BN_{Pre}(X_1) \neq \emptyset$. It means that the anticipation of states from X_1 is rough on the basis of their direct predecessors. The accuracy of anticipation is equal to $\frac{2}{3}$. The point is that there was splitting when the plasmodium moved from one state s_2 to three states s_5, s_6, and s_7.

If we take $X_2 = \{s_5, s_6, s_7, s_8\}$, we will obtain:

- $Pre_*(X_2) = \{s_2, s_3, s_4\}$,
- $Pre^*(X_2) = \{s_2, s_3, s_4\}$.

In this case, $BN_{Pre}(X_2) = \emptyset$. It means that the anticipation of states from X_2 is exact on the basis of their direct predecessors. Obviously, the accuracy of anticipation is equal to 1. It holds, because $BN_{Pre}(U) = \emptyset$ for any set U containing all final states.

Fig. 23.1 An exemplary transition system

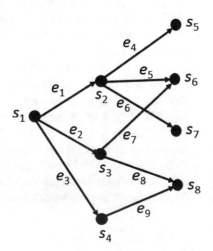

The definitions given above are derived from the original rough set approach. By replacing standard set inclusion with majority set inclusion in the original definition of the lower predecessor anticipation of a set of states in a transition system (cf. [25]), we obtain the following generalized notion of the β-lower predecessor antici-pation. Let $TS(\mathcal{PM}) = (S, E, T, I)$ be a transition system and $X \subseteq S$. The β-lower predecessor anticipation $Pre_*^\beta(X)$ of X is given by

$$Pre_*^\beta(X) = \{s \in S : Post(s) \neq \emptyset \text{ and } Post(s) \overset{\beta}{\subseteq} X\}.$$

The β-lower predecessor anticipation consists of each state from which $TS(\mathcal{PM})$ goes, in most cases (i.e., in terms of majority set inclusion) to the states in X as results of events occurring at these states.

A rough set description of a given transition system enables us to determine whether a given state anticipates, unambiguously or nearly unambiguously, a next state that is one of the states from a distinguished set (for example, consisting of final states).

Let $TS(\mathcal{PM}) = (S, E, T, I)$ be a transition system, $X \subseteq S$, and $0 \leq \beta < 0.5$. If $s \in Pre_*(X)$, then s is said to be a strict anticipator of states from X. If $s \in Pre_*^\beta(X)$, then s is said to be a quasi-anticipator of states from X. A set of all strict anticipators of X will be denoted by $\overline{Ant}(X)$ whereas a set of all quasi-anticipators of X will be denoted by $\widetilde{Ant}(X)$. One can see that for a given X, $\overline{Ant}(X) \subseteq \widetilde{Ant}(X)$.

In concurrent games [1, 2, 18], transition systems are a basic concept in defining games. In case of *Physarum polycephalum* and *Badhamia utricularis* we can extend a transition system $TS(\mathcal{PM}) = (S, E, T, I)$ to a concurrent game defined as follows:

$$\mathcal{G} = (S, \mathcal{P}(S), Agt_{1,2}, Act, Mov, Tab, (\preceq_A)_{A \in Agt_{1,2}}),$$

where:

- S is a set of states of $TS(\mathcal{PM})$ represented by attractants occupied by plasmodia of *Physarum polycephalum* or *Badhamia utricularis*;
- $\mathcal{P}(S)$ is a set of payoffs (states of the game) for *Physarum polycephalum* and *Badhamia utricularis*;
- $Agt_{1,2} = \{Ph, Bu\}$ is a set of two players presented by plasmodia of *Physarum polycephalum* or *Badhamia utricularis* respectively;
- Act is a non-empty set of strategies represented by $\overline{Ant}(X)$ or $\widetilde{Ant}(X)$ for each payoff $X \subseteq S$, an element of $Act^{Agt_{1,2}}$ is called a move;
- $Mov: \mathcal{P}(S) \times Act^{Agt_{1,2}} \to 2^{Act} \setminus \{\emptyset\}$ is a mapping indicating the available sets of actions to a given player in a given set of states;
- $Tab: \mathcal{P}(S) \times Act^{Agt_{1,2}} \to \mathcal{P}(S)$ is the transition table which associates, with a given set of states of the game and a given move of the players, the set of states of the game resulting from that move;

- for each $A \in Agt_{1,2}$, $\preceq_A$ is a preorder (reflexive and transitive relation) over subsets of S, called the *preference relation* of player A, indicating a direction of *Physarum polycephalum* or *Badhamia utricularis* propagation.

Hence, in the bio-inspired game, we deal with a form of concurrent games, where player's strategies are not exclusive (they are intersected) and player's payoffs are not exclusive, too: there are X_1 and X_2 such that $X_1 \cap X_2 \neq$ and $\overline{Ant}(X_1) \cap \overline{Ant}(X_2) \neq$. The last one is the main feature of reflexive games. In these games, we cannot define exclusive strategies at all. The point is that we try to cheat players by false public announcements, since the key task of any reflexive game is to have all the opponent's actions become transparent for us, while our goals and motives remain hidden for the competitor. So, in reflexive games, we must deal with an unlimited hierarchy of cognitive pictures:

- each of the players (say Ph and Bu) can have their own picture about a state of affairs A, let us denote these pictures by $K_{Ph}A$ and $K_{Bu}A$;
- the first-order reflexion is expressed by means of pictures of the second order which are designated by $K_{Bu}K_{Ph}A$ and $K_{Ph}K_{Bu}A$, where $K_{Bu}K_{Ph}A$ are pictures of agent Bu about pictures of agent Ph, and $K_{Ph}K_{Bu}A$ are pictures of agent Ph about pictures of agent Bu, etc.

To sum up, the key task of reflexive games consists in finding the level of reflexion n of the competitor ($n > 0$) to move onto reflexion level n or $n + 1$ and to act on the basis of the given level.

Let us consider the reflexive nature of *Physarum polycephalum* and *Badhamia utricularis* competition. Assume that:

- $\overline{Ant}_{Ph}(X)$ is a strategy to obtain X for the plasmodium of *Physarum polycephalum*,
- $\overline{Ant}_{Bu}(X)$ is a strategy to obtain X for the plasmodium of *Badhamia utricularis*.

Both have started from different transition states. Suppose, *Physarum polycephalum* has started form $Y_{Ph} \subseteq \overline{Ant}_{Ph}(X)$ and *Badhamia utricularis* has started form $Y_{Bu} \subseteq \overline{Ant}_{Bu}(X)$. According to Y_{Ph}, the *Badhamia* plasmodium can guess that the *Physarum* plasmodium is going to occupy X. And according to Y_{Bu}, the *Physarum* plasmodium can guess that the *Badhamia* plasmodium is going to occupy the same X. Hence, we deal with cognitive pictures $K_{Ph}X$ and $K_{Bu}X$, where $K_{Ph}X := Y_{Ph}$ and $K_{Bu}X := Y_{Bu}$.

Let us notice that the strategy game between *Physarum polycephalum* and *Badhamia utricularis* plasmodia can be a first-order reflexive game, when *Physarum polycephalum* behaves (e.g. it can change its strategy) knowing that *Badhamia utricularis* wants to occupy X in accordance with the picture $K_{Bu}X$ and *Badhamia utricularis* behaves (e.g. it can change its strategy) knowing that *Physarum polycephalum* wants to occupy X in accordance with the picture $K_{Ph}X$. Nevertheless, in human simulation of *Physarum* games we can perform any higher-order reflexive game.

Usually, a strategy game is understood as an ordered set

$$\mathcal{G}' = (N, (Str_i)_{i \in N}, Str, (u_i)_{i \in N}),$$

where

- $N = \{1, 2, \ldots, i, \ldots, N\}$ is the set of players;
- for each $i \in N$, Str_i is the set of all possible strategies for agent i;
- Str is the set of strategy profiles, a strategy profile is an n-tuple $(str_1, \ldots, str_N)$, where $str_1 \in Str_1, \ldots, str_N \in Str_N$;
- u_i is the utility for i defined as a mapping from $\prod_{i \in N} str_i$ into $\mathbf{R}$, i.e. it is a payoff for agent i if the strategy profile is $(str_1, \ldots, str_N)$.

However, in the *Physarum* game the set of payoffs cannot have a partition into disjoint subsets $\bigsqcup_{i \in N} u_i$, because there are no partitions $\bigsqcup_{i \in N} Str_i$ and $\bigsqcup_{i \in N} str_i$. Nevertheless, we can define a Nash equilibrium for the *Physarum* game, too.

Let $str_{Agt_{1,2}} := str_{Ph} \times str_{Bu}$ denote a pair of strategies for *Physarum polycephalum* and *Badhamia utricularis* plasmodia. It is called a strategy profile. The outcome if the players $Agt_{1,2}$ follow the strategy $str_{Agt_{1,2}}$ is denoted by

$$Out_G(X_{Ph}, X_{Bu}, str_{Agt_{1,2}}),$$

where $X_{Ph}, X_{Bu} \subseteq S$. All possible outcomes if the players $Agt_{1,2}$ obey $str_{Agt_{1,2}}$ is denoted by

$$Out_G(str_{Agt_{1,2}}) = \bigcup_{X_{Ph}, X_{Bu} \subseteq S} Out_G(X_{Ph}, X_{Bu}, str_{Agt_{1,2}}).$$

Let us take a move $m_{Agt_{1,2}}$ and an action m' for some player $B \in Agt_{1,2}$. Assume, the move $n_{Agt_{1,2}}$ with $n_A = m_A$ when $A \in Agt_{1,2}$ and $A \neq B$, $n_B = m'$ is denoted by $m_{Agt_{1,2}}[B \to m']$. Then a Nash equilibrium is defined as follows. Let G be a bio-inspired *Physarum* game with preference relation $(\preceq_A)_{A \in Agt_{1,2}}$ and let $X \subseteq S$ be a state of G. A Nash equilibrium of G from X is a strategy profile $str_{Agt_{1,2}}$ such that

$$Out(X, X, str_{Agt_{1,2}}[B \to str']) \preceq_B Out(X, X, str_{Agt_{1,2}})$$

for the player $B \in Agt_{1,2}$ and all strategies str' of B. This means in fact that *Physarum polycephalum* leaves some attractants of X to be occupied by *Badhamia utricularis* and *Badhamia utricularis* leaves some attractants of X to be occupied by *Physarum polycephalum*.

23.4 Rough Set Strategy Approximations in Physarum Games

General assumptions for games on *Physarum* machines were presented in [26, 27]. Now, we give an example of a strategy game created within the *Physarum* machine with strategies approximated on the basis of a rough set model based on the VPRSM approach described in Sect. 23.3.

Plasmodium motions, in *Physarum* machines, can be controlled by different topologies of stimuli (attractants or repellents). This control capability can be used to build a strategy game based on the medium of one-cell organisms in their plasmodial stages. In general, strategies of such a game can be approximated on the basis of some additive measures reflecting desired plasmodium transitions and occupation of attractants.

In the simulation of *Physarum* game, we have two players: the first plays for the *Physarum polycephalum* plasmodia, the second for the *Badhamia utricularis* plasmodia. Locations of original points of both plasmodia are randomly generated. The players can control motions of plasmodia via attracting or repelling stimuli. There are two situations which can be defined for the game:

1. Locations of attractants and repellents are a priori generated in the random way. During the game, each player can activate one stimulus (attractant or repellent) at each step.
2. Locations of attractants and repellents are determined by the players during the game. At each step, each player can put one stimulus (attractant or repellent) at any location and this stimulus becomes automatically activated.

The activated attractant A^* causes that the plasmodia propagate protoplasmic veins towards it and feed on it. It means that new transitions are created between the current active points of plasmodia and a new one on the attractant A^*. Propagating protoplasmic veins is possible if the current active points are located in the region of influence (ROI) of A^*. It means that a proper neighborhood of A^* is taken into consideration. From that moment, the activated attractant A^* is occupied by plasmodia. It is worth noting that, as the experiments showed, the attractant occupied by the plasmodium of *Physarum polycephalum* cannot be simultaneously occupied by the plasmodium of *Badhamia utricularis* and vice versa. Moreover, the *Physarum polycephalum* plasmodium grows faster and could grow into branches of *Badhamia utricularis*, while the *Badhamia utricularis* plasmodium could grow over *Physarum polycephalum* veins.

The activated repellent R^* can change the direction of plasmodium motions or can avoid propagating plasmodium protoplasmic veins towards activated attractants. Such influences are possible if plasmodia are in the region of influence (ROI) of R^*.

The control capabilities presented above enable the players to choose, at each step, one of possible tactics:

1. The attractant or repellent activated by the player can help propagation of his/her plasmodia (of either *Physarum polycephalum* or *Badhamia utricularis*).
2. The attractant or repellent activated by the player can disturb propagation of the second player's plasmodia.

The second possibility is worth considering if we adopt the strategy approximations based on the rough set model described in the remaining part of this section. During the game, the players can switch between two possible tactics according to the current game configuration.

At the end of the game, we determine who wins. As it was mentioned earlier, we propose to use a strategy approximation measure based on the rough set model created on the basis of a transition system describing behavior of the *Physarum* machine during the game.

Let a transition system $TS(\mathcal{PM}) = (S, E, T, I)$ describe behavior of the *Physarum* machine $\mathcal{PM} = (P, A, R)$ described by the set $V = \{v_1, v_2, \ldots, v_r\}$ of protoplasmic veins at the end of the game. We can divide the set S of states in $TS(\mathcal{PM})$ into two sets:

- S^1 including all states from S corresponding to attractants from A activated by the first player during the game and occupied by his/her plasmodia;
- S^2 including all states from S corresponding to attractants from A activated by the second player during the game and occupied by his/her plasmodia.

Hence, $S^1 \cap S^2 = \emptyset$ and $S^1 \cup S^2 \subseteq S$. The set S^1 is a payoff for *Physarum polycephalum* and the set S^2 is a payoff for *Badhamia utricularis*.

For each state $s \in S^1$ and $s \in S^2$, we determine the β-lower predecessor anticipation $Pre^{\beta}_*(s)$ of s as it was described in Sect. 23.3.

For each player, we determine the strategy approximation measure as follows:

- for player 1:

$$\sigma^1 = card(Pre^{\beta}_*(S^1)),$$

- for player 2:

$$\sigma^2 = card(Pre^{\beta}_*(S^2)),$$

where $card$ is the cardinality of the set.

It is easy to see that the strategy approximation measure is parameterized. For $\beta = 0$ we have the most rigorous case, i.e., based on the standard rough set approach. In this case, the difficulty level of the game is the highest one.

The goal of each player i, where $i = 1, 2$, is to maximize the cardinality of his/her $Pre^{\beta}_*(S^i)$, i.e., to have as many as possible strict anticipators (or quasi-anticipators in case of relaxed requirements) of states corresponding to attractants activated by him/her. This goal is disturbed if the co-player causes that plasmodia are attracted by his/her stimuli. In this case, a situation that our action is used by the co-player to reach his/her goal or to disperse our efforts can be modeled.

The player whose the strategy approximation measure is greater wins.

As an example, let us consider a transition system (fragment) $TS(\mathcal{PM}) = (S, E, T, I)$ shown in Fig. 23.2, describing behavior of two plasmodia (of *Physarum polycephalum* and of *Badhamia utricularis*) during the game. The original point of two plasmodia is represented by the state s_1. The first player plays for the *Physarum polycephalum*. The second player plays for the *Badhamia utricularis*. States corresponding to attractants activated by the first player are circled. All other states are assumed to be activated by the second player. Hence, $S^1 = \{s_2, s_3, s_5, s_6, s_7\}$ and $S^2 = \{s_4, s_8\}$. So, $card(S^2) < card(S^1)$, but it does not mean that player 1 wins.

Fig. 23.2 A transition
system describing the game,
where player 1 plays for
Physarum polycephalum and
occupies attractants s_2, s_3,
s_5, s_6, s_7 and player 2 plays
for *Badhamia utricularis* and
occupies attractants s_4, s_8

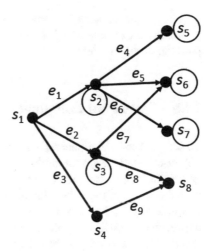

If we assume $\beta = 0$ (i.e., the most rigorous case), then, for player 1, $Pre_*(S^1) = \{s_2\}$ and $\sigma^1 = 1$, and for player 2 $Pre_*(S^2) = \{s_4\}$ and $\sigma^2 = 1$. Hence, nobody wins, because:

- $Post(s_1) = \{s_2, s_3, s_4\}$ and $Post(s_1) \nsubseteq S^1$,
- $Post(s_2) = \{s_5, s_6, s_7\}$ and $Post(s_2) \subseteq S^1$,
- $Post(s_3) = \{s_6, s_8\}$ and $Post(s_3) \nsubseteq S^1$,
- $Post(s_4) = \{s_8\}$ and $Post(s_4) \nsubseteq S^1$,
- $Post(s_1) = \{s_2, s_3, s_4\}$ and $Post(s_1) \nsubseteq S^2$,
- $Post(s_2) = \{s_5, s_6, s_7\}$ and $Post(s_2) \nsubseteq S^2$,
- $Post(s_3) = \{s_6, s_8\}$ and $Post(s_3) \nsubseteq S^2$,
- $Post(s_4) = \{s_8\}$ and $Post(s_4) \subseteq S^2$.

If we assume $\beta = 0.4$ (i.e., more relaxed case), then $Pre_*^{0.4}(S^1) = \{s_1, s_2\}$, $Pre_*^{0.4}(S^2) = \{s_8\}$ and $\sigma^1 = 2, \sigma^2 = 1$. This means that player 1 wins, because:

- $Post(s_1) = \{s_2, s_3, s_4\}$ and $Post(s_1) \overset{0.4}{\subseteq} S^1$,
- $Post(s_2) = \{s_5, s_6, s_7\}$ and $Post(s_2) \overset{0.4}{\subseteq} S^1$,
- $Post(s_3) = \{s_6, s_8\}$ and $Post(s_3) \overset{0.4}{\nsubseteq} S^1$,
- $Post(s_4) = \{s_8\}$ and $Post(s_4) \overset{0.4}{\nsubseteq} S^1$,
- $Post(s_1) = \{s_2, s_3, s_4\}$ and $Post(s_1) \overset{0.4}{\nsubseteq} S^2$,
- $Post(s_2) = \{s_5, s_6, s_7\}$ and $Post(s_2) \overset{0.4}{\nsubseteq} S^2$
- $Post(s_3) = \{s_6, s_8\}$ and $Post(s_3) \overset{0.4}{\nsubseteq} S^2$,
- $Post(s_4) = \{s_8\}$ and $Post(s_4) \overset{0.4}{\subseteq} S^2$.

Thus, a player wins if he/she occupied more predecessors before the end of the game. In the example of Fig. 23.2 the first player won, because this player built more paths in occupying attractants.

23.5   Physarum Language

To program computational tasks for plasmodium transitions, we are developing a new object-oriented programming language [12, 14, 21], called the *Physarum* language. In general, object-oriented programming (OOP) is a programming paradigm based on the concept of the object that has data fields describing the object and associated procedures, called methods, for manipulating the data. We can distinguish two main approaches in OOP programming languages: class-based and prototype-based languages [7]. The prototype-based approach is less common than the class-based one, although, it has a great deal to offer. This model is also called class-less or instance-based programming because prototype-based languages are based upon the idea that objects that represent individuals can be created without reference to class-defining. In this approach, the objects that are manipulated at runtime (the objects that make it an "object-oriented" approach) are the prototypes. JavaScript, the very popular now prototype-based language, has been an inspiration to us and we have implemented a number of its mechanisms in the *Physarum* language. For example, there are inbuilt sets of prototypes corresponding to both the high-level models used for describing behaviour of *Physarum polycephalum* (e.g., ladder diagrams, transition systems, timed transition systems, Petri nets) and the low-level model (distribution of stimuli). High-level models used in programming *Physarum* machines were considered by us in our earlier papers, i.e.:

- ladder diagrams (see [28]),
- transition systems ([12]) and timed transition systems (see [22]),
- Petri nets (see [23]).

According to the prototype-based approach, objects are created by means of a copy operation, called cloning, which is applied to a prototype. Objects can be instantiated (cloning) via the keyword *new* using defined constructors. Methods are used to manipulate features of the objects and create relationships between objects.

A grammar of the language has been described by means of the Java Compiler Compiler (JavaCC) [9] tool. JavaCC is the most popular parser generator for use with Java applications. To describe the syntax of the language, we use the Extended Backus–Naur Form (EBNF) notation, cf. [15]. In the EBNF notation, we use the following nonterminals:

- ID for the identifier,
- STRING for the sequence of characters,
- INT for the integer value.

In the module of simulating *Physarum* games included in the created software tool described in Sect. 23.6, we use the low-level model. The fragment of a grammar describing the *Physarum* language concerning the low-level model has the following EBNF form:

```
LowLevel = "#LOW_LEVEL", {LowLevelExpression, ";"};

LowLevelExpression = ID, (LowLevelCreation
| LowLevelManipulation);

LowLevelCreation = "=", "new", (LayerCreation
| AttractantCreation | RepellentCreation | PhysarumCreation);

LayerCreation = "Layer";

AttractantCreation = "Attractant",
"(", INT, ",", INT, ",", INT, ")";

RepellentCreation = "Repellent",
"(", INT, ",", INT, ",", INT, ")";

PhysarumCreation = "Physarum",
"(", INT, ",", INT, ",", INT, ")";

LowLevelManipulation = ".", (LayerManipulation
| ElementManipulation);

LayerManipulation = "add", "(", STRING, ")";

ElementManipulation = "setSize", "(", INT, ",", INT, ")";
```

23.6 Software Tool

To support research on programming *Physarum* machines and simulating *Physarum* games, we are developing a specialized software tool, called the *Physarum* software system, shortly *PhysarumSoft* (see [24]). The tool was designed for the Java platform. A general structure of this tool is shown in Fig. 23.3. We can distinguish three main parts of *PhysarumSoft*:

- *Physarum* language compiler.
- Module of programming *Physarum* machines.
- Module of simulating *Physarum* games.

Fig. 23.3 A general
structure of *PhysarumSoft*

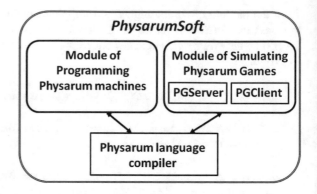

The main features of *PhysarumSoft* are the following:

- Multiplatforming. Thanks to the Java technology, the created tool can be run on various software and hardware platforms. In the future, the tool will be adapted for platforms available in mobile devices and as a service in the cloud.
- User-friendly interface.
- Modularity. The project of *PhysarumSoft* and its implementation covers modularity. It makes the tool to easily extend in the future.

To simulate games on *Physarum* machines, we are developing a special module of *PhysarumSoft* called the *Physarum* game simulator. This module works under the client-server paradigm. A general structure of the *Physarum* game simulator is shown in Fig. 23.4.

The server-side application of the *Physarum* game simulator is called *PGServer*. The main window of *PGServer* is shown in Fig. 23.5. In this window, the user can:

- select the port number on which the server listens for connections,
- start and stop the server,
- set the game strategy:

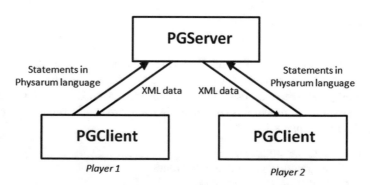

Fig. 23.4 A general structure of the *Physarum* game simulator

Fig. 23.5 The main window of *PGServer*

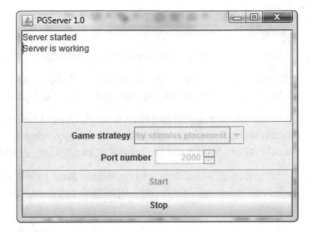

– strategy by stimulus placement,
– strategy by stimulus activation,

• shadow information about actions undertaken.

The client-side application of the *Physarum* game simulator is called *PGClient*. The main window of *PGClient* is shown in Figs. 23.6 and 23.8. In this window, the user can:

• set the server IP address and its port number,
• start participation in the game,

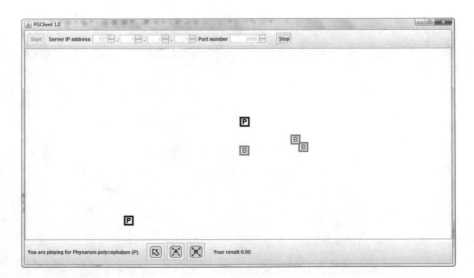

Fig. 23.6 The main window of *PGClient* for the first strategy

- manipulate stimuli (place or activate them) during the game,
- monitor the current result.

In the *Physarum* game simulator, we have two players:

- the first plays for the *Physarum polycephalum* plasmodia,
- the second plays for the *Badhamia utricularis* plasmodia.

Locations of original points of both plasmodia are randomly generated. The players can control motions of plasmodia via attracting or repelling stimuli. There are two strategies, which can be defined for the game, described in Sect. 23.4:

1. Locations of attractants and repellents are a priori generated in the random way.
2. Locations of attractants and repellents are determined by the players during the game.

The client-side main window for the first strategy (locations of attractants and repellents are *a priori* generated in the random way) is shown in Fig. 23.6. At the beginning, original points of *Physarum polycephalum* and *Badhamia utricularis* as well as stimuli are scattered randomly on the plane. The window after several player movements is shown in Fig. 23.7. A box labeled by *P* represents an original point of *Physarum polycephalum*. A box labeled by *B* represents an original point of *Badhamia utricularis*. A single circle denotes an attractant whereas a double circle— a repellent. Different background colors of stimuli differentiate between players.

The client-side main window for the second strategy (locations of attractants and repellents are determined by the players during the game) is shown in Fig. 23.8. At the beginning, original points of *Physarum polycephalum* and *Badhamia utricularis* are scattered randomly on the plane. During the game, players can place stimuli.

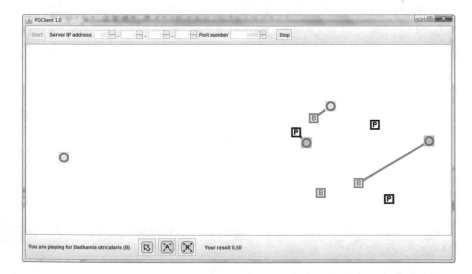

Fig. 23.7 The main window of *PGClient* for the first strategy after several player movements

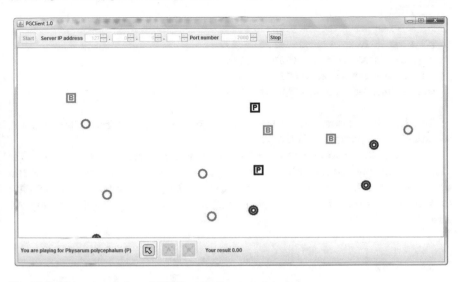

Fig. 23.8 The main window of *PGClient* for the second strategy

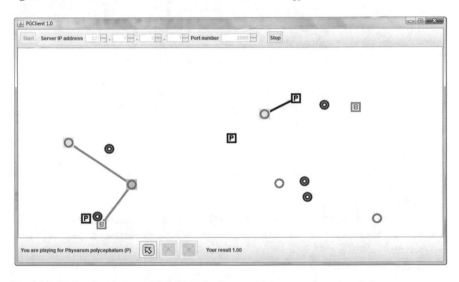

Fig. 23.9 The main window of *PGClient* for the second strategy after several player movements

New veins of plasmodia are created. The window after several player movements is shown in Fig. 23.9.

Communication between clients and the server is realized through text messages containing statements of the *Physarum* language mentioned in Sect. 23.5. The exemplary code responsible for creation of stimuli has the form:

```
p1_a1=new Attractant(195,224,1);
p1_a2=new Attractant(541,310,1);
p1_a1=new Attractant(580,92,2);
p2_r1=new Repellent(452,130,2);
p2_r1=new Repellent(659,327,1);
```

The two first parameters of the stimulus constructors determine the location whereas the last parameter is the player ID.

The server sends to clients information about the current configuration of the *Physarum* machine (localization of original points of *Physarum polycephalum* and *Badhamia utricularis*, localization of stimuli as well as a list of edges, corresponding to veins of plasmodia, between active points) through the XML file. The exemplary XML file has the form:

```
<?xml version="1.0" encoding="UTF-8" standalone="no"?>
<network>
<elements>
<element id="0" player="0" type="0" x="96" y="310"/>
<element id="1" player="0" type="0" x="766" y="178"/>
<element id="2" player="0" type="0" x="566" y="248"/>
<element id="3" player="0" type="3" x="550" y="53"/>
<element id="4" player="0" type="3" x="374" y="534"/>
<element id="5" player="0" type="3" x="746" y="217"/>
<element id="6" player="1" type="1" x="195" y="224"/>
<element id="7" player="1" type="1" x="541" y="310"/>
<element id="8" player="2" type="1" x="580" y="92"/>
<element id="9" player="2" type="2" x="452" y="130"/>
<element id="10" player="1" type="2" x="659" y="327"/>
</elements>
<veins>
<vein createdBy="0" firstNodeID="2" secondNodeID="7"/>
<vein createdBy="3" firstNodeID="3" secondNodeID="8"/>
</veins>
</network>
```

The attribute "player" equal to 0 means that elements are created by the system, in this case, they are original points of plasmodia (*Physarum polycephalum* or *Badhamia utricularis*).

23.7 Conclusions

The Variable Precision Rough Set Model has been used by us to approximate strategy games created on *Physarum* machines. Notice that a variety of approaches gives us a wide range of possibilities for definition strategies of *Physarum* games. The problem is that the strategies used by plasmodia cannot be presented as atoms and their one strategy can be considered as a set containing several members. So, the plasmodium strategies can even be intersected. More interesting approaches for further research are those based on combined rough sets and fuzzy sets in different ways (cf. [8]).

Acknowledgments This research is being fulfilled by the support of FP7-ICT-2011-8.

References

1. Abramsky, S.: Sequentiality vs. concurrency in games and logic. Math. Struct. Comput. Sci. **13**, 531–565 (2003)
2. Abramsky, S., Mellies, P.A.: Concurrent games and full completeness. In: Proceedings of the 14th Symposium on Logic in Computer Science, pp. 431–442. Trento, Italy (1999)
3. Adamatzky, A.: Physarum Machines: Computers from Slime Mould. World Scientific, Singapore (2010)
4. Adamatzky, A.: Slime mould processors, logic gates and sensors. Philos. Trans. R. Soc. A **373**(2046), 20140216 (2015)
5. Adamatzky, A. (ed.): Advances in Physarum Machines: Sensing and Computing with Slime Mould. Springer, Heidelberg (2016)
6. Adamatzky, A., Erokhin, V., Grube, M., Schubert, T., Schumann, A.: Physarum chip project: growing computers from slime mould. Int. J. Unconv. Comput. **8**(4), 319–323 (2012)
7. Craig, I.: Object-Oriented Programming Languages: Interpretation. Springer, London (2007)
8. Dubois, D., Prade, H.: Rough fuzzy sets and fuzzy rough sets. Int. J. Gen. Syst. **17**(2–3), 191–209 (1990)
9. JavaCC. http://java.net/projects/javacc/
10. Lefebvre, V.: Lectures on Reflexive Game Theory. Leaf & Oaks, Los Angeles (2010)
11. Nakagaki, T., Yamada, H., Toth, A.: Maze-solving by an amoeboid organism. Nature **407**, 470–470 (2000)
12. Pancerz, K., Schumann, A.: Principles of an object-oriented programming language for Physarum polycephalum computing. In: Proceedings of the 10th International Conference on Digital Technologies (DT'2014), pp. 273–280. Zilina, Slovak Republic (2014)
13. Pancerz, K., Schumann, A.: Rough set models of Physarum machines. Int. J. Gen. Syst. **44**(3), 314–325 (2015)
14. Pancerz, K., Schumann, A.: Some issues on an object-oriented programming language for Physarum machines. In: Bris, R., Majernik, J., Pancerz, K., Zaitseva, E. (eds.) Applications of Computational Intelligence in Biomedical Technology, Studies in Computational Intelligence, vol. 606, pp. 185–199. Springer International Publishing, Switzerland (2016)
15. Pattis, R.E.: A notation to describe syntax. http://www.cs.cmu.edu/~pattis/misc/ebnf.pdf
16. Pawlak, Z.: Theoretical Aspects of Reasoning about Data Rough Sets. Kluwer Academic, Dordrecht (1991)
17. Pawlak, Z., Skowron, A.: Rudiments of rough sets. Inf. Sci. **177**, 3–27 (2007)
18. Rideau, S., Winskel, G.: Concurrent strategies. In: Proceedings of the Symposium on Logic in Computer Science (LICS'2011), pp. 409–418. Toronto, Ontario, Canada (2011)

19. Schumann, A.: Reflexive games and non-Archimedean probabilities. P-Adic Numbers, Ultra-metric Anal. Appl. **6**(1), 66–79 (2014)
20. Schumann, A., Adamatzky, A.: The double-slit experiment with Physarum polycephalum and p-adic valued probabilities and fuzziness. Int. J. Gen. Syst. **44**(3), 392–408 (2015)
21. Schumann, A., Pancerz, K.: Towards an object-oriented programming language for Physarum polycephalum computing. In: Szczuka, M., Czaja, L., Kacprzak, M. (eds.) Proceedings of the Workshop on Concurrency, Specification and Programming (CS&P'2013), pp. 389–397. Warsaw, Poland (2013)
22. Schumann, A., Pancerz, K.: Timed transition system models for programming Physarum machines: extended abstract. In: Popova-Zeugmann, L. (ed.) Proceedings of the Workshop on Concurrency, Specification and Programming (CS&P'2014), pp. 180–183. Chemnitz, Germany (2014)
23. Schumann, A., Pancerz, K.: Towards an object-oriented programming language for Physarum polycephalum computing: a Petri net model approach. Fundamenta Informaticae **133**(2–3), 271–285 (2014)
24. Schumann, A., Pancerz, K.: Physarumsoft - a software tool for programming Physarum machines and simulating Physarum games. In: Proceedings of the 2015 Federated Conference on Computer Science and Information Systems (FedCSIS'2015), pp. 607–614. Lodz, Poland (2015)
25. Schumann, A., Pancerz, K.: Roughness in timed transition systems modeling propagation of plasmodium. In: Ciucci, D. et al. (eds.) Proceedings of the 10th International Conference on Rough Sets and Knowledge Technology (RSKT'2015), pp. 482–491. Tianjin, China (2015)
26. Schumann, A., Pancerz, K.: Interfaces in a game-theoretic setting for controlling the plasmodium motions. In: Proceedings of the 8th International Conference on Bio-inspired Systems and Signal Processing (BIOSIGNALS'2015), pp. 338–343. Lisbon, Portugal (2015)
27. Schumann, A., Pancerz, K., Adamatzky, A., Grube, M.: Bio-inspired game theory: The case of Physarum polycephalum. In: Proceedings of the 8th International Conference on Bio-inspired Information and Communications Technologies (BICT'2014). Boston, Massachusetts, USA (2014)
28. Schumann, A., Pancerz, K., Jones, J.: Towards logic circuits based on Physarum polycephalum machines: the ladder diagram approach. In: Cliquet Jr., A., Plantier, G., Schultz, T., Fred, A., Gamboa, H. (eds.) Proceedings of the International Conference on Biomedical Electronics and Devices (BIODEVICES'2014), pp. 165–170. Angers, France (2014)
29. Ziarko, W.: Variable precision rough set model. J. Comput. Syst. Sci. **46**(1), 39–59 (1993)

Chapter 24
Computing a Worm: Reverse-Engineering Planarian Regeneration

Daniel Lobo and Michael Levin

Abstract In order to understand and control complex biological systems, we need to unravel the information processing and computations required to regulate their dynamics. The development of a complete organism from a single cell or the restoration of lost structures and body parts after amputations require the coordination of millions of cells exchanging and processing information. Understanding these dynamic processes from the results of biological perturbation experiments represent an outstanding challenge due to the characteristic non-linear dynamics and feed-back loops of their molecular and biophysical regulatory mechanisms—an inverse problem with no analytical or computationally tractable solutions. To bridge the gap between molecular-level mechanistic data and systems-level outcomes, we have developed a computational methodology based on heuristic algorithms to automatically reverse-engineer dynamic regulatory networks directly from experimental results. Using this method, applied to problems of pattern regulation during metazoan regeneration, we inferred the first comprehensive regulatory network of planarian regeneration, capable of explaining the most relevant experiments of anterior-posterior specification during regeneration. Here we summarize our results and study the dynamics of the inferred regulatory model, unraveling the information processing and computations required to regenerate a correct morphology.

D. Lobo (✉)
Department of Biological Sciences, University of Maryland, Baltimore County,
1000 Hilltop Circle, Baltimore, MD 21250, USA
e-mail: lobo@umbc.edu

M. Levin
Department of Biology, Center for Regenerative and Developmental Biology,
Tufts University, 200 Boston Avenue, Suite 4600, Medford, MA 02155, USA
e-mail: michael.levin@tufts.edu

© Springer International Publishing Switzerland 2017
A. Adamatzky (ed.), *Advances in Unconventional Computing*,
Emergence, Complexity and Computation 23,
DOI 10.1007/978-3-319-33921-4_24

637

24.1   Introduction

The biochemical and biophysical mechanisms that control the development of patterns, shapes, and anatomical forms are among the most complex regulatory processes in biology. These mechanisms coordinate the information exchange and processing required by cells to produce, from a fertilized egg made of a single cell, the intricate morphologies and patterns characteristic of biological organisms. In general, these processes and regulations do not stop after maturity, but continue during the entire life-span to maintain a homeostatic state, replacing senescing cells, resisting tumorigenesis, and repairing injuries [66, 67]. Certain organisms can repair themselves to an extraordinary extent. Salamanders can regenerate a complete functional limb from a total amputation [13], also regenerating their eyes, jaws, tails, and portions of the brain. The champions of regeneration are planarian flatworms—complex bilaterians that can regenerate any body part from almost any amputated fragment [1], a process that can include the formation of a complete new brain, eyes, or any other missing organ or structure in the remaining tissue [80]. Understanding the mechanisms governing these regeneration abilities will pave the way for extraordinary biomedical and bioengineering applications.

Currently, there is a vigorous scientific effort to unravel the key components governing the regulation of regeneration. A huge and increasing dataset of perturbation experiments in the literature account for the discovered genes and signaling molecules that are necessary for the organism to regenerate a correct morphology or body part. These experiments show how a certain disturbance of a regulatory component during regeneration can result in an aberrant morphology differing from the wild type one, such as the regeneration of ectopic organs or a complete body region [73]. In particular, in the case of planarians, the perturbation of regulatory key components can produce worms with double, triple, or quadruple heads, each of them including a complete brain, eyes, and any other characteristic [28, 32, 63, 69, 78]. These results serve to identify the building blocks, such as specific genes, that are necessary to regenerate a complete and correct morphology. However, knowing the required proteins is not the same as understanding the dynamics that are sufficient to enable a chunk of tissue to re-scale itself and produce exactly the missing structures in their correct locations and orientations. In particular, the plethora of recent data on stem cell differentiation in planaria does not, in itself, provide information that enables one to directly derive large-scale anatomical patterning. The next step is thus to construct from these perturbation experiments a model of the mechanisms that explains these morphological outcomes—quantitative models that can put the components of the system together to explain and predict the known and unknown morphological outcomes under perturbations [42].

In this context, a mechanism is understood not as the molecular details governing the interactions between specific proteins or genes, but as a model describing the larger picture of how the system can achieve targeted and controlled behaviors with the coordination of all its components [36]. Among modeling approaches, dynamic models based on the mathematical theory of dynamical systems represents an ideal approach for the understanding of the dynamic mechanisms of complex biological processes including regeneration [17, 23, 57, 59]. Indeed, mathematical dynamic models based on differential equations can unambiguously define the relationships and characteristics of every component in the system, that is, the biological computations that occur in the organism [60]. However, despite the clear benefits of defining dynamic models of regulation, there is a lack of mathematical models with predictive ability in the biological sciences, precisely due to the difficulty to construct them from data [42].

Indeed, the reverse-engineering of models directly from biological data represents an inverse problem, for which there are no analytical solutions [49]. Instead, Artificial Intelligence methods have been a successful approach with a long tradition [31]. Computational tools have been proposed and used for the reverse-engineering of mechanistic models from quantitative gene expression profiling data [4, 10–12, 14, 24, 27, 30, 55, 61, 86, 90, 94], time-series concentration data [22, 65, 83], and 1-dimensional gene expression patterns [7, 20, 25, 29, 33, 34, 37, 52, 53, 68, 75–77]. Here, we review a novel method for the reverse-engineering of dynamic models directly from the morphological results of perturbation experiments, and a proof of principle of its capabilities by automatically inferring a comprehensive dynamic regulatory network explaining the mechanisms of planarian regeneration, a fascinating problem intensely studied for over a century [62].

24.2 The Computations of Planarian Regeneration

Planarians are freshwater worms with a fairly complex morphology, including a true brain, two eyes, a nervous system with two ventral nerve cords, a musculature system, a branched stomach, and a pharynx for feeding [1, 74, 81]; moreover, planarians are complex enough to learn basic tasks in the lab [19, 85, 91]. Yet, despite their morphological complexity, planarians are able to regenerate a complete body from almost any amputated piece, including any missing organ such as the brain, eyes, muscles or nerves and with the correct proportions for the new size (Fig. 24.1). Indeed, planarian worms have an extraordinary capacity to remodel their morphologies, proportionally scaling down their bodies when starved, and restoring their original size when fed again [6, 64].

Much research has been done at the bench to understand the regeneration ability of planarians. Knock downs of certain genes by RNAi (RNA interference) or pharmacological interventions blocking the communication channels between cells (gap junctions) can alter the resultant morphology regenerated from an amputated worm fragment [40, 74]. For example, either by knocking down β-catenin or by blocking the gap junctions with octanol, a trunk piece regenerates a double-headed worm: instead of a tail in the posterior end, the worm regenerates an ectopic head [28, 32, 58, 63, 69]. Conversely, knocking down notum in a trunk piece results in the regeneration of a double tail worm: instead of a normal head in the anterior end, the worm regenerates an ectopic tail [71]. Similarly to these extraordinary perturbation experiments,

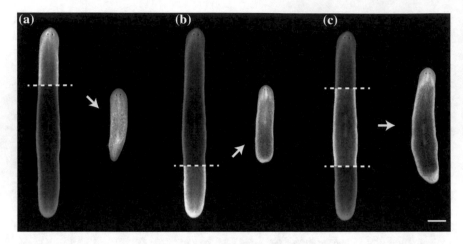

Fig. 24.1 Planarian worms can regenerate a complete body from almost any amputated piece, including **a** a head fragment, **b** a tail fragment, and **c** a trunk fragment. Each fragment regenerates any missing body region or organ and remodel anything left correctly to the new proportions

dozens of genes have been discovered as necessary for the correct regeneration of a wild type morphology; knocking them down individually or in combination with others alter the signaling mechanisms in the worm, producing a specific incorrect morphology (Fig. 24.2).

Despite the discovery of many necessary components for proper regeneration in planaria, a fundamental question still remains of how an amputated fragment is able to determine which body parts and regions are missing and need to be regenerated and which are remaining and need to be remodeled for the new size. Can we use the theory of computation to aid us in the understanding of the mechanisms necessary for regenerating a correct morphology? In contrast and complement to most molecular-genetic studies today, which focus on the material implementation of various model systems, we suggest it is imperative to understand the bigger picture of information flow and control at multiple levels of organization. To this end, Mitchell [60] proposed four questions that need to be answered in any biological computational system:

1. How is information represented in the system?
2. How is information read and written by the system?
3. How is it processed?
4. How does this information acquire function (or purpose, or meaning)?

As an answer to the first question, the information in the planarian system is represented at least at two levels [49]. The first level of information consists in the genetic and epigenetic material derived from the planarian egg, which contains the information to develop a complete worm starting from a single cell. This information is stored as an indirect encoding in the genome: there is no direct correspondence between genes and the resultant morphology, instead a developmental process orchestrated by

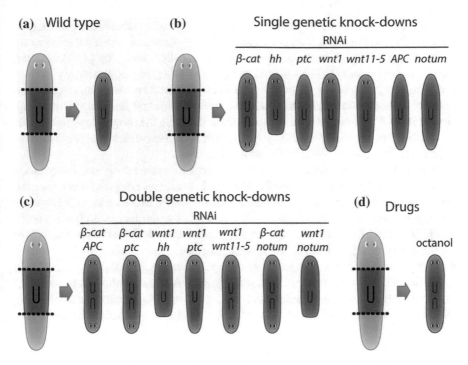

(a) Wild type   **(b)**   Single genetic knock-downs

(c) Double genetic knock-downs   **(d)** Drugs

Fig. 24.2 Perturbation experiments of planarian regeneration to explain. **a** Control experiment (no perturbation). **b** Single genetic knock-downs (RNAi). **c** Double genetic knock-downs. **d** Pharmacological perturbations altering physical properties of the organism

the interactions between genes and their resultant physical products and forces build the emergent morphology [15, 16, 18, 35, 54]. In addition, there exists a second level of information in an adult worm consisting in the maintained patterns of signaling molecules and physical properties (such as bioelectric gradients) that stores information used during regeneration. Indeed, perturbing these secondary patterns can produce different regeneration outcomes, such as double head or double tail worms, without changing the information stored in the genome [5, 40, 63].

These two levels of information stored in the worm are read and written by the molecular machinery in the cell. Signaling molecules can act as transcription factors directly binding into the DNA molecule, as enzymes affecting and changing the state of other proteins, or as intercellular signaling attaching to membrane receptors. In addition, signaling molecules can alter the physical properties in the system, such as the membrane potential in the cell, which in turn can alter the location of other charged molecules [40]. These molecular events also process the information stored in the worm, in the form of analog computations by chemical systems [21, 88].

Finally, the information stored in the genome together with the information stored in the patterns of signaling molecules and physical properties such as bioelectrical gradients acquire purpose by directing the production of specific structural proteins as well as the migration, mitosis, and apoptosis of cells, that, in turn, results in the regeneration of a specific morphology. Importantly, the planarian worm contains along its body a crucial population of adult stem cells, called neoblasts [3, 84, 92, 95]. Neoblasts are the only cells in the worm that can divide, having the ability to auto-renew and to produce any other cell type in any tissue. The neoblasts are required both for regeneration and for morphostatic maintenance and on-going remodeling. Hence, it is clear that the signaling mechanisms dictating the morphology to regenerate needs to communicate with the population of neoblasts, which then are instructed to produce the correct tissues in the correct locations.

All these processes represent a complex information processing machinery based on non-linear interactions and feedback loops. Indeed, this complexity represents a barrier for the inference of the mechanisms that give rise to the correct (or incorrect under perturbations) morphologies from a dataset of experiments, an inverse problem [72]. Indeed, there are no computer program or analytical solution that can solve the general inverse problem [2, 56]. Instead, stochastic and heuristic methods are the usual methods to find solutions to inverse problems [8, 9, 26, 39, 51, 82]. In the next section, we present our heuristic approach for solving the inverse problem in planarian regeneration: an automated reverse-engineering method for discovering mechanistic models of planarian regeneration directly from experimental data.

24.3   Reverse-Engineering Planarian Regeneration

In order to find mechanistic models that can explain a dataset of morphological experiments, we have developed a computational method based on evolutionary computation to reverse-engineer mechanistic regulatory networks directly from morphological data [45]. The computational system consists in three interconnected components: (1) a database based on a mathematical ontology curating the planarian experiments and their morphological outcomes, (2) a systems-level modeling and simulator framework to formulate and test regulatory models within virtual worms, and (3) a heuristic computational method to automatically reverse-engineer regulatory networks. In our system, the heuristic method (3) generates candidate models, which are executed by cells in a simulated worm to which various experimental perturbations are applied in silico (2). The results are tested against outcomes of known experiments in the database (1), providing feedback to guide the revision of candidate models, in a cycle that leads to progressively more correct models. The resultant models can then be interrogated with respect to their structure and dynamics, and used to make testable predictions about novel experiments not yet in the database.

Computational algorithms cannot process yet complex multidimensional scientific data, such as experimental procedures and resultant morphologies, directly from the papers published in the literature. Instead, in order to apply automated inferring methods, we need to curate first all the relevant data with formalisms and ontologies [87]. To this end, we have created ontologies based on mathematical graphs [50] to specifically formalize planarian and limb regeneration experiments and curate centralized databases containing hundreds of experiments from the literature [43, 47]. These databases can describe in an unambiguous and precise manner a large variety of different morphologies, including topological information, region characteristics, shapes, and overall sizes as well as organ locations and properties, and experimental procedures such as surgical amputations, cuts, grafts, as well as genetic and pharmacological treatments (Fig. 24.3). These resources are not only valuable for applying automated methods, but any human scientist can also access, mine, and formalize new experiments with specific user-friendly software tools based on automatically-generated cartoon diagrams and drag-and-drop interfaces [44, 46].

A candidate model of planarian regeneration needs to be tested against the current knowledge of perturbation experiments, to evaluate if its predicted outcomes correspond to those obtained in vivo. For this, we implemented a system-level simulator of planarian regeneration that can take as input a regulatory model and a formalized experiment and perform the same experiment in silico according to the given model. A model is defined as a system of partial differential equations, which define the whole-body dynamics of the planarian worm (Fig. 24.4). The simulator can perform in silico the same perturbations defined in the formal ontology defining the experiments, including surgical amputations, genetic knock-downs, and pharmacological drugs. Each product in the system represents the level of expression or activity, since some genes, like β-catenin, can be widely expressed in the organism, but only be active in the cell nucleus of certain areas of the worm in response to a signaling event. The resultant morphology from the simulation can then be compared with that in the database, and a score can be computed that quantitatively describe how

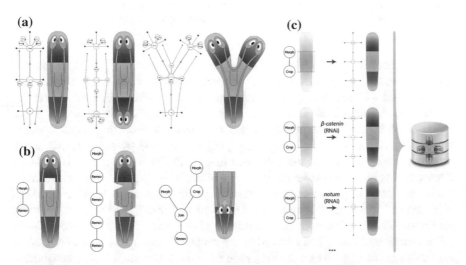

Fig. 24.3 A user- and computer-friendly mathematical ontology for planarian regeneration. **a** A set of morphological outcomes from the literature formalized with the ontology as a graph representation and automatically-generated cartoons. **b** Formalized worm manipulations as described in the literature. **c** A curated database of experiments linking manipulations to morphological outcomes

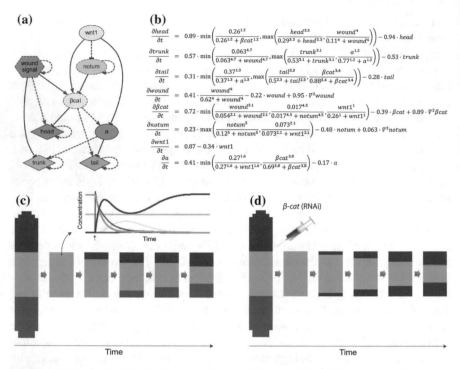

Fig. 24.4 A system-level simulator of planarian regeneration. **a** Schematic representation of a regulatory network model. **b** A complete system of partial differential equations describing a regulatory network model. **c** An in silico experiment showing the dynamics of the system. **d** A perturbation in silico experiment (knocking down the gene β-catenin) showing a different outcome

close the model is in recapitulating the morphologies from the experiments curated in the database. The current version of the virtual worm simulates only biochemical signaling, and will be extended to include bioelectrical gradients in the next version.

We also developed an automated computational method that takes as input a dataset of formalized experiments and returns a mechanistic model, reverse-engineering all the necessary components, regulations, and parameters, that, when simulated, recapitulates all the outcomes defined in the input dataset (Fig. 24.5). The method is based on evolutionary computation, where a population of candidate models are iteratively crossed and mutated to produce new models, which are then evaluated and scored in the simulator to rank them and keep the best ones for the next generation. The initial population is made of random models, including random interactions and parameters. The cross-over and mutation operators create new models from existing ones by combining the products, links, and parameters of two models from the current population, and then probabilistically adding, removing, and altering products, regulatory links, and parameters. The method automatically performs all these operations without any manual intervention. During this process, the population evolves toward candidate models with less and less error, until a model with

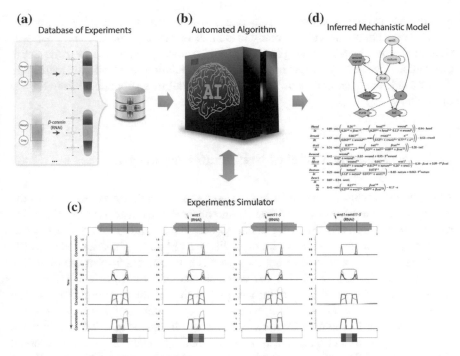

Fig. 24.5 Automated method for the reverse-engineering of regulatory networks directly from morphological experimental data. **a** The input dataset of formalized experiments. **b** The heuristic machine learning algorithm based on evolutionary computation. **c** The in silico experiment simulator used by the algorithm to evaluate candidate models. **d** The output of the algorithm is an inferred regulatory network model defined as a complete system of partial differential equations

zero error is found, which can perfectly recapitulate all the experiments formalized in the database. Thus, while it is possible to seed the initial population with candidate human-derived models, the method is able to generate models with high explanatory power with no human input. This is an important feature, since the (ever-growing) data of planarian regeneration are so numerous and remarkable, that no constructive, comprehensive model had ever been proposed (or specified in sufficient detail to be quantitatively testable) despite over 120 years of focused attention.

The distributed nature of the evolutionary algorithm makes it amenable for a parallel implementation in a high performance computing environment. The simulation and evaluation of each of the new models produced by the method can be performed independently from each other. We designed a parallel implementation of the algorithm based on a master-slave scheme. The master node stores the evolving population, applies the cross-over and mutation operators to create new models, and select the best models to keep in the population. A set of slave nodes receive from the master node the new models, simulate and evaluate them in parallel, and return the model scores to the master node. In addition, the evolving population is divided in subpopulations (called islands), which allows further parallelization (while the master node is busy creating and selecting new candidate models for a subpopulation, others subpopulations are being evaluated in the slave nodes) and preserves the population diversity (a single good model cannot quickly take over the entire population) [93]. Every 250 generations, all subpopulations are randomly paired and their models are randomly shuffled to avoid the stall of a subpopulation in a local optimum.

24.4 A Dynamic Model of Planarian Regeneration

Applying our method to the main head-versus-tail regeneration experiments in the worm, we reverse-engineered the first comprehensive dynamic network of planarian regeneration (Fig. 24.6). First, we encoded with the formal ontology a dataset of in vivo experiments to serve as input to the algorithm. The dataset included the most important experiments of anterior-posterior regeneration from the planarian literature [28, 32, 63, 69–71, 79], including surgical, pharmacological, and genetic (RNAi-mediated knockdown) perturbations and their resulting morphologies. The algorithm ran for 42 hours using 256 cores in a computer cluster (*Stampede* at the Texas Advanced Computing Center), performing billions of in silico experiments, and returning as output a regulatory network model (system of partial differential equations) with all parameters and relationships inferred de novo. This network can recapitulate in the in silico simulator all the resultant morphologies from all the experiments included in the input dataset.

Interestingly, the comprehensive model automatically reverse-engineered by our method is not a hairball of puzzling connections, but a small set of regulatory interactions mostly consistent with our molecular knowledge from the literature that can readily be understood by a human scientist. In addition, the model predicted new regulatory interactions yet-to-be-discovered. Interestingly, the algorithm consistently predicted that notum directly inhibits wnt, an interaction that was molecularly confirmed recently in vivo [38], after the prediction made by our model and contrary to the prevalent hypothesis in the field. Moreover, our method detected as necessary two unknown genes (labeled *a* and *b*, Fig. 6): without them, the algorithm could not find a regulatory network that could recapitulate the input dataset of experiments. These unknown genes can be further characterized by searching for known genes with the same predicted interactions in other organisms curated in specialized databases, such as the STRING protein-protein interaction database [89]. We are currently validating at the bench candidates for such novel genes [48].

Importantly, the mathematical nature of the model allows us to study the dynamic mechanisms of the signaling system (Fig. 24.7). Visualizing the state space during the regeneration of a trunk fragment, we can analyze the dynamics defined by the system. Figure 24.7 shows the state trajectory of different morphological and genetic products in three locations in the trunk fragment: anterior, center, and posterior. In all the products shown, the initial state (the pre-perturbation wild-type morphology) is the same in the three fragment locations, except in the case of notum, which is activated

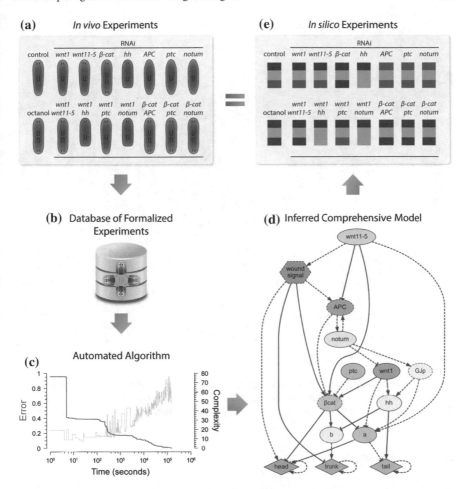

Fig. 24.6 Automatically-inferred comprehensive model of planarian regeneration. **a** Dataset of formalized in vivo experiments. **b** Formalization of the experiments in a database. **c** Evolutionary process run by the automatic method. **d** Schematic of the resultant inferred regulatory network model. **e** The inferred model can recapitulate in silico all the patterning outcomes of the input dataset of in vivo experiments

higher in the anterior side of the fragment that in the posterior side. In the wild type experiment (no treatment), the system defines three clear attractors corresponding to the head, trunk, and tail morphological outcomes, where the trajectories of the anterior, center, and posterior locations of the fragment converge respectively. However, knocking-down β-catenin or notum produces a bifurcation in the dynamic regulatory system, altering the attractors location and hence the trajectories and resulting morphological outcomes.

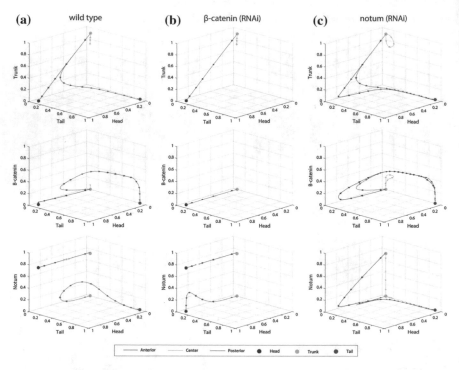

Fig. 24.7 Phase portraits showing the dynamics of the comprehensive model for the **a** control experiment, **b** knock down of β-catenin, and **c** knock down of notum. This analysis reveals the dynamic properties defined by the model, including the trajectories, attractors, and bifurcations

24.5 Discussion

Planarian worms possess an extraordinary ability to regenerate their complete bodies from almost any amputated fragment. The inherent complexity of the biological regulation of patterns, shapes, and forms prevent us to manually derive mechanistic models from the large dataset of perturbation experiments existent in the literature. Instead, it is clear that computational methods are needed to aid us in reverse-engineering mathematical models with the capacity to recapitulate the observed outcomes, provide accurate predictions, and offer a comprehensive view of all the necessary components and relations controlling the system.

Here we reviewed an integrated approach for the reverse engineering of dynamic models of regeneration directly from morphological outcomes. Our methodology is based on a formal ontology to mathematically describe regenerative experiments and morphologies, a systems-level simulator of tissue patterning, and a machine learning heuristic algorithm to automatically infer dynamic regulatory models directly from the data. Together, these components can automate the process of discovering a mechanistic model, based on partial differential equations, that can recapitulate the exact outcomes described in the input dataset of experiments. As a proof of principle, we applied our method to the problem of planarian regeneration, resulting in a comprehensive model able to recapitulate a diverse set of genetic and pharmacological experiments. The inferred model includes all the necessary components, interactions, and parameters to compute the precise morphological outcomes. Precisely, the computational nature of the model, accurately described as a set of biochemical processes, allow us to describe and study the exact dynamics of the system, revealing the attractors and bifurcations responsible for the observed behavior in the different experiments and the inherent controlling mechanisms in the system. Furthermore, the dynamic model can readily be used for predicting the outcome of any novel experimental perturbation of its defined components, which can be then validated by replicating the same experiments in vivo at the bench.

Our method is not restricted to regenerative experiments, but it is general enough to be applicable to any biological system which experiments can be formalized and simulated in silico. Indeed, we have used a similar approach for the reverse engineering of a dynamic regulatory network of a melanoma-like phenotype in *Xenopus* [41]. In this case, the system is stochastic: two simulations with the same conditions can result in different outcomes. Importantly, the reverse-engineered signaling system model could not only recapitulate the observed phenotypes, but also the exact stochastic ratios obtained in vivo. In conclusion, this methodology is widely applicable, and able to unravel complex regulatory systems directly from morphological data. Indeed, the discovered dynamic models represent a mechanistic understanding of the unconventional computations necessary to control these biological systems, offering key insights for finding novel interventions to complex, multidimensional processes such as regeneration and diseases such as cancer.

In addition to the computational approaches to understanding the regenerative properties of extant species of planaria, there is another sense of computation in this field: that performed by the worm itself. Using networks of chemically- and bioelectrically-connected elements, these remarkable animals compute everything necessary to restore a complex 3-dimensional shape despite unpredictable experimental perturbations. It remains to be seen whether the smallest computational units are cells, subcellular organelles (e.g., cytoskeleton), or multi-cellular subnetworks. Nevertheless, in addition to the obvious implications for regenerative medicine and robust artificial constructs, understanding the general dynamics that enable this will surely enrich our understanding of unconventional computing. This novel architecture performs remarkable feats of real-time pattern memory and distributed decision-making. As much as we seek to make computational models of real worm species produced by the biosphere's evolutionary history, the worms are themselves models of fundamental computational principles that once discovered, will have implications well beyond any single material implementation.

Acknowledgments This work was supported by NSF grant EF-1124651, NIH grant GM078484, USAMRMC grant W81XWH-10-2-0058, and the Mathers Foundation. Computation used the Extreme Science and Engineering Discovery Environment (XSEDE), which is supported by NSF grant ACI-1053575, and a cluster computer awarded by Silicon Mechanics.

References

1. Aboukhatwa, E., Aboobaker, A.: An Introduction to Planarians and Their Stem Cells. Wiley, New York (2015). doi:10.1002/9780470015902.a0001097.pub2
2. Aster, R.C., Thurber, C.H.: Parameter Estimation and Inverse Problems, 2nd edn. Academic Press, Waltham (2012)
3. Baguna, J., Saló, E., Auladell, C.: Regeneration and pattern formation in planarians. iii. evidence that neoblasts are totipotent stem-cells and the source of blastema cells. Development **107**(1), 77–86 (1989)
4. Basso, K., Margolin, A.A., Stolovitzky, G., Klein, U., Dalla-Favera, R., Califano, A.: Reverse engineering of regulatory networks in human b cells. Nat. Genet **37**(4), 382–390 (2005). doi:10.1038/ng1532
5. Beane, W.S., Morokuma, J., Adams, D.S., Levin, M.: A chemical genetics approach reveals h, k-atpase-mediated membrane voltage is required for planarian head regeneration. Chem. Biol. **18**(1), 77–89 (2011)
6. Beane, W.S., Morokuma, J., Lemire, J.M., Levin, M.: Bioelectric signaling regulates head and organ size during planarian regeneration. Development **140**(2), 313–22 (2013)
7. Becker, K., Balsa-Canto, E., Cicin-Sain, D., Hoermann, A., Janssens, H., Banga, J.R., Jaeger, J.: Reverse-engineering post-transcriptional regulation of gap genes in drosophila melanogaster. PLOS Comput. Biol. **9**(10), e1003,281 (2013). doi:10.1371/journal.pcbi.1003281
8. Bonabeau, E.: From classical models of morphogenesis to agent-based models of pattern formation. Artif. Life **3**(3), 191–211 (1997)
9. Bongard, J., Lipson, H.: Automated reverse engineering of nonlinear dynamical systems. Proc. Natl. Acad. Sci. U. S. A **104**(24), 9943–9948 (2007). doi:10.1073/pnas.0609476104
10. Bonneau, R., Facciotti, M.T., Reiss, D.J., Schmid, A.K., Pan, M., Kaur, A., Thorsson, V., Shannon, P., Johnson, M.H., Bare, J.C., Longabaugh, W., Vuthoori, M., Whitehead, K., Madar, A., Suzuki, L., Mori, T., Chang, D.E., DiRuggiero, J., Johnson, C.H., Hood, L., Baliga, N.S.: A predictive model for transcriptional control of physiology in a free living cell. Cell **131**(7), 1354–1365 (2007). doi:10.1016/j.cell.2007.10.053
11. Bonneau, R., Reiss, D.J., Shannon, P., Facciotti, M., Hood, L., Baliga, N.S., Thorsson, V.: The inferelator: an algorithm for learning parsimonious regulatory networks from systems-biology data sets de novo. Genome Biol. **7**(5) Artn R36 (2006). doi:10.1186/Gb-2006-7-5-R36
12. Botman, D., Röttinger, E., Martindale, M.Q., de Jong, J., Kaandorp, J.A.: A computational approach towards a gene regulatory network for the developing *Nematostella vectensis*. PLOS ONE **9**(7), e103341 (2014). doi:10.1371/journal.pone.0103341
13. Brockes, J.P., Kumar, A.: Appendage regeneration in adult vertebrates and implications for regenerative medicine. Science **310**(5756), 1919–1923 (2005). doi:10.1126/science.1115200
14. Cantone, I., Marucci, L., Iorio, F., Ricci, M.A., Belcastro, V., Bansal, M., Santini, S., di Bernardo, M., di Bernardo, D., Cosma, M.P.: A yeast synthetic network for in vivo assessment of reverse-engineering and modeling approaches. Cell **137**(1), 172–181 (2009). doi:10.1016/j.cell.2009.01.055
15. Carroll, S.: Endless Forms Most Beautiful: The New Science of Evo Devo and the Making of the Animal Kingdom. W. W. Norton & Company, New York (2005)
16. Carroll, S., Grenier, J., Weatherbee, S.: From DNA to Diversity: Molecular Genetics and the Evolution of Animal Design. Wiley, Blackwell, New York (2004)
17. Chara, O., Tanaka, E.M., Brusch, L.: Mathematical Modeling of Regenerative Processes, **108**, 283–317 (2014). doi:10.1016/B978-0-12-391498-9.00011-5
18. Cohen, I.R., Harel, D.: Explaining a complex living system: dynamics, multi-scaling and emergence. J. Royal Soc. Interface **4**(13), 175–182 (2007). doi:10.1098/rsif.2006.0173
19. Corning, W.C., Freed, S.: Planarian behaviour and biochemistry. Nature **219**(160), 1227–9 (1968)
20. Crombach, A., Wotton, K.R., Cicin-Sain, D., Ashyraliyev, M., Jaeger, J.: Efficient reverse-engineering of a developmental gene regulatory network. PLOS Comput. Biol. **8**(7), e1002589 (2012). doi:10.1371/journal.pcbi.1002589

21. Daniel, R., Rubens, J.R., Sarpeshkar, R., Lu, T.K.: Synthetic analog computation in living cells. Nature **497**(7451), 619–23 (2013). doi:10.1038/nature12148

22. Edwards, J.S., Palsson, B.O.: The escherichia coli mg1655 in silico metabolic genotype: Its definition, characteristics, and capabilities. Proc. Natl. Acad. Sci. U. S. A. **97**(10), 5528–5533 (2000). doi:10.1073/pnas.97.10.5528

23. Ellner, S.P., Guckenheimer, J.: Dynamic Models in Biology. Princeton University Press, Princeton (2006)

24. Faith, J.J., Hayete, B., Thaden, J.T., Mogno, I., Wierzbowski, J., Cottarel, G., Kasif, S., Collins, J.J., Gardner, T.S.: Large-scale mapping and validation of escherichia coli transcriptional regulation from a compendium of expression profiles. PLOS Biol. **5**(1), e8 (2007). doi:10.1371/journal.pbio.0050008

25. Fomekong-Nanfack, Y., Kaandorp, J.A., Blom, J.: Efficient parameter estimation for spatiotemporal models of pattern formation: case study of drosophila melanogaster. Bioinformatics **23**(24), 3356–3363 (2007). doi:10.1093/bioinformatics/btm433

26. Ganguly, N., Sikdar, B.K., Deutsch, A., Canright, G., Chaudhuri, P.: A Survey on Cellular Automata. Report, Centre for High Performance Computing, Dresden University of Technology (2003)

27. Gardner, T.S., di Bernardo, D., Lorenz, D., Collins, J.J.: Inferring genetic networks and identifying compound mode of action via expression profiling. Science **301**(5629), 102–105 (2003). doi:10.1126/science.1081900

28. Gurley, K.A., Rink, J.C.: Sanchez Alvarado, A.: β-catenin defines head versus tail identity during planarian regeneration and homeostasis. Science **319**(5861), 323–327 (2008)

29. Gursky, V.V., Panok, L., Myasnikova, E.M., Manu Samsonova, M.G., Reinitz, J., Samsonov, A.M.: Mechanisms of gap gene expression canalization in the drosophila blastoderm. BMC Syst. Biol. **5**, 118 (2011). doi:10.1186/1752-0509-5-118

30. Hecker, M., Lambeck, S., Toepfer, S., van Someren, E., Guthke, R.: Gene regulatory network inference: Data integration in dynamic models—review. Biosystems **96**(1), 86–103 (2009). doi:10.1016/j.biosystems.2008.12.004

31. Hunter, L.: Artificial intelligence and molecular biology. AI Mag. **11**(5), 27–36 (1990)

32. Iglesias, M., Gomez-Skarmeta, J.L., Saló, E., Adell, T.: Silencing of smed-β-catenin generates radial-like hypercephalized planarians. Development **135**(7), 1215–1221 (2008)

33. Ilsley, G.R., Fisher, J., Apweiler, R., DePace, A.H., Luscombe, N.M.: Cellular resolution models for even skipped regulation in the entire Drosophila embryo, vol. 2 (2013). doi:10.7554/eLife.00522

34. Jaeger, J., Blagov, M., Kosman, D., Kozlov, K.N., Manu, Myasnikova, E., Surkova, S., Vanario-Alonso, C.E., Samsonova, M., Sharp, D.H., Reinitz, J.: Dynamical analysis of regulatory interactions in the gap gene system of drosophila melanogaster. Genetics **167**(4), 1721–1737 (2004). doi:10.1534/genetics.104.027334

35. Jaeger, J., Crombach, A.: Life attractors: understanding developmental systems through reverse engineering and in silico evolution, Adv. Exp. Med. Biol. **751**, 93–119 (2012)

36. Jaeger, J., Sharpe, J.: On the Concept of Mechanism in Development. Oxford University Press, Oxford (2014)

37. Jaeger, J., Surkova, S., Blagov, M., Janssens, H., Kosman, D., Kozlov, K.N., Manu, Myasnikova, E., Vanario-Alonso, C.E., Samsonova, M., Sharp, D.H., Reinitz, J.: Dynamic control of positional information in the early drosophila embryo. Nature **430**(6997), 368–371 (2004). doi:10.1038/Nature02678

38. Kakugawa, S., Langton, P.F., Zebisch, M., Howell, S.A., Chang, T.H., Liu, Y., Feizi, T., Bineva, G., O/'Reilly, N., Snijders, A.P., Jones, E.Y., Vincent, J.P.: Notum deacylates wnt proteins to suppress signalling activity. Nature **519**(7542), 187–192 (2015). doi:10.1038/nature14259

39. Koza, J.: Genetic Programming: On the Programming of Computers by Means of Natural Selection (Complex Adaptive Systems). The MIT Press, Cambridge (1992)

40. Levin, M.: Endogenous bioelectrical networks store non-genetic patterning information during development and regeneration. J. Physiol. **592**(11), 2295–2305 (2014). doi:10.1113/jphysiol.2014.271940

41. Lobikin, M., Lobo, D., Blackiston, D., Martyniuk, C., Tkachenko, E., Levin, M.: Serotonergic regulation of melanocyte conversion: a bioelectrically regulated network for stochastic all-or-none hyperpigmentation. Sci. Signal. **8**(397), ra99 (2015)
42. Lobo, D., Beane, W., Levin, M.: Modeling planarian regeneration: a primer for reverse-engineering the worm. PLOS Comput. Biol. **8**(4), e1002481 (2012). doi:10.1371/journal.pcbi. 1002481
43. Lobo, D., Feldman, E.B., Shah, M., Malone, T.J., Levin, M.: A bioinformatics expert system linking functional data to anatomical outcomes in limb regeneration. Regeneration **1**(2), 37–56 (2014). doi:10.1002/reg2.13
44. Lobo, D., Feldman, E.B., Shah, M., Malone, T.J., Levin, M.: Limbform: a functional ontology-based database of limb regeneration experiments. Bioinformatics **30**(24), 3598–600 (2014). doi:10.1093/bioinformatics/btu582
45. Lobo, D., Levin, M.: Inferring regulatory networks from experimental morphological pheno-types: a computational method reverse-engineers planarian regeneration. PLOS Comput. Biol. **11**(6), e1004295 (2015). doi:10.1371/journal.pcbi.1004295
46. Lobo, D., Malone, T.J., Levin, M.: Planform: an application and database of graph-encoded planarian regenerative experiments. Bioinformatics **29**(8), 1098–1100 (2013). doi:10.1093/bioinformatics/btt088
47. Lobo, D., Malone, T.J., Levin, M.: Towards a bioinformatics of patterning: a computational approach to understanding regulative morphogenesis. Biol. Open **2**(2), 156–169 (2013). doi:10.1242/bio.20123400
48. Lobo D., Morokuma, J., Levin, M.: Computational discovery and in vivo validation of hnf4 as a regulatory gene in planarian regeneration. Bioinformatics, in press. doi: 10.1093/bioinformatics/btw299
49. Lobo, D., Solano, M., Bubenik, G.A., Levin, M.: A linear-encoding model explains the variability of the target morphology in regeneration. J. R. Soc. Interface **11**(92) (2014). doi:10.1098/rsif.2013.0918
50. Lobo, D., Vico, F., Dassow, J.: Graph grammars with string-regulated rewriting. Theor. Comput. Sci. **412**(43), 6101–6111 (2011). doi:10.1016/j.tcs.2011.07.004
51. Lobo, D., Vico, F.J.: Evolutionary development of tensegrity structures. Biosystems **101**(3), 167–176 (2010). doi:10.1016/j.biosystems.2010.06.005
52. Manu, Surkova, S., Spirov, A.V., Gursky, V.V., Janssens, H., Kim, A.R., Radulescu, O., Vanario-Alonso, C.E., Sharp, D.H., Samsonova, M., Reinitz, J.: Canalization of gene expression and domain shifts in the drosophila blastoderm by dynamical attractors. PLOS Comput. Biol. **5**(3), e1000303 (2009). doi:10.1371/journal.pcbi.1000303
53. Manu, Surkova: S., Spirov, A.V., Gursky, V.V., Janssens, H., Kim, A.R., Radulescu, O., Vanario-Alonso, C.E., Sharp, D.H., Samsonova, M., Reinitz, J.: Canalization of gene expression in the drosophila blastoderm by gap gene cross regulation. Plos Biol. **7**(3), 591–603 ARTN e1000049 (2009). doi:10.1371/journal.pbio.1000049
54. Marcus, G.: The Birth Of The Mind: How A Tiny Number of Genes Creates the Complexities of Human Thought. Basic Books, New York (2003)
55. Margolin, A.A., Nemenman, I., Basso, K., Wiggins, C., Stolovitzky, G., Dalla Favera, R., Califano, A.: Aracne: an algorithm for the reconstruction of gene regulatory networks in a mammalian cellular context. BMC Bioinform. **7** Suppl 1, S7 (2006). doi:10.1186/1471-2105-7-S1-S7
56. McCarthy, J.: The Inversion of Functions Defined by Turing Machines, vol. 34, pp. 177–181. Princeton University Press, Princeton (1956)
57. Meinhardt, H.: Models of Biological Pattern Formation. Academic Press, Cambridge (1982)
58. Meinhardt, H.: Beta-catenin and axis formation in planarians. Bioessays **31**(1), 5–9 (2009)
59. Meinhardt, H.: Models for the generation and interpretation of gradients. Cold Spring Harb. Perspect. Biol. **1**(4) (2009). doi:10.1101/cshperspect.a001362
60. Mitchell, M.: Biological computation. Comput. J. **55**(7), 852–855 (2012). doi:10.1093/comjnl/bxs078

61. Molinelli, E.J., Korkut, A., Wang, W., Miller, M.L., Gauthier, N.P., Jing, X., Kaushik, P., He, Q., Mills, G., Solit, D.B., Pratilas, C.A., Weigt, M., Braunstein, A., Pagnani, A., Zecchina, R., Sander, C.: Perturbation biology: Inferring signaling networks in cellular systems. PLOS Comput. Biol. **9**(12), e1003290 (2013). doi:10.1371/journal.pcbi.1003290

62. Morgan, T.: Experimental studies of the regeneration of planaria maculata. Dev. Genes Evol. **7**(2), 364–397 (1898)

63. Oviedo, N.J., Morokuma, J., Walentek, P., Kema, I., Gu, M.B., Ahn, J.M., Hwang, J.S., Gojobori, T., Levin, M.: Long-range neural and gap junction protein-mediated cues control polarity during planarian regeneration. Dev. Biol. **339**(1), 188–199 (2010)

64. Oviedo, N.J., Newmark, P.A.: Sánchez Alvarado, A.: Allometric scaling and proportion regulation in the freshwater planarian schmidtea mediterranea. Dev. Dyn.**226**(2), 326–333 (2003)

65. Patil, K.R., Nielsen, J.: Uncovering transcriptional regulation of metabolism by using metabolic network topology. Proc. Natl. Acad. Sci. U.S.A. **102**(8), 2685–2689 (2005). doi:10.1073/pnas. 0406811102

66. Pearson, B.: Sanchez Alvarado, A.: Regeneration, stem cells, and the evolution of tumor suppression. Cold Spring Harb. Symp. Quant. Biol. **73**, 565–572 (2008). doi:10.1101/sqb.2008. 73.045

67. Pellettieri, J., Sanchez Alvarado, A.: Cell turnover and adult tissue homeostasis: From humans to planarians. Annu. Rev. Genet. **41**, 83–105 (2007). doi:10.1146/annurev.genet.41.110306. 130244

68. Perkins, T.J., Jaeger, J., Reinitz, J., Glass, L.: Reverse engineering the gap gene network of drosophila melanogaster. PLOS Comput. Biol. **2**(5), 417–428 (2006). doi:10.1371/journal. pcbi.0020051

69. Petersen, C.P., Reddien, P.W.: Smed-βcatenin-1 is required for anteroposterior blastema polarity in planarian regeneration. Science **319**(5861), 327–330 (2008)

70. Petersen, C.P., Reddien, P.W.: A wound-induced wnt expression program controls planarian regeneration polarity. Proc. Natl. Acad. Sci. U. S. A. **106**(40), 17061–17066 (2009). doi:10. 1073/pnas.0906823106

71. Petersen, C.P., Reddien, P.W.: Polarized notum activation at wounds inhibits wnt function to promote planarian head regeneration. Science **332**(6031), 852–855 (2011). doi:10.1126/ science.1202143

72. Ramm, A.G.: Inverse Problems. Mathematical and analytical techniques with applications to engineering. Springer, New York (2005)

73. Reddien, P.W., Bermange, A.L., Murfitt, K.J., Jennings, J.R., Alvarado, A.S.: Identification of genes needed for regeneration, stem cell function, and tissue homeostasis by systematic gene perturbation in planaria. Develop. Cell **8**(5), 635–649 (2005). doi:10.1016/j.devcel.2005.02. 014

74. Reddien, P.W., Sanchez Alvarado, A.: Fundamentals of planarian regeneration: Annu. Rev. cell Develop. Biol.**20**, 725–757 (2004)

75. Reinitz, J., Kosman, D., Vanario-Alonso, C.E., Sharp, D.H.: Stripe forming architecture of the gap gene system. Develop. Genet. **23**(1), 11–27 (1998). doi:10.1002/(Sici)1520-6408(1998)23: 1<11:Aid-Dvg2>3.0.Co;2-9

76. Reinitz, J., Mjolsness, E., Sharp, D.H.: Model for cooperative control of positional information in drosophila by bicoid and maternal hunchback. J. Exp. Zool. **271**(1), 47–56 (1995). doi:10. 1002/jez.1402710106

77. Reinitz, J., Sharp, D.H.: Mechanism of eve stripe formation. Mech. Develop. **49**(1–2), 133–158 (1995). doi:10.1016/0925-4773(94)00310-J

78. Reuter, H., Mäz, M., Vogg, M., Eccles, D., Gírfol-Boldú, L., Wehner, D., Owlarn, S., Adell, T., Weidinger, G., Bartscherer, K.: β-catenin-dependent control of positional information along the ap body axis in planarians involves a teashirt family member. Cell Rep. **10**(2), 253–265 (2015). doi:10.1016/j.celrep.2014.12.018

79. Rink, J.C., Gurley, K.A., Elliott, S.A.: Sánchez Alvarado, A.: Planarian hh signaling regulates regeneration polarity and links hh pathway evolution to cilia. Science **326**(5958), 1406–1410 (2009)

80. Roberts-Galbraith, R.H., Newmark, P.A.: On the organ trail: insights into organ regeneration in the planarian. Curr. Opin. Genet. Develop. **32**, 37–46 (2015). doi:10.1016/j.gde.2015.01.009
81. Saló, E., Abril, J.F., Adell, T., Cebricá, F., Eckelt, K., Fernandez-Taboada, E., Handberg-Thorsager, M., Iglesias, M., Molina, M.D.D., Rodrguez-Esteban, G.: Planarian regeneration: achievements and future directions after 20 years of research. Int. J. Develop. Biol. **53**(8–10), 1317–1327 (2009)
82. Schmidt, M., Lipson, H.: Distilling free-form natural laws from experimental data. Science **324**(5923), 81–85 (2009)
83. Schmidt, M.D., Vallabhajosyula, R.R., Jenkins, J.W., Hood, J.E., Soni, A.S., Wikswo, J.P., Lipson, H.: Automated refinement and inference of analytical models for metabolic networks. Phys. Biol. **8**(5), 055011 (2011). doi:10.1088/1478-3975/8/5/055011
84. Scimone, M., Kravarik, K., Lapan, S., Reddien, P.: Neoblast specialization in regeneration of the planarian schmidtea mediterranea. Stem Cell Rep. **3**(2), 339–352 (2014). doi:10.1016/j.stemcr.2014.06.001
85. Shomrat, T., Levin, M.: An automated training paradigm reveals long-term memory in planaria and its persistence through head regeneration. J. Exp. Biol. **216**(20), 3799–3810 (2013). doi:10.1242/jeb.087809
86. Sirbu, A., Ruskin, H., Crane, M.: Comparison of evolutionary algorithms in gene regulatory network model inference. BMC Bioinform. **11**(1), 59 (2010)
87. Soldatova, L., King, R.: An ontology of scientific experiments. J. R. Soc. Interface **3**(11), 795–803 (2006)
88. Solë, R.V., Macia, J.: Expanding the landscape of biological computation with synthetic multicellular consortia. Nat. Comput. 1–13 (2013). doi:10.1007/s11047-013-9380-y
89. Szklarczyk, D., Franceschini, A., Wyder, S., Forslund, K., Heller, D., Huerta-Cepas, J., Simonovic, M., Roth, A., Santos, A., Tsafou, K.P., Kuhn, M., Bork, P., Jensen, L.J., von Mering, C.: String v10: protein-protein interaction networks, integrated over the tree of life. Nucl. Acids Res. **43**(D1), D447–D452 (2015). doi:10.1093/nar/gku1003
90. Tegner, J., Yeung, M.K., Hasty, J., Collins, J.J.: Reverse engineering gene networks: integrating genetic perturbations with dynamical modeling. Proc. Natl. Acad. Sci. U. S. A. **100**(10), 5944–5949 (2003). doi:10.1073/pnas.0933416100
91. Van Oye, P.: Over het geheugen bij de platwormen en andere biologische waarnemingen bij deze dieren. Natuurwet. Tijdschr **2**, 1–9 (1920)
92. Wagner, D.E., Wang, I.E., Reddien, P.W.: Clonogenic neoblasts are pluripotent adult stem cells that underlie planarian regeneration. Science **332**(6031), 811–816 (2011). doi:10.1126/science.1203983
93. Whitley, D., Rana, S., Heckendorn, R.B.: The island model genetic algorithm: on separability, population size and convergence. J. Comput. Inf. Technol. **7**, 33–48 (1999)
94. Yeung, M.K.S., Tegnér, J., Collins, J.J.: Reverse engineering gene networks using singular value decomposition and robust regression. Proc. Natl. Acad. Sci. **99**(9), 6163–6168 (2002). doi:10.1073/pnas.092576199
95. Zhu, S.J., Hallows, S.E., Currie, K.W., Xu, C., Pearson, B.J.: A mex3 homolog is required for differentiation during planarian stem cell lineage development. eLife **4** (2015). doi:10.7554/eLife.07025

Chapter 25
An Integrated *In Silico* Simulation and Biomatter Compilation Approach to Cellular Computation

Savas Konur, Harold Fellermann, Larentiu Marian Mierla, Daven Sanassy, Christophe Ladroue, Sara Kalvala, Marian Gheorghe and Natalio Krasnogor

Abstract Recent advances Synthetic Biology are ushering a new practical computational substrate based on programmable information processing via biological cells. Due to the difficulties in orchestrating complex programmes using myriads of relatively simple, limited and highly stochastic processors such as living cells, robust computational technology to specify, simulate, analyse and compile cellular programs are in demand. We provide the INFOBIOTICS WORKBENCH (IBW) tool, a software platform developed to model and analyse stochastic compartmentalized systems, which permits using various computational techniques, such as *modelling*, *simulation, verification* and *biocompilation*. We report here the details of our work for modelling, simulation and, for the first time, biocompilation, while verification is reported elsewhere in this book. We consider some basic genetic logic gates to illustrate the main features of the IBW platform. Our results show that membrane computing provides a suitable formalism for building synthetic biology models. The software platform we developed permits analysing biological systems through the computational methods integrated into the workbench, providing significant advantages in terms of time, and enhanced understanding of biological functionality.

S. Konur · M. Gheorghe
School of Electrical Engineering and Computer Science, University of Bradford,
Richmond Road, Bradford BD7 1DP, UK

L. Marian Mierla
Faculty of Mathematics and Computer Science, University of Bucharest,
Str. Academiei Nr.14, Sector 1, C.P., 010014 Bucharest, Romania

C. Ladroue · S. Kalvala
Department of Computer Science, University of Warwick, Coventry CV4 7AL, UK

H. Fellermann · D. Sanassy · N. Krasnogor (✉)
School of Computing Science, Newcastle University, Claremont Tower,
Newcastle-upon-Tyne NE1 7RU, UK
e-mail: natalio.krasnogor@newcastle.ac.uk

© Springer International Publishing Switzerland 2017
A. Adamatzky (ed.), *Advances in Unconventional Computing*,
Emergence, Complexity and Computation 23,
DOI 10.1007/978-3-319-33921-4_25

25.1   Introduction

Synthetic biology, a rapidly growing discipline, is concerned with the design and implementation of new biological phenotypes [1]. These "designer" phenotypes do not appear in nature yet can be –in principle– constructed by a judicious combination of natural or synthetic bioparts [9, 28, 43].

Although synthetic biology promises a revolutionary new technology to engineer biological systems, the complex nature of the interactions within genetic and signalling networks makes the design and analysis of these designer phenotypes difficult to implement in the wet lab. Thus, in a similar fashion like systems biology, its elder sibling discipline, synthetic biology depends heavily on sophisticated computation tools for specifying, simulating, verifying and "compiling" biological designs. Building the models of these networks and analysing their behaviour through computational methods can provide significant advantages in terms of time, and enhanced understanding of biological functionality.

Several computational methods and tools (e.g. GEC [39], Eugene [8] and Proto [4]) have emerged in recent years with respect to the synthetic biology design and analysis. However, current state of the art tools provide limited support in terms of analytical techniques available. Most existing tools rely, predominantly, on simulation. This limits the understanding of biological functionality. Also, most of the existing approaches do not consider the *compartmental* nature of biological systems. This is an important limitation considering the fact that biological process are highly organised not only in time but also in space [14].

In this chapter, we study some synthetic biology design problems using the methods and techniques derived from *membrane computing*. Membrane computing [38] is a branch of natural computing inspired by the hierarchical structure of living cells with various compartments inside them, or the network of cells occurring in tissues and organs, and key functions describing molecular interactions of species and macromolecules. Thus, membrane computing presents an explicit account of spatial organisation, providing a natural formalism for modelling (and verifying) synthetic (and systems) biology models. It emphasises the *compartmentalised* nature of biological systems (not supported by most techniques and tools currently used in synthetic biology). Indeed, it has been formally shown [18, 19] that "compartmentalisation" fundamentally alters the way in which biochemical reactions take place and thus its explicit inclusion in biomodelling should be pursued. Membrane computing also supports the study of the computational power, complexity and efficiency of its models, and deals with their applications in various fields.

We base our system on a particular model in membrane computing called a *P system*, consisting of a membrane structure and rewriting rules operating on multisets of objects [38]. P systems mimic chemical reactions and transportation across membranes or cellular division or death processes by repeatedly applying rules. P systems provide a clear mapping of different regions and compartments of a biological systems into membranes, as illustrated in various case studies [2, 3, 7, 17, 52]. Each molecular species is associated with an object in the multiset corresponding to a membrane mapping the region or compartment where the molecules are located. Since P systems are close to biology, they are a suitable formalism for representing biological systems, especially (multi-)cellular systems and molecular interactions taking place in different locations of living cells [20].

Stochastic P (SP) systems [45] are a probabilistic extension of P systems, where reaction rates are derived from elementary rate constants in accordance with the law of mass action kinetics. This modelling technique is ideally suited for biological and especially for genetic systems, where the inherent noise present in dynamics of small copy number of systems cannot be accurately captured by more traditional ordinary differential equation based methods. SP systems are a natural, intuitive and amenable formalism, capturing the stochastic dynamics of biological and chemical systems [44]. They have been used in the analysis of various biological systems [11, 29–31, 48].

The INFOBIOTICS WORKBENCH (IBW) tool is a software platform designed to model and analyse stochastic P systems. IBW permits applying various computational techniques, such as *modelling, simulation, verification* and *biocompilation*. The simulation component includes various simulation algorithms implemented in a high-performance environment, whereas the verification part integrates various model checking strategies, both quantitative and qualitative, with various model checkers. The biocompilation component helps selecting appropriate devices into the designed system. This integrative approach makes IBW the only tool synthesising these different techniques in one tool. Indeed, none of the existing systems and synthetic biology tools facilitates modelling, simulation, verification and biocompilation in one workbench.

In this chapter, we show how IBW utilises these techniques in the analysis of synthetic biological systems. In particular, we illustrate the main capabilities of the IBW platform in the case of some basic genetic logic gates. We present several simulation and performance benchmarking results as well as our initial results on the automatic construction of genetic devices using IBW's biomatter compilation module.

This manuscript extends our previous work [32, 48] by (i) considering a wider range of design approaches for the genetic gates discussed in the previous articles, (ii) providing a more extensive set of experimental results, (iii) discussing some new features and (iv) presenting an extended discussion that helps the reader better understand IBW's features and capabilities.

The chapter is organised as follows: Sect. 25.2 discusses the IBW platform and some of its features. Section 25.3 presents the basic genetic gates and the design approaches considered in our analysis. Section 25.4 presents the results of the experiments, carried out using IBW's different features. Section 25.5 concludes the chapter and discusses our future plans.

25.2 INFOBIOTICS WORKBENCH

The IBW tool [11, 12] enables prototyping synthetic biology programs for multi-cellular computation. The tool provides support for modelling, simulation, verification and optimisation of SP system models.

25.2.1 Modelling

In IBW, system models are constructed using stochastic P (SP) systems, augmented with a *two-dimensional lattice* representation to capture the spatial aspect of a biological system and the structure of membranes distributed geometrically within it. The geometrical representation of membranes permits modelling molecular exchange between adjacent cells. For the sake of keeping the presentation simple, we do not provide the mathematical definitions of SP systems. We refer the interested readers to [22].

IBW provides a dedicated DSL (Domain Specific Langauge) for SP systems, supported by a graphical model editor. The language is very modular in the sense that modules and libraries can be reused by different SP system models, and multiple copies of SP systems can be distributed in different parts of a geometrical lattice, which facilitates the modelling of bacterial colonies containing different types of cells.

The language allows rules to be grouped into higher level units, called modules. For example, a simplified expression of a protein can be defined as

```
simpleProteinExpression(X,Y,c₁,l)  =
    {
    rules:
    r₁:  [ geneₓ ]ₗ -c₁ →  [ geneₓ + mRNAₓ ]ₗ
    r₂:  [ mRNAₓ ]ₗ -c₁ →  [ mRNAₓ + X ]ₗ
    }
```

meaning when at location l, $gene_X$ can react and create $mRNA_X$, which in turn produces X (at location l as well). Similarly, the expression of two proteins is given as

```
constitutiveProteinExpressionTwoGenes(X,Y,c₁,c₂,c₃,l)  =
    {
    rules:
    r₁:  [ gene_XY ]ₗ -c₁ →  [ gene_XY + mRNA_XY ]ₗ
    r₂:  [ mRNA_XY ]ₗ -c₂ →  [ mRNA_XY + X ]ₗ
    r₃:  [ mRNA_XY ]ₗ -c₃ →  [ mRNA_XY + Y ]ₗ
    }
```

where the variables X and Y represent proteins, c_1, c_2 and c_3 represent reaction constants and l represents a membrane. The processes, e.g., protein binding and debinding and protein degradation, can be defined in a similar way.

25.2.2   Simulation

IBW allows simulation of SP systems using stochastic simulation algorithms [24, 25], that sample stochastic trajectories of the continuous time, discrete state Markov process defined by the SP system semantics. Generally, these algorithms generate mathematically exact samples without introducing artefacts. Since this comes to the expense of high computational runtimes [47], different algorithmic variants have been reported that potentially provide large performance improvements.

IBW delegates stochastic simulations to NGSS (next generation stochastic simulator), a stochastic simulator developed by our group [49]. NGSS simulates stochastic models and generates time-series for all the molecular species present in the system. NGSS supports nine different variants of the stochastic simulation algorithms (SSA) employing various optimisations in order to improve computational performance. Eight exact SSA formulations are included. These are Direct Method (DM) [25] and First Reaction Method (FRM) [24], Next Reaction Method (NRM) [23], Optimised Direct Method (ODM) [13], Sorting Direct Method (SDM) [37], Logarithmic Direct Method (LDM) [36], Partial Propensities Direct Method (PDM) [41] and Composition Rejection (CR) [50]. An approximation algorithm, Tau Leaping (TL) [26] is also considered.

In order to obtain tractable simulation time for a particular model, it is important to select the most efficient SSA. A new method to predict the fastest SSA for a given biomodel (using *machine learning* techniques) has been introduced by our group and will be evaluated in our analysis [49]. The technique uses a trained predictor to determine the fastest SSA based on model topological properties. The easy-to-use SSAPREDICT [49] web application implements the prediction functionality by allowing a user to upload biochemical models in the community standard SBML format [27]. Once uploaded, two types of model dependency graphs (reaction and species dependencies) are generated for the model. A fast C++ routine is then called to analyse 32 fast-to-compute graph topological properties using the igraph library [16]. The dependency graph topological property values are then used as inputs to the predictor.

To find the best predictor, we evaluated some well-known methods, e.g. linear regression, logistic regression, support vector classifier (SVC), and a nearest neighbour classifier. For each predictor, we performed a 10-fold cross-validation experiment and measured the mean accuracy and standard deviation of the predictions. Based on the experimental results, a trained linear SVC is the the strongest predictor and returns a prediction of the fastest SSA for the given model [40]. The linear SVC predictor is trained using static topological properties of 380 models taken from the BioModels database [35] (models curated from scientific literature), as well as performance benchmark data for every model evaluated with each algorithm in the NGSS simulator.

25.2.3 Verification

The verification aspects required for the robust design of multicellular computing programs is described in full details in a separate chapter, *Kernel P Systems and Stochastic P Systems for Modelling and Formal Verification of Genetic Logic Gates*, in this book.

25.2.4 Biocompilation

Through simulation and verification, the user is able to design an *in silico* construct that meets their criteria. The next step is then to test the design *in vivo* in an actual organism. We have developed a *biomatter compilation* module, to be integrated into the new version of IBW. It combines known biology, a database of genetic parts, and user knowledge, to automatically build a viable DNA sequence to be used in *in vivo* or *in vitro* experiments.

Once the user has specified the functional parts for the construct (e.g. promoters, protein coding sequences), the biocompiler uses built-in genetics knowledge to add parts not necessary for the design but mandatory for genetic sequences (e.g., ribosome binding sites (RBS), spacers and terminators), as well as restriction enzyme sites for experimental handling. All the parts are then arranged so as to make biological sense, through a mapping to a constraint solving problem. Each part is assigned a position (an integer) and all biological (e.g., non-overlapping devices, parts order within a device given their types) or user-defined requisites are translated into an integer programming problem. We use the Java library JaCoP [33] to solve the integer programming problem, thus determining an optimal arrangement.

The user can also provide in-house knowledge to add constraints on the resulting sequence, e.g., enforcing the direction of a device or the relative position of parts. This is done through a very simple language (ATGC, for Assistant To Genetic Compilation [34]). For example, if the user wants to enforce the relative positions of two promoters, they use the command:

ATGC ARRANGE Promoter2 Promoter1

They can also ask for a device to be implemented in the reverse direction:

ATGC myDevice DIRECTION REVERSE

In this case, the parts are re-arranged so that the device starts with a terminator and ends with a promoter, and the sequences are reverse-complemented.

Figure 25.1 describes the compiler's work flow. From an unstructured set of parts and user-defined constraints (1), the biocompiler completes the devices with extra parts (2) and finds an optimal arrangement (3). The part sequences are found in public databases [9, 28, 43] or provided by the user.

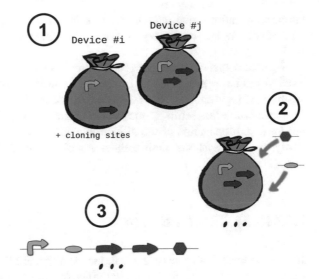

Fig. 25.1 The biocompiler's workflow. From unstructured sets of parts and user requisites, to complete devices to viable DNA sequences

25.3 Two Genetic Logic Gates

Several Synthetic Boolean logic gates have been studied previously, including [5, 51], and [42]. In this article, we will consider two basic logic gates: AND and OR, whose logic diagrams and truth tables are given in Fig. 25.2. Here, we discuss two different design approaches: the genetic parts and designs proposed by Beal et al. [5] and Tamsir et al. [51].

The analysis of the circuits can be done using other synthetic biology tools, e.g. GEC [39], Eugene [8] and Proto [4]. Here, we use the IBW tool, which provides

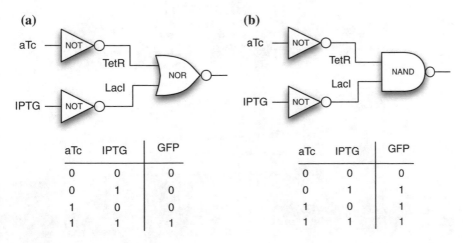

Fig. 25.2 The logic AND and OR gates

support for multicellular modelling and analysis. The tool also facilitates the formal analysis of synthetic biology systems by integrating some third-party model checking tools.

We remark that we model the genetic circuits using the concept of membranes. The necessity of being able to model and analyse multicellular systems might not be very visible in genetic circuits, but as we will see in Sect. 25.3.2, some systems do not work in a *membrane-less* setting. This is also in line with typical behaviour of bacterial strains (e.g. propagation of signalling molecules across cells [11]). Therefore, when analysing such systems, multicellular population behaviour should be taken into account.

25.3.1 Beal et al.'s Design

In this approach, both gates use two inducers, aTc and IPTG, as input and use GFP as output. aTc and IPTG disable the activities of TetR and LacI proteins, respectively.

Figure 25.3a illustrates the genetic design of an AND gate, which receives two input signals: aTc and IPTG. In this system, the transcription factors LacI and TetR are expressed by a gene controlled by the same promoter. The aTc molecules repress TetR, and IPTG molecules repress LacI, to prevent them from inhibiting the production of GFP by binding to the corresponding promoter which up-regulates the

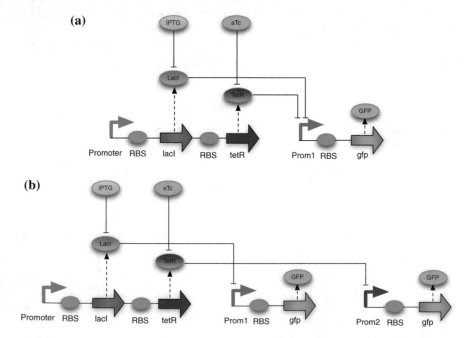

Fig. 25.3 Beal et al.'s genetic devices functioning as an AND and OR gate

expression of GFP. If both IPTG and aTc are set to high, then neither LacI nor TetR can inhibit the GFP production.

Figure 25.3b illustrates a genetic OR gate, comprising two mechanisms. Each mechanism leads to the production of GFP, when it is activated. The first mechanism is repressed by LacI while the second is repressed by TetR. Therefore, GFP can be produced from the former when IPTG is set to high and from the latter when aTc is set to high.

The stochastic model comprises a set of SP system rules, governing the kinetic and stochastic behaviour of the system. The rewriting rules and the kinetic constants (taken from [5]) of the devices are described below. We note that in order to observe the intended behaviours in the steady-state, we have slightly modified the original model defined in [5] such that the inputs are given as continuous inflows to the system, rather than as initial concentration.

(a) AND gate by Beal et al.

Rule		Kinetic rate
r_{0a} :	$\xrightarrow{k_{0a}}$ IPTG	$k_{0a} \in \{0, 1000\}$
r_{0b} :	$\xrightarrow{k_{0b}}$ aTc	$k_{0b} \in \{0, 1000\}$
r_1 :	gene_LacI_TetR $\xrightarrow{k_1}$ gene_LacI_TetR + mLacI_TetR	$k_1 = 0.12$
r_2 :	mLacI_TetR $\xrightarrow{k_2}$ mLacI_TetR + LacI	$k_2 = 0.1$
r_3 :	mLacI_TetR $\xrightarrow{k_3}$ mLacI_TetR + TetR	$k_3 = 0.1$
r_4 :	LacI + IPTG $\xrightarrow{k_4}$ LacI-IPTG	$k_4 = 1.0$
r_5 :	TetR + aTc $\xrightarrow{k_5}$ TetR-aTc	$k_5 = 1.0$
r_{6a} :	gene_GFP + LacI $\xrightarrow{k_{6a}}$ gene_GFP-LacI	$k_{6a} = 1.0$
r_{6b} :	gene_GFP-LacI $\xrightarrow{k_{6b}}$ gene_GFP + LacI	$k_{6b} = 0.01$
r_{7a} :	gene_GFP + TetR $\xrightarrow{k_{7a}}$ gene_GFP-TetR	$k_{7a} = 1.0$
r_{7b} :	gene_GFP-TetR $\xrightarrow{k_{7b}}$ gene_GFP + TetR	$k_{7b} = 0.01$
r_8 :	gene_GFP $\xrightarrow{k_8}$ gene_GFP + GFP	$k_8 = 1.0$
r_9 :	GFP $\xrightarrow{k_9}$	$k_9 = 0.001$
r_{10} :	LacI $\xrightarrow{k_{10}}$	$k_{10} = 0.01$
r_{11} :	TetR $\xrightarrow{k_{11}}$	$k_{11} = 0.01$
r_{12} :	mLacI_TetR $\xrightarrow{k_{12}}$	$k_{12} = 0.001$
(b) OR gate by Beal et al.		
Rule		Kinetic rate
$r_0 - r_5$	same as the rules $r_0 - r_5$ of the AND gate above	
r_{6a} :	gene_GFP1 + LacI $\xrightarrow{k_{6a}}$ gene_GFP1-LacI	$k_{6a} = 1.0$
r_{6b} :	gene_GFP1-LacI $\xrightarrow{k_{6b}}$ gene_GFP1 + LacI	$k_{6b} = 0.01$
r_{7a} :	gene_GFP2 + TetR $\xrightarrow{k_{7a}}$ gene_GFP2-TetR	$k_{7a} = 1.0$
r_{7b} :	gene_GFP2-TetR $\xrightarrow{k_{7b}}$ gene_GFP2 + TetR	$k_{7b} = 0.01$
r_8 :	gene_GFP1 $\xrightarrow{k_8}$ gene_GFP1 + GFP	$k_8 = 1.0$
r_9 :	gene_GFP2 $\xrightarrow{k_9}$ gene_GFP2 + GFP	$k_9 = 1.0$
$r_{10} - r_{13}$	same as the rules $r_9 - r_{12}$ of the AND gate	

If we consider the AND gate, the rules r_1 to r_3 describe the expression the LacI and TetR proteins from gene_LacI_TetR, regulated by the same promoter. r_4 and r_5 describe the binding of LacI and IPTG and TetR and aTc, respectively. r_{6a} and r_{6b} describe the inhibition activity of LacI, i.e. its binding to the promoter that up-regulates the GFP production. r_{7a} and r_{7b} define the same process for TetR. r_8 describes the expression of GFP. r_9 to r_{12} define the degradation process of various molecular species.

The AND and OR gates can be defined using modules for the expression of LacI and TetR, for binding LacI to IPTG, TetR to aTc, LacI to gene_GFP, TetR to gene_GFP, and debinding, the expression of GFP and finally degradation reactions for GFP, LacI, TetR and mLacI_TetR.

We note that we have modelled this circuit using four membranes, each representing a different input combination. Although we can also use the membrane-less modelling approach, the membrane concept allows us to observe and compare the dynamic behaviour of each case at the same time.

25.3.2   Tamsir et al.'s Design

Tamsir et al. [51] implement Boolean gates in colonies of bacterial populations, where each bacterial strain implements one elementary NOT, OR, or NOR gate. Via bacterial quorum sensing, these elemental gates can be wired to implement other, potentially more elaborate circuits. Their AND gate, for example, is implemented according to De Morgan's law as A AND B = NOT (NOT A OR NOT B). Each of the three NOT's is performed by one bacterial strain, while OR is obtained simply by collocating the bacteria in the same medium—see Fig. 25.4 for a schema of the circuit.

The first NOT strain receives the quorum sensing molecule Ara as input. Ara is a transcription factor of the promoter PBAD which, when activated, initiates the transcription of cI. cI, in turn, represses the downstream promoter PCI which initiates transcription of an output quorum sensing molecule, here AHL. The second NOT strain is identical to the first, but uses PLac in place of PBAD, making it susceptible to the initiation factor aTc. Thus, if either Ara or aTc but not both are present in the medium, at least one of the strains will release AHL into the medium. Only if both Ara and aTc are provided, AHL production is turned off in both strands. To transform this behaviour into an AND operation, a third NOT strain is receptive to AHL via the transcription factor RhlI. When activated this transcription factor up-regulates production of cI via the Phrl promoter, which in turn represses transcription of the output gene YFP.

The table below formalises the described behaviour as a set of kinetic rules, including transcription, translation, diffusion and degradation of all molecular species. For simplicity, however, we omit the details of the involved quorum sensing mechanism and couple the gates directly by assuming that RhlI itself is produced in the first two bacterial strands, from where it diffuses out of the bacteria into the medium

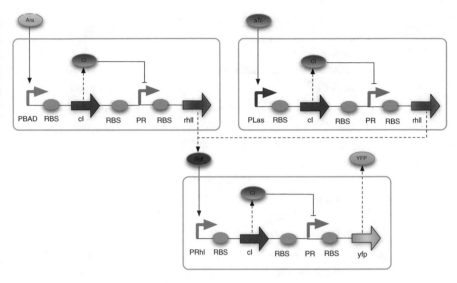

Fig. 25.4 Tamsir et al.'s genetic circuit functioning as an AND gate

and eventually into the third bacterial strand. Doing so, this simplification allows us to focus on the implementation of the Boolean logic, rather than the details of the cell-to-cell communication.

We remark that Tamsir et al.'s design would not work in a membrane-less setting: if the circuits implementing the three different NOT gates where co-located in a single membrane, crosstalk between the common cI protein and the different PR promoters would introduce crosstalks corresponding to short cuts in the wiring of the circuit.

(c) AND gate by Tamsir et al.

Rule		Kinetic rate
r_{0a} :	$\xrightarrow{k_{0a}}$ Ara	$k_{0a} \in \{0, 100\}$
r_{0b} :	$\xrightarrow{k_{0b}}$ aTc	$k_{0b} \in \{0, 100\}$
r_1 :	Ara $\xrightarrow{k_1}$	$k_1 = 0.1$
r_2 :	aTc $\xrightarrow{k_2}$	$k_2 = 0.1$
r_3 :	RhlI $\xrightarrow{k_3}$	$k_3 = 0.1$
r_4 :	Ara $\xrightarrow{k_4}$ notA:Ara	$k_4 = 1$
r_5 :	notA:Ara $\xrightarrow{k_5}$ Ara	$k_5 = 1$
r_6 :	notA:Ara $\xrightarrow{k_6}$	$k_6 = 0.1$
r_{7a} :	notA:PBAD + notA:Ara $\xrightarrow{k_{7a}}$ notA:PBAD-Ara	$k_{7a} = 0.01$
r_{7b} :	notA:PBAD-Ara $\xrightarrow{k_{7b}}$ notA:PBAD + notA:Ara	$k_{7b} = 1$
r_8 :	notA:PBAD-Ara $\xrightarrow{k_8}$ notA:PBAD-Ara + notA:mCI	$k_8 = 10$

Continued from previous page

Rule		Kinetic rate
r_9 :	notA:mCI $\xrightarrow{k_9}$	$k_9 = 5$
r_{10} :	notA:mCI $\xrightarrow{k_{10}}$ notA:mCI + notA:CI	$k_{10} = 100$
r_{11} :	notA:CI $\xrightarrow{k_{11}}$	$k_{11} = 0.1$
r_{12a} :	notA:PCI + notA:CI $\xrightarrow{k_{12a}}$ notA:PCI-CI	$k_{12a} = 0.01$
r_{12b} :	notA:PCI-CI $\xrightarrow{k_{12b}}$ notA:PCI + notA:CI	$k_{12b} = 1$
r_{13} :	notA:PCI $\xrightarrow{k_{13}}$ notA:PCI + notA:mRhlI	$k_{13} = 10$
r_{14} :	notA:mRhlI $\xrightarrow{k_{14}}$	$k_{14} = 5$
r_{15} :	notA:mRhlI $\xrightarrow{k_{15}}$ notA:mRhlI + notA:RhlI	$k_{15} = 100$
r_{16} :	notA:RhlI $\xrightarrow{k_{16}}$	$k_{16} = 0.1$
r_{17} :	notA:RhlI $\xrightarrow{k_{17}}$ RhlI	$k_{17} = 1$
r_{18} :	RhlI $\xrightarrow{k_{18}}$ notA:RhlI	$k_{18} = 1$
r_{19} :	aTc $\xrightarrow{k_{19}}$ notB:aTc	$k_{19} = 1$
r_{20} :	notB:aTc $\xrightarrow{k_{20}}$ aTc	$k_{20} = 1$
r_{21} :	notB:aTc $\xrightarrow{k_{21}}$	$k_{21} = 0.1$
r_{22a} :	notB:PLac + notB:aTc $\xrightarrow{k_{22a}}$ notB:PLac-aTc	$k_{22a} = 0.01$
r_{22b} :	notB:PLac-aTc $\xrightarrow{k_{22b}}$ notB:PLac + notB:aTc	$k_{22b} = 1$
r_{23} :	notB:PLac-aTc $\xrightarrow{k_{23}}$ notB:PLac-aTc + notB:mCI	$k_{23} = 10$
r_{24} :	notB:mCI $\xrightarrow{k_{24}}$	$k_{24} = 5$
r_{25} :	notB:mCI $\xrightarrow{k_{25}}$ notB:mCI + notB:CI	$k_{25} = 100$
r_{26} :	notB:CI $\xrightarrow{k_{26}}$	$k_{26} = 0.1$
r_{27a} :	notB:PCI + notB:CI $\xrightarrow{k_{27a}}$ notB:PCI-CI	$k_{27a} = 0.01$
r_{27b} :	notB:PCI-CI $\xrightarrow{k_{27b}}$ notB:PCI + notB:CI	$k_{27b} = 1$
r_{28} :	notB:PCI $\xrightarrow{k_{28}}$ notB:PCI + notB:mRhlI	$k_{28} = 10$
r_{29} :	notB:mRhlI $\xrightarrow{k_{29}}$	$k_{29} = 5$
r_{30} :	notB:mRhlI $\xrightarrow{k_{30}}$ notB:mRhlI + notB:RhlI	$k_{30} = 100$
r_{31} :	notB:RhlI $\xrightarrow{k_{31}}$	$k_{31} = 0.1$
r_{32} :	notB:RhlI $\xrightarrow{k_{32}}$ RhlI	$k_{32} = 1$
r_{33} :	RhlI $\xrightarrow{k:33}$ notB:RhlI	$k_{33} = 1$
r_{34} :	RhlI $\xrightarrow{k_{34}}$ notC:RhlI	$k_{34} = 1$
r_{35} :	notC:RhlI $\xrightarrow{k_{35}}$ RhlI	$k_{35} = 1$
r_{36a} :	notC:Prhl + notC:RhlI $\xrightarrow{k_{36a}}$ notC:Prhl-RhlI	$k_{36a} = 0.01$
r_{36b} :	notC:Prhl-RhlI $\xrightarrow{k_{36b}}$ notC:Prhl + notC:RhlI	$k_{36b} = 1$
r_{37} :	notC:Prhl $\xrightarrow{k_{37}}$ notC:Prhl + notC:mYFP	$k_{37} = 10$
r_{38} :	notC:mYFP $\xrightarrow{k_{38}}$	$k_{38} = 5$
r_{39} :	notC:mYFP $\xrightarrow{k_{39}}$ notC:mYFP + notC:YFP	$k_{39} = 100$

25.4   Analysis

In this section, we present the results of various analyses, which can be used to infer whether the devices function according to their desired behaviour. The source files of models and experimental results can be accessed online [21].

25.4.1   Simulation

25.4.1.1   Beal et al.'s Design

The trajectories of both gate dynamics for the four different input combinations of *low* and *high* aTc and IPTG are shown in Fig. 25.5. The gates quickly approach a steady state with output concentrations, which implements the desired Boolean logics. During the short transient period, GFP is produced in marginal quantities even in the absence of input signals, but this expression is suppressed once LacI and TetR repress the respective promoters and the present GFP degrades.

Figure 25.6 shows the transfer functions (gate output for varying input values) of the AND and OR gates. In principle, the genetic AND and OR devices closely implement the requested transfer functions and express *high* GFP amounts under the presence of both (AND gate) or either of the two inputs (OR gate). Yet, the simulations also reveal that the gate outputs follow their inputs more or less linearly

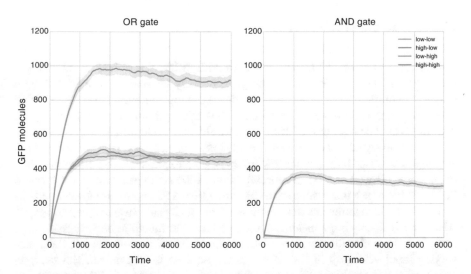

Fig. 25.5 GFP expression in the AND (*left*) and (OR) gate over time for the aTc/IPTG input combinations *low-low*, *low-high*, *high-low*, and *high-high*. The error bands denote a confidence interval of 95 % for 100 statistically independent samples

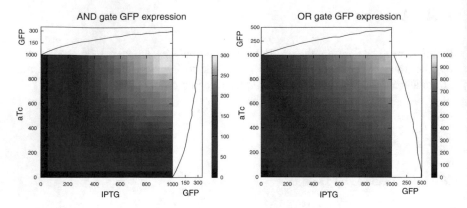

Fig. 25.6 Heat map visualisations of the AND and OR gate transfer functions obtained by stochastic simulation. Colors indicate GFP expression for different aTc, IPTG input values. For the AND gate (OR gate, resp.), the *top* (*bottom*, resp.) inlay shows the steady-state response of the gate for varying IPTG amounts under constant $aTc - 1000$ ($aTc = 0$, resp.), the *right* (*left*, resp.) inlay shows the gate response for varying aTc under constant $IPTG = 1000$ ($IPTG = 0$, resp.)

and do not implement a clear switching behaviour where the output concentration would drastically change around some critical threshold input value. Depending on the application, the observed linear behaviour can cause problems by accumulating errors when complicated circuits are composed by feeding the output of one gate into other downstream gates.

25.4.1.2 Tamsir et al.'s Design

The trajectories of the AND gate dynamics with different aTc and Ara input combinations are shown in Fig. 25.7. The gates quickly approach a steady state with output concentrations that implement the desired Boolean logics. However, the YFP expression in the *low-low* input configuration is not negligible, at nearly 20 % of the *high-high* input configuration. Ideally, one would expect a logic gate implementation to have output signals that vary more greatly between on and off states. The *low-high* and *high-low* states have similar low levels of YFP expression, slightly higher than the *low-low* states. For a better representation of the AND gate behaviour it would be preferable to see these three states have identical YFP expression levels.

Figure 25.8 shows the transfer functions of the AND gates. In principle, the genetic AND device closely implements the requested transfer functions and expresses high YFP amounts under the presence of both inputs. The figure on the left illustrates varying input values up to 100, which is the on state for the model. However, an inflow of 100 is too high, causing a relatively high amounts of YFP expression as shown in Fig. 25.7 for the *low-low*, *high-low* and *low-high* input states. We therefore lower the input value to vary up to 20, and show the corresponding results in the

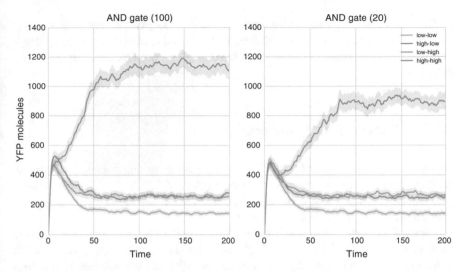

Fig. 25.7 YFP expression in the AND gate over time for the aTc/Ara input combinations *low-low*, *low-high*, *high-low*, and *high-high*. The bands denote a confidence interval of 95 % for 100 statistically independent samples. The figures on the *left* and *right* shows the results under the constant inflow of aTc and Ara with a rate of 100 and 20, respectively

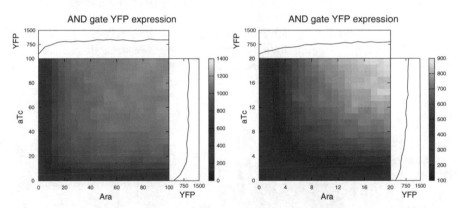

Fig. 25.8 Heat map visualisations of the AND gate transfer functions obtained by stochastic simulation. Colors indicate YFP expression for different aTc, Ara input values. For each heat map, the *top* inlay shows the steady-state response of the gate for varying inflow of Ara amounts under constant inflow of $aTc = 100$ ($aTc = 20$, resp.), the *right* inlay shows the gate response for varying inflow of aTc under constant inflow of $Ara = 100$ ($Ara = 20$, resp.)

heat map on the right. The behaviour then becomes more similar to what one would expect from an AND gate, e.g. similar to left heat map in Fig. 25.6.

Again, the simulations indicate a linear relationship between gate inputs and outputs, rather than a sharp switching behaviour. In this case, however, these smooth transition functions are less of a problem, as the gate outputs are rectified by the quorum-sensing mechanism among bacterial strains (which has not been taken into account in our model).

Using this analysis, we could inform and refine the design of this synthetic Boolean gate to overcome these predicted deficiencies. Repeated *in silico* hypothesis testing is key to minimising costly web lab work when designing complex synthetic biological systems.

25.4.2 Performance Benchmarking

To assess the implications of algorithmic performance for this model, we have also benchmarked the performance of each of the mentioned SSA variants for the Boolean gate models. As a proof of concept, we present our results on the AND gate, constructed according to Beal et al.'s approach.

For each algorithm, 100 runs were performed and each simulation completed to 6000 seconds of simulation time. The metric for measuring performance used is *reactions per second (rps)*. Rps is calculated by dividing the number of reactions executed by the amount of computational (process) time required. For each model we tried four different configurations of gate inputs aTc and IPTG *(high-high, high-low, low-high and low-low)* where low is zero molecules and high is 1000 molecules. Our benchmarking experiments were performed on a single core of an Intel Core i7-3770 CPU @ 3.40GHz with 8GB RAM.

Figure 25.9 shows the algorithmic performance for the AND gate model in the *high-high* and *low-low* configurations. These results demonstrate that small differences in a model (in this case, the concentrations of two species) may result in large

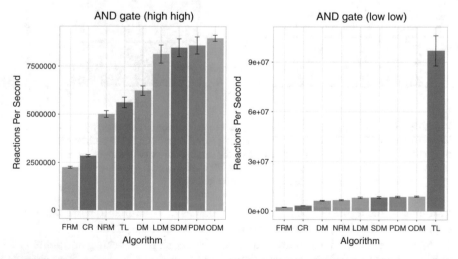

Fig. 25.9 Algorithm benchmark performance results in rps of each algorithm for the AND gate with aTc and IPTG in *high-high* (constant 1000 1000) and *low-low* (constant 0 0) input configuration. Each algorithm's performance was evaluated as the mean of a total of 100 runs

differences in algorithmic performance profiles. We can see that for the *low low* configuration TL is the fastest performing simulation algorithm, and outperforms others by an order of magnitude. However, in all other configurations ODM is the better algorithmic selection and strongly outperforms TL (see Fig. 25.9). Whilst many SSA runs can be distributed over nodes of a HPC cluster, the minimum possible simulation time is still bounded by the time required for one run. Therefore, the selection of the fastest algorithm for simulating a given model results in a performance increase irrespective of the use of HPC facilities.

To complement the performance benchmark result shown in Fig. 25.9, we evaluated the efficacy of SSAPREDICT tool [49] using these models. The predictor recommended ODM as the fastest algorithm for these models. While ODM is indeed the fastest algorithm for the gate in *high-high* configuration, the algorithm is outperformed by TL in the gate's *low-low* configuration. The SSAPREDICT tool currently analyses the topological properties of the model and does not consider species amounts which explains this inaccuracy. Note, however, that ODM is the fastest performing *exact* algorithm also for the *low-low* configuration.

25.4.3   Biocompilation

Figure 25.10 shows the result of the biocompilation from the specifications decided upon during the design stage. The only functional parts specified by the user are promoters and genes. No extra constraints were required. The biocompiler automatically completed the devices with RBS and terminators and found a viable arrangement for the parts. The sequences for the parts were looked up in the biobricks database. The RBS's were calculated so as to produce similar initiation translation rates (1000 A.U). The promoter A90 was custom-built by [15] through a combinatorial exploration of excerpts of promoters. It results into an AND operator for IPTG and aTc. Since it is not found in a common repository, we specify its whole sequence with the ATGC directive:

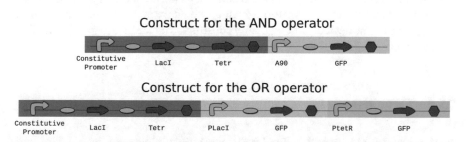

Fig. 25.10 Constructs resulting from the biocompilation of Beal et al.'s design. Each operator is made of two or three devices. RBS and terminators have been added to complete the devices, and the sequences were found from the BioBricks database

Construct for the AND operator

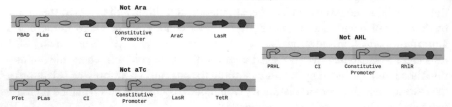

Fig. 25.11 Constructs resulting from the biocompilation of Tamsir et al.'s construct. The design involves 3 cell types, each of which features a distinct NOR gate (this follows the original specification by Tamsir et al. [51], not the shortened specification of Sect. 25.3.2)

ATGC PART A90 PROMOTER TACAACGTCGTGTTAGCTGCCTTTCGTC(...)

The list of specified parts for these constructs is:

- Custom-built promoter A90: sequence: TACAACGTCGTGTTAGCTGCCTTT CGT CTTCAATAATTCTTGACATCCCTATCAGTGATAGAGATACTTTGTGG AATTGT GAGCGGATAACATTTCACACAG
- Generic constitutive promoter: (biobricks entry: BBa_J23100)
- LacI: lacI repressor from E. coli (biobricks entry: BBa_C0012)
- Tetr: tetracycline repressor from transposon Tn10 (biobricks entry: BBa_C0040)
- PLacI: lacI regulated promoter (biobricks entry: BBa_R0010)
- PtetR: TetR repressible promoter (biobricks entry: BBa_R0040)
- GFP: green fluorescent protein derived from jellyfish Aequeora victoria wild-type (biobricks entry: BBa_E0040)

Tamsir's Design

Tamsir et al.'s design [51] for an AND operator is made of 3 cell types, communicating through an intermediate signal. The AND operator is decomposed into two NOT operators for each of Ara and aTc, the results of which are fed to a NOT operator.

Promoters, genes and terminators are all specified in the supplementary material of [51], Sect. 6. The devices, automatically constructed by the biocompiler, are shown in Fig. 25.11.

25.5   Conclusions

In this chapter we have described how synthetic biology can be used to programme multicellular computing systems. We have also presented the INFOBIOTICS WORKBENCH, a software platform for modelling and analysis of biological systems expressed in stochastic P systems using various computational techniques, e.g. simulation and verification. We have also presented our initial results on the automatic construction of genetic devices using a biomatter compilation module. We have illustrated our approach on two basic genetic devices: the AND and OR gates. For our analysis, we have considered two different design approaches: Beal et al. [5] and Tamsir et al. [51]. We are planning to analyse more complex gates discussed in various articles, e.g. [42, 51]. Biomatter compilation (or biocompilation for short) is a rapidly advancing area of research. For example in [10] we have described an algorithm for planning multistage assembly of DNA libraries (which can be derived from our Biocompiler) with shared intermediates that greedily attempts to maximize DNA reuse as to minimise errors and cost of DNA synthesis. Although the problem of reuse is formally NP-hard and APX-hard by reduction from vertex cover, the heuristic we developed provides solutions of near-minimal stages and thanks to almost instantaneous planning of DNA libraries it can be used as a metric of "manufacturability" to guide DNA library design as we have shown both theoretically and empirically that it runs in linear time. Furthermore, we also complemented our computational work with a microfluidic device [6] that can take biological programmes and carry out de novo synthesis and cell-free cloning of custom DNA libraries in sub-microliter reaction droplets using programmable digital microfluidics. Taken together, the work described here illustrates a close loop where multicellular programmes can be prototyped in silico and implemented via microfluidic devices which, with further research, will enable in the near future rapid real-world and practical multicellular programming. We are currently working on a new version of IBW, which will incorporate more methods for specifying, modelling, testing and simulating biological systems, and will facilitate these processes for biologists. In addition, the new version will also include the biocompilation module that we are currently developing. The future version of the biocompiler will have an optimisation module, with an integrated RBS calculator [46] and codon optimisation [53], with the overall aim to be as close to the kinetic parameters decided upon in the simulation stage.

Acknowledgments SKo and MG acknowledge the EPSRC (EP/I031812/1) support. CL and SKa are supported by EPSRC (EP/I03157X/1). HF, DS and NK's work is supported by EPSRC (EP/I031642/1, EP/I031642/2, EP/J004111/1, EP/J004111/2, EP/L001489/1 and EP/L001489/2).

References

1. Andrianantoandro, E., Basu, S., Karig, D.K., Weiss, R.: Synthetic biology: new engineering rules for an emerging discipline. Mol. Syst. Biol. **2**(1), 14 (2006)
2. Bakir, M.E., Ipate, F., Konur, S., Mierlă, L., Niculescu, I.: Extended simulation and verification platform for kernel P systems. In: 15th International Conference on Membrane Computing. LNCS, vol. 8961, pp. 158–168. Springer, Switzerland (2014)
3. Bakir, M.E., Konur, S., Gheorghe, M., Niculescu, I., Ipate, F.: High performance simulations of kernel P systems. In: Proceedings of the 2014 IEEE 16th International Conference on High Performance Computing and Communication. HPCC '14, pp. 409–412. France, Paris (2014)
4. Beal, J., Lu, T., Weiss, R.: Automatic compilation from high-level biologically-oriented programming language to genetic regulatory networks. PLoS ONE **6**(8), e22,490 (2011)

5. Beal, J., Phillips, A., Densmore, D., Cai, Y.: High-level programming languages for biomolecular systems. In: Design and Analysis of Biomolecular Circuits, pp. 225–252. Springer, New York (2011)
6. Ben-Yehezkel, T., Rival, A., Raz, O., Cohen, R., Marx, Z., Camara, M., Dubern, J.F., Koch, B., Heeb, S., Krasnogor, N., Delattre, C., Shapiro, E.: Synthesis and cell-free cloning of dna libraries using programmable microfluidics. Nucleic Acids Research pp. first published online October 19, 2015 doi:10.1093/nar/gkv1087 (2015)
7. Bernardini, F., Gheorghe, M., Krasnogor, N., Muniyandi, R.C., Perez-Jimenez, M.J., Romero-Campero, F.J.: On P systems as a modelling tool for biological systems. In: Membrane Computing, Lecture Notes in Computer Science, vol. 3850, pp. 114–133. Springer, Heidelberg (2006)
8. Bilitchenko, L., Liu, A., Cheung, S., Weeding, E., Xia, B., Leguia, M., Anderson, J.C., Densmore, D.: Eugene—a domain specific language for specifying and constraining synthetic biological parts, devices, and systems. PLoS ONE 6(4), e18,882 (2011)
9. Biofab: International Open Facility Advancing Biotechnology. http://biofab.synberc.org
10. Blakes, J., Raz, O., Feige, U., Bacardit, J., Widera, P., Ben-Yehezkel, T., Shapiro, E., Krasnogor, N.: Heuristic for maximizing dna reuse in synthetic dna library assembly. ACS Synth. Biol. 8(3), 529–542 (2014)
11. Blakes, J., Twycross, J., Konur, S., Romero-Campero, F.J., Krasnogor, N., Gheorghe, M.: Infobiotics workbench: A p systems based tool for systems and synthetic biology. In: Applications of Membrane Computing in Systems and Synthetic Biology. Emergence, Complexity and Computation, vol. 7, pp. 1–41. Springer International Publishing, Switzerland (2014)
12. Blakes, J., Twycross, J., Romero-Campero, F.J., Krasnogor, N.: The infobiotics workbench: an integrated in silico modelling platform for systems and synthetic biology. Bioinformatics 27(123), 3323–3324 (2011)
13. Cao, Y., Li, H., Petzold, L.: Efficient formulation of the stochastic simulation algorithm for chemically reacting systems. J. Chem. Phys. 121(9), 4059–4067 (2004)
14. Chen, K.H., Boettiger, A.N., Moffitt, J.R., Wang, S., Zhuang, X.: Spatially resolved, highly multiplexed RNA profiling in single cells. Science 348(6233) (2015)
15. Cox, R.S., Surette, M.G., Elowitz, M.B.: Programming gene expression with combinatorial promoters. Mol. Syst. Biol. 3(1) (2007)
16. Csardi, G., Nepusz, T.: The igraph software package for complex network research. Int. J. Complex Syst. 1695 (2006). http://igraph.sf.net
17. Dragomir, C., Ipate, F., Konur, S., Lefticaru, R., Mierlă, L.: Model checking kernel P systems. In: 14th International Conference on Membrane Computing, LNCS, vol. 8340, pp. 151–172. Springer, Berlin, Heidelberg (2013)
18. Fellermann, H., Hadorn, M., Füchslin, R.M., Krasnogor, N.: Formalizing modularization and data hiding in synthetic biology. J. Emerg. Technol. Comput. Syst. 11(3), 24:1–24:20 (2014)
19. Fellermann, H., Krasnogor, N.: Chemical production and molecular computing in addressable reaction compartments. In: Language. Life, Limits, Lecture Notes in Computer Science, vol. 8493, pp. 173–182. Springer International Publishing, Switzerland (2014)
20. Frisco, P., Gheorghe, M., Pérez-Jiménez, M.J. (eds.): Applications of Membrane Computing in Systems and Synthetic Biology. Springer, Switzerland (2014)
21. Genetic Boolean Gates. http://www.scim.brad.ac.uk/skonur/models/gates
22. Gheorghe, M., Manca, V., Romero-Campero, F.J.: Deterministic and stochastic P systems for modelling cellular processes. Nat. Comput. 9(2), 457–473 (2010)
23. Gibson, M.A., Bruck, J.: Efficient exact stochastic simulation of chemical systems with many species and many channels. J. Phys. Chem. A 104(9), 1876–1889 (2000)
24. Gillespie, D.: A general method for numerically simulating the stochastic time evolution of coupled chemical reactions. J. Comput. Phys. 22(4), 403–434 (1976)
25. Gillespie, D.T.: Exact stochastic simulation of coupled chemical reactions. J. Phys. Chem. 81(25), 2340–2361 (1977)
26. Gillespie, D.T.: Approximate accelerated stochastic simulation of chemically reacting systems. J. Chem. Phys. 115(4), 1716–1733 (2001)

27. Hucka, M., et al.: The systems biology markup language (SBML): a medium for representation and exchange of biochemical network models. Bioinformatics **19**(4), 524–531 (2003)
28. iGem: Parts Registry. http://partsregistry.org/
29. Konur, S., Gheorghe, M.: A property-driven methodology for formal analysis of synthetic biology systems. IEEE/ACM Trans. Comput. Biol. Bioinf. **12**, 360–371 (2015)
30. Konur, S., Gheorghe, M., Dragomir, C., Ipate, F., Krasnogor, N.: Conventional verification for unconventional computing: a genetic XOR gate example. Fundamenta Informaticae **134**, 97–110 (2014)
31. Konur, S., Gheorghe, M., Dragomir, C., Mierla, L., Ipate, F., Krasnogor, N.: Qualitative and quantitative analysis of systems and synthetic biology constructs using P systems. ACS Synth. Biol. **4**(1), 83–92 (2015)
32. Konur, S., Ladroue, C., Fellermann, H., Sanassy, D., Mierla, L., Ipate, F., Kalvala, S., Gheorghe, M., Krasnogor, N.: Modeling and analysis of genetic boolean gates using Infobiotics Workbench. In: Verification of Engineered Molecular Devices and Programs, pp. 26–37. Vienna, Austria (2014)
33. Kuchcinski, K.: Constraints-driven scheduling and resource assignment. ACM Trans. Des. Autom. Electron. Syst. **8**(3), 355–383 (2003)
34. Ladroue, C., Kalvala, S.: Constraint-based genetic compilation. In: Algorithms for Computational Biology, LNBI, vol. 9199. Springer International, Heidelberg (2015)
35. Li, C., Donizelli, M., Rodriguez, N., Dharuri, H., Endler, L., Chelliah, V., Li, L., He, E., Henry, A., Stefan, M.I., Snoep, J.L., Hucka, M., Le Novère, N., Laibe, C.: BioModels database: an enhanced, curated and annotated resource for published quantitative kinetic models. BMC Syst. Biol. **4**, 92 (2010)
36. Li, H., Petzold, L.: Logarithmic direct method for discrete stochastic simulation of chemically reacting systems. Technical report, Department of Computer Science, University of California: Santa Barbara (2006)
37. McCollum, J.M., Peterson, G.D., Cox, C.D., Simpson, M.L., Samatova, N.F.: The sorting direct method for stochastic simulation of biochemical systems with varying reaction execution behavior. Comput. Biol. Chem. **30**(1), 39–49 (2006)
38. Păun, G.: Computing with membranes. J. Comput. Syst. Sci. **61**(1), 108–143 (2000)
39. Pedersen, M., Phillips, A.: Towards programming languages for genetic engineering of living cells. J. R. Soc. Interface **6**(Suppl 4), S437–S450 (2009)
40. Pedregosa, F., Varoquaux, G., Gramfort, A., Michel, V., Thirion, B., Grisel, O., Blondel, M., Prettenhofer, P., Weiss, R., Dubourg, V., Vanderplas, J., Passos, A., Cournapeau, D., Brucher, M., Perrot, M., Duchesnay, E.: Scikit-learn: machine learning in Python. J. Mach. Learn. Res. **12**, 2825–2830 (2011)
41. Ramaswamy, R., Gonzalez-Segredo, N., Sbalzarini, I.F.: A new class of highly efficient exact stochastic simulation algorithms for chemical reaction networks. J. Chem. Phys. **130**(24), 244104 (2009)
42. Regot, S., Macia, J., Conde, N., Furukawa, K., Kjellen, J., Peeters, T., Hohmann, S., de Nadal, E., Posas, F., Sole, R.: Distributed biological computation with multicellular engineered networks. Nature **469**(7329), 207–211 (2011)
43. Roberts, R.J., Vincze, T., Posfai, J., Macelis, D.: REBASE–a database for DNA restriction and modification: enzymes, genes and genomes. Nucleic Acids Res. **38**(Database issue), D234–D236 (2010)
44. Romero-Campero, F.J., Twycross, J., Camara, M., Bennett, M., Gheorghe, M., Krasnogor, N.: Modular assembly of cell systems biology models using P systems. Int. J. Found. Comput. Sci. **20**(3), 427–442 (2009)
45. Romero-Campero, F.J., Twycross, J., Cao, H., Blakes, J., Krasnogor, N.: A multiscale modeling framework based on P systems. In: Membrane Computing, LNCS, vol. 5391, pp. 63–77. Springer, Heidelberg (2009)
46. Salis, H.M.: The ribosome binding site calculator. In: Synthetic Biology. Part B Computer Aided Design and DNA Assembly, Methods in Enzymology, vol. 498, pp. 19–42. Academic Press, USA (2011)

47. Sanassy, D., Blakes, J., Twycross, J., Krasnogor, N.: Improving computational efficiency in stochastic simulation algorithms for systems and synthetic biology. In: SynBioCCC. 11th European Conference on Artificial Life (2011)
48. Sanassy, D., Fellermann, H., Krasnogor, N., Konur, S., Mierlă, L., Gheorghe, M., Ladroue, C., Kalvala, S.: Modelling and stochastic simulation of synthetic biological Boolean gates. In: 16th IEEE International Conference on High Performance Computing and Communications, HPCC '14, pp. 404–408. Paris, France (2014)
49. Sanassy, D., Widera, P., Krasnogor, N.: Meta-stochastic simulation of biochemical models for systems and synthetic biology. ACS Synth. Biol. 4(1), 39–47 (2015)
50. Slepoy, A., Thompson, A.P., Plimpton, S.J.: A constant-time kinetic Monte Carlo algorithm for simulation of large biochemical reaction networks. J. Chem. Phys. 128(20), 205101 (2008)
51. Tamsir, A., Tabor, J.J., Voigt, C.A.: Robust multicellular computing using genetically encoded NOR gates and chemical 'wires'. Nature 469(7329), 212–215 (2011)
52. Twycross, J., Band, L., Bennett, M., King, J., Krasnogor, N.: Stochastic and deterministic multiscale models for systems biology: an auxin-transport case study. BMC Syst. Biol. 4(1), 34 (2010)
53. Welch, M., Villalobos, A., Gustafsson, C., Minshull, J.: You're one in a googol: optimizing genes for protein expression. J. R. Soc. Interface/R. Soc. 6(Suppl 4), S467–S476 (2009)

Chapter 26
Plant Roots as Excellent Pathfinders: Root Navigation Based on Plant Specific Sensory Systems and Sensorimotor Circuits

Ken Yokawa and František Baluška

Abstract Roots are underground plant organs hidden in the soil and coping with many environmental challenges. The root system forms ultimately complex networks of roots with numerous root apices at the distal ends of all roots. All these root apices move away from the plant body, being pushed via the elongation region in which cells rapidly elongate. Each root apex acts as an autonomous sensory organ receiving information from numerous sensory systems feeding into the root apex transition zone. The latter is acting as command center navigating growing root apices through very complex underground environment. New root apices are formed continuously behind the growth zone in endogenous manner, initiated at the stele-cortex interface via cell divisions in the pericycle and endodermis. All this allows exploratory root systems to effectively explore and exploit large areas of heterogeneous soil. In order to find out the underlying biological mechanisms, root behavior can be observed and manipulated in laboratory. Roots use their plant-specific cognition and problem-solving apparatus which allows them to exploit heterogeneous soil for water and mineral nutrition. Plant-specific memory and processing of sensory information are discussed also from the perspective of plant-specific unconventional computing. We hope that our better understanding of root behavior will be relevant for the bio-inspired robotics.

K. Yokawa (✉) · F. Baluška
IZMB, University of Bonn, 53115 Bonn, Germany
e-mail: yokawa@uni-bonn.de

F. Baluška
e-mail: Baluska@uni-bonn.de

K. Yokawa
Department of Biological Sciences, Tokyo Metropolitan University,
192-039 Tokyo, Japan

© Springer International Publishing Switzerland 2017
A. Adamatzky (ed.), *Advances in Unconventional Computing*,
Emergence, Complexity and Computation 23,
DOI 10.1007/978-3-319-33921-4_26

677

26.1   Introduction

In natural ecosystems, organisms are facing numerous challenges from extremely heterogeneous environments that represent ever-changing complex systems. Organisms living on the Earth can survive such environments via their abilities to solve continuously problems and challenges. In addition, plants have a sessile nature and spend their whole life in a fixed place after seed germination. Since plants normally show no quick movements like animals, it looks like sessile plants are passively facing these environmental insults. However, in fact, plants are making excellent use of their sensory and physiological abilities to cope with these environmental challenges. In contrast to general views, plants are not passive organisms devoid of any movements. For example, plant organ movements, known as tropisms, are directing the parts of plant body towards favorable and away of unfavorable locations. With these intrinsic mechanisms, plants and their organs can search for light, free space, water, nutrients, oxygen; or avoid physical barrier and toxic soil patches. Roots are underground organs which are known to be the most active part of plant body showing many tropisms. Root tropism was firstly described in the book 'The Power of Movement in Plants' [13]. In the last paragraph of this book, they concluded with a statement that root tip acts like the brain of one of the lower animals seated at the anterior pole of the plant body. This shocking proposal was first heavily criticized, then ignored, and finally revived as the Darwin's 'root-brain' theory [5, 25]. Plant roots are considered as semi-autonomous organs, which are capable of exploring and exploiting of complex environments. Importantly, this process is essential for the plant survival. Using newly designed simple experimental apparatus [44], we have recently reported, that growing maize roots show elaborate sensing ability to volatile compounds and light. In this chapter, we discuss our recent findings regarding root behavior that open new possibilities to assemble living root-based unconventional computer. We are discussing decision-making ability of maize roots based on our results using Y-maze system and U-turn behavior of roots growing within thin glass capillaries and controlled by light. We will argue that root systems, consisting of numerous root apices, are operating as parallel computing units supported by their sensory root apices. This new view of plant roots is relevant for future technologies based on unconventional computing.

26.2   Y-Maze: Binary Decision-Making of Maize Roots

Mazes have been used for many psychological or cognitive tests [29]. Especially, they have the advantage to observe animal cognitive behaviors online because mazes consist of sequences of branching paths, which abstract properties of the actual environment available for living. Basically, an essential goal (food, exit etc.) can be set somewhere in the maze, and then the process of the solution (behavior) is observed by the examiners. Animals solving the maize require continuous on-line decision-making and problem-solving abilities based on their memory, learning, cognition, and intelligence. Although there are diverse variants of these experimental mazes, the Y-maze (T-maze) was introduced as more simplified system for examining decision-making of rats, namely left or right choice. As rewards, food can be placed in one of the end, which emits specific odor attracting rats. After first several times of learning training, the working memory can then be tested by reversing the reward place between left and right ends. It was also reported that the behavior and learning of zebra fish can be explored using Y-shape water tank [2, 12]. With such Y-maze system, examiners can observe the two different stages of behavior of animals, (1) decision-making in unknown environment guided by senses or cognitions; and (2) learning based on the trial-and-error learning process.

As shown in the Fig. 26.1, we have developed the first Y-maze test system for plant roots [44]. It was firstly used to find out if maize roots can determine the right way according to the gradient of odor, namely volatile compounds. It is well-known that plants communicate among individuals by emitting and receiving diverse volatile chemicals [14, 22, 23]. For example, ethylene is airborne plant stress hormone regulating plant development, growth, stress adaptation and fruits ripening. Likewise to aerial organs, also underground roots use volatile compounds for many purposes [15, 16, 30].

Maize roots were placed in the top of 12 cm squire-shaped Petri dish equipped with Y-maze inside (Fig. 26.1). Reagents and solvents as a control are applied in the each end of branch. Only root parts of three to four days-after germination seedlings are set in the entrance to the Y-maze. The Petri dish is tightly sealed with parafilm in order to keep moisture inside the maze. Since light input has a big impact on root behavior and growth [45], the growth experiment in Y-maze must be conducted in darkness. As discussed above, plants have elaborate abilities to regulate their growth as well as direction by perceiving many airborne volatile molecules. Roots are attracted or repelled in response to certain chemical species. In this Y-maze system, roots

Fig. 26.1 Y-maze for maize roots

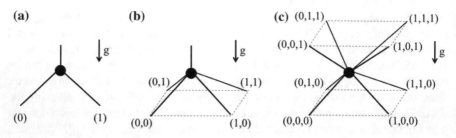

Fig. 26.2 Graphs connected with a vertex. The vertex, shown in *black circle*, indicates the place where roots determine the paths. **a** Binary tree, **b** several targets on a plane surface and **c** sterical array of targets. The *arrow* indicates the gravity vector

decide which way to grow by sensing the gradient of volatile compounds emitted from the source placed at the end of one branch. We have previously reported that ethylene and diethyl ether attracts, whereas methyl jasmonate repels maize roots [44]. There are enormous combinations of the chemicals and concentrations, which enable to control root behaviors in these maze paths. These results open the attractive possibility of using roots to solve the problem of binary partition tree in an acyclic graph structure as shown in the Fig. 26.2a. Furthermore, extending the paths of Y-maze (Fig. 26.2b) may improve the system coping with more complex problems or situations. In addition, steric design of paths; as shown in the Fig. 26.2c, can also be possible because the direction of root growth is not much restricted by gravity vector as long as it grows in darkness [9].

It is known that the information of external input (environmental stresses) is transduced rapidly through the plant body via cellular signaling factors such as calcium and reactive oxygen species [20], as well as electric and hydraulic signals [3, 18]. Therefore, further studies are highly essential to investigate whether complex root networks once formed in designed paths intermediate the exchange of information among other roots, like wiring of electric circuits. In fact, root system has more advantage than conventional circuits because living roots are capable, due to their inherent phenotypic plasticity, of changing the aspects of network in response to variable environments.

26.3 Light Controls U-Turn Behavior of Maize Roots

Roots normally grow in the soil to fix the plant body and to obtain water as well as mineral nutrients. Unlike the aerial part of plants, it is well known that roots grow downward along gravity vector. Root apices sense gravity and direct their growth accordingly. Since Charles and Francis Darwin had described this phenomenon more than 100 years ago [13], many plant physiologists have been intensively studying this sensory aspect of plant sciences. This ability is of course essential for many plants to grow the roots downward. However, is gravity only one environmental factor that provokes the root behavior going back into soil? It has already been shown that roots accomplish many tropisms and have sophisticated abilities to integrate numerous incoming environmental inputs into adaptive motoric responses [6]. Among those environmental factors, light is one of physical factors that strongly promote avoidance behavior of roots [40, 42, 43, 45, 46]. This behavior of roots includes negative phototropism or escaping phototropism. Light induced U-turn behavior of maize roots growing within thin glass capillaries upwards against gravity is particularly relevant in this respect [9]. As shown in the Fig. 26.3, maize roots suddenly turn downward upon illumination from above, whereas maize roots keep growing upwards in darkness even if they emerge out from the capillaries into the free space (Yokawa and Baluška, submitted). This root behavior suggests that light is perceived as dangerous signal by maize roots which recognize sensitively the source of light and try to escape from illuminated space [9]. We have previously reported that very short illumination (several seconds) causes rapid generation of superoxide in *Arabidopsis* root apices which trigger quick behavioral response of negative phototropism [42]. There are also other reports that roots quickly change their physiological conditions and respond to light incidents immediately [34, 35]. In addition to illumination period, plant roots also detect very dim light [17]. These findings reveal that roots possess very sensitive sensory systems which record external light. Obviously, these sensory systems feed perceptual information into root-specific motoric system which drive root avoidance and escape tropisms. If roots would wait until gene expression and protein transcription are modified after receiving light information, then such dangerous situations would be prolonged and potentially very dangerous. Roots are plant organs optimized for their growth and functioning underground must get back into the soil as

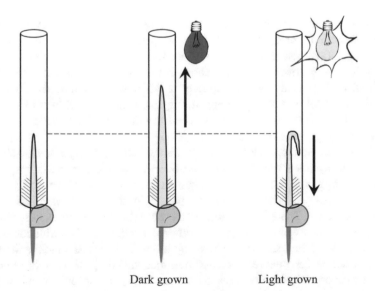

Dark grown   Light grown

Fig. 26.3 Light-induced U-turn behavior of inverted maize roots in a glass capillary

quickly as possible. Recently, the inhibition of feeding behavior of nematode caused by light-induced hydrogen peroxide generation was reported. It was revealed that hydrogen peroxide functions as primary messenger in this phenomenon [8]. This reaction was observed readily upon illumination, and it was proposed that photosensitive molecule such as flavins in vivo might be the source of hydrogen peroxide [8]. This is very relevant to our findings that *Arabidopsis* roots generate reactive oxygen species already after 10 sec of illumination [42]. It is rather surprising that different species, from different kingdoms, such as plant roots and nematodes; utilize same signaling molecules for the photophobic behavior [41].

What makes roots so extraordinary sensitive to environmental factors? Charles and Francis Darwin had firstly performed an elegant experiment that if root cap of maize was ablated, roots lost the response to gravity [13]. Since then, many plant physiologists have been studying the function of root cap [21]. As described, U-turn behavior of maize roots, induced by light illumination is based on intact root cap [9]. Root cap can be considered as sensory organ of plants which, together with transition zone, integrate the external information and direct the root growth [4]. In real plants living outside in the nature, thousands of root apices exploring the environment and commanding the whole plant represent and enormous underground parallel computing units. From the view of bioengineering, this root escape behavior controlled by light in thin capillaries is probably applicable, in combination with other systems, to construct future computing units based on living organisms.

26.4 Anticipation and Memory in Plants

In addition to plant tropisms, some other abilities must be necessary for survival in such a harsh environment as soil. For example, animals approach or evade the favorable or unfavorable places by using their senses. However animals cannot keep targeting just by following the information released from the targets. Sometimes, they 'anticipate' the challenges to increase the efficiency of their actions. In order to anticipate or expect something to happen, very important ability, 'memory', should be utilized as much as possible. Experience from the past gives us a chance to cut down waste energy and avoid coming dangers.

Do plants use memory? It has been discussed whether plants posses the ability of memory. This argument caused many controversial situations because many scientists know that plant have no brain and equivalent nervous system found in animals. However, the biological details about the memory even in our human brain are not fully understood yet. Therefore, the term of memory should be used for describing the ability to store sensory information for a certain period. In fact, it is well documented that plant can keep the memory of stress situations [24] such as attack by herbivore [19, 33] based on gene expressions and hormone signaling. It helps to reduce the expensive cost or avoidable damages in advance. This plant-specific memory resembles in some feed-forward, feed-back and delay circuits. Furthermore, it was reported that memory based memristors operate in plants [27, 38, 39]. Memristor, a resistor with memory, was firstly presumed as forth circuit element in 1971 [11]. Since then, engineers have intensively been studying it because if materialized, the ability of conventional circuits will be drastically improved in terms of memory capacity. It is a very intriguing fact that plants already acquired this sophisticated ability driven by the complex action of ion channels and signaling networks. Interestingly, recent studies have been reporting that slime molds *Physarum polycephalum*, brainless organisms, can memorize and distinguish the area they visited once before [31] using secretions from their body [32]. It was shown that slime molds even capable of making decision and of problem-solving [1, 7, 26]. Plant roots are well-known to secrete large amounts of slime from the root cap secrete cells, and chemical compounds in the secreted slime could be used as similar externalized memory for navigation and path-finding of plants roots exploring complex environment of soil. Taken together, likewise animals, plants also have evolved ability of storing memories of past situations for the necessity to survive the ever-changing environments [10]. However, our understanding of plant memory is only rudimentary and further works and new experimental designs are required.

26.5 Possible Use of Roots for Unconventional Computing

How can we use abilities of living roots for computing? Roots behavior guided by chemical compounds (volatiles), light, gravity, etc. can of course be tracked by image sensor such as camera, and the results of movement (solution) are also processed by appropriate image processing methods. Another possibility is to detect and analyze the intensity of certain chemical information of propagating signaling (e.g., calcium wave, electric potentials) at a point in plant body with an electrode.

Besides membrane potential and cellular signaling, some other components are also capable of transducing the mediator of information. In this respect, there are several reports showing the possibility that plant roots can behave a kind of conducting cable. As an example, boron is known to be important component cross-linking of RGII pectins in plant cell walls. Tanada was first to propose that boron can be a factor to generate the bioelectric field in plants [37]. Boron in roots might also play a role for propagating the bioelectric field because roots also contain abundant boron in cell wall [28]. Tanada also found out the interesting phenomena that illuminated maize root stuck to the negative-charged surface of glass because electron potential of root cell membrane was changed upon illumination. This phenomenon was coined as 'Tanada effect' [35, 36, 47] and it could be useful for future development of unconventional computing based on living roots functioning as sensors, circuit elements and processors.

Acknowledgments Ken Yokawa was supported by the JSPS (Japanese Society for the Promotion of Science) Postdoctoral Fellowship. This work was supported in part by JSPS KAKENHI, Grant-in-Aid for JSPS fellows, No. 261654.

References

1. Adamatzky, A.: Slime mould processors, logic gates and sensors. Philos. Trans. Math. Phys. Eng. Sci. **373**, 2046 (2015)
2. Aoki, R., Tsuboi, T., Okamoto, H.: Y-maze avoidance: an automated and rapid associative learning paradigm in zebrafish. Neurosci. Res. **91**, 69–72 (2015)
3. Baluška, F. (ed.): Long-Distance Systemic Signaling and Communication in Plants. Springer, Berlin (2013)
4. Baluška, F., Mancuso, S.: Root apex transition zone as oscillatory zone. Front. Plant Sci. **4**, 354 (2013)
5. Baluška, F., Mancuso, S., Volkmann, D., et al.: The 'root-brain' hypothesis of Charles and Francis Darwin: revival after more than 125 years. Plant Signal Behav. **4**, 1121–1127 (2009)
6. Baluška, F., Mancuso, S., Volkmann, D., et al.: Root apex transition zone: a signalling-response nexus in the root. Trends Plant Sci. **15**, 402–408 (2010)
7. Beekman, M., Latty, T.: Brainless but multi-headed: decision making by the acellular slime mould *Physarum polycephalum*. J. Mol. Biol. **427**, 3734–3743 (2015)
8. Bhatla, N., Horvitz, H.R.: Light and hydrogen peroxide inhibit C. elegans feeding through gustatory receptor orthologs and pharyngeal neurons. Neuron **85**, 804–818 (2015)
9. Burbach, C., Markus, K., Zhang, Y., et al.: Photophobic behavior of maize roots. Plant Signal Behav. **7**, 874–878 (2012)
10. Calvo, P., Baluška, F.: Conditions for minimal intelligence across eukaryota: a cognitive science perspective. Front. Psychol. **6**, 1329 (2015)
11. Chua, L.O.: Memristor-the missing circuit element. IEEE Trans. Circuits Syst. **18**, 507–519 (1971)
12. Cognato Gde, P., Bortolotto, J.W., Blazina, A.R., et al.: Y-Maze memory task in zebrafish (Danio rerio): the role of glutamatergic and cholinergic systems on the acquisition and consolidation periods. Neurobiol. Learn. Mem. **98**, 321–328 (2012)
13. Darwin, C., Darwin, F.: The power of movement in plants. John Murray (1880)
14. Dicke, M., Baldwin, I.T.: The evolutionary context for herbivore-induced plant volatiles: beyond the 'cry for help'. Trends Plant Sci. **15**, 167–175 (2010)
15. Falik, O., Mordoch, Y., Quansah, L., et al.: Rumor has it..: relay communication of stress cues in plants. PloS One **6**, e23625 (2011)
16. Falik, O., Mordoch, Y., Ben-Natan, D., et al.: Plant responsiveness to root-root communication of stress cues. Ann. Bot. **110**, 271–280 (2012)
17. Feldman, L.J., Briggs, W.R.: Light-regulated gravitropism in seedling roots of maize. Plant Physiol. **83**, 241–243 (1987)
18. Fromm, J., Lautner, S.: Electrical signals and their physiological significance in plants. Plant Cell Environ. **30**, 249–257 (2007)
19. Galis, I., Gaquerel, E., Pandey, S.P., et al.: Molecular mechanisms underlying plant memory in JA-mediated defence responses. Plant Cell Environ. **32**, 617–627 (2009)
20. Gilroy, S., Suzuki, N., Miller, G., et al.: A tidal wave of signals: calcium and ROS at the forefront of rapid systemic signaling. Trends Plant Sci. **19**, 623–630 (2014)
21. Hahn, A., Zimmermann, R., Wanke, D., et al.: The root cap determines ethylene-dependent growth and development in maize roots. Mol. Plant **1**, 359–367 (2008)
22. Heil, M., Karban, R.: Explaining evolution of plant communication by airborne signals. Trends Ecol. Evol. **25**, 137–144 (2010)
23. Holopainen, J.K., Blande, J.D.: Molecular plant volatile communication. Adv. Exp. Med. Biol. **739**, 17–31 (2012)
24. Jaskiewicz, M., Conrath, U., Peterhänsel, C.: Chromatin modification acts as a memory for systemic acquired resistance in the plant stress response. EMBO Rep. **12**, 50–55 (2011)
25. Kutschera, U., Briggs, W.R.: From Charles Darwin's botanical country-house studies to modern plant biology. Plant Biol. **11**, 785–795 (2009)
26. Latty, T., Beekman, M.: Food quality and the risk of light exposure affect patch-choice decisions in the slime mold Physarum polycephalum. Ecology **91**, 22–27 (2010)

27. Markin, V.S., Volkov, A.G., Chua, L.: An analytical model of memristors in plants. Plant Signal. Behav. **9**, e972887 (2014)
28. Matoh, T.: Boron in plant cell walls. Plant Soil. **193**, 59–70 (1997)
29. Olton, D.S.: Mazes, maps, and memory. Am. Psychol. **34**, 583–596 (1979)
30. Rasmann, S., Kollner, T.G., Degenhardt, J., et al.: Recruitment of entomopathogenic nematodes by insect-damaged maize roots. Nature **434**, 732–737 (2005)
31. Reid, C.R., Latty, T., Dussutour, A., et al.: Slime mold uses an externalized spatial memory to navigate in complex environments. Proc. Natl. Acad. Sci. U. S. A. **109**, 17490–17494 (2012)
32. Reid, C.R., Beekman, M., Latty, T., et al.: Amoeboid organism uses extracellular secretions to make smart foraging decisions. Behav. Ecol. **24**, 812–818 (2013)
33. Stork, W., Diezel, C., Halitschke, R., et al.: An ecological analysis of the herbivory-elicited JA burst and its metabolism: plant memory processes and predictions of the moving target model. PloS One **4**, e4697 (2009)
34. Suzuki, T., Tanaka, M., Fujii, T.: Function of light in the light-induced geotropic response in Zea roots. Plant Physiol. **67**, 225–228 (1981)
35. Tanada, T.: A rapid photoreversible response of barley root tips in the presence of 3-indoleacetic acid. Proc. Natl. Acad. Sci. U. S. A. **59**, 376 (1968)
36. Tanada, T.: Substances essential for a red, far-red light reversible attachment of mung bean root tips to glass. Plant Physiol. **43**, 2070 (1968)
37. Tanada, T.: Boron-induced bioelectric field change in mung bean hypocotyl. Plant Physiol. **53**, 775–776 (1974)
38. Volkov, A.G., Reedus, J., Mitchell, C.M., et al.: Memristors in the electrical network of Aloc vera L. Plant Signal. Behav. **9**, e29056 (2014a)
39. Volkov, A.G., Tucket, C., Reedus, J., et al.: Memristors in plants. Plant Signal. Behav. **9**, 28152 (2014b)
40. Wan, Y., Jasik, J., Wang, L., et al.: The signal transducer NPH3 integrates the phototropin1 photosensor with PIN2-based polar auxin transport in Arabidopsis root phototropism. Plant Cell **24**, 551–565 (2012)
41. Yokawa, K., Baluška, F.: *C. elegans* and *Arabidopsis thaliana* show similar behavior: ROS induce escape tropisms both in illuminated nematodes and roots. Plant Signal Behav. **10**, e1073870 (2015)
42. Yokawa, K., Kagenishi, T., Kawano, T., et al.: Illumination of Arabidopsis roots induces immediate burst of ROS production. Plant Signal Behav **6**, 1460–1464 (2011)
43. Yokawa, K., Kagenishi, T., Baluška, F.: Root photomorphogenesis in laboratory-maintained Arabidopsis seedlings. Trends Plant Sci. **18**, 117–119 (2013)
44. Yokawa, K., Derrien-Maze, N., Mancuso, S., et al.: Binary decisions in maize root behavior: Y-maze system as tool for unconventional computation in plants. Int. J. Unconv. Comput. **10**, 381–390 (2014)
45. Yokawa, K., Fasano, R., Kagenishi, T., et al.: Light as stress factor to plant roots - case of root halotropism. Front. Plant Sci. **5**, 718 (2014)
46. Yokawa, K., Koshiba, T., Baluška, F.: Light-dependent control of redox balance and auxin biosynthesis in plants. Plant Signal. Behav. **9**, e29522 (2014)
47. Yunghans, H., Jaffe, M.: Phytochrome controlled adhesion of mung bean root tips to glass: a detailed characterization of the phenomenon. Physiol. Plant. **23**, 1004–1016 (1970)

Chapter 27
Soft Plant Robotic Solutions: Biological Inspiration and Technological Challenges

B. Mazzolai, V. Mattoli and L. Beccai

Abstract Plants have a sessile lifestyle, and, as a consequence, their structures have a modular organization to ensure surviving in case of environmental damage or predation. Moreover, they developed strategies for efficiently use the resources available in their surroundings, and a well-organized sensing system that allows them to explore the environment and react rapidly to potentially dangerous circumstances. In particular plant roots behaviour emerges from the complex and dynamic interaction between their morphology, sensory-motor control, and environment. Despite the richness of behaviours, mechanisms and features shown, only in recent years scientists and engineers started to consider plants as a possible source of inspiration for developing new technological solutions. In this chapter we highlight how plants and plant roots represent a new model in bioinspired soft robotics and technologies, reporting on few examples of solutions inspired by plants, including growing robots, osmosis-based actuators, controllable hygromorphic materials, and mechanoperceptive systems.

27.1 Introduction

Plants have a sessile lifestyle, and, as a consequence of this primordial decision, their structures have a modular organization to ensure that in case of environmental damage or predation, some modules of the body can survive and regenerate the individual. For these reasons, the specialization of tissues and cells in plants is minimized, compared to most animals, to limit predatory damage.

B. Mazzolai (✉) · V. Mattoli · L. Beccai
Center for Micro-BioRobotics, Istituto Italiano di Tecnologia,
Viale Rinaldo Piaggio 34, 56025 Pontedera, Italy
e-mail: barbara.mazzolai@iit.it

V. Mattoli
e-mail: virgilio.mattoli@iit.it

L. Beccai
e-mail: lucia.beccai@iit.it

© Springer International Publishing Switzerland 2017
A. Adamatzky (ed.), *Advances in Unconventional Computing*,
Emergence, Complexity and Computation 23,
DOI 10.1007/978-3-319-33921-4_27

687

Moreover, because plants must efficiently use the resources available in their surroundings, as they cannot physically move to another place, they exhibit a well-organized sensing system that allows them to explore the environment and react rapidly to potentially dangerous circumstances. Below ground, roots can sense a multitude of abiotic and biotic signals, enabling the appropriate responses. The root apices drive the root growth in search of nutrients and water to feed the whole plant body. Plant roots show efficient exploration capabilities, adapting themselves morphologically to the environment to explore, penetrating the environment with a number of sensorized tips, and accomplishing a capillary exploration of the whole volume of soil. Interestingly, modularity, movement, evolved sensing systems and distributed control are among the most important problems of contemporary robotics. Actually, the exploration capability of plant roots emerges from the complex and dynamic interaction between their morphology, sensory-motor control, and environment. This adaptive behaviour is the basic principle of what in robotics is called "embodied intelligence", which has shifted robotic research away from the traditional view that reduces adaptive behaviour to control and computation [1, 2]. This principle has been adopted in a wide range of current approaches to the development of intelligent artefacts [3, 4]. Many tasks become much easier if embodiment is taken into account. Nature, in fact, gives wide demonstrations of this principle. As already stated, it is well-established that plants are able to show considerable plasticity in their morphology and physiology in response to variability within their environment, particularly as regards foraging for resources [5, 6]. The mechanical properties of the plant roots and the morphology of their structure can be considered for developing the first level of control embedded in the mechanical structures of plant-inspired artefacts [7–9].

More in detail, there are many plant features that can be exploited in technological solutions. First, their movement capabilities offer important cues for implementing new strategies in robotics: nastic (movement that is independent of the spatial direction of a stimulus) or tropic (the response of a plant is influenced by the direction of a stimulus); active (live plant cells activate and control the response by moving ions and changing the permeability of membranes based on potential actions) or passive (movements that are based on dead tissue that is suitable to undergo predetermined modifications upon changes in environmental conditions); reversible or irreversible. This implies the development of new actuation solutions for steering or elongating robotic parts. Other movements take advantage of the water flux among cells, explained as an osmotic phenomenon. Plants use osmosis for driving "slow" movements or for triggering "fast" movements. Interestingly, a new generation of actuators, characterized by low power consumption and high energy density, can be derived from this fundamental plant property [10, 11].

Sensing capabilities are fundamental properties for robots that must move in unstructured environments. In this scenario, soft robotic approaches point towards a new generation of robots capable of soft movements and soft and safe interaction with the environment and humans [12–17]. Compliance is one of the key requirements to accomplish delicate and new tasks in the real world, together with the design of soft sensing systems, especially artificial tactile sensors, which convey information not only about the environment but also about the robot's movements (e.g., interaction forces, bending). In this context, not only animals but also plants can offer new ideas for developing innovative sensing systems that incorporate, at the microscale, several transduction mechanisms for detecting several parameters, which are intelligently "interpreted" for efficient behaviour [18].

Plants, which have recently been considered as a new model in bioinspired and soft robotics [7, 9, 19], must address "problems" that are common also in animals, such as, for example, squid, cuttlefish, and, especially, octopus, which represents the paradigm of soft robotics and embodied intelligence [20–22]. These include distributed control to manage the infinite degrees of freedom of their body, high flexibility, the capability of growing and/or elongating their extremities, and distributed sensing capabilities. Plants have also recently been considered as a revolutionary concept for developing mathematical models of swarm intelligence [23] or *emergent behaviour* given by the coordination of the plant roots or root apices of the whole organism towards optimal targets. This research can lead to the development of novel methods of collective decision making in decentralized structures with local computation and simple communication.

In the following, we highlight a few examples of solutions inspired by plants, including growing robots, materials, actuations, and sensing systems.

27.2 Roots—Growing Robots and Emerging Behaviour

Unlike animals, for which growth is related to development, the distinctive feature of roots is the link between their growth and their movement [24]. Plant roots are able to find low-resistance pathways and exploit cracks in the soil, overcoming soil penetration resistance [25–27]. Their exploration capability arises from the root apex [28–30], which is able to move and penetrate soil by growing at the apical region [31, 32], following or escaping from environmental stimuli. The apex growth process enables roots to adapt their morphology and organ development to environmental conditions such as the soil texture and mechanical impedance because the cell division and morphology are directly influenced by the interaction with the surrounding environment [33–35]. Root development is driven by two continuous processes in the apex, cell division and cell elongation, which occur in the meristematic and elongation regions, respectively (see Fig. 27.1a). We called the growing process "elongation from the tip" (EFT) [8]. Newly generated cells move from the meristematic region to the elongation region, where they expand axially because of the water absorbed by osmosis and the directional loosening of the cell wall [36, 37]. This action allows the root to penetrate the soil with only a small part of its structure (the apex), while the remainder of the structure is stationary and in contact with the soil (the mature zone). This process provides the pressure required for the forward advancement of the root. The penetration may be straight or curved, depending on whether the cell growth is symmetric or differential. EFT in a plant root represents an efficient solution for soil penetration, and the preliminary results reported in [38] were considered a starting

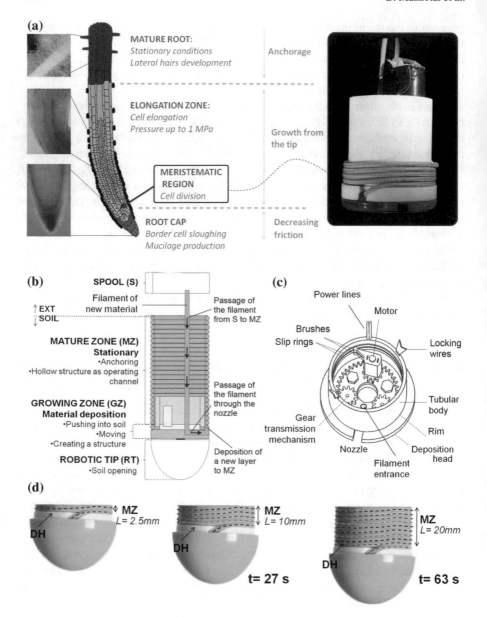

Fig. 27.1 **a** Plant root structure and relative functions. The bioinspired artefact mimics the meristematic region, releasing a filament that constitutes the body of the artificial root. **b** Schematic representation of the cross section of the *root-like* device. **c** 3D design of the growing zone. **d** From left to right, a sequence of images showing how the filament deposition implements the growth process. Images adapted with permission from [8]

point and a source of inspiration for designing and developing an innovative robotic soil exploration system. Another strategy to reduce friction between the root and soil is the continual production of cells in the meristematic region. These cells move to the root cap and then slough off from its outer surface while producing mucus [39]. In this way, the cap cells create an interface between the soil and the root apex. The root cap protects the delicate cells in the meristematic region, and the mucus promotes root penetration by reducing friction [39–42]. Mature cells, which are situated behind the apex, are strongly anchored to the soil and thus allow the apex to move forward. This anchorage, also called root-soil adhesion [43], is achieved by root hairs, secondary roots, and the root architecture [43–45].

We recently developed robotic systems that are inspired both by the "elongation from the tip (growing) process" [8] and by the sloughing cells in the apical root region [46], described previously. These represent a new generation of devices that are able to grow and construct their own body and are inspired by the growth and soil penetration behaviours in plant roots. Specifically, the proposed root-like device grows through a monotonic process that continuously adds new material. This device can efficiently penetrate soil by taking advantage of root penetration behaviours, in particular, EFT. Moreover, because the design is based on plant root principles, the system described represents a useful platform for the experimental validation of theories and hypotheses concerning actual plant root systems. The device is composed of a growing zone and a stationary mature zone. The mature zone consists of a hollow, tubular structure that allows the transfer of new material from a spool (external to the robotic root) and power to the growing zone. The growing zone is a customized additive layering mechanism that generates a force for penetration into the soil and consists of: (a) a rotating deposition head, (b) a guiding nozzle on the deposition head, and (c) a motor and a transmission mechanism (i.e., a gearset) (Fig. 27.1b,c). The material, which is in the form of a filament (polypropylene, PP), can be easily transferred from the external spool to the nozzle at the periphery of the deposition head. The "growing" capability at the tip region is achieved with an additive manufacturing technique similar to fusion deposition modelling (FDM) [47]. The layer-by-layer deposition creates the mature zone, which uses the tubular body as a support structure. The rotation of the deposition head is driven by a motor coupled to the transmission mechanism. The tubular body moves axially inside the mature zone in a passive manner, and any rotational movement between them is prevented by locking wires installed on the body. The axial slipping movement induced by the addition of filament material allows the new material to be automatically distributed on the surface of the deposition head, in the growing zone (Fig. 27.1b). To prevent any twisting between the filament and the motor power cables, the power is transferred by two annular electrodes (i.e., slip rings) that are fixed to the internal surface of the tubular body and are always in contact with the brushes connected to the rotating motor. The cables are connected to the electrodes through an axial hole in the tubular body (Fig. 27.1c). The rotary motion of the deposition head is converted into a linear motion at the tip that provides the motive force for overcoming soil resistance (Fig. 27.1d).

Another important lesson that we can take from the study of plant roots is related to the "control" strategies they seem to implement. Despite their lack of a central nervous system, if we look at the growing and explorative behaviour of roots in the soil, we cannot escape the recognition that the growth pattern of the root apparatus is not chaotic at all. On the contrary, it is coordinated and efficiently shaped to exploit soil resources and avoid hazards. Plant roots exhibit an exceptional capacity to sense very weak oxygen, water, temperature and nutrient gradients in the soil [48]. However, how they manage to navigate towards a source of a resource without being distracted by local variations remain unclear. An interesting analogy can be found in the collective behaviour adopted by birds during long-range migration. The navigation follows a very weak gradient, and due to local variations, it is an almost impossible task for an individual bird, whereas a flock acting collectively overcomes this obstacle by working as an integrated array of sensors [49]. Consequently, it is tempting to propose that a collective behaviour of root tips could emerge even in plant roots from the individual activity of a single root apex. In the same way as birds, an individual root apex could adopt a collective behaviour that minimizes the influence of local fluctuations for exploration purposes. The "swarm rules" come out as simple rules for different organisms: information left by the chemical traces produced by every single apex and instruction of local density may be considered the basis of such rules for roots.

27.3   Movements—Osmotic and Hygromorphyc Actuators

Plants did not evolve contractile proteins on the basis of animal muscular motion. Nevertheless, plants are able to move in a variety of ways to perform all the activities needed to spread their genetic materials, to explore their environment and to search for nutrients and energy [50]. Some of them also developed remarkably complex defensive behaviours.

The question of how plants move without muscles has attracted the interest of many scientists since the pioneering work of the Darwins [24, 51]. At the macroscopic level, the most obvious difference between animal and plant motions is their timescales. Even if several examples of fast plant movements can be found in nature, more typical movements (e.g., growing or tropic movements) occur in a temporal window of hours, several orders of magnitude slower than that in animals. This is because these movements are based on relatively slow mechanisms, such as osmosis for the active movements (controlled by plant metabolism) or the vapour/water uptake equilibrium in the case of passive movements (mostly based on dead tissues, suitably structured at the microscale level). A schematic representation of the different plant movements, following the classification proposed by [52], is presented in Fig. 27.2.

The tropic growth of roots and the opening/closing of stomata in plant leaves are typical active osmotic-driven slow movements (purely water-driven movements) based on living cells that change volume by modulating the permeability of their

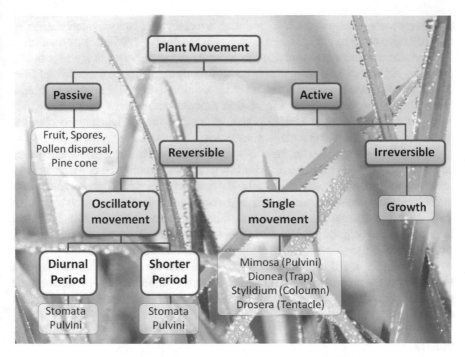

Fig. 27.2 Classification of plant movements according to [52]

membranes, thus controlling ion movements on the basis of metabolic stimuli. In the case of fast active movements, the basic osmotic cellular water uptake mechanism is often coupled with mechanical instabilities and/or leverage mechanisms that amplify the capacity to move. This is, for example, the case of the snap-buckling mechanism acting in *Dionaea muscipula* [53], the pollen release of the *Stylidium* flower [52] and the stimulus-responsive closure of the leaves in *Mimosa pudica* [54].

From an engineering point of view, plant movements show several remarkable features that could be successfully translated into engineering mechanisms and principles. These include characteristics such as energy efficiency, high actuation force, and a rich motion repertoire (acquired during hundreds of millions of years of evolution processes).

Starting from such a biomimetic perspective, our group recently introduced two complementary approaches for plant-inspired actuation, one based on an osmosis mechanism and another based on passive mechanisms exploiting water uptake from environmental humidity.

27.3.1   Osmosis-Based Actuator

The term "osmosis" originates from the Greek "ωσμoς" (osmos), which means a "push or impulsion". Osmosis is a chemo-physical entropy-driven process that relies on solvent transport through a semi-permeable membrane (osmotic membrane) able to block (also partially) the passage of the solute (osmolyte). When two solutions at different concentrations are separated by an osmotic membrane, a pressure (osmotic pressure) is generated across this membrane, driving the passage of the solvent from the less concentrated solution to the more concentrated one. This process is present in every living system, performing several fundamental roles at the physiological level. As previously introduced, in plants, osmosis is the fundamental mechanism that drives both slow and fast active movements.

To turn the osmosis phenomenon into an actuation mechanism, four elements are necessary: an osmotic membrane, a rigid structure, an osmotic power reservoir (two solutions with different concentrations) and a compliant transducer, which transforms the solvent flow into a movement. Plants are complex hydraulic systems with a high level of compartmentalization, in which the fundamental osmotic unit is the cell that exploits the osmotic process through its stiff walls, composed of hierarchically structured cellulose microfibrils embedded in a pectin matrix [55, 56].

The hardness of the cells can be finely modulated by plants by regulating in a selective way the turgor pressure ($\Delta\Pi$), defined as the difference between the osmotic pressures (also called osmotic potentials) outside and inside the cell (respectively, $\Delta\Pi_o$ and $\Delta\Pi_i$) [57].

The osmotic potential Π is given to a first approximation by the Van't Hoff equation [13]

$$\Pi = \Phi i M R T \tag{27.1}$$

where M is the molar concentration of the osmolyte in solution, R the universal gas constant, T the temperature (in K) and i the Van't Hoff coefficient, which accounts for the particle effect in the presence of electrolytes [58]. Φ is an empirical factor that takes into account the non-ideal behaviour of solutes. In the case of strong common electrolytes, deviations from this law are generally less than 2 %.

The water flow ($\dot{q}$) induced across the membrane by the osmotic pressure depends on the osmotic membrane surface (S_{OM}) and osmotic membrane permeability (α_{OM}) following the simple equation

$$\dot{q} = S_{OM}\alpha_{OM}(\sigma\Delta\Pi - \Delta P) \tag{27.2}$$

where σ is an osmolyte rejection coefficient ($s \sim 1$ for an ideal membrane), $\Delta\Pi$ the osmotic pressure difference, and ΔP the pressure difference between the two sides of the osmotic membrane. If $\Delta P > \Delta\Pi$, the water flow reverses its direction and flows from the more concentrated to the more dilute solution (reverse osmosis) [59].

The osmotic principle, which is used by plants for their movements, has been considered for actuating artificial systems. An osmotic actuator was proposed to steer the tip of a mechatronic system inspired by the apex of plant roots [19], with applications in soil exploration and monitoring. This type of actuator was based on electro-osmosis because of its potential for reversibility. In particular, the steering concept was based on three cells that are separated by pairs of semi-permeable osmotic membranes and ion-selective membranes and individually coupled with a piston mechanism. Technical issues primarily related to the degradation of the ion-selective membranes during the electro-osmotic process (lead acetate was used as a salt) encouraged the development of an alternative actuation strategy based on forward osmosis.

A first parametric modelling of a forward-osmosis actuator inspired by plants was carried out by [10], with the aim of providing useful guidelines to develop a system able to reach a targeted performance in terms of the actuation time and force. The actuator was modelled with a system composed of an osmotic membrane, an actuation chamber containing the osmotic solution with a rigid frame and a deformable elastic membrane (bulging membrane), and a reservoir containing the solute (see Fig. 27.3a).

The parametric equations express the actuator characteristics of interest (in particular, the actuation time, maximum force, and power density) in terms of the geometric and physical parameters of the system (e.g., dimensions, elastic modulus of membrane, and osmotic membrane permeability). One of the most interesting results

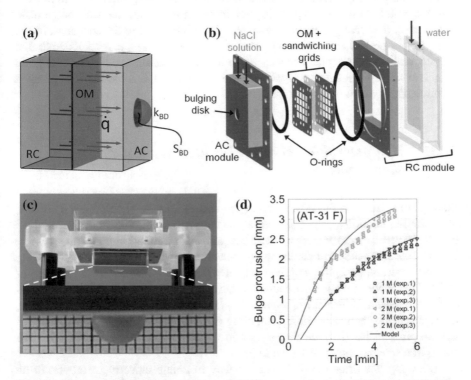

Fig. 27.3 a Osmotic actuator concepts: the solvent flows from the reservoir chamber (RC) to the actuation chamber (AC) through the osmotic membrane (OM): the actuation work is transduced through the elastic bulging deformation of a disk-membrane. **b** Exploded view of the main components of the actuator. **c** Assembled prototype of the actuator; in the inset: bulging of the elastic membrane (scale bar 1 mm). **d** Typical bulge protrusion vs. time for different samples at two different concentrations. Solid line is the theoretical prediction. Images adapted with permission from [11]

is the demonstration that relatively fast osmotic actuation (on the order of a few minutes) is achievable with a centimetre-scale actuator having the proper geometry.

Based on these results and on the evaluation of practical issues such as materials selection (membranes and osmolytes) and mechanical constraints, a fully working osmotic actuator prototype was developed by [11]. The actuator is characterized by fast dynamics (on the order of minutes) and by high energy and power density. The system is composed of an actuation chamber made of AISI 316 stainless steel containing a semi-permeable forward-osmosis membrane (specifically designed for operation with NaCl, with a rejection coefficient in the range of 0.95–0.97) kept in position by 2 metal grids and an elastomeric bulging membrane. Sodium chloride (NaCl) was chosen as the osmolyte. The dimensions of the osmotic membrane were fixed to 10×10 mm^2, while the diameter hole for the bulging disk protrusion was 5 mm (see Fig. 27.3b). With a solution of 2 M NaCl, the actuator is able to produce a bulge protrusion of several mm in a few minutes, well agreeing with the theoretical model, resulting in the fastest forward osmosis-based actuator in the literature to date [60–64] (see Fig. 27.3c, d). Measurement of the blocking force showed that the actuator is able to produce up to 20 N of force on the 5 mm diameter bulge disk. In this case, the force generated results closely proportional to the concentration of osmolyte, in perfect agreement with the model [11]. Finally, it is remarkable that the timescale of that actuator is very close to the relaxation time of an ideal giant plant cell with the same typical size as the actuator, thus supporting the pursued bioinspired approach (see Fig. 27.4).

27.3.2 Hygromorphic Actuation Inspired by Passive Movements

Some plants show the remarkable ability to convert simple environmental stimuli, in particular, humidity, into a reversible mechanical motion [65], even using dead tissues. Well-known examples of these movements are found in pine cones, which release their ripe seeds by opening their scales in a dry environment (instead closing in a wet environment) [66], and the seeds of wild wheat, in which the alternation of wet and dry environments in circadian cycles results in propulsion into the soil [67].

Even if several different implementations of this principle are present in nature, the basic mechanism relies on the coupling of bilayered composite structures with anisotropic structured tissue that swells in different specific directions upon the absorption of environmental moisture [68].

Following this principle, the simplest way to obtain anisotropic motion in an artificial actuator is to couple a humidity-responsive material (active layer) with an elastic material insensitive to environmental humidity (passive layer). The main function of the passive layer is to convert the water-driven swelling of the active layer into a bending actuation.

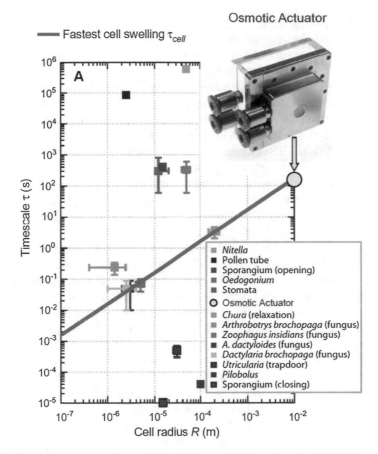

Fig. 27.4 The osmotic actuator timescale matches the relaxation time of an ideal giant plant cell of the same typical size, i.e., cell radius R. Image adapted with permission from [11]

Several polymeric materials able to react to environmental stimuli (commonly called smart polymers) have been studied so far, with the specific aim of targeting biomimetic applications. In this regard, hydrogels are a particular class of active materials, composed by a network of hydrophilic polymeric chains containing a water-based solution, able to react to several stimuli (pH, temperature, and biochemical processes), producing macroscopic actuation through reversible swelling and shrinking due to water exchange with the environment. In this sense, they are interesting candidates for mimicking the swelling and shrinking behaviour of plant cells. Moreover, complex motion could be, in principle, obtained by inducing some anisotropy at the material level, e.g., embedding into the hydrogel matrix some oriented stiffer fibres. Nevertheless, hydrogels show some limitations, mainly related to their poor mechanical properties and their poor (or null) controllability via electrical stimuli in environments with moderate humidity, such air in typical conditions (highly desirable in bioinspired robotics applications).

As an alternative to develop actuation bending mechanisms based on reversible adsorption and desorption of environmental humidity, a new approach was recently proposed [69], also combining the possibility of achieving passive (humidity-driven) and active (electrically driven) actuation with a single composite material.

In this case, the moisture sensitive material is a well-known conjugated conductive polyelectrolyte complex named poly(3,4-ethylenedioxythiophene):poly(styrenesulfonate) (PEDOT:PSS), having the peculiar capacity to absorb water due to its hydrophilic PSS part. This capability makes the polymer a very good candidate for the development of water-sensitive systems, such as humidity sensors [70], as well as humidity-driven actuators. Following the bilayer approach, a thin PEDOT:PSS film is coupled with a thick humidity-inert layer of a passive elastomeric layer, such as poly-dimethylsiloxane (PDMS), acting as a structural material.

In a pure passive mechanism, when the bilayer is subjected to a high-humidity environment, the PEDOT:PSS layer swells, increasing its volume (isotropically), following the water vapour absorption, generating the bending of the bilayer due to the constraints of the passive layer. The original equilibrium is quickly recovered when the system is exposed to an environment with lower humidity.

Even more interestingly, the system can also be controlled by applying an electrical stimulus, in particular, an electric current. As PEDOT:PSS is a conductive polymer with a reasonable conductivity (on the order of hundreds of Ω/square), a low voltage can generate enough current to induce a localized Joule-heating effect, sufficient to release the fraction of water absorbed at the equilibrium with air having a certain humidity level. The drying of the active part results in a shrinking of the same that induces a bending movement in the reverse direction with respect to the one induced by an increase of humidity. Additionally, in this case, the actuation is perfectly reversible, and upon removing the stimulus (electrical current), the system quickly comes back to the original equilibrium state (see Fig. 27.5a for a schematic representation of the working principle of the hygromorphic actuator).

The good performance of the actuator in terms of response time (faster than other systems based on similar mechanisms) is mainly due to the thickness of the active layer, which is on the order of a few hundred nanometres, against a thickness of about one hundred microns for the passive layer. This thickness and the consequent high surface-to-volume ratio allow a faster water adsorption/desorption process with respect to the bulk polymer. Interestingly, even if the thickness of the active layer is really thin and the maximum contraction in PEDOT:PSS due to humidity desorption is only on the order of 2% [71], due to the mechanical properties of the active and passive layers (with elastic moduli of approximately 1 GPa and 2 MPa) and the actuator design (mainly aspect ratio and active/passive material thickness ratio), the bilayer is capable of a large bending displacement.

Figure 27.5c-b shows an example of a cantilever-shaped actuator in which the bending is induced by the variation of the environmental moisture level and a flower-shaped hygromorphic actuator controlled by a current flowing through the petals. Due to the straight fabrication process and the possibility of easily customizing

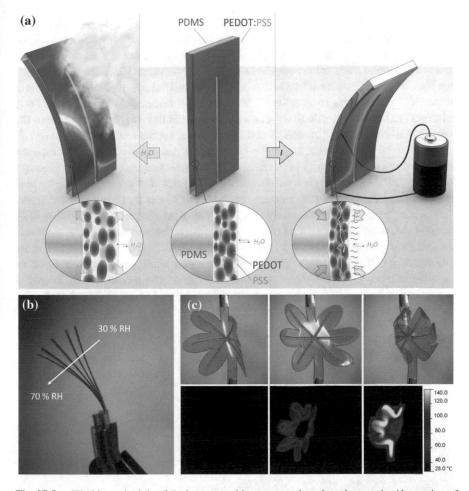

Fig. 27.5 a Working principle of the hygromorphic actuators, based on the sorption/desorption of water from environmental humidity: beginning from the equilibrium position (*central*), an electric current induces a Joule-heating effect that results in a desorption of water from the PEDOT:PSS layer that shrinks, thus producing a bending (*right*); conversely, increasing the environmental humidity, the actuator bends from the equilibrium position in the reverse direction due to the expansion of the PEDOT:PSS layer due to the sorption of water until the establishment of a new equilibrium (*left*). **b** Example of a *cantilever-shaped* actuator in which the bending is induced by the variation in the environmental moisture level. **c** *Flower-shaped* hygromorphic actuator controlled by Joule effect: visible (*top*) and thermal (*bottom*) camera pictures. Images adapted with permission from [69]

the shape of the material via laser cutting (making it in a single step shape and conductive patterns for Joule heating), this active material can find applications in several bioinspired and soft robotics applications, such as soft grippers, manipulators, and active elements on the millimetre scale.

Nevertheless, the most striking feature of this bilayer material is its capability to work as an integrated multifunctional system, acting at the same time as a functional material, able to perform actuation and sensing, and as structural material. This is a point of particular interest because the search for integrated multifunctional materials has been targeted by material scientists and engineers in recent decades with the aim of providing a new way to develop smarter and more integrated devices and robotic artefacts [72]. This is particularly true for bioinspired devices, trying to mimic the features and behaviour of materials and mechanisms encountered in nature.

An interesting demonstrator of the capability of the multifunctional composite and its potential applicability in the field of biomimetic and soft robotics is given by the "artificial" *Mimosa pudica*. Among different species able to move on the scale of seconds, the *Mimosa* is a paradigmatic examples of plants that react to mechanical stimuli (i.e., touch by an external subject) by closing their leaves for protection [73]. To imitate the behaviour of this plant, we exploited the variation of the electrical conductivity of the PEDOT:PSS layer when it is subjected to a mechanical stress (piezoresistive characteristic). The artificial *Mimosa pudica* prototype consists of a six-leaf-shaped structure with typical dimensions of some centimetres, in such a shape that the leaves close when a current is applied (see Fig. 27.6). The detection of

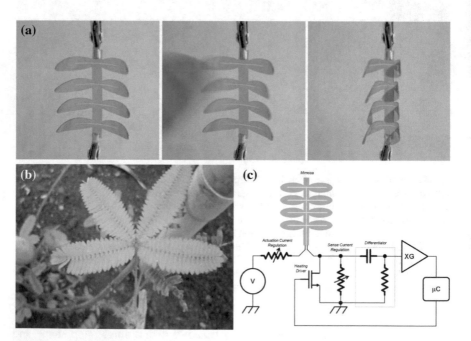

Fig. 27.6 a A leaf-shaped multifunctional sensing/actuator system inspired by the Mimosa pudica plant (*left*); when the leaf is gently touched by a finger (*central*), the system reacts to the touch stimulus by closing its leaves (*right*). Images adapted with permission from [69]. **b** Picture of a real Mimosa pudica. **c** Scheme of conditioning electronics for the leaf-shaped actuator inspired by Mimosa pudica, used to implement the reactive behaviour

the touch stimulus is performed by monitoring the resistance value of the conductive path on the *Mimosa* with a suitable conditioning system and an analogue filter (see Fig. 27.6c). When a fast variation in the resistivity is detected, a current is applied to the structure, and the leaves close, as in their natural counterpart.

27.4 Mechanoperception—Soft Bending and Force Sensing

New tactile sensing strategies and technology can be inspired by the world of plants because, to grow and develop, they need to adapt and respond to the mechanical stresses that the environment presents (i.e., 'exogenous'), as well as those related to their inner structure (i.e., 'endogenous'). In the first case, the stimuli can be, for example, wind, soil constraints and mechanical barriers, or passing animals, whereas in the second case, they are related to the turgor pressure driving cell expansion. In any case, plants perceive the caused stresses, i.e., they are characterized by the so called 'mechanoperception' [74], and as a consequence their cells deform. We took inspiration from these mechanisms to design and develop a soft sensing cylindrical body that can discriminate between its convex and concave sides, in addition to discriminating between bending and indentation stimuli [18]. In particular, we were inspired by two observations of plant biologists in *Arabidopsis* plants [75, 76], as follows: (i) upon root bending, a stretching of the cells on the root convex side occurs, while cells on the concave side are compressed; and (ii) convex side cells have changes (increase) in cytosolic Ca^{2+} levels, while on the concave walls, cells do not show a detectable response. Starting from this, we used soft and flexible materials, such as elastomers and conductive textiles, to build two capacitive sensing elements that intimately conform to the shape of the soft body and follow its deformations, and we combined the responses of the sensing sites to retrieve information about bending and force. Figure 27.7 shows a flexible and soft body prototype in polydimethylsiloxane (PDMS) and two embedded capacitive sensing elements (namely S1 and S2) at its opposite walls, i.e., 180 °C from each other. When the module is bent, the sensing elements are under tension and compression solicitation, i.e., on convex and concave, respectively, just like in the plant root. This happens because their constituent materials allow an adaptation to the reversible change in the deformation produced by the bending movement, i.e., two stretchable conductive fabric parallel electrodes and as a dielectric, a silicone elastomer film (Ecoflex 0010, Smooth-On, 300 µm thick). The active area of each sensing element is 5×5 mm^2. It is possible to correlate the normalized capacitance variation ($\Delta C / C_0$ with C_0 being the is the capacitance value when no stimulation is applied) of the sensing elements S1 and S2 to the different mechanical stimulations applied to the soft body, such as bending and/or an external force. In Fig. 27.8, the typical mechanical configurations of a cantilever and a beam clamped at both extremities subjected to both force and bending are shown.

It is worth emphasizing that the proposed strategy is trivial from the computational point of view because given a particular mechanical configuration, the applied stimulation is retrieved by looking at the sign of the responses of sensing elements

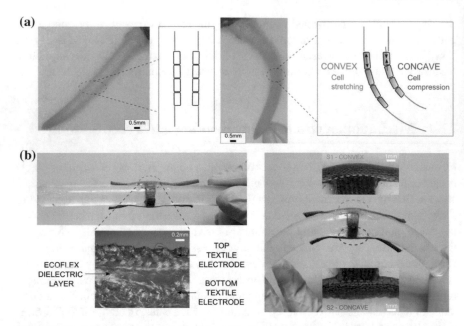

Fig. 27.7 The plant root inspired soft sensing body (composed figure adapted from [18]. **a** Images of *Zea mays* roots and drawings depicting the deformations induced in the root cells by bending (*right*), i.e., cells are stretched and compressed at convex and concave side, respectively. **b** (*Left, top*) images showing the soft body with two opposite sensing sites in rest configuration and (*left, bottom*) cross-section of one capacitive tactile sensor; (*Right*) image of the soft body subjected to bending and inset photographs of the two solicited sensing elements

S1 and S2. In the cantilever beam configuration, the capacitance variation of the concave sensing side is negative, while the response on the convex side is positive. In addition, the latter can be used to quantitatively determine the maximum bending angle of the cantilever. Instead, when the body is clamped at both extremities, both responses present a positive variation. However, the signal on the concave side is always much larger in amplitude (due to the force applied directly on it) with respect to the convex side. This allows distinguishing between the two sides and measuring simultaneously the externally applied force and the maximum deflection of the beam.

In the literature, very few studies address the integration of soft sensing systems in soft bodies, while this is a crucial topic in soft robotics. In bioinspired arms and flexible beams, pressure detection is achieved or the movement of the arm is used to reconstruct their spatial configuration [77, 78]. However, these systems are not able to distinguish between the convex and concave sides or to discriminate between bending and force solicitations, as they are only capable of a single sensing function (i.e., force or strain sensing). Therefore, starting from the study of plant-root mechanoperception, it is possible to define a strategy to reveal bending and applied force in a soft body with only two sensing elements of the same kind and a null computational effort. This is in contrast to traditional approaches, where the

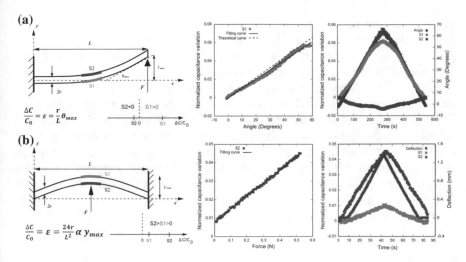

Fig. 27.8 A schematic of two mechanical configurations of the *cylindrical* soft body studied in [18] and of the related experimental results obtained when soliciting the system by an external force F: **a** cantilever beam configuration; **b** body clamped at both extremities. *On the left*, in both cases the normalized capacitance variation $\Delta C / C_0$ is shown as function of (**a**) the bending angle Θ, and (**b**) the maximum beam deflection y_{max}, and the signs of the S1 and S2 output signals are shown. At the centre, graphs show: (**a**) the dependence of $\Delta C / C_0$ relative to S2 on the bending angle; (**b**) the dependence of $\Delta C / C_0$ relative to S1 on force F. *On the right*, graphs show the dependence of $\Delta C / C_0$ relative to S1 and S2, on: (**a**) Θ; (**b**) y_{max}. Images are adapted from [18]

integration of single units, each addressing a specific functionality, is needed, with the result of a dramatic increase in the complexity of the robotic design (both mechanical and algorithmic) [13].

27.5 Conclusions

Plant-inspired robots and technologies involve a variety of scientific and technological objectives due to the challenges of developing plant root-like artefacts, and these objectives cannot be met with current technologies. The vision of this research is to explore a new concept of robotic artefact and behavioural modelling that allows a distributed, extensive and intensive monitoring of the soil. Plant-inspired robots or PLANTOIDS [79] distributed on the ground surface and able to penetrate and advance in soils have the potential to become a very appealing solution to a variety of different problems in soil exploration and environment protection. By imitating the plant root strategy, the robotic artefact should move slowly, efficiently exploring the environment, coordinating its artificial roots (without a "brain") and showing high actuation forces despite low power consumption. An important question to address is whether plants exhibit intelligent behaviour. A simple definition of plant intelligence could be adaptively variable growth and development during the lifetime of the individual. This research should advance the state of the art in intelligent systems, particularly in robotics and ICT, as well as in other disciplines, such as biology and agronomy. The approach proposed for the development of some of the considered technologies is visionary, thanks to the inspiration from plants, and it should bring essential contributions for achieving robotic systems that are more natural and safer. Exploiting adaptive abilities in plants could lead to the development of smart devices not only with the ability to sense, but with the capability to follow stimuli and take decisions to accomplish the required tasks, limiting computational costs. Applications for such technologies inspired by plants include soil monitoring and exploration for contamination or mineral deposits, whether on earth or other planets, but could also include medical and surgical applications, like new flexible endoscopes, able to steer and grow in delicate human organs.

References

1. Pfeifer, R., Scheier, C.: Understanding Intelligence. MIT Press, Cambridge (1999)
2. Pfeifer, R., Bongard, J.: How The Body Shapes The Way We Think: A New View of Intelligence. MIT Press, Cambridge (2007)
3. Pfeifer, R., Lungarella, M., Iida, F.: Self-organization, embodiment, and biologically inspired robotics. Science **318**, 1088 (2007)
4. Cianchetti, M., Arienti, A., Follador, M., Mazzolai, B., Dario, P., Laschi, C.: Design concept and validation of a robotic arm inspired by the octopus. Mater. Sci. Eng. C **31**, 1230–1239 (2011)
5. Hodge, A.: Root decisions. Plant Cell Environ. **32**, 7–9 (2009)
6. Trewavas, A.: What is plant behaviour? Plant Cell Environ. **32**, 606–616 (2009)
7. Mazzolai, B., Laschi, C., Dario, P., Mugnai, S., Mancuso, S.: The plant as a biomechatronic system. Plant Signal. Behav. **5**(2), 1 (2010)
8. Sadeghi, A., Tonazzini, A., Popova, L., Mazzolai, B.: A novel growing device inspired by plant root soil penetration behaviors. PLoS ONE **9**(2), e90139 (2014)
9. Mazzolai, B., Beccai, L., Mattoli, V.: Plants as model in biomimetics and biorobotics: new perspectives. Front. Bioeng. Biotechnol. Bionics. Biomim. **2**, 2 (2014)
10. Sinibaldi, E., Puleo, G., Mattioli, F., Mattoli, V., Di Michele, F., Beccai, L., Tramacere, F., Mancuso, S., Mazzolai, B.: Osmotic actuation modelling for innovative biorobotic solutions inspired by the plant kingdom. Bioinspiration Biomim. **8**, 025002 (2013)
11. Sinibaldi, E., Argiolas, A., Puleo, G.L., Mazzolai, B.: Another lesson from plants: the forward osmosis-based actuator. PLoS ONE **9**(7), e102461 (2014)
12. Iida, F., Laschi, C.: Soft robotics: challenges and perspectives. Procedia Comput. Sci. **7**, 99–102 (2011)
13. Kim, S., Laschi, C., Trimmer, B.: Soft robotics: a bioinspired evolution in robotics. Tren. Biotechnol. **31**, 287–294 (2013)
14. Pfeifer, R., Lungarella, M., Iida, F.: The challenges ahead for bio-inspired 'soft' robotics. Commun. ACM **55**, 76–87 (2012)
15. Trivedi, D., Rahn, C.D., Kier, W.M., Walker, I.D.: Soft robotics: biological inspiration, state of the art, and future research. Appl. Bionics Biomech. **5**, 99–117 (2008)
16. Trimmer, B.: Soft robots. Curr. Biol. **23**, R639–R641 (2013)
17. Lipson, H.: Challenges and opportunities for design, simulation, and fabrication of soft robots. SoRo **1**, 21–27 (2013)

18. Lucarotti, C., Totaro, M., Sadeghi, A., Mazzolai, B., Beccai, L.: Revealing bending and force in a soft body through a plant root inspired approach. Sci. Rep. **5**, 8788 (2015)
19. Mazzolai, B., Mondini, A., Corradi, P., Laschi, C., Mattoli, V., Sinibaldi, E., Dario, P.: A miniaturized mechatronic system inspired by plant roots. IEEE Trans. Mech. **16**(2), 201–212 (2011)
20. Laschi, C., Mazzolai, B., Cianchetti, M., Mattoli, V., Bassi-Luciani, L., Dario, P.: Design of a biomimetic robotic octopus arm. Bioinspiration Biomim. **4**, I015006 (2009)
21. Margheri, L., Laschi, C., Mazzolai, B.: Soft robotic arm inspired by the octopus.I. from biological functions to artificial requirements. Bioinspiration Biomim. **7**, 025004 (2012)
22. Mazzolai, B., Margheri, L., Cianchetti, M., Dario, P., Laschi, C.: Soft robotic arm inspired by the octopus. II. from artificial requirements to innovative technological solutions. Bioinspiration Biomim. **7**, 025005 (2012)
23. Ciszak, M., Comparini, D., Mazzolai, B., Baluska, F., Arecchi, T.-F., Vicsek, T., Mancuso, S.: Swarming behavior in plant roots. Plos One **7**(1), e29759 (2012)
24. Darwin, C.: The Power of Movement in Plants. John Murray, London (1880)
25. Inoue, N., Arase, T., Hagiwara, M., Amano, T., Hayashi, T., et al.: Ecological significance of root tip rotation for seedling establishment of Oryza sativa L. Ecol. Res. **14**, 31–38 (1999)
26. Brown, A.H.: Circumnutations: from Darwin to space flights. Plant Physiol. **101**, 345–348 (1993)
27. Vollsnes, A., Futsaether, C., Bengough, A.: Quantifying rhizosphere particle movement around mutant maize roots using time lapse imaging and particle image velocimetry. Eur. J. Soil Sci. **61**, 926–939 (2010)
28. Verbelen, J.-P., De Cnodder, T., Le, J., Vissenberg, K., Baluska, F.: Root apex of Arabidopsis thaliana consists of four distinct zones of growth activities: meristematic zone, transition zone, fast elongation zone, and growth terminating zone. Plant Sig. Behav. **1**, 296–304 (2006)
29. Dexter, A.: Mechanics of root growth. Plant Soil **98**, 303–312 (1987)
30. Baluska, F., Mancuso, S., Volkmann, D., Barlow, P.: Root apices as plant command centres: the unique 'brain-like' status of the root apex transition zone. Biologia **59**, 7–19 (2004)
31. Ishikawa, H., Evans, M.L.: Specialized zones of development in roots. Plant Physiol. **109**, 725 (1995)
32. Barlow, P.W.: The root cap: cell dynamics, cell differentiation and cap function. J. Plant Growth Reg. **21**, 261–286 (2002)
33. Barlow, P.W.: The response of roots and root systems to their environment? An interpretation derived from an analysis of the hierarchical organization of plant life. Env. Exp. Bot. **33**, 1–10 (1993)
34. Unger, P.W., Kaspar, T.C.: Soil compaction and root growth: a review. Agronomy J. **86**, 759–766 (1994)
35. Iijima, M., Morita, S., Barlow, P.W.: Structure and function of the root cap. Plant Prod. Sci. **11**, 17–27 (2008)
36. Schopfer, P.: Biomechanics of plant growth. Am. J. Bot. **93**, 1415–1425 (2006)
37. Croser, C., Bengough, A.G., Pritchard, J.: The effect of mechanical impedance on root growth in pea (Pisum sativum). II. cell expansion and wall rheology during recovery. Physiol. Plantarum **109**, 150–159 (2000)
38. Tonazzini, A., Sadeghi, A., Popova, L., Mazzolai, B.: Plant root strategies for robotic soil penetration. Biomim. Biohybrid Syst. **8064**, 447–449 (2013)
39. Hawes, M.C., Bengough, G., Cassab, G., Ponce, G.: Root caps and rhizosphere. J. Plant Growth Reg. **21**, 352–367 (2002)
40. Bengough, A., McKenzie, B.: Sloughing of root cap cells decreases the frictional resistance to maize (Zea mays L.) root growth. J. Exp. Bot. **48**, 885 (1997)
41. Iijima, M., Higuchi, T., Barlow, P.W., Bengough, A.G.: Root cap removal increases root penetration resistance in maize (Zea mays L.). J. Exp. Bot. **54**, 2105 (2003)
42. Bengough, A., Kirby, J.: Tribology of the root cap in maize (Zea mays) and peas (Pisum sativum). New phytol. **142**, 421–425 (1999)

43. Goodman, A., Ennos, A.: The effects of soil bulk density on the morphology and anchorage mechanics of the root systems of sunflower and maize. Annal. Bot. **83**, 293–302 (1999)
44. Bengough, A., McKenzie, B., Hallett, P., Valentine, T.: Root elongation, water stress, and mechanical impedance: a review of limiting stresses and beneficial root tip traits. J. Exp. Bot. **62**, 59–68 (2011)
45. Czarnes, S., Hiller, S., Dexter, A., Hallett, P., Bartoli, F.: Root-soil adhesion in the maize rhizosphere: the rheological approach. Plant Soil **211**, 69–86 (1999)
46. Sadeghi, A., Tonazzini, A., Popova, L., Barbara, M.: Innovative robotic mechanism for soil penetration inspired by plant roots. Proc. IEEE ICRA (2013)
47. Crump, S.S.: Apparatus and method for creating three-dimensional objects. US Patent 5,121,329 1992
48. Gilroy, S., Trewavas, A.: Signal processing and transduction in plant cells: the end of the beginning? Nat. Rev. Mol. Cell Biol. **2**, 307–314 (2001)
49. Couzin, I.: Collective minds. Nature **445**, 715 (2007)
50. Dumais, J., Forterre, Y.: "Vegetable Dynamicks": The role of water in plant movements. Annu. Rev. Fluid Mech. **44**, 453–478 (2012)
51. Darwin, C.: Insectivorous Plants. Murray, London (1875)
52. Hill, B.S., Findlay, G.P.: The power of movement in plants: the role of osmotic machines. Q Rev. Biophys. **14**, 173–222 (1981)
53. Forterre, Y., Skotheim, J.M., Mahadevan, Dumais J.L.: How the Venus fly-trap snaps. Nature **433**, 421–425 (2005)
54. Samejima, M., Sibaoka, T.: Changes in the extracellular ion concentration in the main pulvinus of Mimosa pudica during rapid movement and recovery. Plant Cell Physiol. **21**(3), 467–479 (1980)
55. Taiz, L., Zeiger, E.: Plant Physiology. Sinauer Associates, Sunderland (2003)
56. Baskin, T.I.: Anisotropic expansion of the plant cell wall. Annu. Rev. Cell Dev. Biol. **21**, 203–322 (2005)
57. Bengough, A.G., Croser, C., Pritchard, J.: A biophysical analysis of root growth under mechanical stress. Plant Soil **189**, 155–164 (1997)
58. Atkins, P.W., de Paula, J.: Phys. Chem., 8th edn. Oxford University Press, New York (2006)
59. Cath, T.Y., Childress, A.E., Elimelech, M.: Forward osmosis: principles, applications, and recent developments. J. Membr. Sci. **281**, 70–87 (2006)
60. Li, Y.-H., Su, Y.-C.: Miniature osmotic actuators for controlled maxillofacial distraction osteogenesis. J. Micromech. Microeng. **20**, 1–8 (2010)
61. Piyasena, M.E., Newby, R., Miller, T.J., Shapiro, B., Smela, E.: Electroosmotically driven microfluidic actuators. Sens. Actuator. B **141**, 263–269 (2009)
62. Wang, C., Wang, L., Zhu, X., Wang, Y., Xue, J.: Low-voltage electroosmotic pumps fabricated from track-etched polymer membranes. Lab. Chip **12**, 1710–1716 (2012)
63. Sudaresan, V.B., Leo, D.J.: Modeling and characterization of a chemomechanical actuator using protein transporter. Sens. Actuator. B **131**, 384–393 (2008)
64. Herrlich, S., Spieth, S., Messner, S., Zengerle, R.: Osmotic micropumps for drug delivery. Adv. Drug Deliv. Rev. **64**, 1617–1627 (2012)
65. Burgert, I., Fratzl, P.: Actuation systems in plants as prototypes for bioinspired devices. Philos. Trans. Math. Phys. Eng. Sci. **367**, 1541 (2009)
66. Dawson, C., Vincent, J.F.V., Rocca, A.-M.: How pine cones open. Nature **290**, 668 (1997)
67. Elbaum, R., Zaltzman, L., Burgert, I., Fratzl, P.: The role of wheat awns in the seed dispersal unit. Science **316**, 884 (2007)
68. Reyssat, E., Mahadevan, L.: Hygromorphs: from pine cones to biomimetic bilayers. J. R Soc. Interface **6**, 951 (2009)
69. Taccola, S., Greco, F., Sinibaldi, E., Mondini, A., Mazzolai, B., Mattoli, V.: Toward a new generation of electrically controllable hygromorphic soft actuators. Adv. Mater. **27**(10), 1668–1675 (2015)

70. Taccola, S., Greco, F., Zucca, A., Innocenti, C., de Julin, Fernndez C., Campo, G., Sangregorio, C., Mazzolai, B., Mattoli, V.: Characterization of free-standing PEDOT:PSS/iron oxide nanoparticles composite thin films and application as conformable humidity sensors. ACS Appl. Mater. Interface. **5**(13), 6324–6332 (2013)
71. Okuzaki, H., Suzuki, H., Ito, T.: Electrically driven PEDOT/PSS actuators. Synth. Met. **159**, 2233 (2009)
72. Gibson, R.F.: A review of recent research on mechanics of multifunctional composite materials and structures. Compos. Struct. **92**, 2793 (2010)
73. Braam, J.: In touch: Plant responses to mechanical stimuli. New Phytol. **165**, 373 (2005)
74. Monshausen, G.B., Haswell, E.S.: A force of nature: molecular mechanisms of mechanoperception in plants. J. Exp. Bot. **64**, 4663–4680 (2013)
75. Monshausen, G.B., Bibikova, T.N., Weisenseel, M.H., Gilroy, S.: Ca2+regulates reactive oxygen species production and pH during mechanosensing in Arabidopsis roots. Plant Cell **21**, 2341–2356 (2009)
76. Monshausen, G.B., Gilroy, S.: Feeling green: mechanosensing in plants. Trend. Cell Biol. **19**, 228–235 (2009)
77. Cianchetti, M., Renda, F., Licofonte, A., Laschi, C.: Sensorization of continuum soft robots for reconstructing their spatial configuration. Proc. IEEE BioRob (2012)
78. Shapiro, Y., Kosa, G., Wolf, A.: Shape tracking of planar hyper-flexible beams via embedded PVDF deflection sensors. IEEE/ASME Trans. Mechatron. **19**, 1260–1267 (2014)
79. http://www.plantoidproject.eu

Chapter 28
Thirty Seven Things to Do with Live Slime Mould

Andrew Adamatzky

Abstract Slime mould *Physarum polycephalum* is a large single cell capable for distributed sensing, concurrent information processing, parallel computation and decentralised actuation. The ease of culturing and experimenting with Physarum makes this slime mould an ideal substrate for real-world implementations of unconventional sensing and computing devices. In the last decade the Physarum became a swiss knife of the unconventional computing: give the slime mould a problem it will solve it. We provide a concise summary of what exact computing and sensing operations are implemented with live slime mould. The Physarum devices discussed range from morphological processors for computational geometry to experimental archeology tools, from self-routing wires to memristors, from devices approximating a shortest path to analog physical models of space exploration.

28.1 Introduction

Acellular slime mould *P. polycephalum* has quite sophisticated life cycle [114], which includes fruit bodied, spores, single-cell amoebas, syncytium. At one phase of its cycle the slime mould becomes a plasmodium. The plasmodium is a coenocyte: nuclear divisions occur without cytokinesis. It is a single cell with thousands of nuclei. The plasmodium is a large cell. It grows up to tens centimetres when conditions are good. The plasmodium consumes microscopic particles and bacteria. During its foraging behaviour the plasmodium spans scattered sources of nutrients with a network of protoplasmic tubes. The plasmodium optimises it protoplasmic network to cover all sources of nutrients, stay away from repellents and minimise transportation of metabolites inside its body. The plasmodium's ability to optimise its shape [96] attracted attention of biologists, then computer scientists [8] and engineers. Thus the field of slime mould computing was born.

A. Adamatzky (✉)
Unconventional Computing Centre, University of the West of England, Bristol, UK
e-mail: andrew.adamatzky@uwe.ac.uk

© Springer International Publishing Switzerland 2017
A. Adamatzky (ed.), *Advances in Unconventional Computing*,
Emergence, Complexity and Computation 23,
DOI 10.1007/978-3-319-33921-4_28

So far, the plasmodium is the only useful for computation stage of *P. poly-cephalum*'s life cycle. Therefore further we will use word 'Physarum' when referring to the plasmodium. Most computing and sensing devices made of the Physarum explore one or more key features of the Physarum's physiology and behaviour:

- the slime mould senses gradients of chemo attractants and repellents [56, 102, 128]; it responds to chemical or physical stimulation by changing patterns of electrical potential oscillations [81, 104] and protoplasmic tubes contractions [120, 133];
- it optimises its body to maximise its protoplasm streaming [54]; and,
- it is made of hundreds, if not thousands, of biochemical oscillators [79] with varied modes of coupling [63].

Here we offer very short descriptions of actual working prototypes of Physarum based sensors, computers, actuators and controllers. Details can be found in pioneer book on Physarum machines [8] and the 'bible' of slime mould computing [21].

28.2 Optimisation and Graphs

28.2.1 Shortest Path and Maze

Given a maze we want to find a shortest path between the central chamber and an exit. This was the first ever problem solved by Physarum. There are two Physarum processors which solve the maze. First prototype [96] works as follows. The slime mould is inoculated everywhere in a maze. The Physarym develops a network of protoplasmic tubes spanning all channels of the maze. This network represents all possible solutions. Then oat flakes are placed in a source and a destination site. Tube lying along the shortest (or near shortest) path between two sources of nutrients develop increased flow of cytoplasm. This tube becomes thicker. Tubes branching to sites without nutrients become smaller due to lack of cytoplasm flow. They eventually collapse. The sickest tube represents the shortest path between the sources of nutrients. The selection of the shortest protoplasmic tube is implemented via interaction of propagating bio-chemical, electric potential and contractile waves in the plasmodium's body, see mathematical model in [121]. The approach is not efficient because we must literally distribute the computing substrates everywhere in the physical representation of the problem. A number of computing elements would be proportional to a sum of lengths of the maze's channels.

Second prototype of the Physarum maze solver is based on Physarum' chemoattraction [13]. An oat flake is placed in the central chamber. The Physarum is inoculated somewhere in a peripheral channel. The oat flake releases chemoattractants. The chemoattractants diffuse along the maze's channels. The Physarum explores its vicinity by branching out protoplasmic tubes into opening of nearby channels. When a wave-front of diffusing attractants reaches Physarum, the Physarum halts lateral exploration. Instead it develops an active growing zone propagating along gradient of the attractants' diffusion. The sickest tube represents the shortest path between the sources of nutrients. The approach is efficient because a number of computing elements would be proportional to a length of the shortest path.

28.2.2 Towers of Hanoi

Given n discs, each of unique size, and three pegs, we want to move the entire stack to another peg by moving one top disk at a time and not placing a disk on top of smaller disc. The set of all possible configurations and moves of the puzzle forms a planar graph with $3n$ vertices. To solve the puzzle one must find shortest paths between configurations of pegs with discs on the graph [66, 67, 105]. Physarum solves shortest path (Sect. 28.2.1) therefore it can solve Tower of Hanoi puzzle. This is experimentally demonstrated in [103]. Sometimes Physarum does not construct an optimal path initially. However, if its protoplasmic networks are damaged and then allowed to regrow the path closer to optimal then before develops [103].

28.2.3 Travelling Salesman Problem

Given a graph with weighted edges, find a cyclic route on the graph, with a minimum sum of edge weights, spanning all nodes, where each node is visited just once. Commonly, weight of an edge is an Euclidean length of the edge. Physarum is used as component of an experimental device approximating the shortest cyclic route [138]. A map of eight cities is considered. A set of solutions is represented by channels arranged in a star graph. The channels merge in a central chamber. There are eight channels for each city. Each channel encodes a city and the step when the city appears in the route. There are sixty four channels. Physarum is inoculated in the central chamber. The slime mould then propagates into the channels. A node of the data graph is assumed to be visited when its corresponding channel is colonised by Physarum. Growth of the Physarum in the star-shape is controlled by a recurrent neural network. The network takes ratios of colonizations of the channels as input and produces patterns of illumination, projected onto the channels, as output. The network is designed to prohibit revisiting of already visited nodes and simultaneous visits to multiple nodes. The Physarum propagates into channels and pulls back. Then propagates to other channels and pulls back from some of them. Eventually the system reaches a stable solution where no propagation occurs. The stable solution represents the minimal distance cyclic route on the data graph [138].

A rather more natural approximate algorithm of solving the travelling salesman problem is based on a construction of an α-shape (Sect. 28.3.3). This how humans solve the problem visually [87]. An approximate solution of the travelling salesman problem by a shrinking blob of simulated Physarum is proposed in [73]. The

Physarum is inoculated all over the convex hull of the data set. The Physarum's blob shrinks. It adapts morphologically to the configuration of data nodes. The shrinkage halts when the Physarum no longer covers all data nodes. The algorithm is not implemented with real Physarum.

28.2.4 Spanning Tree

A spanning tree of a finite planar set is a connected, undirected, acyclic planar graph, whose vertices are points of the planar set. The tree is a minimal spanning tree where sum of edge lengths is minimal [98]. As algorithm for computing a spanning tree of a finite planar set based on morphogenesis of a neuron's axonal tree was initially proposed [1]: planar data points are marked by attractants (e.g. neurotrophins) and a neuroblast is placed at some site. Growth cones sprout new filopodia in the direction of maximal concentration of attractants. If two growth cones compete for the same site of attractants then a cone with highest energy (closest to previous site or branching point) wins. Fifteen years later we implemented the algorithm with Physarum [4].

Degree of Physarum branching is inversely proportional to a quality of its substrate. Therefore to reduce a number of random branches we cultivate Physarum not on agar but just humid filter paper. Planar data set is represented by a configuration of oat flakes. Physarum is inoculated at one of the data sites. Physarum propagates to a virgin oat flake closest to the inoculation site. Physarum branches, if there are several virgin flakes nearby. It colonises next set of flakes. The propagation goes on until all data sites are spanned by a protoplasmic network. The protoplasmic network approximates the spanning tree. The resulted tree does not remain static though. Later cycles can be formed and the tree is transformed to one of proximity graphs, e.g. relative neighbourhood graph or Gabriel graph [5].

28.2.5 Approximation of Transport Networks

Motorway networks are designed with an aim of efficient vehicular transportation of goods and passengers. Physarum protoplasmic networks evolved for efficient intracellular transportation of nutrients and metabolites. To uncover similarities between biological and human-made transport networks and to project behavioural traits of biological networks onto development of vehicular transport networks we conducted an evaluation and approximation of motorway networks by Physarum in fourteen geographical regions: Africa, Australia, Belgium, Brazil, Canada, China, Germany, Iberia, Italy, Malaysia, Mexico, The Netherlands, UK, and USA [11].

We represented each region with an agar plate, imitated major urban areas with oat flakes, inoculated Physarum in a capital, and analysed structures of protoplasmic networks developed. We found that, the networks of protoplasmic tubes grown by Physarum match, at least partly, the networks of human-made transport arteries. The shape of a country and the exact spatial distribution of urban areas, represented by sources of nutrients, play a key role in determining the exact structure of the plasmodium network. In terms of absolute matching between Physarum networks and motorway networks the regions studied can be arranged in the following order of decreasing matching: Malaysia, Italy, Canada, Belgium, China, Africa, the Netherlands, Germany, UK, Australia, Iberia, Mexico, Brazil, USA. We compared the Physarum and the motorway graphs using such measures as average and longest shortest paths, average degrees, number of independent cycles, the Harary index, the Π-index and the Randić index. Using these measures we find that motorway networks in Belgium, Canada and China are most affine to Physarum networks. With regards to measures and topological indices we demonstrated that the Randić index could be considered as most bio-compatible measure of transport networks, because it matches very well the slime mould and man-made transport networks, yet efficiently discriminates between transport networks of different regions [31].

Many curious discoveries have been made. Just few of them are listed below. All segments of trans-African highways not represented by Physarum have components of non-paved roads [24]. The east coast transport chain from the Melbourne urban area in the south to the Mackay area in the north, and the highways linking Alice Springs and Mount Isa and Cloncurry, are represented by the slime mould's protoplasmic tubes in almost all experiments on approximation of Australian highways [26]. If the two parts of Belgium were separated with Brussels in Flanders, the Walloon region of the Belgian transport network would be represented by a single chain from Tournai in the north-west to the Liege area in the north-east and down to southernmost Arlon; motorway links connecting Brussels with Antwerp, Tournai, Mons, Charleroi and Namur, and links connecting Leuven with Liege and Antwerp with Genk and Turnhout, are redundant from the Physarum's point of view [30]. The protoplasmic network forms a subnetwork of the man-made motorway network in the Netherlands; a flooding of large area around Amsterdam will lead to substantial increase in traffic at the boundary between flooded and non-flooded area, paralysis and abandonment of the transport network and migration of population from the Netherlands to Germany, France and Belgium [33]. Physarum imitates the 1947 year separation of Germany into East Germany and West Germany [27].

28.2.6 Mass Migration

People migrate towards sources of safe life and higher income. Physarum migrates into environmentally conformable areas and towards source of nutrients. In [25] we explored this analogy to imitate Mexican migration to USA. We have made a 3D Nylon terrain of USA and placed oat flakes, to act as sources of attractants and nutrients, to ten areas with highest concentration of migrants: New York, Jacksonville, Chicago, Dallas, Houston, Denver, Albuquerque, Phoenix, Los Angeles, San Jose. We inoculated Physarum in a locus between Ciudad Juárez and Nuevo Laredo, allowed it to colonise the template for five to ten days, and analysed routes of migration. From results of laboratory experiments we extracted topologies of migratory routes, and highlighted a role of elevations in shaping the human movement networks.

28.2.7 Experimental Archeology

Experimental archeology uses analytical methods, imaginative experiments, or transformation of a matter [37, 49, 70] in a context of human activity in the past. Knowing that Physarum is fruitful substrate for simulating transport routes (Sect. 28.2.5) and migration (Sect. 28.2.6) we explored foraging behaviour of the Physarum to imitate development of Roman roads in the Balkans [58]. Agar plates were cut in a shape of Balkans area. We placed oat flakes in seventeen areas, corresponding to Roman provinces and major settlements and inoculated Physarum in Thessaloniki. We found that Physarum imitates growth of Roman roads to a larger extent. For example, the propagation of Physarum from Thessaloniki towards the area of Scopje and Stoboi matches the road aligned with the Valley of Axios, which is a key communication artery between the Balkan hinterland and Aegean area from Bronze age. A range of historical scenarios was uncovered [58], including movement along via Diagonalis, the long diagonal axis that crossed central Balkan; propagation to the East towards Byzantium, towards the North along the coast of Euxeinus Pontus, and from Thessaloniki to Dyrrachion, along the western part of via Egnatia [58].

28.2.8 Evacuation

Evacuation is a rapid but temporary removal of people from the area of danger. Physarum moves away from sources of repellents or areas of uncomfortable environmental conditions. Would Physarum be able to find a shortest route of evacuation in geometrically constrained environment? To find out we undertook a series of experiments [77]. We made a physical, scaled down, model of a whole floor of the real office building and inoculated Physarum in one of the rooms. In first scenario we placed a crystal of a salt in the room with Physarum in a hope that a gradient of sodium chloride diffusion would repel Physarum towards exist of the building along the shortest path. This did not happen. Physarum got lost in the template. By placing attractants near the exit we rectified the mishap and allowed Physarum to find a shortest route away from the 'disaster' [77]. Evacuation is one of few problems where computer models of Physarum find better solution than the living Physarum [76].

28.2.9 Space Exploration

We employed the foraging behaviour of the Physarum to explore scenarios of future colonisation of the Moon and the Mars [34]. We grown Physarum on three-dimensional templates of these planet and analysed formation of the exploration routes, dynamical reconfiguration of the transportation networks as a response to addition of hubs. The developed infrastructures were explored using proximity graphs and Physarum inspired algorithms of supply chain designs. Interesting insights about how various lunar missions will develop and how interactions between hubs and landing sites can be established are given in [34].

28.3 Geometry

28.3.1 Voronoi Diagram

Let $\mathbf{P}$ be a non-empty finite set of planar points. A planar Voronoi diagram of the set $\mathbf{P}$ is a partition of the plane into such regions that, for any element of $\mathbf{P}$, a region corresponding to a unique point p contains all those points of the plane which are closer to p than to any other node of $\mathbf{P}$. A unique region $vor(p) = \{z \in \mathbf{R}^2 : d(p, z) < d(p, m) \forall m \in \mathbf{R}^2, m \neq z\}$ assigned to the point p is called a Voronoi cell of the point p. The boundary of the Voronoi cell of the point p is built of segments of bisectors separating pairs of geographically closest points of the given planar set $\mathbf{P}$. A union of all boundaries of the Voronoi cells determines the planar Voronoi diagram: $VD(\mathbf{P}) = \cup_{p \in \mathbf{P}} \partial vor(p)$ [101].

The basic concept of constructing Voronoi diagrams with reaction-diffusion systems is based on an intuitive technique for detecting the bisector points separating two given points of the set $\mathbf{P}$. If we drop reagents at the two data points the diffusive waves, or phase waves if the computing substrate is active, travel outwards from the drops. The waves travel the same distance from the sites of origin before they meet one another. The points where the waves meet are the bisector points [29].

Plasmodium growing on a nutrient substrate from a single site of inoculation expands circularly as a typical diffusive or excitation wave. When two plasmodium waves encounter each other, they stop propagating. To approximate a Voronoi diagram with Physarum [8], we physically map a configuration of planar data points by inoculating plasmodia on a substrate. Plasmodium waves propagate circularly from each data point and stop when they collide with each other. Thus, the plasmodium waves approximate a Voronoi diagram, whose edges are the substrate's loci not occupied by plasmodia. Time complexity of the Physarum computation is proportional to a maximal distance between two geographically neighbouring data points, which is capped by a diameter of the data planar set, and does not depend on a number of the data points.

28.3.2   Delaunay Triangulation

Delaunay triangulation is a dual graph of Voronoi diagram. A Delaunay triangulation of a planar set is a triangulation of the set such that a circumcircle of any triangle does not contain a point of the set [50]. There are two ways to approximate the Delaunay triangulation with Physarum. First is based on the setup of Voronoi diagram processor (Sect. 28.3.1). Previously, we wrote that when propagating fronts of Physarum meet they stop. This is true. But what happens after they stop is equally interesting. Physarum forms a bridge—a single protoplasmic tube—connecting the stationary Physarum fronts. Such protoplasmic tubes typically span geographically neighbouring sites of inoculation and they cross sites of first contacts of growing wave fronts. These tubes represent edges of the Delaunay triangulation. This is why we proposed in [112] that the Voronoi diagram and the Delaunay triangulation are constructed simultaneously by Physarum growing on a nutrient agar.

Second method of approximating the Delaunay triangulation is implemented on a non-nutrient agar. We represent planar data points by inoculation sites. The inoculants propagate and form a planar proximity graph spanning all inoculation sites. At the beginning of such development a Gabriel graph [59] is formed. Then additional protoplasmic tubes emerge and the Gabriel graph is transformed to the Delaunay triangulation [5].

28.3.3   Concave Hull

The α-shape of a planar set $\mathbf{P}$ is an intersection of the complement of all closed discs of radius $1/\alpha$ that includes no points of $\mathbf{P}$ [57]. A concave hull is a connected α-shape without holes. This is a non-convex polygon representing area occupied by $\mathbf{P}$. Given planar set $\mathbf{P}$ represented by physical objects Physarum must approximate concave hull of $\mathbf{P}$ by its thickest protoplasmic tube. We represent data points by somniferous pills placed directly on a non-nutrient agar. The pills emits attractants to 'pull' Physarum towards $\mathbf{P}$ but they also emit repellents preventing Physarum from growing inside $\mathbf{P}$ [14]. The combination of long-distance attracting forces and short-distance ('short' is $O(D)$, where D is a diameter of $\mathbf{P}$) repelling forces allows us to implement Jarvis wrapping algorithm [71]. We select a starting point which is extremal point of $\mathbf{P}$. We pull a rope to other extremal point. We continue until the set $\mathbf{P}$ is wrapped completely. We tested feasibility of the idea in laboratory experiments [14]. In each experiment we arranged several half-pills in a random fashion near centre of a Petri dish and inoculated an oat flake colonised by Physarum few centimetres away from the set $\mathbf{P}$. Physarum propagates towards set $\mathbf{P}$ and starts enveloping the set with its body and the network of protoplasmic tubes. The plasmodium does not propagate inside configuration of pills. The plasmodium completes approximation of a shape by entirely enveloping $\mathbf{P}$ in a day or two.

28.4   Computing Circuits

28.4.1   Attraction-Based Logical Gates

When two growing zones of separate Physarum cells meet they repel if there is a free space to deviate to. If there is no opportunity to deviate the cells merge. This feature is employed in the construction of Boolean logical gates—NOT, OR and AND—in [125]. The gates are made of segments of agar gel along which the Physarum propagates. To implement input '1' (TRUE) in channel x a piece of Physarum is inoculated in x otherwise the input is considered to be '0' (FALSE). Attractants are placed in the end of the output channels to stimulate growth of the Physarum towards outputs. The Physarum propagates towards closest source of attractants along a shortest path.

The gate OR is a $\searrow_\downarrow\nearrow$ junction. Physarum placed in one of the inputs propagates towards the output. If each input contains the Physarum, the propagating cells merge and appear in the output as if they were a single cell. Even if both inputs are '1' the Physarum cells have no space to avoid collision and therefore the merge and propagate into the output channel.

The gate AND looks like distorted 'H': $\underset{\top}{\rule{1.2cm}{0.4pt}}\overset{\downarrow}{\underset{\top}{}}\downarrow$ When only one input is '1' the Physarum propagates towards closes attractant and exits along right output channel. When both inputs are '1' the Physarum from the right input channel propagates into the right output channel. The Physarum from the left input channel avoids merging with another Physarum and propagates towards left output channel. The left output channel realises AND.

28.4.2   Ballistic Logical Gates

In designs of ballistic gates [9] we employ inertia of the Physarum growing zones. On a non-nutrient substrate the plasmodium propagates as a traveling localisation, as a compact wave-fragment of protoplasm. The plasmodium-localisation travels in its originally predetermined direction for a substantial period of time even when no gradient of chemo-attractants is present. We explore this feature of Physarum localisations to design a two-input two-output Boolean gates. The gate realising AND on one output and OR on another output look like horizontally flipped 'K': $\rangle_\downarrow^\downarrow$. When left input is '1' the Physarum propagates inertially along the vertical (on the right) output channel. The same happens when right input is '1'. If both inputs are '1' then the Physarum from the right input propagates along vertical output channel but the Physarum from the left input repels from the right-input-Physarum and moves into the left output channel. The left output channel realises AND and the right output channel realises OR.

The gate NOT is an asymmetric cross junction: $_\rightarrow \downarrow _\rightarrow$. Vertical input channel is twice as long as horizontal input channel. Vertical input is constant TRUE: Physarum is always inoculated their. Horizontal input is a variable. If variable input is '0' then Physarum from the constant TRUE vertical input propagates into the vertical output. If variable input is '1' then Physarum from the input channel propagates into the horizontal output channel and blocks path of the Physarum representing constant TRUE. Both ballistic gates work very well without attractants. However they work ever better when attractants are placed into output channels. Cascading of the gates into a binary adder is demonstrated in [9].

28.4.3 Opto-Electronics Logical Gates

In prototypes of repellent gates [90], active growing zones of slime mould representing different inputs interact with each other by electronically switching light inputs and thus invoking photo avoidance. The gates NOT and BAND are constructed using this feature.

The gate NOT is made of two electrodes. The electrodes are connected to a power supply. There is a green LED (with it is independent power supply) on one electrode. The Physarum is inoculated on another electrode. Input '1' is represented by LED's light on. Output is represented by the Physarum closing the circuit between two electrodes. When LED is illuminated (input '1') the Physarum does not propagate between electrodes, thus output '0' is produced. When LED is off (input '0') the Physarum closes the circuit by propagating between the electrodes.

The gate NAND is implemented with two LEDs. When both inputs are '0' LEDs are off and the Physarum closes the circuit between its inoculation electrode and one of the LED electrodes, chosen at random. When both inputs are '1', both LEDs are on, they repel Physarum. The Physarum does not propagate to any electrodes. When only one input is '1', one LED is on and another LED is off, the Physarum propagates toward the electrode with non-illuminating LED and closes the circuit.

28.4.4 Frequency Based Logical Gates

The Physarum responds to stimulation with light, nutrients and heating by changing frequency of its electrical potential oscillations. We represent TRUE and FALSE values by different types of stimuli and apply threshold operations to frequencies of the Physarum oscillations. We represent Boolean inputs as intervals of oscillation frequency [130]. Thus we experimentally implement OR, AND, NOT, NOR, NAND, XOR and XNOR gates, see details in [130].

28.4.5 Micro-Fluidic Logical Gates

When a fragment of protoplasmic tube is mechanically stimulated, e.g. gently touched by a hair, a cytoplasmic flow in this fragment halts and the fragment's resistivity to the flow dramatically increases. The cytoplasmic flow is then directed via adjacent protoplasmic tubes. A basic gate looks like a 'Y'-junction of two protoplasmic tubes x and y with a horizontal bypass z between them

Segments x and y are inputs. Segment z is output. When both input segments are intact and there is a flow of cytoplasm between them there is no flow of cytoplasm via z. If one of the input tubes is mechanically stimulated the flow through this tube stops and the flow is diverted via the output tube z. If both input tubes are stimulated there is no flow via input or output tubes. Thus we implement XOR gate [28]. A mechanically stimulated fragment restores its flow of cytoplasm in one minute: the gate is reusable.

By adding one more output (bypass) tube to XOR gate we produce a gate with two inputs and two outputs: z and p:

The output z represents XOR. The tube p represents NOR because a cytoplasmic flow is directed via p if both tubes x and y are blocked. More complicated gates and memory devices can be found in [28].

28.4.6 Intra-Cellular Collision Based Computing

The paradigm of a collision-based computing originates from the computational universality of the Game of Life, Fredkin–Toffoli conservative logic and the billiard-ball model with its cellular-automaton implementation [2]. A collision-based computer employs mobile localisations, e.g. gliders in Conway's Game of Life cellular automata, to represent quanta of information in active non-linear media. Information values, e.g. truth values of logical variables, are given by either the absence or presence of the localizations or by other parameters such as direction or velocity. The localizations travel in space and collide with each other. The results of the collisions are interpreted as computation. Physarum ballistic gates (Sect. 28.4.2) are also based on collisions, or interactions, between active growing zones of Physarum.

However, signals per se, represented by growing body of Physarum, are not localised. Intra-cellular vesicles—up 100 nm droplets of liquid encapsulated by a lipid bilayer membrane—could be convenient representations of localised signals.

In [91] we outlined pathways towards collision-based computing with vesicles inside the Physarum cell. The vesicles travel along actin and tubulin network and collide with each other. The colliding vesicles may reflect, fuse or annihilate. The vesicles reflect in over half of the collisions observed. The vesicles fuse in one of seven collisions. The vesicles annihilate and unload their cargo in one of ten collisions. The vesicles becomes paired and travel as a single object in one of ten collisions. Based on the experimental observations, we derive soft spheres collision [89] gates and also gates NOT and FAN- OUT.

28.4.7 Kolmogorov–Uspensky Machine

In 1950s Kolmogorov outlined a concept of an algorithmic process, an abstract machine, defined on a dynamically changing graph [82]. The structure later became known as Kolmogorov–Uspensky machine [83]. The machine is a computational process on a finite undirected connected graph with distinctly labelled nodes [65]. A computational process travels on the graph, activates nodes and removes or adds edges. A program for the machine specifies how to replace the neighbourhood of an active node with a new neighbourhood, depending on the labels of edges connected to the active node and the labels of the nodes in proximity to the active node [40]. The Kolmogorov–Uspensky machine is more flexible than a Turning machine because it recognises in real time some predicates not recognisable in real time by the Turing machine [64], it is stronger than any model of computation that requires $\Omega(n)$ time to access its memory [48, 113]. "Turing machines formalize computation as it is performed by a human. Kolmogorov–Uspensky machines formalize computation as it performed by a physical process" [40].

We implement the Kolmogorov–Uspensky machine in the Physarum as follows [3]. Stationary nodes are represented by sources of nutrients. Dynamic nodes are assigned to branching site of the protoplasmic tubes. The stationary nodes are labelled by food colourings because Physarum exhibits an hierarchy of preferences to different colourings. An active zone in the storage graph is selected by inoculating the Physarum on one of the stationary nodes. An edge of the Kolmogorov–Uspensky machine is a protoplasmic tube connecting the nodes.

Program and data are represented by the spatial configuration of stationary nodes. Results of the computation over a stationary data node are represented by the configuration of dynamic nodes and edges. The initial state of a Physarum machine (Physarum implementation of Kolmogorov–Uspensky machine) includes part of an input string (the part which represents the position of Physarum relative to stationary nodes), an empty output string, the current instruction in the program and the storage structure consisting of one isolated node. The physical graph structure developed by Physarum is the result of its computation. The Physarum machine halts when all data nodes are utilised. At every step of the computation there is an active node and an active zone (nodes neighbouring the active node). The active zone has a limited complexity: all elements of the zone are connected by some chain of edges to the initial node. The size of the active zone may vary depending on the computational task. In the Physarum machine, an active node is a trigger of contraction/excitation waves, which spread all over the plasmodium tree and cause pseudopodia to propagate, the shape to change and protoplasmic veins to annihilate. The active zone comprises stationary and/or dynamic nodes connected to an active node with tubes of protoplasm. Instructions of the Physarum machine are INPUT, OUTPUT, GO, HALT, ADD NODE, REMOVE NODE, ADD EDGE, REMOVE EDGE, IF. The INPUT is done via distribution of sources of nutrients. The OUTPUT is recorded optically. The SET instruction causes pointers to redirect. It is realised by placing a fresh nutrient source in the experimental container. When a new node is created, all pointers from the old node point to the new node [3].

28.5 Electronics

28.5.1 Wires

Protoplasmic tubes of the Physarum are conductive and therefore can be used as wires. A resistivity of Physarum protoplasmic tubes is of the same rank as resistivity of a cardiac and skeletal muscles of dogs and humans [62]. By using 1–5 cm protoplasmic tube as a wire we can light up a LED, and keep it illuminated for days. We can power up a piezo audio transducer [15] using the Physarum wire. Due to high, comparing to conventional conductors, resistivity of the Physarum we must apply a high voltage to power loads. For example, to operate the LED array we should apply 10 V and 3.9 μA direct current. To produce a 30 dB sound with buzzer we need to apply 8 V.

The Physarum wires are self-growing. To connect two pins of a circuit with a wire we inoculate Physarum at one pin and place a source of attractants near another pin. The Physarum grows a protoplasmic tube connecting two pins. The Physarum propagates well, and relatively fast, 1–5 mm/h, on a bare surface of electronic boards. Growing Physarum circuits can be controlled by white and blue light, chemical and thermal gradients, and electrical fields. Physarum wires can be robustly routed with a wide range of organic volatiles [52].

Physarum wires can self-repair after a substantial damage. After part of a protoplasmic tube is removed (1–2 mm segment) the tube restores its integrity in 6–9 h. Typically, a cytoplasm from cut-open ends spills out on a substrate. Each spilling of cytoplasm becomes covered by a cell wall and starts growing. In few hours growing parts of the tube meet with each other and merge. Restoration of tubes conductivity was confirmed by electrical measurements [15].

Physarum wire can transfer analogue signals below 19 KHz without distortion [132]. Physarum wires perform well in digital communication systems. For example, the protoplasmic tubes are used to establish communication between Arduino Mega and Digital 3-axis compass using I2C protocol. Valid magnometric data was confirmed by movement/rotation of the manometer and subsequent change in data receive alone the Physarum wire [132].

Physarum wires are not immortal. To make them last longer we cover them with conducting organic polymer polypyrrole [53]. A localised section of protoplasmic tube is treated with ferric chloride and exposed to vapour of pyrrole monomer. A 1 cm section of the treated tube has resistance 100 kOhm. The treatment is selective therefore we can produce live and functionalised Physarum wires in the same computing circuit.

28.5.2 Low Pass Filter

Physarum protoplasmic tubes are conductors. Do they modify analog or digital signals passing through them? Signal propagation in Physarum's protoplasmic tubes was tested using a frequency response network analyser [132]. Sinusoidal voltage waveforms are sent via the protoplasmic tubes at frequencies from 10 Hz to 4 MHz. The transfer function of voltage waveform passing through the tube is frequency dependent. Signals of higher frequencies are dramatically reduced in strength while low frequency waveforms remain largely unaffected. The magnitude-frequency profile matches a low pass filter. In most cases there was some attenuation of the voltage through the wire at the pass-band frequency range with a mean attenuation of −6 dB. The cut off frequency was defined as −3 dB from the pass-band magnitude; the mean cut off frequency was 19 KHz. The phase-frequency response showed the 45° phase shift to be aligned very closely to the cut off frequency in all protoplasmic tubes [132].

28.5.3 Oscillators

An electronic oscillator is a device that produces periodic electronic signal. Electronic Physarum chips need oscillators to have a source of regularly spaced pulses. The experimental electronic oscillator, which converts direct current to alternating current signal, is made with Physarum as follows [18]. We made two electrodes setup— Physarum spans two electrodes with a single protoplasmic tube, apply direct current potential in a range 2–15 V and measure electrical potential difference between the electrodes. When we apply an electrical potential to a protoplasmic tube we observe oscillations of the output electrical potential. The oscillations of the output potential are caused by periodic changes in resistance of the protoplasmic tube connecting the electrodes. Average resistance of a 10 mm protoplasmic tube is 3 MOhm.

Resistance of the protoplasmic tube exhibits oscillatory behaviour with highly pronounced dominating frequency 0.014 Hz. The resistance oscillations have average amplitude 0.59 MOhm, minimum amplitude of resistance oscillations observed was 0.11 MOhm and maximum amplitude 1 MOhm. Oscillations in resistance observed are due to peristaltic contractions of the protoplasmic tube [116]. Average output potential and average amplitude of output potential oscillations grow linearly with the increase of the input potential. Frequency of oscillations remains almost constant. Physarum oscillator produces the same frequency oscillations at 2 and 15 V applied potential. A ratio of average amplitude of output potential oscillations to average output potential decreases by a power low with increase of input potential.

28.5.4 Tactile Sensor

A tactile sensor is a device that responds to a physical contact between the device and an object. When a segment of a glass capillary is placed across protoplasmic tube, which spans reference and recording electrodes, Physarum demonstrates two types of responses to application of this load: an immediate response with a high-amplitude impulse and a prolonged response with changes in its oscillation pattern [16]. The immediate response is a high-amplitude spike: its amplitude is 12.33 mV and its duration is 150 s. The prolonged response is an envelop of increased amplitude oscillations. For example, an average amplitude of oscillations before stimulation is 2.3 mV and duration of each wave was 120 s. The amplitude of waves in the prolonged response to stimulation is 5.29 mV with a period of a wave increased to 124 s. Tactile sensor developed in [16] is non-reusable. While the load rests on the protoplasmic tube Physarum starts colonising the load. Removal of the load damages the protoplasmic tube. To rectify this deficiency we designed Physarum tactile bristle [19].

To make a tactile bristle with Physarum we stuck a bristle in the agar blob on the recording electrode [19]. In a couple of days after inoculation of Physarum to an agar blob on a references electrode the Physarum propagates to and colonises agar blob on a recording electrode. Physarum climbs up the bristle and occupies one third to a half of the bristle's length. The sensor works by deflecting the bristle. A sensed object does not come into direct contact with Physarum but only with a tip of the bristle not-colonised by Physarum. A typical response of Physarum to deflection of the bristle is comprised of an immediate response—a high-amplitude impulse, and a prolonged response. High-amplitude impulse is always well pronounced, prolonged response oscillations can sometimes be distorted by other factors, e.g. growing branches of a protoplasmic tube or additional strands of plasmodium propagating between the agar blobs. Responses are repeatable not only in different experiments but also during several rounds of stimulation in the same experiment [19].

28.5.5 Colour Sensor

A colour sensor is a device that gives wavelength-dependent response when illuminated. Physarum is photo-sensitive. It changes pattern of electrical potential oscillatory activity when illuminated [41, 134]. Moreover the Physarum distinguishes the colour of illumination [17]. We placed Physarum between two electrodes and illuminated it with red, green, blue or white light. We also illuminated Physarum with white light via transparent lens. Amount of light on the blob was 80–120 LUX for each colour. We say the Physarum recognises a colour of the light if it reacts to illumination with the colour by a unique changes in amplitude and periods of oscillatory activity. We found that Physarum recognises when red and blue light are switched on and when red light is switched off. Red and blue illuminations decrease frequency of oscillations. Red light increases amplitude of oscillations but blue light decreases the amplitude. Physarum does not differentiate between green and white lights. Switching off red light leads to increase of period and decrease of amplitude of oscillations [17].

28.5.6 Chemical Sensor

A chemical sensor is a device that gives a selective response when exposed to a target chemical substance. Physarum senses and responds to volatile aromatic substances [10, 12, 22]. We studied Physarum's binary preferences to various volatile chemicals [51] and derived an experimental mapping between subset of chemoattractants and chemorepellents: farnesene, tridecane, s(-)limonene, cis-3-hexenylacetate, geraniol, benzyl alcohol, linalool, nonanal and amplitude and frequency of electrical potential oscillation of Physarum [131]. Physarum increases frequency of electrical potential oscillations when exposed to strongest attractants—farnesene, tridecane, s(-)limonene, cis-3-hexenylacetate. Exposure to repellents—linlool, benzyl alcohol, nonanal—leads to a decrease of oscillation frequency, and, for linlool and benzyl alcohol, increase of the oscillation amplitude. Physarum chemical sensor discriminates individual chemicals by changing amplitude and frequency of its electrical potential oscillations; it can detect chemical from a distance of several centimetres [131].

28.5.7 Memristors

A memristor is resistor with memory, which resistance depends on how much current had flown through the device [46, 115]. Memristor is a material implication $\rightarrow$, a universal Boolean logical gate. In laboratory experiments [60] we demonstrated that protoplasmic tubes of Physarum show current versus voltage profiles consistent with memristive system. Experimental laboratory studies shown pronounced hysteresis and memristive effects exhibited by the slime mould [60]. Being a memristive element the slime mould's protoplasmic tube can also act a low level sequential logic element [60] operated with current spikes, or current transients. In such a device logical input bits are temporarily separated. Memristive properties of the slime mould's protoplasmic tubes gives us a hope that a range of 'classical' memristor-based neuromorphic architectures can be implemented with Physarum. Memristor is an analog of a synaptic connection [99]. Being the living memristor each protoplasmic tube of Physarum may be seen as a synaptic element with memory, which state is modified depending on its pre-synaptic and post-synaptic activities. Therefore a network of Physarum's protoplasmic tubes is an associate memory network. A memristor can be also made from Physarum bio-organic electrochemical transistor (Sect. 28.5.10) by removing a drain electrode [47].

28.5.8 Schottky Diodes

A diode is a two-terminal passive non-linear device which conducts mainly in one direction. The diodes are used as current rectifiers to change alternating current to direct current. A forward voltage drop is a difference between electrical potentials of anode and cathode. A Schottky diode is a diode which has low forward voltage drop, comparing to other families of diodes. A device showing some resemblance to Schottky diode can be made of Physarum [47]. In this device, the Physarum spans two asymmetrical junction electrodes—gold and indium. Cyclic voltage-current characteristics are measured. The measurements reveal suppression of conductivity for low voltage values in the direction of the positive bias and rectification features, pronounced more for the low values of the bias voltage [47]. Physarum per se is not a diode: the rectifying properties emerge due to combination of features of the asymmetric electrode junction with electrochemical activity of the Physarum [47].

28.5.9 Voltage Divider

Voltage divider is a circuit that produces a given fraction of an input voltage as an output voltage [68]. A 10 mm protoplasmic tube acts as a simple divider: the output voltage is c. 0.9 of the input voltages, voltages tested up to 20 V [15]. A typical voltage divider has two resistances, represented by two protoplasmic tubes. The Physarum divider produces output with 12 % accuracy. This error is linear and might be due to differences in the protoplasmic tubes' resistances. Resistors in Physarum voltage divider can be made adjustable by applying illumination, heat, or loading the protoplasmic tubes with functional nano particles [92, 132]. For example, loading Physarum with magnetite lower resistance of protoplasmic tube to 10–20 KOhm, making it compatible by value with common resistors [92]. One of the tube constituting the divider can be transformed to a potentiometer by making an output electrode a conductive micro needle [132].

28.5.10   Transistors

A transistor is a three-terminal active device that power amplifies input signal. The additional power comes from an external source of power. An organic electrochemical transistor is a semiconducting polymer channel in contact with an electrolyte. Its functioning is based on the reversible doping of the polymer channel. A hybrid Physarum bio-organic electrochemical transistor is made by interfacing an organic semiconductor, poly-3,4-ethylenedioxythiophene doped with poly-styrene sulfonate, with the Physarum [118]. Physarum is used instead of electrolyte. Electrical measurements in three-terminal mode uncover characteristics similar to transistor operations. The device operates in a depletion mode similarly to standard electrolyte-gated transistors. The Physarum transistor works well with platinum, golden and silver electrodes. If the drain electrode is removed and the device becomes two-terminal, it exhibits cyclic voltage-current characteristics similar to memristors [118].

28.6   Robotics

28.6.1   Robot Controllers

Physarum responds to stimuli by changing pattern of its electrical potential oscillations and cytoplasm shuttling. By interfacing the Physarum with actuators we can make the slime mould controller for robots. Two prototypes of such robotic controllers are made: controller for a hexapod robot [126] and controller for a robotic android head [61].

A Physarum controller for hexapod robot is made of a star shaped template [126]. It has six circular wells connected by the channels that meet at a single point. The Physarum grows inside the template. Physarum in each well acts as cytoplasm shuttle streaming oscillators. The Physarum oscillators in the wells coupled via Physarum body colonising the channels between the wells. Blue light used as a stimulus. The shuttle streaming of cytoplasm is measured via light absorbance. Oscillations of shuttle streaming in the wells, as a response to a stimulation with light, modulate phase and frequency of the robot legs's movement and cause the robot to change its direction of movement.

An electrical activity of Physarum in response to stimulations is converted to affective state in the design of Physarum emotional controller [61]. A Physarum is inoculated on a multi-electrode array and stimulated with nutrients (attractant) and light (repellent). Extracellular electrical potential is recorded. The recorded data is split into chunks. We employed a circumplex model of affect, where emotions are plotted in two-dimensions determined by the polarity and arousal level. The chunks are assigned polarity. Potential recorded during stimulation with attractant is given positive polarity. Data obtained during illumination of Physarum is assigned negative polarity. A level of arousal is proportional to amplitude of the electrical potential. Emotions are assigned to the data chunks, based on the polarity and the arousal of chunks and fed into an android robot. The data activate the motors placed in the positions matching sites of real muscles in a human face. Actuation of the motors causes movements of an artificial skin. The movements are expressed as affective facial expression of the android [61].

28.6.2 Actuators

Physarum contracts its body in a phase with oscillation of calcium waves. By placing a column of water on one side of the Physarum dumbbell shape we calculated that the Physarum weighting 5 mg, can lift up a load 36 times heavier than its own weight [127]. In [7] we studied how the Physarum can manipulate on a water surface. To make the Physarum propelling a 'boat' we take a small piece of a plastic foam, inoculate the Physarum on the foam and place this floater on a water surface. When Physarum is illuminated, it increases its peristaltic which transferred into a movement of the boat. If we allow Physarum to develop its protoplasmic tubes outside the float, on the water surface, then periodic contractions of tubes, stimulated by light, will make the Physarum boat propel away from the source of light. Another way to move the floater, is to place a stationary floater (anchor) with an attractants nearby the Physarum boat. Then Physarum develops protoplasmic tree towards the attractants and colonises the anchorr. Then Physarum straightens its tubes thus pulling the boat towards the anchor [7].

28.6.3 Nervous System

The Physarum senses tactile, chemical and optical stimuli and converts the stimuli into characteristic patterns of its electrical potential oscillations. The electrical responses to stimuli may propagate along protoplasmic tubes for distances exceeding tens of centimetres, like impulses in neural pathways do. The Physarum makes decision about its propagation direction based on information fusion from thousands of spatially extended protoplasmic loci, similarly to a neurone collecting information from its dendritic tree. When growing on a non-nutrient substrate Physarum develops shapes resembling body of a single neuron. It looks like neuron—can it be behave as one? In [20] we speculate on whether an alternative—would-be—nervous systems can be developed and practically implemented from the slime mould. We uncover analogies between the slime mould and neurons, and demonstrate that the slime mould can play a role of primitive mechanoreceptors, photoreceptors, chemoreceptors; we also show how the Physarum neural pathways develop [20].

Physarum neural networks do not have synapses represented as discrete structural elements. Synapse-like morphological contacts could not be formed. When two pieces of Physarum are inoculated at a distance form each other, they start exploring space around them and form branching networks of protoplasmic tubes. When two networks, grown from different sites of inoculation come into contact they usually fuse forming a single united network. However, there is a functional analog of synapses and an intrinsic feature of Physarum protoplasmic tubes which makes literally any loci of Physarum network a synapse. This is a memristive property (Sect. 28.5.7).

We explored the analogy between behaviour of neuron growth cones and Physarum active growing zones [20]. To test if Physarum can develop information pathways we conducted several experiments on one to one scale models of human skull and brain. We used real scale models for the following reasons. First, to show that information pathways made of protoplasmic tubes can be tens of centimetres length and thus match lengths of neural pathways. Second, to demonstrate that—when propagating inside human skull—the plasmodium follows general anatomical trajectories of ocular and olfactory nerves. We found that morphology of information pathways developed by Physarum on a human skull matches well anatomy of the real nervous pathways. Impressive results we also obtained in imitation of sensorial innervation of the front scalp: we inoculated Physarum on the frontal bone above glabella and placed few oat flakes on the pariental bone. In two days Physarum developed an extensively branching tree of protoplasmic tubes. The tree spanned substantial part of the frontal lobe, even covering its lateral parts, crossed coronal sutura and developed actively branching growing zones moving towards the target site on the parietal bone [20].

28.6.4   Illusions

A configuration of three Pac-Man shapes positioned at a vertices of an imaginary triangle and looking towards the centre of the triangle is a famous illusory contour [78]. When we look at this configuration of the shapes we are getting an impression of a white triangle defined by the Pac-Man shapes. The illusory contour disappears when the Pac-Man shapes are facing away from each other. Physarum shows tendency to 'experience' the same illusion as humans do [117]. The Pac-Man shapes are made of a nutrient rich agar and placed on a non-nutrient agar. The Physarum is inoculated in the centre of each Pac-Man. The Physarum develops protoplasmic networks spanning the configuration of Pac-Man shapes. When spanning the configuration, when Pac-Man shapes are facing each other, the Physarum formed a network matching a contour of the illusory triangle in 4/5 of experiments [117]. In the scenarios of away facing Pan-Man shapes the Physarum matched the illusory triangle only in the half of experiments and constructed spanning tree and other graphs in another half of experiments. Conclusion was that Physarum shows behaviour which mimic illusory impressions of humans [117]. Exact mechanisms of such behaviour of Physarum are different from humans, however there may be some subtle analogies between how we visually scan pictures [135] and how the Physarum perceives its environment.

28.7 Energy Production

28.7.1 Modulation of Energy Generation

A microbial fuel cell is a biological electrochemical device which uses microorganisms to convert energy of organic substates into an electrical energy. The microorganisms colonise electrodes, metabolise organic material and donate electrons to the electrodes [69]. A basic fuel cell is made of two electrodes and an ion selective membrane. Electrodes are made of folded carbon veil. Two scenarios are explored in [119] for anode and cathode sites of Physarum inoculation. When Physarum is housed on anode no significant difference between the test and the control fuel cells is observed. In experiments with Physarum inoculated on cathode a statistically significant power increase is observed. A peak open circuit voltage of Physarum fuel cell and control fuel cell was around 0.6 V [119]. When 9.4 kOhm external load is connected, the voltage drops to 0.2 V in the control fuel cell and 0.25 V in the Physarum fuel cell. During 7 days of experiments power generated by the Physarum fuel cell was 12.5–15 μW and power generated by the control fuel cell 10–11 V [119].

28.7.2 Biodiesel Production

This is not about computation, sensing, actuation or analog physical modelling. But we included this section because future unconventional devices will need energy. Biodiesel is a liquid fuel made of vegetable oil or animal fat and used in compression-ignition engines. It is produced by a conversion of a carboxylic acid ester into a different carboxylic acid ester by reacting lipids with an alcohol producing fatty acid esters [88]. Biodiesel is advantageous to fossil fuels because it is biodegradable and has low toxicity. However, biodiesel is expensive to produce: cost of raw material is 2/3 of production costs. Slime moulds have relatively high concentration of lipids in their bodies. Thus they could be a good alternative to existing technologies of biomass production. *P. polycephalum* is found to be a champion, amongst *Myxomecetes*, in lipid production [122]. Physarum produces equivalent of 65 g of dry biomass and over 7 g of lipids per litre of culture medium in four days [122]. The biomass production rate exceeds that of algae [45]. Tran et al. [123] show that cultivation of Physarum on 37.5 g per litre rice brans yields 7.5 g dry biomass and 0.9 g lipid in five days.

28.8 Arts

28.8.1 Music Generation

Physarum expresses its physiological states in patterns of its electrical potential oscillators [23]. In [6] 5–10 days recordings were processed and converted to sound track by mapping parameters of electrical potential oscillation to pitch, attack and duration of tones. First ever sound track was produced from Physarum's electrical activity in [6]. The music reflected a physiological transition of Physarum from a comfortable foraging state to a state of active search for disappearing nutrients to decision making state to transformation to sclerotium. When the track played to auditorium at various presentations, majority of people were getting a feeling of a dramatic development in the 'life of Physarum'. Later the transformation of Physarum activity recording into sounds has been taken at a more professional, from music point of view, level in [94]. There electrical activity of Physarum was converted to parameters of sinusoidal oscillators; the rhythmic behaviour of Physarum was shown to produce different timbres.

Memristive properties of Physarum (Sect. 28.5.7) are used to generate musical responses in [43]. A vocabulary notes are assigned voltage values. The Physarum current-voltage response to electrical stimulation is recorded. Discrete voltages are converted to notes. The notes are fed into a MIDI keyboard. During interactive music performance between human composer and Physarum, a feedback to the Physarum is implemented. Parts of well-know melody: Elgar's Nimrod and Beethoven's Für Elise were generated for live performances with Physarum in [43].

28.8.2 Modelling Creativity

Creativity is manifested by divergent thinking and lack of lateral inhibition, making remote associations between ideas and concepts, switching between ideation, generating ideas of actualities, risk taking, nonconformity [84]. In [32] we show how to use live Physarum to analyse many scenarios of an individual creativity genesis. The divergent thinking is expressed by Physarum in its simultaneous reaction to several sources of attractants and repellents, and parallel implementation of sensorial fusion. Functional non-conformity of Physarum is manifested in its self-avoidance. Activity in both hemispheres and exchange of activities between hemispheres are considered to be attributes of human creativity [84]. In Physarum the hemispheres' activity is represented by simultaneous oscillatory activity, with biochemical oscillators located to distant parts of Physarum body and oscillating with different frequencies and amplitudes. Interaction between 'hemispheres' is instantiated by waves of calcium waves and waves of contractile activity propagating along protoplasmic tubes.

Cognitive control of divergent thinking is a requisite of creativity [84]. A person with extremely divergent thinking who is unable to control these associations would be potentially classified as mentally ill. However, those who can fit their high schizotypy (a range of personality characteristics ranging from normal to schizophrenia) traits into rigorous cognitive frameworks may be classified as gifted or even genius. Thus creativity could be positioned together with autism and schizophrenia in the same phase space. Physarum imitates cognitive control by tuning regularity of its protoplasmic network. A degree of branching of a Physarum network may be considered as a representation of a degree of schizotypy. Then 'mathematical savant' slime mould grows a low branching highly symmetrical protoplasmic networks and severely 'autistic' Physarum develops highly asymmetric low branching networks [32].

28.9   Things Inspired by Physarum but Never Done with a Real One

1. A non-quantum implementation of Shor's factorisation algorithm [39] is inspired by Physarum ability to retain the time-periods of stimulation, the anticipatory behaviour [106].
2. Single electron circuit solving maze problem [111] is based on a cellular automaton imitation of Physarum.
3. Particle based model of Physarum approximates moving average and low-pass filters in one-dimensional data sets and spatial computation of splines in two-dimensional data set [74].
4. Physarum logic is developed by interpreting basic features of the Physarum foraging behaviour in terms of process calculi and spatial logic without modal operators [109].
5. Soft amoeboid robots [100]—models of coupled-oscillator-based robots [129] are based on general principles of Physarum behaviour, especially coordination of distant parts of its cell.
6. Physarum concurrent games are proposed in [110]. In these games rules can change, players update their strategies and actions, resistance points are reduced to payoffs. At the time these games were proposed, we were unaware about the lethal reaction followed a fusion between Physarum of two different strains [44]: a degree of lethality depends on the position and size of invasion between strains, which supports the ideas developed in [110].
7. Physarum is interfaced with a field-programmable array in [93]. The hybrid system performs predefined arithmetic operations derived by digital recognition of membrane potential oscillations.
8. A Physarum-inspired algorithm for solution of the Steiner tree problem is proposed in [86] and applied to optimise the minimal exposure problem and worst-case coverage in wireless sensor networks [86, 124].

9. An abstract implementation of reversible logical gates with Physarum is proposed in [108].
10. Mechanisms of Physarum foraging behaviour are employed in robotics algorithm for simultaneous localisation and mapping [75].
11. A range of algorithms for network optimisation is derived from a model of Physarum shortest path formation [42, 121]. Most of these algorithms are based on a feedback between traffic throughout a tube and the tube's capacity. They include dynamical shortest path algorithms [136], optimal communication paths in wireless sensor networks [55, 137], supply chains design [137], shortest path tree problem [137], design of fault tolerant graphs [38], and multi-cast routing [85]. Behaviour of Physarum solver [121] was also applied to deriving a shortest path on Riemann surface [95].
12. Physarum-inspired algorithm for learning Bayesian network structure from data is designed in [107].
13. Attraction based two-input two-output gate realising AND and OR and three-input two-output gate realising conjunction of three inputs and negation of one input with disjunction of two other inputs are constructed in particle-based model of Physarum [72]; the gates are cascaded into a one-bit adder.
14. A nano-device aimed to solve Boolean satisfiability problem, inspired by optimisation of protoplasmic networks by Physarum, is designed in [35, 36]. The device works on fluctuations generated from thermal energy in nanowires, electrical Brownian ratchets.
15. Solution of a the 'exploration versus exploitation' dilemma by Physarum making a choice between colonising nutrients and escaping illuminated areas is used in a tug of war model [80]: a parallel search of a space by collectives of locally correlated agents and decision making in situations of uncertainty. The ideas are developed further in a design of experimental device where a single photon solves the multi-armed bandit problem [97].

28.10 Post Coitum Omne Animal Triste Est

Excitement of being able to make computing, sensing and actuating devices with live Physarum eventually fades down. Let us wake up now. Is there any real use of the slime mould? Is it fast? No. Physarum is a very slow creature. It might take the Physarum several days to compute Voronoi diagram or a shortest path. Is it quick in responding to stimuli? Not really. Is it robust? No two experiments are the same. Physarum always behaves differently. Results of a computation performed by Physarum are only valid when at least dozen of experiments are done cause a single trial might mean nothing. Are the Physarum computing devices reliable? Failure rate is near 30%. The slime mould is not a miraculous computing substrate. It is just a user-friendly non-demanding living creature which changes its form and shape to stay comfortable in the fields of attractants and repellents. Why did we love Physarum then? Because by taking myriad of different shapes and exhibiting fantastically rich spectrum of electrical potential oscillations Physarum makes a unique fruitful material for interpretations of its behaviour in terms of purposeful transformation of data to results. The unconventional computing is an art of interpretation. Physarum feeds out phantasy and fuels our unconventional computing dreams. Will we ever see the Physarum in commercial computing or sensing devices? Not tomorrow. In no way Physarum can win over the silicon technology which has been optimised non-stop for decades and decades. But a success depends on many factors, not just technological ones. Success is in finding a vacant niche and flourishing there. More likely applications of Physarum computers will be in disposable hybrid processing devices used for sensing and decision-making in environments and situations where speed does not matter but being energy efficient, adaptable and self-healing is important.

References

1. Adamatzky, A.: Neural algorithm for constructing minimal spanning tree. Neural Netw. World **6**, 335–339 (1991)
2. Adamatzky, A.: Collision-Based Computing. Springer, London (2002)
3. Adamatzky, A.: Physarum machine: implementation of a Kolmogorov-Uspensky machine on a biological substrate. Parallel Process. Lett. **17**(04), 455–467 (2007)
4. Adamatzky, A.: Growing spanning trees in plasmodium machines. Kybernetes **37**(2), 258–264 (2008)
5. Adamatzky, A.: Developing proximity graphs by Physarum polycephalum: does the plasmodium follow the Toussaint hierarchy? Parallel Process. Lett. **19**(01), 105–127 (2009)
6. Adamatzky, A.: Physarum music. https://youtu.be/mvBSkt6LhJE (2010)
7. Adamatzky, A.: Physarum boats: if plasmodium sailed it would never leave a port. Appl. Bionics Biomech. **7**(1), 31–39 (2010)
8. Adamatzky, A.: Physarum Machines: Computers from Slime Mould, vol. 74. World Scientific, Singapore (2010)
9. Adamatzky, A.: Slime mould logical gates: exploring ballistic approach. arXiv preprint arXiv:1005.2301 (2010)
10. Adamatzky, A.: On attraction of slime mould Physarum polycephalum to plants with sedative properties. Nat. Proc. **10**, 3334 (2011)
11. Adamatzky, A.: Bioevaluation of World Transport Networks. World Scientific, Singapore (2012)
12. Adamatzky, A.: Simulating strange attraction of acellular slime mould Physarum polycephaum to herbal tablets. Math. Comput. Model. **55**(3), 884–900 (2012)
13. Adamatzky, A.: Slime mold solves maze in one pass, assisted by gradient of chemo-attractants. IEEE Trans. NanoBioscience **11**(2), 131–134 (2012)
14. Adamatzky, A.: Slime mould computes planar shapes. Int. J. Bio-Inspired Comput. **4**(3), 149–154 (2012)
15. Adamatzky, A.: Physarum wires: self-growing self-repairing smart wires made from slime mould. Biomed. Eng. Lett. **3**(4), 232–241 (2013)
16. Adamatzky, A.: Slime mould tactile sensor. Sens. Actuators B: Chem. **188**, 38–44 (2013)
17. Adamatzky, A.: Towards slime mould colour sensor: recognition of colours by Physarum polycephalum. Org. Electron. **14**(12), 3355–3361 (2013)
18. Adamatzky, A.: Slime mould electronic oscillators. Microelectron. Eng. **124**, 58–65 (2014)
19. Adamatzky, A.: Tactile bristle sensors made with slime mold. IEEE Sens. J. **14**(2), 324–332 (2014)

20. Adamatzky, A.: A would-be nervous system made from a slime mold. Artif. Life **21**(1), 73–91 (2015)
21. Adamatzky, A. (ed.): Advances in Physarum Machines: Sensing and Computing with Slime Mould. Springer, Berlin (2016)
22. Adamatzky, A., De Lacy Costello, B.: Physarum attraction: why slime mold behaves as cats do? Commun. Integr. Biol. **5**(3), 297–299 (2012)
23. Adamatzky, A., Jones, J.: On electrical correlates of physarum polycephalum spatial activity: can we see physarum machine in the dark? Biophys. Rev. Lett. **6**(01n02), 29–57 (2011)
24. Adamatzky, A., Kayem, A.V.D.M.: Biological evaluation of Trans-African highways. Eur. Phys. J. Spec. Top. **215**(1), 49–59 (2013)
25. Adamatzky, A., Martinez, G.J.: Bio-imitation of Mexican migration routes to the USA with slime mould on 3D terrains. J. Bionic Eng. **10**(2), 242–250 (2013)
26. Adamatzky, A., Prokopenko, M.: Slime mould evaluation of Australian motorways. Int. J. Parallel Emerg. Distrib. Syst. **27**(4), 275–295 (2012)
27. Adamatzky, A., Schubert, T.: Schlauschleimer in Reichsautobahnen: slime mould imitates motorway network in Germany. Kybernetes **41**(7/8), 1050–1071 (2012)
28. Adamatzky, A., Schubert, T.: Slime mold microfluidic logical gates. Mater. Today **17**(2), 86–91 (2014)
29. Adamatzky, A., De Lacy Costello, B., Asai, T.: Reaction-Diffusion Computers. Elsevier, Amsterdam (2005)
30. Adamatzky, A., De Baets, B., Van Dessel, W.: Slime mould imitation of Belgian transport networks: redundancy, bio-essential motorways, and dissolution. Int. J. Unconv. Comput. **8**(3), 235–261 (2012)
31. Adamatzky, A., Akl, S., Alonso-Sanz, R., Van Dessel, W., Ibrahim, Z., Ilachinski, A., Jones, J., Kayem, A.V.D.M., Martínez, G.J., De Oliveira, P., et al.: Are motorways rational from slime mould's point of view? Int. J. Parallel Emerg. Distrib. Syst. **28**(3), 230–248 (2013)
32. Adamatzky, A., Armstrong, R., Jones, J., Gunji, Y.-P.: On creativity of slime mould. Int. J. Gen. Syst. **42**(5), 441–457 (2013)
33. Adamatzky, A., Lees, M., Sloot, P.: Bio-development of motorway network in the Netherlands: a slime mould approach. Adv. Complex Syst. **16**(02n03), 1250034 (2013)
34. Adamatzky, A., Armstrong, R., De Lacy Costello, B., Deng, Y., Jones, J., Mayne, R., Schubert, T., Sirakoulis, G.C., Zhang, X.: Slime mould analogue models of space exploration and planet colonisation. J. Br. Interplanet. Soc. **67**, 290–304 (2014)
35. Aono, M., Kim, S.-J., Zhu, L., Naruse, M., Ohtsu, M., Hori, H., Hara, M.: Amoeba-inspired sat solver. In: Proceedings of the NOLTA, pp. 586–589 (2012)
36. Aono, M., Kasai, S., Kim, S.J., Wakabayashi, M., Miwa, H., Naruse, M.: Amoeba-inspired nanoarchitectonic computing implemented using electrical brownian ratchets. Nanotechnology **26**(23), 234001 (2015)
37. Ascher, R.: Experimental archeology. Am. Anthropol. **63**(4), 793–816 (1961)
38. Becker, M., Kromker, M., Szczerbicka, H.: Evaluating heuristic optimization, bio-inspired and graph-theoretic algorithms for the generation of fault-tolerant graphs with minimal costs. In: Information Science and Applications, pp. 1033–1041. Springer, Berlin (2015)
39. Blakey, Ed.: Towards non-quantum implementations of shor's factorization algorithm. Int. J. Unconv. Comput. **10**(4), 317–338 (2014)
40. Blass, A., Gurevich, Y.: Algorithms: a quest for absolute definitions. Bull. EATCS **81**, 195–225 (2003)
41. Block, I., Wohlfarth-Bottermann, K.E.: Blue light as a medium to influence oscillatory contraction frequency in Physarum. Cell Biol. Int. Rep. **5**(1), 73–81 (1981)
42. Bonifaci, V., Mehlhorn, K., Varma, G.: Physarum can compute shortest paths. J. Theor. Biol. **309**, 121–133 (2012)
43. Braund, E., Sparrow, R., Miranda, E.: Physarum-based memristors for computer music. In: Adamatzky, A. (ed.) Advances in Physarum Machines. Springer, Heidelberg (2016)
44. Carlile, M.J.: The lethal interaction following plasmodial fusion between two strains of the myxomycete physarum polycephalum. J. Gen. Microbiol. **71**(3), 581–590 (1972)

45. Chisti, Y.: Biodiesel from microalgae. Biotechnol. Adv. **25**(3), 294–306 (2007)
46. Chua, L.O.: Memristor-the missing circuit element. IEEE Trans. Circuit Theory **18**(5), 507–519 (1971)
47. Cifarelli, A., Dimonte, A., Berzina, T., Erokhin, V.: Non-linear bioelectronic element: Schottky effect and electrochemistry. Int. J. Unconv. Comput. **10**(5–6), 375–379 (2014)
48. Cloteaux, B., Ranjan, D.: Some separation results between classes of pointer algorithms. DCFS **6**, 232–240 (2006)
49. Coles, J.: Experimental Archaeology. Academic Press, London (1979)
50. Delaunay, B.: Sur la sphere vide. Izv. Akad. Nauk SSSR, Otdelenie Matematicheskii i Estestvennyka Nauk **7**(793–800), 1–2 (1934)
51. De Lacy Costello, B., Adamatzky, A.I.: Assessing the chemotaxis behavior of Physarum polycephalum to a range of simple volatile organic chemicals. Commun. Integr. Biol. **6**(5), e25030 (2013)
52. de Lacy Costello, B., Adamatzky, A.I.: Routing of physarum polycephalum "signals" using simple chemicals. Commun. Integr. Biol. **7**(3), e28543 (2014)
53. de Lacy Costello, B.J.P., Mayne, R., Adamatzky, A.: Conducting polymer-coated Physarum polycephalum towards the synthesis of bio-hybrid electronic devices. Int. J. Gen. Syst. **44**(3), 409–420 (2015)
54. Dietrich, M.R.: Explaining the "pulse of protoplasm": the search for molecular mechanisms of protoplasmic streaming. J. Integr. Plant Biol. **57**(1), 14–22 (2015)
55. Dourvas, N., Tsompanas, M.-A., Ch Sirakoulis, G., Tsalides, P.: Hardware acceleration of cellular automata Physarum polycephalum model. Parallel Process. Lett. **25**(01), 1540006 (2015)
56. Durham, A.C., Ridgway, E.B.: Control of chemotaxis in Physarum polycephalum. J. Cell Biol. **69**(1), 218–223 (1976)
57. Edelsbrunner, H., Kirkpatrick, D.G., Seidel, R.: On the shape of a set of points in the plane. IEEE Trans. Inf. Theory **29**(4), 551–559 (1983)
58. Evangelidis, V., Tsompanas, M.-A., Ch Sirakoulis, G., Adamatzky, A.: Slime mould imitates development of Roman roads in the Balkans. J. Archaeol. Sci. Rep. **2**, 264–281 (2015)
59. Gabriel, K.R., Sokal, R.R.: A new statistical approach to geographic variation analysis. Syst. Biol. **18**(3), 259–278 (1969)
60. Gal, E., Adamatzky, A., de Lacy Costello, B.: Slime mould memristors. BioNanoScience **5**(1), 1–8 (2013)
61. Gale, E., Adamatzky, A.: Translating slime mould responses: a novel way to present data to the public. In: Adamatzky, A. (ed.) Advances in Physarum Machines. Springer, Heidelberg (2016)
62. Geddes, L.A., Baker, L.E.: The specific resistance of biological material—a compendium of data for the biomedical engineer and physiologist. Med. Biol. Eng. **5**(3), 271–293 (1967)
63. Grebecki, A., Cieślawska, M.: Plasmodium of Physarum polycephalum as a synchronous contractile system. Cytobiologie **17**(2), 335–342 (1978)
64. Grigor'ev, DYu.: Kolmogoroff algorithms are stronger than turing machines. J. Math. Sci. **14**(5), 1445–1450 (1980)
65. Gurevich, Y.: Kolmogorov machines and related issues. Bull. EATCS **35**, 71–82 (1988)
66. Hinz, A.M.: The Tower of Hanoi. Enseign. Math. **35**(2), 289–321 (1989)
67. Hinz, A.M.: Shortest paths between regular states of the Tower of Hanoi. Inf. Sci. **63**(1), 173–181 (1992)
68. Horowitz, P., Hill, W.: The Art of Electronics. Cambridge University Press, Cambridge (1989)
69. Ieropoulos, I.A., Greenman, J., Melhuish, C., Hart, J.: Comparative study of three types of microbial fuel cell. Enzym. Microb. Technol. **37**(2), 238–245 (2005)
70. Ingersoll, D., Yellen, J.E., Macdonald, W.: Experimental Archaeology. Columbia University Press, New York (1977)
71. Jarvis, R.A.: On the identification of the convex hull of a finite set of points in the plane. Inf. Process. Lett. **2**(1), 18–21 (1973)
72. Jones, J., Adamatzky, A.: Towards Physarum binary adders. BioSystems **101**(1), 51–58 (2010)

73. Jones, J., Adamatzky, A.: Computation of the travelling salesman problem by a shrinking blob. Nat. Comput. **13**(1), 1–16 (2014)
74. Jones, J., Adamatzky, A.: Material approximation of data smoothing and spline curves inspired by slime mould. Bioinspir. Biomim. **9**(3), 036016 (2014)
75. Kalogeiton, V.S., Papadopoulos, D.P., Ch Sirakoulis, G.: Hey Physarum! can you perform slam? Int. J. Unconv. Comput. **10**(4), 271–293 (2014)
76. Kalogeiton, V.S., Papadopoulos, D.P., Georgilas, I.P., Ch Sirakoulis, G., Adamatzky, A.I.: Biomimicry of crowd evacuation with a slime mould cellular automaton model. In: Computational Intelligence, Medicine and Biology, pp. 123–151. Springer, Berlin (2015)
77. Kalogeiton, V.S., Papadopoulos, D.P., Georgilas, I.P., Ch Sirakoulis, G., Adamatzky, A.I.: Cellular automaton model of crowd evacuation inspired by slime mould. Int. J. Gen. Syst. **44**(3), 354–391 (2015)
78. Kanizsa, G.: Subjective contours. Sci. Am. **234**(4), 48–52 (1976)
79. Kauffman, S., Wille, J.J.: The mitotic oscillator in Physarum polycephalum. J. Theor. Biol. **55**(1), 47–93 (1975)
80. Kim, S.-J., Aono, M., Hara, M.: Tug-of-war model for the two-bandit problem: nonlocally-correlated parallel exploration via resource conservation. BioSystems **101**(1), 29–36 (2010)
81. Kishimoto, U.: Rhythmicity in the protoplasmic streaming of a slime mood, Physarum polycephalum. i. a statistical analysis of the electrical potential rhythm. J. Gen. Physiol. **41**(6), 1205–1222 (1958)
82. Kolmogorov, A.N.: On the concept of algorithm. Uspekhi Mat. Nauk **8**(4), 175–176 (1953)
83. Kolmogorov, A.N., Uspenskii, V.A.: On the definition of an algorithm. Uspekhi Matematicheskikh Nauk **13**(4), 3–28 (1958)
84. Kuszewski, A.M.: The genetics of creativity: a serendipitous assemblage of madness (2009)
85. Liang, M.X., Gao, C., Liu, Y.X., Tao, L., Zhang, Z.L.: A new Physarum network based genetic algorithm for bandwidth-delay constrained least-cost multicast routing. In: Advances in Swarm and Computational Intelligence, pp. 273–280. Springer, Berlin (2015)
86. Liu, L., Song, Y., Zhang, H., Ma, H., Vasilakos, A.V.: Physarum optimization: a biology-inspired algorithm for the steiner tree problem in networks. IEEE Trans. Comput. **64**(3), 819–832 (2015)
87. MacGregor, J.N., Ormerod, T.: Human performance on the traveling salesman problem. Percept. Psychophys. **58**(4), 527–539 (1996)
88. Ma, F., Hanna, M.A.: Biodiesel production: a review. Bioresour. Technol. **70**(1), 1–15 (1999)
89. Margolus, N.: Universal cellular automata based on the collisions of soft spheres. In: Collision-Based Computing, pp. 107–134. Springer, London (2002)
90. Mayne, R., Adamatzky, A.: Slime mould foraging behaviour as optically coupled logical operations. Int. J. Gen. Syst. **44**(3), 305–313 (2015)
91. Mayne, R., Adamatzky, A.: On the computing potential of intracellular vesicles. PloS one **10**(10), e0139617 (2015)
92. Mayne, R., Adamatzky, A.: Toward hybrid nanostructure-slime mould devices. Nano LIFE **5**(01), 1450007 (2015)
93. Mayne, R., Tsompanas, M.-A., Ch Sirakoulis, G., Adamatzky, A.: Towards a slime mould-FPGA interface. Biomed. Eng. Lett. **5**(1), 51–57 (2015)
94. Miranda, E.R., Adamatzky, A., Jones, J.: Sounds synthesis with slime mould of Physarum polycephalum. J. Bionic Eng. **8**(2), 107–113 (2011)
95. Miyaji, T., Ohnishi, I.: Physarum can solve the shortest path problem on Riemannian surface mathematically rigourously. Int. J. Pure Appl. Math. **47**(3), 353–369 (2008)
96. Nakagaki, T., Yamada, H., Toth, A.: Path finding by tube morphogenesis in an amoeboid organism. Biophys. Chem. **92**(1), 47–52 (2001)
97. Naruse, M., Berthel, M., Drezet, A., Huant, S., Aono, M., Hori, H., Kim, S.-J.: Single-photon decision maker. Sci. Rep. **5**, 13253 (2015)
98. Nešetřil, J., Milková, E., Nešetřilová, H.: Otakar Borůvka on minimum spanning tree problem translation of both the 1926 papers, comments, history. Discret. Math. **233**(1), 3–36 (2001)

99. Pershin, Y.V., Di Ventra, M.: Experimental demonstration of associative memory with memristive neural networks. Neural Netw. **23**(7), 881–886 (2010)
100. Piovanelli, M., Fujie, T., Mazzolai, B., Beccai, L.: A bio-inspired approach towards the development of soft amoeboid microrobots. In: 4th IEEE RAS & EMBS International Conference on Biomedical Robotics and Biomechatronics (BioRob), pp. 612–616. IEEE (2012)
101. Preparata, F.P., Shamos, M.I.: Computational Geometry: An Introduction. Springer, Berlin (1985)
102. Rakoczy, L.: Application of crossed light and humidity gradients for the investigation of slime-molds. Acta Soc. Bot. Pol. **32**(2), 393–403 (1963)
103. Reid, C.R., Beekman, M.: Solving the Towers of Hanoi - how an amoeboid organism efficiently constructs transport networks. J. Exp. Biol. **216**(9), 1546–1551 (2013)
104. Ridgway, E.B., Durham, A.C.: Oscillations of calcium ion concentrations in Physarum polycephalum. J. Cell Biol. **69**(1), 223–226 (1976)
105. Romik, D.: Shortest paths in the Tower of Hanoi graph and finite automata. SIAM J. Discret. Math. **20**(3), 610–622 (2006)
106. Saigusa, T., Tero, A., Nakagaki, T., Kuramoto, Y.: Amoebae anticipate periodic events. Phys. Rev. Lett. **100**(1), 018101 (2008)
107. Schön, T., Stetter, M., Tomé, A.M., Puntonet, C.G., Lang, E.W.: Physarum learner: a bio-inspired way of learning structure from data. Expert Syst. Appl. **41**(11), 5353–5370 (2014)
108. Schumann, A.: Conventional and unconventional reversible logic gates on Physarum polycephalum. Int. J. Parallel Emerg. Distrib. Syst. **17**, 1–14 (2015)
109. Schumann, A., Adamatzky, A.: Physarum spatial logic. New Math. Nat. Comput. **7**(03), 483–498 (2011)
110. Schumann, A., Pancerz, K., Adamatzky, A., Grube, M.: Bio-inspired game theory: the case of Physarum polycephalum. In: Proceedings of the 8th International Conference on Bioinspired Information and Communications Technologies, pp. 9–16. ICST (Institute for Computer Sciences, Social-Informatics and Telecommunications Engineering) (2014)
111. Shinde, Y., Oya, T.: Design of single-electron "slime-mold" circuit and its application to solving optimal path planning problem. Nonlinear Theory Appl. IEICE **5**(1), 80–88 (2014)
112. Shirakawa, T., Adamatzky, A., Gunji, Y.-P., Miyake, Y.: On simultaneous construction of Voronoi diagram and Delaunay triangulation by Physarum polycephalum. Int. J. Bifurc. Chaos **19**(09), 3109–3117 (2009)
113. Shvachko, K.V.: Different modifications of pointer machines and their computational power. In: Mathematical Foundations of Computer Science, pp. 426–435. Springer, Berlin (1991)
114. Stephenson, S.L., Stempen, H., Hall, I.: Myxomycetes: A Handbook of Slime Molds. Timber Press, Portland (1994)
115. Strukov, D.B., Snider, G.S., Stewart, D.R., Williams, R.S.: The missing memristor found. Nature **453**(7191), 80–83 (2008)
116. Sun, T., Tsuda, S., Zauner, K.-P., Morgan, H.: Single cell imaging using electrical impedance tomography. In: 4th IEEE International Conference on Nano/Micro Engineered and Molecular Systems, NEMS, pp. 858–863. IEEE (2009)
117. Tani, I., Yamachiyo, M., Shirakawa, T., Gunji, Y.-P.: Kanizsa illusory contours appearing in the plasmodium pattern of Physarum polycephalum. Front. Cell. Infect. Microbiol. **4**, 10 (2014)
118. Tarabella, G., D'Angelo, P., Cifarelli, A., Dimonte, A., Romeo, A., Berzina, T., Erokhin, V., Iannotta, S.: A hybrid living/organic electrochemical transistor based on the Physarum polycephalum cell endowed with both sensing and memristive properties. Chem. Sci. **6**(5), 2859–2868 (2015)
119. Taylor, B., Adamatzky, A., Greenman, J., Ieropoulos, I.: Physarum polycephalum: towards a biological controller. Biosystems **127**, 42–46 (2015)
120. Teplov, V.A., Romanovsky, Y.M., Latushkin, O.A.: A continuum model of contraction waves and protoplasm streaming in strands of Physarum plasmodium. Biosystems **24**(4), 269–289 (1991)

121. Tero, A., Kobayashi, R., Nakagaki, T.: Physarum solver: a biologically inspired method of road-network navigation. Phys. A Stat. Mech. Appl. **363**(1), 115–119 (2006)
122. Tran, H.T.M., Stephenson, S.L., Chen, Z., Pollock, E.D., Goggin, F.L.: Evaluating the potential use of myxomycetes as a source of lipids for biodiesel production. Bioresour. Technol. **123**, 386–389 (2012)
123. Tran, H., Stephenson, S., Pollock, E.: Evaluation of Physarum polycephalum plasmodial growth and lipid production using rice bran as a carbon source. BMC Biotechnol. **15**(1), 67 (2015)
124. Tsompanas, M.-A.I., Mayne, R., Ch Sirakoulis, G., Adamatzky, A.I.: A cellular automata bioinspired algorithm designing data trees in wireless sensor networks. Int. J. Distrib. Sens. Netw. **501**, 471045 (2015)
125. Tsuda, S., Aono, M., Gunji, Y.-P.: Robust and emergent Physarum logical-computing. Biosystems **73**(1), 45–55 (2004)
126. Tsuda, S., Zauner, K.-P., Gunji, Y.-P.: Robot control: from silicon circuitry to cells. Biol. Inspired Approaches Adv. Inf. Technol. **3853**, 20–32 (2006)
127. Tsuda, S., Jones, J., Adamatzky, A.: Towards Physarum engines. Appl. Bionics Biomech. **9**(3), 221–240 (2012)
128. Ueda, T., Muratsugu, M., Kurihara, K., Kobatake, Y.: Chemotaxis in Physarum polycephalum: effects of chemicals on isometric tension of the plasmodial strand in relation to chemotactic movement. Exp. Cell Res. **100**(2), 337–344 (1976)
129. Umedachi, T., Idei, R., Ito, K., Ishiguro, A.: A fluid-filled soft robot that exhibits spontaneous switching among versatile spatiotemporal oscillatory patterns inspired by the true slime mold. Artif. Life **19**(1), 67–78 (2013)
130. Whiting, J.G.H., de Lacy Costello, B.P.J., Adamatzky, A.: Slime mould logic gates based on frequency changes of electrical potential oscillation. Biosystems **124**, 21–25 (2014)
131. Whiting, J.G.H., de Lacy Costello, B.P.J., Adamatzky, A.: Towards slime mould chemical sensor: mapping chemical inputs onto electrical potential dynamics of Physarum polycephalum. Sens. Actuators B Chem. **191**, 844–853 (2014)
132. Whiting, J.G.H., de Lacy Costello, B.P.J., Adamatzky, A.: Transfer function of protoplasmic tubes of Physarum polycephalum. Biosystems **128**, 48–51 (2015)
133. Wohlfarth-Bottermann, K.E.: Oscillatory contraction activity in Physarum. J. Exp. Biol. **81**(1), 15–32 (1979)
134. Wohlfarth-Bottermann, K.E., Block, I.: The pathway of photosensory transduction in Physarum polycephalum. Cell Biol. Int. Rep. **5**(4), 365–373 (1981)
135. Yarbus, A.L.: Eye Movements During Perception of Complex Objects. Springer, New York (1967)
136. Zhang, X., Wang, Q., Adamatzky, A., Chan, F.T.S., Mahadevan, S., Deng, Y.: An improved Physarum polycephalum algorithm for the shortest path problem. Sci. World J. **2014**, 487069 (2014)
137. Zhang, X., Mahadevan, S., Deng, Y.: Physarum-inspired applications in graph-optimization problems. Parallel Process. Lett. **25**(01), 1540005 (2015)
138. Zhu, L., Aono, M., Kim, S.-J., Hara, M.: Amoeba-based computing for traveling salesman problem: long-term correlations between spatially separated individual cells of Physarum polycephalum. Biosystems **112**(1), 1–10 (2013)

Chapter 29
Experiments in Musical Biocomputing: Towards New Kinds of Processors for Audio and Music

Eduardo Reck Miranda and Edward Braund

Abstract The emerging field of Unconventional Computing is developing new algorithms and computing architectures inspired by or implemented in biological, physical and chemical systems. We are investigating how Unconventional Computing may benefit the future of the music industry and related audio engineering technologies. In this chapter, after a brief introduction to Unconventional Computing, we present our research into harnessing the behaviour of a slime mould called *Physarum polycephalum* to build new kinds of processors for audio and music. The plasmodium of *Physarum polycephalum* is a large single cell with a myriad of diploid nuclei, which moves like a giant amoeba in its pursuit for food. The organism is amorphous, and although without a brain or any serving centre of control, can respond to the environmental conditions that surround it. As our research progressed, we have successfully harnessed the organism to implement a sound synthesiser and a musical sequencer, grow biological audio wires, and build an interactive biocomputer that can listen and produce musical responses in real-time.

29.1 Introduction

Over the course of the last 60 years, digital computing technology has played a pivotal part in the development of the music industry. Advances in computer science have had a significant impact on both the way we consume and produce music. For example, today we have digital audio workstations to compose music on and mass distribution of music on the internet. As such, the music industry is likely to progress in tandem with the development of new computing technologies.

E.R. Miranda (✉) · E. Braund
Interdisciplinary Centre for Computer Music Research (ICCMR), Plymouth University,
Plymouth, UK
e-mail: eduardo.miranda@plymouth.ac.uk

E. Braund
e-mail: edward.braund@students.plymouth.ac.uk

© Springer International Publishing Switzerland 2017 739
A. Adamatzky (ed.), *Advances in Unconventional Computing*,
Emergence, Complexity and Computation 23,
DOI 10.1007/978-3-319-33921-4_29

To cite this chapter, let us first look to the development of today's conventional computer. The term 'computer' originally referred to large groups of people who followed strict sets of rules to solve a mathematical or logic based problem [33]. In the 1930s, English mathematician and wartime code-breaker Alan Turing formalised the behaviour of these people-based computers to create the first stored-program computing model: the Turing Machine [35]. Shortly after, Hungarian-born mathematician John von Neumann developed a stored-program computing architecture [36]. These two developments are considered the predecessors to today's commercial computers. During the past eight decades, what we consider to be conventional or standard computation has advanced at a rapid speed. There is a growing consensus amongst computer scientists that we will soon reach the limit of today's conventional computing systems. This is a result of our ever-growing need for new kinds of computers to address problems that are arduous to address with current computing technologies. An example being self-organisation in non-equilibrium systems, to cite but one.

Research into new Unconventional Computing (UC) schemes develops new concepts, algorithms, models and computing architectures inspired by or implemented in, biological, physical and chemical systems. Computer scientists are developing UC technologies, which harness basic processes from natural phenomena because of their naturally efficient approach to solving certain types of complex problems. Current UC paradigms include, but are not limited to, quantum computing, DNA computing, reaction-diffusion computing and slime mould computing [6].

It is our belief that if research into UC mirrors the level of development made to date in today's classical computation, then future computers are likely to be unrecognisably different from those we use today.

We are investigating ways in which UC modes of computation may be used to enhance or go beyond the current technological offering in computer music, and how such advances may provide innovative pathways for the future of music. The objective of this chapter is to offer an insight into the potential of UC for music technology and report on the outcomes of the work we have been developing at Plymouth University's Interdisciplinary Centre for Computer Music Research (ICCMR).

This chapter's structure is as follows. Firstly, we present the field of UC and briefly introduce our research. Next, we review two projects aimed at harnessing a biological organism called *Physarum polycephalum*: a step sequencer and a granular synthesiser. Then, we introduce an experiment aimed at growing biological audio wires, followed by an introduction to an experimental interacting musical biocomputer. The chapter concludes with a reflection on the exciting progress we have made so far and an indication of what we are doing next.

29.2 Background

29.2.1 The Beginning: Cellular Automata Simulations

In computer music, there is a tradition of experimenting with emerging technologies. Until recent years, developments put forward by the field of unconventional computation have been left unexploited, which is likely due to the field's heavy theoretical nature, complexity and lack of accessible prototypes. As Adamatzky explains: *"unconventional computing is chock full of theoretical stuff. There are just a handful of experimental laboratory prototypes. They are outstanding but difficult for non-experts to play with."* [2]. These constraints are likely the reasons why cellular automata (CA) models were the first methods used to bridge the two fields: CA are capable of simulating some aspects of biological and physical media that have been explored in UC; e.g., chemical reaction-diffusion.

Cellular automata (CA) are computer tools that can be programmed to model the evolution of a system over time. Normally, a cellular automaton is implemented on a computer as a grid of cells. Every cell can exist in a defined quantity of states, which are conventionally represented by an integer, and displayed on the computer screen by colours. By way of evolving the model, rules are applied to the cells informing them to change state according to the state of their neighbourhood. Typically, these rules remain the same throughout the model, but this is not always the case. Initially, at time $t = 0$, each cell is assigned its starting state. The model can then produce a new generation ($t = 1$) of the grid by applying the rules. This process can continue for an infinite amount of generations.

The first known composer who used CA in his work was Iannis Xenakis in the mid-1980s [38]. Xenakis's piece 'Horos' used a CA to create an evolution of orchestral clusters. Following the release of this composition, a handful of projects using CA started to appear. One example is Miranda's application of a modelled reaction-diffusion computer to control a granular synthesiser, which he developed in 1992 on a Connection CM-200 parallel computer at Edinburgh Parallel Computing Centre [27]. Here, the CA grid was divided into a number of sections, which were each assigned to a sine wave oscillator. The automaton was programmed to model the behaviour of a network of oscillating neurones. Upon the automaton's grid refreshing, the mean state of each oscillator's area was calculated and mapped to control the frequency and amplitude of a sound partial (synthesised by the respective oscillator). All partials were then combined to create a short burst of sound, lasting for a few milliseconds, referred to as a sound granule. Hence the synthesis technique: granular synthesis, where a rapid succession of sound granules produces a continuous sound. Each of these granules represented the entire automaton's grid at the respective refresh point (Fig. 29.1). Miranda used this system to generate sounds for several pieces of electroacoustic music. Such pieces included 'Olivine Tress', composed in 1993, which is believed to be the first piece of music composed on a parallel computer [26].

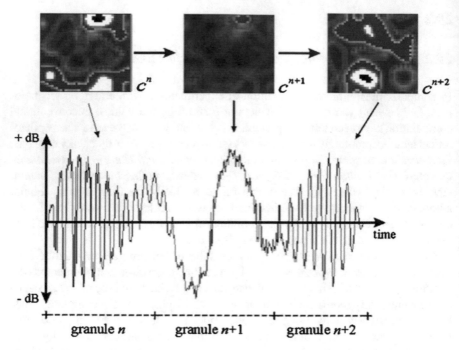

Fig. 29.1 At each time-step, as the cellular automaton evolves it generates parameters to synthesise a sound grain

CA modelling is an inviting way for computer musicians to experiment with using unconventional computing schemes, but they do have their limitations. For instance, real-time use can be problematic due to the time the grid takes to regenerate and modelling certain systems with accuracy may be problematic. From his various experiments with CA in music, Miranda has suggested that *"cellular automata are more suitable for sound synthesis than for musical composition"* [29]. The reason for this is likely due to musical composition involving aspects of culture and convention, whereas sound synthesis lends itself to the complexity of CA. For other examples where CA have been used in music see [8, 28] and for an excellent reference on computing in CA collectives see [1].

29.2.2 Genuine and Hybrid Schemes

The continued research in unconventional computing is tightening the coupling between silicon and biological machines. Biological computers harness abstractions derived from biological systems to perform calculations by processing, storing and retrieving data. The first implementation of computational technology based on biological concepts is believed to have been carried out by Adleman in 1994 [7]. Since then, there has been a huge amount of interest in biological computing across disciplines. From a musical perspective, biological computing has some very attractive possibilities, as we shall demonstrate with examples below.

One early project that investigates the feasibility of employing a hybrid wetware-silicon device in computer music is Miranda et al.'s *"Sound synthesis with In Vitro Neuronal Networks"* [30]. In this project, the authors were interested in harnessing the spiking interactions between neurones to produce sound. Here, brain cells were acquired from a seven-day-old Hen embryo and cultured in vitro. Culturing brain cells in an in vitro environment encourages them to form synapses—a structure that allows a nerve cell to transmit a chemical or electrical signal to another cell. Once grown, they positioned the culture into a MEA (multi-electrode array) in such a way that at least two electrodes made a connection into the neuronal network. One electrode was then arbitrarily chosen as the input into the system while another, as the output. The input is then used to stimulate the network with electrical impulses while the output is used to monitor and record the subsequent spiking behaviour.

With the recorded behaviour, the team developed and experimented with a number of sonification methods using additive and granular synthesis techniques to convey the neuronal network's behaviour. In one of these experiments, nine oscillators made up the additive synthesis framework, with the first having its amplitude and frequency controlled directly by the recorded behaviour. The other eight oscillator's parameters are multiples of the first's parameters. Initially, the authors used the gathered behavioural data in its raw, uncompressed form, which produced excessively long sounds. To circumvent this, they implemented a data compression algorithm that retained the target behaviour while removing uneventful data.

The main aim of the project was to create a sound synthesis tool/instrument with a good level of control and repeatability. To achieve this, the authors considered a machine-learning algorithm aimed at controlling the spiking behaviour of the network. Currently, this part of the project is still in its infancy, but their initial results have shown that they can control spiking behaviour in about one-third of cases.

Another example of a genuine unconventional computing scheme being developed for music is Kirke et al.'s [22] quantum computing system, *Q-Muse*. Quantum computing takes advantage of quantum mechanics to perform computations; e.g., quantum superposition, entanglement and tunnelling. For instance, quantum superposition involves photons being in multiple physical states, e.g. being in multiple locations or representing multiple binary states simultaneously. This superposition ability is a key element behind the promised speed-up of quantum over classical computing.

29.3   Towards Musical Biocomputing: *Physarum Polycephalum*

For most, the computing schemes such as the aforementioned, may seem out of reach for the average computer musician. It is worth noting that each of the discussed uses of genuine unconventional computing schemes for music were developed in

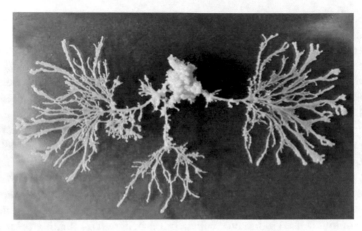

Fig. 29.2 *Physarum polycephalum* is an accessible biological substrate for research into UC (From George Barron's online collection of fungi images, with permission)

collaboration between computer music research departments and computer science departments. Such working partnerships, although fitting for the interdisciplinary nature of unconventional computing for music, can be limiting from the perspective of the computer musician. As such, we chose to experiment with the biological computing substrate *Physarum polycephalum* for our investigations into how unconventional modes of computing may be used for music. Uniquely, *Physarum polycephalum* (Fig. 29.2) requires comparatively fewer resources than most other unconventional computing substrates. The organism is cheap, openly obtainable, considered safe to use and has a robustness that allows for ease of application. It is for these reasons that we believe *Physarum polycephalum* to be an appropriate substrate for our research. Moreover, we believe that by selecting such an accessible computing medium, we can encourage computer musicians who are interested but hesitant to explore the potential of unconventional computing paradigms in their works, due to the difficulty of finding schemes to adapt to their needs.

Physarum polycephalum, henceforth known as *P. polycephalum*, is a plasmodial slime mould belonging to the order *Physarales*, subclass *Myxogastromycetidae*, class *Myxomycete*. *P. polycephalum* exhibits a complex lifecycle and is naturally found in cool, moist and dark environments. During its vegetative plasmodium phase, it exists as an amorphous single cell visible via the human eye, with a multitude of diploid nuclei. The plasmodium, although without a brain or any serving centre of control, can respond with natural parallelism to the environmental conditions that surround it (hence '*polycephalum*', which literally means 'many heads'). The slime mould propagates along gradients of stimuli while building a route-efficient network of protoplasmic veins connecting foraging efforts and areas of colonisation. It moves towards chemoattractants (a substance or agent that induces an organism to migrate towards it) and away from chemorepellents (a substance or agent that has the opposite effect of a chemoattractant). Upon discovery of food (e.g., source

Fig. 29.3 Photograph of a Petri dish containing plasmodium of *P. polycephalum*, showing: (A) the place where it has been inoculated, (B) protoplasmic network connecting areas of colonisation, (C) colonised region containing nutrients (in this case an oat flake), and (D) extending pseudopods forming a search front along a gradient towards another oat flake, marked by (E)

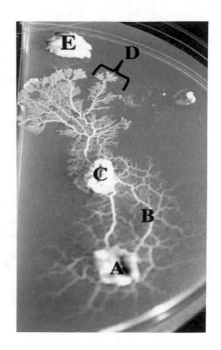

of nutrients such as oat flakes), the plasmodium surrounds it and feeds through the process of phagocytosis, ingesting nutrients that are distributed across the organism through the protoplasmic veins via shuttle streaming (Fig. 29.3). Conversely, the area is avoided if the slime mould discovers matter that does not entice its appetite. It also reacts to other phenomena, such as light, but these are beyond the scope of this chapter.

The plasmodium's topology is a network of biochemical oscillators: waves of contraction or relaxation, which collide inducing shuttle streaming. This intracellular activity produces fluctuating levels of electrical potential as pressure within the cell changes. Typically this is in the range of ±50 mV, displaying oscillations at periods of approximately 50–200 s with amplitudes of ±5–10 mV [24], dependent on the organism's physiological state and environmental conditions. Research has been put forward demonstrating that such patterns can be used to denote behaviour accurately [4]. This electrical behaviour is just one of the characteristics of *P. polycephalum* that renders it attractive for research into UC.

A natural characteristic of the slime mould is the time it can take to span an environment. Depending on how the experimental environment is set up, it can take several hours to exhibit substantial growth and exhaust available sources of nutrients. Much research is being conducted worldwide to utilise more instantaneous behavioural aspects of the organism; e.g., using intracellular activity as real-time logic gates [5]. In the meantime, one solution adopted by a number of research laboratories is to work with computer models of the slime mould. For instance, Jones [21] developed a model, which simulates fairly realistically the dynamics of the organism.

Another approach, which is the one we have adopted for some of the works presented in this chapter, is to record the behaviour of the organism and subsequently use the data off-line. This approach is more attractive because it enables us to experiment directly with the biological substrate. Moreover, as we will see later, speed to span an environment is not necessary an issue for the kinds of systems we are interested in building.

We have been conducting a number of experiments investigating ways to harness the behaviour of the plasmodium of *P. polycephalum* with a view on building audio and music systems, and ultimately a musical biocomputer. To begin our research, we conducted an initial proof of concept study with *P. polycephalum*. Here, we were not concerned with building computing systems; rather, we were interested to study the behaviour that renders the organism interesting to computer scientists is interesting and usable for computer musicians. In this study, a foraging environment was constructed with electrodes embedded in areas containing oat flakes. Then, as the organism navigated the foraging environment, we recorded electrical readings from each of the electrodes. Such readings were subsequently rendered as frequency and amplitude values for a bank of oscillators forming an additive synthesiser. From this project, we could confirm that the organism's behaviour is indeed interesting for computer music. For more details about this preliminary study, please refer to [31].

29.3.1 Culturing Methods and Materials

When conducting our research with the plasmodium of *P. polycephalum* at ICCMR BioMusic Labs, we maintain a farm that adopts techniques from [2]. Here, plasmodium is farmed in the dark at room temperature on a moist, porous substrate in plastic containers. The farm is fed daily with oat flakes, moistened every other day and replanted onto a new substrate weekly. To inoculate plasmodium into an experimental arena, colonised oat flakes or heaps of pseudopods are removed from the farm and positioned as desired.

In order to record the behaviour of the plasmodium, we use a combination of time-lapse imagery and/or electrical potential data by means of bare-wire electrodes. We introduce these methods where appropriate as we present our experiments and systems below; see [9] for more detail.

29.3.2 Step Sequencer

Step sequencers are devices that loop through a defined quantity of steps at given time intervals. Each of these steps can normally exist in one of two states: active or inactive. When active, a given sound event will be triggered as the sequencer reaches its respective position in the loop. No sound is produced when the reached position is inactive.

Fig. 29.4 Step sequencer
architecture

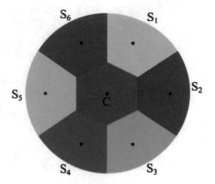

P. polycephalum's ability to solve mazes, find shortest paths and develop networks linking sources of nutrients informed the development of a step sequencer where the organism's behaviour controlled step activation and regulated sound event triggering. We theorised such a device when reviewing sets of time-lapse images with correlating electrical activities. We noticed how the organism oscillates protoplasm around a network of veins to colonised regions, and how this related to the architecture of a musical step sequencer.

29.3.2.1 Methods

In order to implement the step sequencer [12], we designed an environment that represents a step sequencer's architecture as schematically shown in Fig. 29.4. It consisted of a Petri dish divided into six electrode zones, representing sequencing steps ($S_1 \ldots S_6$), arranged in a circular fashion (representing the sequencer loop) with a central inoculation area (C).

To entice propagation and facilitate colonisation, each of the electrode zones and the inoculation area were furnished with oat flakes (Fig. 29.5, right-hand side). In order to record the electrical behaviour of the organism, we used two forms of hardware: a USB manual focus camera and an ADC-20 high-resolution data logger manufactured by Pico Technology, UK. Within each Petri dish, bare-wire electrodes were placed through small holes in the base with wiring underneath secured using adhesive tack (Fig. 29.5, left-hand side). Electrode arrays consisted of one reference electrode and six measurement electrodes, one for each step of the sequencer. The reference resided in the centre (i.e., in the inoculation area) and gives a ground potential for each of the measurement electrodes. Electrodes were coated with a 2 % non-nutrient agar, which kept humidity high for the plasmodium. Due to the agar substrate being liquid-based and thus a conductor, a non-conductive plastic wall isolates electrode zone's area, allowing the data logger to take readings without interference. By way of limiting the light intensity and promoting growth, a black enclosure encapsulated the experimental arrangement. White LEDs periodically illuminated the enclosure for image capture.

Fig. 29.5 Photographs showing the construction of the environment for the step sequencer. Shown on the *left hand side* is the bare-wire electrode array wired into position. The *right hand side* shows the completed growth environment with each electrode embedded within blocks of agar

To begin recording behaviour, we took pseudopods from the *P. polycephalum* farm and inoculated them into the experimental arena. The inoculation source was put through a period of starvation beforehand to speed up initial propagation speed.

We programmed the data logger to sample each electrode 100 times per second, and then averaged the readings to produce a single value for every second. This level of recording detail was necessary to capture the natural gradients exhibited with various progressions, some of which are fairly prompt. In tandem, we took image snapshots at intervals of 5 min with the LEDs turning on 5 s before and staying active for 10 s.

29.3.2.2   Results

The recording process took five days to complete, with us terminating the experiment as the plasmodium entered its dormant Sclerotium phase, which occurred as a result of a drop in humidity and lack of fresh food sources. Figure 29.6 shows line plots of the electrical activity on each measurement electrode.

It took the organism just under 12 h to propagate to a step. When the plasmodium arrives at an electrode, a fast change in voltage is registered, which in our results is an increase ranging from ±5–15 mV. Propagation to each step came equally from both the central node as well as neighbouring steps. Activities and conditions within colonised regions cause differing intracellular activity, which in turn can result in electrical impulses across the organism. Such impulses are typically registered in regions connected directly, but may spread a distance across the organism if conditions are favourable (electrodes 3, 4 and 5 in Fig. 29.6, marked by the rectangle). When arriving at two electrodes at the same time, electrical readings exhibit synchronised patterns. In our experiment, this is the case for electrodes 1 and 2, as shown by the triangle in Fig. 29.6. The Sclerotium phase is characterised by an increase in voltage, marked by a circle on electrode 1 in Fig. 29.6.

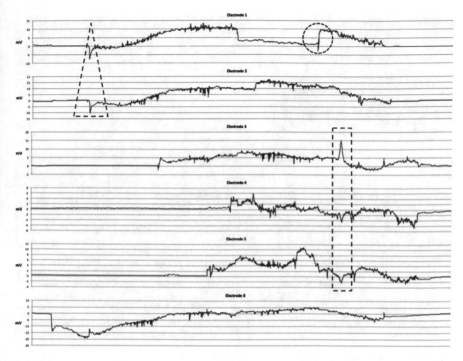

Fig. 29.6 Line plots depicting the electrical behaviour of the plasmodium within the experimental arrangement shown in Fig. 29.5

29.3.2.3 The Step Sequencer

We implemented our step sequencer application in Max, manufactured by Cycling74, USA, with some data handling operations being dealt with in Java. The application functions by recalling the recorded behaviour at a user-defined frequency into the device. Readings are adjusted to become absolute values. The device then steps through each electrode's measurement taking a sample at a user-defined speed in terms of beat per minutes. A step only becomes active within the sequence once it becomes colonised by the organism. Otherwise, no reading is taken. Once activated, the system looks for a level of change in electrical potential in order to retire steps from triggering sounds when the organism is no longer active. This is achieved by storing readings over a short period and reviewing any oscillatory behaviour. By way of forging a connection between the user of the device and the source of the recorded behaviour, the images are played back in the user interface with perfect synchrony to the electrical readings.

The user interface (Fig. 29.7) provides an interactive graph showing the combined electrical potential readings (on the top right hand side). This allows the user to change the current position of the data being recalled, creating means to restructure the output of the sequencer.

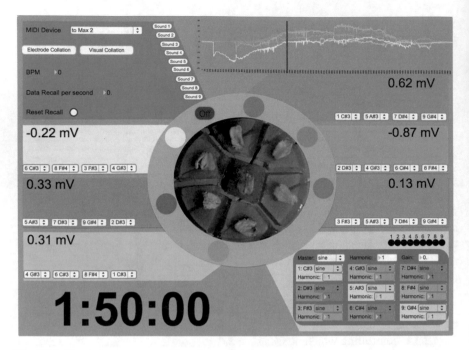

Fig. 29.7 The step sequencer user interface

We developed different versions of the sequencer to probe its usability in realistic musical production. One of these versions looked at harnessing the plasmodium's behaviour to extend the functionality of a conventional step sequencer by triggering different sounds as a function of each step's electrical potential readings. The system allows us to associate four sounds to each step of the sequencer. Readings are used to trigger one of the four sound samples; e.g., sound A would be triggered with voltage values between 0 and 15 mV, and so on.

Another version took the form of a MIDI instrument. Here, the recorded behaviour is put through a musification model where the user can change the parameters. In this version, readings taken by the sequencer are used to trigger a set of nine MIDI notes that are programmed in by the user. All steps are allocated a set of four notes from the nine available, which are then each assigned to a voltage trigger range. When a note is triggered, its velocity is produced through scaling the step's current electrical potential value to the MIDI data range. In order to determine the duration of a note, the system calculates the current mean potential of all other steps (active and non-active) to produce an average potential difference value. The higher this value, the more significant the note duration will be within the sequence, with a maximum duration being four beats. The sequencer is limited to only allow six notes to sound at a time; if a note is triggered but is unavailable due to being made active by another step, the note with the closest value in the step's priority list will sound. To make this version of the sequencer versatile, the user can alter each step's note priority order in real-time.

29.3.2.4 Discussions

The output of our device in both scenarios produced a variety of interesting arrangements, which can be used by musicians in several different ways. In the sample-triggering scenario, we found that by allocating sounds to a relative voltage range—for example, higher velocity sounds to elevating voltages—we can achieve a naturally progressive output. However, at certain points, the sample arrangement can become repetitive. From a musical perspective, we believe the MIDI scenario gave more useful and interesting results. This version of the device outputs a progressive arrangement of notes, which correspond morphologically to the recorded foraging behaviour. As a musification of behaviour, the output does convey an auditory representation of voltage levels quite accurately. This is because each note's velocity is produced directly by the respective step's voltage—a parameter that directly relates to energy. Moreover, by producing each note's duration with a potential difference value, it is possible to compare activity at each step through listening. From a musical perspective, having their velocity controlled this way is slightly dull due to voltage levels also controlling which notes are triggered: each note is played with a similar velocity every time. As a compositional tool, we found this version of the sequencer useful for arranging and applying interesting variations over time to a given set of notes.

We are currently making headway in advancing the notion of a *P. polycephalum* step sequencer by tracking the intracellular movement of the organism under a microscope in real-time. Here, we are harnessing research put forward by Cifarelli et al. [14], which demonstrates that it is possible to load the organism with micro-particles that are then distributed by shuttle streaming. We are exploring ways to use the movement of such micro-particles to trigger sounds as they pass through certain areas of the organism.

29.3.3 Granular Synthesis

Granular synthesis is the rapid succession of short sound events (typically 1–100 millisecond-long-sounds), referred to as grains that together form a larger sound object. This method of sound synthesis is inspired by Dennis Gabor's acoustical quanta [15], which proposed that complex sound events are made up of a myriad of simple sonic grains. There is a wide span of approaches to harnessing granular synthesis for music. Examples include Miranda's before mentioned application of a modelled reaction-diffusion computer to control a granular synthesiser. In this section, we report on an approach to granular synthesis harnessing the plasmodium of *P. polycephalum* [11].

   This project was an extension of our discussed preliminary work, where the organism's electrical readings were scaled to control the frequencies and amplitudes of a group of oscillators. The results of this preliminary project were interesting and created a reasonable auditory representation of the organism's behaviour. However, results were hampered by the quantity of data generated by recording the organism's electrical activity. To render such large data useable, compression stages were implemented to the detriment of the relationship between the sound and the behaviour of the plasmodium. As such, we developed a different approach to sound synthesis with *P. polycephalum*, where we take advantage of the organism's oscillatory behaviour by regarding it as a granular audio oscillator whose output can be controlled via various methods of interacting with the organism.

29.3.3.1   Methods

Due to *P. polycephalum* taking several days to span an environment, we designed custom software that implements our granular synthesis approach over the duration of the organism being active in accordance with composer-defined parameters. Here, we take 100 samples from each electrode every second, which are then averaged to produce a single reading per second. The software then transcribes these readings into audio buffers. At composer-defined intervals, each buffer is addressed to produce a sound grain, which are sequenced together in ascending order according to their running electrical potential average. Grain lengths are determined by scaling each electrode's potential difference value against the current average of all other electrodes, to a composer defined minimum and maximum grain length range. Each electrode's buffer is only addressed for grain creation if the organism is active in the respective area. Our software achieves this by reviewing each incoming measurement for oscillatory activity. Once initiated, the software automates the granular synthesis composition until all food sources have been exhausted and the plasmodium starts to fructify or progress into its dormant Sclerotium phase. Upon conditions being met, the system halts and renders the resulting audio file. Currently, we have designed the system to accommodate up to eight electrode inputs.

29.3.3.2   Results

We have experimented extensively with this approach to granular synthesis. Shown in Fig. 29.8 is an example of an environment that we created for the synthesiser. Here, four electrodes are arranged on the vertices of a square, with a reference electrode positioned in the centre. In the context of this project, this arrangement represents four grain oscillators.

   Shown in Fig. 29.9 is a cochleogram of the granular piece produced by our system from the setup in Fig. 29.8. In this example, the system generated 174 s worth of audio material from five days worth of plasmodium electrical activity. Notice the dark lines

Fig. 29.8 The slime environment for the granular synthesiser

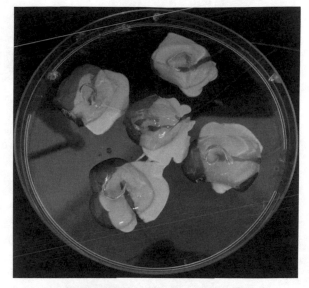

Fig. 29.9 Cochleogram of the result of our granular synthesis approach

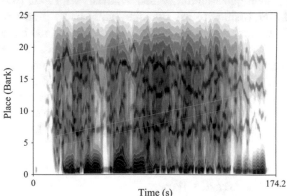

in Fig. 29.9. These relate morphologically to the electrical plots of the experiment shown in Fig. 29.10.

29.3.3.3 Discussions

Composing with granular synthesis can be an extensive and sonically detailed process. Compositions that conform to conventional musical theory have a temporal hierarchy of structure. Part of the compositional process is managing the interaction between structures on different time scales—from individual note level to the topmost level of a complete composition. When composing with granular synthesis there are additional levels that go below note level to grain level. At grain level, there

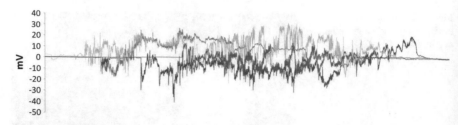

Fig. 29.10 Graphs depicting the electrical activity within the growth environment shown in Fig. 29.8. Each *line plot* corresponds to the activity recorded on one of the electrodes

is a massive quantity of control data required to advance to higher perspective sound levels. For example, if each grain has n quantity of parameters (these often exceed double digits), and there is q amount of grains in a second long sound object, then n multiplied by q equals the amount of data needed to produce a second of audio. As such, composers wishing to adopt granular synthesis in their works often need algorithms that produce grains in accordance with global parameters.

Our approach to granular synthesis with *P. polycephalum* is useful for the composer wishing to use granular synthesis as it automates the production of grains. Furthermore, the sonic result can be fine tuned by taking advantage of the organism's innate taxis response to various stimuli. Composers can also create different audio densities and output lengths by creating different electrode arrangements.

At this point of our research, we were able to confirm that *P. polycephalum* exhibits properties that can be harnessed to implement systems for generative audio and music. However, we were conscious that we needed to move on to develop and study systems beyond musification and sonification in order to progress with our research.

29.3.4 Audio Wires

Adamatzky investigated the possibility of using the protoplasmic tubes of *P. polycephalum* as self-programming conductive pathways for electronic devices [3]. His experiments demonstrated that the protoplasmic tubes remain functioning and conductive when applied with potentials in excess of 15 V. Furthermore, he showed that if a protoplasmic tube becomes broken, it repairs itself and becomes conductive once more. These results are exciting because they suggest that it might be possible to build audio signal wires and hardware with such protoplasmic tubes.

We are currently investigating how to use the protoplasmic tubes to conduct audio signals. For our investigations, we have been growing protoplasmic tubes at lengths of 10, 25, 50, 75 and 100 mm, and testing their ability to transmit an audio signal by comparing the amplitude, phase and frequency spectra of the audio before and after transmission.

Fig. 29.11 50 mm protoplasmic tube grown between two electrodes

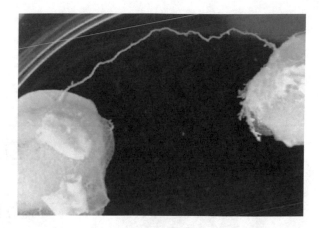

29.3.4.1 Methods

In order to grow the tubes, we position two electrodes spaced at a certain distance from each other. Electrodes are coated with a 2 % non-nutrient agar, with each of their terminals connected to a 1/4 in TRS socket. To grow the audio wire, we take a colonised oat flake from our farm and place it on one of the electrodes while placing a fresh oat flake on the other. This arrangement causes the plasmodium to propagate along a chemical gradient to the fresh oat, resulting in a protoplasmic tube linking the two electrodes (Fig. 29.11). The TRS sockets' ground terminals are connected using a short piece of wire.

29.3.4.2 Preliminary Results

Although we are currently in the midst of our experimentation, preliminary tests suggest that the notion of developing biological audio wires with *P. polycephalum* is viable, as shown by the amplitude statistics in Table 29.1 and frequency analysis in Fig. 29.12, which show no change in frequency spectra and a small reduction in amplitude. These experiments gave us good insights on how one might deploy the slime mould as an active component on electronic circuit boards. While we are currently conducting more experiments with audio wires, we also began to look into the possibility of developing analogue musical hardware that encompasses components grown from *P. polycephalum*.

Table 29.1 An overview of the amplitude statistics of an audio signal before and after transmission through a 100 mm protoplasmic tube

	Before transmission	After transmission
Peak amplitude	−4.80 dB	−5.94 dB
True peak amplitude	−4.78 dB	−5.92 dB
Average RMS amplitude	−27.52 dB	−28.66 dB
DC offset	−1.12 %	−0.98 %
Loudness	−23.58 dB	−36.69 dB
Perceived loudness	−16.00 dB	−17.15 dB

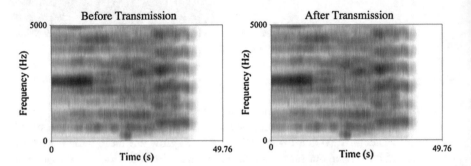

Fig. 29.12 A spectrogram of an audio signal before and after transmission through a 100 mm protoplasmic tube

29.4 Biocomputing for Music with *P. Polycephalum*

In this section, we introduce our current research into developing musical analogue hardware-wetware with *P. polycephalum*. First, we give an overview to underlying concepts behind our work, the first of which is the memristor.

The material base of computing architectures has revolved around the three fundamental passive circuit components: capacitor, inductor and resistor. In 1971, Chua [13] theorised a fourth fundamental component: the memristor, but it was not developed on any physical medium until 2008 [34]. This component is exciting due to its potential to revolutionise the material basis of computation. The memristor relates magnetic flux linkage and charge and, unlike the other three passive circuit components, is nonlinear. A memristor alters its resistance as a function of the previous charge that has flown through it. The current versus voltage characteristic of a memristor, when applied with an AC voltage, is a pinched hysteresis loop—a Lissajous figure formed by two perpendicular oscillations. Hysteresis is where the output of a system is dependent on both its current input and history of previous inputs. In an ideal memristor, this figure is observed as a figure of 8 where the centre intersection is at zero volts and current (Fig. 29.13). For an excellent introduction to memristors see [20].

Fig. 29.13 Example of hysteresis in an ideal memristor (arbitrary values used)

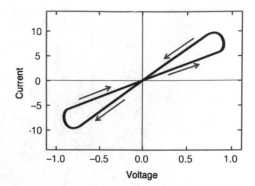

We believe that the memristor's nonlinear ability to alter its resistance as a function of both its current input, and history of previous inputs, holds potential for music generation. Unfortunately, until recently, we have been unable to explore such potential due to the component not yet being commercially available. There has, however, been one investigation using a simulation of a memristor network under a DC voltage [18] to generate music [17]. Other labs have also been restricted in their memristor interests due to this lack of accessibility. As a result, researchers have been looking past conventional electrical engineering approaches and have found that a selection of organic systems exhibit memristive characteristics. Examples include human blood [23], human skin [19] and Aloe Vera plants [37].

In 2009, Pershin et al. [32] published a paper that described the plasmodium's adaptive learning behaviour in terms of a memristive model. Shortly after, Gale et al. [16] demonstrated in laboratory experiments that the protoplasmic tube of *P. polycephalum* showed I–V profiles consistent with memristive systems. These discoveries have conveniently allowed us to coincide our musical research interests in memristors and unconventional computing with *P. polycephalum*. Moreover, it has gifted us with an early opportunity to investigate how some of the memristor's characteristics may provide new pathways for music before they are made widely accessible. At ICCMR's BioMusic Lab, we are following up from the research progresses to build analogue hardware-wetware systems that encompasses components grown from *P. polycephalum*. In the subsequent section, we give an overview of one such system we developed for the piece *BioComputer Music*, which premièred on the 1st of March 2015 at the Peninsula Arts Contemporary Music Festival, Plymouth University, UK.

29.4.1 BioComputer Music

In *BioComputer Music* [10], we experiment with the plasmodium's memristive characteristics: its nonlinear ability to alter its resistance as a function of the previous input current to generate musical accompaniments. As an overview, we have designed a

Fig. 29.14 The prototype biocomputer

system that comprises analogue hardware-wetware (Fig. 29.14) and a basic software communication platform.

Our system translates the performance of a pianist into MIDI note information. The result of such translation is scaled into voltage values to create a list, which the system uses to fabricate a discretized symmetrical AC voltage waveform with a step dwell time of 2-s. These voltage values are within the magnitude of ±250 mV, which we had chosen as in our experimentation we found that it produced the best results in terms of hysteresis and did not damage the organism. Once fabricated, the waveform is passed through a *P. polycephalum* protoplasmic tube . We then measure the instantaneous resistance at each voltage step and, using mapping/sonification techniques, transcribe the evolved output to musical notes, which are played through electromagnets arranged above the piano's strings (Fig. 29.15). A recording of a rehearsal of the *BioComputer Music* piece can be found at [25].

We observed that *P. polycephalum* hysteresis loops vary heavily in magnitude from organism to organism. This might not be desirable in specific electrical engineering situations, but it is ideal for generative music: it results in a slightly varied output each time the system is used. This is an interesting behaviour, which is often sought after in interactive music systems.

As a summary, the musical result of the *BioComputer Music* system in its current implementation is as follows. If a repetitive sequence of notes is input, then the fabricated waveform may become increasingly directional, resulting in decreased changes in resistance. Moreover, if only a single note is input, after the initial change in resistance, the organism over time will essentially become an ordinary resistor. Thus, the resistance measurement will remain the same, resulting in the same note playing with no variation. Conversely, under an extremely dynamic input, changes of resistance are likely to be higher, resulting in a wide span of notes being output.

Fig. 29.15 The biocomputer plays the piano through electromagnets placed on the strings

At this stage of implementation, due to current resource constraints, the *BioComputer Music* system only uses the *P. polycephalum* components to generate lists of notes. To apply rhythmic structure, we use a conventional second order Markov chain. To discern note durations, we use a Gaussian distribution, which operates within the performer-defined minimum and maximum range. We are currently looking into feasible methods of generating rhythm from *P. polycephalum* components and anticipate that we will have an autonomous *P. polycephalum* music generation system in the near future.

29.5 Concluding Remarks

This chapter introduced our research into UC and music, focusing our work with the slime mould *P. polycephalum*. Before we ventured ourselves into harnessing the organism to build a biocomputer, we developed a number of case studies to gain a better understanding of the organism. As part of these studies, we developed a number of musification and sonification methods, two of which were introduced in this paper: the step sequencer and the granular sound synthesiser. The outcomes from these studies were interesting in their own right, but our ultimate goal is not musification or sonification of organism. Rather, our aim is to build musical biocomputers and develop new approaches to music with them. Nevertheless, our studies enabled us to amass a significant practical knowledge about *P. polycephalum*. They paved the way towards the implementation of a prototype biocomputer with slime moulds acting as memristors. This has opened an exciting new ground for our research musically and scientifically.

To the best of our knowledge, *BioComputer Music* is the first musical composition that harnesses biological circuit components to generate music. In regards to our research into music with unconventional computing, the *BioComputer* system marks the beginning of a new avenue. Although the system behind the piece is our first implementation, the musical result is both engaging and interesting. Of course, there are several areas we need to develop. Foremost, we intend to review our method of transcribing notes into voltages. Here, we are researching and experimenting with the most meaningful way of taking advantage of the *P. polycephalum* component's nonlinear conductance profile. Also, we are working towards developing feasible methods of generating note duration and rhythmic structure with *P. polycephalum* components.

As we move on to work with living matter in computing technology, essentially we will be harnessing the intelligence of such organisms to compose music with. Undoubtedly, new forms of music making will emerge from Unconventional Computing. *BioComputer Music* is only a glimpse of what is to come.

References

1. Adamatzky, A.: Computing in Nonlinear Media and Automata Collectives. CRC Press, Boca Raton (2001)
2. Adamatzky, A.: Physarum Machines: Computers from Slime Mould, vol. 74. World Scientific, Singapore (2010)
3. Adamatzky, A.: Physarum wires: self-growing self-repairing smart wires made from slime mould. Biomed. Eng. Lett. **3**(4), 232–241 (2013)
4. Adamatzky, A., Jones, J.: On electrical correlates of Physarum polycephalum spatial activity: Can we see Physarum Machine in the dark? Biophys. Rev. Lett. **6**(01n02), 29–57 (2011)
5. Adamatzky, A., Schubert, T.: Slime mold microfluidic logical gates. Mater. Today **17**(2), 86–91 (2014)
6. Adamatzky, A., Teuscher, C.: From Utopian to Genuine Unconventional Computers. Luniver Press, Frome (2006)
7. Adleman, L.M.: Molecular computation of solutions to combinatorial problems. Science **266**(5187), 1021–1024 (1994)
8. Beyls, P.: Cellular automata mapping procedures. In: Proceedings of the ICMC. Citeseer (2004)
9. Braund, E., Miranda, E.: Music with Unconventional Computing: A System for Physarum Polycephalum Sound Synthesis. In: Aramaki, M., Derrien, O., Kronland-Martinet, R., Ystad, S.I (eds.) Sound, Music, and Motion. Lecture Notes in Computer Science, pp. 175–189. Springer International Publishing, Heidelberg (2014)
10. Braund, E., Miranda, E.: BioComputer music: generating musical responses with Physarum polycephalum-based memristors. In: Computer Music Multidisciplinary Research (CMMR): Music, Mind, and Embodiment. Plymouth, UK (2015)
11. Braund, E., Miranda, E.: Music with unconventional computing: granular synthesis with the biological computing substrate Physarum polycephalum. In: Computer Music Multidisciplinary Research (CMMR): Music, Mind, and Embodiment. Plymouth, UK (2015)
12. Braund, E., Miranda, E.: Music with unconventional computing: towards a step sequencer from plasmodium of physarum polycephalum. In: Johnson, C., Carballal, A., Correia, J. (eds.) Evolutionary and Biologically Inspired Music, Sound, Art and Design. Lecture Notes in Computer Science, vol. 9027, pp. 15–26. Springer International Publishing, Heidelberg (2015)
13. Chua, L.O.: Memristor-the missing circuit element. IEEE Trans. Circuit Theory **18**(5), 507–519 (1971)

14. Cifarelli, A., Dimonte, A., Berzina, T., Erokhin, V.: On the loading of slime mold Physarum polycephalum with microparticles for unconventional computing application. BioNanoScience 4(1), 92–96 (2014)
15. Gabor, D.: Acoustical quanta and the theory of hearing. Nature 159(4044), 591–594 (1947)
16. Gale, E., Adamatzky, A., de Lacy Costello, B.: Slime mould memristors. BioNanoScience 5, 1–8 (2013)
17. Gale, E., Matthews, O., de Costello, B.L., Adamatzky, A.: Beyond Markov Chains, Towards Adaptive Memristor Network-based Music Generation. arXiv preprint arXiv:1302.0785 (2013)
18. Gale, E., de Lacy Costello, B., Adamatzky, A.: Emergent spiking in non-ideal memristor networks. Microelectron. J. 45(11), 1401–1415 (2014)
19. Grimnes, S., Lütken, C.A., Martinsen, O.G.: Memristive properties of electro-osmosis in human sweat ducts. In: World Congress on Medical Physics and Biomedical Engineering, 7–12 Sept 2009, Munich, Germany, pp. 696–698. Springer, Heidelberg (2009)
20. Johnsen, G.K.: An introduction to the memristor-a valuable circuit element in bioelectricity and bioimpedance. J. Electr. Bioimpedance 3(1), 20–28 (2012)
21. Jones, J.: The emergence and dynamical evolution of complex transport networks from simple low-level behaviours. Ijuc 6(2), 125–144 (2010)
22. Kirke, A., Shadbolt, P., Neville, A., Antoine, A., Miranda, E.: Q-Muse: A quantum computer music system designed for a performance for orchestra, electronics and live internet-connected photonic quantum computer. In: Conference on Interdisciplinary Musicology (CIM). Berlin (2014)
23. Kosta, S.P., Kosta, Y.P., Bhatele, M., Dubey, Y.M., Gaur, A., Kosta, S., Gupta, J., Patel, A., Patel, B.: Human blood liquid memristor. Int. J. Med. Eng. Inform. 3(1), 16–29 (2011)
24. Meyer, R., Stockem, W.: Studies on microplasmodia of physarum polycephalum V: electrical activity of different types of microplasmodia and macroplasmodia. Cell Biol. Int. Rep. 3(4), 321–330 (1979)
25. Miranda, E.: Biocomputer music. http://tinyurl.com/kszgm3r
26. Miranda, E.R.: Cellular automata music: an interdisciplinary project. J. New Music Res. 22(1), 3–21 (1993)
27. Miranda, E.R.: Granular synthesis of sounds by means of a cellular automaton. Leonardo 28, 297–300 (1995)
28. Miranda, E.R.: Evolving cellular automata music: from sound synthesis to composition. In: Proceedings of 2001 Workshop on Artificial Life Models for Musical Applications (2001)
29. Miranda, E.R.: Computer Sound Design: Synthesis Techniques and Programming, vol. 1. Taylor & Francis, UK (2002)
30. Miranda, E.R., Bull, L., Gueguen, F., Uroukov, I.S.: Computer music meets unconventional computing: towards sound synthesis with in vitro neuronal networks. Comput. Music J. 33(1), 9–18 (2009)
31. Miranda, E.R., Adamatzky, A., Jones, J.: Sounds synthesis with slime mould of physarum polycephalum. J. Bionic Eng. 8(2), 107–113 (2011)
32. Pershin, Y.V., La Fontaine, S., Di Ventra, M.: Memristive model of amoeba learning. Phys. Rev. E 80(2), 21,926 (2009)
33. Stepney, S.: Programming unconventional computers: dynamics, development, self-reference. Entropy 14(10), 1939–1952 (2012)
34. Strukov, D.B., Snider, G.S., Stewart, D.R., Williams, R.S.: The missing memristor found. Nature 453(7191), 80–83 (2008)
35. Turing, A.M.: On computable numbers, with an application to the Entscheidungsproblem. J. Math. 58(345–363), 5 (1936)
36. von Neumann, J.: First Draft of a Report on the EDVAC (1945)
37. Volkov, A., Reedus, J., Mitchell, C.M., Tucket, C., Forde-Tuckett, V., Volkova, M.I., Markin, V.S., Chua, L.: Memristors in the electrical network of Aloe vera L. Plant Signal. Behav. 9(4), e29,056 (2014)
38. Xenakis, I.: Formalized Music: Thought and Mathematics in Composition, vol. 6. Pendragon, United States (1992)

Chapter 30
Immunocomputing and Baltic Indicator of Global Warming

Alexander O. Tarakanov and Alla V. Borisova

Abstract Unconventional computing of sea surface temperature (SST) was once featured by NASA as a *unique merger of science and art*. Our approach led to a discovery that just one geographical point could be sufficient to track global anomalies of SST based on El Niño Southern Oscillation (ENSO). Such single point in the Pacific Ocean off of the island of Isabella in the Galapagos Islands was named the *Galapagos indicator*. Now we show that a single point in the Baltic Sea off of the coast of Göteborg could be also sufficient to track ENSO. We propose to name it the *Baltic indicator*. We also demonstrate that two crisis falls of oil price in 2008 and 2014 followed just after the local maximums of Baltic indicator. However, Baltic and Galapagos indicators do not confirm any settled global warming from the beginning of this century.

30.1 Introduction

Immunocomputing (IC) is based on the principles (especially—mathematical models) of information processing by proteins and immune networks and it has been generalized as a new way to intelligent signal processing [8]. Coupling with modern brain research [1], our approach has led to amazing discovery of deep biomolecular similarities between marine sponges, immune system and human brain [11]. On the other hand, the forecast of hydrophysical fields [12, 13] *might not seem to harbor an artistic statement, but a novel application of data simulation has led to a unique merger of science and art* [7]. Namely, rather simple mathematical model of global dynamics of sea surface temperature (SST) has been developed and implemented in so called SST-simulator [14]. This software has been applied to track global anomalies of SST [6].

A.O. Tarakanov (✉) · A.V. Borisova
St. Petersburg Institute for Informatics and Automation,
Russian Academy of Sciences, Saint Petersburg, Russia
e-mail: tar@iias.spb.su

© Springer International Publishing Switzerland 2017   763
A. Adamatzky (ed.), *Advances in Unconventional Computing*,
Emergence, Complexity and Computation 23,
DOI 10.1007/978-3-319-33921-4_30

Global anomalies of SST associated with El Niño Southern Oscillation (ENSO) have been known to disturb weather, environment, economy, and human lives worldwide. Our approach has led to a discovery that just one geographical point which we have named the *Galapagos indicator* could be sufficient to track ENSO [14]. Now we show that a single point in the Baltic Sea could be also sufficient to track ENSO. We propose to name it the *Baltic indicator*. We also demonstrate that two crisis falls of oil price in 2008 and 2014 followed just after the local maximums of both Baltic and Galapagos indicators. However, these indicators do not confirm any settled global warming during 2000–2014.

30.2 Data

The El Niño can be defined as a severe warm anomaly of SST of the Peruvian coastal upwelling system [14]. The ENSO events show that SST anomalies in this local area can be important for global environmental changes such as the hurricanes, floods, droughts and so on all over the world and may be even the global warming.

The Southern Oscillation Index (SOI) is usually used to track ENSO as a measure of the difference in sea level air pressure between Tahiti and Darwin, Australia [4]. The most common index of ENSO is also average SST over the large area of Pacific Ocean (10°S–0°S, 90°W–80°W) named Niño 1+2 [4].

Global SST data can be obtained from NASA Giovanni system for sea surface areas all over the world for years beginning from 2000. This study has been utilized data from the Moderate Resolution Imaging Spectroradiometer (MODIS), which is generated by the Ocean Biology Processing Group (OBPG). The MODIS SST data have been acquired from the Goddard Earth Sciences Data and Information Services Center (GES DISC) Giovanni system [3].

Europe Brent oil price has been obtained from US EIA [5].

30.3 Methods

Our approach has been implemented in so called SST-simulator intended for the modeling of the daily evolution of SST based on the monthly SST data of NASA [14]. This simulator in general is a numerical model which includes temporal cellular automa (CA) and spatial interpolation. The states of CA are discrete values of SST (0.1 °C digitization). The cells of CA are points of a spatial grid (from 0.25° to 1° by latitude and longitude depending on size of an area). The values of SST over the grid are computed by CA for every time step (1 day) and then interpolated by splines for any point of the area.

Our SST-simulator now covers 18 different areas (Table 30.1), including three large equatorial oceanic areas (Pacific ENSO, Atlantic and Indian), four large non-equatorial oceanic areas (North Pacific, Drake Passage, Bermuda Triangle and

South-East Australia), seven seas (Baltic, Black, Caspian, Barents, North, Norwegian and Mediterranean) and four other areas of special interest (Gulf of Mexico, Hawaii, Crete and South West Europe together with North West Africa).

This SST-simulator provides fast computing of SST value in any point of any of the above area for any date from 1 January 2000. If given date is within 2000–2014 then SST is computed using monthly data of NASA for given year. These data are stored for any month as a matrix of monthly SST over a fixed spatial grid of area by Latitude and Longitude. Such matrix is considered as the values of SST on 15th day of the corresponding month. The corresponding matrices for other days of any month are computed using linear interpolation by days between two nearest months. Then SST in any point of the area is computed using spline interpolation between nearest points of the spatial grid. The animated results in colors can be viewed in the corresponding YouTube videos for all these areas (see Table 30.1).

Table 30.1 Baltic, Galapagos and most anomalous indicators of global SST (Sea Surface Temperature)

Geo point	SD[a]	Indicator	Area	Simulation of daily SST (YouTube clip)
57.75N, 11.5E	1.4	Baltic	Göteborg Sweden	http://youtu.be/JIng8MAXTsQ
58N, 9E	1.9	MAI[b]	Baltic Sea	http://youtu.be/JIng8MAXTsQ
1S, 91W	1.2	Galapagos	Isabella Island	http://youtu.be/jNRmuIykFzo
3S, 84W	1.4	MAI	Equatorial Pacific	http://youtu.be/jNRmuIykFzo
41N, 172E	1.8	MAI	North Pacific	http://youtu.be/ybdXV85k-hs
44.3N, 29E	1.8	MAI	Black Sea	http://youtu.be/UFe5M44vk5U
29N, 92W	1.4	MAI	Gulf of Mexico	http://youtu.be/PSW3XiljffE
55.5S, 67W	1.4	MAI	Drake Passage	http://youtu.be/f6zXQEWo1Js
39N, 50.2E	1.3	MAI	Caspian Sea	http://youtu.be/yzyzreBg09Y
74.5N, 16E	1.3	MAI	Barents Sea	http://youtu.be/sXAtZwDqY9Q
5N, 2E	1.3	MAI	Equatorial Atlantic	http://youtu.be/_E3d-YFk6ek
36.5N, 8W	1.2	MAI	South West Europe	http://youtu.be/oMDF-0KQc4o
56N, 8E	1.2	MAI	UK and Ireland	http://youtu.be/rmmgnUvkyzA
69.5N, 2W	1.1	MAI	Norwegian Sea	http://youtu.be/1m0_OhE4QeI
33.5N, 76.5W	1.1	MAI	Bermuda Triangle	http://youtu.be/eP8WpF6nvAg
34.5N, 12E	1.0	MAI	Mediterranean Sea	http://youtu.be/M8OpKEXsVP8
18S, 40E	0.9	MAI	Indian Ocean	http://youtu.be/jYYQ7iKgEFA
35.5S, 154E	0.9	MAI	South East Australia	http://youtu.be/NzrNaSUakIc
19N, 155W	0.9	MAI	Hawaii	http://youtu.be/E2et_vmC0UY
34.1N, 26.7E	0.8	MAI	Crete	http://youtu.be/TqC008dgFlw

[a] Standard Deviation of monthly SST anomaly (Celsius degree)
[b] Most Anomalous Indicator over the region

If given date is later than 31 December 2014 then SST is forecasted just as a mean value of SST over its 15 known values on this date for the years 2000–2014. For example, Fig. 30.1 demonstrates such forecast for the Baltic Sea on 22 December 2015.

SST-simulator can also compute SST anomaly on any date in any point of any of the 18 areas. Therefore, monthly anomalies can be computed in all points of the grid and compared by given criteria. For example, Galapagos indicator is such a point where monthly anomaly of SST is maximal (+2.6 °C) over all points of ENSO area in December 2002 when the El Niño event occured [14]. Another example is so called most anomalous indicator (MAI) as a point where standard deviation of mothly anomaly during 2000–2014 is maximal over all points of given area [14]. Such singular point of MAI can be determined for any area of the world ocean (sea, lake) and thus all these areas can be compared and tracked just by their singular points.

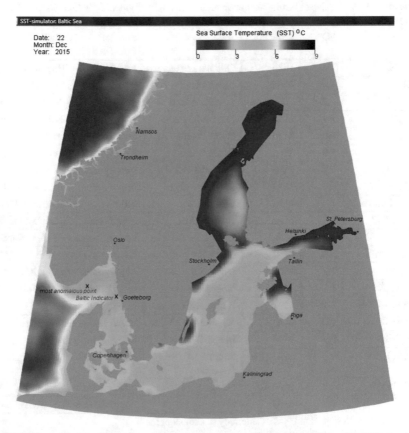

Fig. 30.1 Forecast of the Baltic Sea surface temperature on 22 December 2015 using SST-simulator

30.4   Results

Table 30.1 contains 20 singular points of 18 areas all over the world ocean, including two points of special interest (Baltic and Galapagos indicators) and 18 most anomalous points of the areas. Two geographical coordinates of every point and the standard deviation (SD) of SST anomaly have been computed using methods of Sect. 30.3 and data of Sect. 30.2 for 15 years 2000–2014.

Note that Galapagos indicator (1°S, 91°W) is located rather far from the point of MAI (3°S, 84°W) of the same Equatorial Pacific area of ENSO, while 180 monthly SST anomalies (time series over the years 2000–2014) in these points are rather close: SD is 1.2 °C versus 1.4 °C (see Table 30.1) and correlation coefficient is rather high (0.86, not shown in Table 30.1).

The proposing Baltic indicator (57.75°N, 11.5°E) is located not so far from the corresponding point of MAI (58°N, 9°E) of the Baltic Sea while SST anomalies in

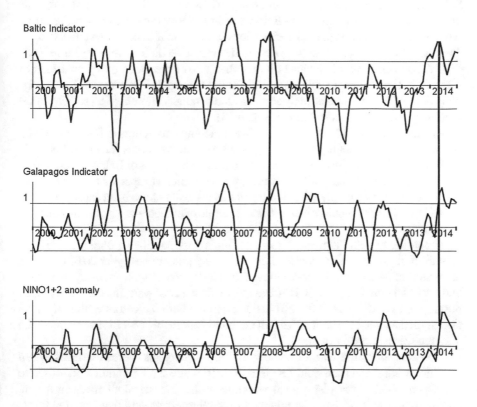

Fig. 30.2 Three-months running mean of monthly SST anomalies in Baltic and Galapagos indicators of El Niño over Niño 1+2 regional index anomaly. Note that two local maximums of Galapagos indicator and Niño 1+2 in 2008 and 2014 follow just after the corresponding local maximums of Baltic indicator (*bold vertical lines*)

these points are not so close by their SD (1.4 vs. 1.9 in Table 30.1) but correlation is also high (0.84, not shown in Table 30.1).

Figure 30.2 shows the SST anomalies in the points of Baltic and Galapagos indicators together with Niño 1+2 regional index anomaly for years 2000–2014 (three-months running mean of 180 monthly SST anomalies).

Although these time series look rather different, they have three clear coincided maximums in 2006, 2008 and 2014 when both Baltic and Galapagos indicators show montly anomaly of SST greater than +1°C. Note that two local maximums of Galapagos indicator and Niño 1+2 in 2008 and 2014 follow just after the corresponding local maximums of Baltic indicator (see bold vertical lines in Fig. 30.2).

30.5 Discussion

Baltic indicator is located off of the coast of Göteborg, Sweden not so far from the most anomalous point of the Baltic Sea in the Skagerrak strait (see Fig. 30.1). Amazingly, this point is the most anomalous not only in the Baltic Sea but also in all 18 areas around the world (note its maximal standard deviation 1.9 °C in Table 30.1).

Galapagos indicator is located on the other side of the globe off of the island of Isabella in the Galapagos Islands in the Pacific Ocean. It is not surprising that the behaviour of SST anomalies in these points looks rather different (e.g., see Fig. 30.3) and independent (e.g., correlation is just 0.13).

Nevertheless, Fig. 30.3 shows three clear matching maximums of Baltic and Galapagos indicators in 2006, 2008 and 2014. Moreover, two crisis falls of oil price in 2008 and 2014 follow just after the corresponding maximums of Baltic indicator (see bold vertical lines in Fig. 30.3). Of course, we don't intend to speculate that such falls of Brent crude price or any other factors in this price may have been caused or influenced by El Niño events. We just indicate a couple of evident correspondence in their extremums.

According to both Baltic and Galapagos indicators, so called *global warming* can be indicated just from the year 2013 up to the beginning of next year 2014 when the anomalies of SST in the opposite sides of the globe showed their sharp, synchronous and extremal raise in Fig. 30.3. However, no evident global warming can be indicated during the past 15 years 2000–2014. For example, both indicators in this Fig. 30.3 clear indicate three cold anomalies (where mothly anomaly of SST is less than −1°C) after the warm anomaly in 2008.

Moreover, after the warm anomaly in 2008, Galapagos indicator shows warm El Niño in 2009 and then cold La Niña in 2010 while the SST indicators of Crete and Cyprus both show just a local warming in the Eastern Mediterranean during 2009–2010 (Fig. 30.4). Surprisingly, before the Syrian uprising that began in 2011, the greater Fertile Crescent experienced the most severe drought in the instrumental record [2].

As for unconventional computing, it is worth highlighting that SST-simulator provides fast computing of SST on any date of 2000–2014 in any point of 18 large

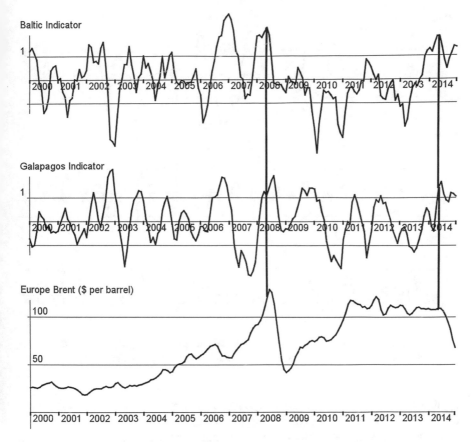

Fig. 30.3 Baltic and Galapagos indicators of El Niño over oil price. Note that two crisis falls of the price in 2008 and 2014 follow just after the corresponding local maximums of Baltic indicator (*bold vertical lines*)

areas all over the globe. Mean value of such SST in a fixed point and on a fixed date of all 15 years can be considered as a reasonable forecast of SST in this point and on this date of 2015 (e.g., see Fig. 30.1). Such *forecast by mean* then can be made more exact by the following way.

Based on the mean values, SST-simulator also provides fast computing of daily anomalies of SST and their visualizing similar to daily SST in Fig. 30.1. Specifically, SST-simulator computes monthly anomalies as time series (*signal*) in given point. For example, Fig. 30.3 shows such time series in two points of Baltic and Galapagos indicators during 2000–2014. Therefore, the forecast by mean can be made more exact by the forecasting of the anomalies of SST.

Mathematically, the task of forecasting of such anomalies is similar to the forecast of any time series or signal and its solution can be based on immunocomputing

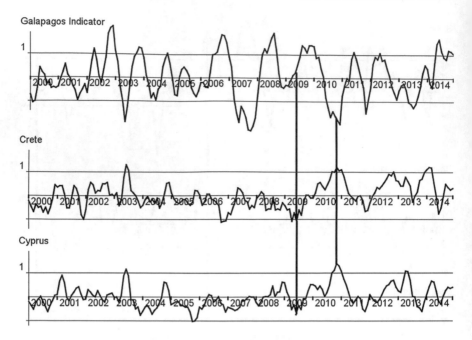

Fig. 30.4 Galapagos indicator shows El Niño–La Niña events in 2009–2010 (*vertical lines*) when both Crete and Cyprus indicators show a local warming in the Eastern Mediterranean

approach [8]. This approach can be applied at least by two different ways that can be defined as *forecast by automata* and *forecast by recognition.*

Forecast by automata consists in representing the signal as a type of cellular automata and its identification by immunocomputing [12]. If such identification exists, it gives a rigorous mathematical model of the signal which future state can be predicted unambiguously by its previous states. An example of such forecast of monthly index of solar activity (sun spots number) can be found in [9].

If such identification does not exist for a reasonable class of cellular automata or the obtained automat is too primitive for the actual signal then the forecast can be provided without cellular automata using general immucomputing approach to intelligent signal processing [8]. This forecast by recognition includes supervised learning of the obtained temporal anomalies of SST in one or several spatial points and recognition of future anomalies in given point by a similar spatio-temporal situation in the past. An example of such spatio-temporal forecast of SST anomalies of the Caspian Sea, Black Sea and Barents Sea can be found in [10].

Of course, we treat our forecast reasonable for one year only and such year (e.g. 2015) should be the next to the data of NASA (e.g. 2000–2014). So, we don't consider an overyear prediction (e.g. 2016) without the previous year data (e.g. 2015) and without a comparison of our forecast with these data.

30.6 Conclusion

The SST anomalies of Baltic indicator correspond fine to those of Galapagos indicator and Baltic indicator does not confirm any settled global warming from the beginnig of this century.

We can also speculate that anomality of SST in both Baltic and Galapagos indicators may be caused by strong currents in both areas. Such currents may arise where the Baltic Sea meets the North Sea in the area of Baltic indicator. Strong currents may also arise in the area of Galapagos indicator where cold Peruvian coastal upwelling system meets with warm waters from Mexican coast and Equatorial Pacific.

References

1. Goncharova, L.B., Tarakanov, A.O.: Molecular networks of brain and immunity. Brain Res. Rev. **55**, 155–166 (2007)
2. Kelley, C.P., Mahtadi, S., Cane, M.A., Seager, R., Kushnir, Y.: Climate change in the Fertile Crescent and implications of the recent Syrian drought. Proc. Nat. Acad. Sci. U. S. A. **112**, 3241–3246 (2015)
3. NASA Giovanni-Interactive Visualization and Analysis (2015). http://disc.sci.gsfc.nasa.gov/giovanni
4. Monthly atmospheric and SST indices. In: NOAA National Weather Service. Climate Prediction Center. http://www.cpc.ncep.noaa.gov/data/indices/
5. Petroleum and other liquids. In: US Energy Information Administration (2015). http://www.eia.gov/petroleum/
6. Research Highlight: Galapagos indicator of El Niño using monthly SST from NASA Giovanni system. In: NASA GES DISC News (Nov 22, 2013). http://disc.sci.gsfc.nasa.gov/gesNews/giovanni_news_oct_nov_2013_online
7. Russian scientist creates simulation of daily sea surface temperatures in the Caspian Sea. In: NASA GES DISC News (Jan 14, 2011). http://disc.sci.gsfc.nasa.gov/giovanni/gesNews/2011
8. Tarakanov, A.: Immunocomputing for intelligent intrusion detection. IEEE Comput. Intell. M. **3**, 22–30 (2008)
9. Tarakanov, A.: Immunocomputing for geoinformation fusion and forecast. In: Popovich, V., Schrenk, M., Claramunt, C., Korolenko, K. (eds.) Lecture Notes in Geoinformation and Cartography, pp. 125–134. Springer, Berlin (2009)
10. Tarakanov, A.: Immunocomputing for spatio-temporal forecast. In: Mo, H. (ed.) Handbook of Research on Artificial Immune Systems and Natural Computing: Applying Complex Adaptive Technologies, pp. 241–261. IGI Global, Hershey, PA (2009)
11. Tarakanov, A.O., Fuxe, K.G.: Integrin triplets of marine sponges in the murine and human MHCI-CD8 interface and in the interface of human neural receptor heteromers and subunits. SpringerPlus **2**, 128 (2013)
12. Tarakanov, A., Prokaev, A.: Identification of cellular automata by immunocomputing. J. Cell. Autom. **2**, 39–45 (2007)
13. Tarakanov, A., Prokaev, A., Varnavskikh, E.: Immunocomputing of hydroacoustic fields. Int. J. Unconv. Comput. **3**, 123–133 (2007)
14. Tarakanov, A.O., Borisova, A.V.: Galapagos indicator of El Niño using monthly SST from NASA Giovanni system. Environ. Model. Softw. **50**, 12–15 (2013)

Chapter 31
Experimental Architecture
and Unconventional Computing

Rachel Armstrong

Abstract This chapter examines aspects of unconventional computing from a design perspective through the practice of architecture. It reflects on how non-scientific forms of investigation may help develop cultural and economic frameworks for design thinking and scientific innovation, by building public and commercial interest in the field.

31.1 Introduction

This chapter takes a first principles view of the discipline of architecture, which is engaged with the choreography of space. In other words, architecture explores how matter and information are channeled and organized, within a site. It engages with form, functionality and transformation. It is also concerned with how these designed spaces are inhabited.

Conventionally, this practice results in the production of a building around which an existing industry of construction, dwelling and property ownership exists. Yet, such a manifestation is only a subset of the study of architecture, which also influences the production of space. For example, architecture is also concerned with material, economic, political, environmental and cultural parameters that shape our encounters with, and inhabitation of, a site.

Yet, these classical principles of understanding and shaping space are being questioned. Our cities produce 40 % of our carbon footprint—which is greater than that of transport (Green Building Council, Not dated)—and therefore how we shape our living spaces is contributing to our environmental impact. Small changes in how we produce the built environment can therefore have potentially significant impacts on our biospherical relationship. While current industrial approaches seek to increase efficiencies in current methods of production, these incremental changes are not

R. Armstrong (✉)
School of Architecture, Planning and Landscape, Newcastle University,
Newcastle upon Tyne, England
e-mail: Rachel.armstrong3@ncl.ac.uk

© Springer International Publishing Switzerland 2017
A. Adamatzky (ed.), *Advances in Unconventional Computing*,
Emergence, Complexity and Computation 23,
DOI 10.1007/978-3-319-33921-4_31

going to give us a reversal of the paradigm that underpins our civilizations—whereby making a building has a negative impact on our environment.

But how do we completely turn around the impact of our developmental paradigm on the world around us so that the way we make things and how we inhabit spaces can actually increase the liveliness of the planet—rather than reduce it?

While no formal method for achieving this has been established, architects are exploring the possibilities of an expanded portfolio of tools that may enable our buildings to perform more compatibly with the natural world and even enhance the ecosystems in which our homes exist. Rather than 'mimicking' life's processes, it may be possible to directly embody them and use the creativity of the material realm in surprising new ways.

In the last 30 years developments in biotechnology have enabled us to work directly with the dynamic processes of natural systems by harnessing the properties of lively things, which exceed the performance of a natural biological portfolio. This extends into the realm of self-organising chemistries. Our rapidly increasing discoveries in these fields are providing us with toolset that enable designers to directly work with these systems as a technical platform [29, 30]. While these systems are not formalised they inspire designers to imagine how they may shape space in new ways and so, give rise to alternative aesthetics, politics, economies, material transformations and social relationships that possess a different quality of purchase on the world around us.

Although a mature operating system that achieves this is not already available—an overlapping set of scientific practices such as unconventional computing, are generating the conditions in which architects and designers no longer represent natural systems but may discover ways of shaping space by coordinating the functions of some of nature's material processes.

31.2 Experimental Architecture

My practice is situated within a field of design investigation called experimental architecture. Peter Cook introduced the term into the design lexicon in 1970 in a book by a title of that name [16], which critiqued the avant-garde of modernism as a way of opening up more experimental forms of architecture that resisted an increasing functional approach to the development of cities. This was perceived to taking place at the expense of creativity and community, leading to uninspired living spaces and bland metropolitan expanses. While Cook's critique was theoretical, Lebbeus Woods championed the idea of experimental architecture as a practical form of research through drawing. He challenged established disciplinary tropes and raised new issues for architectural consideration that were beyond the reach of established knowledge canons such as, massive scale events like, war and natural disaster [55].

My own contribution to the field of experimental architecture is to move this research field into a laboratory space, where models, prototypes and installations can be realized. In this context, experimental architecture becomes a design practice that tackles the choreography of space by producing a set of design tactics and spatial programs that are realized and explored through a system of prototypes. My research specifically embodies an ecological approach to architectural construction that is predicated on shifting away from the traditional view of architecture as a static, form-giving subject, to develop other methods of producing architecture [7]. In this context, architecture is not simply about making a building, but in developing strategic approaches to organizing matter in time and space—for which computation is a necessary practice.

31.3 Multidisciplinary Method

Experimental architecture offers a multi disciplinary platform for the investigation of how emerging technologies may influence design practice. Since its boundaries are expansive, rather than reductive, it fundamentally aims to create partnerships between practices and knowledge sets by sharing the risk of experimentation between stakeholders. Experimental architecture also promotes creativity by identifying common themes of interest and bringing designers together with cutting edge scientific research fields and model systems. Collaborations include an integration of dynamic chemistries with cybernetics [8], architectural applications of dynamic droplets [29, 30] and 3D printing with lifelike chemistries [7]. These explorations developed a founding set of concepts for further technical development of the emerging technical systems towards wider applications, exemplified in Cronin's 'chemputer' that proposes to 'print any drug' [4]. Experimental architecture explores such possibilities in social and commercial contexts, which has resulted in collaborations with Autodesk (agent modeling of dynamic droplets) [37] and Unilever (Living Kitchen) [56]. The value of these explorations is in developing meaningful scenarios in which society, commerce and research are collectively invested in breaking new ground through emerging technical platforms.

Experimental architecture therefore becomes a practical engagement and research conversation between collaborators. Yet, it does not seek to build hierarchies of truth, or generate power differentials. Rather, it takes a constructivist approach that is consistent with Isabelle Stengers' notion of building 'an ecology of practices' where different knowledge sets, or participants, create their own new understanding of a subject by working alongside each other. Rather than proposing endless critiques of the various methods used by the contributing disciplines, Stengers looks to the power of synthesis between participants to produce ideas that can reinform us about reality by reinforcing the commonalities between disciplines. Potentially, by coming into contact with new views, collaborations may challenge fundamental assumptions or viewpoints [43]. These new perspectives, in turn, feed back into methods of production across the contributory disciplines, to further refine professional and commercial practices [39].

The value of multidisciplinary collaborations in producing novelty has been recognized in an National Science Foundation (NSF) report [40], p. 9, which proposed that these approaches have the potential to bring significant economic and human benefits, particularly through the emergence of new, combined technologies. This so-called NBIC (Nano Bio Info Cogno) convergence has led to funded 'sandpits' both by the NSF and European Union (EU), where multidisciplinary practitioners, predominantly scientists, collaboratively addressed 'grand' challenges such as artificial photosynthesis, and has resulted in projects such as the cyberplasm robot.[1] Convergence between the Two Cultures is also gathering support as a method of innovation from central funding sources, and currently the UK government is supporting STEAM (Science, Technology, Engineering, Arts and Mathematics) in which the contributory role of arts in innovation, is supported and recognized [23].

31.4   Design-Led Prototyping

Finding ways to engage the imagination to extend possibilities and take risks beyond obvious, or likely outcomes that characterize a conventional scientific hypothesis, are key to creative practices. Design-led experiment generally differs from scientific experiment in that it aims to produce better questions, rather than provide definitive answers to a challenge. Design-led experiments encourage blue sky (that pushes back the limits of possibility)—and even 'black' sky (which deals with unknowns), thinking [24]. As such, frequent use is made of propositions that are conceptually plausible, yet experimentally unproven. Taking a propositional approach to working with emergent systems helps keep solution spaces open for exploration and therefore do not already suppose outcomes. This enables access to a broad range of possible applications that may not have been entirely predicted from the outset. A range of design-led methods are used in experimental architecture projects including, Design Fiction [44], Science Fiction [25], 'fictionalism'[2] and 'post-normal' scientific research [26]. The 'prototypes' that arise out of these practices are therefore not the first iteration of a final product but a manifestation of a way of thinking. Yet, in experimental architecture, these prototypes do not stand alone as imaginary constructs but are reflected back on research findings and design explorations to reinform new possibilities. Within this probabilistic terrain, technical and cultural issues are therefore identified that are underpinned by real questions raised by emerging technical systems. These are further evaluated through the production of drawings, models, prototypes and installations.

Indeed, architectural research widely employs experimental approaches and has been used in the European Commission-funded VISIONS project [26], in architectural prototyping [20] and also Bio Design [36]. Moreover, the National Endowment for Science, Technology and the Arts (NESTA) proposes that there is a link between

---

[1]The 'cyberplasm' robot is an example of the kind of projects that have arisen from the NBIC 'sandpits', which is a melange of biological and mechanical systems [19].

[2]Fictionalism is the philosophical view that a serious intellectual inquiry need not aim at truth [32].

science fiction and innovation [48], where speculative approaches build mutual relationships between scientific and design practices. Yet the aim is not to predict the future, but to increase the probability that desired outcomes will come true, since storytelling and working through how alternative systems may work, prepares societies for change. Even with no ready means of testing the proposals, speculative explorations can be used as comparative models—or to explore incomplete knowledge sets. This allows increasingly more reasoned trajectories to be developed, which may reach a point where they are experimentally testable and gain the status of scientific hypotheses.

31.5 Sustainability

Explorations in experimental architecture have been most marked in the field of sustainability. The current definition proposes a form of human development that can meet the needs of today's populations without compromising those of future generations [53]. Yet, sustainable agendas have been dominated by industrial paradigms that privilege solutions that reduce energy and resource consumption. However, when thinking about the environment in a broader context, an ecologically engaged culture must also nurture other possibilities for prosperity beyond the possibilities available in current modern industrial approaches. Most particularly, if we are to reverse the developmental paradigm on which our own survival depends then the production of a building must benefit, not damage our natural world. Experimental architecture therefore seeks to explore how new technologies may establish alternative approaches towards making our living spaces and underpin the construction of cities that promote, rather than diminish, the liveliness of our planet.

One way of potentially achieving this is to work alongside—rather than against—natural imperatives. Biomimicry [12] proposes that Nature has all the solutions we need for sustainable development. However, despite much investment in organic-looking buildings, contemporary architecture applies these potential solutions using the same principles and manufacturing platforms that gave rise to the Industrial Age. Inevitably, these have similar environmental impacts. While industrial technologies *per se* are not 'bad', having no realistic manufacturing alternatives to working with digitized machines raises significant issues from the global scale and intensity at which they are applied. Even more local and distributed technical platforms such as, 3D printing techniques do not address the consumptive paradigm, or materiality at the heart of manufacturing platforms [5]. Moreover, these issues drive a hierarchy of order where machines 'feed' on natural resources to provide carbon-based metabolisms, in which living systems are sacrificed for the propagation of fundamentally inert structures. Think of how living trees are transformed into wooden planks before we build with them. With access to fossil fuel reserves, carbon stores from the ground are being unlocked and pumped into our atmosphere as side effects of modern life, which is contributing to a global shift in meteorological events—a phenomenon called climate change.

31.6   Unconventional Computing

Experimental architecture proposes to counter the global impacts of industrialization by diversifying the technical portfolio that underpins human development.[3]

Nature itself offers an alternative production platform with environmentally enlivening material solutions and was widely harnessed before the industrial age to provide technical solutions. For example, horses were used for transport and agrarian land management systems, like irrigation, were developed for increasing food production. Biotechnology has enabled living systems to be manipulated in ways where life's processes themselves can be thought of in a different kind of technical capacity—not just as a resource for consumption by machines, but as an operating system that can perform useful work. Yet, for these material processes to reach a broader sphere of relevance, a toolset is needed that enables designers and engineers to compute with and directly manipulate them.

Unconventional computing provides such an opportunity as it seeks forms of computation that cannot be framed by the conventions of Turing or conventional architectures [14]. While there is no formal definition, it embraces a range of overlapping practices from the digital modeling of biological systems, to chemical computing [1]. These techniques are of great relevance to experimental architecture. The aim is to find ways of directly using physics and chemistry of the natural realm as a way of choreographing space in according to human-centered ethics. In other words its aim is to direct natural events that are in keeping with human interests but share the same impacts as natural events—a programmable technology that performs as nature does. As such, experimental architecture takes a first principles approach to computation that is set within a value landscape—where the practice of computing is considered as the art and science of sorting and ordering the world in the quest to acquire new knowledge. At the heart of this knowledge acquisition system is a form of counting—or computation—that is based on a particular philosophy of numbers. Today's counting systems are founded on a common (Western) theory of numbers that have been around for a few hundred years. 'Zero' was invented to symbolize empty space in 1598, while John Wallis introduced the infinity symbol a little later in 1655.

While digital computing is not possible without the concept of zero, it can operate quite happily without infinity. Modern computing deals with finitudes and drawing limits around a solution space so that the computer knows where to search and when to stop its calculations, while natural systems are unbounded. These iterations are based on different kinds of excursions based in physics and chemistry. Some of these drivers such as, Jeremy England's notion of 'adaptive dissipation' [54] shape evolutionary processes and occupy what Stuart Kauffman describes as 'the adjacent possible' [33]. So, while digital systems can be applied to rationalize natural phenomena, its solutions are always abstractions of its material potential. This abstraction also

---

[3]Such platforms may have soft bodies, be recyclable or use completely different kinds of resources than modern industrial systems.

extends to the structure of DNA, which can be readily represented as a series of ones and zeros. This has given rise to a host of genetic algorithms and digital breeding projects, which produce algorithms that describe new combinations of morphology and behavior, which are then likened to biological systems.

While these are informative, unconventional computing practices may be valuable in observing and shaping the outcomes of biological systems in new ways—not as abstractions—but as functions of material processes. Since number theory is an evolving discipline within the field of mathematics, it is important to ensure that we do not only limit ourselves to using one model, or simulation of the world [15]. Potentially then, unconventional computing can unlock more of the creative potential of the natural realm by 'counting' directly with the material realm through its spontaneously occurring iterative processes.

31.7   Dissipative Systems

Lifelike materials do not create abstractions of the natural realm as they works directly with metabolism, and orchestrate the massive parallel processing potential of matter. This facilitates change and leakage in architectures, which are composed of agile materialities and soft scaffoldings that act as the sites for new kinds of syntheses. We can even invent compounds have never existed before in the history of the cosmos [34]. Recently, the origins of life sciences have been paying close attention to the role of non-classical objects in lifelike systems such as, dissipative structures, with unique programs that embody the transition from inert to living matter. They include a range of examples across a variety of scales from—crystal growth, dynamic droplets, whirlpools, the birth of galaxies, minimal kinds of proto-life and even life itself.

Organisms themselves may be considered as computational systems that can literally be applied to solving complex challenges as in the case of slime moulds [3]. Much attention has been given to DNA as a structural and chemical code that shapes material decisions in living processes but lately, the origins of life sciences is paying close attention to dissipative structures as a kind of minimal proto-life and includes self-assembling chemical systems such as dynamic droplets [9]. Dissipative structures were chosen as a computational substrate for experimental architecture as they offer both a material and also a model system for life's processes. Their behavior is being explored from a first principles perspective, to consider the possibility of working with iterative processes that underpin natural systems and therefore may be considered as an alternative design strategy than symbolic (mathematical) forms of representation, when thinking about the characteristics of living systems.

Dissipative structures were first described by Nobel prize-winning chemist Ilya Prigogine in the1960s and are paradoxical systems that are spontaneously formed by the massive flow of energy and matter through a space. Fundamentally they consist of an oscillator, excitable medium and an open system [2]. Dissipative systems exhibit emergent properties like pattern formation and is possible as they are 'far from equilibrium' [42]. Prigogine proposed that dissipative structures underpinned the organization of living systems [38] and the dissipation-driven adaptation of matter [54] has recently been proposed to support the emergence of lifelike phenomena, such as self-replication [31]. By considering dissipative structures as an iterative process that underpins self-organizing systems, they can be read as computational programs that shape natural events. Experimental architecture aims to establish a portfolio of unconventional computing principles that examine the points of intervention within the progression of dissipative systems so that these structures may be directly interrogated and influenced by a range of strategies. The aim is to directly provoke change in the system rather than represent it as information and then re-translate the outcome into material events.

These principles are based on the iterations of these physical systems, which can be thought of as an oscillator and do not have complete equivalence with mathematical numbers. Rather than representing patterns, they directly produce maps of the computational system they embody that can be read through time-based events, such as the Belousov Zhabotinsky reaction [49]. Through experimental architecture these outcomes are given value, meaning and can be interpreted—although these are relational and personal assessments, not objective, universal, or empirical. Nevertheless they have value in a design context in shaping an understanding of highly complex phenomena. For example, within the practice of the built environment, the morphological growth of cities has been likened to fractal formation. By recognising these patterns, an understanding of the complex phenomena can be made related to ideas about networks and interactions that sustain cities through their transport. By understanding the organizational principles that underpin such systems it becomes possible to propose the processes of movement and mobility that can give rise to the diffusion and segregation of different spatial activities, which ultimately give rise to physical morphologies. Ultimately, such ideas can then be transposed and embedded into models that have potential applications to inform urban policy [10].

In the case of dissipative structures, recognising the patterning portfolio of the system helps designers understand that these phenomena are spontaneously arising from overlapping fields of activity. These produce waves of observable activity and material events that directly feedback into the outputs of the system as they arise, so that new spaces and possibilities can be characterised within a portfolio for design.

31.7.1 Experimentally Generating Dissipative Structures

Dissipative systems form spontaneously at the intersections between potentiating fields of activity. This can be demonstrated using simple kitchen ingredients that provoke dynamic interactions. For example, water affinity can be used to produce a very simple dissipative system by harnessing the energetic and material potential that exists between substances that have very different relationships with water namely, glycerin that strongly attracts water and olive oil, which repels water. A base layer of glycerin and a top layer of olive oil create an interface at which droplets of food colouring are introduced using a hand-held pipette. Adding rock salt to the droplets

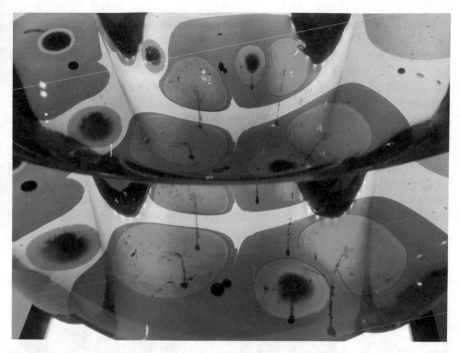

Fig. 31.1 Photograph by Rachel Armstrong, Academy of Arts and Sciences, Belgrade November 2013. Oil and glycerin interface with table salt saturated droplets of food colouring that leave crystalline trails in the transient dissipative structures that emerge from the overlapping fields that are competing for water availability

increases their affinity for water, which provides further complexity and tension in the system. The hygroscopic properties of glycerin eventually overcome the osmotic pressure of the food color droplets, at which point they explode downwards leaving structural trails in their wake that outline the resultant transient dissipative structure. Oil soluble coloring spreads sideways through the olive oil interface as a secondary interaction providing a residue of molecular activity—or, three-dimensional painting. Below is an installation for the 'On Architecture' exhibition at the Department of Arts and Sciences in Belgrade where food colour droplets have discharged into the base layer of glycerin leaving spidery threads of salt crystals behind (see Fig. 31.1).

31.8 Assimilation and Growth

Although dissipative structures spontaneously emerge as a complete body, they are capable of growth through the assimilation of different materials. Such processes can be seen at work in highly organized dissipative structures that are giant cells, which can take up foreign particles into their substance, so that they can continue to thrive.

For example, the protoplasm of Bryopsis plumosa [51], a hairy, giant cell seaweed that grows to around 15 cm in diameter that blooms in early spring and late summer around European shores. It has developed an incredible ability to regenerate from cell fragments that possess several nuclei and have access to light and nutrients, which has allowed it to flourish under the harshest conditions along the shoreline. If is it completely destroyed say for example, after being battered by the environment, or cut up with a razor blade—this hairy green seaweed is able to completely regenerate. In its regenerative state Bryopsis incorporates foreign particles into its substance during the recovery phase. If magnetic particles are added to the culture medium of the regenerating plant, it can be encouraged to grow into new configurations by applying a magnetic field across the rejuvenating tissues. The implications for ecological beings are that their parts, but their wholes do not define them. Possessing a magnetic particle does not turn the Bryopsis into a magnet; rather it enables new

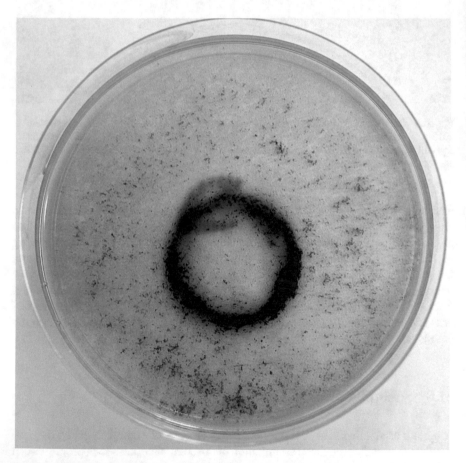

Fig. 31.2 Photograph by Rachel Armstrong, Cantacuzino Institute, Bucharest June 2009. Magnetite particles ingested by the healing Bryopsis giant cell can be magnetically manipulated into specific configurations—in this case—an ouroboros

properties to be conferred on the organism. Despite its resilience there are also limits to how much magnetite the regenerating body can assimilate and its powers of regeneration, from which it can only recover once.

Below a Bryopsis organism has been mechanically destroyed and its protoplasts regenerated with a solution containing magnetite in which a magnetic field has been applied. The recovering body has taken on the morphological configuration of the magnetic force, which in this case, is the alchemical symbol of an ouroboros (see Fig. 31.2).

31.9 Dynamic Droplets

Dynamic droplets are specific examples of dissipative structures that arise at the interface between oil and water [9].

31.9.1 Bütschli System

Zoologist Otto Bütschli described a particular dynamic droplet system based on a simple soap-making recipe in 1892, which bestows the system with a simple, material-producing metabolism that contributes to the complexity of the system's outputs. He added alkali (potash—potassium hydroxide) to a field of olive oil (oleic acid) [13] in an attempt to dispel the 'myths' of vitalism, which argued that 'life' was conferred on matter through a special essence. Bütschli argued that this experiment demonstrated that chemistry alone could account for living phenomena based on similarities in the structure and movement between the two systems. The recipe was updated for a modern laboratory practice using 3M sodium hydroxide solution and extra virgin olive oil. When a drop of alkali was added to the oil field it spread out and broke up into tiny mobile droplets about a millimeter in diameter (see Fig. 31.3). Using a low power setting on a binocular microscope at times four magnification, this 'Bütschli' preparation was studied using modern microscopy for the first time in over a hundred years.

The droplets possessed strikingly lifelike behaviours but did not meet the criteria for being fully living, particularly as they lacked a central biological program, such as DNA. They could move around the environment, sense it and even interact with each other on a population scale basis. Remarkably, this system produced a range of characteristic morphologies and movements, which were documented by conducting over 300 experiments under identical conditions.

Using the principles of unconventional computing, each Bütschli droplet is an iterative agent that operates in a particular way. This perspective may enable us to work with low-level infrastructures, which produce material programs that directly promote life without reducing or abstracting them. Owing to their physicality, Bütschli droplets are leaky objects that are environmentally sensitive. They exist within a

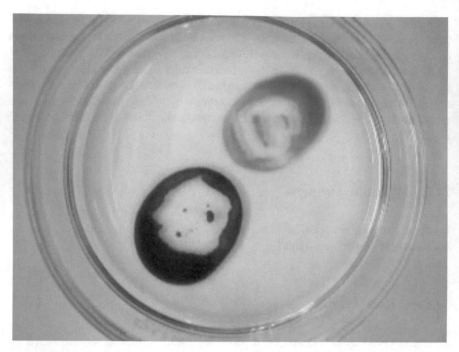

Fig. 31.3 Movie still by Martin Hanczyc, Southern University of Denmark, Odense June 2009. 3M sodium hydroxide fields stained with blue and red water-soluble dyes producing Bütschli droplets in a field of olive oil

range of limits and are sensitive to their environment, eventually succumbing to the forces of disorder, or entropy, at which point they lose their lively characteristics. Dissipative structures also interact beyond their boundaries through active fields. Perhaps the best way to imagine this effect is to think of a storm chaser that can feel the presence of a twister long before they ever reach the eye of the storm.

From a design perspective, is not necessary to understand 'how' each Bütschli droplet operates in order for it to be a design tool—just in the same way that it is not necessary to know 'how' a 3D printer system works, in order to understand the contexts in which its design applications may be appropriate. However, it is important that a design 'language' is developed in which value can be attributed to the emergent events as they exhibit a range of unique characteristics. Bütschli system iterations are continuous, exist against background activity and 'count' spatiotemporally in parallel and never really reach a 'zero' quantity. Their computational sequence grows exponentially from material iterations that arise from some (not yet established) fundamental baseline position that reaches a peak state (which defines the iteration) and falls off again back to the background level, which is above a zero level and starts to rise again back to another peak value (which defines the next iteration). The rise, collapse and coupling of Bütschli droplets may account for how natural phenomena appear to cluster, establishing the initial conditions for another in its locality.

31.9.2   Spatiotemporal counting

The progression of Bütschli system iterations are differentially distributed over the droplet interface, being influenced by the production of matter, which generates further complexity. While the fundamental chemical program remains constant—the endothermic (heat absorbing) interaction of alkali and oil produces soap crystals (sodium oleate)—the spatialised local conditions shape the computational process—some diffusion waves will be reflected, others quenched in the production of sodium oleate (soap). The kinetics of this process are extremely complex but from a design perspective, it is sufficient to recognize the range of possible structures and movements and how they may be further influenced in developing new architectural tactics and programs.

Below is a droplet that is shaped by the differential interactions of the basic chemical program of the Bütschli system over its interface, which is shaped through a complex ecology of physical and chemical relationships that include microscale disturbances in molecular flow, variations in chemical equilibrium and the physical properties of the droplet (see Fig. 31.4).

Notably, the Bütschli droplets are polarized from the outset of their production owing to mass molecular flow centered on disturbances at the interface, which generate an anterior and posterior pole to the droplets. Between these reference points the specific outcomes of the chemical program varies. The droplet moves from the anterior end, where there is maximum interface and surface area to volume ratio for the most efficient interaction between oil and alkali. As crystals are produced they block the reaction interface and are swept backwards towards the posterior pole where other reaction products (impurities in the oleic acid) accumulate. The anterior end therefore absorbs more energy than the posterior pole, its reagents being replaced by an internal current of alkali—perhaps by the Marangoni effect, which regenerates the interface [30]. This process continues until a critical surface area to volume ratio is reached, when the reagent replenishment rate falls behind crystal production. At this point a transition in the movement of the droplet occurs, where crystalline deposits are rapidly deposited at the interface. This increased precipitation occurs away from the direction of travel, so soap crystals are rapidly deposited at the posterior pole. This asymmetrical extension alters the centre of gravity of the droplet, so it begins to move erratically through the medium and becomes quiescent when crystalline deposits completely occlude the droplet interface. The differential effect of the temporal and spatial position of the interacting species is evident from the range of morphologies and movements of the osmotic structures produced by Bütschli droplets [7].

Examples of 3D osmotic structures produced by the Bütschli system are shown in the graphic below (see Fig. 31.5).

A illustrative diagram was constructed to graphically represent a morphological range of osmotic species and reflect on their relationships as a function of time, based on findings documented over the course of 300 replicate experiments [9]. The graphic is centered at time zero from which concentric circles radiate, signifying

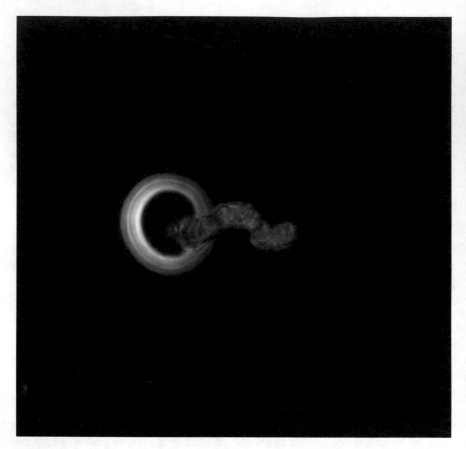

Fig. 31.4 Movie still by Rachel Armstrong, Southern University of Denmark, Odense June 2009. Bütschli droplet producing an osmotic structure

an exponentially increasing series of time intervals and encapsulates the intense self-organizing activity that happens early on in the chemical reaction, which falls off rapidly with the passage of time (see Fig. 31.6). An estimated ninety per cent of chemical activity is completed within five minutes of activation of the system, although individual droplets have been observed to be active as long as an hour after their genesis. A spiral that represents complexity that also radiates from the origin and depicts the high frequency of events around the start of the reaction that becomes less frequent as time unfolds. The various morphologies and behaviours that indicate change in the system are subjectively grouped based on a personalized interpretation of the typologies observed. These are not absolute categories but are nonetheless distinctive forms that share many overlapping characteristics. For example, the complex oyster chains are striking in appearance but only differ by degree, from the complex marine landscapes. Specifically, 'oysters' produce a large mass of material, which anchors them and their soft bodies bulge from their material shell-like tethers.

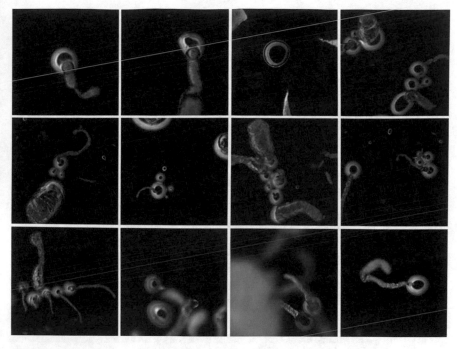

Fig. 31.5 Collage of movie stills by Rachel Armstrong, Southern University of Denmark, Odense, June 2009. Range of osmotic structures produced by the Bütschli system

31.9.3 Cellular Automata

The osmotic structures produced by the Bütschli system possess a range of recognizable typologies that reflect the local limits of the adaptive fitness landscape. Intriguingly, these bear some morphological similarities to Stephan Rafler's Smooth Life, which is a 'continuous' form of cellular automaton that uses floating point values instead of integers, which are an approximation of real numbers that can support a trade-off between range and precision [22]. Notionally, perhaps like the material peaks that form the iterative system of dissipative structures, the Smooth Life program produces an evolving set of recognisable events with associated graphics. However, Rafler's model does not demonstrate agent or material transformation as the Bütschli system does—which lends weight to the idea that the system is influenced by its embodiment and deposition of material in generating highly complex outputs. This warrants further research and study but is beyond the scope of this essay.

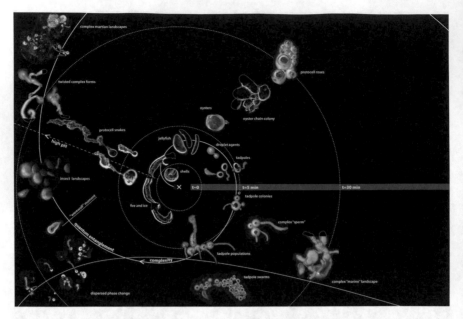

Fig. 31.6 Diagram designed by Rachel Armstrong and drawn by Simone Ferracina, July 2012. This diagram depicts dynamic droplets as 'actors' that operate within the many variable influences encountered in their oil field as an ontolological 'map' of events. While the diagram is drawn as a 2D topology, the field events are manifold and open up multidimensional spaces through their continuous interactions that shape the evolution of the system

31.10 Computational Strategies

The Bütschli droplets are transformers that complexify adaptive landscapes, which means it is not possible to exactly predict their outputs. Such an approach does not attempt to replace classical forms of computing but generate a broader portfolio of materials and techniques available for the choreography of space. Their applications are far from formalized and are best considered as a 'graphical' system that enables designers to directly view the dynamic potential in matter at far from equilibrium states. The design strategy taken to deal with this uncertainty is to work 'along with' these structures and exercise forms of soft control to influence and shape the computational process. We are already familiar with using these kinds of techniques as artisan practices over the millennia. They range from cooking, gardening and agriculture—to working with small children, or herding cats.

However, the production of dissipative systems alone is not sufficient for a formal computing practice. A nurturing (open), excitable medium in which computational operations can take place is integral to the development of a 'dissipative hardware and software'. Moreover, a set of strategies through which these excitable media can be orchestrated is needed. From an experimental architectural perspective, the objective in working with a dissipative computational platform is not to manufacture 'objects' but to provoke the conditions in which desired events are likely—although not inevitable. Dissipative systems, like the Bütschli droplets, are embedded in their matrix, so computational stratagems to alter their performance may be targeted at their internal and external composition. These approaches are context sensitive and therefore vary depending on the composition of the local environment.

A series of Bütschli droplet experiments were conducted using an experimental architectural approach to explore how dissipative structures could be used to modify and distribute matter in time and space.

31.10.1 Internal Droplet Modification

Buütschli droplets can be designed to create a range of different products by adding different chemistries to their internal environment, or droplet body. For example, aqueous inorganic salts may be added to a field of Buütschli droplets to produce 'secondary' forms on contact with the alkaline droplet body, such as microstructures, which are deposited at the oil/water interface. For example, insoluble, magnetic 'magnetite' crystals can be produced by adding iron 2/iron 3 salts [11] that produce

Fig. 31.7 Photograph Rachel Armstrong, Southern University of Denmark, Odense, June 2009. Modified Bütschli droplets produce magnetite inclusion bodies when they come in contact with iron II and iron III salts in a ratio of 1:2

inclusion bodies in the droplets that may also be deposited within the olive oil matrix (see Fig. 31.7).

31.10.2 External Environment Modification

Changing the external conditions of the medium can also alter the behaviour of the Bütschli system based on physical and chemical changes, such as surface tension and chemotaxis [46]. For example, adding ethanol, or 'alcohol' to the olive oil field, produces a rather a dramatic effect characterised by the population-scale, sudden

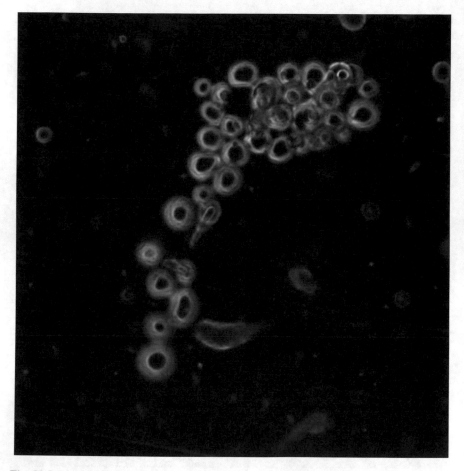

Fig. 31.8 Movie still Rachel Armstrong, Southern University of Denmark, Odense, June 2009. Bütschli droplets respond to the presence of ethanol by clustering around the site of its addition to the olive oil medium

movement of the Bütschli droplets towards the alcohol source (see Fig. 31.8). This may be explained by changes in surface tension that promote movement of the droplet dynamics, but also by reducing the viscosity of the olive oil [7].

31.11   Dissipative Systems in Unconventional Computing

31.11.1   Questions of Scale

Droplets are easier to see and can be easily worked with by hand if they are made bigger by adding an extract of their waste products to their body, which also enables them to last longer. Although these droplets are attenuated, they still possess the fundamental properties of dissipative systems. For example, they can produce microstructures following the strategic adding salt solutions that transform soluble carbon dioxide into an insoluble carbonate precipitate (see Fig. 31.9).

Further constraints placed on the system can provoke a range of phenomena. For example Turing bands can be produced by reducing the physical dimensions of the reaction space to 2 cm, which approximates with the maximum diameter possible

Fig. 31.9 Photograph Rachel Armstrong, Southern University of Denmark, February 2010. Modified Bütschli droplets produce complex structures in the presence of mineral solutions

Fig. 31.10 Photograph Rachel Armstrong, Natural History Museum, Vienna, April 2011. Modified Bütschli droplets exhibit Turing band pattern formation

for a modified droplet. These undulating configurations are the consequence of reaction/diffusion systems, which Alan Turing believed underpinned pattern formation in animals during morphogenesis, like dappling [47]. This transformational ability was clearly demonstrated in this installation in Vienna at the Natural History Museum (see Fig. 31.10).

31.11.2 Population Scale Interactions

Perhaps surprisingly, despite being able to generate highly complex outputs, Bütschli droplets are remarkably resilient, robust and therefore predictable, as they operate within 'limits' of possibility, which are imposed by the properties of the system and its context. However, they can behave unpredictably at tipping points and give rise to novel, emergent, complex events that cannot be deduced from their characteristic behaviours, such as, striking changes in structure and movement.

In the following micrographs, a population of Bütschli droplets undergoes spontaneous phase change in morphology and behaviour when an uncharacterised chemical tipping point in the system is reached. Figure 31.11 shows the configuration

Fig. 31.11 Movie still Rachel Armstrong, Southern University of Denmark, Odense, June 2009. Spontaneous assemblage of Bütschli droplets producing osmotic structures

of droplets before this event, where droplets are observed producing microstructures. Figure 31.12 shows the same droplet population following a conjunction with second population of droplets that have been produced at the same time under identical conditions join the assemblage. After thirty seconds of intra-action the droplets simultaneously separate and produce microstructures. The trigger for this unexpected event has not been identified and the phenomenon requires further research.[4]

---

[4]This video can be seen at: https://www.youtube.com/watch?v=gB6MKMqbLIM.

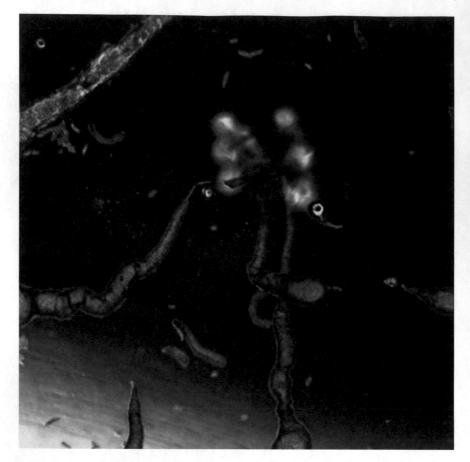

Fig. 31.12 Movie still Rachel Armstrong, Southern University of Denmark, Odense, June 2009. Spontaneous phase change in morphology and behaviour in an assemblage of dynamic droplets that reach an unknown chemical 'tipping point' in the system

31.12   Elemental Infrastructures

31.12.1   Enabling Flow

A fundamental requirement for the production of dissipative structures is the availability of life-promoting media—namely, they are open and excitable. These are essential in facilitating flow, which enables the dynamic exchanges that allow dissipative structures to persist and evade the decay towards a disordered state. Such media may be thought of as elemental systems namely: air, heat, water and earth.

The importance of elemental systems in facilitating the performance of attenuated Bütschli droplets was demonstrated in Philip Beesley's Hylozoic Ground installation for the 2010 Venice Architecture Biennale. This jungle-like, room-sized, cybernetic, interactive mechanical matrix detected gallery visitors through an array of sensors that were coupled to effectors through a primitive neutral network. On activation the system shivered and raised friendly rubbery fronds of spikes as a somewhat disconcerting but entirely harmless greeting.

Modified Bütschli droplets were installed as the dissipative system of choice, which could persist for the full 3-month duration of this exhibit. Sustenance was provided through an arrangement of flasks containing elemental infrastructures. Here, Bütschli droplets were entangled with the cybernetic system and its environmental interactions by responding to the presence of carbon dioxide through the production of tiny, brightly coloured structures about the size of a little fingernail (see Fig. 31.13). Walking through this installation might be likened to exploring inside a giant nose—where the plastic fronds act as sinus hairs and the 'golden apple-like' flasks are 'smart' snot glands that can smell and taste the presence of carbon dioxide [8].

Fig. 31.13 Photograph Rachel Armstrong, Canadian pavilion, Architecture Biennale, Venice, August 2010. The Hylozoic Ground installation is a cybernetic matrix that integrates a range of different dynamic chemistries to aesthetically and functionally complement the dynamic processes that inform architect Philip Beesley's installation. Here, chemical organs containing modified Bütschli droplets are open to the air and produce solid carbonate microsculptures in response to the presence of carbon dioxide

The Hylozoic Ground flasks were open to the air and so, enabled the on-going exchange of carbon dioxide across the air/water interface. Yet, increasing the flow of resources generates more potential in unlocking the synthetic capacity of the modified Bütschli droplets. The capacity for architectural scale synthesis of a dissipative structure based fabric is speculatively and experimentally explored in the Future Venice project [36]. The long-term survival of the city is addressed by programming dissipative structures to literally transform and empower its foundations with lifelike qualities, which enable the urban fabric to literally fight back against the damaging effects of natural elements in a struggle for survival—and therefore secure its longevity (see Fig. 31.14).

A series of dynamic droplets with a range of different programs were experimentally designed to chemically converse with the lagoon environment. These exchanges may be triggered by specific environmental conditions to produce a variety of synthetic actions that result in an accretion process that is mediated through dynamic droplet interactions. The watery infrastructure of the city provides the specially engineered droplets with an abundant flow of nutrients such as, dissolved carbon dioxide and minerals. This enables the collective action of the droplets to form an artificial garden reef underneath the city's foundations and creates a solid structure that spreads

Fig. 31.14 Photograph Rachel Armstrong, Canalside Venice, August 2010. Stromatolite-like formations in the Venetian canals are shaped over time to produce mineralized materials

Venice's point load over a much broader base. Consequently Venice is prevented from sinking so quickly into the soft mud that it has been founded on.

A natural version of this accretion process is observed around the lagoon side and the canals, which is orchestrated by the native marine wildlife. Potentially, dynamic droplets could work alongside these organisms to co-construct an architecture that is mutually beneficial to the marine ecology and the city. Importantly, should the environmental conditions change and the lagoon dries out—say for example, Pietro Tiatini and his colleagues succeed in anthropogenically lifting the city by pumping seawater into its deflated aquifers [45], or if when the MOSES gates are raised in 2014 [50] the native ecology reaches a catastrophic tipping point—then the dissipative structures within the Venice waterways may work alongside the smart droplets, to follow a different program and re-appropriate their intra-actions. Specifically, as the waters subside droplets coat the woodpiles in a downwards direction with a protective layer of 'biocrete' that stops them rotting when they are exposed to the air—instead of spreading outwards to form a reef.

Potentially, the computational processes associated with dissipative structures could be applied to the whole bioregion of Venice. Facilitated by flowing water, the right kinds of metabolisms and spatial programs could give rise to tactics that generate, new relationships between natural and artificial systems. These may become the bedrock for forging life-promoting, synthetic ecologies. In this context the manipulation of dissipative structures enables the development of spatial programs and design tactics to generate a constant flux between fabric, space, structure and location. The outputs of the system do not imitate Nature but work as an alternative kind of Nature that operate at 'low level' organizing principles, which influence the performance of dissipative systems. These are then realized using a common language based in physics and chemistry that enables coupling between different material bodies, which is shared with the natural world.

Yet, these potential unconventional computing strategies do not propose a comprehensive solution to Venice's precarious future—or indeed our legacy our environmental woes. It does not attempt to 'solve' the inevitable changes that accompany a lively environment but enables us to imagine new solutions and respond differently to environmental events and challenges.

Such dynamic practices generate new possibilities for the city of Venice—which can now be imagined in terms of its organizing clusters—which are formations produced by dynamic processes that convert inert materials into active ecologies. They enable designers and engineers to regard the fabric of city in a new way—not as a collection of decorative objects—but as a pulsating, vibrating, transforming, flowing material in which multiple, collaborating dissipative bodies are immersed and work alongside human communities to produce new kinds of experiences and spaces for inhabitation.

31.13 Urban Applications

We are now living at the time of the Anthropocene [18] an era in Earth's history where human activity is considered to be shaping geological events. Since our cities are so large—occupying around 3 % of the Earth's land surface [41]—even small changes in their environmental performance may bring significant benefits to our ecosystems. Through experimental architectural methods, we can consider how a lifelike materiality for cities conferred by using dissipative structures may provide a different platform for human development than industrial processes.

The 21st century is the age of the megacity. These are metropolitan edifices have already sprung that now house more than 10 million inhabitants and appear 'endless' as they stretch over hundreds of square kilometers. Yet, through our own design and engineering practices, the materiality of our living spaces is in conflict with the needs of their inhabitants. Scarce drinking water is used to flush excrement. Fertile soils are being scorched by intense agricultural practices and the produce transported hundreds of miles to provide the illusion of abundance within our urban deserts, which offer little or no fertility of their own, since they are largely devoid of fertile soils. Above these metropolitan colossuses the skies are choking with invisible toxins and beyond them, their waste is stretched out into oceans of particulate plastics.

This scenario demands practical and theoretical research on living materials and technology that builds towards a scenario where architecture not only deals with the manufacturing of objects but also choreographs the flow of matter through our living spaces. This may be concentrated in technologized sites that operate as architectural 'organs', such as algae bioreactors, which possess unique physiologies that shape the character and performance of spaces. If living materials can be practically realized within our buildings, the fabric of our homes and cities may be enabled to perform the equivalent work to machines by orchestrating the molecular flow of matter through our most intimate environments.

For example, architectural organs may provide biomass for food and fuel—or produce low-level lighting through bioluminescence. They could be situated in under used and under imagined spaces such as cavity walls within our homes, or be on proud display like the designs as in Philip's Microbial home, which is fuelled by the actions of microorganisms [35]. In this scenario, our cities will not be imagined as Le Corbusier's machines for living in [17]—but as ecologies for thriving in— where each city possesses a unique urban metabolism that arises from its natural and architectural physiologies.

Working with the programmability of dissipative structures brings Nature right into our living spaces in ways that could make our homes more resilient and attractive. Yet this is not fiction—such systems are already being developed for installation in the modern built environment. We're already tapping into this relationship between natural systems and machines using renewables—and now, unconventional computing creates the possibility of orchestrating the performance of micro agricultures that may help us live in densely populated spaces in new ways. Indeed, in the near future we are likely to see the rise of 'smart green' cities where—Nature and digital

Fig. 31.15 Photograph Colt International. Arup, SSC Gmbh. BIQ house that houses microalgae in façade panels containing water. They feed on sunlight and carbon dioxide to produce biomass that is collected, dried and burned back in the building to produce energy that offsets carbon emissions

information systems are brought together and work together do 'more' with fewer resources—so that we can meet the needs of our urban populations differently.

A range of ambitious proposals is currently being tested in experimental contexts and in bespoke building installations that are setting new benchmarks for the next generation of 'sustainable' building designs. For example, Astudio architects and Sustainable Now Technologies are constructing a bioprocessor for a 6th form college in Twickenham, as a next-generation ecological architecture. The bioreactor system provided a focus for the school curriculum by generating outputs that could be used by the students. For example, the biomass could be used as a fertilizer for green roofs and walls.

For example, the BIQ (Intelligent building) in Hamburg opened in November 2014. This apartment block has been constructed with watery facades that house microalgae, which produce biomass from sunlight and carbon dioxide and reduce energy consumption within the residential apartments by virtue of the thermalsolar effect (see Fig. 31.15).

While explorations such as living facades, are still very much at the experimental stage, they facilitate the convergence of different media so that the technological systems on which we currently rely, do not inevitably damage our ecosystems, but may actually strengthen them. Ultimately, by combining the self-organizing principles of dissipative structures with conventional technical platforms [40] may produce different kinds of environmental impacts in which the built and living environments mutually reinforce each other to produce a livelier planet.

31.14 Conclusion

Cross-disciplinary experimental platforms such as, experimental architecture may contribute to the development, pedagogy, social engagement and commercialization of unconventional computing. In doing so, a future portfolio of unconventional computers may have significant impacts in underpinning new ways of living and working. For example, it may be possible to use some of the strategies of the natural realm to directly address environmental challenges so that we no longer 'mimic' nature's processes but directly apply its organizational ruleset. In this way we may be able to go beyond performance limits of biological systems and generate new kinds of materials and spaces that have not yet been encountered. For the lifelike qualities of materials at non-equilibrium states to reach a broader sphere of relevance to design and engineering practices, a toolset is needed that enables designers and engineers to compute with and directly manipulate them. Experimental architecture works synergistically with unconventional computing to explore these possibilities. While the relationships and the practice are not formalized, the juxtaposition between design-led and scientific experiments has the potential to create propositions and contexts in which new kinds of computing and technical systems can be explored—both speculatively and practically. These projects are not simply fictions but can be hypotheses that are iteratively examined and mapped out through drawings, models and prototypes as potential trajectories. These explorations may inform further design-led projects or even subsequent scientific experiments.

Since unconventional computing is an exploratory and creative practice, there are many points of synergy with experimental architecture. Yet design disciplines do not propose to produce scientific outcomes—or focus on empirical data—but nonetheless add to the field of knowledge by unfolding new possibilities and provoking further questions for exploration that may lead to new applications for science and technology. Specifically, developments in the fields of physics and chemistry point towards the importance of dissipative structures in better understanding the organizational principles of living systems and how these may directly be applied to the practice of the built environment. These have been discussed in this chapter as a potential platform for unconventional computing by applying experimental architecture to create an environment in which incomplete knowledge can be worked with intuitively and re-informed through design-led experiment. Such explorations are complementary to the field of unconventional computing and opens up a space for articulating ideas and developing language protocols that are not dominated by industrial agendas and near to market objectives. Instead researchers from across a spectrum of disciplines are enabled to engage imaginatively with the philosophy, theory and practice of knowledge acquisition in unconventional computing to develop its cultural relevance and continue to expand its frontiers in new directions.

The field of unconventional computing may ultimately wield significant influence on the fabrics that are unique to this planet and even collectively comprise a force capable of geoengineering-scale impacts. We may see the advent of new kinds of Nature that are deeply entangled with the NBIC convergence. Rather than the current practice of layering green vegetation over grey concrete [21], p. 230 to achieve a new kind of urban 'efficiency'—the 'grey' digital technologies of smart cities may be directly threaded into the green fibres of 'sustainable' living materials. Out of these convergences alternative intelligences, gastrulating architectures, and production platforms for human development may begin to emerge with qualitatively different kinds of impacts that are similar to the natural realm. Within these placental enfoldings new architectural strategies, programs and tactics are made available to experimental architects and urban designers, who use iterative, persistent experimentation to explore these spaces. Rather than seeking atomic control of events, they embrace risk as a condition of existence and develop a broad palette of lively, multi materialities that incessantly coalesce to provoke new spatial experiences. Unconventional computing may equip these designers with a new range of materials and technologies that can help them achieve these aims—fuzzy surfaces, cloudy vistas, fragile details, quantum logic, soft scaffoldings and all kinds of teratogenic in-betweens. Such transgressive technical systems may infiltrate the spandrels between the mineralized bones of industrial construction [27]. Yet these nascent landscapes and complex, fertile substrates do not claim to provide totalizing solutions to the constantly unfolding multiplicities and challenges that we are facing. Rather, they catalyse new opportunities for invention by providing an emerging palette of new possibilities and paradoxes from which we may birth new kinds of architectures. In this way the built environment shares a common project with the natural realm that can be shaped by human ethics through the production of life's poetry and our mutual, continued survival into an ever-unfolding adjacent possible that is full of surprises.

References

1. Adamatzky, A., De Lacy Costello, B.: Reaction-diffusion path planning in a hybrid chemical and cellular-automaton processors. Chaos, Solitons Fract. **16**, 727–736 (2003)
2. Adamatzky, A.De, Lacy Costello, B., Asai, T.: Reaction-Diffusion Computers. Elsevier Science, London (2005)
3. Adamatzky, A., Armstrong, R., Jones, J., Gunji, Y.K.: On creativity of slime mould. Int. J. Gen. Syst. **42**(5), 441–457 (2013)
4. Adams, T.: The chemputer that could print out any drug. The Observer (2012). www.guardian.co.uk/science/2012/jul/21/chemputer-that-prints-out-drugs. Accessed 19 Feb 2015
5. Armstrong, R.: 3D printing will destroy the world unless it tackles the issue of materiality. Architectural Review (2014). www.architectural-review.com/home/products/3d-printing-will-destroy-the-world/8658346.article. Accessed 28 Feb 2015
6. Armstrong, R.: Designing with protocells: applications of a novel technical platform. Life **4**(3), 457–490 (2014)
7. Armstrong, R.: Vibrant Architecture: Material Realm as a Codesigner of Living Spaces. De Gruyter Open, Berlin (2015)
8. Armstrong, R., Beesley, P.: Soil and protoplasm: the Hylozoic ground project. Archit. Des. **81**(2), 78–89 (2011)
9. Armstrong, R., Hanczyc, M.M.: Bütschli dynamic droplet system. Artif. Life J. **19**(3–4), 331–346 (2013)
10. Batty, M.: Building a science of cities, Working papers series. 170 (2011). www.bartlett.ucl.ac.uk/casa/pdf/paper170.pdf. Accessed 10 Feb 2015
11. Berger, P., Adelman, N.B., Beckman, K.J., Campbell, D.J., Ellis, A.B., Lisensky, G.C.: Preparation and properties of an aqueous ferrofluid. J. Chem. Educ. **76**, 943–948 (1999)
12. Beynus, J.: Biomimicry: Innovation inspired by Nature. William Morrow, New York (1997)
13. Bütschli, O.: Untersuchungen ueber microscopische Schaume und das Protoplasma, Leipzig (1892)
14. Centre for Nonlinear Studies. Unconventional Computing Conference, Los Alamos (2007). http://cnls.lanl.gov/uc07/. Accessed 29 Oct 2015
15. Chatelin, F.: Qualitative computing: A computational journey into nonlinearity. World Scientific, Singapore (2012)
16. Cook, P.: Experimental Architecture. Universe Books, New York (1970)
17. Corbusier, L.: (C.E. Jeanneret) Toward an Architecture. Getty Research Institute, Los Angeles (1923, reprint 2007)
18. Crutzen, P.J., Stoermer, E.F.: The anthropocene. Glob. Change Newsl. **41**, 17–18 (2000)
19. Cyberplasm Team. Cyberplasm: a micro-scale biohybrid robot developed using principles of synthetic biology (2010). http://cyberplasm.net/. Accessed 18 Feb 2015
20. Davies, K., Vercruysse, E.: Incisions in the haze. In: Sheil, B. (ed.) Manufacturing the Bespoke: Making and prototyping architecture AD reader, pp. 162–174. Wiley, London (2012)
21. Dean, P.J.: Delivery without Discipline: Architecture in the age of design. Ann Arbor: proquest, Umi Dissertation Publishing (2011)
22. Doctorow, C.: Game of Life with floating point operations: beautiful SmoothLife. Boing Boing (11 Oct 2012). http://boingboing.net/2012/10/11/game-of-life-with-floating-poi.html. Accessed 20 Feb 2015
23. Else, L.: From STEM to STEAM. Culture Lab, New Scientist (8 Aug 2012). www.newscientist.com/blogs/culturelab/2012/08/john-maeda-steam.html. Accessed 19 Feb 2015

24. Evans, D.: Black Sky Think. Lib. **3**, 22–25 (2013)
25. Fuller, S.: Humanity 2.0. Basingstoke: Palgrave Macmillan (2011)
26. Funtowicz, S.O., Ravetz, J.R.: Three types of risk assessment and the emergence of post-normal science. In: Krimsky, S., Golding, D. (eds.) Social Theories of Risk, pp. 251–273. Praeger, Westport (1992)
27. Gould, S.J., Lewontin, R.C.: The spandrels of San Marco and the Panglossian paradigm: a critique of the adaptionist programme. Proc. R. Soc. Lond. Ser. B **205**(1161), 581–598 (1979)
28. Green Building Council. Not dated. Buildings and climate change. www.eesi.org/files/climate.pdf. Accessed 29 Oct 2015
29. Hanczyc, M.M., Ikegami, T.: Protocells as smart agents for architectural design. Techno. Arts **7**(2), 117–120 (2009)
30. Hanczyc, M.M., Toyota, T., Ikegami, T., Packard, N.H., Sugawara, T.: Fatty acid chemistry at the oil-water interface: self-propelled oil droplets. J. Am. Chem. Soc. **129**(30), 9386–9391 (2007)
31. Jarzynski, C.: Nonequilibrium equality for free energy differences. Phys. Rev. Lett. **78**, 2690–2693 (1997)
32. Kalderon, M.E.: Fictionalism in Metaphysics. Clarendon Press, Broadbridge (2005)
33. Kauffman, S.A.: Reinventing the Sacred: A new view of science, reason, and religion. First trade, paper edn. Basic Books, New York (2008)
34. Lehn, J.M.: Perspectives in supramolecular chemistry-from molecular recognition towards molecular information processing and self-organization. Angewandte Chem. Int. Ed. **27**(11), 89–121 (1988)
35. McGuirk, J.: Philips microbial home takes kitchen design back to the future. Art and Design, The Guardian (21 November 2011). www.guardian.co.uk/artanddesign/2011/nov/21/philips-kitchen-design-microbial-home. Accessed 19 Feb 2015
36. Myers, W., Antonelli, P.: Bio Design: Nature, Science, Creativity. London: Thames & Hudson/New York: MOMA (2013)
37. Olguin, C.: Programming matter across domains and scales. J. B. Interplanet. Soc. **67**(7,8,9), 354–358 (2014)
38. Prigogine, I., Stengers, I.: Order Out of Chaos: Man's New Dialogue with Nature. Heinemann, London (1984)
39. Resnick, L.B.: Introduction. In: Resnick, L.B. (ed.) Knowing, Learning and Instruction: Essays in honor of Robert Glaser, pp. 1–24. Erlbaum, Hillsdale (1989)
40. Roco, M.C., Bainbridge, W.S.: Converging Technologies for Improving Human Performance, Nanotechnology, Biotechnology. Information Technology and Cognitive Science. NSF/DOC-sponsored report. Springer, Dordrecht (2003)
41. Schirber, M.: Cities cover more of earth than realized. Live Science (11 Mar 2005). www.livescience.com/6893-cities-cover-earth-realized.html. Accessed 8 Jan 2015
42. Schrödinger, E.: What is life? The physical aspect of the living cell. Based on lectures delivered under the auspices of the Dublin Institute for Advanced Studies at Trinity College, Dublin, in February 1943 (1944). http://whatislife.stanford.edu/LoCo_files/What-is-Life.pdf. Accessed 16 Feb 2015
43. Stengers, I.: God's heart and the stuff of life. Pli **9**, 86–118 (2000)
44. Sterling, B.: Design fiction: Growth assembly. Sascha Pohflepp and Daisy Ginsberg. Beyond the Beyond. Wired (3 Jan 2012). www.wired.com/beyond_the_beyond/2012/01/design-fiction-sascha-pohflepp-daisy-ginsberg-growth-assembly/. Accessed 18 Feb 2015
45. Teatini, P., Castelletto, N., Ferronato, M., Gambolati, G., Tosi, L.: A new hydrogeologic model to predict anthropogenic uplift of Venice. Water Resour. Res. **47**(12), W12507 (2011)
46. Toyota, T., Maru, N., Hanczyc, M.M., Ikegami, T., Sugawara, T.: Self-propelled oil droplets consuming fuel surfactant. J. Am. Chem. Soc. **131**(14), 5012–5013 (2009)
47. Turing, A.M.: The chemical basis of morphogenesis. Philos. Trans. R. Soc. Lond. Ser. B, Biolog. Sci. **237**(641), 37–72 (1952)
48. Turney, J.: Imagining Technology. NESTA working paper (2013). www.nesta.org.uk/library/documents/Imagining_Technology.pdf. Accessed 19 Feb 2015

49. Tyson, J.T.: What everyone should know about the Belousov–Zhabotinsky reaction. In: Levin, S.A. (ed.) Frontiers in Mathematical Biology, pp. 569–587. Springer, New York (1994)
50. United Nations Office for Disaster Risk Reduction. US$6.7 billion floodgates to protect Venice in 2014 (21 Dec 2012). www.unisdr.org/archive/30174. Accessed 19 Feb 2015
51. Vladimirescu, A.: Can We Re-build a Cell? Bryopsis- An Experimental Model! Artificial Life XI. MIT Press, Cambridge (2008)
52. WETFab : Event booklet. School of Chemistry, University of Glasgow (24–25 Jan 2011). http://www.chem.gla.ac.uk/cronin/files/news/WETFAB.pdf. Accessed 26 Feb 2015
53. World Commission on Environment and Development. *Our Common Future*. Report of the World Commission on Environment and Development. Published as Annex to General Assembly document A/42/427 (1987). www.un-documents.net/our-common-future.pdf. Accessed 25 Feb 2015
54. Wolchover, N.: A new physics theory of life. Quanta (22 Jan 2014). www.quantamagazine.org/20140122-a-new-physics-theory-of-life/. Accessed 10 Feb 2015
55. Woods, L.: War and Architecture. Princeton Architectural Press, New York (1996)
56. Youde, K.: Unilever employees get taste of the future at Royal Society event (5 Nov 2013). www.citmagazine.com/article/1219480/unilever-employees-taste-future-royal-society-event. Accessed 28 Feb 2015

Index

© Springer International Publishing Switzerland 2017
A. Adamatzky (ed.), *Advances in Unconventional Computing*,
Emergence, Complexity and Computation 23,
DOI 10.1007/978-3-319-33921-4